biomedicines

Roles of NF-κB in Cancer and Their Therapeutic Approaches

Edited by
Véronique Baud

Printed Edition of the Special Issue Published in *Biomedicines*

www.mdpi.com/journal/biomedicines

MDPI

Roles of NF-κB in Cancer and Their Therapeutic Approaches

Roles of NF-κB in Cancer and Their Therapeutic Approaches

Special Issue Editor

Véronique Baud

MDPI • Basel • Beijing • Wuhan • Barcelona • Belgrade

MDPI

Special Issue Editor
Véronique Baud
University Paris Descartes
France

Editorial Office
MDPI
St. Alban-Anlage 66
Basel, Switzerland

This edition is a reprint of the Special Issue published online in the open access journal *Biomedicines* (ISSN 2227-9059) from 2017–2018 (available at: http://www.mdpi.com/journal/biomedicines/special_issues/NF-kB).

For citation purposes, cite each article independently as indicated on the article page online and as indicated below:

LastName, A.A.; LastName, B.B.; LastName, C.C. Article Title. *Journal Name* **Year**, *Article number*, Page Range.

ISBN 978-3-03897-117-7 (Pbk)
ISBN 978-3-03897-118-4 (PDF)

Cover image courtesy of Thierry Molina. Pathology department, Hôpital Necker, Paris, France and Research unit NF-κB, Differentiation and Cancer, EA7324, Université Paris Descartes, Paris, France.

Contents

About the Special Issue Editor

Véronique Baud, Ph.D., is a Research Director at the INSERM (National Institute for Health and Medical Research), and head of the team "NF-κB, Differentiation and Cancer" at the University Paris Descartes, Paris, France. After having carried out a Ph.D. in Molecular Genetics in the laboratory of Thomas Tursz, Institut Gustave Roussy, Villejuif, France, she performed a post-doctoral training in the laboratory of Michael Karin, University of California, San Diego, CA, USA. Back to France, she obtained a permanent position at the INSERM. Her group works on the interface between signal transduction and cancer with a focus on the alternative NF-κB signaling pathway, how it is regulated, and its contributions towards tumor development and resistance to conventional cancer therapies.

.

Preface to "Roles of NF-κB in Cancer and Their Therapeutic Approaches"

Although mortality rates have declined in recent years, the majority of cancers are still difficult to treat and the medical need for better cancer treatment is evident. The current anticancer armamentarium includes many active agents that are applied across tumor types. However, most of these broadly active anticancer drugs have a small therapeutic index and barely discriminate between malignant and normal cells. In recent years, the focus has shifted to the development of rationally designed, molecularly-targeted therapy for the treatment of a specific cancer, therefore offering the promise of greater specificity coupled with reduced systemic toxicity. The NF-κB transcription factor family has emerged as a promising target for cancer therapy.

More than thirty years after their discovery, it is evident that the importance of the NF-κB transcription factors is continuing to grow due to their central role in many situations of cellular and organismal challenge. NF-κB plays a critical role in the immune and inflammatory responses, but also controls cell growth and protects cell from apoptosis. More recently, the relevance of NF-κB in tumor maintenance, tumor development and possibly even in tumor initiation is becoming more evident. It has also been shown that NF-κB plays an important role in inducing resistance to conventional anti-cancer treatments.

The seventeen articles gathered in this volume explore the routes from NF-κB basic research, cancer research and oncogenomics into the development of NF-κB-based cancer therapeutics and biomarkers.

We hope you will enjoy reading this Special Issue and learn more about the exciting era that we are experiencing in the field of NF-κB and its application in cancer research.

The editor and all participants look forward to meeting the NF-κB community in Grasse in October 2018 for the 4th edition of the European NF-κB subunit workshop and discussing further developments in this rapidly moving field.

Véronique Baud
Special Issue Editor

Review

Unlocking the NF-κB Conundrum: Embracing Complexity to Achieve Specificity

Federica Begalli, Jason Bennett, Daria Capece, Daniela Verzella, Daniel D'Andrea, Laura Tornatore and Guido Franzoso *

Centre for Cell Signalling and Inflammation, Department of Medicine, Imperial College London, London W12 0NN, UK; f.begalli@imperial.ac.uk (F.B.); j.bennett@imperial.ac.uk (J.B.); d.capece@imperial.ac.uk (D.C.); d.verzella@imperial.ac.uk (D.V.); d.dandrea@imperial.ac.uk (D.D.); l.tornatore@imperial.ac.uk (L.T.)
* Correspondence: g.franzoso@imperial.ac.uk; Tel.: +44-(0)20-3313-8421

Received: 30 June 2017; Accepted: 10 August 2017; Published: 22 August 2017

Abstract: Transcription factors of the nuclear factor κB (NF-κB) family are central coordinating regulators of the host defence responses to stress, injury and infection. Aberrant NF-κB activation also contributes to the pathogenesis of some of the most common current threats to global human health, including chronic inflammatory diseases, autoimmune disorders, diabetes, vascular diseases and the majority of cancers. Accordingly, the NF-κB pathway is widely considered an attractive therapeutic target in a broad range of malignant and non-malignant diseases. Yet, despite the aggressive efforts by the pharmaceutical industry to develop a specific NF-κB inhibitor, none has been clinically approved, due to the dose-limiting toxicities associated with the global suppression of NF-κB. In this review, we summarise the main strategies historically adopted to therapeutically target the NF-κB pathway with an emphasis on oncology, and some of the emerging strategies and newer agents being developed to pharmacologically inhibit this pathway.

Keywords: nuclear factor κB; NF-κB inhibitors; cancer; IκB kinase; Gadd45β; ubiquitin

1. Introduction

Just over 30 years ago, Ranjan Sen and David Baltimore discovered a protein complex that bound to a conserved immunoglobulin regulatory DNA sequence in activated B lymphocytes and that they referred to as "Nuclear Factor binding to the κ-Light-chain-enhancer B site" or, in short, nuclear factor κB (NF-κB) [1]. Decades of research into this factor have subsequently elucidated the multiple mechanisms of action of the signalling pathways NF-κB embodies, its ubiquitous presence in tissues and the pivotal roles it plays in governing a myriad of physiological processes ranging from the regulation of the immune system and inflammatory response to the regulation of autophagy, senescence, cell survival, cell proliferation, and cell differentiation [2–5].

Considering the multitude of stimuli that are capable of inducing NF-κB and the vast spectrum of functions NF-κB plays in normal tissues, it is perhaps unsurprising that these key features of the NF-κB pathway are frequently hijacked by cancer [6]. Indeed, aberrant NF-κB activation is a hallmark of most human neoplasias, where it drives oncogenesis, disease recurrence and therapy resistance, largely by inducing a transcriptional programme that suppresses apoptosis of cancer cells and orchestrates the inflammatory reaction in the tumour microenvironment (TME) [7]. There is also evidence that NF-κB can promote tumour cell proliferation, local invasion, metastatic dissemination, metabolism reprogramming and the epithelial-mesenchymal transition (EMT), thereby contributing to essentially all hallmarks of cancer [8–10]. In certain cancer types, such as multiple myeloma, diffuse large B-cell lymphoma (DLBCL), Hodgkin's lymphoma and glioblastoma multiforme (GBM), NF-κB is frequently stably activated by recurrent genetic alterations of upstream regulators of its pathway [6,11–20]. More commonly, however, constitutive NF-κB activation in cancer ensues from inflammatory stimuli and other cues emanating from the TME or oncogenic alterations lying outside the traditional NF-κB pathway, such as *RAS* and *PTEN* mutations. These mechanisms are often responsible for inducing stable NF-κB activation in various types of solid malignancy, including colorectal, pancreatic, ovarian and breast carcinoma [11]. Indeed, the wealth of available genetic, biochemical and clinical evidence provides a compelling rationale for therapeutically targeting NF-κB in a wide range of human malignancies. Nevertheless, developing a specific and clinically useful NF-κB inhibitor has proven a seemingly insurmountable problem. Historically, the challenge with conventional NF-κB-targeting strategies has been to achieve contextual, selective inhibition of the NF-κB pathogenetic activities, given the pleiotropic physiological functions and ubiquitous nature of the NF-κB pathway.

This review will examine the main strategies historically adopted to therapeutically target the NF-κB pathway in cancer, illustrating the principal classes of synthetic compounds and natural products that have been developed to inhibit oncogenic NF-κB signalling, and focusing on some of the more promising emerging approaches being developed to overcome the historical limitations of conventional NF-κB-targeting therapeutics. For a more general overview of the NF-κB pathway and its regulation and functions throughout the course of oncogenesis, we refer to the excellent reviews that have already extensively covered these topics [21–29].

2. The Nuclear Factor κB (NF-κB) Pathway

In mammals, NF-κB comprises a family of five proteins, namely RelA/p65, RelB, p50/NF-κB1 (p105), p52/NF-κB2 (p100), and c-Rel, which form virtually all possible combinations of homo- and hetero-dimeric NF-κB complexes [5,30]. The members of this family are characterised by the presence of a highly conserved 300-amino acid N-terminal region known as the Rel-homology domain (RHD), which is responsible for the dimerization, DNA binding and nuclear translocation of NF-κB subunits, as well as their interaction with IκB regulatory proteins [4]. In resting cells, NF-κB complexes are normally held inactive in the cytoplasm by binding to members of the IκB family of proteins, including IκBα, IκBβ and IκBε. These proteins all contain a so-called ankyrin repeat domain (ARD), which interacts with NF-κB dimers and blocks their nuclear import by masking their nuclear localization signals (NLS) [4]. IκB proteins can also prevent nuclear NF-κB complexes from binding to DNA and can shuttle them out of the nucleus by means of their nuclear export signal (NES) [4]. The C-termini of the p105 and p100 precursor proteins also contain IκB-like ankyrin repeats, which must be degraded in order to generate the mature p50 and p52 subunits, respectively [8,28].

NF-κB can be activated from these latent cytoplasmic pools in response to a large variety of stimuli capable of causing the phosphorylation of IκB proteins on conserved serine residues by the IκB kinase (IKK) complex [31,32]. The site-specific IκB phosphorylation by IKK in turn creates a destruction motif, which is recognised by the SKP1-Cullin 1-F-box protein (SCF) E3 ubiquitin-protein ligase complex, $SCF^{\beta TrCP}$, comprising the core subunits, SKP1 and Cullin 1 (CUL1), the RING component, RING-box protein 1 (RBX1; also known as ROC1/HRT1), and the F-box protein, β-transducin repeat-containing protein (βTrCP), in conjunction with a member of the Ubc4/5 family of E2 ubiquitin-conjugating enzymes, leading to the K48-linked polyubiquitination of IκBs at conserved lysine residues and their subsequent proteolytic degradation by the 26S proteasome [4,21,25,26,33–38].

Following the removal of IκBs, the released NF-κB complexes are free to enter the nucleus [4], where they bind to distinctive decameric DNA elements, known as κB sites, and regulate transcription of a diverse array of genes, encoding numerous inflammatory mediators, immunoregulators, apoptosis inhibitors, developmental factors and other genes responsible for moulding the host defence responses to stress, injury and infection [4,5]. Notably, the outcome of NF-κB activation is the induction of diverse and tightly controlled transcriptional programmes, which exhibit a wide degree of tissue- and context-specificity. Precisely how ubiquitous NF-κB complexes achieve this transcriptional diversity in tissues is not completely understood, but appears to hinge at least in part upon the specific composition of the NF-κB dimers being activated, their post-translational modification state, their interaction with other transcription factor pathways, and the specific configuration of the chromatin [5].

2.1. The Canonical NF-κB Pathway

The pathways of NF-κB activation are broadly classified as the canonical or classical pathway and the non-canonical or alternative pathway, depending upon the nature of the IKK complexes and IκB proteins involved. In the canonical NF-κB pathway, exposure to inflammatory cytokines such as tumour necrosis factor-α (TNF-α) and interleukin 1β (IL-1β), antigen or other immune signals, and microbial products such as pathogen-associated molecular patterns (PAMPs), leads to the rapid activation of an IKK complex typically comprising the two homologous catalytic subunits, IKKα and IKKβ, and the regulatory scaffold subunit, IKKγ (also known as NEMO) [32]. Upon engagement of their cognate receptors, these ligands trigger a signalling cascade that culminates with the K63-linked polyubiquitination of IKKγ/NEMO [39], a reaction that, unlike K48-linked ubiquitination, does not target proteins for degradation by the proteasome, but rather causes post-translational modifications, which enable IKKγ/NEMO to interact with other signalling effectors, leading to the downstream activation of IKKβ by site-specific phosphorylation on serine residues, S177 and S181, within the activation loop (Figure 1) [4,21,27,34,38,40,41].

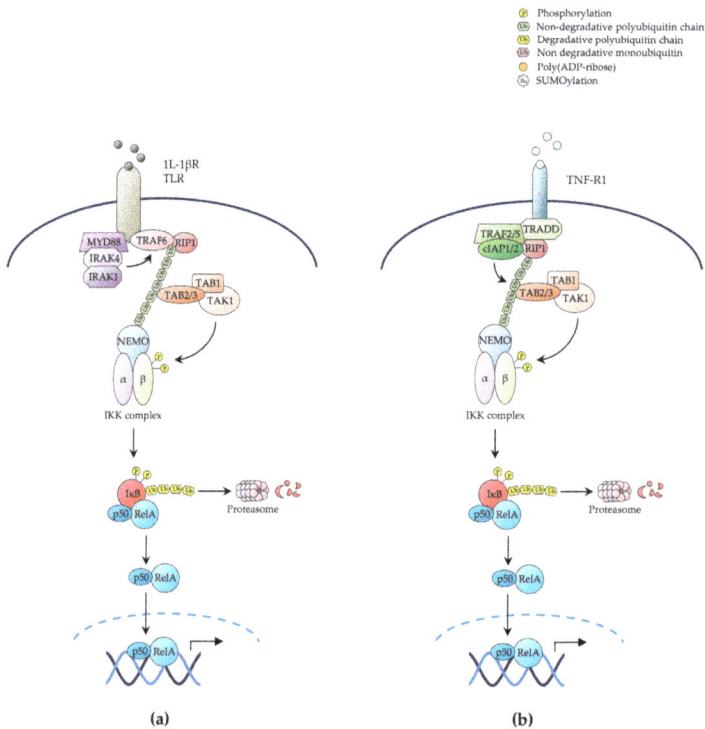

Figure 1. *Cont.*

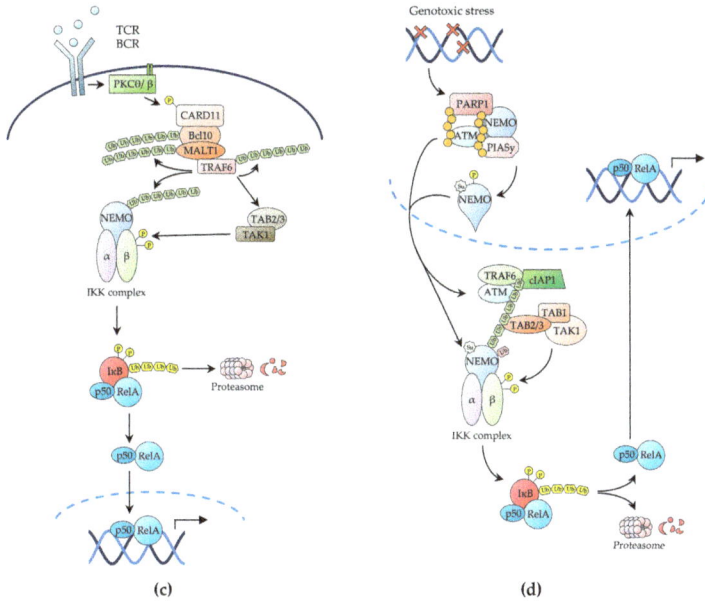

Figure 1. The canonical NF-κB pathway. Schematic representation of the three main upstream pathways of canonical NF-κB activation through: (**a**) IL-1βR and TLR receptors; (**b**) TNF-R1 receptor; (**c**) the antigen receptors, in B and T cells; (**d**) genotoxic stress. As results of the exposure to activating stimuli, NEMO is recruited to the respective proximal signalling complexes, bound to the receptors, through non-degradative polyubiquitin chains. The recruitment of the IKK complex via NEMO then allows the TAK1 kinase to activate IKKβ by the phosphorylation on specific T-loop serine residues, leading to IκBs' phosphorylation and degradation, thereby freeing the active NF-κB dimers. NF-κB, nuclear factor κB; TLR, toll-like receptor; NEMO, NF-κB essential modulator; IKK, IκB kinase; TAK1, Transforming growth factor β-activated kinase 1.

The main physiological function of canonical IKK activation is to enable the IKKβ-mediated phosphorylation of IκBα on S32 and S36, thereby targeting the inhibitor for SCF$^{\beta TrCP}$-dependent K48-linked ubiquitination at conserved lysine residues, K21 and K22, and subsequent proteolysis by the proteasome [8,34]. Biochemical and genetic studies have shown that IKKβ is both necessary and sufficient for IκBα phosphorylation through a mechanism that strictly depends upon IKKγ/NEMO. By contrast, IKKα is not involved in IκBα phosphorylation, but rather contributes to canonical NF-κB-mediated transcriptional responses by directly modulating the activity of NF-κB subunits, histones, transcriptional co-activators/co-repressors, and other non-IκB substrates [21,28]. While both IKKα and IKKβ have the ability to phosphorylate NF-κB subunits, such as RelA and c-Rel, and engage in crosstalk with other signalling pathways, such as the mammalian target of rapamycin (mTOR) and mitogen-activated protein kinase (MAPK) pathways, only IKKα is capable of accumulating in the nucleus and phosphorylating nuclear substrates, such as chromatin components [21,23,42,43].

Conversely, in addition to targeting IκBα, activated IKKβ, but not IKKα, phosphorylates the other canonical IκB proteins, IκBβ and IκBε, which, like IκBα, contain a conserved N-terminal signal-responsive motif, comprising two serine residues serving as target sites for IKKβ-mediated phosphorylation and one or two lysine residues that are targeted for inducible K48-linked ubiquitination by SCF$^{\beta TrCP}$ and Ubc4/5-family E2 ligases [34]. Notably, whilst upon stimulation, IκBα is subject to rapid proteolytic degradation and resynthesis, the signal-induced proteolysis and resynthesis of IκBβ and IκBε occur with characteristically delayed kinetics [34]. Indeed, canonical IκB proteins are also functionally distinct in other important ways. For instance, unlike IκBα, nuclear IκBβ can associate with DNA-bound NF-κB complexes at specific promoter κB sites to prolong transcription of inflammatory genes, such as TNF-α, in response to bacterial lipopolysaccharide (LPS) [44]. Moreover, the expression and function of IκBε appears to be distinctively restricted to the haematopoietic lineage [4]. Irrespective of the mechanisms involved in the regulation and function of different IκBs, the primary outcome of canonical NF-κB activation, which is the predominant form of NF-κB signalling, is the release of RelA- and c-Rel-containing NF-κB dimers, the most ubiquitous and abundant of which is the RelA/p50 heterodimer [4,8,21,25,29,45]. Upon nuclear translocation, these dimers bind to DNA to coordinate the expression of inducible transcriptional programmes responsible for orchestrating immune and inflammatory responses, cell survival programmes and other host defence mechanisms, through a process that is characterised as being rapid and self-limiting [4,8,21,25,29,45].

2.1.1. NF-κB Activation by IL-1β Receptor (IL-1βR) and Pattern Recognition Receptors (PRRs)

Multiple upstream signalling pathways, including those evoked by cell-surface receptors such as IL-1β receptor (IL-1βR), toll-like receptors (TLRs), TNF receptor 1 (TNF-R1) and T-cell receptor (TCR), cytoplasmic receptors such as nucleotide-binding oligomerisation domain (NOD)-like and RIG-I-like receptors, and stress signals such as genotoxic stress, converge on the IKK complex to induce canonical NF-κB activation [21]. Once activated, these receptors and their associated protein complexes coalesce into intracellular signalling networks that utilise adaptor protein interactions, protein phosphorylation, non-degradative ubiquitination and other signal-transducing mechanisms as means to attain the downstream activation of IKK. These signalling processes hinge on the signal-induced proximity and correct positioning of diverse multimeric units, and involve common signalling intermediates, such as TNF receptor-associated factor (TRAF)-family ubiquitin E3 ligases and the IKK kinase, transforming growth factor β-activated kinase 1 (TAK1) (Figure 1a) [21,24,27].

In the case of IL-1βR and TLRs, cognate ligand recognition triggers the recruitment to the receptor of the adaptor protein, myeloid differentiation primary response protein 88 (MYD88), and the protein kinases, interleukin-1 receptor-associated kinase (IRAK)1 and IRAK4, which in turn recruit TRAF6, thereby enabling it to oligomerise and manifest its ubiquitin E3 ligase activity [21,23,24,46]. Activated TRAF6 then catalyses its own K63-polyubiquitination as well as the K63-polyubiquitination of its target substrates, including IKKγ/NEMO and IRAK1, by operating in conjunction with a ubiquitin E2 ligase complex comprising Ubc13 and the Ubc-like protein, Uev1A [21,24,47]. The polyubiquitin chains of IRAK1 in turn form a binding scaffold that interacts, on the one hand, with the IKKγ/NEMO ubiquitin-binding domain (UBD) to recruit the IKK complex and, on the other hand, with the UBD of the TAK1-associated adaptor proteins, TAB2/3, to recruit the TAK1 kinase complex, also comprising TAB1, thereby enabling TAK1 to phosphorylate IKKβ on its T-loop serine residues, S177 and S181, resulting in the activation of IKK (Figure 1a) [34,48,49]. The MAPKKK, MEKK3, has also been shown to serve as an IKK kinase downstream of IL-1βR and TLRs, as well as of TNF-R1 [21,24].

The innate immunity pathways of NF-κB activation initiated by the engagement of intracellular NOD-like receptors (NLRs) and RIG-I-like receptors (RLRs) share an overall similar organisation and common signalling intermediates, such as TRAF proteins, with the NF-κB activation pathways triggered by IL-1βR or TLR stimulation [35]. NLRs are evolutionarily conserved cytoplasmic receptors activated in response to PAMPs, such as intracellular bacterial peptidoglycans, and damage-associated molecular patterns (DAMPs) such as cellular stress products [35]. Upon ligand engagement, NOD1 and NOD2—the best-characterised NLRs—recruit the protein kinase, RIP2 (also known as RICK), which in turn associates with TRAF2, TRAF5 or TRAF6, enabling TRAF proteins to catalyse their own K63-polyubiquitination as well as that of their target proteins, including RIP2 and IKKγ/NEMO, in conjunction with the Ubc13/Uev1A E2 ligase [35]. Polyubiquitinated RIP2 then recruits the TAK1 kinase complex via TAB2/3-mediated interaction, leading to IKKβ activation by TAK1-catalysed phosphorylation.

Likewise, cytoplasmic RIG-I-like receptors (RLRs), comprising a family of RNA helicases (RLHs), serve as sensors of viral RNA produced during viral replication in infected cells [35]. Upon exposure to their ligands, the RLR receptors, retinoic acid-inducible gene-I (RIG-I) and melanoma differentiation-associated gene 5 (MDA5), associate through CARD domain-mediated interaction with the adaptor protein, mitochondrial antiviral signalling protein (MAVS; also known as IPS-1), triggering the formation of detergent-resistant, prion-like MAVS aggregates [35,50]. Upon polymerisation, MAVS then recruits multiple TRAF proteins, including TRAF2, TRAF3 and TRAF6, which catalyse the K63-linked polyubiquitination of their protein substrates, thereby creating docking sites for UBD-containing proteins [35]. The resulting ubiquitin-mediated protein–protein interactions then enable the assembly of multimeric signalling complexes, leading to the downstream activation of IKK, with the release of NF-κB dimers from IκBs, and the TRAF3-dependent activation of the non-canonical IKK kinases, TANK-binding kinase (TBK1; also known as NAK) and IκB kinase ε (IKKε; also known as IKK-i). Both TBK1 and IKKε then phosphorylate the IRF-family transcription factors, IRF3 and IRF7, thereby enabling them to cooperate with NF-κB in triggering antiviral immune responses through the induction of type-1 interferons and other antiviral molecules [35,51,52].

2.1.2. NF-κB Activation by TNF-R1

Accumulating evidence reveals an increasingly greater complexity in the ubiquitin-mediated signalling mechanisms that lead to inducible IKK activation in response to a variety of NF-κB-inducing signals, whereby parallel, sequential, alternative or even hybrid types of ubiquitin linkage and ubiquitin-like proteins are involved in signal propagation (reviewed in [21]. An example of this complexity is provided by the NF-κB-inducing pathway emanating from TNF-R1 (Figure 1b). TNF-α recognition causes this receptor to trimerise, triggering the recruitment of the adaptor proteins, TNF receptor-associated death domain protein (TRADD), and the ubiquitin E3 ligases, TRAF2, TRAF5, cellular inhibitor of apoptosis (c-IAP)1 and c-IAP2, which in turn recruit the protein kinase, receptor-interacting protein 1 (RIP1) [35]. Upon ligand-induced signalling complex assembly at the receptor intracellular tail, c-IAP1 and c-IAP2 catalyse the K63-linked polyubiquitination of RIP1 on K377, operating in conjunction with a ubiquitin E2 ligase complex utilising UbcH5. This polyubiquitination event, but interestingly not the RIP1 kinase activity, is then responsible for TNF-α-induced NF-κB activation, which it mediates by enabling the RIP1 polyubiquitin chains to interact with the TAK1 and IKK kinase complexes through the UBDs of TAB2/3 and IKKγ/NEMO, respectively, leading to activation of IKKβ by TAK1-mediated phosphorylation (Figure 1b) [3,21,25,48,53–59].

Various signalling pathways involving alternative forms of non-degradative ubiquitination and ubiquitin signalling molecules have also been shown to play an important role in NF-κB activation by TNF-R1 (reviewed in [21]). One such ubiquitin-dependent pathway of NF-κB activation involves the linear M1-linked polyubiquitination of RIP1 and IKKγ/NEMO by the linear ubiquitination assembly complex (LUBAC), consisting of HOIP, HOIL-1L and Sharpin, which is recruited to the activated TNF-R1 in a TRADD-, TRAF2- and c-IAP1/2-dependent manner, resulting in IKKβ activation by trans-autophosphorylation [59–62]. More recently, LUBAC-mediated M1-linked ubiquitination has been shown to also contribute to NF-κB activation by other receptors, including IL-1βR and the TNF-R-family receptors, CD40 and transmembrane activator and CAML interactor (TACI; also known as TNFRSF13B) [21]. Additionally, TNF-α was shown to induce both the K63- and K11-linked polyubiquitination of RIP1, suggesting that distinct TNF-α-induced populations of ubiquitinated RIP1 and, possibly, RIP1 molecules with mixed polyubiquitin chains, contribute to regulate NF-κB signalling [21,41]. A similar situation entailing hybrid K63- and M1-linked polyubiquitin chains has also been reported for IRAK1, IRAK4 and MYD88 in the context of IL-1βR or TLR stimulation [21]. Furthermore, in vitro, IKKγ/NEMO has been shown to be a substrate for K6- and K27-linked polyubiquitination by c-IAP1 and tripartite motif protein 23 (TRIM23), an E3 ubiquitin ligase involved in NF-κB activation by RIG-I-like receptors [21,41]. Indeed, since the N-terminal methionine and any of the seven lysine residues of ubiquitin can form ubiquitin linkages, and both ubiquitination and deubiquitination enzymes (further discussed below) have evolved distinct ubiquitin-linkage specificities, it is conceivable that the precise contribution of the remarkable conformational and functional diversity of polyubiquitin chains to the regulation of NF-κB signalling has just begun to be unravelled.

2.1.3. NF-κB Activation by Antigen Receptors

Upon antigen recognition, in the context of cognate MHC molecules, the TCR initiates a signalling cascade that results in the activation of the serine-threonine kinase, protein kinase C (PKC)θ, within lipid raft microdomains of the immunological synapse [35,63,64]. Activated PKCθ then phosphorylates the molecular scaffold, caspase recruitment domain-containing protein 11 (CARD11; also known as CARMA1), triggering a conformational change, which enables CARD11 to recruit the adaptor protein, BCL10, and the paracaspase, mucosa-associated lymphoid tissue lymphoma translocation protein 1 (MALT1), thereby forming the CARD11-BCL10-MALT1 (CBM) signalling complex (Figure 1c) [35,64]. Genetic evidence demonstrates that each component of the CBM complex is essential for IKKγ/NEMO ubiquitination and IKK activation by TCR stimulation [35,64]. Interestingly, multiple chromosomal and genetic abnormalities affecting each constituent of the CBM complex or anyhow resulting in constitutive CBM complex activation have been reported in various types of lymphoma, including DLBCL and mucosa-associated lymphoid tissue (MALT) lymphoma, where they drive NF-κB activation and oncogenesis [16,17,35,64]. Upon CBM complex formation, BCL10 and MALT1 undergo oligomerisation, thereby enabling MALT1 to bind to TRAF6, causing it to also oligomerise with activation of its E3 ligase activity [35]. MALT1-associated TRAF6 then catalyses its self-polyubiquitination as well as the K63-linked polyubiquitination of its target proteins, including MALT1 and IKKγ/NEMO, via a process that involves the E2 ligase Ubc13/Uev1A (Figure 1c) [35]. BCL10 is also inducibly polyubiquitinated in response to TCR stimulation, although it is unclear whether TRAF6 catalyses this reaction, and both the BCL10 and MALT1 K63-polyubiquitin chains then serve as docking sites that contribute to recruit the IKK complex via the IKKγ/NEMO UBD [35,63].

While TAK1 appears to be recruited to the CBM scaffold via polyubiquitinated TRAF6-mediated interaction with the TAB2/3 UBD, TAK1 activation in response to TCR stimulation appears to also require a further, CBM-independent step (Figure 1c) [22]. Recent studies have suggested a role for the serine-threonine kinase, phosphoinoisitide-dependent kinase (PDK)1, and the molecular adaptor, adhesion- and degranulation-promoting adaptor protein (ADAP), in enabling this additional signalling step by promoting the PKCθ-mediated recruitment of CARD11 and TAK1 to the PKCθ signalosome, thereby resulting in TAK1 activation and IKKβ phosphorylation [22]. There is also evidence of a role for caspase-8, an initiator caspase involved in apoptosis and lymphocyte activation, in positive regulation of NF-κB signalling downstream of the TCR, and both caspase-8 and its regulator, cellular FLICE-like inhibitory protein (c-FLIP), have been found in association with MALT1 in TCR-activated T cells [22]. Notwithstanding, the precise role of caspase-8 in TCR-induced NF-κB activation and the molecular details underlying this role remain to be determined [22,63,65,66]. A similar signalling pathway, utilising many of the same signalling intermediates also involved in TCR-induced NF-κB signalling, including the CBM complex, has been shown to mediate NF-κB activation by B-cell receptor (BCR) stimulation in B lymphocytes (reviewed in [63]).

2.1.4. NF-κB Activation by Genotoxic Stress

Genotoxic stress induced by ionising radiation or chemotherapeutic drugs results in canonical NF-κB activation, which enables cells to survive while the DNA damage is being repaired. This pathway of NF-κB activation involves the trafficking of signalling intermediates between the nucleus and the cytoplasm (reviewed in [35,67]). The NF-κB-inducing signal is initiated in the nucleus by the DNA damage sensors, ataxia telangiectasia mutated (ATM) kinase and poly(ADP-ribose)-polymerase-1 (PARP-1), which, upon binding to DNA strand breaks, synthesizes poly(ADP-ribose) (PAR) chains attached to itself and other target proteins [21,29,35,67,68]. Both enzymes are essential for DNA damage-induced canonical NF-κB signalling [21,29,35,67,68]. Following dissociation from damaged DNA sites, PAR-conjugated PARP-1 proteins recruit IKKγ/NEMO (devoid of any IKK catalytic subunit) to the nucleus to assemble a nuclear signalosome comprising activated ATM, IKKγ/NEMO, and the E3 small ubiquitin-like modifier (SUMO)-protein ligase, protein inhibitor of activated STAT protein y (PIASy) [21,35,67]. PARP-1 signalosome assembly then enables the PIASy-mediated IKKγ/NEMO SUMOylation on K277 and K309 via a process that requires the E2 ligase, Ubc9, the PAR-mediated PIASy modification, and an involvement of the death domain-containing proteins, p53-induced death domain protein (PIDD)1 and RIP1 [35,67]. Importantly, PARP-1 signalosome assembly also enables the ATM-dependent phosphorylation of IKKγ/NEMO on S85 [35,67].

In addition to mediating this essential nuclear function, DNA damage-activated ATM is exported from the nucleus to the cytoplasm, where it binds to TRAF6 via its TRAF-interaction motif. This binding then activates the TRAF6 E3 ubiquitin ligase activity, leading to TRAF6-mediated self-polyubiquitination, in a reaction catalysed in conjunction with the E2 ubiquitin-conjugating enzyme, Ubc13 [35,41,67]. The newly formed TRAF6 polyubiquitin chains then trigger the recruitment of c-IAP1, resulting in the assembly of the ATM-TRAF6-c-IAP1 signalling complex, which in turn recruits the TAK1 and IKK kinase complexes via TAB2- and IKKγ/NEMO-mediated interactions, respectively, resulting in TAK1 activation by trans-autophosphorylation [69]. However, in addition to these cytoplasmic signalling events, efficient IKKβ phosphorylation requires the nuclear export of SUMOylated IKKγ/NEMO, generated in the PARP-1 signalosome [67,70]. Upon integration in the cytoplasmic IKK kinase complex, SUMOylated IKKγ/NEMO is monoubiquitinated on K285 by the c-IAP1 ubiquitin E3 ligase bound to the ATM-TRAF6 signalling complex, leading to TAK1-mediated IKKβ phosphorylation and NF-κB activation [69,71]. Therefore, IKKγ/NEMO monoubiquitination appears to integrate the nuclear signalling pathway driven by the PARP-1 signalosome, with the cytoplasmic ATM-driven pathway induced in response to genotoxic stress [69]. Precisely how IKKγ/NEMO monoubiquitination enables IKKβ phosphorylation is unclear. However, it is possible that it induces a conformational change in the IKK complex that makes IKKβ more accessible to TAK1 [34,35,72,73].

Although canonical NF-κB activation by ionising radiation and chemotherapy is transient, it has nonetheless been shown to contribute to both radio- and chemo-resistance in cancer [74]. This ability of NF-κB to promote tumour-cell survival following treatment with DNA-damaging agents has been extensively demonstrated in cell lines and primary tissues, both in vitro and in vivo, in a wide range of tumour types, including breast carcinoma, squamous cell carcinoma and thyroid carcinoma [75–78]. Likewise, most, if not all, genotoxic therapeutic agents, including paclitaxel, vinblastine, vincristine, doxorubicin, 5-fluorouracil, cisplatin and tamoxifen, have been shown to induce effective NF-κB activation [75,77]. The products of various NF-κB target genes, most notably *cyclin D1*, *BCL-2*, *Bcl-X$_L$*, *survivin* and *XIAP*, have been implicated, each in diverse oncological contexts, as mediators of NF-κB-dependent tumour-cell survival leading to radiotherapy and/or chemotherapy resistance [79,80]. Therefore, the therapeutic inhibition of the NF-κB pathway has also been pursued in order to enhance the clinical effects of radio- and chemo-therapy in cancer (further discussed below).

2.2. The Non-Canonical NF-κB Pathway

The non-canonical or alternative NF-κB pathway is induced by a distinct group of TNF-family ligands, including lymphotoxin β (LTβ), CD40 ligand (CD40L), B-cell activating factor (BAFF), receptor activator of NF-κB ligand (RANKL), TNF-related weak inducer of apoptosis (TWEAK) and tumor necrosis factor superfamily member 14 (TNFSF14; also known as LIGHT) [4,45], and governs developmental and immune processes, such as secondary lymphoid organogenesis, B-cell survival and maturation, bone morphogenesis and dendritic-cell activation [81–89]. In contrast to canonical NF-κB signalling, which is subject to rapid and transient activation, the non-canonical NF-κB pathway is activated with distinctively slower kinetics. Additionally, it does not require IKKβ or IKKγ/NEMO, nor does it involve the proteolysis of canonical IκBs. Instead, it exclusively relies on IKKα and the signal-induced processing of p100, which releases RelB/p52 heterodimers (Figure 2) [4,21,25,34].

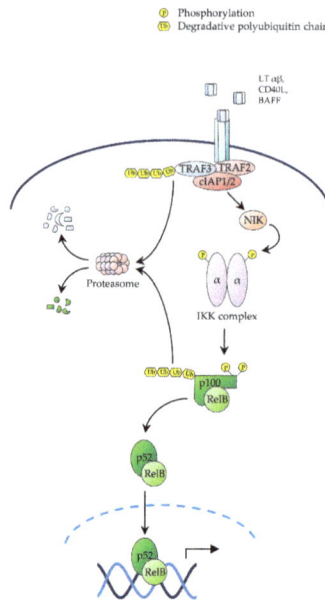

Figure 2. The non-canonical NF-κB pathway. Schematic representation of the non-canonical pathway of NF-κB activation. Activation of this pathway depends on the signal-induced stabilisation of NIK and subsequent NIK-mediated phosphorylation of IKKα on T-loop serine residues. In the absence of receptor stimulation, NIK is constitutively degraded via the ubiquitin-proteasome pathway controlled by the activity of the E3 ubiquitin ligase, c-IAP1/2. Upon ligand-mediated receptor engagement, TRAF3 is degraded with subsequent stabilisation of NIK, which in turn phosphorylates IKKα, leading to C-terminal ubiquitination and proteasome-mediated processing p100. NIK, NF-κB-inducing kinase.

Receptor engagement by ligands activating non-canonical NF-κB signalling triggers a signalling cascade that results in the stabilisation of NF-κB-inducing kinase (NIK), which in turn, upon accumulation in cells, phosphorylates IKKα on T-loop serine residues, S176 and S180, leading to IKKα activation [4,34]. Activated IKKα then phosphorylates p100 on C-terminal serine residues, S866 and S870, thereby creating a docking site for SCF$^{\beta\,\text{TrCP}}$, similar to the destruction motif generated by signal-induced phosphorylation of IκBs [25,34]. Upon recruitment to p100, SCF$^{\beta\,\text{TrCP}}$ catalyses the K48-linked polyubiquitination of p100 on lysine residue, K856, in conjunction with a Ubc4/5 ubiquitin E2 ligase, thereby targeting the C-terminal, IκB-like domain of the precursor for partial proteolysis by the proteasome [25,34]. A proteasomal stop signal, comprising a glycine rich region (GRR) present between the RHD and ARD domains of p100, protects the p100 N-terminal domain from proteolysis, thereby enabling the generation of the mature p52 subunit, which together with RelB forms RelB/p52 heterodimers, which translocate to the nucleus to regulate the transcriptional programme governed by non-canonical NF-κB signalling (Figure 2) [4,45,90].

In unstimulated cells, NIK is constitutively bound to a complex comprising TRAF2, TRAF3 and c-IAP1 or c-IAP2, whereby TRAF3 serves as an adaptor interacting directly with both NIK and TRAF2, which in turn binds to c-IAP1/2, enabling it to catalyse the NIK K48-linked polyubiquitination, resulting in constitutive NIK degradation by the proteasome [4,21,25,27]. Upon receptor engagement, the NIK-associated TRAF2/TRAF3/c-IAP1/2 complex is recruited to the receptor via TRAF3-mediated interaction, enabling TRAF2 to catalyse the non-degradative K63-polyubiquitination of c-IAP1/2. This redirects the c-IAP1/2 E3 ligase activity from NIK to TRAF3, causing the c-IAP1/2-mediated K48-polyubiquitination of TRAF3 [4,21,25,27]. The resulting proteosomal degradation of TRAF3 then destabilizes the TRAF2/TRAF3/c-IAP1/2 complex, leading to NIK dissociation from c-IAP1/2, with consequent NIK stabilization and accumulation of newly synthesized NIK [4,21,25,27,91]. As the NIK kinase domain adopts an intrinsically active conformation, accumulated NIK does not require any additional phosphorylation event for activation, and consequently binds to and phosphorylates IKKα, leading to p100 processing and NF-κB activation (Figure 2) [4,21,25,27]. Interestingly, a number of genetic alterations promoting NIK stabilization and p100 processing, such as *c-IAP1/2* and *TRAF3* deletions and *NIK* amplifications, have been reported in multiple myeloma, where they drive constitutive NF-κB activation and multiple myeloma cell survival [21,24,26,28,34,68,92].

In addition to their roles as NF-κB precursors, both p100 and p105 can serve as IκB inhibitory proteins by means of their C-terminal IκB-like ARD domains [4,8,25,33]. Degradative ubiquitination of the precursors may therefore serve a dual purpose: On the one hand, it may promote their partial proteolysis to produce active NF-κB subunits, which can form homodimers or heterodimeric complexes with other NF-κB-family proteins; on the other hand, it may result in their complete degradation, thereby liberating sequestered NF-κB dimers from their interaction with inhibitory C-terminal IκB-like domains [4,8,25,33]. Notably, in contrast to the signal-induced processing of p100, which is tightly regulated by the non-canonical NF-κB pathway via IKKα-mediated p100 phosphorylation, with minimal basal processing in the absence of stimulation, the processing of p105 to p50 is largely a constitutive process that may occur either co- or post-translationally [25]. Notwithstanding, in addition to being subject to signal-induced partial proteolysis, controlled by $SCF^{\beta TrCP}$-dependent ubiquitination, p100 is constitutively targeted for complete degradation, which facilitates efficient, stimulus-induced non-canonical NF-κB activation, via a process that depends on the glycogen synthase kinase 3 (GSK3)-mediated phosphorylation of p100 at C-terminal serine residue, S707, and subsequent ubiquitination catalysed by an SCF complex containing the F-box protein, FBXW7α [24,25,33].

2.3. Termination of NF-κB Signalling

The feedback control and timely termination of the NF-κB response are essential to ensure the restoration of homeostasis and prevent excessive inflammation, tissue damage and the onset of neoplasias. To maintain transience and ensure the prompt cessation of NF-κB signalling, multiple negative feedback mechanisms have evolved to control the NF-κB pathway at various levels. The first discovered and best characterised attenuation mechanism consists in the resynthesis of IκB proteins following NF-κB activation [4,25,27,28]. All *IκB* genes, including those encoding canonical IκBs and NF-κB precursors, contain κB sites in their promoters [72]. An essential negative feedback mechanism is mediated by IκBα, which upon its resynthesis, helps to terminate the NF-κB response by virtue of its ability to enter the nucleus, dissociate NF-κB complexes from DNA, and export them into the cytoplasm [25,28,72].

A number of additional downstream inhibitory mechanisms are mediated by various types of post-translational modification of NF-κB subunits, including site-specific phosphorylation, acetylation and ubiquitination, which attenuate the NF-κB response by directly affecting the protein stability, DNA-binding affinity and/or transcriptional activity of nuclear NF-κB dimers and/or their interactions with transcriptional cofactors [25,27,28]. For instance, in LPS-stimulated macrophages, nuclear IKKα has been shown to phosphorylate the C-terminal domains of RelA and c-Rel within DNA-bound NF-κB complexes, thereby accelerating NF-κB protein turnover with consequent downregulation of inflammatory gene expression [93]. Nuclear RelA has also been reported to undergo proteosomal degradation, which contributes to the termination of NF-κB-dependent transcription, through a mechanism mediated by the ubiquitin E3 ligases, suppressor of cytokine signalling-1 (SOCS-1) and PDZ and LIM domain protein 2 (PDLIM2) [25,27,28]. Several other mechanisms participate in the cessation of NF-κB signalling, including RelA acetylation, which diminishes the DNA-binding affinity of RelA-containing complexes and affect their interaction with both histone acetyltransferases (HATs) and histone deacetylases (HDACs) [28]. Likewise, site-specific RelA dephosphorylation by wild-type p53-induced phosphatase 1 (WIP1) has been shown to weaken the RelA interaction with the transcriptional coactivator, p300, resulting in an attenuation of RelA-dependent gene transcription [29]. Moreover, the oxidation of redox-sensitive cysteine residues in the DNA-binding domains of NF-κB subunits has been suggested to dampen the NF-κB response [28,34].

In addition to these downstream processes, multiple feedback mechanisms regulate NF-κB signalling components operating either at the level or upstream of the IKK complex. One such mechanism relies on the intrinsic self-limiting capacity of the IKK complex, dependent upon the IKKβ-mediated phosphorylation of IKKγ/NEMO on serine residue, S68, and the IKKβ auto-phosphorylation on C-terminal serine residues, resulting in a disruption of essential structural motifs and interactions of IKK subunits [21]. IKKβ further catalyses the BCL10 phosphorylation, which has been shown to downregulate TCR-induced NF-κB activation [21]. Additionally, a number of phosphatases, including protein phosphatase (PP)2A and PP2C, contribute to inhibit NF-κB activation by dephosphorylating T-loop serines in the catalytic IKK subunits [21,23].

Notably, accumulating evidence demonstrates the importance of negative feedback mechanisms affecting the ubiquin system. Recent studies have shown that the NEMO-like adaptor, optineurin, plays an important role in the negative regulation of TNF-α-induced NF-κB activation by competing with NEMO for binding to the polyubiquitin chains of several IKKγ/NEMO-interacting proteins [21,27]. Additionally, several deubiquitination enzymes (DUBs), including A20 (also known as tumor necrosis factor α-induced protein 3; TNFAIP3) and the tumour suppressor, CYLD (also known as cylindromatosis), implicated in familial cylindromas, have been shown to control critical signalling steps upstream of the IKK complex [4,5,35]. Both A20 and CYLD are direct transcriptional targets of NF-κB and, as such, are rapidly induced upon NF-κB activation, thereby providing essential feedback mechanisms promoting the cessation of the NF-κB response. A20 and CYLD both mediate this function, at least in part, by hydrolysing the K63-polyubiquitin chains of a number of signalling molecules involved in IKK activation, including RIP1/2, TRAF proteins, NOD2, MALT1 and IKKγ/NEMO itself [4,21,35,65]. Interestingly, A20 may further limit NF-κB signalling by virtue of its ubiquitin-editing function, whereby upon removing the RIP1 K63-polyubiquitin chains via the DUB activity mediated by in its N-terminal domain, it can catalyse the RIP1 K48-linked polyubiquitination via the ubiquitin E3 ligase activity in its C-terminal domain, thereby targeting RIP1 for proteosomal degradation [5,35]. Several other DUBs have been implicated as negative regulators of canonical NF-κB signalling. These include OTU domain-containing protein 7B (OTUD7B; also known as cezanne), ubiquitin-specific peptidase 11 (USP11), USP15 and USP21, each exhibiting distinct ubiquitin-linkage specificity [41]. Interestingly, some of the more recently identified DUBs antagonising NF-κB activation have been shown to target M1-type ubiquitin linkages. For instance, otulin (also known FAM105B) has been reported to HOIP-mediated interaction and specifically remove M1 linear polyubiquitin chains [5,21]. Moreover, in addition to cleaving K63 ubiquitin linkages, CYLD has been shown to degrade M1-linked polyubiquitin chains from various components of the TNF-R1 and NOD2 signalling complexes, including RIP1. Curiously, A20 also has the ability to bind to M1 polyubiquitin chains, but appears to have the opposite effect on ubiquitin chain stability, by preventing their removal [41].

An additional mechanism of NF-κB inhibition that is being viewed with increasing interest is mediated by the tumour suppressor, WW domain-containing oxidoreductase (*WWOX*), which is frequently inactivated by gene mutation, deletion or chromosomal translocation in multiple types of haematological and solid cancer [94–102]. WWOX has been shown to negatively regulate canonical NF-κB signalling by means of its ability to directly bind to IκBα and, thereby, impede proteasome-mediated IκBα proteolysis [94,97,98]. This inhibitory activity of WWOX has additionally been implicated in the suppression of canonical NF-κB activation by the HTLV-1 protein, Tax, in adult T-cell leukaemia (ATL) [94]. Therefore, genetic WWOX inactivation in cancer contributes to constitutive NF-κB activation. Interestingly, in addition to the aforementioned genetic mechanisms of inhibition, *WWOX* is negatively regulated by non-canonical NF-κB signalling and, therefore, appears to mediate a mechanism of cross-amplification of canonical NF-κB activity by the non-canonical NF-κB pathway [94].

Far fewer negative feedback mechanisms have been reported for the non-canonical NF-κB pathway. An important checkpoint in this respect is mediated by TRAF3. In addition to being transcriptionally upregulated by NF-κB to promote NIK degradation, leading to p100 stabilization and attenuation of non-canonical NF-κB signalling, TRAF3 is subject to post-translational stabilisation by means of OTUD7B, which targets TRAF3 for deubiquitination of its K48-linked polyubiquitin chains, thereby preventing NIK accumulation and the processing of p100 [21,103]. IKKα-mediated NIK phosphorylation has also been reported to accelerate NIK turnover, thereby contributing to the feedback regulation of non-canonical NF-κB signalling [41]. Additionally, microRNA-146a may negatively regulate both canonical and non-canonical NF-κB activation by downmodulating the expression of IRAK1, TRAF6 and RelB [104].

3. Therapeutic Targeting of the NF-κB Pathway in Cancer

The central role that NF-κB plays in a vast range of human malignant and non-malignant pathologies has catalysed an intensive effort by the pharmaceutical industry and academic laboratories over the past two and a half decades to develop a specific NF-κB inhibitor for clinical indication in these diseases [2,3,11,105]. In 2006, a survey had already counted no fewer than 750 candidate therapeutics designed to target the NF-κB pathway, and this number is certain to have grown considerably over the past decade [3]. These compounds comprise a disparate variety of chemical classes, including peptidomimetics, small molecules, small interfering RNAs and microbial products and their derivatives. Many of them function as general inhibitors of NF-κB signalling, while for many others the mode of action is poorly understood [3,21,105]. Yet, notwithstanding the tremendous progress made in recent years towards unravelling the intricate signalling networks governing the NF-κB pathway and its functions, and the intensive efforts and mighty investments committed to develop a specific, clinically useful NF-κB inhibitor, the output of pharmacological interventions has been disappointingly scant, with a continuing dismaying absence of a specific NF-κB or IKK inhibitor in the clinical anticancer armamentarium. The insurmountable challenge, having precluded the clinical success of these NF-κB-targeting approaches, has been to achieve contextual selective inhibition of the NF-κB pathogenetic activity, while preserving the essential physiological functions of NF-κB. By contrast, the efforts so far have resulted in drug candidates often causing the global suppression of NF-κB and its pleiotropic and ubiquitous functions, leading to severe dose-limiting toxicities.

This review examines the main approaches utilised to therapeutically target NF-κB, with a focus on oncology, illustrating for each approach the underlying rationale and most representative classes of compounds, while emphasising emerging strategies and some of the most promising future directions. For ease of discussion, we have broadly classified NF-κB-targeting therapeutics on the basis of their mode of action, depending upon the level of the NF-κB signalling pathway at which they operate (Figure 3). Given the breadth of the basic and translational research in this area, it has not been possible to cover all relevant aspects of the preclinical and clinical pharmacology of the vast number of molecules generated to inhibit NF-κB. For further information on these molecules and their effects, we therefore refer to the excellent reviews that have previously covered these topics [2,3,105–107].

The canonical pathway   The non-canonical pathway

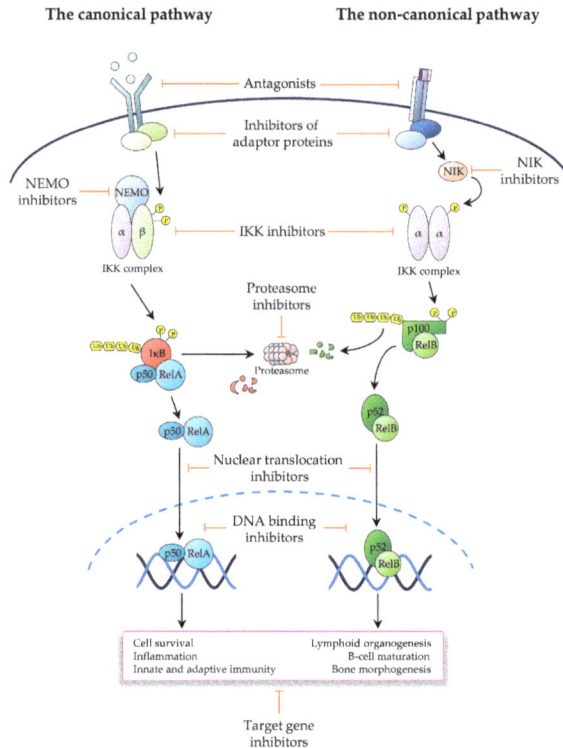

Figure 3. Schematic representation of the main strategies utilised to therapeutically inhibit the NF-κB signalling pathway. Depicted are the canonical (**left**) and non-canonical (**right**) pathways of NF-κB activation. Also depicted are the main therapeutic strategies utilised to inhibit these NF-κB signalling pathways in the oncological context.

3.1. Inhibitors Operating Upstream of the IKK Complex

NF-κB can be activated by a multitude of stimuli, which initiate distinct signalling pathways that converge on the IKK complex (Figures 1 and 2). From a clinical standpoint, this paradigm provides a significant opportunity for therapeutic interventions aimed at interfering with pathogenic NF-κB activation (Figure 3). An important area of drug discovery in this context has been the TNF-R superfamily, which governs diverse physiological processes, including inflammation, cell survival and lymphoid organogenesis, and drives the pathogenesis of chronic inflammatory diseases as well as multiple cancer types [108,109]. TNF-Rs comprise a family of 29 structurally-related receptors, which are bound by 19 ligands of the TNF superfamily [108]. Due to the important roles that TNF-Rs play in widespread human pathologies, there has been a great deal of interest over the past few decades in developing therapeutics targeting these receptors or their ligands [110]. The first such a therapeutic has been infliximab, a TNF-α-specific neutralising antibody, which was approved in 1998 for the treatment of Crohn's disease, followed by the approval in the same year of etanercept [108]. Both drugs are currently in clinical use for the treatment of rheumatoid arthritis, psoriasis, psoriatic arthritis, and other forms of chronic arthritis, and infliximab is also in clinical use for Chron's disease and ulcerative colitis [108]. However, patients treated with these drugs often experience significant side effects, including fevers, chills, nausea, shortness of breath, tachycardia and hypotension. Indeed, the dose-limiting toxicities and immunosuppressive activities of TNF-R signalling inhibitors have precluded the broader clinical development of these agents beyond chronic inflammatory diseases, in particular in the area of oncology [109]. Notwithstanding, a few molecules interfering with TNF-R signalling have found indication in niche areas of oncology. For instance, the human TNF-α analogues, tasonermin (Beromun), has been clinically approved as an adjunct therapy to surgery for sarcoma to prevent or delay amputation and to treat unresectable soft-tissue sarcoma of the limbs [111]. Likewise, brentuximab (Vedotin, Adcetris), a toxin-conjugated chimeric antibody targeting the TNF-R-family receptor CD30 (also known as TNFRSF8), is approved for the treatment of Hodgkin's lymphoma and anaplastic large-cell lymphoma (ALCL) (Table 1), two cancer types that express particularly high surface levels of CD30 [112].

Table 1. Selection of inhibitors of the NF-κB signalling pathway that are either in clinical development or have been clinically approved.

Compound	Molecular Target	Cancer Type	Ongoing Clinical Trials
Upstream IKKs complex			
Brentuximab (Vedotin)	CD30	HL, Anaplastic large cell lymphoma, etc.	NCT01657331, NCT02462538, NCT01807598, NCT02939014, NCT03007030, NCT02169505, NCT01900496, etc.
Ibrutinib (PCI-32765)	BTK	MCL, CLL, WM, DLBCL, FL, MM, and NSCLC, etc.	NCT02801578, NTC0275689, NCT02943473, NCT02321540, NCT02558816, NCT02420912, NCT02315768, NCT02451111, NCT02356458, etc.
IMO-8400	TLR 7, 8, and 9	WM, DLBCL	NCT02252146
LCL-161	cIAPs	Ovarian cancer, MM	NCT02649673, NCT02890069, NCT01955434
Birinapant (TL32711)	cIAPs	Solid tumours and high grade serous carcinomas	NCT02587962, NCT02756130
Ubiquitin proteasome pathway			
Bortezomib	Proteasome	AML, lymphoma, MDS, neuroblastoma, ALL, etc.	NCT02308280, NCT02535806, NCT01736943, NCT01534260, NCT02613598, NCT02356458, NCT01241708, NCT03016988, NCT02139397, NCT02237261, etc.
Carfizomib	Proteasome	MM, neuroendocrine cancer, NHL, DLBCL, MCL, FL, peripheral T-cell lymphoma, HL, T-cell NHL, solid tumours, leukaemia, etc.	NCT02302495, NCT02572492, NCT02318784, NCT02142530, NCT02867618, NCT01738594, NCT02512926, etc.
Ixazomib (MNL-9708)	Proteasome	Glioblastoma, MM, lymphoma, amyloidosis, solid tumours, B-cell lymphoma, lymphoma, etc.	NCT02630030, NCT02924272, NCT02942095, NCT02312258, NCT02477215, NCT02898259, etc.
MLN4924 (Pevonedistat)	NEDD8	AML, solid tumours, chronic myelomonocytic leukaemia, MDS	NCT01814826, NCT02782468, NCT02610777, NCT03009240, NCT03057366
NF-κB target genes			
DTP3	Gadd45β/MKK7	MM	MR/L005069/1
ABT-199	BCL-2	CLL, WM, MCL, AML, NHL, DLBCL, FL, MM, MDS, etc.	NCT02677324, NCT02471391, NCT02558816, NCT02203773, NCT02055820, NCT03136497, NCT03128879, NCT02427451, etc.

In B-cell lymphoma and leukaemia, the NF-κB pathway can be inhibited by therapeutic agents targeting proximal signalling events downstream of the BCR. Upon antigen engagement, the BCR triggers a signalling cascade that involves receptor-associated CD79A/B heterodimers, SRC-family protein tyrosine kinases and Burton tyrosine kinase (BTK), which, as in the case of TCR-induced signalling, leads to the downstream the assembly of a PKC signalosome (mediated by PKCβ, rather than PKCθ, as in T cells) and the CBM signalling complex (Figure 1c) [65]. BTK is an essential signalling intermediate in the BCR-induced pathway leading to NF-κB activation and B-cell survival [113]. Notably, while BTK is expressed in virtually all cells of the haematopoietic lineage, except for T cells and plasma cells, its functions in NF-κB activation and cell survival appear to be dispensable outside of B-cell lineage [114]. Consequently, BTK has been developed into an effective therapeutic target upstream of IKK in various types of B-cell malignancy, including chronic lymphocytic leukaemia (CLL), mantle-cell lymphoma (MCL), follicular lymphoma (FL), DLBCL and acute lymphoblastic leukaemia (ALL) [113]. The first-in-class oral BTK inhibitor, ibrutinib (PCI-32765), has demonstrated impressive clinical responses in clinical trials as a single agent and has been subsequently approved by the FDA for the treatment of refractory MCL (November 2013), CLL (February 2014) and Waldenström's macroglobulinemia (WM; January 2015) [115]. Ibrutinib irreversibly binds to cysteine residue, C481, in the active site of BTK, thereby inhibiting BTK phosphorylation on T223 and resulting in loss of BTK function, NF-κB inhibition and induction of tumour-cell apoptosis [114]. Owing to BTK's restricted pattern of expression, ibrutinib is generally relatively well tolerated, causing for the most part only transient adverse effects, such as diarrhoea, nausea, vomiting, hypertension, urinary and upper respiratory tract infections, fatigue, arthralgia, pyrexia, and peripheral oedema [116]. However, secondary resistance is almost inevitable, and tumours with oncogenic NF-κB-pathway alterations affecting signalling events downstream of BTK, such *CARD11* mutations and *A20* mutations and deletions, are naturally refractory to this agent, thereby limiting its clinical utility, especially in patients with aggressive NF-κB-pathway mutated lymphomas (Table 1) [113].

The gene encoding the adaptor protein, MYD88, which mediates NF-κB and MAPK activation downstream of all TLRs, with the exception of TLR3 [117], is recurrently mutated in haematological malignancies, such as DLBCL, WM and CLL, where it induces constitutive NF-κB and STAT3 activation, thereby promoting cancer-cell survival and oncogenesis [117,118]. These findings provide a strong rationale for therapeutically targeting TLR signalling in certain types of lymphoma and leukaemia. Congruently, preclinical studies have demonstrated that the antisense oligonucleotide TLR inhibitor, IMO-8400, which specifically targets TLR7, TLR8 and TLR9, is effective in diminishing the growth of WM and DLBCL xenografts, driven by gain-of-function MYD88 mutations [118]. A phase I/II trial of IMO-8400 is ongoing in patients in WM and DLBCL (Table 1), and second generation TLR 7/TLR 8/TLR9 inhibitors are currently in development. Another strategy aimed at therapeutically targeting TLR signalling in cancer is directed at the downstream protein kinases, IRAK1/4. This strategy has demonstrated preclinical efficacy in tumours harbouring MYD88 mutations, and preclinical studies of IRAK1/4 small-molecule inhibitors have reported encouraging preliminary results in melanoma [119], myelodysplastic syndrome (MDS) [120] and T-cell acute lymphoblastic leukaemia (T-ALL) [121,122].

The non-canonical NF-κB pathway plays an important role in the pathogenesis of multiple myeloma, where it is constitutively activated by recurrent genetic alterations, including *NIK* amplifications and *c-IAP1/2* and *TRAF3* deletions [10,12,14]. Recent studies suggest an additional role for non-canonical NF-κB signalling in other types of malignancy, such as DLBCL [45]. Within this signalling pathway, NIK is an especially attractive target, owing to its central role in controlling IKKα phosphorylation and p100 processing (Figure 3). Two NIK small-molecule inhibitors, which were recently developed by Amgen, AM-0216 and AM-0561 [123], have demonstrated significant therapeutic activity in multiple myeloma cells, in vitro [124–126]. However, further studies are required to evaluate their potential therapeutic efficacy, in vivo.

c-IAP proteins play an important role in tipping the balance between canonical NF-κB activation and inhibition of non-canonical signalling. Upon TNF-R1 stimulation, receptor-associated c-IAP proteins contribute to recruit the LUBAC complex by ubiquitinating various signalling intermediates, resulting in NF-κB activation [127]. By contrast, in the non-canonical pathway, c-IAP-mediated ubiquitination reactions result in constitutive NIK degradation, thereby dampening non-canonical NF-κB activation [45]. Underscoring these opposing functions of c-IAPs in canonical and non-canonical NF-κB activation, *c-IAP* genes are subject to both amplification and deletion in human cancer [127]. Consequently, the aim of the therapeutic interventions targeting c-IAP proteins in cancer is to inhibit canonical NF-κB activation, without interfering with non-canonical NF-κB signalling. The endogenous c-IAP antagonist, second mitochondria-derived activator of caspases (SMAC), provides an attractive target for achieving this aim. In response to apoptosis-inducing stimuli, SMAC is released from mitochondria into the cytoplasm, where it binds to the conserved c-IAP domain, baculovirus IAP repeat (BIR), via its N-terminal AVPI tetrapeptide, thereby neutralising the prosurvival activity of c-IAPs [128]. This paradigm has served as a reference point for the development of drug mimetics that act as selective c-IAP inhibitors. At least two such compounds, i.e., LCL161 and birinapant (TL32711), have now entered clinical trials in patients with solid cancers, such as ovarian serous carcinoma (Table 1) [129]. Although these compounds have been shown to promote cancer-cell death by stimulating an increase in cytokine production, they have also been reported to have severe dose-limiting toxicities, in particular the onset of cytokine-release syndrome, which significantly restrict their clinical application [130]. Other common, but less severe, side effects include vomiting, nausea, fatigue, and anorexia [131].

Therefore, although targeting upstream NF-κB signalling components has yielded tangible clinical results in the treatment of certain cancers and represents an attractive therapeutic strategy from the standpoint of achieving a degree of tissue- and context-specificity, this approach has so far been limited by the onset of dose-limiting adverse effects, inherent cancer recalcitrance and/or an early onset of secondary drug resistance [109,113,130]. Nevertheless, as the clinical efforts in this area are relatively new and the underlying research is constantly advancing the understanding the intricate upstream signalling networks leading to IKK activation, targeting these networks as a means of therapeutic intervention holds promise for the future development of safe and effective anticancer therapeutics. While strictly speaking, most therapies interfering with upstream signalling intermediates will lack NF-κB-selective specificity, as they will inevitably also affect pathways beyond the NF-κB pathway, this limitation may be overcome at least in part by targeting protein-protein interactions involved in IKK activation, an area that remains largely unexplored and is likely to deliver effective and more specific NF-κB-targeting therapeutics. Although developing molecules that affect protein–protein interactions presents clear challenges, the heavy reliance of NF-κB activation pathways upon adaptor molecules and inducible formation of multimeric signalling complexes, coupled with the flourishing progress being made in unravelling new interactions and their regulation, is certain to catalyse the translational research efforts to develop NF-κB inhibitors in the future.

3.2. IKK Inhibitors

Owing to its central role as the signal integration hub for NF-κB activation pathways (Figures 1 and 2), the IKK complex has been the focus of significant drug discovery efforts since its discovery in 1996. However, while a vast spectrum of inhibitors has been developed throughout the years, only a few of these agents have ever been entered into clinical trials, and none has been clinically approved (Figure 3). A seminal paper by Michael Karin and colleagues in 2007 irreversibly tempered the initial enthusiasm over these agents as candidate NF-κB-targeting therapeutics [132]. This paper demonstrated that pharmacological IKKβ inhibition results in elevated systemic levels of IL-1β, owing to increased pro-IL-1β processing and IL-1β secretion by macrophages and neutrophils upon bacterial infection or exposure to endotoxin, leading to overt systemic inflammation and lethality in mice [132]. These severe on-target toxicities of IKKβ inhibitors have exposed the serious consequences of long-term IKK inhibition, which combined with the associated immunodeficiency and increased risks of malignancies arising from the liver, skin and other tissues [9,11,107,132–134], have irrevocably undermined any research efforts to clinically develop IKKβ-targeting therapeutics.

Despite their amino acidic sequence similarity, IKKα and IKKβ play largely distinct roles in NF-κB activation [107,135], whereby canonical NF-κB signalling strictly relies upon IKKβ-mediated phosphorylation of IκBs, while non-canonical NF-κB activation exclusively relies on IKKα. Nonetheless, IKKα also contributes to canonical NF-κB-dependent transcriptional responses by modulating the nuclear activities of RelA, histones and various transcriptional co-activators and co-repressors [2,136]. While the focus of the translational research has been on developing specific

IKKβ inhibitors, these molecules often also target IKKα [3,136], owing to the high degree of similarity between these kinases.

IKKα/IKKβ inhibitors can be broadly classified into three major groups on the basis of their mode of action: ATP analogues, allosteric modulators, and agents interfering with the kinase activation loops [3,136]. ATP analogues are the largest group and comprise both natural products such as β-carboline, and small-molecule inhibitors such as SPC-839 (Celgene). SPC-839 is a synthetic quinazoline analogue, which exhibits an approximately 200-fold selectivity for IKKβ over IKKα [2,3,105,137]. The imidazoquinoxaline derivative, BMS-345541, binds to an allosteric pocket present on both IKKα and IKKβ and is an example of an allosteric modulator, having a 10-fold higher selectivity for IKKβ than IKKα. Curiously, while BMS-345541 does not interfere with ATP binding to IKKβ, it disrupts ATP binding to IKKα [2,3,11,105,137,138]. BMS-345541 has so far mainly been tested in vitro, and demonstrated some efficacy against collagen-induced arthritis in mouse models [2,3,11,105,137–141]. Tiol-reactive compounds, including parthenolide, arsenite and epoxyquinoids, inhibit IKKβ by interacting with T-loop cysteine residue, C179 [3,105]. Although their exact mode of action is not well understood, this class of compounds appears to induce post-translational modifications that curtail IKKβ activity [3,105]. Parthenolide displays poor bioavailability and therefore has limited therapeutic application [142]. A new IKKβ inhibitor more recently developed by Sanofi-Aventis, SAR-113945, has been investigated in four different clinical trials in non-oncological patients, with initial promising results [143]. However, SAR-113945 has failed to demonstrate clinical efficacy in follow-on clinical trials, underscoring the challenge of striking an acceptable balance between efficacy and adverse side-effects [143]. Overall, the disappointing clinical performance of IKKβ inhibitors has considerably dampened the interest in this class of agents, and this trend is likely to continue in the future, as shown by the dramatic decline in the number of patent applications filed on these agents in recent years [107].

A significant effort has also been made towards developing molecules that target the IKKγ/NEMO scaffold (Figure 3), using one of three main targeting strategies aimed at perturbing either the IKKγ/NEMO interaction with catalytic IKK subunits, IKKγ/NEMO dimerization, or IKKγ/NEMO ubiquitination. The IKKγ/NEMO interactions with IKKα and IKKβ involve the kinase C-terminal NEMO-binding domain (NBD), which consists of the hexapeptide amino acid sequence, LDWSWL [135,136]. Peptidomimetics of this sequence have been shown to disrupt the IKKγ/NEMO–IKKα/IKKβ interaction and accordingly inhibit NF-κB activation by various upstream signals. However, the therapeutic utility of these agents is limited by their poor bioavailability and significant instability, in vivo [135]. Notwithstanding, small-molecule inhibitors of the IKKγ/NEMO–IKKα/IKKβ interaction may have the potential to provide useful agents to pharmacologically target the NF-κB pathway [136]. Following publication of the crystal structure of the IKKγ/NEMO–IKKβ interface, four phenothiazine derivatives were developed and demonstrated to have inhibitory effects on NF-κB activation in macrophages, in vitro [135]. However, further studies are required to determine whether these agents retain their inhibitory activity, in vivo, and crucially whether they are suitable for further development, especially in consideration of the safety concerns raised by IKKβ inhibitors [135]. Peptidomimetics were also developed to interfere with IKKγ/NEMO dimerisation [144]. However, to our knowledge, no significant preclinical development has ever been reported on any of these agents. Moreover, there has been significant interest in generating agents targeting the IKKγ/NEMO UBAN (i.e., UBD in the ABIN proteins and NEMO) domain, which transduces IKK activation signals by mediating the inducible interactions between IKKγ/NEMO and the polyubiquitin scaffolds of other signalling components [21,145]. Several agents have been shown to disrupt these ubiquitin-dependent IKKγ/NEMO interactions, including peptidomimetics [57], as well as small molecules such as anthraquinone derivatives of emodin [145]. However, despite the ability of some IKKγ/NEMO-targeting agents to modulate NF-κB activation without interfering with the IL-1β release, there appear to have been no clinical trials ever initiated to investigate the clinical safety and efficacy of these molecules.

The IKK-related kinase, TBK1 was originally discovered due to its involvement in NF-κB activation by PKCε-mediated signals [146]. As well as IKKε, TBK1 primarily regulates the activation of IRF-family factors, such as IRF3, IRF5 and IRF7, by RLRs and other receptors [147–149], but also phosphorylates RelA to enhance its transcriptional activity [136,146,150]. TBK1 and IKKε share a 64% sequence identity, but display only 27% identity with classical IKKs, and are both involved in inflammatory responses, oncogenesis, and insulin resistance [107]. Moreover, IKKε has been found overexpressed in breast and ovarian carcinoma, while TBK1 cooperates with RAS in promoting malignant transformation, in vitro, and is overexpressed in the carcinoma of the lung, colon and breast [151–156]. The antiinflammatory agents, BX765 and CYT387 (momelotinib), were originally developed as inhibitors of 3-phosphoinositide-dependent protein kinase 1 (PDK1) [157] and Janus kinases (JAKs), respectively, but were subsequently shown to also have potent inhibitory activity against TBK1 and IKKε [158]. Consequently, BX765 and CYT387 have been used as starting points for the molecular design of new TBK1 and IKKε antagonists. However, similar to these prototypes, the resulting compounds also demonstrated broad kinase target specificity [159]. Domainex and Myrexis have since generated more selective TBK1 and IKKε inhibitors, such as DMXD-011 and MPI-0485520, respectively [159]. DMXD-011 displays good drug-like properties, is orally bioavailable and has shown promising therapeutic activity in in vivo inflammatory disease models [160]. Likewise, MPI-0485520 displays good oral bioavailability and has been investigated in mouse models of systemic lupus erythematous, rheumatoid arthritis and other autoimmune disorders, with significant therapeutic activity demonstrated, especially in the context of especially in the context of rheumatoid arthritis [161]. However, further investigations are required to determine the potential clinical benefit resulting from these agents and whether their preclinical efficacy translates to the oncological context [159].

3.3. Ubiquitin and Proteasome Pathway Inhibitors

The ubiquitin pathway provides a highly versatile and tightly regulated system of signal propagation and protein degradation conserved throughout the eukarya dominion [162–164]. Over the past decade, this system has witnessed a burgeoning interest, not only because of its central importance in signal transduction and protein degradation, but also due to its potential for serving as a treasure trove of drug targets for therapeutic intervention in a wide range of malignant and inflammatory pathologies [162,165]. The ubiquitin pathway consists of a three-step enzymatic process, catalysed by as many protein complexes, whereby a ubiquitin-activating enzyme (E1) first binds to and activates a ubiquitin molecule in an ATP-dependent reaction, followed by the sequential transfer of the activated ubiquitin molecule to the active site of a ubiquitin-conjugating enzyme (E2) and, finally, to the target protein through the site-specific recognition and correct positioning by a ubiquitin-protein ligase (E3), which operates in conjunction with the E2 ubiquitin-conjugating enzyme to catalyse the covalent attachment of an 8-kDa ubiquitin molecule to the target site(s) of the protein substrate [25,33,162,165–167]. In addition to catalysing the attachment of the first ubiquitin molecule onto the protein substrate, the same three-step enzymatic process conjugates further ubiquitin molecules to form polyubiquitin chains [167].

Ubiquitin-mediated signalling pathways play a central role in the regulation of both canonical and non-canonical NF-κB activation (Figures 1 and 2) [21,24]. They also regulate oncogenesis by participating in either tumour suppression or tumour promotion [21,24,35,162]. Therefore, interfering with the ubiquitin system can affect cancer development and progression in many different ways (Figure 3). Notwithstanding, tumour cells that depend on constitutive NF-κB signalling for survival often display sensitivity to inhibitors of the ubiquitin-proteasome pathway (UPP), owing to the essential role of this pathway in the proteolytic degradation of IκB proteins [6,25,162]. The active proteasome consists of a large 2.4-MDa protein complex, which comprises a 20S catalytic core of a cylindrical shape and two regulatory 19S components forming a lid-like structure at both extremities of the 20S cylinder, and catalyses the proteolysis of substrate proteins via an ATP-dependent process [167].

The first proteasome inhibitors developed were molecules of the class of peptide aldehydes and were subsequently extensively used for preclinical research [168]. The best characterised of these molecules is MG132, which also served as a prototype for the generation of the next classes of proteasome inhibitors that later found common use in the clinical practice [169]. The first proteasome inhibitor ever tested in humans was bortezomib (velcade; formerly known as PS-341), a reversible boronic-acid inhibitor of the 20S catalytic subunit [137,162]. Bortezomib received the approval of the FDA in 2003 for the treatment of refractory multiple myeloma, and is currently in clinical use as a front-line therapy in combination with other agents for the treatment of multiple myeloma and mantle cell lymphoma (MCL) [162]. A number of clinical trials are ongoing to assess the efficacy of bortezomib in further oncological indications, including solid cancers (Table 1). An especially promising area of development for bortezomib and other proteasome inhibitors is their use as part of combination therapies aimed at overcoming radio- and chemo-resistance in cancer. Indeed, bortezomib has been shown to have a strong synergistic activity when used in combination with radiotherapy and/or chemotherapy in various types of haematological and solid cancer [170–175]. The second-generation proteasome inhibitor, carfilzomib (Kyprolis), was approved in 2012 as a single agent and in 2016 in combination with dexamethasone, with or without lenalidomide, for the treatment of patients with relapsed or refractory multiple myeloma who have received at least one line of prior therapy. Carfilzomib is an epoxyketone compound, acting as an irreversible proteasome inhibitor to afford prolonged therapeutic inhibition [162], and is currently being investigated in multiple myeloma in combination with other agents, as well as other oncological indications (Table 1). Since both bortezomib and carfilzomib can be only administered intravenously or subcutaneously, the boronic proteasome inhibitor, ixazomib (MLN-9708), was recently developed as an oral therapy and was approved by the FDA in 2015 in combination with lenalidomide and dexamethasone for the treatment of patients with multiple myeloma who have received at least one line of prior therapy [176]. Ixazomib is currently being evaluated in patients with other cancer types and as part of other combination regimens (Table 1).

The human genome encodes two E1 ubiquitin-activating enzymes, UBA1 and UBA5, an estimated fifty E2 ubiquitin-conjugating (UBC) enzymes, and more than six hundred E3 ligases, conferring a large degree of substrate specificity to the ubiquitin cascade [11,25,166]. It follows that therapeutically targeting individual E3s, which selectively bind to the recognition sites of protein substrates, will achieve maximal target specificity, as compared to targeting any E1s or E2s. The inhibitor of the ubiquitin-like protein, neural precursor cell-expressed developmentally down-regulated 8 (NEDD8), MLN4924 (pevonedistat), was developed to inhibit Cullin-RING E3 ubiquitin ligases, the largest family of E3s, which require activation by E1/E2-mediated NEDDylation [165]. As a result of this mode of action inhibiting multiple E3 ligases, MLN4924 has broad target specificity. Although most of its adverse effects are mild or moderate, several higher-grade toxicities been reported, including febrile neutropenia, thrombocytopenia and elevated circulating levels of aspartate transaminase [165,166,177]. Notwithstanding, clinical trials of MLN4924 in combination with 5-azacytidine are underway in patients with acute myeloid leukaemia (AML) and are yielding promising initial results (Table 1) [177].

The F-box protein, β-TrCP, is involved in the signal-dependent recognition and degradative K48-linked ubiquitination of canonical IκBs and p100, as well as in the recognition and ubiquitination of a large group of other target substrates, including β-catenin, Snail, Emi1, Wee1, Cdc25A and Claspin [178,179]. From a theoretical standpoint at least, targeting NF-κB signalling via β-TrCP would represent a more specific and, possibly, safer alternative to proteasome inhibition [178,179], although a limitation of the approach would remain the accumulation of protein substrates outside the NF-κB pathway, such as β-catenin, which contributes to colorectal carcinogenesis [178,179]. Notwithstanding this caveat, small-molecule inhibitors of β-TrCP, such as GS143, have been developed and shown to inhibit signal-induced IκBα ubiquitination [180]. However, little information is available on the mode of action of this agent, although it seemingly involves an interaction with both β-TrCP and IκBα, or its further characterisation [165].

While proteasome inhibitors, such as bortezomib, can provide significant clinical benefit in multiple myeloma and a few other indications, these agents inhibit NF-κB and many other essential cellular pathways that rely on the proteasome function [6,162,166], and are therefore by no means specific for the NF-κB pathway. Moreover, proteasome inhibitors target these pathways in normal and cancer cells alike, thereby resulting in a low therapeutic index and significant dose-limiting toxicities [137,166,181,182]. Indeed, despite their indisputable commercial and clinical success, proteasome inhibitors as a therapy present several limitations, which remain largely unaddressed, including their broad cellular activities, dose-limiting side effects and the relatively rapid onset of secondary drug resistance [6,162,165,166]. In particular, treatment with bortezomib is often associated with peripheral neuropathy, which is a dose-limiting adverse effect, as well as trombocytopenia, neutropenia, nausea, diarrhea, and fatigue [162]. Furthermore, drug resistance is inevitable and generally develops within a year from the start of treatment [166]. While carfilzomib administration is associated with different side effects and the onset of peripheral neuropathy is significantly less frequent than with bortezomib, patients can nonetheless develop renal impairment and cardiovascular complications [162]. Likewise, treatment with Ixazomib can induce peripheral neuropathy, gastrointestinal adverse effects and skin rash [162].

Importantly, from a mechanistic standpoint, it is also unclear that the clinical response to proteasome inhibitors in patients with multiple myeloma and other B-cell malignancies results from the inhibition of NF-κB signalling, as it is becoming increasingly clear that the cellular accumulation of undigested proteins in these immunoglobulin-producing tumours can activate the unfolded protein response (UPR), thereby accelerating tumour-cell death [162]. Therefore, there remains a need for novel therapeutic strategies capable of selectively targeting the NF-κB pathway in these oncological indications [137,166,179]. While targeting upstream components of the UPP, such as NF-κB-activating E3 ligases, would be a preferable strategy than targeting the proteasome, as it could mitigate at least some of the limitations of proteasome inhibition, and, more broadly, represents a highly promising area for future drug discovery and development in oncology, this strategy would nevertheless be unlikely to ever yield a specific NF-κB-targeting agent [137,162,165,166].

3.4. Inhibitors of NF-κB Nuclear Activities

Once liberated from IκBs, NF-κB dimers migrate into the nucleus where they regulate gene expression. This activation step offers a significant opportunity for therapeutic intervention, since, at least from a theoretical standpoint, a number of nuclear activities of NF-κB could be targeted with drug agents, including the post-translational modification of NF-κB proteins and their ability to dimerise, translocate into the nucleus, bind to DNA and interact with chromatin components, coactivators and corepressors and other transcription factors [3,105]. While the drug development output in this area has been relatively limited, a few examples of candidate therapeutics targeting the nuclear translocation and DNA-binding activity of NF-κB complexes are worthy of consideration (Figure 3). In particular, several strategies have been developed to inhibit the NF-κB nuclear translocation, including small peptidomimetics, such as SN-50, which encompasses the NLS of p50 [3,105,183]. At high concentrations, SN-50 has been shown to saturate the transport machinery importing p50-containing dimers into the nucleus [3,105,183]. However, apart from the high peptide concentrations required to achieve this effect, a drawback of SN-50 is its non-specific inhibition of transcription factors other than NF-κB complexes, such as AP-1-family factors [3,105,183]. Additionally, dehydroxymethylepoxyquinomicin (DHMEQ), a derivative of the antibiotic epoxyquinomicin C, isolated from *Amycolatopsis*, has been shown to exhibit a potent and specific inhibitory activity on NF-κB nuclear import by means of its ability to directly bind to NF-κB dimers [3,105,183,184]. Studies in mouse models have demonstrated a therapeutic effect of DHMEQ in prostate carcinoma and an immunomodulatory effect in ovarian carcinoma. Despite these promising preclinical results, to our knowledge, no clinical trials have been initiated to evaluate DHMEQ in cancer patients [185–187].

Furthermore, sesquiterpene lactone (SL) compounds have been shown to inhibit the DNA-binding activity of RelA-containing NF-κB dimers by interacting with cysteine residue, C38, within RelA's DNA-binding loop 1 (L1) [3,105]. By binding to homologous cysteine residues within p50 and c-Rel, certain SL compounds have been shown to have the additional capacity to inhibit NF-κB complexes containing these subunits [3,105]. Interestingly, some SLs, such as parthenolide, also have the ability to inhibit IKKβ [3,105]. However, while this dual effect of parthenolide on the NF-κB pathway has attracted some interest, the poor bioavailability of this agent has precluded any further drug development effort [142,188]. However, the search for a functionally equivalent compound has resulted in the generation of the amino acid-analogue, dimethylaminoparthenolide (DMAPT), exhibiting enhanced bioavailability [189]. A phase I clinical trial of DMAPT was initiated in 2009 in patients with AML, but was then suspended later in the same year [189,190]. Another class of agents designed to interfere with the NF-κB DNA-binding activity consists of decoy oligodeoxynucleotides, which can compete for binding to NF-κB complexes with κB DNA sites on specific gene promoters [2]. Most therapeutic decoy oligodeoxynucleotides were further modified to increase their in vivo stability, as well as their affinity for NF-κB complexes [3,105]. A clinical trial of the NF-κB decoy oligodeoxynucleotide, anesiva, was initiated in 2005 to test its efficacy as a local ointment for the treatment of atopic dermatitis, but its clinical use was subsequently discontinued in 2014 (NCT00125333, [190]).

3.5. Inhibitors of NF-κB Downstream Effectors

Since NF-κB induces transcriptional programmes that affect all hallmarks of cancer [79,191], an attractive alternative to therapeutically targeting NF-κB in malignant disease would be to inhibit the non-redundant, cancer cell-specific downstream effectors of the NF-κB oncogenic functions (Figure 3). Given that the insurmountable challenge with conventional NF-κB or IKKβ inhibitors has been to achieve cancer-cell specificity, due to the pleiotropic and ubiquitous functions of NF-κB [11], agents targeting these effectors, having functional restriction to cancer cells or their microenvironment, could provide safer and more selective anticancer therapeutics, lacking the dose-limiting toxicities of global NF-κB inhibitors.

Our group recently sought to obtain proof-of-concept for this principle in multiple myeloma, the paradigm of NF-κB-driven malignant diseases. Since a key pathogenetic function of NF-κB in multiple myeloma is to upregulate genes that block apoptosis and, despite its ubiquitous nature, NF-κB signalling induces highly tissue- and context-specific transcriptional programs [6,11,12,14,79,192], we targeted an essential downstream effector of this pathogenically critical activity of NF-κB, in order to achieve cancer cell-selective therapeutic specificity and thereby circumvent the limitations of conventional IKK/NF-κB-targeting drugs. Several years ago, we identified the immediate-early gene, growth arrest and DNA damage 45B (*GADD45B*), as a novel transcriptional target of NF-κB and effector of the NF-κB-dependent inhibitory activity on JNK signalling and apoptosis in response to TNF-α and other cues [193,194]. GADD45β is a member of the GADD45 family of proteins, also comprising GADD45α and GADD45γ, which play distinct roles in multiple cellular functions, including cell-cycle regulation, DNA repair, apoptosis, senescence and DNA demethylation [195–197]. Interesting, GADD45β is the only member of this family that is largely regulated downstream of NF-κB signalling [193,194]. A number of studies, including some from our own group, have demonstrated that GADD45β and other members of this family play many of their important biological roles by regulating the activity MAPK pathways, such as the JNK and p38 pathway [195,198,199]. Indeed, more recently, we have identified the complex formed by GADD45β and the JNK kinase, MKK7, as a functionally critical survival module downstream of NF-κB and novel therapeutic target in multiple myeloma [194,198,200–203]. As most normal cells do not constitutively express *GADD45B* [204], and, unlike mice lacking *RelA* or any *IKK* subunit [11], *Gadd45b*$^{-/-}$ mice are viable, fertile and die of old age [205,206], we reasoned that, in contrast to global NF-κB blockade, pharmacological GADD45β inhibition would be well tolerated in vivo.

We therefore developed the D-tripeptide inhibitor of the GADD45β/MKK7 complex, DTP3, which effectively disrupts this complex at nanomolar concentrations, in vitro, by binding to MKK7, and as a result, selectively kills multiple myeloma cells by inducing MKK7/JNK-dependent apoptosis, without toxicity to normal tissues [200,201]. We showed that, due to this target-selective mode of action, DTP3 displays an excellent cancer-cell selective specificity in multiple myeloma cell lines and primary cells from patients, in vitro, and eradicates multiple myeloma xenografts in mice, with excellent tolerability and no apparent side effects at the therapeutic doses [200,201]. The first-in-human phase I/IIa clinical study of DTP3 has recently been initiated in patients with refractory or relapsed multiple myeloma, and initial results from this study preliminarily demonstrate the clinical safety of this agent, alongside a cancer-selective pharmacodynamic response (Table 1).

Further investigations will be required to determine the potential clinical benefit resulting from this approach in multiple myeloma patients, as well as its potential side-effects, propensity to develop drug resistance, and therapeutic efficacy in combination with other agents [207,208]. Nevertheless, the available body of preclinical data and the encouraging initial clinical results preliminarily demonstrate that cancer-selective inhibition of the NF-κB pathway is possible and promises to provide an effective therapeutic strategy, with no preclusive toxicity, that could profoundly benefit patients with multiple myeloma and, potentially, other cancers where NF-κB is a driver of pathogenesis. Indeed, the same principle we developed of targeting an axis of the NF-κB pathway with cancer-restricted function, rather than NF-κB globally, could be similarly applied to also selectively inhibit NF-κB oncogenic functions beyond the suppression of cancer-cell apoptosis [200,201], such as functions in governing tumour-associated inflammation.

Several antiapoptotic genes, in addition to *GADD45B*, have been shown to be transcriptionally regulated by NF-κB, including those encoding various BCL-2-family members such as B-cell lymphoma-extra large (*Bcl-X$_L$*), myeloid cell leukaemia sequence 1 (*MCL1*), *A1* (also known as BFL-1) and, at least in certain tissue contexts, B-cell lymphoma 2 (*BCL-2*) itself [191,209]. These proteins are also frequently deregulated in certain cancer types, often as a result of chromosomal translocations or other genetic abnormalities, and promote oncogenesis by means of their ability to suppress cancer-cell apoptosis [210–213]. BCL-2-family members have also been shown to contribute to NF-κB-dependent radio- and chemo-resistance in cancer [80]. Accordingly, the therapeutic inhibition of these proteins has also been pursued as a combination therapy with radiation and chemotherapeutic drugs to overcome resistance to these agents in recalcitrant forms of malignancy [214]. Congruently, certain NF-κB-dependent multiple myeloma cell lines express low levels of GADD45β and are completely refractory to DTP3-induced killing [200], confirming the existence of GADD45β-independent mechanisms for NF-κB-dependent survival in certain subtypes of multiple myeloma, and such mechanisms are certain to also exist in other types of malignancy. The BCL-2 family of proteins comprises an evolutionarily conserved group of 20 members, which share one or more of four so-called BCL-2 homology (BH) domains, referred to as BH1, BH2, BH3 and BH4, and are involved in either suppressing or promoting apoptosis [211–213]. The members of this family can be classified into three functionally and structurally distinct groups: prosurvival proteins such as BCL-X$_L$, MCL1, A1/BFL-1 and BCL-2 itself; multidomain proapoptotic effectors such as BCL-2-associated X protein (BAX) and BCL-2 antagonist/killer (BAK); and BH3-only proteins, which convey apoptosis-initiating signals, such as BCL-2-interacting mediator of cell death (BIM), BH3 interacting-domain death agonist (BID), Bcl-2-associated death promoter (BAD), and p53 upregulated modulator of apoptosis (PUMA) [211–213]. Historically, the main strategy utilised to inhibit the antiapoptotic activity of BCL-2-family members has been to generate cell-permeable drug mimetics of the BH3 domains of proapoptotic BCL-2-like proteins, which neutralise prosurvival BCL-2 proteins by binding to their surface hydrophobic groove [211–213,215]. However, many of these agents have shown limited therapeutic activity as single agents and/or significant side-effects due to their low specificity [211,215]. A notable exception in this group of molecules has been ABT-199 (venetoclax, RG7601, GDC-0199), a potent and selective first-in-class BCL-2 inhibitor [211,213,215]. ABT-199 has demonstrated significant anticancer activity in various models of lymphoma and leukaemia, both in vitro and in vivo, and has entered clinical trials in patients with non-Hodgkin's lymphoma, CLL, AML and T-ALL, both as a single agent and in combination with other drugs [211,216]. ABT-199 was subsequently granted breakthrough status designation by the FDA in 2016 for the treatment of patients with relapsed or refractory CLL with 17p deletion (Table 1).

4. Conclusions

Given the central role of aberrant NF-κB activation in the pathogenesis of the large majority of human diseases, the therapeutic targeting of the NF-κB pathway has been aggressively pursued by the pharmaceutical industry and academic laboratories for over two and a half decades. However, this goal has proven thus far an insurmountable challenge, due to the severe dose-limiting toxicities associated with the global suppression of NF-κB, resulting in the dismaying absence of a specific NF-κB or IKK inhibitor in the current anticancer armamentarium. Nevertheless, the past three decades, since NF-κB was first discovered in 1986, have witnessed tremendous advances in the understanding of the intricate signalling networks governing the NF-κB pathway and its multiple functions, and these efforts are now bearing some of the long-awaited fruits in the field of translational medicine, by pointing toward potential safer alternatives to therapeutic NF-κB inhibition. Undoubtedly, the targeting of upstream signalling mechanisms and protein-protein interactions governing the contextual and tissue-specific activation of NF-κB signalling and, at the opposite end of the spectrum, the non-redundant effectors of the diverse tissue-specific transcriptional programmes that NF-κB activates in order to exert its biological functions are amongst the most attractive strategies being developed to achieve contextual, cancer-cell selective therapeutic inhibition of the NF-κB pathway, and thereby circumvent the preclusive limitations of conventional NF-κB inhibitors. Perhaps, the most valuable lesson to be learnt from these initial therapeutic attempts is that the complexity of the intricate signalling networks governing the NF-κB regulation and function appears to hold the key to untangle the NF-κB therapeutic riddle and translate it into concrete clinical benefits. Indeed, while significant ground remains to be covered and the limited clinical successes obtained so far in select experimental clinical contexts have yet to transform into healthcare benefit for the broader patient population, the basic and translational knowledge unravelled by these studies on the biological complexity in the NF-κB pathway is providing tangible new opportunities for cancer-selective therapeutic intervention, which are certain to attract growing interest in the future and be further capitalised upon.

Acknowledgments: The work was supported in part by Cancer Research UK programme grant A15115, Medical Research Council (MRC) Biomedical Catalyst grant MR/L005069/1 and Bloodwise project grant 15003 to Guido Franzoso.

Conflicts of Interest: The authors declare no conflict of interest.

Abbreviations

A1	BCL2A1—BCL2-related protein A1
A20	TNFAIP3 —Tumour necrosis factor α-induced protein 3
ABIN-1	A20 binding inhibitor of NF-κB
ADAP	Adhesion- and degranulation-promoting adaptor protein
ALCL	Anaplastic large-cell lymphoma
ALL	Acute lymphoblastic leukaemia
AML	Acute myeloid leukaemia
ARD	Ankyrin repeat domain
ATM	Ataxia telangiectasia mutated
BAD	Bcl-2-associated death promoter
BAFF	B-cell activating factor
BAK	BCL-2 antagonist/killer
BAX	BCL-2-associated X protein
BCL10	B-cell CLL/lymphoma 10
BCL-2	B-cell lymphoma 2
Bcl-XL	B-cell lymphoma-extra large

BCR	B-cell receptor
BH	BCL-2 homology
BID	BH3 interacting-domain death agonist
BIM	BCL-2-interacting mediator of cell death
BIR	Baculovirus IAP repeat
CAML	Calcium modulating ligand
CARD	Caspase recruitment domain
CBM	CARD11-BCL10-MALT1
CD40L	CD40 ligand
Cdc25A	Cell division cycle 25A
c-FLIP	Cellular FLICE/caspase8-like inhibitory protein
c-IAP	Cellular inhibitor of apoptosis
CLL	Chronic lymphocytic leukaemia
CYLD	Cylindromatosis lysine 63 deubiquitinase
DAMP	Damage-associated molecular patterns
DHMEQ	Dehydroxymethylepoxyquinomicin
DLBCL	Diffuse large B-cell lymphoma
DMAPT	Dimethylaminoparthenolide
DUB	Deubiquitination enzyme
Emi1-FBXO 5	F-box protein 5
EMT	Epithelial-mesenchymal transition
FBXW	F-box and WD repeat domain containing
FDA	US Food and Drug Administration
FL	Follicular lymphoma
GADD45β	Growth arrest and DNA damage inducible β
GBM	Glioblastoma multiforme
GRR	Glycine rich region
GSK3	Glycogen synthase kinase 3
HAT	Histone acetyltransferase
HDAC	Histone deacetylases
HL	Hodgkin's lymphoma
HOIL-1L	RBCK1—RanBP-type and C3HC4-type zinc finger containing 1
HOIP	RNF31—Ring finger protein 31
IKK	IκB kinase
IL-1β	Interleukin 1 β
IL-1β	Interleukin 1β
IL-1βR	IL-1β receptor
IRAK	Interleukin-1 receptor-associated kinase
IκB	Nuclear factor of κ light polypeptide gene enhancer in B-cells inhibitor
JAK	Janus kinase
LPS	Lipopolysaccharide
LTβ	Lymphotoxin β
LUBAC	Linear ubiquitination assembly complex
MALT	Mucosa-associated lymphoid tissue
MALT1	Mucosa-associated lymphoid tissue lymphoma translocation protein 1
MAP	Mitogen activated protein
MAPKKK	Mitogen activated protein kinase kinase kinase
MAVS	Mitochondrial antiviral signalling protein
MCL	Mantle-cell lymphoma
MCL1	Myeloid cell leukaemia sequence 1
MDS	Myelodysplastic syndrome
MEKK	Mitogen activated protein kinase kinase
MM	Multiple myeloma
mTOR	Mechanistic target of rapamycin
MYD88	Myeloid differentiation primary response protein 88
NEDD8	Neural precursor cell-expressed developmentally down-regulated 8
NES	Nuclear export signal
NF-κB	Nuclear factor κ B (Nuclear Factor binding to the κ-Light-chain-enhancer B site)
NHL	Non-Hodgkin's lymphoma
NIK	NF-κB-inducing kinase
NLR	NOD-like receptor
NLS	Nuclear localization signals
NOD	Nucleotide-binding oligomerisation domain
NSCLC	Non-Small Cell Lung Cancer

OTUD7B	OTU domain-containing protein 7B
PAMP	Pathogen-associated molecular pattern
PAR	Poly(ADP-ribose)
PARP-1	Poly(ADP-ribose)-polymerase-1
PDK	Phosphoinoisitide-dependent kinase
PDK1	3-Phosphoinositide-dependent protein kinase 1
PDLIM2	PDZ and LIM domain protein 2
PIASy	Protein inhibitor of activated STAT protein y
PIDD	p53-induced death domain protein
PKC	Protein kinase C
PP	Protein phosphatase
PTEN	Phosphatase and tensin homolog
PUMA	p53 upregulated modulator of apoptosis
RANKL	Receptor activator of NF-κB ligand
RAS	Rat sarcoma virus oncogene
RBX1	RING-box protein 1
RHD	Rel-homology Domain
RIP	Receptor interacting protein
RLH	RNA helicase
RLR	RIG-I-like receptor
SCF	SKP1-Cullin 1-F-box protein
Sharpin	SHANK-associated RH domain interacting protein
SL	Sesquiterpene lactone
SMAC	Second mitochondria-derived activator of caspases
SOCS-1	Suppressor of cytokine signalling-1
STAT	Signal transducer and activator of transcription
SUMO	Small ubiquitin-like modifiers
TAB	TAK1-associated binding protein
TACI	TNFRSF13B—TNF receptor superfamily member 13B
TAK1	Transforming growth factor β-activated kinase 1
T-ALL	T-cell acute lymphoblastic leukaemia
TBK1	TANK-binding kinase
TCR	T-cell receptor
TLR	Toll-like receptor
TME	Tumour microenvironment
TNF-R	TNF receptor
TNFSF	Tumour necrosis factor superfamily member
TNF-α	Tumour necrosis factor-α
TRADD	TNF receptor-associated death domain protein
TRAF	TNF receptor-associated factor
TRIM23	Tripartite motif protein 23
TWEAK	TNF-related weak inducer of apoptosis
UBA	Ubiquitin-activating enzyme, E1
UBC	Ubiquitin-conjugating, E2
Ubc/Uev	Ubiquitin-conjugating enzyme/ubiquitin E2 variant
UBD	Ubiquitin-binding domain
UPP	Ubiquitin-proteasome pathway
UPR	Unfolded protein response
USP	Ubiquitin-specific peptidase
WIP1	Wild-type p53-induced phosphatase 1
WM	Waldenström's macroglobulinemia
WWOX	WW domain-containing oxidoreductase
βTrCP	β-transducin repeat-containing protein

References

1. Sen, R.; Baltimore, D. Multiple nuclear factors interact with the immunoglobulin enhancer sequences. *Cell* **1986**, *46*, 705–716. [CrossRef]
2. Karin, M.; Yamamoto, Y.; Wang, Q.M. The IKK NF-κB system: A treasure trove for drug development. *Nat. Rev. Drug Discov.* **2004**, *3*, 17–26. [CrossRef] [PubMed]
3. Gilmore, T.D.; Herscovitch, M. Inhibitors of NF-κB signaling: 785 and counting. *Oncogene* **2006**, *25*, 6887–6899. [CrossRef] [PubMed]

4. Hayden, M.S.; Ghosh, S. NF-κB, the first quarter-century: Remarkable progress and outstanding questions. *Genes Dev.* **2012**, *26*, 203–234. [CrossRef] [PubMed]

5. Zhang, Q.; Lenardo, M.J.; Baltimore, D. 30 Years of NF-κB: A blossoming of relevance to human pathobiology. *Cell* **2017**, *168*, 37–57. [CrossRef] [PubMed]

6. Staudt, L.M. Oncogenic activation of NF-κB. *Cold Spring Harb. Perspect. Biol.* **2010**, *2*, a000109. [CrossRef] [PubMed]

7. Xia, Y.; Shen, S.; Verma, I.M. NF-κB, an active player in human cancers. *Cancer Immunol. Res.* **2014**, *2*, 823–830. [CrossRef] [PubMed]

8. Hoesel, B.; Schmid, J.A. The complexity of NF-κB signaling in inflammation and cancer. *Mol. Cancer* **2013**, *12*, 86. [CrossRef] [PubMed]

9. Grivennikov, S.I.; Greten, F.R.; Karin, M. Immunity, inflammation, and cancer. *Cell* **2010**, *140*, 883–899. [CrossRef] [PubMed]

10. Karin, M. NF-κB as a critical link between inflammation and cancer. *Cold Spring Harb. Perspect. Biol.* **2009**, *1*, a000141. [CrossRef] [PubMed]

11. DiDonato, J.A.; Mercurio, F.; Karin, M. NF-κB and the link between inflammation and cancer. *Immunol. Rev.* **2012**, *246*, 379–400. [CrossRef] [PubMed]

12. Annunziata, C.M.; Davis, R.E.; Demchenko, Y.; Bellamy, W.; Gabrea, A.; Zhan, F.; Lenz, G.; Hanamura, I.; Wright, G.; Xiao, W.; et al. Frequent engagement of the classical and alternative NF-κB pathways by diverse genetic abnormalities in multiple myeloma. *Cancer Cell* **2007**, *12*, 115–130. [CrossRef] [PubMed]

13. Demchenko, Y.N.; Glebov, O.K.; Zingone, A.; Keats, J.J.; Bergsagel, P.L.; Kuehl, W.M. Classical and/or alternative NF-κB pathway activation in multiple myeloma. *Blood* **2010**, *115*, 3541–3552. [CrossRef] [PubMed]

14. Keats, J.J.; Fonseca, R.; Chesi, M.; Schop, R.; Baker, A.; Chng, W.J.; van Wier, S.; Tiedemann, R.; Shi, C.X.; Sebag, M.; et al. Promiscuous mutations activate the noncanonical NF-κB pathway in multiple myeloma. *Cancer Cell* **2007**, *12*, 131–144. [CrossRef] [PubMed]

15. Roschewski, M.; Staudt, L.M.; Wilson, W.H. Diffuse large B-cell lymphoma-treatment approaches in the molecular era. *Nat. Rev. Clin. Oncol.* **2014**, *11*, 12–23. [CrossRef] [PubMed]

16. Pasqualucci, L.; Neumeister, P.; Goossens, T.; Nanjangud, G.; Chaganti, R.S.; Küppers, R.; Dalla-Favera, R. Hypermutation of multiple proto-oncogenes in B-cell diffuse large-cell lymphomas. *Nature* **2001**, *412*, 341–346. [CrossRef] [PubMed]

17. Pasqualucci, L.; Dominguez-Sola, D.; Chiarenza, A.; Fabbri, G.; Grunn, A.; Trifonov, V.; Kasper, L.H.; Lerach, S.; Tang, H.; Ma, J.; et al. Inactivating mutations of acetyltransferase genes in B-cell lymphoma. *Nature* **2011**, *471*, 189–195. [CrossRef] [PubMed]

18. Pasqualucci, L.; Dalla-Favera, R. SnapShot: Diffuse large B cell lymphoma. *Cancer Cell* **2014**, *25*, 132.e1. [CrossRef] [PubMed]

19. Bredel, M.; Scholtens, D.M.; Yadav, A.K.; Alvarez, A.A.; Renfrow, J.J.; Chandler, J.P.; Yu, I.L.; Carro, M.S.; Dai, F.; Tagge, M.J.; et al. NFKBIA deletion in glioblastomas. *N. Engl. J. Med.* **2011**, *364*, 627–637. [CrossRef] [PubMed]

20. Cahill, K.E.; Morshed, R.A.; Yamini, B. Nuclear factor-κB in glioblastoma: Insights into regulators and targeted therapy. *Neuro-Oncology* **2015**, *18*, 329–339. [CrossRef] [PubMed]

21. Hinz, M.; Scheidereit, C. The IκB kinase complex in NF-κB regulation and beyond. *EMBO Rep.* **2014**, *15*, 46–61. [CrossRef] [PubMed]

22. Paul, S.; Schaefer, B.C. A new look at T cell receptor signaling to nuclear factor-κB. *Trends Immunol.* **2013**, *34*, 269–281. [CrossRef] [PubMed]

23. Liu, F.; Xia, Y.; Parker, A.S.; Verma, I.M. IKK biology. *Immunol. Rev.* **2012**, *246*, 239–253. [CrossRef] [PubMed]

24. Chen, Z.J. Ubiquitin signalling in the NF-κB pathway. *Nat. Cell Biol.* **2005**, *7*, 758–765. [CrossRef] [PubMed]

25. Collins, P.E.; Mitxitorena, I.; Carmody, R.J. The Ubiquitination of NF-κB subunits in the control of transcription. *Cells* **2016**, *5*, 23. [CrossRef] [PubMed]

26. Wertz, I.E.; Dixit, V.M. Signaling to NF-κB: Regulation by ubiquitination. *Cold Spring Harb. Perspect. Biol.* **2010**, *2*, a003350. [CrossRef] [PubMed]

27. Sun, S.C.; Ley, S.C. New insights into NF-κB regulation and function. *Trends Immunol.* **2008**, *29*, 469–478. [CrossRef] [PubMed]

28. Oeckinghaus, A.; Ghosh, S. The NF-κB family of transcription factors and its regulation. *Cold Spring Harb. Perspect. Biol.* **2009**, *1*, a000034. [CrossRef] [PubMed]

29. Perkins, N.D. The diverse and complex roles of NF-κB subunits in cancer. *Nat. Rev. Cancer* **2012**, *12*, 121–132. [CrossRef] [PubMed]

30. Smale, S.T. Dimer-specific regulatory mechanisms within the NF-κB family of transcription factors. *Immunol. Rev.* **2012**, *246*, 193–204. [CrossRef] [PubMed]

31. NF-κB. 2017. Available online: www.nf-kb.org (accessed on 27 January 2017).

32. Häcker, H.; Karin, M. Regulation and function of IKK and IKK-related kinases. *Sci. STKE* **2006**, *2006*, re13. [CrossRef] [PubMed]

33. Busino, L.; Millman, S.E.; Pagano, M. SCF-mediated degradation of p100 (NF-κB2): Mechanisms and relevance in multiple myeloma. *Sci. Signal.* **2012**, *5*, pt14. [CrossRef] [PubMed]

34. Perkins, N.D.; Gilmore, T.D. Good cop, bad cop: The different faces of NF-κB. *Cell Death Differ.* **2006**, *13*, 759–772. [CrossRef] [PubMed]

35. Chen, Z.J.; Sun, L.J. Nonproteolytic functions of ubiquitin in cell signaling. *Mol. Cell* **2009**, *33*, 275–286. [CrossRef] [PubMed]

36. Winston, J.T.; Strack, P.; Beer-Romero, P.; Chu, C.Y.; Elledge, S.J.; Harper, J.W. The SCFβ-TRCP-ubiquitin ligase complex associates specifically with phosphorylated destruction motifs in IκBα and β-catenin and stimulates IκBα ubiquitination in vitro. *Genes Dev.* **1999**, *13*, 270–283. [CrossRef] [PubMed]

37. Spencer, E.; Jiang, J.; Chen, Z.J. Signal-induced ubiquitination of IκBα by the F-box protein Slimb/β-TrCP. *Genes Dev.* **1999**, *13*, 284–294. [CrossRef] [PubMed]

38. Wei, D.; Sun, Y. Small RING Finger Proteins RBX1 and RBX2 of SCF E3 Ubiquitin Ligases: The Role in Cancer and as Cancer Targets. *Genes Cancer* **2010**, *1*, 700–707. [CrossRef] [PubMed]

39. Laplantine, E.; Fontan, E.; Chiaravalli, J.; Lopez, T.; Lakisic, G.; Véron, M.; Agou, F.; Israël, A. NEMO specifically recognizes K63-linked poly-ubiquitin chains through a new bipartite ubiquitin-binding domain. *EMBO J.* **2009**, *28*, 2885–2895. [CrossRef] [PubMed]

40. Delhase, M.; Hayakawa, M.; Chen, Y.; Karin, M. Positive and negative regulation of IκB kinase activity through IKKβ subunit phosphorylation. *Science* **1999**, *284*, 309–313. [CrossRef] [PubMed]

41. Schmukle, A.C.; Walczak, H. No one can whistle a symphony alone—How different ubiquitin linkages cooperate to orchestrate NF-κB activity. *J. Cell Sci.* **2012**, *125*, 549–559. [CrossRef] [PubMed]

42. Lee, D.F.; Kuo, H.P.; Chen, C.T.; Hsu, J.M.; Chou, C.K.; Wei, Y.; Sun, H.L.; Li, L.Y.; Ping, B.; Huang, W.C.; et al. IKK β suppression of TSC1 links inflammation and tumor angiogenesis via the mTOR pathway. *Cell* **2007**, *130*, 440–455. [CrossRef] [PubMed]

43. Huang, W.C.; Ju, T.K.; Hung, M.C.; Chen, C.C. Phosphorylation of CBP by IKKα promotes cell growth by switching the binding preference of CBP from p53 to NF-κB. *Mol. Cell* **2007**, *26*, 75–87. [CrossRef] [PubMed]

44. Rao, P.; Hayden, M.S.; Long, M.; Scott, M.L.; West, A.P.; Zhang, D.; Oeckinghaus, A.; Lynch, C.; Hoffmann, A.; Baltimore, D.; et al. IκBβ acts to inhibit and activate gene expression during the inflammatory response. *Nature* **2010**, *466*, 1115–1119. [CrossRef] [PubMed]

45. Cildir, G.; Low, K.C.; Tergaonkar, V. Noncanonical NF-κB Signaling in Health and Disease. *Trends Mol. Med.* **2016**, *22*, 414–429. [CrossRef] [PubMed]

46. Lin, S.C.; Lo, Y.C.; Wu, H. Helical assembly in the MyD88-IRAK4-IRAK2 complex in TLR/IL-1R signalling. *Nature* **2010**, *465*, 885–890. [CrossRef] [PubMed]

47. Deng, L.; Wang, C.; Spencer, E.; Yang, L.; Braun, A.; You, J.; Slaughter, C.; Pickart, C.; Chen, Z.J. Activation of the IκB kinase complex by TRAF6 requires a dimeric ubiquitin-conjugating enzyme complex and a unique polyubiquitin chain. *Cell* **2000**, *103*, 351–361. [CrossRef]

48. Xu, M.; Skaug, B.; Zeng, W.; Chen, Z.J. A ubiquitin replacement strategy in human cells reveals distinct mechanisms of IKK activation by TNFα and IL-1β. *Mol. Cell* **2009**, *36*, 302–314. [CrossRef] [PubMed]

49. Wang, C.; Deng, L.; Hong, M.; Akkaraju, G.R.; Inoue, J.; Chen, Z.J. TAK1 is a ubiquitin-dependent kinase of MKK and IKK. *Nature* **2001**, *412*, 346–351. [CrossRef] [PubMed]

50. Gack, M.U.; Shin, Y.C.; Joo, C.H.; Urano, T.; Liang, C.; Sun, L.; Takeuchi, O.; Akira, S.; Chen, Z.; Inoue, S.; et al. TRIM25 RING-finger E3 ubiquitin ligase is essential for RIG-I-mediated antiviral activity. *Nature* **2007**, *446*, 916–920. [CrossRef] [PubMed]

51. Le Negrate, G. Viral interference with innate immunity by preventing NF-κB activity. *Cell. Microbiol.* **2012**, *14*, 168–181. [CrossRef] [PubMed]

52. Ikeda, F.; Hecker, C.M.; Rozenknop, A.; Nordmeier, R.D.; Rogov, V.; Hofmann, K.; Akira, S.; Dötsch, V.; Dikic, I. Involvement of the ubiquitin-like domain of TBK1/IKK-i kinases in regulation of IFN-inducible genes. *EMBO J.* **2007**, *26*, 3451–3462. [CrossRef] [PubMed]
53. Ea, C.K.; Deng, L.; Xia, Z.P.; Pineda, G.; Chen, Z.J. Activation of IKK by TNFα requires site-specific ubiquitination of RIP1 and polyubiquitin binding by NEMO. *Mol. Cell* **2006**, *22*, 245–257. [CrossRef] [PubMed]
54. Haas, T.L.; Emmerich, C.H.; Gerlach, B.; Schmukle, A.C.; Cordier, S.M.; Rieser, E.; Feltham, R.; Vince, J.; Warnken, U.; Wenger, T.; et al. Recruitment of the linear ubiquitin chain assembly complex stabilizes the TNF-R1 signaling complex and is required for TNF-mediated gene induction. *Mol. Cell* **2009**, *36*, 831–844. [CrossRef] [PubMed]
55. Poyet, J.L.; Srinivasula, S.M.; Lin, J.H.; Fernandes-Alnemri, T.; Yamaoka, S.; Tsichlis, P.N.; Alnemri, E.S. Activation of the IκB kinases by RIP via IKKgamma /NEMO-mediated oligomerization. *J. Biol. Chem.* **2000**, *275*, 37966–37977. [CrossRef] [PubMed]
56. Scheidereit, C. IκB kinase complexes: Gateways to NF-κB activation and transcription. *Oncogene* **2006**, *25*, 6685–6705. [CrossRef] [PubMed]
57. Chiaravalli, J.; Fontan, E.; Fsihi, H.; Coic, Y.M.; Baleux, F.; Véron, M.; Agou, F. Direct inhibition of NF-κB activation by peptide targeting the NOA ubiquitin binding domain of NEMO. *Biochem. Pharmacol.* **2011**, *82*, 1163–1174. [CrossRef] [PubMed]
58. Bertrand, M.J.; Milutinovic, S.; Dickson, K.M.; Ho, W.C.; Boudreault, A.; Durkin, J.; Gillard, J.W.; Jaquith, J.B.; Morris, S.J.; Barker, P.A. cIAP1 and cIAP2 facilitate cancer cell survival by functioning as E3 ligases that promote RIP1 ubiquitination. *Mol. Cell* **2008**, *30*, 689–700. [CrossRef] [PubMed]
59. Gerlach, B.; Cordier, S.M.; Schmukle, A.C.; Emmerich, C.H.; Rieser, E.; Haas, T.L.; Webb, A.I.; Rickard, J.A.; Anderton, H.; Wong, W.W.; et al. Linear ubiquitination prevents inflammation and regulates immune signalling. *Nature* **2011**, *471*, 591–596. [CrossRef] [PubMed]
60. Ikeda, F.; Deribe, Y.L.; Skånland, S.S.; Stieglitz, B.; Grabbe, C.; Franz-Wachtel, M.; van Wijk, S.J.; Goswami, P.; Nagy, V.; Terzic, J.; et al. SHARPIN forms a linear ubiquitin ligase complex regulating NF-κB activity and apoptosis. *Nature* **2011**, *471*, 637–641. [CrossRef] [PubMed]
61. Tokunaga, F.; Nakagawa, T.; Nakahara, M.; Saeki, Y.; Taniguchi, M.; Sakata, S.; Tanaka, K.; Nakano, H.; Iwai, K. SHARPIN is a component of the NF-κB-activating linear ubiquitin chain assembly complex. *Nature* **2011**, *471*, 633–636. [CrossRef] [PubMed]
62. Tokunaga, F.; Sakata, S.; Saeki, Y.; Satomi, Y.; Kirisako, T.; Kamei, K.; Nakagawa, T.; Kato, M.; Murata, S.; Yamaoka, S.; et al. Involvement of linear polyubiquitylation of NEMO in NF-κB activation. *Nat. Cell Biol.* **2009**, *11*, 123–132. [CrossRef] [PubMed]
63. Thome, M.; Charton, J.E.; Pelzer, C.; Hailfinger, S. Antigen receptor signaling to NF-κB via CARMA1, BCL10, and MALT1. *Cold Spring Harb. Perspect. Biol.* **2010**, *2*, a003004. [CrossRef] [PubMed]
64. Turvey, S.E.; Durandy, A.; Fischer, A.; Fung, S.Y.; Geha, R.S.; Gewies, A.; Giese, T.; Greil, J.; Keller, B.; McKinnon, M.L.; et al. The CARD11-BCL10-MALT1 (CBM) signalosome complex: Stepping into the limelight of human primary immunodeficiency. *J. Allergy Clin. Immunol.* **2014**, *134*, 276–284. [CrossRef] [PubMed]
65. Afonina, I.S.; Elton, L.; Carpentier, I.; Beyaert, R. MALT1—A universal soldier: Multiple strategies to ensure NF-κB activation and target gene expression. *FEBS J.* **2015**, *282*, 3286–3297. [CrossRef] [PubMed]
66. Cheng, J.; Montecalvo, A.; Kane, L.P. Regulation of NF-κB induction by TCR/CD28. *Immunol. Res.* **2011**, *50*, 113–117. [CrossRef] [PubMed]
67. McCool, K.W.; Miyamoto, S. DNA damage-dependent NF-κB activation: NEMO turns nuclear signaling inside out. *Immunol. Rev.* **2012**, *246*, 311–326. [CrossRef] [PubMed]
68. Hayden, M.S.; Ghosh, S. Signaling to NF-κB. *Genes Dev.* **2004**, *18*, 2195–2224. [CrossRef] [PubMed]
69. Hinz, M.; Stilmann, M.; Arslan, S.Ç.; Khanna, K.K.; Dittmar, G.; Scheidereit, C. A cytoplasmic ATM-TRAF6-cIAP1 module links nuclear DNA damage signaling to ubiquitin-mediated NF-κB activation. *Mol. Cell* **2010**, *40*, 63–74. [CrossRef] [PubMed]
70. Huang, T.T.; Wuerzberger-Davis, S.M.; Wu, Z.H.; Miyamoto, S. Sequential modification of NEMO/IKKgamma by SUMO-1 and ubiquitin mediates NF-κB activation by genotoxic stress. *Cell* **2003**, *115*, 565–576. [CrossRef]
71. Lee, M.H.; Mabb, A.M.; Gill, G.B.; Yeh, E.T.; Miyamoto, S. NF-κB induction of the SUMO protease SENP2: A negative feedback loop to attenuate cell survival response to genotoxic stress. *Mol. Cell* **2011**, *43*, 180–191. [CrossRef] [PubMed]

72. Perkins, N.D. Integrating cell-signalling pathways with NF-κB and IKK function. *Nat. Rev. Mol. Cell Biol.* **2007**, *8*, 49–62. [CrossRef] [PubMed]
73. Schmitz, M.L.; Kracht, M.; Saul, V.V. The intricate interplay between RNA viruses and NF-κB. *Biochim. Biophys. Acta* **2014**, *1843*, 2754–2764. [CrossRef] [PubMed]
74. Deorukhkar, A.; Krishnan, S. Targeting inflammatory pathways for tumor radiosensitization. *Biochem. Pharmacol.* **2010**, *80*, 1904–1914. [CrossRef] [PubMed]
75. Wang, C.Y.; Cusack, J.C., Jr.; Liu, R.; Baldwin, A.S., Jr. Control of inducible chemoresistance: Enhanced anti-tumor therapy through increased apoptosis by inhibition of NF-κB. *Nat. Med.* **1999**, *5*, 412–417. [PubMed]
76. Ahmed, K.M.; Zhang, H.; Park, C.C. NF-κB regulates radioresistance mediated by β1-integrin in three-dimensional culture of breast cancer cells. *Cancer Res.* **2013**, *73*, 3737–3748. [CrossRef] [PubMed]
77. Godwin, P.; Baird, A.M.; Heavey, S.; Barr, M.P.; O'Byrne, K.J.; Gately, K. Targeting nuclear factor-κB to overcome resistance to chemotherapy. *Front. Oncol.* **2013**, *3*, 120. [CrossRef] [PubMed]
78. Starenki, D.; Namba, H.; Saenko, V.; Ohtsuru, A.; Yamashita, S. Inhibition of nuclear factor-κB cascade potentiates the effect of a combination treatment of anaplastic thyroid cancer cells. *J. Clin. Endocrinol. Metab.* **2004**, *89*, 410–418. [CrossRef] [PubMed]
79. Baud, V.; Karin, M. Is NF-κB a good target for cancer therapy? Hopes and pitfalls. *Nat. Rev. Drug Discov.* **2009**, *8*, 33–40. [CrossRef] [PubMed]
80. Li, F.; Sethi, G. Targeting transcription factor NF-κB to overcome chemoresistance and radioresistance in cancer therapy. *Biochim. Biophys. Acta* **2010**, *1805*, 167–180. [CrossRef] [PubMed]
81. Bonizzi, G.; Karin, M. The two NF-κB activation pathways and their role in innate and adaptive immunity. *Trends Immunol.* **2004**, *25*, 280–288. [CrossRef] [PubMed]
82. Shih, V.F.; Tsui, R.; Caldwell, A.; Hoffmann, A. A single NFκB system for both canonical and non-canonical signaling. *Cell Res.* **2011**, *21*, 86–102. [CrossRef] [PubMed]
83. Dejardin, E.; Droin, N.M.; Delhase, M.; Haas, E.; Cao, Y.; Makris, C.; Li, Z.W.; Karin, M.; Ware, C.F.; Green, D.R. The lymphotoxin-β receptor induces different patterns of gene expression via two NF-κB pathways. *Immunity* **2002**, *17*, 525–535. [CrossRef]
84. Feng, B.; Cheng, S.; Hsia, C.Y.; King, L.B.; Monroe, J.G.; Liou, H.C. NF-κB inducible genes BCL-X and cyclin E promote immature B-cell proliferation and survival. *Cell Immunol.* **2004**, *232*, 9–20. [CrossRef] [PubMed]
85. Li, Q.; Verma, I.M. NF-κB regulation in the immune system. *Nat. Rev. Immunol.* **2002**, *2*, 725–734. [CrossRef] [PubMed]
86. Rescigno, M.; Martino, M.; Sutherland, C.L.; Gold, M.R.; Ricciardi-Castagnoli, P. Dendritic cell survival and maturation are regulated by different signaling pathways. *J. Exp. Med.* **1998**, *188*, 2175–2180. [CrossRef] [PubMed]
87. Chang, J.; Wang, Z.; Tang, E.; Fan, Z.; McCauley, L.; Franceschi, R.; Guan, K.; Krebsbach, P.H.; Wang, C.Y. Inhibition of osteoblastic bone formation by nuclear factor-κB. *Nat. Med.* **2009**, *15*, 682–689. [CrossRef] [PubMed]
88. Jimi, E.; Aoki, K.; Saito, H.; D'Acquisto, F.; May, M.J.; Nakamura, I.; Sudo, T.; Kojima, T.; Okamoto, F.; Fukushima, H.; et al. Selective inhibition of NF-κB blocks osteoclastogenesis and prevents inflammatory bone destruction in vivo. *Nat. Med.* **2004**, *10*, 617–624. [CrossRef] [PubMed]
89. Dejardin, E. The alternative NF-κB pathway from biochemistry to biology: Pitfalls and promises for future drug development. *Biochem. Pharmacol.* **2006**, *72*, 1161–1179. [CrossRef] [PubMed]
90. Moorthy, A.K.; Savinova, O.V.; Ho, J.Q.; Wang, V.Y.; Vu, D.; Ghosh, G. The 20S proteasome processes NF-κB1 p105 into p50 in a translation-independent manner. *EMBO J.* **2006**, *25*, 1945–1956. [CrossRef] [PubMed]
91. Zarnegar, B.J.; Wang, Y.; Mahoney, D.J.; Dempsey, P.W.; Cheung, H.H.; He, J.; Shiba, T.; Yang, X.; Yeh, W.C.; Mak, T.W.; et al. Noncanonical NF-κB activation requires coordinated assembly of a regulatory complex of the adaptors cIAP1, cIAP2, TRAF2 and TRAF3 and the kinase NIK. *Nat. Immunol.* **2008**, *9*, 1371–1378. [CrossRef] [PubMed]
92. Liao, G.; Zhang, M.; Harhaj, E.W.; Sun, S.C. Regulation of the NF-κB-inducing kinase by tumor necrosis factor receptor-associated factor 3-induced degradation. *J. Biol. Chem.* **2004**, *279*, 26243–26250. [CrossRef] [PubMed]
93. Lawrence, T.; Bebien, M.; Liu, G.Y.; Nizet, V.; Karin, M. IKKα limits macrophage NF-κB activation and contributes to the resolution of inflammation. *Nature* **2005**, *434*, 1138–1143. [CrossRef] [PubMed]
94. Fu, J.; Qu, Z.; Yan, P.; Ishikawa, C.; Aqeilan, R.I.; Rabson, A.B.; Xiao, G. The tumor suppressor gene WWOX links the canonical and noncanonical NF-κB pathways in HTLV-I Tax-mediated tumorigenesis. *Blood* **2011**, *117*, 1652–1661. [CrossRef] [PubMed]

95. Hong, Q.; Hsu, L.J.; Schultz, L.; Pratt, N.; Mattison, J.; Chang, N.S. Zfra affects TNF-mediated cell death by interacting with death domain protein TRADD and negatively regulates the activation of NF-κB, JNK1, p53 and WOX1 during stress response. *BMC Mol. Biol.* **2007**, *8*, 50. [CrossRef] [PubMed]

96. Li, M.Y.; Lai, F.J.; Hsu, L.J.; Lo, C.P.; Cheng, C.L.; Lin, S.R.; Lee, M.H.; Chang, J.Y.; Subhan, D.; Tsai, M.S.; et al. Dramatic co-activation of WWOX/WOX1 with CREB and NF-κB in delayed loss of small dorsal root ganglion neurons upon sciatic nerve transection in rats. *PLoS ONE* **2009**, *4*, e7820. [CrossRef] [PubMed]

97. Huang, S.S.; Su, W.P.; Lin, H.P.; Kuo, H.L.; Wei, H.L.; Chang, N.S. Role of WW Domain-containing Oxidoreductase WWOX in Driving T Cell Acute Lymphoblastic Leukemia Maturation. *J. Biol. Chem.* **2016**, *291*, 17319–17331. [CrossRef] [PubMed]

98. Chen, S.J.; Huang, S.S.; Chang, N.S. Role of WWOX and NF-κB in lung cancer progression. *Transl. Respir. Med.* **2013**, *1*, 15. [CrossRef] [PubMed]

99. Karras, J.R.; Schrock, M.S.; Batar, B.; Huebner, K. Fragile Genes That Are Frequently Altered in Cancer: Players Not Passengers. *Cytogenet. Genome Res.* **2016**, *150*, 208–216. [CrossRef] [PubMed]

100. Jahid, S.; Sun, J.; Gelincik, O.; Blecua, P.; Edelmann, W.; Kucherlapati, R.; Zhou, K.; Jasin, M.; Gümüş, Z.H.; Lipkin, S.M. Inhibition of colorectal cancer genomic copy number alterations and chromosomal fragile site tumor suppressor FHIT and WWOX deletions by DNA mismatch repair. *Oncotarget* **2017**. [CrossRef]

101. Smida, J.; Xu, H.; Zhang, Y.; Baumhoer, D.; Ribi, S.; Kovac, M.; von Luettichau, I.; Bielack, S.; O'Leary, V.B.; Leib-Mösch, C.; et al. Genome-wide analysis of somatic copy number alterations and chromosomal breakages in osteosarcoma. *Int. J. Cancer* **2017**, *141*, 816–828. [CrossRef] [PubMed]

102. Yang, W.; Wang, X.M.; Yuan, H.Y.; Liu, Z.H.; Gao, S.; Peng, L. Exploring the mechanism of WWOX growth inhibitory effects on oral squamous cell carcinoma. *Oncol. Lett.* **2017**, *13*, 3198–3204. [CrossRef] [PubMed]

103. Hu, H.; Brittain, G.C.; Chang, J.H.; Puebla-Osorio, N.; Jin, J.; Zal, A.; Xiao, Y.; Cheng, X.; Chang, M.; Fu, Y.X.; et al. OTUD7B controls non-canonical NF-κB activation through deubiquitination of TRAF3. *Nature* **2013**, *494*, 371–374. [CrossRef] [PubMed]

104. Etzrodt, M.; Cortez-Retamozo, V.; Newton, A.; Zhao, J.; Ng, A.; Wildgruber, M.; Romero, P.; Wurdinger, T.; Xavier, R.; Geissmann, F.; et al. Regulation of monocyte functional heterogeneity by miR-146a and Relb. *Cell Rep.* **2012**, *1*, 317–324. [CrossRef] [PubMed]

105. Gupta, S.C.; Sundaram, C.; Reuter, S.; Aggarwal, B.B. Inhibiting NF-κB activation by small molecules as a therapeutic strategy. *Biochim. Biophys. Acta* **2010**, *1799*, 775–787. [CrossRef] [PubMed]

106. Kwak, J.H.; Jung, J.K.; Lee, H. Nuclear factor-κB inhibitors; a patent review (2006–2010). *Expert Opin. Ther. Pat.* **2011**, *21*, 1897–1910. [CrossRef] [PubMed]

107. Llona-Minguez, S.; Baiget, J.; Mackay, S.P. Small-molecule inhibitors of IκB kinase (IKK) and IKK-related kinases. *Pharm. Pat. Anal.* **2013**, *2*, 481–498. [CrossRef] [PubMed]

108. Croft, M.; Benedict, C.A.; Ware, C.F. Clinical targeting of the TNF and TNFR superfamilies. *Nat. Rev. Drug Discov.* **2013**, *12*, 147–168. [CrossRef] [PubMed]

109. Sedger, L.M.; McDermott, M.F. TNF and TNF-receptors: From mediators of cell death and inflammation to therapeutic giants—Past, present and future. *Cytokine Growth Factor Rev.* **2014**, *25*, 453–472. [CrossRef] [PubMed]

110. Mody, R. Targeting death receptors: Is this trail still hot? *Transl. Pediatr.* **2013**, *2*, 66–69. [CrossRef] [PubMed]

111. Walsh, G. Biopharmaceutical benchmarks. *Nat. Biotechnol.* **2014**, *32*, 992–1000. [CrossRef] [PubMed]

112. Khan, N.; Moskowitz, A.J. Where Do the New Drugs Fit in for Relapsed/Refractory Hodgkin Lymphoma? *Curr. Hematol. Malig. Rep.* **2017**, *12*, 227–233. [CrossRef] [PubMed]

113. Zhang, S.Q.; Smith, S.M.; Zhang, S.Y.; Lynn Wang, Y. Mechanisms of ibrutinib resistance in chronic lymphocytic leukaemia and non-Hodgkin lymphoma. *Br. J. Haematol.* **2015**, *170*, 445–456. [CrossRef] [PubMed]

114. Novero, A.; Ravella, P.M.; Chen, Y.; Dous, G.; Liu, D. Ibrutinib for B cell malignancies. *Exp. Hematol. Oncol.* **2014**, *3*, 4. [CrossRef] [PubMed]

115. FDA Website. 2017. Available online: www.fda.gov (accessed on 10 March 2017).

116. Wiestner, A. The role of B-cell receptor inhibitors in the treatment of patients with chronic lymphocytic leukemia. *Haematologica* **2015**, *100*, 1495–1507. [CrossRef] [PubMed]

117. Wang, J.Q.; Jeelall, Y.S.; Ferguson, L.L.; Horikawa, K. Toll-like receptors and cancer: MYD88 mutation and inflammation. *Front. Immunol.* **2014**, *5*, 367. [CrossRef] [PubMed]

118. Brenner, L.; Arbeit, R.D.; Sullivan, T. IMO-8400, an Antagonist of Toll-like Receptors 7, 8, and 9, in Development for Genetically Defined B-Cell Lymphomas: Safety and Activity in Phase 1 and Phase 2 Clinical Trials. *Blood* **2014**, *124*, 3101.

119. Srivastava, R.; Geng, D.; Liu, Y.; Zheng, L.; Li, Z.; Joseph, M.A.; McKenna, C.; Bansal, N.; Ochoa, A.; Davila, E. Augmentation of therapeutic responses in melanoma by inhibition of IRAK-1,-4. *Cancer Res.* **2012**, *72*, 6209–6216. [CrossRef] [PubMed]

120. Rhyasen, G.W.; Bolanos, L.; Fang, J.; Jerez, A.; Wunderlich, M.; Rigolino, C.; Mathews, L.; Ferrer, M.; Southall, N.; Guha, R.; et al. Targeting IRAK1 as a therapeutic approach for myelodysplastic syndrome. *Cancer Cell* **2013**, *24*, 90–104. [CrossRef] [PubMed]

121. Li, Z.; Younger, K.; Gartenhaus, R.; Joseph, A.M.; Hu, F.; Baer, M.R.; Brown, P.; Davila, E. Inhibition of IRAK1/4 sensitizes T cell acute lymphoblastic leukemia to chemotherapies. *J. Clin. Investig.* **2015**, *125*, 1081–1097. [CrossRef] [PubMed]

122. Cheng, B.Y.L.; Ng, I.O.L.; Lee, T.K.W. Revisiting the Role of TLR/IRAK Signaling and its Therapeutic Potential in Cancer. *J. Liver* **2015**, *5*, 1. [CrossRef]

123. Li, K.; McGee, L.R.; Fisher, B.; Sudom, A.; Liu, J.; Rubenstein, S.M.; Anwer, M.K.; Cushing, T.D.; Shin, Y.; Ayres, M.; et al. Inhibiting NF-κB-inducing kinase (NIK): Discovery, structure-based design, synthesis, structure-activity relationship, and co-crystal structures. *Bioorg. Med. Chem. Lett.* **2013**, *23*, 1238–1244. [CrossRef] [PubMed]

124. De Leon-Boenig, G.; Bowman, K.K.; Feng, J.A.; Crawford, T.; Everett, C.; Franke, Y.; Oh, A.; Stanley, M.; Staben, S.T.; Starovasnik, M.A.; et al. The crystal structure of the catalytic domain of the NF-κB inducing kinase reveals a narrow but flexible active site. *Structure* **2012**, *20*, 1704–1714. [CrossRef] [PubMed]

125. Demchenko, Y.N.; Brents, L.A.; Li, Z.; Bergsagel, L.P.; McGee, L.R.; Kuehl, M.W. Novel inhibitors are cytotoxic for myeloma cells with NFκB inducing kinase-dependent activation of NFκB. *Oncotarget* **2014**, *5*, 4554–4566. [CrossRef] [PubMed]

126. Ren, X.; Li, X.; Jia, L.; Chen, D.; Hou, H.; Rui, L.; Zhao, Y.; Chen, Z. A small-molecule inhibitor of NF-κB-inducing kinase (NIK) protects liver from toxin-induced inflammation, oxidative stress, and injury. *FASEB J.* **2017**, *31*, 711–718. [CrossRef] [PubMed]

127. Gyrd-Hansen, M.; Meier, P. IAPs: From caspase inhibitors to modulators of NF-κB, inflammation and cancer. *Nat. Rev. Cancer* **2010**, *10*, 561–574. [CrossRef] [PubMed]

128. Benetatos, C.A.; Mitsuuchi, Y.; Burns, J.M.; Neiman, E.M.; Condon, S.M.; Yu, G.; Seipel, M.E.; Kapoor, G.S.; Laporte, M.G.; Rippin, S.R.; et al. Birinapant (TL32711), a bivalent SMAC mimetic, targets TRAF2-associated cIAPs, abrogates TNF-induced NF-κB activation, and is active in patient-derived xenograft models. *Mol. Cancer Ther.* **2014**, *13*, 867–879. [CrossRef] [PubMed]

129. Wang, S.; Bai, L.; Lu, J.; Liu, L.; Yang, C. Targeting Inhibitors of Apoptosis Proteins (IAPs) For New Breast Cancer Therapeutics. *J. Mammary Gland Biol. Neoplasia* **2012**, *17*, 217–228. [CrossRef] [PubMed]

130. Fulda, S. Targeting inhibitor of apoptosis proteins for cancer therapy: A double-edge sword? *J. Clin. Oncol.* **2014**, *32*, 3190–3191. [CrossRef] [PubMed]

131. Infante, J.R.; Dees, E.C.; Olszanski, A.J.; Dhuria, S.V.; Sen, S.; Cameron, S.; Cohen, R.B. Phase I dose-escalation study of LCL161, an oral inhibitor of apoptosis proteins inhibitor, in patients with advanced solid tumors. *J. Clin. Oncol.* **2014**, *32*, 3103–3110. [CrossRef] [PubMed]

132. Greten, F.R.; Arkan, M.C.; Bollrath, J.; Hsu, L.C.; Goode, J.; Miething, C.; Göktuna, S.I.; Neuenhahn, M.; Fierer, J.; Paxian, S.; et al. NF-κB is a negative regulator of IL-1β secretion as revealed by genetic and pharmacological inhibition of IKKβ. *Cell* **2007**, *130*, 918–931. [CrossRef] [PubMed]

133. Hsu, L.C.; Enzler, T.; Seita, J.; Timmer, A.M.; Lee, C.Y.; Lai, T.Y.; Yu, G.Y.; Lai, L.C.; Temkin, V.; Sinzig, U.; et al. IL-1β-driven neutrophilia preserves antibacterial defense in the absence of the kinase IKKβ. *Nat Immunol.* **2011**, *12*, 144–150. [CrossRef] [PubMed]

134. Luedde, T.; Schwabe, R.F. NF-κB in the liver—Linking injury, fibrosis and hepatocellular carcinoma. *Nat. Rev. Gastroenterol. Hepatol.* **2011**, *8*, 108–118. [CrossRef] [PubMed]

135. De Falco, F.; Di Giovanni, C.; Cerchia, C.; De Stefano, D.; Capuozzo, A.; Irace, C.; Iuvone, T.; Santamaria, R.; Carnuccio, R.; Lavecchia, A. Novel non-peptide small molecules preventing IKKβ/NEMO association inhibit NF-κB activation in LPS-stimulated J774 macrophages. *Biochem. Pharmacol.* **2016**, *104*, 83–94. [CrossRef] [PubMed]

136. Gamble, C.; McIntosh, K.; Scott, R.; Ho, K.H.; Plevin, R.; Paul, A. Inhibitory κB Kinases as targets for pharmacological regulation. *Br. J. Pharmacol.* **2012**, *165*, 802–819. [CrossRef] [PubMed]

137. Pal, S.; Bhattacharjee, A.; Ali, A.; Mandal, N.C.; Mandal, S.C.; Pal, M. Chronic inflammation and cancer: Potential chemoprevention through nuclear factor κB and p53 mutual antagonism. *J. Inflamm.* **2014**, *11*, 23. [CrossRef] [PubMed]

138. Burke, J.R.; Pattoli, M.A.; Gregor, K.R.; Brassil, P.J.; MacMaster, J.F.; McIntyre, K.W.; Yang, X.; Iotzova, V.S.; Clarke, W.; Strnad, J.; et al. BMS-345541 is a highly selective inhibitor of I κB kinase that binds at an allosteric site of the enzyme and blocks NF-κB-dependent transcription in mice. *J. Biol. Chem.* **2003**, *278*, 1450–1456. [CrossRef] [PubMed]

139. McIntyre, K.W.; Shuster, D.J.; Gillooly, K.M.; Dambach, D.M.; Pattoli, M.A.; Lu, P.; Zhou, X.D.; Qiu, Y.; Zusi, F.C.; Burke, J.R. A highly selective inhibitor of IκB kinase, BMS-345541, blocks both joint inflammation and destruction in collagen-induced arthritis in mice. *Arthritis Rheum.* **2003**, *48*, 2652–2659. [CrossRef] [PubMed]

140. Berger, A.; Quast, S.A.; Plötz, M.; Kammermeier, A.; Eberle, J. Sensitization of melanoma cells for TRAIL-induced apoptosis by BMS-345541 correlates with altered phosphorylation and activation of Bax. *Cell Death Dis.* **2013**, *4*, e477. [CrossRef] [PubMed]

141. MacMaster, J.F.; Dambach, D.M.; Lee, D.B.; Berry, K.K.; Qiu, Y.; Zusi, F.C.; Burke, J.R. An inhibitor of IκB kinase, BMS-345541, blocks endothelial cell adhesion molecule expression and reduces the severity of dextran sulfate sodium-induced colitis in mice. *Inflamm. Res.* **2003**, *52*, 508–511. [CrossRef] [PubMed]

142. Atta-ur-Rahman. *Studies in Natural Products Chemistry*; Elsevier B.V.: Amsterdam, The Netherlands, 2017; Volume 54, pp. 1–414.

143. Grothe, K.; Flechsenhar, K.; Paehler, T.; Ritzeler, O.; Beninga, J.; Saas, J.; Herrmann, M.; Rudolphi, K. IκB kinase inhibition as a potential treatment of osteoarthritis—Results of a clinical proof-of-concept study. *Osteoarthr. Cartil.* **2017**, *25*, 46–52. [CrossRef] [PubMed]

144. Agou, F.; Courtois, G.; Chiaravalli, J.; Baleux, F.; Coïc, Y.M.; Traincard, F.; Israël, A.; Véron, M. Inhibition of NF-κB activation by peptides targeting NF-κB essential modulator (NEMO) oligomerization. *J. Biol. Chem.* **2004**, *279*, 54248–54257. [CrossRef] [PubMed]

145. Vincendeau, M.; Hadian, K.; Messias, A.C.; Brenke, J.K.; Halander, J.; Griesbach, R.; Greczmiel, U.; Bertossi, A.; Stehle, R.; Nagel, D.; et al. Inhibition of Canonical NF-κB Signaling by a Small Molecule Targeting NEMO-Ubiquitin Interaction. *Sci. Rep.* **2016**, *6*, 18934. [CrossRef] [PubMed]

146. Clément, J.F.; Meloche, S.; Servant, M.J. The IKK-related kinases: From innate immunity to oncogenesis. *Cell Res.* **2008**, *18*, 889–899. [CrossRef] [PubMed]

147. Maelfait, J.; Beyaert, R. Emerging role of ubiquitination in antiviral RIG-I signaling. *Microbiol. Mol. Biol.* **2012**, *76*, 33–45. [CrossRef] [PubMed]

148. Pomerantz, J.L.; Baltimore, D. NF-κB activation by a signaling complex containing TRAF2, TANK and TBK1, a novel IKK-related kinase. *EMBO J.* **1999**, *18*, 6694–6704. [CrossRef] [PubMed]

149. Peters, R.T.; Liao, S.M.; Maniatis, T. IKKε is part of a novel PMA-inducible IκB kinase complex. *Mol. Cell* **2000**, *5*, 513–522. [CrossRef]

150. Adli, M.; Baldwin, A.S. IKK-i/IKKepsilon controls constitutive, cancer cell-associated NF-κB activity via regulation of Ser-536 p65/RelA phosphorylation. *J. Biol. Chem.* **2006**, *281*, 26976–26984. [CrossRef] [PubMed]

151. Shen, R.R.; Hahn, W.C. Emerging roles for the non-canonical IKKs in cancer. *Oncogene* **2011**, *30*, 631–641. [CrossRef] [PubMed]

152. Olefsky, J.M. IKKepsilon: A bridge between obesity and inflammation. *Cell* **2009**, *138*, 834–836. [CrossRef] [PubMed]

153. Barbie, D.A.; Tamayo, P.; Boehm, J.S.; Kim, S.Y.; Moody, S.E.; Dunn, I.F.; Schinzel, A.C.; Sandy, P.; Meylan, E.; Scholl, C.; et al. Systematic RNA interference reveals that oncogenic KRAS-driven cancers require TBK1. *Nature* **2009**, *462*, 108–112. [CrossRef] [PubMed]

154. Kim, J.Y.; Welsh, E.A.; Oguz, U.; Fang, B.; Bai, Y.; Kinose, F.; Bronk, C.; Remsing Rix, L.L.; Beg, A.A.; Rix, U.; et al. Dissection of TBK1 signaling via phosphoproteomics in lung cancer cells. *Proc. Natl. Acad. Sci. USA* **2013**, *110*, 12414–12419. [CrossRef] [PubMed]

155. Jiang, Z.; Liu, J.C.; Chung, P.E.; Egan, S.E.; Zacksenhaus, E. Targeting HER2(+) breast cancer: The TBK1/IKKε axis. *Oncoscience* **2014**, *1*, 180–182. [CrossRef] [PubMed]

156. Barbie, T.U.; Alexe, G.; Aref, A.R.; Li, S.; Zhu, Z.; Zhang, X.; Imamura, Y.; Thai, T.C.; Huang, Y.; Bowden, M.; et al. Targeting an IKBKE cytokine network impairs triple-negative breast cancer growth. *J. Clin. Investig.* **2014**, *124*, 5411–5423. [CrossRef] [PubMed]

157. Feldman, R.I.; Wu, J.M.; Polokoff, M.A.; Kochanny, M.J.; Dinter, H.; Zhu, D.; Biroc, S.L.; Alicke, B.; Bryant, J.; Yuan, S.; et al. Novel small molecule inhibitors of 3-phosphoinositide-dependent kinase-1. *J. Biol. Chem.* **2005**, *280*, 19867–19874. [CrossRef] [PubMed]

158. Clark, K.; Plater, L.; Peggie, M.; Cohen, P. Use of the pharmacological inhibitor BX795 to study the regulation and physiological roles of TBK1 and IκB kinase epsilon: A distinct upstream kinase mediates Ser-172 phosphorylation and activation. *J. Biol. Chem.* **2009**, *284*, 14136–14146. [CrossRef] [PubMed]

159. Yu, T.; Yang, Y.; Yin, D.Q.; Hong, S.; Son, Y.J.; Kim, J.H.; Cho, J.Y. TBK1 inhibitors: A review of patent literature (2011–2014). *Expert Opin. Ther. Pat.* **2015**, *25*, 1385–1396. [CrossRef] [PubMed]

160. Domainex. 2017. Available online: http://www.domainex.co.uk/drug-discovery-pipeline/inflammatory-diseases (accessed on 8 March 2017).

161. Richards, B.; Cronin, M.; Seager, N.; Niederjohn, J.; Baichwal, V.; Chan, A.; Robinson, R.; Hess, M.; Davis, T.; Papac, D.; et al. Cellular and In Vivo Properties of MPI-0485520, a Novel and Potent Small Molecule Inhibitor of IKKe. *FASEB J.* **2010**, *24*. [CrossRef]

162. Manasanch, E.E.; Orlowski, R.Z. Proteasome inhibitors in cancer therapy. *Nat. Rev. Clin. Oncol.* **2017**, *14*, 417–433. [CrossRef] [PubMed]

163. Glickman, M.H.; Ciechanover, A. The ubiquitin-proteasome proteolytic pathway: Destruction for the sake of construction. *Physiol. Rev.* **2002**, *82*, 373–428. [CrossRef] [PubMed]

164. Schwartz, A.L.; Ciechanover, A. Targeting proteins for destruction by the ubiquitin system: Implications for human pathobiology. *Annu. Rev. Pharmacol. Toxicol.* **2009**, *49*, 73–96. [CrossRef] [PubMed]

165. Bulatov, E.; Ciulli, A. Targeting Cullin-RING E3 ubiquitin ligases for drug discovery: Structure, assembly and small-molecule modulation. *Biochem. J.* **2015**, *467*, 365–386. [CrossRef] [PubMed]

166. Morrow, J.K.; Lin, H.K.; Sun, S.C.; Zhang, S. Targeting ubiquitination for cancer therapies. *Future Med. Chem.* **2015**, *7*, 2333–2350. [CrossRef] [PubMed]

167. Adams, J. The proteasome: A suitable antineoplastic target. *Nat. Rev. Cancer* **2004**, *4*, 349–360. [CrossRef] [PubMed]

168. Kisselev, A.F.; Goldberg, A.L. Proteasome inhibitors: From research tools to drug candidates. *Chem. Biol.* **2001**, *8*, 739–758. [CrossRef]

169. Goldberg, A.L. Development of proteasome inhibitors as research tools and cancer drugs. *J. Cell Biol.* **2012**, *199*, 583–588. [CrossRef] [PubMed]

170. Russo, S.M.; Tepper, J.E.; Baldwin, A.S., Jr.; Liu, R.; Adams, J.; Elliott, P.; Cusack, J.C., Jr. Enhancement of radiosensitivity by proteasome inhibition: Implications for a role of NF-κB. *Int. J. Radiat. Oncol. Biol. Phys.* **2001**, *50*, 183–193. [CrossRef]

171. Goel, A.; Dispenzieri, A.; Greipp, P.R.; Witzig, T.E.; Mesa, R.A.; Russell, S.J. PS-341-mediated selective targeting of multiple myeloma cells by synergistic increase in ionizing radiation-induced apoptosis. *Exp. Hematol.* **2005**, *33*, 784–795. [CrossRef] [PubMed]

172. Goktas, S.; Baran, Y.; Ural, A.U.; Yazici, S.; Aydur, E.; Basal, S.; Avcu, F.; Pekel, A.; Dirican, B.; Beyzadeoglu, M. Proteasome inhibitor bortezomib increases radiation sensitivity in androgen independent human prostate cancer cells. *Urology* **2010**, *75*, 793–798. [CrossRef] [PubMed]

173. O'Neil, B.H.; Raftery, L.; Calvo, B.F.; Chakravarthy, A.B.; Ivanova, A.; Myers, M.O.; Kim, H.J.; Chan, E.; Wise, P.E.; Caskey, L.S.; et al. A phase I study of bortezomib in combination with standard 5-fluorouracil and external-beam radiation therapy for the treatment of locally advanced or metastatic rectal cancer. *Clin. Colorectal Cancer* **2010**, *9*, 119–125. [CrossRef] [PubMed]

174. Salem, A.; Brown, C.O.; Schibler, J.; Goel, A. Combination chemotherapy increases cytotoxicity of multiple myeloma cells by modification of nuclear factor (NF)-κB activity. *Exp. Hematol.* **2013**, *41*, 209–218. [CrossRef] [PubMed]

175. Zhao, Y.; Foster, N.R.; Meyers, J.P.; Thomas, S.P.; Northfelt, D.W.; Rowland, K.M., Jr.; Mattar, B.; Johnson, D.B.; Molina, J.R.; Mandrekar, S.J.; et al. A phase I/II study of bortezomib in combination with paclitaxel, carboplatin, and concurrent thoracic radiation therapy for non-small-cell lung cancer: North Central Cancer Treatment Group (NCCTG)-N0321. *J. Thorac. Oncol.* **2015**, *10*, 172–180. [CrossRef] [PubMed]

176. Shirley, M. Ixazomib: First Global Approval. *Drugs* **2016**, *76*, 405–411. [CrossRef] [PubMed]

177. Swords, R.T.; Coutre, S.; Maris, M.B.; Zeidner, J.F.; Foran, J.M.; Cruz, J.C.; Erba, H.P.; Berdeja, J.G.; Tam, W.; Vardhanabhuti, S.; et al. Results of a clinical study of pevonedistat (pev), a first-in-class NEDD8-activating enzyme (NAE) inhibitor, combined with azacitidine (aza) in older patients (pts) with acute myeloid leukemia (AML). In Proceedings of the 58th American Society of Hematology Annual Meeting & Exposition, San Diego, CA, USA, 3 December 2016.

178. Kanarek, N.; London, N.; Schueler-Furman, O.; Ben-Neriah, Y. Ubiquitination and degradation of the inhibitors of NF-κB. *Cold Spring Harb. Perspect. Biol.* **2010**, *2*, a000166. [CrossRef] [PubMed]

179. Kanarek, N.; Ben-Neriah, Y. Regulation of NF-κB by ubiquitination and degradation of the IκBs. *Immunol. Rev.* **2012**, *246*, 77–94. [CrossRef] [PubMed]

180. Nakajima, H.; Fujiwara, H.; Furuichi, Y.; Tanaka, K.; Shimbara, N. A novel small-molecule inhibitor of NF-κB signaling. *Biochem. Biophys. Res. Commun.* **2008**, *368*, 1007–1013. [CrossRef] [PubMed]

181. Jiang, X.; Chen, Z.J. The role of ubiquitylation in immune defence and pathogen evasion. *Nat. Rev. Immunol.* **2011**, *12*, 35–48. [CrossRef] [PubMed]

182. Richardson, P.G. Improving the therapeutic index in myeloma. *Blood* **2010**, *116*, 4733–4734. [CrossRef] [PubMed]

183. Horie, R. Molecularly-targeted Strategy and NF-κB in lymphoid malignancies. *J. Clin. Exp. Hematop.* **2013**, *53*, 185–195. [CrossRef] [PubMed]

184. Brassesco, M.S.; Roberto, G.M.; Morales, A.G.; Oliveira, J.C.; Delsin, L.E.; Pezuk, J.A.; Valera, E.T.; Carlotti, C.G., Jr.; Rego, E.M.; De Oliveira, H.F.; et al. Inhibition of NF-κB by Dehydroxymethylepoxyquinomicin Suppresses Invasion and Synergistically Potentiates Temozolomide and γ-Radiation Cytotoxicity in Glioblastoma Cells. *Chemother. Res. Pract.* **2013**, *2013*, 593020. [CrossRef] [PubMed]

185. Kozakai, N.; Kikuchi, E.; Hasegawa, M.; Suzuki, E.; Ide, H.; Miyajima, A.; Horiguchi, Y.; Nakashima, J.; Umezawa, K.; Shigematsu, N.; et al. Enhancement of radiosensitivity by a unique novel NF-κB inhibitor, DHMEQ, in prostate cancer. *Br. J. Cancer* **2012**, *107*, 652–657. [CrossRef] [PubMed]

186. Nishio, H.; Yaguchi, T.; Sugiyama, J.; Sumimoto, H.; Umezawa, K.; Iwata, T.; Susumu, N.; Fujii, T.; Kawamura, N.; Kobayashi, A.; et al. Immunosuppression through constitutively activated NF-κB signalling in human ovarian cancer and its reversal by an NF-κB inhibitor. *Br. J. Cancer* **2014**, *110*, 2965–2974. [CrossRef] [PubMed]

187. Verzella, D.; Fischietti, M.; Capece, D.; Vecchiotti, D.; Del Vecchio, F.; Cicciarelli, G.; Mastroiaco, V.; Tessitore, A.; Alesse, E.; Zazzeroni, F. Targeting the NF-κB pathway in prostate cancer: A promising therapeutic approach? *Curr. Drug Targets.* **2016**, *17*, 311–320. [CrossRef] [PubMed]

188. Ghantous, A.; Gali-Muhtasib, H.; Vuorela, H.; Saliba, N.A.; Darwiche, N. What made sesquiterpene lactones reach cancer clinical trials? *Drug Discov. Today* **2010**, *15*, 668–678. [CrossRef] [PubMed]

189. Shanmugam, R.; Kusumanchi, P.; Appaiah, H.; Cheng, L.; Crooks, P.; Neelakantan, S.; Peat, T.; Klaunig, J.; Matthews, W.; Nakshatri, H.; et al. A water soluble parthenolide analogue suppresses in vivo tumor growth of two tobacco associated cancers, lung and bladder cancer, by targeting NF-κB and generating reactive oxygen species. *Int. J. Cancer* **2011**, *128*, 2481–2494. [CrossRef] [PubMed]

190. Adis Insight. 2017. Available online: http://adisinsight.springer.com/search (accessed on 25 January 2017).

191. NF-kB Target Genes. 2017. Available online: https://www.bu.edu/nf-kb/gene-resources/target-genes/ (accessed on 25 January 2017).

192. Bennett, J.; Moretti, M.; Thotakura, A.K.; Tornatore, L.; Franzoso, G. The Regulation of the JNK Cascade and Programmed Cell Death by NF-κB: Mechanisms and Functions. In *Trends in Stem Cell. Proliferation and Cancer Research*; Springer: Dordrecht, The Netherlands, 2013.

193. Jin, R.; De Smaele, E.; Zazzeroni, F.; Nguyen, D.U.; Papa, S.; Jones, J.; Cox, C.; Gelinas, C.; Franzoso, G. Regulation of the GADD45β promoter by NF-κB. *DNA Cell Biol.* **2002**, *21*, 491–503. [CrossRef] [PubMed]

194. De Smaele, E.; Zazzeroni, F.; Papa, S.; Nguyen, D.U.; Jin, R.; Jones, J.; Cong, R.; Franzoso, G. Induction of GADD45β by NF-κB downregulates pro-apoptotic JNK signalling. *Nature* **2001**, *414*, 308–313. [CrossRef] [PubMed]

195. Salvador, J.M.; Brown-Clay, J.D.; Fornace, A.J., Jr. GADD45 in stress signaling, cell cycle control, and apoptosis. *Adv. Exp. Med. Biol.* **2013**, *793*, 1–19. [CrossRef] [PubMed]

196. Schäfer, A. GADD45 proteins: Key players of repair-mediated DNA demethylation. *Adv. Exp. Med. Biol.* **2013**, *793*, 35–50. [CrossRef] [PubMed]

197. Zhang, L.; Yang, Z.; Liu, Y. GADD45 proteins: Roles in cellular senescence and tumor development. *Exp. Biol. Med.* **2014**, *239*, 773–778. [CrossRef] [PubMed]

198. Papa, S.; Zazzeroni, F.; Bubici, C.; Jayawardena, S.; Alvarez, K.; Matsuda, S.; Nguyen, D.U.; Pham, C.G.; Nelsbach, A.H.; Melis, T.; et al. GADD45 β mediates the NF-κB suppression of JNK signalling by targeting MKK7/JNKK2. *Nat. Cell Biol.* **2004**, *6*, 146–153. [CrossRef] [PubMed]

199. Larsen, C.M.; Døssing, M.G.; Papa, S.; Franzoso, G.; Billestrup, N.; Mandrup-Poulsen, T. Growth arrest- and DNA-damage-inducible 45β gene inhibits c-Jun N-terminal kinase and extracellular signal-regulated kinase and decreases IL-1β-induced apoptosis in insulin-producing INS-1E cells. *Diabetologia* **2006**, *49*, 980–989. [CrossRef] [PubMed]

200. Tornatore, L.; Sandomenico, A.; Raimondo, D.; Low, C.; Rocci, A.; Tralau-Stewart, C.; Capece, D.; D'Andrea, D.; Bua, M.; Boyle, E.; et al. Cancer-selective targeting of the NF-κB survival pathway with GADD45β/MKK7 inhibitors. *Cancer Cell* **2014**, *26*, 495–508. [CrossRef] [PubMed]

201. Tornatore, L.; Acton, G.; Adams, N.; Campbell, E.A.; Kelly, J.; Szydlo, R.M.; Tarbit, M.; Bannoo, S.; D'Andrea, D.; Capece, D.; et al. Cancer-Selective Targeting of the NF-κB Survival Pathway in Multiple Myeloma with the GADD45β/MKK7 Inhibitor, DTP3. *Blood* **2015**, *126*, 868.

202. Tang, G.; Minemoto, Y.; Dibling, B.; Purcell, N.H.; Li, Z.; Karin, M.; Lin, A. Inhibition of JNK activation through NF-κB target genes. *Nature* **2001**, *414*, 313–317. [CrossRef] [PubMed]

203. Papa, S.; Monti, S.M.; Vitale, R.M.; Bubici, C.; Jayawardena, S.; Alvarez, K.; De Smaele, E.; Dathan, N.; Pedone, C.; Ruvo, M.; et al. Insights into the structural basis of the GADD45β-mediated inactivation of the JNK kinase, MKK7/JNKK2. *J. Biol. Chem.* **2007**, *282*, 19029–19041. [CrossRef] [PubMed]

204. Zhang, N.; Ahsan, M.H.; Zhu, L.; Sambucetti, L.C.; Purchio, A.F.; West, D.B. NF-κB and not the MAPK signaling pathway regulates GADD45β expression during acute inflammation. *J. Biol. Chem.* **2005**, *280*, 21400–21408. [CrossRef] [PubMed]

205. Papa, S.; Zazzeroni, F.; Fu, Y.X.; Bubici, C.; Alvarez, K.; Dean, K.; Christiansen, P.A.; Anders, R.A.; Franzoso, G. Gadd45β promotes hepatocyte survival during liver regeneration in mice by modulating JNK signaling. *J. Clin. Investig.* **2008**, *118*, 1911–1923. [CrossRef] [PubMed]

206. Lu, B.; Ferrandino, A.F.; Flavell, R.A. GADD45β is important for perpetuating cognate and inflammatory signals in T cells. *Nat. Immunol.* **2004**, *5*, 38–44. [CrossRef] [PubMed]

207. Karin, M. Whipping NF-κB to Submission via GADD45 and MKK7. *Cancer Cell* **2014**, *26*, 447–449. [CrossRef] [PubMed]

208. Vaccarezza, M.; Vitale, M. Harnessing downstream NF-κB signalling to achieve apoptosis-inducing anti-cancer-specific activity. *Cell Death Dis.* **2016**, *7*, e2306. [CrossRef] [PubMed]

209. Catz, S.D.; Johnson, J.L. Transcriptional regulation of bcl-2 by nuclear factor κB and its significance in prostate cancer. *Oncogene* **2001**, *20*, 7342–7351. [CrossRef] [PubMed]

210. Gomez-Bougie, P.; Amiot, M. Apoptotic machinery diversity in multiple myeloma molecular subtypes. *Front. Immunol.* **2013**, *4*, 467. [CrossRef] [PubMed]

211. Cang, S.; Iragavarapu, C.; Savooji, J.; Song, Y.; Liu, D. ABT-199 (venetoclax) and BCL-2 inhibitors in clinical development. *J. Hematol. Oncol.* **2015**, *8*, 129. [CrossRef] [PubMed]

212. Vogler, M. Targeting BCL2-Proteins for the Treatment of Solid Tumours. *Adv. Med.* **2014**, *2014*, 943648. [CrossRef] [PubMed]

213. Pandey, M.K.; Prasad, S.; Tyagi, A.K.; Deb, L.; Huang, J.; Karelia, D.N.; Amin, S.G.; Aggarwal, B.B. Targeting Cell Survival Proteins for Cancer Cell Death. *Pharmaceuticals* **2016**, *9*, 11. [CrossRef] [PubMed]

214. Seyfried, F.; Demir, S.; Debatin, K.M.; Meyer, L.H. Synergistic Activity of ABT-199 with Conventional Chemotherapy and Dinaciclib in B-Cell Precursor Acute Lymphoblastic Leukemia. *Blood* **2015**, *126*, 2631.

215. Fox, J.L.; MacFarlane, M. Targeting cell death signalling in cancer: Minimising "Collateral damage". *Br. J. Cancer* **2016**, *115*, 5–11. [CrossRef] [PubMed]

216. Levy, M.A.; Claxton, D.F. Therapeutic inhibition of BCL-2 and related family members. *Expert Opin. Investig. Drugs* **2017**, *26*, 293–301. [CrossRef] [PubMed]

biomedicines

MDPI

Review

Noncanonical NF-κB in Cancer

Matthew Tegowski [1] **and Albert Baldwin** [2,*]

1   Curriculum of Genetics and Molecular Biology, The University of North Carolina at Chapel Hill, Chapel Hill, NC 27599, USA; tegowski@email.unc.edu
2   Lineberger Comprehensive Cancer Center, The University of North Carolina at Chapel Hill, Chapel Hill, NC 27599, USA
*   Correspondence: abaldwin@med.unc.edu; Tel.: +1-919-966-3652

Received: 10 May 2018; Accepted: 4 June 2018; Published: 5 June 2018

Abstract: The NF-κB pathway is a critical regulator of immune responses and is often dysregulated in cancer. Two NF-κB pathways have been described to mediate these responses, the canonical and the noncanonical. While understudied compared to the canonical NF-κB pathway, noncanonical NF-κB and its components have been shown to have effects, usually protumorigenic, in many different cancer types. Here, we review noncanonical NF-κB pathways and discuss its important roles in promoting cancer. We also discuss alternative NF-κB-independent functions of some the components of noncanonical NF-κB signaling. Finally, we discuss important crosstalk between canonical and noncanonical signaling, which blurs the two pathways, indicating that understanding the full picture of NF-κB regulation is critical to deciphering how this broad pathway promotes oncogenesis.

Keywords: noncanonical NF-κB; cancer; cellular signaling; inflammation; tumor initiating cells; NIK; RelB; p52

1. Introduction

The Nuclear factor kappa-light-chain-enhancer of activated B cells (NF-κB) pathway is an important regulator of innate and adaptive immune responses, where it regulates responses to pathogens, as well as T and B cell activation [1]. NF-κB regulates immune responses by promoting the transcription of proinflammatory and antiapoptotic genes. Additionally, diverse stimuli such as UV radiation, DNA damage, cytokines, growth factors, and reactive oxygen species have all been shown to lead to NF-κB activation [2]. Activation of NF-κB subunits leads to their nuclear translocation and activation of transcription, and the NF-κB pathway is known to regulate the transcription of many genes including proinflammatory cytokines and chemokines (e.g., IL-6 [3]), cell cycle genes (e.g., cyclin D1 [4]), antiapoptotic genes (e.g., bcl-2 [5]), and extracellular proteases (e.g., MMP3 [6]). Chronic inflammation and DNA damage have long been associated with the development of cancer, and dysregulation of the NF-κB pathway has important effects in cancer [7].

NF-κB pathway activation leads to transcription regulation by dimers of 5 related transcription factors (RelA/p65, RelB, c-Rel, NFKB1/p105, and NFKB2/p100). NFKB1/p105 and NFKB2/p100 subunits require posttranslational proteolytic processing before they can support transcription activation. NFKB1/p105 is thought to be constitutively processed into the active p50 subunit concurrent with translation [8], whereas NFKB2/p100 remains unprocessed until noncanonical pathway activation induces its proteasome-dependent processing into the active p52 subunit (Figure 1) [9]. Although many combinations of dimers have been observed, the most widely studied dimers are the RelA-p50 dimer, which is primarily activated by canonical NF-κB signaling, and the RelB-p52 dimer, which is activated by noncanonical NF-κB signaling.

Activation of canonical NF-κB is dependent on a kinase complex that contains the scaffold protein NF-κB essential modifier (NEMO) and the inhibitor of NF-κB kinase β (IKKβ) and IKKα.

Upon activation, IKKβ phosphorylates the inhibitor of NF-κB α (IκBα) that binds to and inhibits RelA-p50 dimers, confining the dimers to the cytoplasm. Phosphorylated IκBα is rapidly targeted for degradation by the proteasome, with subsequent accumulation of nuclear RelA-p50 [10,11]. Noncanonical NF-κB, however, is dependent on the stabilization of a labile kinase, NF-κB-inducing kinase (NIK) and the catalytic activity of IKKα.

While the canonical NF-κB pathway is rapidly inducible and can be activated by inflammatory cytokines and other stimuli, the noncanonical NF-κB pathway is primarily activated by a set of cytokine/receptor pairs in the tumor necrosis factor receptor superfamily including BAFF receptor (BAFFR), CD40, lymphotoxin B receptor (LTβR), Fn14, and receptor activator of nuclear factor kappa-B (RANK) [12–16]. In the absence of a stimulus, noncanonical NF-κB is kept inactive by the continual ubiquitination and proteasomal degradation of the critical upstream kinase NIK. TNF-receptor associated factor 3 (TRAF3) is critical for maintaining low basal NIK levels, and cells with inactivating TRAF3 mutations have upregulated NIK protein levels [17–19]. TRAF3, however does not have ubiquitinase activity. Instead, TRAF3 recruits NIK to a degradation complex containing TRAF2 and cellular inhibitors of apoptosis 1 (cIAP1) and cIAP2 [20–22]. All components of this complex are required, although cIAP1 and cIAP2 seem to have redundant functions and NIK stabilization does not occur unless both cIAP1 and cIAP2 are reduced [20]. Upon complex formation, NIK protein is marked for proteasome-mediated degradation with K48-linked ubiquitin chains [17]. The continual degradation of this critical kinase in the absence of a stimulus keeps the noncanonical NF-κB pathway inactive.

Upon receptor stimulation, the NIK-degradation complex is recruited to the active receptor complex. Instead of marking NIK for degradation, cIAP1 and cIAP2 ubiquitylate TRAF2 and TRAF3, which are then rapidly degraded (Figure 1) [17,20,23]. Without TRAF3 recruiting NIK for destruction, NIK protein rapidly accumulates. NIK phosphorylates residues on p100 and the downstream kinase IKKα [9,24,25]. Phosphorylation of IKKα activates its kinase activity and IKKα phosphorylates several residues on the C-teminus of p100 [9]. Although NIK phosphorylates p100, IKKα activity is required for p100 processing, and mutating IKKα phosphorylation sites on p100 prevents processing and pathway activation (Figure 1) [9,25]. Phosphorylated p100 is recognized by the E3 ubiquitin ligase βTRCP and processed into the p52 subunit in a proteasome-dependent manner [26]. Although IKKα activity is required for p100 processing, and IKKβ activity is not, it is unclear what the IKKα-containing complex that phosphorylates p100 consists of. Recent work from the Ghosh lab suggests that IKKα can form a hexamer in vitro, and IKKα mutated at residues predicted to be important for hexamerization fails to transduce noncanonical signaling [27]. However, size exclusion chromatography shows that overexpressed IKKα exists predominately as dimers in cells [27]. This suggests that if the hexadimeric complexes exist, they may be transient upon noncanonical receptor stimulation. After processing, the IκB-like inhibitory properties of p100 (see below) are lost and the RelB-p52 dimer can promote transcription of target genes (Figure 1). Although noncanonical and canonical NF-κB regulate different target genes, there is significant overlap in regulated genes [28]. While there is evidence that these differences arise from small variations in the DNA sequence of κB sites [29,30], there is still a lot unknown about how dimer specificity is determined and what the biological relevance may be [28]. In fact, although the noncanonical pathway has been elucidated as quite separate from the canonical NF-κB pathway, there is significant crossregulation between the components of each pathway, emphasizing the importance of the NF-κB system as a single, highly complex system with disease relevance in many types of cancer.

The processing of p100 is a critical event in the noncanonical NF-κB pathway, in order to generate the functional p52 transcription factor. This processing removes C-terminal ankyrin repeats, which are similar to those in the IκB proteins that sequester and inhibit the canonical RelA-p50 dimers. Therefore, p100 binds and inhibits RelB in the absence of an activating signal, and p100 processing is a critical event in the activation of the noncanonical NF-κB pathway [31–33]. Further, p100 has been shown to inhibit the canonical pathway by binding and inhibiting RelA, an activity known as IκBδ, through its ankyrin repeats [28,34]. In this review, we explore what is currently known about the impact of noncanonical NF-κB components on cancer initiation, growth, and survival. We also explore functions of noncanonical NF-κB kinases outside of their traditional roles in regulating this pathway. Further, we explore interesting crossregulation mechanisms linking canonical NF-κB components with noncanonical NF-κB.

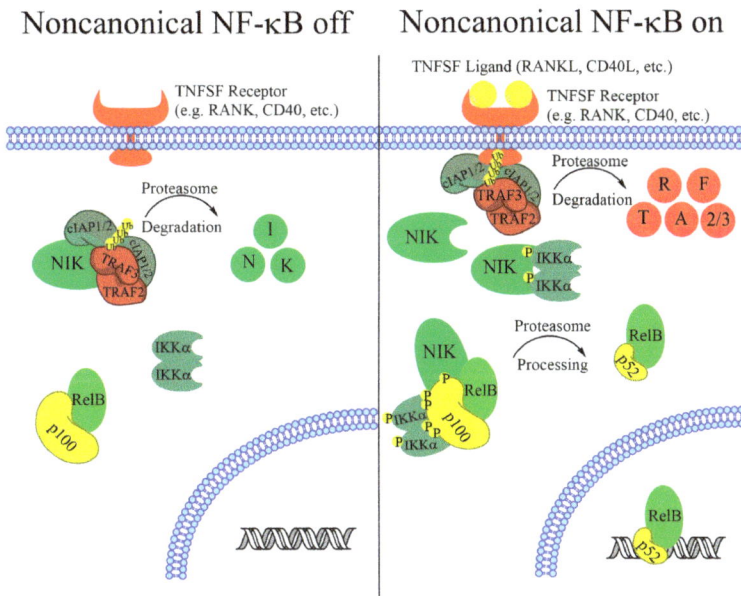

Figure 1. Overview of the Noncanonical NF-κB signaling pathway. In the absence of a stimulus the critical kinase NIK is constitutively targeted for degradation by a ubiquitination complex containing TRAF2, TRAF3, cIAP1, and cIAP2. Upon receptor activation, NIK is stabilized, leading to IKKα activation and p100 processing to p52. RelB-p52 dimers then translocate to the nucleus and activate transcription.

2. Noncanonical NF-κB Activation in Multiple Myeloma

Noncanonical NF-κB is critical for B-cell homeostasis as well as lymph node and germinal center establishment [35–39]. Multiple myeloma (MM) is caused by uncontrolled proliferation of bone marrow plasma cells, which are post germinal center mature B cells that secrete antibodies into the serum [40]. Since both canonical and noncanonical NF-κB are critical for B-cell development and germinal center formation [41], it is perhaps expected that NF-κB activation would be an important development in MM. MM cell adhesion to bone marrow stromal cells induces NF-κB activity and drug resistance [42,43]. Further, production of the NF-κB target gene RANKL in the tumor cells activates osteoclasts, leading to bone resorption at the sites of lesions [16,44].

Although dysregulation of the NF-κB pathway occurs in many tumor types, mutations in NF-κB pathway regulators are rare in solid tumors [7]. However, NF-κB-activating mutations are estimated to occur in about 20% of multiple myeloma patients [45,46]. Mutations in CD40, LTβR, NIK, TRAF2, TRAF3, cIAP1, cIAP2, CYLD, NFKB1/p105, and NFKB2/p100 have been observed, with TRAF3 inactivation being the most frequent alteration. NIK protein is stabilized in TRAF3 mutant MM cell lines, which induces constitutive processing of p100 to p52 and hyperactive noncanonical NF-κB [45,46].

Both canonical and noncanonical NF-κB depend on proteasome activity for activation. Canonical NF-κB requires the degradation of the IκB proteins to release the RelA-p50 dimers, whereas noncanonical NF-κB requires the proteasome for p100 processing to p52. The glucocorticoid dexamethasone, which can block NF-κB signaling [47,48], and the proteasome inhibitor bortezomib are impactful treatments for MM in the clinic [49]. Interestingly, patients with TRAF3 mutations, and therefore hyperactive noncanonical NF-κB, are resistant to dexamethasone treatment. However, TRAF3 mutant tumors were found to be exquisitely responsive to proteasome inhibition with bortezomib [45]. This heightened sensitivity to proteasome inhibition in tumors with hyperactive NF-κB can informs treatment decisions. Further, these lessons may apply to tumor types other than MM where NF-κB is persistently activated, even if no mutations in the pathway are present.

Even though primarily noncanonical-regulating genes are mutated in MM, activation of canonical NF-κB is observed in many MM cells. Further, inhibition of both canonical and noncanonical pathways is effective in reducing growth and promoting cell death in many MM cell lines [46,50], indicating that canonical NF-κB is also critical for cell survival in many MM cells. This suggests that while MM has a special dependence on activation of the noncanonical pathway, significant crosstalk between the NF-κB pathways occurs, and both are often required for MM cell survival and proliferation. Importantly, targeting NF-κB activity by inhibiting IKKs or the proteasome induces cytotoxicity in MM cells [46,49]. Finding new methods of targeting NF-κB in multiple myeloma cells, as well as preventing resistance to these therapies would be important to improving treatment success in this disease.

3. Noncanonical NF-κB in Other Cancers

Noncanonical NF-κB has been shown to regulate mammary gland development [51,52]. It is, therefore, intriguing that activity of the noncanonical NF-κB transcription factors RelB and p100/p52 (NFKB2) have been implicated in promoting breast cancer. It has been known for decades that breast cancer cells show increased NF-κB activity, which supports growth and survival [53]. More recently, the expression of RelB and NFKB2 was shown to be elevated in estrogen receptor negative (ER$^-$) breast tumors, compared to ER$^+$ tumors [54]. ER$^-$ tumors are generally more aggressive and have a worse prognosis than ER$^+$ tumors. Additionally, estrogen receptor has been shown to repress NF-κB activity [55], and FoxA1, a cofactor critical for estrogen receptor transcriptional activity [56], also directly represses RelB expression [57]. RelB in turn has been shown to inhibit estrogen receptor expression in breast cancer cells by upregulating a transcriptional repressor Blimp1 [58]. RelB expression in breast cancer cells promotes an epithelial-to-mesenchymal transition (EMT), and supports the self-renewal of tumor initiating cells [55,59,60]. Further, patients with higher expression of RelB and NFKB2 had decreased disease free survival as well as decreased overall survival [54]. The noncanonical NF-κB transcription factors are upregulated in ER$^-$ breast cancers, and they effect important processes such as EMT and self-renewal. However, since the critical upstream kinase NIK, has not been found to be widely stabilized in breast cancer cells, more work needs to be done to uncover the mechanisms by which RelB and NFKB2 expression and activity are promoted in breast cancers.

The activity of another hormone receptor, androgen receptor (AR), is intimately linked with noncanonical NF-κB signaling. Many prostate tumors are driven by AR activity, and can be effectively treated with surgical or chemical androgen deprivation therapy [61]. Some prostate cancers will acquire resistance and become castration-resistant tumors, often as a result of AR ligand-independent activity [62]. Work from the Gao laboratory demonstrated that p52 expression could support tumor growth of androgen-dependent LNCaP prostate cancer cells in castrated mice [63]. Further, p52 could induce the nuclear localization and DNA binding activity of AR in the absence of ligand. The expression of p52 is high in a castration-resistant subline of LNCaP cells, and importantly, knockdown of p52 in these cells reduces AR activity and DNA binding [63]. How noncanonical NF-κB is activated in castration-resistant prostate cancer needs to be elucidated to determine whether inhibition of noncanonical NF-κB may be a druggable pathway in castration-resistant prostate cancer.

Activation of noncanonical NF-κB, along with NIK stabilization and constitutive p100 processing, has also been observed in pancreatic cancer cell lines [64,65]. Interestingly, multiple mechanisms for noncanonical NF-κB activation in pancreatic cancer cells have been discovered. First, work from the Storz lab has shown that NIK stabilization is critical for proliferation and tumorigenicity of pancreatic cells [66]. Further, expression of a negative regulator of NIK, TRAF2, was shown to be limited as a consequence of constitutive degradation [66]. Work from our lab has also shown that GSK3α activity activates both canonical and noncanonical NF-κB signaling, and that GSK3α activity supports the proliferation and survival of pancreatic cancer cells [67]. In addition to described roles in hematologic malignancies, prostate cancer, breast cancer, and pancreatic cancer, the noncanonical NF-κB pathway and its components have been shown to support the proliferation, survival, or tumor initiating cells of glioma [68,69], ovarian [70], and endometrial cancers [71].

Recently, RelB-p52 dimers were shown to regulate a group of nucleic acid editing enzymes (APOBECs) that are implicated in tumor progression [72]. A growing body of evidence suggests that APOBEC enzymes are responsible for increasing tumorigenesis by inducing mutations at cytosine residues immediately preceded by a thymine [72], especially in breast and ovarian cancers [73]. APOBEC enzymes are a family of enzymes that deaminate cytosines on viral RNA to leave an abundance of uracil residues. This activity behaves in practice as hypermutation of viral RNA [74]. Although they edit RNA residues, APOBEC proteins have been shown to mutate DNA in cancer, where the resulting genomic hypermutation can drive tumor growth. Given its roles in the innate immune response, a possible connection between APOBEC expression and NF-κB activity seems likely. Indeed, the noncanonical NF-κB transcription factors RelB and p52 were recently shown to induce APOBEC3B expression in some cancer cells [75]. Leonard et al. demonstrated that protein kinase C (PKC) activation leads to increased APOBEC3B expression by inducing the binding of RelB and p52 to several sites in the APOBEC3B promoter. Further, this activity was sensitive to both IKK and proteasome inhibitors, indicating that IKK activation and p100 processing are required for this activity [75].

TERT is the catalytic subunit for the telomerase enzyme, which maintains the protective telomeres at the ends of chromosomes that would otherwise be lost progressively every cell division [76]. Maintenance of telomeres is required for the indefinite proliferative potential of cancer cells and TERT activity can be measured in 80–90% of cancers [76,77]. In some cancers, TERT upregulation occurs as a result of point mutations at hotspots in the TERT promoter, which allow for the binding of new transcription factors, such as the Myc or Ets transcription factors [78,79]. Activation of noncanonical NF-κB, but not canonical NF-κB, was shown to increase TERT expression in glioblastoma cells containing one of these hotspot mutations (C250T) [80]. Ets1/2 transcription factors were found to be recruited to the mutated TERT promoter by p52 in order to induce TERT expression (Figure 2), and reversion of the C250T promoter mutation blocked p52 binding, and noncanonical NF-κB-induced TERT expression [80]. Further, noncanonical NF-κB was required for in vivo tumor growth [80]. This shows that activation of noncanonical NF-κB can support tumor growth of some glioblastoma by upregulating TERT expression.

Although p52 has generally been shown to promote noncanonical NF-κB and oncogenic functions, its precursor, p100, has been demonstrated to act as a tumor suppressor in an NF-κB-independent mechanism [81]. p100, but not p52, was shown to inhibit anchorage-independent growth by directly interacting with ERK2, and via inhibition of c-Jun/AP-1, inducing the downregulation of miR-494. PTEN expression, normally inhibited by microRNA (miRNA) miR-494, increased [81]. Another study from the same group demonstrated that p100, but not p52, could limit the proliferation of bladder cancer cells by promoting miR-302 production via CREB activity. Additionally, miR-302 was demonstrated to suppress cyclin D1 expression, leading to decreased proliferation [82].

The activity of RelB and NFKB2/p52 has been demonstrated in numerous cancer types, and they regulate a diverse set of genes. When activated, these noncanonical NF-κB transcription factors promote tumor initiation, growth, and survival. Inhibition of their activity could yield advances in the treatment of numerous cancers, including breast, brain, ovarian, and prostate.

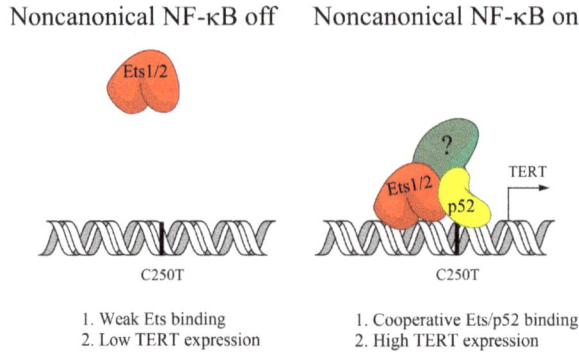

Figure 2. Cooperation of Ets and p52 in promoting TERT expression when promoter is mutated. The C250T mutation in the TERT promoter creates a p52 binding site. When noncanonical NF-κB is activated in C250T mutant cells, p52 can recruit Ets transcription factors to the mutated TERT promoter and drive transcription.

4. RelB-p52 and EZH2 Cooperation in Cancer

As discussed earlier, activation of the noncanonical NF-κB transcription factors, namely RelB and p52, can support the growth of several cancer types. Recently, a body of evidence has emerged linking RelB-p52 activity with enhancer of zeste homolog 2 (EZH2). EZH2 is a histone methyltransferase that, as the catalytic subunit of the polycomb repressive complex 2 (PRC2), trimethylates H3K27 (H3K27me3) and leads to gene repression [83–86]. EZH2, through its catalytic activity, has been shown to be a tumor promoter, because it can repress tumor suppressors such as p16^{Ink4a} and p14^{ARF} [87]. Iannetti et al. showed that noncanonical NF-κB promotes the expression of EZH2 in chronic lymphocytic leukemia (CLL) cells. Active RelB-p52 promoting EZH2 expression supports a blockade on p53-mediated entry into senescence by downregulating p53 target genes (Figure 3). ChIP analysis in human fibroblasts revealed that RelB and p52 bind to the EZH2 promoter [88], indicating direct transcriptional regulation of EZH2 by noncanonical NF-κB. A supporting study in melanoma cells suggests that noncanonical NF-κB upregulation of EZH2 may be a general mechanism to bypass p53-induced senescence [89]. Interestingly, this study showed that NIK protein was stabilized in melanoma cells and is critical for EZH2 expression. Canonical NF-κB subunits were not found at the EZH2 promoter, and knockdown of RelA had no effect on EZH2 expression [89]. This suggests that activation of noncanonical NF-κB can support EZH2 expression and avoid senescence, without significant crosstalk with canonical NF-κB pathway members.

Although EZH2 is a well-established transcriptional repressor, functioning through its catalytic activity to methylate H3K27, a growing body of evidence has shown that EZH2 has PRC2-independent roles as a transcriptional activator. SWI/SNF mutant tumors were shown to require EZH2 for growth. Interestingly, EZH2 depletion in some SWI/SNF mutant tumors was rescued with a catalytically inactive mutant, suggesting alternative functions of EZH2 [90]. Further, EZH2 was found to act as a coactivator to support the transcriptional activation of target genes controlled by Notch [91], Wnt [92,93], estrogen receptor [92], and androgen receptor [94]. Interestingly, EZH2 was also found to support NF-κB target gene expression in triple-negative breast cancer cells by acting as a coactivator in a complex containing RelA and RelB (Figure 3) [95]. This was further supported by a study from our group, which showed that EZH2 and RelA localized to the RelB promoter and supported transcription of RelB in triple-negative breast cancer cells (Figure 3). This activity was critical for the cancer cells to maintain tumor initiating cells [59]. Intriguingly, this NF-κB-supporting activity in triple-negative breast cancer cells is independent of its methyltransferase activity, and PRC2 complex members, like SUZ12 [60,95]. Since EZH2 enzymatic activity is dispensable for this regulatory mechanism, more work needs to be done to elucidate therapeutic mechanisms to target this noncanonical function of EZH2.

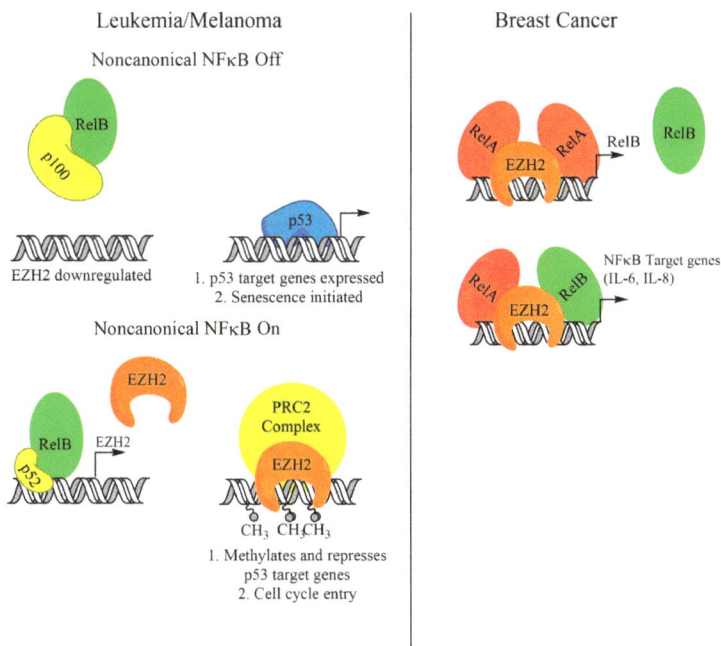

Figure 3. Crosstalk between EZH2 and noncanonical NF-κB components in CLL, melanoma, and breast cancer. Work in CLL and melanoma shows that RelB-p52 dimers can promote the transcription of EZH2. EZH2 can then promote a senescence bypass by repressing p53 target genes.

5. Noncanonical IκB Proteins

Originally discovered to inhibit the NF-κB response, IκB proteins are generally considered to be cytoplasmic molecules that bind and inhibit the transcriptionally active RelA/p65 subunits. However, there are several nontraditional IκB proteins that reside in the nucleus and generally act as transcriptional activators. Two of the most well-studied nuclear IκBs are Bcl-3 and IκBζ. Bcl-3 contains several ankyrin repeat domains, like all other IκB proteins. Unlike other IκB proteins, Bcl-3 preferentially binds p50 and p52, rather than the transcriptionally active RelA, RelB, or c-Rel [96,97]. Since Bcl-3 was originally shown to inhibit the DNA binding of p50 homodimers, this led to the hypothesis that Bcl-3 inhibits NF-κB activity [98]. Further Bcl-3 can blunt the NF-κB-dependent expression of TNFα in response to LPS stimulation [99]. Bcl-3 has also been shown to recruit repressors to viral genes and repress transcription [100]; however Bcl-3 has generally been observed to promote transcriptional activation. Additionally, Bcl-3 expression is induced by NF-κB activity [101]. In response to NF-κB activation, Bcl-3 binds p50 and p52 and acts as a transcriptional activator, promoting the transcription of anti-apoptotic and proliferation genes, such as cyclin D1 [102–105]. In this regard, Bcl-3 has also been shown to inhibit p53 by promoting the transcription of its negative regulator, hdm2 in breast cancer cells [104]. Completing the regulatory circle, p53 has been shown to decrease cyclin D1 by inhibiting Bcl-3. In the absence of Bcl-3, p52 associates with HDACs at the cyclin D1 promoter, inhibiting transcription [106]. Increased expression of Bcl-3 was observed in breast cancer [103], skin squamous cell carcinoma [107], endometrial cancer [108], nasopharyngeal cancer [109], and some lymphomas [110]. Although established as an oncogene, there are still many questions as to how Bcl-3 activity is regulated. It has been known to be modulated by phosphorylation nearly since its discovery [96], and both IKKα and IKKβ have been shown to rapidly phosphorylate Bcl-3 and stimulate Bcl-3-p50 complexes to initiate transcription [111]. Recently, work from the Ghosh lab has given a glimpse into the complex regulation of Bcl-3 by phosphorylation. They showed that Akt phosphorylates Bcl-3 to increase its stabilization and nuclear localization, while IKKs and ERK phosphorylates Bcl-3 to promote its transcriptional activity [112]. Still, there is much that is unknown as to how the oncogenic functions of Bcl-3 are regulated.

Another nuclear IκB protein, IκBζ, promotes the transcription of inflammation-associated genes like IL-6 [113,114]. In general, IκBζ mRNA expression is kept silent and its mRNA is unstable. Upon stimulation of toll-like receptors (TLRs) or IL-1β, but not TNFα, IκBζ mRNA is stabilized and its transcription is initiated in an NF-κB-dependent manner [115,116]. IκBζ is both a target gene of NF-κB, and a coactivator, as it is critical to proper NF-κB target gene expression downstream of TLRs and IL-1β [113,117,118]. Like Bcl-3, IκBζ preferentially interacts with p50 and p52, but not RelA [119,120]. p50 homodimers, which are normally considered to be transcriptionally inactive, activate the transcription of NF-κB target genes when associated with IκBζ [118]. Recently, Nogai and colleagues showed that IκBζ expression is critical for activated B cell-like diffuse large cell B cell lymphoma (ABC-DLBCL) cells, but not germinal cell B cell-like DLBCL (GC-DLBCL) or multiple myeloma cells. Not only did they show that IκBζ expression is critical to maintain the expression of several NF-κB target genes, but that knockdown of IκBζ caused cytotoxicity in ABC-DLBCL cells, emphasizing its important roles in ABC-DLBCL [120]. Often overlooked in NF-κB research, these nontraditional IκB transcription factors have important roles in supporting NF-κB responses in immunity and disease and deserve more research.

6. Alternative Functions of the Noncanonical Kinases

6.1. Other Functions of IKKα

As the critical kinase in mediating activation of the canonical NF-κB pathway, IKKβ has been more highly studied than IKKα. IKKα does not play a major role in controlling IκBα degradation in response to TNFα [121,122]. However, IKKα does play a role in regulating NF-κB target gene expression [123–125]. Interesting evidence that IKKα may shuttle into the nucleus provided evidence of potential alternative functions [126]. Further clarification came when IKKα was found to phosphorylate serine 10 on histone 3 (H3S10), and loss of IKKα caused defects in NF-κB-dependent transcription [127,128]. Much of this mechanistic work was done in MEFs and did not establish whether nuclear IKKα has any consequences in cancer.

Subsequently, nuclear IKKα was observed in a prostate cancer cell line. IKKα, but not IKKβ, was found to phosphorylate the SMRT corepressor in the prostate cancer cell line DU145 in response to laminin attachment. This phosphorylation leads to SMRT and the associated repressor HDAC3 to be exported from the nucleus, leading to optimal activation of NF-κB target genes like cIAP1/2 and IL-8. This mechanism appears to be independent of IKKα-mediated H3S10 phosphorylation [129]. Additionally, interesting work in colorectal cancer cells has implicated IKKα-mediated phosphorylation of SMRT with changes in NF-κB-independent transcription. Nuclear IKKα and phospho-SMRT were observed in colon cancer cell lines as well as patient samples. Upon further investigation, IKKα was observed to phosphorylate SMRT, leading to its exclusion from the nucleus, similar to the prostate cancer cells. Additionally, IKKα presence at Notch target genes in colon cancer cells was identified using ChIP analysis (Figure 4). Further, IKKα was shown to promote Notch-dependent transcription and tumorigenesis in colon cancer [130]. More recently, a truncated p45-IKKα was observed in colorectal cancer cells. This fragment, generated by cathepsin protease activity, was shown to be critical for SMRT phosphorylation as well as expression of cIAP1/2 and several Notch target genes (Figure 4) [131].

Additionally, IKKα has also been shown to suppress gene expression [132]. IKKα activity was found to be associated with decreased expression of a metastasis-associated gene Maspin. Further, this is dependent on both IKKα kinase activity and nuclear localization, but it does not require IKKβ activity [132]. This fascinating result has potentially exciting consequences, as targeting this IKKα activity in prostate cancer with small molecule inhibitors could reduce metastatic burden. However, the mechanism by which IKKα mediates gene repression is not known, and IKKα has not yet been reported to bind at the Maspin locus using ChIP. It is possible these results indicate that activation of noncanonical NF-κB mediated by IKKα could be repressing gene expression as shown in multiple myeloma [133–135]. These mechanisms show that in addition to acting as a primary mediator for noncanonical NF-κB pathway activation, nuclear IKKα directly regulates gene transcription through the regulation of histones and chromatin-associated factors.

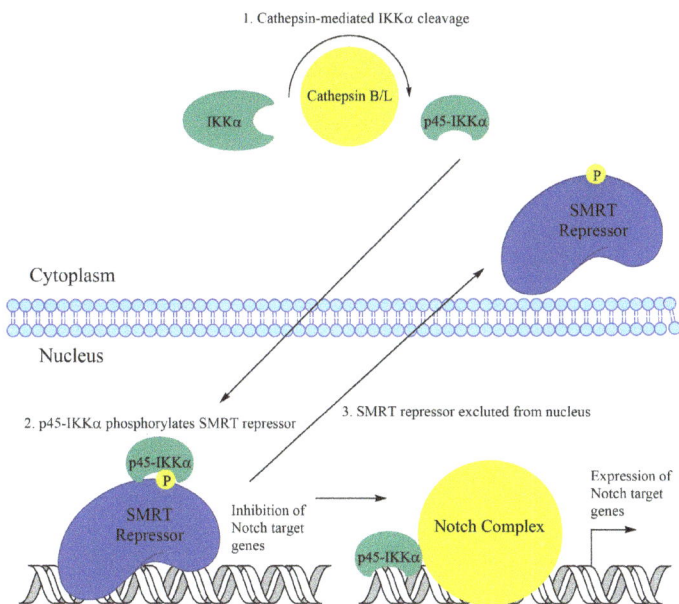

Figure 4. Nuclear IKKα promotes the expression of Notch target genes in colorectal cancer cells. A truncated p45-IKKα has been shown to phosphorylate the SMRT repressor, leading to its exclusion from the nucleus, leading to the expression of Notch target genes.

6.2. Alternative Roles for NIK

NIK is almost exclusively studied for its critical role in promoting noncanonical NF-κB. As described earlier, NIK is expressed at low levels due to constitutive degradation until a stimulus induces its stabilization, and as a consequence, NIK is not stabilized in many cell types. Nonetheless, intriguing functions for NIK have been uncovered. Birbach et al. showed that NIK has a nucleolar targeting sequence that is required for nuclear localization [136]. Interestingly, they show that a mutant of NIK that preferentially accumulates in the nucleolus is less efficient at inducing noncanonical NF-κB activity, suggesting that nucleolar NIK is either less efficient at activating NF-κB, or more interestingly, that it has alternative functions there. Further, nuclear localized NIK can be observed by immunofluorescence in the triple-negative breast cancer cell line MDA-MB-231 [136]. This interesting finding further suggests that nuclear localized NIK may have functions in cancer cells. Several other triple-negative breast cancer cells have been shown to have stabilized NIK, and NIK may promote breast cancer stemness [137,138]. The potential impact of NIK stabilization has not been thoroughly studied in breast cancer. Since EZH2 promotes RelB production and tumor initiating cells in triple-negative breast cancer [60], it is interesting to consider what effects NIK stabilization may have on EZH2-RelB crossregulation.

As discussed earlier, several reports indicate that NIK stabilization and subsequent noncanonical NF-κB activation generally supports the growth and survival of tumor cells. However, a recent report unexpectedly shows that NIK stabilization in certain contexts can act as a tumor suppressor. Canonical NF-κB activity is increased in AML, and it supports survival [139]. Xiu et al. show that NIK stabilization in AML cells induces noncanonical NF-κB, but nuclear RelA was reduced, which led to decreased tumor growth [140]. Interestingly, RelA overexpression promoted tumor growth, whereas RelB overexpression decreased tumor growth [140]. This has interesting implications on NF-κB biology. As described below there is significant crosstalk between canonical and noncanonical NF-κB, and this study suggests that in AML, there is an antagonistic relationship between canonical and noncanonical NF-κB, and activation of canonical NF-κB is ultimately critical to tumor survival.

NIK has many interesting functions due to its role as a noncanonical NF-κB activator. However, its activity has also been shown to regulate processes completely independent of NF-κB activity. Recently, Jung et al. have shown that NIK is localized to mitochondria in a glioma cell line [141]. Not only is NIK localized to mitochondria, but it promotes mitochondrial division and invasion. NIK does this by promoting the mitochondrial localization of Drp1, a GTPase required for mitochondrial fission. Interestingly, this activity is not dependent on NF-κB or IKKα, but can be stimulated by the TNF related ligand TWEAK [141]. TWEAK is a known noncanonical NF-κB inducing agent that can stimulate NIK and IKKα activity [15]. It is interesting that TWEAK induces NIK to support mitochondrial fission, but not NF-κB activation [141]. This intriguing study shows that NIK has alternative functions in some cancer cells that have important effects on metabolism and invasion. Interestingly, Drp1 activity has also been found to support glioblastoma tumor initiating cells [142] and pancreatic tumor growth in vivo [143], highlighting the potential impact of targeting this pathway. NIK is an understudied kinase that has complex effects on development and cancer, and more work needs to be done to identify potential targeting agents against this interesting kinase.

7. Crosstalk between Canonical and Noncanonical NF-κB

The NF-κB pathway involves numerous regulators and co-regulators, many of which have additional known functions outside of NF-κB regulation. Additionally, NF-κB activation is achieved by a variety of stumuli, and the induced target genes can vary by stimulus and cell type, thus the NF-κB response exhibits significant complexity. The noncanonical and canonical NF-κB pathways do not function in isolation, instead, co-regulation and crosstalk are abundant. Importantly, stimulation of canonical NF-κB with TNFα or LPS leads to rapid nuclear accumulation of canonical RelA-p50 dimers, but not RelB [144]. However, increased transcription of RelB in response to TNFα or LPS is dependent on RelA (Figure 5), and this leads to RelB-dependent changes in gene expression, which requires synthesis of new RelB protein [144]. Therefore, canonical NF-κB activation can promote noncanonical target gene expression via induction of RelB transcription. However, RelB has also been shown to suppress RelA DNA binding activity and sequester RelA in the cytoplasm in response to canonical stimuli such as TNFα (Figure 5) [145–147], potentially helping to blunt the strong inflammatory response of the canonical NF-κB pathway. Interestingly, RelA-RelB heterodimerization is dependent on contacts in the rel homology domains of each protein, and the interaction is promoted by phosphorylation of S276 on RelA, but inhibited by phosphorylation of RelB on S368 [148,149]. This intriguing antagonism is cell type dependent as the effect was observed in fibroblasts but not in macrophages [145]. Additionally, the RelA-RelB heterodimer also inhibits RelB-mediated transcription after TNFα treatment [148]. The antagonistic relationship between RelA and RelB also has disease relevance, as was explored earlier in a discussion on acute myeloid leukemia. Canonical upregulation of noncanonical NF-κB subunits followed by mutual antagonism between RelA and RelB introduces multiple layers of complexity to the NF-κB response in cells and the regulation of the RelA-RelB heterodimer as well as the S276 residue on RelA needs to be further elucidated to unravel how the NF-κB response is regulated in different cell types.

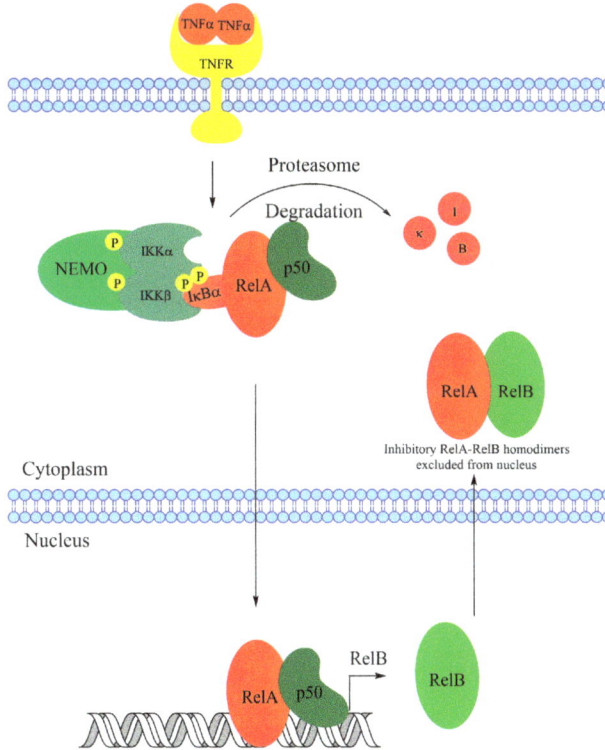

Figure 5. Canonical NF-κB activity stimulates RelB expression. Activation of RelA-p50 dimers with TNFα induces increased expression of RelB. RelB induction serves as negative feedback by binding RelA and sequestering it in the cytoplasm.

There is an additional system for inhibition of NF-κB mediated by the full-length precursor molecules p105 (termed IκBγ) and p100 (termed IκBδ). These molecules were observed to form high molecular weight complexes (IκBsomes) that can bind RelA [31,150]. Further, stimulation of canonical NF-κB can induce the expression of p100, acting as a negative feedback mechanism [151,152]. Upon LPS stimulation, IκBsomes dissociated transiently before eventually reforming [150], supporting the hypothesis that IκBsomes can act as a mechanism to limit canonical NF-κB activation. Since the IκBsome inhibitory complexes were found to exist in resting cells, they may also function as reservoirs that can activate both canonical and noncanonical pathway simultaneously upon stimulation. The existence and function of IκBsomes indicates that the NF-κB pathway is complex and that canonical and noncanonical components contribute to the NF-κB response.

As mentioned previously, RelA promotes RelB activity upon stimulation with TNFα. Work from the Baud laboratory showed that RelB activity downstream of TNFα stimulation is mediated through the canonical IKK complex containing IKKα, IKKβ, and NEMO, which phosphorylates RelB after TNFα treatment at S476 [6]. Phosphorylation at S472 dissociates RelB-containing complexes from IκBα, leading to NF-κB activation. Interestingly, phosphorylation at S472 promotes migration in fibroblasts by inducing a RelB-mediated upregulation of matrix metalloprotease 3 (MMP3), indicating an important biological consequence of this pathway [6,153].

The distinction between the two NF-κB pathways is blurred further when the array of subunit dimers that have been discovered is considered. There are 15 possible NF-κB dimer combinations, and 12 of them have been identified [154,155]. Supershift analysis of electrophoretic mobility shift assays (EMSA) provided evidence that alternative dimer combinations occur, even in response to canonical stimuli like TNF [156]. Additionally, stimulation of noncanonical-activating receptors like LTβR induces DNA-binding activity of both canonical RelA-p50 dimers and noncanonical RelB-p52 dimers [157]. The canonical RelA-p50 and noncanonical RelB-p52 dimers are probably the most commonly observed because of binding stability [154]. In particular, structural analysis of the RelA-p50 dimer has shown contacts create a highly stable dimer [154,158]. On the other hand, the RelB dimerization domain differs from that of RelA, leading it to form more stable complexes with both p50 and p52 (see below) [159]. RelA, however, is generally not observed to interact with p52, although it has been seen in RelB-deficient cells [160]. While the RelA-p50 and RelB-p52 heterodimers appear to be the most abundant within cells, other important dimer combinations have been studied.

One of the most well-studied "alternative" dimers is the RelB-p50 dimer, in which the noncanonical RelB molecule binds with the canonical p50. The DNA-binding activity of the RelB-p50 dimer was observed in response to viral proteins [161], bacterial products (LPS) [162], and cytokines [163,164]. The observed activity of the RelB-p50 dimers is usually delayed, being observed hours after a canonical stimulus. However, in dendritic cells, which have high levels of RelB activity, RelB-p50 dimers were found to respond rapidly to canonical stimuli, such as LPS [165]. Interestingly, although RelB is not usually bound to IκB proteins because of the inhibitory action of p100, low levels of p100 in dendritic cells led to RelB association with p50 as well as to IκB proteins. These studies highlight the complexity of the NF-κB pathway, indicating that the two NF-κB pathways are not district entities, but rather two branches of the same pathway with many nodes of cross-regulation.

8. Concluding Remarks

Activation of the noncanonical NF-κB pathway has important effects in tissue development as well as disease. Multiple myeloma in particular has a strong dependence on noncanonical NF-κB activation. Activation of noncanonical pathway components such as NIK, RelB, or IKKα have been shown to support tumor progression in various cancers, such as multiple myeloma, DLBCL, glioblastoma, breast, prostate, ovarian, and colon cancers, and inhibitors of NIK or IKKα could have therapeutic benefit for patients with these tumors. While much of the NF-κB-related research focuses on IKKβ/RelA, the noncanonical NF-κB pathway has proven to be equally critical in cancer and other diseases. More work needs to be done to elucidate how canonical and noncanonical NF-κB work to drive tumorigenesis, and determine if they can be effectively targeted.

Author Contributions: M.T.: Planned, wrote, and revised the review. A.B.: Planned and revised the review.

Acknowledgments: We are funded by the National Institutes of Health (NIH). The content is solely the responsibility of the authors and does not necessarily represent the official views of the National Institutes of Health.

Conflicts of Interest: The authors declare no conflict of interest.

References

1. Ghosh, S.; Hayden, M.S. New regulators of NF-kappaB in inflammation. *Nat. Rev. Immunol.* **2008**, *8*, 837–848. [CrossRef] [PubMed]
2. Hayden, M.S.; Ghosh, S. Signaling to NF-kappaB. *Genes Dev.* **2004**, *18*, 2195–2224. [CrossRef] [PubMed]
3. Libermann, T.A.; Baltimore, D. Activation of interleukin-6 gene expression through the NF-kappa B transcription factor. *Mol. Cell. Biol.* **1990**, *10*, 2327–2334. [CrossRef] [PubMed]

4. Guttridge, D.C.; Albanese, C.; Reuther, J.Y.; Pestell, R.G.; Baldwin, A.S. NF-kappaB controls cell growth and differentiation through transcriptional regulation of cyclin D1. *Mol. Cell. Biol.* **1999**, *19*, 5785–5799. [CrossRef] [PubMed]

5. Catz, S.D.; Johnson, J.L. Transcriptional regulation of bcl-2 by nuclear factor κB and its significance in prostate cancer. *Oncogene* **2001**, *20*, 7342–7351. [CrossRef] [PubMed]

6. Authier, H.; Billot, K.; Derudder, E.; Bordereaux, D.; Rivière, P.; Rodrigues-Ferreira, S.; Nahmias, C.; Baud, V. IKK phosphorylates RelB to modulate its promoter specificity and promote fibroblast migration downstream of TNF receptors. *Proc. Natl. Acad. Sci. USA* **2014**, *111*, 14794–14799. [CrossRef] [PubMed]

7. Xia, Y.; Shen, S.; Verma, I.M. NF-κB, an active player in human cancers. *Cancer Immunol. Res.* **2014**, *2*, 823–830. [CrossRef] [PubMed]

8. Lin, L.; DeMartino, G.N.; Greene, W.C. Cotranslational Biogenesis of NF-κB p50 by the 26S Proteasome. *Cell* **1998**, *92*, 819–828. [CrossRef]

9. Senftleben, U.; Cao, Y.; Xiao, G.; Greten, F.R.; Krähn, G.; Bonizzi, G.; Chen, Y.; Hu, Y.; Fong, A.; Sun, S.C.; et al. Activation by IKKalpha of a Second, Evolutionary Conserved, NF-kappa B Signaling Pathway. *Science* **2001**, *293*, 1495–1499. [CrossRef] [PubMed]

10. Zhang, Q.; Lenardo, M.J.; Baltimore, D. 30 Years of NF-κB: A Blossoming of Relevance to Human Pathobiology. *Cell* **2017**, *168*, 37–57. [CrossRef] [PubMed]

11. Hayden, M.S.; Ghosh, S. Shared principles in NF-kappaB signaling. *Cell* **2008**, *132*, 344–362. [CrossRef] [PubMed]

12. Claudio, E.; Brown, K.; Park, S.; Wang, H.; Siebenlist, U. BAFF-induced NEMO-independent processing of NF-κB2 in maturing B cells. *Nat. Immunol.* **2002**, *3*, 958–965. [CrossRef] [PubMed]

13. Dejardin, E.; Droin, N.M.; Delhase, M.; Haas, E.; Cao, Y.; Makris, C.; Li, Z.-W.; Karin, M.; Ware, C.F.; Green, D.R. The Lymphotoxin-β Receptor Induces Different Patterns of Gene Expression via Two NF-κB Pathways. *Immunity* **2002**, *17*, 525–535. [CrossRef]

14. Coope, H.J.; Atkinson, P.G.P.; Huhse, B.; Belich, M.; Janzen, J.; Holman, M.J.; Klaus, G.G.B.; Johnston, L.H.; Ley, S.C. CD40 regulates the processing of NF-kappaB2 p100 to p52. *EMBO J.* **2002**, *21*, 5375–5385. [CrossRef] [PubMed]

15. Saitoh, T.; Nakayama, M.; Nakano, H.; Yagita, H.; Yamamoto, N.; Yamaoka, S. TWEAK induces NF-kappaB2 p100 processing and long lasting NF-kappaB activation. *J. Biol. Chem.* **2003**, *278*, 36005–36012. [CrossRef] [PubMed]

16. Novack, D.V.; Yin, L.; Hagen-Stapleton, A.; Schreiber, R.D.; Goeddel, D.V.; Ross, F.P.; Teitelbaum, S.L. The IkappaB function of NF-kappaB2 p100 controls stimulated osteoclastogenesis. *J. Exp. Med.* **2003**, *198*, 771–781. [CrossRef] [PubMed]

17. Liao, G.; Zhang, M.; Harhaj, E.W.; Sun, S.-C. Regulation of the NF-kappaB-inducing kinase by tumor necrosis factor receptor-associated factor 3-induced degradation. *J. Biol. Chem.* **2004**, *279*, 26243–26250. [CrossRef] [PubMed]

18. De Jong, S.J.; Albrecht, J.-C.; Giehler, F.; Kieser, A.; Sticht, H.; Biesinger, B. Noncanonical NF- B Activation by the Oncoprotein Tio Occurs Through a Nonconserved TRAF3-Binding Motif. *Sci. Signal.* **2013**, *6*, ra27. [CrossRef] [PubMed]

19. He, J.Q.; Zarnegar, B.; Oganesyan, G.; Saha, S.K.; Yamazaki, S.; Doyle, S.E.; Dempsey, P.W.; Cheng, G. Rescue of TRAF3-null mice by p100 NF-κB deficiency. *J. Exp. Med.* **2006**, *203*, 2413–2418. [CrossRef] [PubMed]

20. Zarnegar, B.J.; Wang, Y.; Mahoney, D.J.; Dempsey, P.W.; Cheung, H.H.; He, J.; Shiba, T.; Yang, X.; Yeh, W.; Mak, T.W.; et al. Noncanonical NF-κB activation requires coordinated assembly of a regulatory complex of the adaptors cIAP1, cIAP2, TRAF2 and TRAF3 and the kinase NIK. *Nat. Immunol.* **2008**, *9*, 1371–1378. [CrossRef] [PubMed]

21. Vince, J.E.; Wong, W.W.-L.; Khan, N.; Feltham, R.; Chau, D.; Ahmed, A.U.; Benetatos, C.A.; Chunduru, S.K.; Condon, S.M.; McKinlay, M.; et al. IAP Antagonists Target cIAP1 to Induce TNFα-Dependent Apoptosis. *Cell* **2007**, *131*, 682–693. [CrossRef] [PubMed]

22. Varfolomeev, E.; Blankenship, J.W.; Wayson, S.M.; Fedorova, A.V.; Kayagaki, N.; Garg, P.; Zobel, K.; Dynek, J.N.; Elliott, L.O.; Wallweber, H.J.A.; et al. IAP Antagonists Induce Autoubiquitination of c-IAPs, NF-κB Activation, and TNFα-Dependent Apoptosis. *Cell* **2007**, *131*, 669–681. [CrossRef] [PubMed]

23. Brown, K.D.; Hostager, B.S.; Bishop, G.A. Differential signaling and tumor necrosis factor receptor-associated factor (TRAF) degradation mediated by CD40 and the Epstein-Barr virus oncoprotein latent membrane protein 1 (LMP1). *J. Exp. Med.* **2001**, *193*, 943–954. [CrossRef] [PubMed]

24. Liang, C.; Zhang, M.; Sun, S.-C. β-TrCP binding and processing of NF-κB2/p100 involve its phosphorylation at serines 866 and 870. *Cell. Signal.* **2006**, *18*, 1309–1317. [CrossRef] [PubMed]

25. Xiao, G.; Fong, A.; Sun, S.-C. Induction of p100 processing by NF-kappaB-inducing kinase involves docking IkappaB kinase alpha (IKKalpha) to p100 and IKKalpha-mediated phosphorylation. *J. Biol. Chem.* **2004**, *279*, 30099–30105. [CrossRef] [PubMed]

26. Fong, A.; Sun, S.-C. Genetic evidence for the essential role of beta-transducin repeat-containing protein in the inducible processing of NF-kappa B2/p100. *J. Biol. Chem.* **2002**, *277*, 22111–22114. [CrossRef] [PubMed]

27. Polley, S.; Passos, D.O.; Huang, D.-B.; Mulero, M.C.; Mazumder, A.; Biswas, T.; Verma, I.M.; Lyumkis, D.; Ghosh, G. Structural Basis for the Activation of IKK1/α. *Cell Rep.* **2016**, *17*, 1907–1914. [CrossRef] [PubMed]

28. Shih, V.F.-S.; Tsui, R.; Caldwell, A.; Hoffmann, A. A single NFκB system for both canonical and non-canonical signaling. *Cell Res.* **2011**, *21*, 86–102. [CrossRef] [PubMed]

29. Wong, D.; Teixeira, A.; Oikonomopoulos, S.; Humburg, P.; Lone, I.; Saliba, D.; Siggers, T.; Bulyk, M.; Angelov, D.; Dimitrov, S.; et al. Extensive characterization of NF-κB binding uncovers non-canonical motifs and advances the interpretation of genetic functional traits. *Genome Biol.* **2011**, *12*, R70. [CrossRef] [PubMed]

30. Leung, T.H.; Hoffmann, A.; Baltimore, D. One Nucleotide in a κB Site Can Determine Cofactor Specificity for NF-κB Dimers. *Cell* **2004**, *118*, 453–464. [CrossRef] [PubMed]

31. Tao, Z.; Fusco, A.; Huang, D.-B.; Gupta, K.; Kim, D.Y.; Ware, C.F.; Van Duyne, G.D.; Ghosh, G. p100/IκBδ sequesters and inhibits NF-κB through kappaBsome formation. *Proc. Natl. Acad. Sci. USA* **2014**, *111*, 15946–15951. [CrossRef] [PubMed]

32. Cildir, G.; Low, K.C.; Tergaonkar, V. Noncanonical NF-κB Signaling in Health and Disease. *Trends Mol. Med.* **2016**, *22*, 414–429. [CrossRef] [PubMed]

33. Guo, F.; Tänzer, S.; Busslinger, M.; Weih, F. Lack of nuclear factor-kappa B2/p100 causes a RelB-dependent block in early B lymphopoiesis. *Blood* **2008**, *112*, 551–559. [CrossRef] [PubMed]

34. Basak, S.; Kim, H.; Kearns, J.D.; Tergaonkar, V.; O'Dea, E.; Werner, S.L.; Benedict, C.A.; Ware, C.F.; Ghosh, G.; Verma, I.M.; et al. A Fourth IκB Protein within the NF-κB Signaling Module. *Cell* **2007**, *128*, 369–381. [CrossRef] [PubMed]

35. Miyawaki, S.; Nakamura, Y.; Suzuka, H.; Koba, M.; Shibata, Y.; Yasumizu, R.; Ikehara, S. A new mutation, aly, that induces a generalized lack of lymph nodes accompanied by immunodeficiency in mice. *Eur. J. Immunol.* **1994**, *24*, 429–434. [CrossRef] [PubMed]

36. Koike, R.; Nishimura, T.; Yasumizu, R.; Tanaka, H.; Hataba, Y.; Watanabe, T.; Miyawaki, S.; Miyasaka, M. The splenic marginal zone is absent in alymphoplasticaly mutant mice. *Eur. J. Immunol.* **1996**, *26*, 669–675. [CrossRef] [PubMed]

37. Franzoso, G.; Carlson, L.; Poljak, L.; Shores, E.W.; Epstein, S.; Leonardi, A.; Grinberg, A.; Tran, T.; Scharton-Kersten, T.; Anver, M.; et al. Mice deficient in nuclear factor (NF)-kappa B/p52 present with defects in humoral responses, germinal center reactions, and splenic microarchitecture. *J. Exp. Med.* **1998**, *187*, 147–159. [CrossRef] [PubMed]

38. Caamaño, J.H.; Rizzo, C.A.; Durham, S.K.; Barton, D.S.; Raventós-Suárez, C.; Snapper, C.M.; Bravo, R. Nuclear factor (NF)-kappa B2 (p100/p52) is required for normal splenic microarchitecture and B cell-mediated immune responses. *J. Exp. Med.* **1998**, *187*, 185–196. [CrossRef] [PubMed]

39. De Silva, N.S.; Anderson, M.M.; Carette, A.; Silva, K.; Heise, N.; Bhagat, G.; Klein, U. Transcription factors of the alternative NF-κB pathway are required for germinal center B-cell development. *Proc. Natl. Acad. Sci. USA* **2016**, *113*, 9063–9068. [CrossRef] [PubMed]

40. Victora, G.D.; Nussenzweig, M.C. Germinal Centers. *Annu. Rev. Immunol.* **2012**, *30*, 429–457. [CrossRef] [PubMed]

41. Kaileh, M.; Sen, R. NF-κB function in B lymphocytes. *Immunol. Rev.* **2012**, *246*, 254–271. [CrossRef] [PubMed]

42. Chauhan, D.; Uchiyama, H.; Akbarali, Y.; Urashima, M.; Yamamoto, K.; Libermann, T.; Anderson, K. Multiple myeloma cell adhesion-induced interleukin-6 expression in bone marrow stromal cells involves activation of NF-kappa B. *Blood* **1996**, *87*, 1104–1112. [PubMed]

43. Landowski, T.H.; Olashaw, N.E.; Agrawal, D.; Dalton, W.S. Cell adhesion-mediated drug resistance (CAM-DR) is associated with activation of NF-κB (RelB/p50) in myeloma cells. *Oncogene* **2003**, *22*, 2417–2421. [CrossRef] [PubMed]

44. Croucher, P.I.; Shipman, C.M.; Lippitt, J.; Perry, M.; Asosingh, K.; Hijzen, A.; Brabbs, A.C.; van Beek, E.J.; Holen, I.; Skerry, T.M.; et al. Osteoprotegerin inhibits the development of osteolytic bone disease in multiple myeloma. *Blood* **2001**, *98*, 3534–3540. [CrossRef] [PubMed]

45. Keats, J.J.; Fonseca, R.; Chesi, M.; Schop, R.; Baker, A.; Chng, W.-J.; Van Wier, S.; Tiedemann, R.; Shi, C.-X.; Sebag, M.; et al. Promiscuous mutations activate the noncanonical NF-kappaB pathway in multiple myeloma. *Cancer Cell* **2007**, *12*, 131–144. [CrossRef] [PubMed]

46. Annunziata, C.M.; Davis, R.E.; Demchenko, Y.; Bellamy, W.; Gabrea, A.; Zhan, F.; Lenz, G.; Hanamura, I.; Wright, G.; Xiao, W.; et al. Frequent Engagement of the Classical and Alternative NF-κB Pathways by Diverse Genetic Abnormalities in Multiple Myeloma. *Cancer Cell* **2007**, *12*, 115–130. [CrossRef] [PubMed]

47. Scheinman, R.I.; Cogswell, P.C.; Lofquist, A.K.; Baldwin, A.S. Role of transcriptional activation of I kappa B alpha in mediation of immunosuppression by glucocorticoids. *Science* **1995**, *270*, 283–286. [CrossRef] [PubMed]

48. Auphan, N.; DiDonato, J.A.; Rosette, C.; Helmberg, A.; Karin, M. Immunosuppression by glucocorticoids: Inhibition of NF-kappa B activity through induction of I kappa B synthesis. *Science* **1995**, *270*, 286–290. [CrossRef] [PubMed]

49. Raab, M.S.; Podar, K.; Breitkreutz, I.; Richardson, P.G.; Anderson, K.C. Multiple myeloma. *Lancet* **2009**, *374*, 324–339. [CrossRef]

50. Fabre, C.; Mimura, N.; Bobb, K.; Kong, S.-Y.; Gorgun, G.; Cirstea, D.; Hu, Y.; Minami, J.; Ohguchi, H.; Zhang, J.; et al. Dual Inhibition of Canonical and Noncanonical NF- B Pathways Demonstrates Significant Antitumor Activities in Multiple Myeloma. *Clin. Cancer Res.* **2012**, *18*, 4669–4681. [CrossRef] [PubMed]

51. Cao, Y.; Bonizzi, G.; Seagroves, T.N.; Greten, F.R.; Johnson, R.; Schmidt, E.V.; Karin, M. IKKalpha provides an essential link between RANK signaling and cyclin D1 expression during mammary gland development. *Cell* **2001**, *107*, 763–775. [CrossRef]

52. Demicco, E.G.; Kavanagh, K.T.; Romieu-Mourez, R.; Wang, X.; Shin, S.R.; Landesman-Bollag, E.; Seldin, D.C.; Sonenshein, G.E. RelB/p52 NF- B Complexes Rescue an Early Delay in Mammary Gland Development in Transgenic Mice with Targeted Superrepressor I B- Expression and Promote Carcinogenesis of the Mammary Gland. *Mol. Cell. Biol.* **2005**, *25*, 10136–10147. [CrossRef] [PubMed]

53. Sovak, M.A.; Bellas, R.E.; Kim, D.W.; Zanieski, G.J.; Rogers, A.E.; Traish, A.M.; Sonenshein, G.E. Aberrant nuclear factor-kappaB/Rel expression and the pathogenesis of breast cancer. *J. Clin. Investig.* **1997**, *100*, 2952–2960. [CrossRef] [PubMed]

54. Rojo, F.; González-Pérez, A.; Furriol, J.; Nicolau, M.J.; Ferrer, J.; Burgués, O.; Sabbaghi, M.; González-Navarrete, I.; Cristobal, I.; Serrano, L.; et al. Non-canonical NF-κB pathway activation predicts outcome in borderline oestrogen receptor positive breast carcinoma. *Br. J. Cancer* **2016**, *115*, 322–331. [CrossRef] [PubMed]

55. Wang, X.; Belguise, K.; Kersual, N.; Kirsch, K.H.; Mineva, N.D.; Galtier, F.; Chalbos, D.; Sonenshein, G.E. Oestrogen signalling inhibits invasive phenotype by repressing RelB and its target BCL2. *Nat. Cell Biol.* **2007**, *9*, 470–478. [CrossRef] [PubMed]

56. Hurtado, A.; Holmes, K.A.; Ross-Innes, C.S.; Schmidt, D.; Carroll, J.S. FOXA1 is a key determinant of estrogen receptor function and endocrine response. *Nat. Genet.* **2011**, *43*, 27–33. [CrossRef] [PubMed]

57. Naderi, A.; Meyer, M.; Dowhan, D.H. Cross-regulation between FOXA1 and ErbB2 signaling in estrogen receptor-negative breast cancer. *Neoplasia* **2012**, *14*, 283–296. [CrossRef] [PubMed]

58. Wang, X.; Belguise, K.; O'Neill, C.F.; Sanchez-Morgan, N.; Romagnoli, M.; Eddy, S.F.; Mineva, N.D.; Yu, Z.; Min, C.; Trinkaus-Randall, V.; et al. RelB NF- B Represses Estrogen Receptor Expression via Induction of the Zinc Finger Protein Blimp1. *Mol. Cell. Biol.* **2009**, *29*, 3832–3844. [CrossRef] [PubMed]

59. Kendellen, M.F.; Bradford, J.W.; Lawrence, C.L.; Clark, K.S.; Baldwin, A.S. Canonical and non-canonical NF-κB signaling promotes breast cancer tumor-initiating cells. *Oncogene* **2014**, *33*, 1297–1305. [CrossRef] [PubMed]

60. Lawrence, C.L.; Baldwin, A.S. Non-Canonical EZH2 Transcriptionally Activates RelB in Triple Negative Breast Cancer. *PLoS ONE* **2016**, *11*, e0165005. [CrossRef] [PubMed]

61. Tan, M.E.; Li, J.; Xu, H.E.; Melcher, K.; Yong, E. Androgen receptor: Structure, role in prostate cancer and drug discovery. *Acta Pharmacol. Sin.* **2015**, *36*, 3–23. [CrossRef] [PubMed]

62. Harris, W.P.; Mostaghel, E.A.; Nelson, P.S.; Montgomery, B. Androgen deprivation therapy: Progress in understanding mechanisms of resistance and optimizing androgen depletion. *Nat. Rev. Urol.* **2009**, *6*, 76. [CrossRef] [PubMed]

63. Nadiminty, N.; Lou, W.; Sun, M.; Chen, J.; Yue, J.; Kung, H.-J.; Evans, C.P.; Zhou, Q.; Gao, A.C. Aberrant activation of the androgen receptor by NF-kappaB2/p52 in prostate cancer cells. *Cancer Res.* **2010**, *70*, 3309–3319. [CrossRef] [PubMed]

64. Prabhu, L.; Mundade, R.; Korc, M.; Loehrer, P.J.; Lu, T. Critical role of NF-κB in pancreatic cancer. *Oncotarget* **2014**, *5*, 10969–10975. [CrossRef] [PubMed]

65. Wharry, C.E.; Haines, K.M.; Carroll, R.G.; May, M.J. Constitutive non-canonical NFkappaB signaling in pancreatic cancer cells. *Cancer Biol. Ther.* **2009**, *8*, 1567–1576. [CrossRef] [PubMed]

66. Döppler, H.; Liou, G.-Y.; Storz, P. Downregulation of TRAF2 Mediates NIK-Induced Pancreatic Cancer Cell Proliferation and Tumorigenicity. *PLoS ONE* **2013**, *8*, e53676. [CrossRef] [PubMed]

67. Bang, D.; Wilson, W.; Ryan, M.; Yeh, J.J.; Baldwin, A.S. GSK-3α promotes oncogenic KRAS function in pancreatic cancer via TAK1-TAB stabilization and regulation of noncanonical NF-κB. *Cancer Discov.* **2013**, *3*, 690–703. [CrossRef] [PubMed]

68. Lee, D.W.; Ramakrishnan, D.; Valenta, J.; Parney, I.F.; Bayless, K.J.; Sitcheran, R. The NF-κB RelB Protein Is an Oncogenic Driver of Mesenchymal Glioma. *PLoS ONE* **2013**, *8*, e57489. [CrossRef] [PubMed]

69. Ohtsu, N.; Nakatani, Y.; Yamashita, D.; Ohue, S.; Ohnishi, T.; Kondo, T. Eva1 Maintains the Stem-like Character of Glioblastoma-Initiating Cells by Activating the Noncanonical NF-κB Signaling Pathway. *Cancer Res.* **2016**, *76*, 171–181. [CrossRef] [PubMed]

70. Uno, M.; Saitoh, Y.; Mochida, K.; Tsuruyama, E.; Kiyono, T.; Imoto, I.; Inazawa, J.; Yuasa, Y.; Kubota, T.; Yamaoka, S. NF-κB Inducing Kinase, a Central Signaling Component of the Non-Canonical Pathway of NF-κB, Contributes to Ovarian Cancer Progression. *PLoS ONE* **2014**, *9*, e88347. [CrossRef] [PubMed]

71. Ge, Q.-L.; Liu, S.-H.; Ai, Z.-H.; Tao, M.-F.; Ma, L.; Wen, S.-Y.; Dai, M.; Liu, F.; Liu, H.-S.; Jiang, R.-Z.; et al. RelB/NF-κB links cell cycle transition and apoptosis to endometrioid adenocarcinoma tumorigenesis. *Cell Death Dis.* **2016**, *7*, e2402–e2402. [CrossRef] [PubMed]

72. Nik-Zainal, S.; Alexandrov, L.B.; Wedge, D.C.; Van Loo, P.; Greenman, C.D.; Raine, K.; Jones, D.; Hinton, J.; Marshall, J.; Stebbings, L.A.; et al. Breast Cancer Working Group of the International Cancer Genome Consortium Mutational processes molding the genomes of 21 breast cancers. *Cell* **2012**, *149*, 979–993. [CrossRef] [PubMed]

73. Alexandrov, L.B.; Nik-Zainal, S.; Wedge, D.C.; Aparicio, S.A.J.R.; Behjati, S.; Biankin, A.V.; Bignell, G.R.; Bolli, N.; Borg, A.; Børresen-Dale, A.-L.; et al. Signatures of mutational processes in human cancer. *Nature* **2013**, *500*, 415–421. [CrossRef] [PubMed]

74. Harris, R.S.; Liddament, M.T. Retroviral restriction by APOBEC proteins. *Nat. Rev. Immunol.* **2004**, *4*, 868–877. [CrossRef] [PubMed]

75. Leonard, B.; McCann, J.L.; Starrett, G.J.; Kosyakovsky, L.; Luengas, E.M.; Molan, A.M.; Burns, M.B.; McDougle, R.M.; Parker, P.J.; Brown, W.L.; et al. The PKC/NF-κB signaling pathway induces APOBEC3B expression in multiple human cancers. *Cancer Res.* **2015**, *75*, 4538–4547. [CrossRef] [PubMed]

76. Shay, J.W.; Wright, W.E. Role of telomeres and telomerase in cancer. *Semin. Cancer Biol.* **2011**, *21*, 349–353. [CrossRef] [PubMed]

77. Kim, N.W.; Piatyszek, M.A.; Prowse, K.R.; Harley, C.B.; West, M.D.; Ho, P.L.; Coviello, G.M.; Wright, W.E.; Weinrich, S.L.; Shay, J.W. Specific association of human telomerase activity with immortal cells and cancer. *Science* **1994**, *266*, 2011–2015. [CrossRef] [PubMed]

78. Huang, F.W.; Hodis, E.; Xu, M.J.; Kryukov, G.V.; Chin, L.; Garraway, L.A. Highly recurrent TERT promoter mutations in human melanoma. *Science* **2013**, *339*, 957–959. [CrossRef] [PubMed]

79. Horn, S.; Figl, A.; Rachakonda, P.S.; Fischer, C.; Sucker, A.; Gast, A.; Kadel, S.; Moll, I.; Nagore, E.; Hemminki, K.; et al. TERT promoter mutations in familial and sporadic melanoma. *Science* **2013**, *339*, 959–961. [CrossRef] [PubMed]

80. Li, Y.; Zhou, Q.-L.; Sun, W.; Chandrasekharan, P.; Cheng, H.S.; Ying, Z.; Lakshmanan, M.; Raju, A.; Tenen, D.G.; Cheng, S.-Y.; et al. Non-canonical NF-κB signalling and ETS1/2 cooperatively drive C250T mutant TERT promoter activation. *Nat. Cell Biol.* **2015**, *17*, 1327–1338. [CrossRef] [PubMed]

81. Wang, Y.; Xu, J.; Gao, G.; Li, J.; Huang, H.; Jin, H.; Zhu, J.; Che, X.; Huang, C. Tumor-suppressor NFκB2 p100 interacts with ERK2 and stabilizes PTEN mRNA via inhibition of miR-494. *Oncogene* **2016**, *35*, 4080–4090. [CrossRef] [PubMed]

82. Xu, J.; Wang, Y.; Hua, X.; Xu, J.; Tian, Z.; Jin, H.; Li, J.; Wu, X.-R.; Huang, C.; Xu, J.; et al. Inhibition of PHLPP2/cyclin D1 protein translation contributes to the tumor suppressive effect of NF-κB2 (p100). *Oncotarget* **2016**, *7*, 34112–34130. [CrossRef] [PubMed]

83. Cao, R.; Wang, L.; Wang, H.; Xia, L.; Erdjument-Bromage, H.; Tempst, P.; Jones, R.S.; Zhang, Y. Role of histone H3 lysine 27 methylation in Polycomb-group silencing. *Science* **2002**, *298*, 1039–1043. [CrossRef] [PubMed]

84. Müller, J.; Hart, C.M.; Francis, N.J.; Vargas, M.L.; Sengupta, A.; Wild, B.; Miller, E.L.; O'Connor, M.B.; Kingston, R.E.; Simon, J.A. Histone Methyltransferase Activity of a Drosophila Polycomb Group Repressor Complex. *Cell* **2002**, *111*, 197–208. [CrossRef]

85. Kuzmichev, A.; Nishioka, K.; Erdjument-Bromage, H.; Tempst, P.; Reinberg, D. Histone methyltransferase activity associated with a human multiprotein complex containing the Enhancer of Zeste protein. *Genes Dev.* **2002**, *16*, 2893–2905. [CrossRef] [PubMed]

86. Czermin, B.; Melfi, R.; McCabe, D.; Seitz, V.; Imhof, A.; Pirrotta, V. Drosophila Enhancer of Zeste/ESC Complexes Have a Histone H3 Methyltransferase Activity that Marks Chromosomal Polycomb Sites. *Cell* **2002**, *111*, 185–196. [CrossRef]

87. Bracken, A.P.; Kleine-Kohlbrecher, D.; Dietrich, N.; Pasini, D.; Gargiulo, G.; Beekman, C.; Theilgaard-Mönch, K.; Minucci, S.; Porse, B.T.; Marine, J.-C.; et al. The Polycomb group proteins bind throughout the INK4A-ARF locus and are disassociated in senescent cells. *Genes Dev.* **2007**, *21*, 525–530. [CrossRef] [PubMed]

88. Iannetti, A.; Ledoux, A.C.; Tudhope, S.J.; Sellier, H.; Zhao, B.; Mowla, S.; Moore, A.; Hummerich, H.; Gewurz, B.E.; Cockell, S.J.; et al. Regulation of p53 and Rb Links the Alternative NF-κB Pathway to EZH2 Expression and Cell Senescence. *PLoS Genet.* **2014**, *10*, e1004642. [CrossRef] [PubMed]

89. De Donatis, G.M.; Pape, E.L.; Pierron, A.; Cheli, Y.; Hofman, V.; Hofman, P.; Allegra, M.; Zahaf, K.; Bahadoran, P.; Rocchi, S.; et al. NF-kB2 induces senescence bypass in melanoma via a direct transcriptional activation of EZH2. *Oncogene* **2016**, *35*, 2735–2745. [CrossRef] [PubMed]

90. Kim, K.H.; Kim, W.; Howard, T.P.; Vazquez, F.; Tsherniak, A.; Wu, J.N.; Wang, W.; Haswell, J.R.; Walensky, L.D.; Hahn, W.C.; et al. SWI/SNF-mutant cancers depend on catalytic and non-catalytic activity of EZH2. *Nat. Med.* **2015**, *21*, 1491–1496. [CrossRef] [PubMed]

91. Gonzalez, M.E.; Moore, H.M.; Li, X.; Toy, K.A.; Huang, W.; Sabel, M.S.; Kidwell, K.M.; Kleer, C.G. EZH2 expands breast stem cells through activation of NOTCH1 signaling. *Proc. Natl. Acad. Sci. USA* **2014**, *111*, 3098–3103. [CrossRef] [PubMed]

92. Shi, B.; Liang, J.; Yang, X.; Wang, Y.; Zhao, Y.; Wu, H.; Sun, L.; Zhang, Y.; Chen, Y.; Li, R.; et al. Integration of Estrogen and Wnt Signaling Circuits by the Polycomb Group Protein EZH2 in Breast Cancer Cells. *Mol. Cell. Biol.* **2007**, *27*, 5105–5119. [CrossRef] [PubMed]

93. Chang, C.-J.; Yang, J.-Y.; Xia, W.; Chen, C.-T.; Xie, X.; Chao, C.-H.; Woodward, W.A.; Hsu, J.-M.; Hortobagyi, G.N.; Hung, M.-C. EZH2 Promotes Expansion of Breast Tumor Initiating Cells through Activation of RAF1-β-Catenin Signaling. *Cancer Cell* **2011**, *19*, 86–100. [CrossRef] [PubMed]

94. Xu, K.; Wu, Z.J.; Groner, A.C.; He, H.H.; Cai, C.; Lis, R.T.; Wu, X.; Stack, E.C.; Loda, M.; Liu, T.; et al. EZH2 Oncogenic Activity in Castration-Resistant Prostate Cancer Cells Is Polycomb-Independent. *Science* **2012**, *338*, 1465–1469. [CrossRef] [PubMed]

95. Lee, S.T.; Li, Z.; Wu, Z.; Aau, M.; Guan, P.; Karuturi, R.K.M.; Liou, Y.C.; Yu, Q. Context-Specific Regulation of NF-κB Target Gene Expression by EZH2 in Breast Cancers. *Mol. Cell* **2011**, *43*, 798–810. [CrossRef] [PubMed]

96. Nolan, G.P.; Fujita, T.; Bhatia, K.; Huppi, C.; Liou, H.C.; Scott, M.L.; Baltimore, D. The bcl-3 proto-oncogene encodes a nuclear I kappa B-like molecule that preferentially interacts with NF-kappa B p50 and p52 in a phosphorylation-dependent manner. *Mol. Cell. Biol.* **1993**, *13*, 3557–3566. [CrossRef] [PubMed]

97. Kerr, L.D.; Duckett, C.S.; Wamsley, P.; Zhang, Q.; Chiao, P.; Nabel, G.; McKeithan, T.W.; Baeuerle, P.A.; Verma, I.M. The proto-oncogene bcl-3 encodes an I kappa B protein. *Genes Dev.* **1992**, *6*, 2352–2363. [CrossRef] [PubMed]

98. Hatada, E.N.; Nieters, A.; Wulczyn, F.G.; Naumann, M.; Meyer, R.; Nucifora, G.; McKeithan, T.W.; Scheidereit, C. The ankyrin repeat domains of the NF-kappa B precursor p105 and the protooncogene bcl-3 act as specific inhibitors of NF-kappa B DNA binding. *Proc. Natl. Acad. Sci. USA* **1992**, *89*, 2489–2493. [CrossRef] [PubMed]

99. Wessells, J.; Baer, M.; Young, H.A.; Claudio, E.; Brown, K.; Siebenlist, U.; Johnson, P.F. BCL-3 and NF-κB p50 Attenuate Lipopolysaccharide-induced Inflammatory Responses in Macrophages. *J. Biol. Chem.* **2004**, *279*, 49995–50003. [CrossRef] [PubMed]

100. Hishiki, T.; Ohshima, T.; Ego, T.; Shimotohno, K. BCL3 acts as a negative regulator of transcription from the human T-cell leukemia virus type 1 long terminal repeat through interactions with TORC3. *J. Biol. Chem.* **2007**, *282*, 28335–28343. [CrossRef] [PubMed]

101. Brasier, A.R.; Lu, M.; Hai, T.; Lu, Y.; Boldogh, I. NF-kappa B-inducible BCL-3 expression is an autoregulatory loop controlling nuclear p50/NF-kappa B1 residence. *J. Biol. Chem.* **2001**, *276*, 32080–32093. [CrossRef] [PubMed]

102. Fujita, T.; Nolan, G.P.; Liou, H.C.; Scott, M.L.; Baltimore, D. The candidate proto-oncogene bcl-3 encodes a transcriptional coactivator that activates through NF-kappa B p50 homodimers. *Genes Dev.* **1993**, *7*, 1354–1363. [CrossRef] [PubMed]

103. Cogswell, P.C.; Guttridge, D.C.; Funkhouser, W.K.; Baldwin, A.S. Selective activation of NF-κB subunits in human breast cancer: Potential roles for NF-κB2/p52 and for Bcl-3. *Oncogene* **2000**, *19*, 1123–1131. [CrossRef] [PubMed]

104. Kashatus, D.; Cogswell, P.; Baldwin, A.S. Expression of the Bcl-3 proto-oncogene suppresses p53 activation. *Genes Dev.* **2006**, *20*, 225–235. [CrossRef] [PubMed]

105. Westerheide, S.D.; Mayo, M.W.; Anest, V.; Hanson, J.L.; Baldwin, A.S. The Putative Oncoprotein Bcl-3 Induces Cyclin D1 To Stimulate G1 Transition. *Mol. Cell. Biol.* **2001**, *21*, 8428–8436. [CrossRef] [PubMed]

106. Rocha, S.; Martin, A.M.; Meek, D.W.; Perkins, N.D. p53 represses cyclin D1 transcription through down regulation of Bcl-3 and inducing increased association of the p52 NF-kappaB subunit with histone deacetylase 1. *Mol. Cell. Biol.* **2003**, *23*, 4713–4727. [CrossRef] [PubMed]

107. Budunova, I.V.; Perez, P.; Vaden, V.R.; Spiegelman, V.S.; Slaga, T.J.; Jorcano, J.L. Increased expression of p50-NF-κB and constitutive activation of NF-κB transcription factors during mouse skin carcinogenesis. *Oncogene* **1999**, *18*, 7423–7431. [CrossRef] [PubMed]

108. Pallares, J.; Martínez-Guitarte, J.L.; Dolcet, X.; Llobet, D.; Rue, M.; Palacios, J.; Prat, J.; Matias-Guiu, X. Abnormalities in the NF-κB family and related proteins in endometrial carcinoma. *J. Pathol.* **2004**, *204*, 569–577. [CrossRef] [PubMed]

109. Thornburg, N.J.; Pathmanathan, R.; Raab-Traub, N. Activation of nuclear factor-kappaB p50 homodimer/Bcl-3 complexes in nasopharyngeal carcinoma. *Cancer Res.* **2003**, *63*, 8293–8301. [PubMed]

110. Canoz, O.; Rassidakis, G.Z.; Admirand, J.H.; Medeiros, L.J. Immunohistochemical detection of BCL-3 in lymphoid neoplasms: A survey of 353 cases. *Mod. Pathol.* **2004**, *17*, 911–917. [CrossRef] [PubMed]

111. Heissmeyer, V.; Krappmann, D.; Wulczyn, F.G.; Scheidereit, C. NF-kappa B p105 is a target of Ikappa B kinases and controls signal induction of Bcl-3-p50 complexes. *EMBO J.* **1999**, *18*, 4766–4778. [CrossRef] [PubMed]

112. Wang, V.Y.-F.; Li, Y.; Kim, D.; Zhong, X.; Du, Q.; Ghassemian, M.; Ghosh, G. Bcl3 Phosphorylation by Akt, Erk2, and IKK Is Required for Its Transcriptional Activity. *Mol. Cell* **2017**, *67*, 484–497. [CrossRef] [PubMed]

113. Kitamura, H.; Kanehira, K.; Okita, K.; Morimatsu, M.; Saito, M. MAIL, a novel nuclear I kappa B protein that potentiates LPS-induced IL-6 production. *FEBS Lett.* **2000**, *485*, 53–56. [CrossRef]

114. Willems, M.; Dubois, N.; Musumeci, L.; Bours, V.; Robe, P.A. IκBζ: An emerging player in cancer. *Oncotarget* **2016**, *7*, 66310–66322. [CrossRef] [PubMed]

115. Yamazaki, S.; Muta, T.; Matsuo, S.; Takeshige, K. Stimulus-specific induction of a novel nuclear factor-kappaB regulator, IkappaB-zeta, via Toll/Interleukin-1 receptor is mediated by mRNA stabilization. *J. Biol. Chem.* **2005**, *280*, 1678–1687. [CrossRef] [PubMed]

116. Eto, A.; Muta, T.; Yamazaki, S.; Takeshige, K. Essential roles for NF-κB and a Toll/IL-1 receptor domain-specific signal(s) in the induction of IκB-ζ. *Biochem. Biophys. Res. Commun.* **2003**, *301*, 495–501. [CrossRef]

117. Yamamoto, M.; Yamazaki, S.; Uematsu, S.; Sato, S.; Hemmi, H.; Hoshino, K.; Kaisho, T.; Kuwata, H.; Takeuchi, O.; Takeshige, K.; et al. Regulation of Toll/IL-1-receptor-mediated gene expression by the inducible nuclear protein IκBζ. *Nature* **2004**, *430*, 218–222. [CrossRef] [PubMed]
118. Trinh, D.V.; Zhu, N.; Farhang, G.; Kim, B.J.; Huxford, T. The Nuclear IκB Protein IκBζ Specifically Binds NF-κB p50 Homodimers and Forms a Ternary Complex on κB DNA. *J. Mol. Biol.* **2008**, *379*, 122–135. [CrossRef] [PubMed]
119. Yamazaki, S.; Muta, T.; Takeshige, K. A novel IkappaB protein, IkappaB-zeta, induced by proinflammatory stimuli, negatively regulates nuclear factor-kappaB in the nuclei. *J. Biol. Chem.* **2001**, *276*, 27657–27662. [CrossRef] [PubMed]
120. Nogai, H.; Wenzel, S.-S.; Hailfinger, S.; Grau, M.; Kaergel, E.; Seitz, V.; Wollert-Wulf, B.; Pfeifer, M.; Wolf, A.; Frick, M.; et al. IκB-ζ controls the constitutive NF-κB target gene network and survival of ABC DLBCL. *Blood* **2013**, *122*, 2242–2250. [CrossRef] [PubMed]
121. Hu, Y.; Baud, V.; Delhase, M.; Zhang, P.; Deerinck, T.; Ellisman, M.; Johnson, R.; Karin, M. Abnormal Morphogenesis But Intact IKK Activation in Mice Lacking the IKK Subunit of IB Kinase. *Science* **1999**, *284*, 316–320. [CrossRef] [PubMed]
122. Delhase, M.; Hayakawa, M.; Chen, Y.; Karin, M. Positive and negative regulation of IkappaB kinase activity through IKKbeta subunit phosphorylation. *Science* **1999**, *284*, 309–313. [CrossRef] [PubMed]
123. Li, X.; Massa, P.E.; Hanidu, A.; Peet, G.W.; Aro, P.; Savitt, A.; Mische, S.; Li, J.; Marcu, K.B. IKKα, IKKβ, and NEMO/IKKγ Are Each Required for the NF-κB-mediated Inflammatory Response Program. *J. Biol. Chem.* **2002**, *277*, 45129–45140. [CrossRef] [PubMed]
124. Merkhofer, E.C.; Cogswell, P.; Baldwin, A.S. Her2 activates NF-kappaB and induces invasion through the canonical pathway involving IKKalpha. *Oncogene* **2010**, *29*, 1238–1248. [CrossRef] [PubMed]
125. Adli, M.; Merkhofer, E.; Cogswell, P.; Baldwin, A.S. IKKα and IKKβ Each Function to Regulate NF-κB Activation in the TNF-Induced/Canonical Pathway. *PLoS ONE* **2010**, *5*, e9428. [CrossRef] [PubMed]
126. Birbach, A.; Gold, P.; Binder, B.R.; Hofer, E.; de Martin, R.; Schmid, J.A. Signaling Molecules of the NF-κB Pathway Shuttle Constitutively between Cytoplasm and Nucleus. *J. Biol. Chem.* **2002**, *277*, 10842–10851. [CrossRef] [PubMed]
127. Yamamoto, Y.; Verma, U.N.; Prajapati, S.; Kwak, Y.-T.; Gaynor, R.B. Histone H3 phosphorylation by IKK-α is critical for cytokine-induced gene expression. *Nature* **2003**, *423*, 655–659. [CrossRef] [PubMed]
128. Anest, V.; Hanson, J.L.; Cogswell, P.C.; Steinbrecher, K.A.; Strahl, B.D.; Baldwin, A.S. A nucleosomal function for IκB kinase-α in NF-κB-dependent gene expression. *Nature* **2003**, *423*, 659–663. [CrossRef] [PubMed]
129. Hoberg, J.E.; Yeung, F.; Mayo, M.W. SMRT Derepression by the IκB Kinase α: A Prerequisite to NF-κB Transcription and Survival. *Mol. Cell* **2004**, *16*, 245–255. [CrossRef] [PubMed]
130. Fernández-Majada, V.; Aguilera, C.; Villanueva, A.; Vilardell, F.; Robert-Moreno, A.; Aytés, A.; Real, F.X.; Capella, G.; Mayo, M.W.; Espinosa, L.; et al. Nuclear IKK activity leads to dysregulated notch-dependent gene expression in colorectal cancer. *Proc. Natl. Acad. Sci. USA* **2007**, *104*, 276–281. [CrossRef] [PubMed]
131. Margalef, P.; Fernández-Majada, V.; Villanueva, A.; Garcia-Carbonell, R.; Iglesias, M.; López, L.; Martínez-Iniesta, M.; Villà-Freixa, J.; Mulero, M.C.; Andreu, M.; et al. A Truncated Form of IKKα Is Responsible for Specific Nuclear IKK Activity in Colorectal Cancer. *Cell Rep.* **2012**, *2*, 840–854. [CrossRef] [PubMed]
132. Luo, J.-L.; Tan, W.; Ricono, J.M.; Korchynskyi, O.; Zhang, M.; Gonias, S.L.; Cheresh, D.A.; Karin, M. Nuclear cytokine-activated IKKα controls prostate cancer metastasis by repressing Maspin. *Nature* **2007**, *446*, 690–694. [CrossRef] [PubMed]
133. Puto, L.A.; Reed, J.C. Daxx represses RelB target promoters via DNA methyltransferase recruitment and DNA hypermethylation. *Genes Dev.* **2008**, *22*, 998–1010. [CrossRef] [PubMed]
134. Croxton, R.; Puto, L.A.; de Belle, I.; Thomas, M.; Torii, S.; Hanaii, F.; Cuddy, M.; Reed, J.C. Daxx represses expression of a subset of antiapoptotic genes regulated by nuclear factor-kappaB. *Cancer Res.* **2006**, *66*, 9026–9035. [CrossRef] [PubMed]
135. Vallabhapurapu, S.D.; Noothi, S.K.; Pullum, D.A.; Lawrie, C.H.; Pallapati, R.; Potluri, V.; Kuntzen, C.; Khan, S.; Plas, D.R.; Orlowski, R.Z.; et al. Transcriptional repression by the HDAC4–RelB–p52 complex regulates multiple myeloma survival and growth. *Nat. Commun.* **2015**, *6*, 8428. [CrossRef] [PubMed]

136. Birbach, A.; Bailey, S.T.; Ghosh, S.; Schmid, J.A. Cytosolic, nuclear and nucleolar localization signals determine subcellular distribution and activity of the NF-kappaB inducing kinase NIK. *J. Cell Sci.* **2004**, *117*, 3615–3624. [CrossRef] [PubMed]

137. Yamaguchi, N.; Ito, T.; Azuma, S.; Ito, E.; Honma, R.; Yanagisawa, Y.; Nishikawa, A.; Kawamura, M.; Imai, J.; Watanabe, S.; et al. Constitutive activation of nuclear factor-κB is preferentially involved in the proliferation of basal-like subtype breast cancer cell lines. *Cancer Sci.* **2009**, *100*, 1668–1674. [CrossRef] [PubMed]

138. Vazquez-Santillan, K.; Melendez-Zajgla, J.; Jimenez-Hernandez, L.E.; Gaytan-Cervantes, J.; Muñoz-Galindo, L.; Piña-Sanchez, P.; Martinez-Ruiz, G.; Torres, J.; Garcia-Lopez, P.; Gonzalez-Torres, C.; et al. NF-kappaB-inducing kinase regulates stem cell phenotype in breast cancer. *Sci. Rep.* **2016**, *6*, 37340. [CrossRef] [PubMed]

139. Guzman, M.L.; Neering, S.J.; Upchurch, D.; Grimes, B.; Howard, D.S.; Rizzieri, D.A.; Luger, S.M.; Jordan, C.T. Nuclear factor-kappaB is constitutively activated in primitive human acute myelogenous leukemia cells. *Blood* **2001**, *98*, 2301–2307. [CrossRef] [PubMed]

140. Xiu, Y.; Dong, Q.; Li, Q.; Li, F.; Borcherding, N.; Zhang, W.; Boyce, B.; Xue, H.-H.; Zhao, C. Stabilization of NF-κB-Inducing Kinase Suppresses MLL-AF9-Induced Acute Myeloid Leukemia. *Cell Rep.* **2018**, *22*, 350–358. [CrossRef] [PubMed]

141. Jung, J.-U.; Ravi, S.; Lee, D.W.; McFadden, K.; Kamradt, M.L.; Toussaint, L.G.; Sitcheran, R. NIK/MAP3K14 Regulates Mitochondrial Dynamics and Trafficking to Promote Cell Invasion. *Curr. Biol.* **2016**, *26*, 3288–3302. [CrossRef] [PubMed]

142. Xie, Q.; Wu, Q.; Horbinski, C.M.; Flavahan, W.A.; Yang, K.; Zhou, W.; Dombrowski, S.M.; Huang, Z.; Fang, X.; Shi, Y.; et al. Mitochondrial control by DRP1 in brain tumor initiating cells. *Nat. Neurosci.* **2015**, *18*, 501–510. [CrossRef] [PubMed]

143. Kashatus, J.A.; Nascimento, A.; Myers, L.J.; Sher, A.; Byrne, F.L.; Hoehn, K.L.; Counter, C.M.; Kashatus, D.F. Erk2 phosphorylation of Drp1 promotes mitochondrial fission and MAPK-driven tumor growth. *Mol. Cell* **2015**, *57*, 537–551. [CrossRef] [PubMed]

144. Bren, G.D.; Solan, N.J.; Miyoshi, H.; Pennington, K.N.; Pobst, L.J.; Paya, C.V. Transcription of the RelB gene is regulated by NF-κB. *Oncogene* **2001**, *20*, 7722–7733. [CrossRef] [PubMed]

145. Xia, Y.; Pauza, M.E.; Feng, L.; Lo, D. RelB regulation of chemokine expression modulates local inflammation. *Am. J. Pathol.* **1997**, *151*, 375–387. [PubMed]

146. Xia, Y.; Chen, S.; Wang, Y.; Mackman, N.; Ku, G.; Lo, D.; Feng, L. RelB modulation of IkappaBalpha stability as a mechanism of transcription suppression of interleukin-1alpha (IL-1alpha), IL-1beta, and tumor necrosis factor alpha in fibroblasts. *Mol. Cell. Biol.* **1999**, *19*, 7688–7696. [CrossRef] [PubMed]

147. Marienfeld, R.; May, M.J.; Berberich, I.; Serfling, E.; Ghosh, S.; Neumann, M. RelB forms transcriptionally inactive complexes with RelA/p65. *J. Biol. Chem.* **2003**, *278*, 19852–19860. [CrossRef] [PubMed]

148. Jacque, E.; Tchenio, T.; Piton, G.; Romeo, P.-H.; Baud, V. RelA repression of RelB activity induces selective gene activation downstream of TNF receptors. *Proc. Natl. Acad. Sci. USA* **2005**, *102*, 14635–14640. [CrossRef] [PubMed]

149. Maier, H.J.; Marienfeld, R.; Wirth, T.; Baumann, B. Critical role of RelB serine 368 for dimerization and p100 stabilization. *J. Biol. Chem.* **2003**, *278*, 39242–39250. [CrossRef] [PubMed]

150. Savinova, O.V.; Hoffmann, A.; Ghosh, G. The Nfkb1 and Nfkb2 proteins p105 and p100 function as the core of high-molecular-weight heterogeneous complexes. *Mol. Cell* **2009**, *34*, 591–602. [CrossRef] [PubMed]

151. Legarda-Addison, D.; Ting, A.T. Negative regulation of TCR signaling by NF-kappaB2/p100. *J. Immunol.* **2007**, *178*, 7767–7778. [CrossRef] [PubMed]

152. Sun, S.C.; Ganchi, P.A.; Béraud, C.; Ballard, D.W.; Greene, W.C.; Ware, C.F.; Van Duyne, G.D.; Ghosh, G. Autoregulation of the NF-kappa B transactivator RelA (p65) by multiple cytoplasmic inhibitors containing ankyrin motifs. *Proc. Natl. Acad. Sci. USA* **1994**, *91*, 1346–1350. [CrossRef] [PubMed]

153. Baud, V.; Collares, D. Post-Translational Modifications of RelB NF-κB Subunit and Associated Functions. *Cells* **2016**, *5*, 22. [CrossRef] [PubMed]

154. Huxford, T.; Ghosh, G. A Structural Guide to Proteins of the NF- B Signaling Module. *Cold Spring Harb. Perspect. Biol.* **2009**, *1*, a000075. [CrossRef] [PubMed]

155. Hoffmann, A.; Natoli, G.; Ghosh, G. Transcriptional regulation via the NF-κB signaling module. *Oncogene* **2006**, *25*, 6706–6716. [CrossRef] [PubMed]

156. Beg, A.A.; Baldwin, A.S. Activation of multiple NF-kappa B/Rel DNA-binding complexes by tumor necrosis factor. *Oncogene* **1994**, *9*, 1487–1492. [PubMed]

157. Yilmaz, Z.B.; Weih, D.S.; Sivakumar, V.; Weih, F. RelB is required for Peyer's patch development: Differential regulation of p52-RelB by lymphotoxin and TNF. *EMBO J.* **2003**, *22*, 121–130. [CrossRef] [PubMed]

158. Huang, D.B.; Huxford, T.; Chen, Y.Q.; Ghosh, G. The role of DNA in the mechanism of NFkappaB dimer formation: Crystal structures of the dimerization domains of the p50 and p65 subunits. *Structure* **1997**, *5*, 1427–1436. [CrossRef]

159. Huang, D.-B.; Vu, D.; Ghosh, G. NF-κB RelB Forms an Intertwined Homodimer. *Structure* **2005**, *13*, 1365–1373. [CrossRef] [PubMed]

160. Lo, J.C.; Basak, S.; James, E.S.; Quiambo, R.S.; Kinsella, M.C.; Alegre, M.-L.; Weih, F.; Franzoso, G.; Hoffmann, A.; Fu, Y.-X. Coordination between NF-kappaB family members p50 and p52 is essential for mediating LTbetaR signals in the development and organization of secondary lymphoid tissues. *Blood* **2006**, *107*, 1048–1055. [CrossRef] [PubMed]

161. Jiang, H.Y.; Petrovas, C.; Sonenshein, G.E. RelB-p50 NF-kappa B complexes are selectively induced by cytomegalovirus immediate-early protein 1: Differential regulation of Bcl-x(L) promoter activity by NF-kappa B family members. *J. Virol.* **2002**, *76*, 5737–5747. [CrossRef] [PubMed]

162. Hofer, S.; Rescigno, M.; Granucci, F.; Citterio, S.; Francolini, M.; Ricciardi-Castagnoli, P. Differential activation of NF-kappa B subunits in dendritic cells in response to Gram-negative bacteria and to lipopolysaccharide. *Microbes Infect.* **2001**, *3*, 259–265. [CrossRef]

163. Derudder, E.; Dejardin, E.; Pritchard, L.L.; Green, D.R.; Körner, M.; Baud, V. RelB/p50 Dimers Are Differentially Regulated by Tumor Necrosis Factor-α and Lymphotoxin-β Receptor Activation. *J. Biol. Chem.* **2003**, *278*, 23278–23284. [CrossRef] [PubMed]

164. Schjerven, H.; Tran, T.N.; Brandtzaeg, P.; Johansen, F.-E. De novo synthesized RelB mediates TNF-induced up-regulation of the human polymeric Ig receptor. *J. Immunol.* **2004**, *173*, 1849–1857. [CrossRef] [PubMed]

165. Shih, V.F.-S.; Davis-Turak, J.; Macal, M.; Huang, J.Q.; Ponomarenko, J.; Kearns, J.D.; Yu, T.; Fagerlund, R.; Asagiri, M.; Zuniga, E.I.; et al. Control of RelB during dendritic cell activation integrates canonical and noncanonical NF-κB pathways. *Nat. Immunol.* **2012**, *13*, 1162–1170. [CrossRef] [PubMed]

biomedicines

MDPI

Review

The Direct and Indirect Roles of NF-κB in Cancer: Lessons from Oncogenic Fusion Proteins and Knock-in Mice

Tabea Riedlinger [1], Jana Haas [1], Julia Busch [1], Bart van de Sluis [2], Michael Kracht [3] and M. Lienhard Schmitz [1,*]

[1] Institute of Biochemistry, Justus-Liebig-University, D-35392 Giessen, Germany;
    Tabea.Riedlinger@biochemie.med.uni-giessen.de (T.R.); Jana.Haas@biochemie.med.uni-giessen.de (J.H.);
    Julia.Busch@biochemie.med.uni-giessen.de (J.B.)
[2] Department of Pediatrics, Molecular Genetics Section, University of Groningen, University Medical Center
    Groningen, Antonius Deusinglaan 1, 9713 AV, Groningen, The Netherlands; a.j.a.van.de.sluis@umcg.nl
[3] Rudolf-Buchheim-Institute of Pharmacology, Justus-Liebig-University, D-35392 Giessen, Germany;
    Michael.Kracht@pharma.med.uni-giessen.de
*   Correspondence: Lienhard.Schmitz@biochemie.med.uni-giessen.de;
    Tel.: +49-641-994-7570; Fax: +49-641-994-7589

Received: 28 February 2018; Accepted: 17 March 2018; Published: 19 March 2018

Abstract: NF-κB signaling pathways play an important role in the regulation of cellular immune and stress responses. Aberrant NF-κB activity has been implicated in almost all the steps of cancer development and many of the direct and indirect contributions of this transcription factor system for oncogenesis were revealed in the recent years. The indirect contributions affect almost all hallmarks and enabling characteristics of cancer, but NF-κB can either promote or antagonize these tumor-supportive functions, thus prohibiting global NF-κB inhibition. The direct effects are due to mutations of members of the NF-κB system itself. These mutations typically occur in upstream components that lead to the activation of NF-κB together with further oncogenesis-promoting signaling pathways. In contrast, mutations of the downstream components, such as the DNA-binding subunits, contribute to oncogenic transformation by affecting NF-κB-driven transcriptional output programs. Here, we discuss the features of recently identified oncogenic RelA fusion proteins and the characterization of pathways that are regulating the transcriptional activity of NF-κB by regulatory phosphorylations. As NF-κB's central role in human physiology prohibits its global inhibition, these auxiliary or cell type-specific NF-κB regulating pathways are potential therapeutic targets.

Keywords: NF-κB; transcription; cancer

1. The NF-κB System

The NF-κB transcription factor system represents an archetypal signaling pathway that is evolutionary conserved, with core components occurring in insects, cnidarians, and even some unicellular species [1]. One of its key functions is the induction of the immune system in response to pathogens and inflammatory stimuli, but NF-κB is also activated by further adverse stimuli, such as DNA damage and hypoxia [2,3]. The great variety of input signals is reflected by a multitude of target genes that do not only represent inducers and effectors of the immune system, but also genes that impact on cell survival, proliferation, and differentiation [4]. A characteristic feature of NF-κB is the large variety of inducers that utilize a limited set of signal transduction molecules to create transcriptional output programs that are specifically tailored to suit the particular requirements in a specific tissue or organ. This conversion of relatively unspecific input signals to highly specific output programs is not really understood and employs components of the NF-κB core signaling pathways in conjunction with multiple auxiliary systems that confer specificity.

2. The Core NF-κB Activation Pathways

Five different NF-κB DNA-binding subunits share a N-terminal NF-κB/Rel homology domain (RHD), which mediates DNA-binding, dimerization, and the interaction with the inhibitory IκB proteins, as schematically shown in Figure 1. The C-terminal regions of RELA (also known as p65), RELB, and REL (also known as c-Rel) contain transactivation domains (TADs), which associate with further regulatory proteins to trigger mRNA synthesis. The family members NF-κB1 (also known as p105) and NF-κB2 (also known as p100) are precursor proteins that lack TADs but contain Ankyrin repeats in their C-terminal regions [5]. These allow for them to function as IκB proteins and to mediate protein/protein interactions. Either during translation or through phosphorylation-induced partial proteolysis, the precursors are processed to yield their DNA-binding forms p50 and p52, respectively [6]. As p50 and p52 also lack TADs, they cannot trigger transcription unless being associated with a TAD-containing NF-κB family member, suggesting that constitutive binding of p50 or p52 homodimers to κB sites represses transcription [7]. Alternatively, p50 and p52 can activate transcription upon association with IκB family members, such as Bcl3.

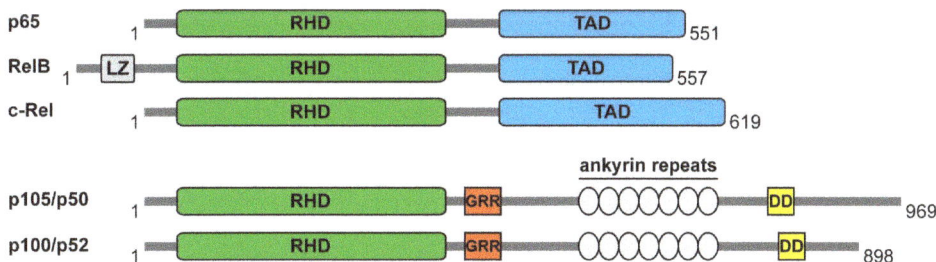

Figure 1. DNA-binding subunits of NF-κB. The functional domains of the five DNA-binding subunits, including the leucine zipper (LZ), the glycine-rich region (GRR), and the death domain (DD) are shown. The number of amino acids is given for the human proteins.

Activation of NF-κB can proceed by canonical, noncanonical, and the atypical NF-κB activation pathway [8]. All of the different NF-κB activating pathways ultimately result in the generation of active, DNA-binding dimers, which were beforehand retained in the cytosol of unstimulated cells by association with inhibitory IκB proteins.

In the canonical pathway, membrane anchored or cytosolic receptors sense cytokines, pathogen-associated molecular patterns (PAMPs), or damage-associated molecular patterns (DAMPs) [9]. Bound receptors self-associate to trigger signaling cascades, which proceed via posttranslational modifications (PTMs), including phosphorylation and ubiquitination. A variety of polyubiquitin chains are formed to mediate NF-κB activation including K63-branched and linear polyubiquitin chains, which are recognized and bound by chain readers such as TAB2 (TGFβ activated kinase 1/MAP3K7 binding protein 2) and NEMO (NF-κB essential modulator), respectively [10,11]. These events lead to activation of the so-called IKK (IκB kinase) complex, which is composed of the catalytic subunits IKKα and IKK in association with the scaffold protein NEMO. The activated IKKs in turn phosphorylate IκBα in order to enable its subsequent tagging with K48-branched ubiquitin chains. This allows rapid IκBα degradation by the proteasome, allowing nuclear entry, DNA-binding and transcriptional activity of the dimerized NF-κB DNA-binding subunits [8].

The noncanonical NF-κB pathway is activated in response to a distinct class of stimuli, particularly in B-cells. The activation of this pathway depends on NF-κB-inducing kinase (NIK) and IKKα, which mediate C-terminal processing of NF-κB2/p100, thereby allowing for the generation of p52/RelB dimers [12]. The noncanonical pathway requires NF-κB-inducing kinase (NIK) and IKKα, and proceeds with much slower kinetics [13].

The atypical NF-κB pathway is activated by various adverse stimuli, including DNA damage, and depends on the inducible modification of NEMO by the ubiquitin-related peptide SUMO-1 (small ubiquitin-like modifier 1). This modification process depends on the SUMO E3 ligase PIASy (protein inhibitor of activated STAT y) [14]. In addition, DNA damage leads to ATM (Ataxia Telangiectasia mutated)-mediated TRAF6 (TNF receptor associated factor 6) activation and cIAP1 (cellular inhibitor of apoptosis protein-1) recruitment, which catalyzes monoubiquitination of NEMO [15]. These modifications ultimately lead to induction of IKK activity, IκBα degradation and the release of the DNA-binding subunits.

3. Auxiliary NF-κB Regulating Mechanisms

Any given NF-κB activating stimulus does not only activate NF-κB, but it will necessarily trigger further signaling pathways. These additional pathways show extensive crosstalk to the NF-κB system at all levels, as elaborated in a number of excellent reviews [16,17]. Furthermore, any NF-κB activating pathophysiological situation, such as an infection, will elicit a complex mixture of NF-κB activating agents and receptors, thus leading to NF-κB activation amidst multiple co-regulated pathways that serve to dampen, trigger, or shape the NF-κB response. As a result of NF-κB activation, the DNA-binding subunits, released from their inhibitor, will bind to their cognate κB genomic binding site, which is conventionally a GGGRNNYYCC (R = purine, N = any nucleotide, Y = pyrimidine) motif [18]. Binding of NF-κB dimers to the genomic κB site is not only dictated by the accessibility of the site and by interaction with other transcription factors, such as AP1, but also by the composition of the κB binding sequence. The extensive characterization of NF-κB binding sites by EMSA-seq (electrophorectic mobility shift assays coupled to sequencing) experiments uncovered the preferred motifs for the p65/p50, p65/p65, and p65/p52 complexes [19]. For example, p65/p50 dimers preferentially bind the motif GGGGRTTTCC, while the p65/p52 heterodimer binds to the sequence NNNGGGGRYTT. Binding specificity is also achieved by the interaction of nuclear NF-κB subunits with further transcription factors, such as AP1 family members [20], Erg [21], and E2F1 [22].

Moreover, the DNA-binding capacity of NF-κB DNA-binding subunits can also be regulated in order to achieve a further level of regulation. For example, acetylation of p65 at Lys 221 causes a conformation change that favours κB DNA binding [23]. Conversely, the DNA-binding activity can be inhibited upon association of p65 with PIAS1 [24] or by PTMs, such as nitration of Tyr 66 and Tyr 152, or asymmetric dimethylation of Arg 30 [25,26]. The recent years have witnessed the identification of further PTMs at the DNA-binding subunits. The PhosphositePlus® database lists 64 different modifications for the p65 subunit and comparable numbers of PTMs for p105 (61), p100 (49), c-Rel (22), and RelB (20). The DNA-binding subunits are modified by many different PTMs, including phosphorylation, ubiquitination, SUMO modification, O-GlcNAcylation, and various types of methylation. This strong increase in PTM site identification is contrasted by our limited understanding of their physiological function. Several knock-in mouse models have been generated that started to unravel the physiological roles of the individual PTMs.

4. Functional Analysis of p65 Phosphorylations In Vivo

The importance of individual p65 modification sites has been investigated in various knock-in mouse models, which are summarized in Figure 2. Published data are available for five different p65 knock-in mouse models and the observed range of phenotypes is remarkably broad. None of these mice that are mutated in p65 phosphorylation sites recapitulate the phenotype of RelA-deficient mice, which show embryonic lethality that is caused by massive liver degeneration with hepatocyte apoptosis [27]. The first knock-in mouse model was published almost 10 years ago by the laboratory of Sankar Ghosh [28]. In this animal, the phosphorylated Ser 276 was changed to a non-phosphorylatable Ala (p65 S276A). Cells isolated from these mice display a significant reduction of NF-κB-dependent transcription. Most of the p65 S276A knock-in embryos die at different embryonic days due to variegated developmental abnormalities in limb or eye formation, which are due to HDAC (histone deacetylase)-dependent interference with gene expression. Further experiments showed that p50/p65 S276A heterodimers cannot efficiently bind to CBP/p300 and instead recruit HDACs, resulting in aberrant repression of non-NF-κB-regulated genes through epigenetic mechanisms and the direct repression of a subset of NF-κB target genes [28]. The recruitment of HDACs requires the DNA-binding activity of p65, as demonstrated by the creation of a further knock-in mouse model where the p65 S276A mutant was additionally changed at Arg 274 to Ala, thus creating a non-DNA-binding p65 S276A/R274A double mutant [28]. These p65 S276A/R274A mice show no variegation of the phenotype, but die during embryogenesis due to massive hepatocyte apoptosis, thus resembling the phenotype of p65-deficient mice [27]. While the p65 S276A mice are phosphorylation-deficient, the p65 S276D knock-in mice express a phosphomimetic form of p65. These animals are viable, but consistent with an increased transcriptional activity of p65- show elevated expression of proinflammatory cytokines and chemokines, resulting in a systemic hyperinflammatory phenotype leading to death 8–20 days after birth. The lethality of the p65 S276D animals could be rescued by crossing with a strain lacking the TNF receptor 1 (TNFR1) [29], similar to the rescue of $RelA^{-/-}/Tnfr^{-/-}$ double knockout animals [30]. However, upon aging, the p65 S276D/$Tnfr^{-/-}$ mice develop chronic keratitis together with elevated expression of inflammatory cytokines in the cornea, leading to the development of a phenotype that resembles a human disease called keratoconjunctivitis sicca ("dry eyes") [29]. In addition, p65 S276D mice show hyperproliferation and dysplasia of the mouse epidermis [31]. Several mouse mutants were generated for the analysis of phosphorylated key residues in the C-terminal TADs. Mice harboring the T505A mutation develop normally but exhibit increased hepatocyte proliferation following damage resulting from carbon tetrachloride treatment or liver partial hepatectomy [32]. The p65 T505A mice also show an earlier onset of tumorigenesis in the N-nitrosodiethylamine model of hepatocellular carcinoma, as consistent with previous data from reconstituted cells, suggesting that Thr 505 phosphorylation functions to suppress the tumour-promoting functions of p65 [33]. A further knock-in model addressed the relevance of Ser 534 phosphorylation, a site corresponding to the well-studied Ser 536 in human p65. The p65 S534A mice are born at normal Mendelian ratios and are healthy. Following the injection of lipopolysaccharide (LPS) they display slightly increased expression of selective NF-κB target genes and a strongly increased sensitivity to LPS-induced death [34]. The inhibitory activity of p65 Ser534 phosphorylation on NF-κB activity may also be due to a decreased half-life of the p65 protein, which is a feature that can be regulated by phosphorylation at Ser 536 and also at Ser 468 [35–37]. A regulatory effect of Ser 468 phosphorylation on the total amount of p65 was also observed in mice where Ser 467 (the mouse homolog of human p65 Ser 468) was mutated to Ala. This substitution causes reduced de novo synthesis of the p65 protein by an unexplored mechanism. The reduced p65 levels likely contribute to diminished TNF-induced expression of a selected group of NF-κB dependent target genes and also increased TNF-triggered apoptosis [38]. A serendipitous observation was the reduced weight gain in male p65 S467A mice. Feeding a high fat diet resulted in a strong body weight increase in wildtype animals, while male p65 S467A mice were partially protected from an increase in body weight. This phenotype is likely due to the elevated locomotor activity of these animals, but it is currently completely unclear why this phenotype only occurs in male mice [38]. It will be also very interesting to study the mechanisms that are underlying this increased locomotor activity, which might be attributable to NF-κB's role for the brain [39].

In summary, most of the reports show a stunning biological variability of the p65 knock-in phenotypes rather than a uniform modulation of NF-κB-mediated transcription. We anticipate that one DNA-binding subunit can be simultaneously modified at several positions, but the typical modification patterns must be revealed in future studies. The analysis of further p65 variants that are simultaneously mutated at several sites will probably reveal a more complete picture of the regulatory potential of the auxiliary NF-κB regulating pathways.

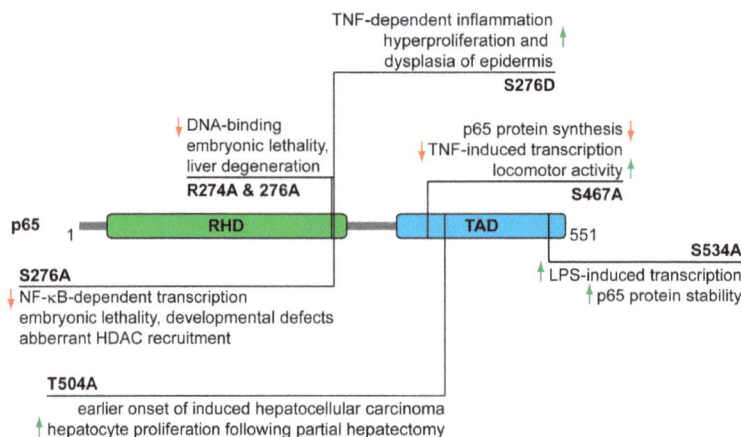

Figure 2. Schematic summary of the phenotypes of the p65 (RelA) knock-in mice mutated at the indicated phosphorylation sites.

5. The Multiple Roles of NF-κB in Cancer

The central position of NF-κB as a signaling hub in human physiology also involves this network in several human ailments, including cancer. A Pubmed search with the keywords "NF-κB" and "cancer" revealed >17,200 Pubmed entries for this topic. The complex role of NF-κB in cancer can be attributed to changes in the canonical and noncanonical pathways, and to indirect and direct mechanisms.

The direct contribution of the NF-κB system to cancer was revealed by a number of genetic changes of NF-κB regulators detected in cancer cells [40]. These changes include mutation, amplification, and fusion of NF-κB regulators. Constitutive or chronic NF-κB activation can be achieved by gain-of-function (GOF) mutations that lead to continuous signaling to the IKK complex. Alternatively, the same effect can be achieved by loss-of-function (LOF) mutations disrupting negative regulators, such as CYLD or A20, which can lead to the inappropriate termination of the NF-κB response. The indirect contribution of NF-κB is attributable to the fact that this transcription factor contributes to the regulation of most (but not all) classical hallmarks of cancer, as defined by the Hanahan and Weinberg review: sustained proliferative signaling, resistance to cell death, replicative immortality, induction of angiogenesis, activation of invasion and metastasis, and reprogramming of energy metabolism [41]. But, also enabling characteristics, such as the acquisition of genome instability and tumor-promoting inflammation, are processes in which NF-κB-is involved.

5.1. The Indirect Roles of NF-κB in Cancer

In this section we will discuss the indirect roles of NF-κB in cancer by describing its contribution to the classical hallmarks and enabling characteristics, as schematically depicted in Figure 3. Many apoptotic stimuli, including the cytokine TNF, ionizing radiation, and chemotherapeutic agents, such as daunorubicine, trigger the anti-apoptotic functions of NF-κB [42–44]. This mechanism is clinically important as many chemotherapeutical agents induce NF-κB, which consequently protects cancer cells from cell death. This phenomenon has been described in numerous studies, for example in cisplatin and camptothecin-treated lung cancer [45] and anthracycline- and taxane-treated breast cancer [46]. These results led to the suggestion that targeting the anti-apoptotic role of NF-κB helps to overcome therapy resistance [47]. Proof-of-concept studies showed that selective NF-κB inhibition enhances the response to chemotherapy in gastric cancer [48] and IKK inhibition sensitizes melanoma cells to doxorubicin-induced cell death [49]. Of note, in some settings, NF-κB can also trigger cell death, for example in response to oxidative stress [50] or in the execution of TRAIL- and CD95-mediated apoptosis [51].

The role of NF-κB for replicative immortality is not well established, but an extensive crosstalk between the telomerase catalytic subunit (TERT) and NF-κB has been described. TERT activity ensures telomere elongation and thus helps to evade tumors from the fate of replicative senescence. NF-κB p65 was found to modulate TNF-induced nuclear translocation as well as telomerase activity of TERT in multiple myeloma cells [52]. The expression of the *Tert* gene is also directly upregulated by NF-κB [53]. But, this regulation is mutual, as vice versa TERT directly regulates NF-κB-dependent gene expression by binding to DNA-bound p65, potentially leading to hyperexpression of an extremely tumor promoting set of genes [54].

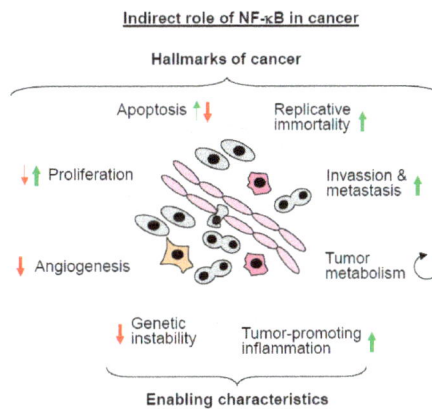

Figure 3. Schematic summary of the indirect roles of NF-κB for the hallmarks and enabling characteristics of cancer. The relative thickness of the arrows indicates the estimated relative contribution of NF-κB for the support (arrow up) or antagonism (arrow down) for the indicated processes.

Also, the activation of invasion and metastasis involves NF-κB-dependent processes. NF-κB plays a significant role in the regulation of epithelial-mesenchymal transition (EMT), which is an early event in metastasis [55]. Inhibition of NF-κB signaling by expression of a dominant negative IκBα prevents EMT in Ras-transformed epithelial cells. This study also showed that NF-κB activation increases the transition to a mesenchymal phenotype, while its inhibition in mesenchymal cells causes even a reversal of EMT [56]. Collectively, these data suggest that NF-κB contributes both to the induction and also the maintenance of EMT. TNF-triggered EMT depends on NF-κB-mediated upregulation of Twist1, which is one of the key transcription factors modulating EMT [57]. In breast cancer cells, NF-κB activation contributes not only to expression of Twist1, but also to further EMT-regulating transcription factors, such as SLUG and SIP1 (Smad interacting protein 1) [58]. In addition, many cell adhesion molecules, such as integrins, selectins, ICAM-1 (intercellular adhesion molecule 1), E-selectin, and VCAM-1 (vascular cell adhesion molecule 1) are directly regulated by NF-κB [59]. These molecules contribute to cancer cell extravasation, but NF-κB activity is also important in the non-tumorigenic cells at the remote sites, which are colonized by tumor cells. This was shown in a mouse model of metastasis where the injection of lung carcinoma cells results in a reduced formation of metastatic foci in the livers of animals with a liver-specific deletion of IKKβ [60].

More recent evidence showed a contribution of NF-κB for the remodeling of tumor metabolism. Tumor cells often (but not always) prefer to use glycolysis instead of oxidative phosphorylation, as initially reported in 1927 by Otto Warburg [61]. The preferential use of glycolysis leads to a reduced ATP production, but might allow tumor cells to survive under hypoxic conditions and also to deliver substrates and intermediates for various anabolic pathways [62]. The shift from oxidative phosphorylation to glycolysis is mainly mediated by p53 and hypoxia inducible factor (HIF)-1, but also NF-κB participates in this regulation. Knockdown of NF-κB p65 in mouse embryonic fibroblasts causes cellular reprogramming to aerobic glycolysis, thus recapitulating the Warburg effect [63]. After glucose starvation, NF-κB p65 triggers p53 expression, which in turn leads to the mitochondrial upregulation of cytochrome c oxidase 2 [63], which is a crucial component of the electron transport chain. Another report showed that loss of p53 leads to p65-triggered expression of the glucose transporter GLUT3 (glucose transporter 3) in tumor cells, resulting in increased glucose consumption and anaerobic glycolysis [64]. The p65 protein can also be transported into the mitochondria in the absence of p53. Mitochondrial p65 can associate with the mitochondrial genome to repress mitochondrial gene expression and oxidative phosphorylation [65], and it will be interesting to study the molecular mechanisms allowing p65 entry into the mitochondria.

NF-κB also sustains proliferative signaling, but this is seen only in specific cell types, such as lymphocytes. An example for a direct and cell-specific effect of NF-κB is provided by mice that are lacking the DNA-binding subunit c-Rel, which display defects in B-cell proliferation [66,67]. Similarly, also RelB-deficient mice show defective proliferative responses in B cells [68]. In addition, *Rel*$^{-/-}$ mice show reduced keratinocyte proliferation and epidermal thickness [69]. Growth inhibitory roles of NF-κB have also been reported in mice expressing a dominant-negative IκBα protein. These mice show hyperplasia of the epidermal epithelium, suggesting that NF-κB restricts the proliferation of this cell type in vivo [70]. In addition to its importance for proliferation of specific cell types, NF-κB might also be relevant for proliferation of cancer cells, although this aspect is not fully explored. For example, RelB is required for proliferation of endometrioid adenocarcinoma [71] and overexpression of dominant-negative IκBα interferes with the proliferation of cervix carcinoma cells [72].

Many solid tumors induce angiogenesis in order to ensure the supply of nutrients and oxygen as well as to evacuate carbon dioxide and metabolic wastes [41]. This process is antagonized by NF-κB, as revealed in a study using mice expressing a dominant-negative IκBα protein exclusively in endothelial cells, resulting in the selective repression of NF-κB in endothelial cells. Inoculated tumors grow faster and more aggressive in these transgenic mice, concomitant with a striking increase in tumor vascularization [73]. Therefore, angiogenesis repression by NF-κB challenges the paradigm that systemic NF-κB inhibition can serve as a universal anti-cancer strategy.

But, NF-κB also regulates two cancer enabling characteristics of cancer cells. The contribution of NF-κB for tumor-promoting inflammation is very well documented and also covered by a number of excellent reviews [74,75]. In order avoid repetition and redundancy, we refer to these articles, which convincingly summarize how cells of the immune system create a tumor-promoting microenvironment by the NF-κB-dependent production of growth and survival factors.

The enabling characteristics of cancer include the acquisition of genetic instability, which confers selective advantage on cell subclones to enable their outgrowth and succession of clonal expansions [41]. NF-κB is activated by DNA damage, for example, by an ATM-mediated pathway in response to DNA double strand breaks induced by oxidative stress. In addition, NF-κB also contributes to the repair of double-strand breaks by homologous recombination. DNA damage triggers the association of p65 with the CtIP-BRCA1 complex and thus stimulates DNA repair [76]. A role of NF-κB for genome stability was revealed by the analysis of p65-deficient mouse and human cells, which show all of the signs of genomic instability, including high frequencies of gene deletions, DNA mutations, and chromosomal translocations [77]. The molecular mechanisms explaining the contribution of p65 for genomic stability are not clear and need to be elucidated in the future.

In summary, numerous studies have revealed an important contribution of NF-κB to many aspects of cancer biology. Although the majority of features are consistent with a tumor-promoting function of this transcription factor, a global inhibition of NF-κB is not possible, as it would foster angiogenesis and genetic instability.

5.2. The Direct Roles of NF-κB in Cancer

The direct contribution of the NF-κB system to cancer is seen by the direct mutation of NF-κB regulatory proteins [40]. Many of these mutations lead to lymphomas, which likely have its cause in the important role of the NF-κB pathway in normal lymphocyte development and activation. A review by Neil Perkins makes the important point that single mutations in NF-κB DNA-binding subunits are relatively rare, as they cannot adequately mimic the diverse nature and complexity of the NF-κB response [78]. Thus, the majority of NF-κB-driven cancers are induced by gain-of-function (GOF) mutations in the upstream activators of NF-κB. While GOF mutations of upstream activators lead to tumor formation that is dependent on the simultaneous induction of further signaling pathways, mutations of the DNA-binding subunits can trigger oncogenesis without the requirement for additional signals, as summarized in Figure 4.

The contribution of NF-κB for tumorigenesis by upstream activators is exemplified by mutations that are found in components of the so-called CBM complex, which is essential for NF-κB activation in lymphocytes by activation of the T-cell receptor (TCR) or B-cell receptor (BCR) [79]. The CBM complex is composed of the scaffold protein CARMA1/CARD11 (CARD-MAGUK1/caspase recruitment domain family member 11), the adaptor protein BCL10 (B-Cell CLL/lymphoma 10) and MALT1 (mucosa-associated lymphoid tissue lymphoma translocation protein 1), which has a dual role as a scaffold protein and a paracaspase [80,81].

Direct role of NF-κB in cancer

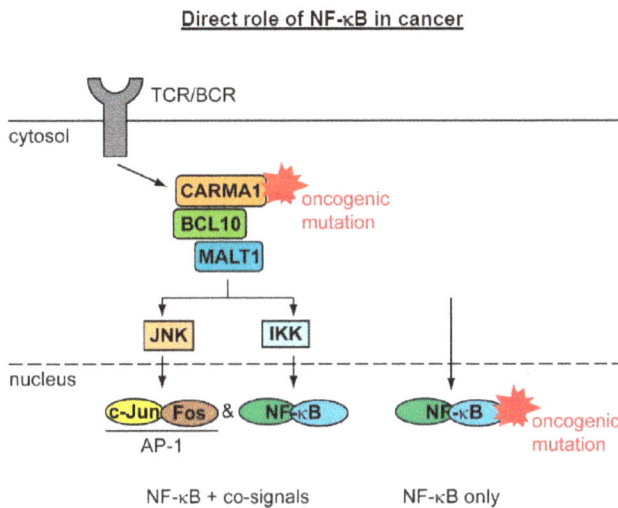

Figure 4. Schematic display of mutations occurring at upstream or downstream components of the NF-κB system. Mutations of upstream regulators are exemplified by the gain-of-function (GOF) mutations in the CBM complex that typically lead to activation of further tumor-promoting signaling pathways.

Receptor activation leads to the recruitment of CARMA1 to lipid rafts, followed by its phosphorylation, which induces a conformational change, thus allowing for its oligomerization, the formation of BCL10 filaments and the attachment of MALT1 [82]. The BCL10/MALT1 filaments are further decorated by the ubiquitin E3 ligase TRAF6, which likely results in the all-or-none activation of the downstream pathways [82]. Active TRAF6 mediates the attachment of K63-branched polyubiquitin chains at the C-terminus of MALT1, which allows for further protein/protein interactions and the activation of NF-κB and JNK/AP1 (c-Jun N-terminal kinase/activator protein 1) signaling. The MALT1 protein has proteolytic activity and cleaves its substrate proteins after a positively charged arginine residue in the P1 position [80,81]. Mutations leading to the constitutive activation of the CBM complex are frequently found in lymphoid malignancies. GOF mutations of CARMA1 often occur in diffuse large B-cell lymphoma, B-cell lymphoproliferative disorders, and also in T-cell leukemia/lymphomas [83,84]. But, also the other members of the CBM complex are frequently found to be mutated, overexpressed, or affected by chromosomal translocations (MALT1-API2 or c-IAP2-MALT1). The GOF mutations of CBM members promote lymphomagenesis and MALT1 inhibitors have shown to allow for new treatment options for patients with refractory t(11;18)-positive MALT1 lymphoma [85–87]. The constitutively active CBM complex leads to constant activation of NF-κB and also to the permanent induction of the JNK/AP1 pathway and expression of RNA-binding proteins (Roquin-1 and -2, Regnase-1), which mediate the stabilization of several NF-κB-dependent transcripts. Of note, constitutive activation of NF-κB alone is not sufficient for lymphomagenesis and requires further JNK-derived signals. This was revealed in a mouse model where the transgenic expression of the constitutively active CARMA1 mutant CARMA1 (L225LI) is sufficient to trigger aggressive B-cell lymphoproliferation. Inhibition of either NF-κB or JNK blocks proliferation of the CARMA1 (L225LI)-expressing B-cells, showing that only cooperative NF-κB and JNK activation drives the malignant growth [88].

While mutation of the upstream activators of NF-κB necessarily affects the co-regulatory pathways, there is also evidence for an oncogenic function of the DNA-binding subunits, as summarized in Figure 5. The DNA-binding subunits represent the endpoint of the NF-κB cascade and their ability to mediate oncogenic transformation shows the importance of NF-κB-dependent transcriptional programs. Soon after cloning of the avian retroviral protein v-Rel it became clear that retroviruses encoding v-Rel are highly oncogenic and transform chicken cells in vitro and in vivo [89,90]. Transgenic mice expressing v-Rel in T-cells develop immature, multicentric aggressive T-cell leukemia/lymphomas, showing that ectopic expression of v-Rel alone is sufficient for oncogenic transformation [91]. Constitutive p50/v-Rel DNA-binding is induced upon v-Rel epression, but the transforming activity also occurs in transgenic thymocytes lacking p50. The ability of v-Rel to trigger or repress gene expression depends on the cell type [92], but v-Rel-mediated induction of gene expression seems not to be as potent as gene induction by p65 [93]. Transformed chicken spleen cells show v-Rel localization in the nucleus and also in the cytoplasm. [94], but a threshold nuclear level of the v-Rel oncoprotein is required for the transformation of avian lymphocytes [95]. In the nucleus, v-Rel cooperates with many transcription factors, such as the AP1 subunits c-Fos and c-Jun, IRF4, and SP1 to mediate oncogenic transformation [96–98].

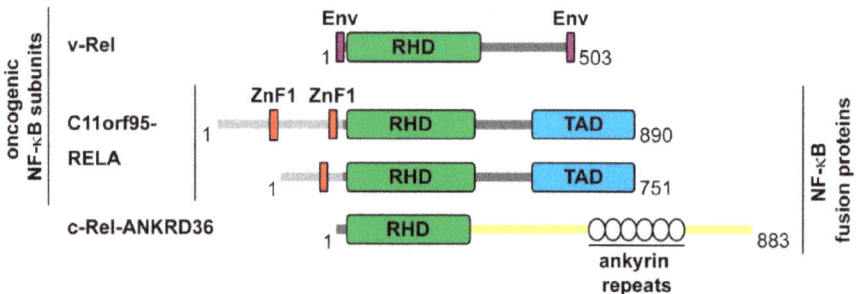

Figure 5. Schematic summary of oncogenic NF-κB DNA-binding subunits. The color code is similar to Figure 1, the v-Rel protein contains short sequence remnants from the REV env (envelope) gene at the N- and C-termini. Different parts of the C11ORF95 proteins, which contain zinc fingers (ZnF) are fused to the N-termini of p65.

From all five different DNA-binding subunits, only the v-Rel homologue c-Rel has the ability to transform avian lymphoid cells dependent from the presence of both C-terminal TADs [99,100]. However, the removal of either of the 2 C-terminal TADs from c-Rel augments its transforming activity, suggesting that a chronic low-level of transcriptional activation is optimal for oncogenic transformation [101]. The c-Rel encoding *REL* locus is frequently altered (rearranged, amplified, mutated) in a variety of B- and T-cell malignancies. The *REL* locus at the chromosomal position 2p16.1-15 is amplified in Hodgkin's lymphoma (~46%) and diffuses large B cell lymphoma (DLBCL) (~15%) [101]. Also, T-cell lymphomas (natural killer, peripheral T cells, anaplastic large cells) show the frequent amplification of the *REL* locus [102]. Increased c-Rel expression has been observed in Hodgkin lymphoma where the *REL* locus was found to be translocated to a position near the light chain enhancer [103], or alternatively by integration of an EBV genome near to the c-Rel encoding region [104]. Experimental proof for a causative role of c-Rel overexpression in the process of tumorigenesis has been obtained in a mouse model. Overexpression of c-Rel under the control of a hormone-responsive mouse mammary tumour virus promoter led to the induction of mammary tumors, but these develop with a long latency, suggesting the requirement of secondary events [105]. Also, nuclear localisation of c-REL is frequently observed in cancer cells.

Furthermore, oncogenic fusion proteins have been described for c-Rel and p65. A chimeric c-REL-ANKRD36 protein occurs in the RC-K8 DLBCL cell line due to a deletion on chromosome 2 [106]. Interestingly, the fusion partner ANKRD36 has several ankyrin repeats, which are also found in p100, p105 and the IκB proteins. However, it is unclear whether this fusion protein is responsible for the proliferation of the RC-K8 cells. Another oncogenic fusion protein, C11orf95-RELA, is expressed in more than two-thirds of supratentorial ependymomas, a rare brain tumor, where C11orf95-RELA accumulates in the nuclei [107]. A causative role of C11orf95-RELA for tumorigenesis was shown in experiments where the expression of the fusion protein in mouse forebrain neural stem cells converted them to brain tumor cells, probably by driving an aberrant transcription program [107]. The various C11orf95-RELA fusion proteins contain one or two zinc finger domains, raising the possibility that this fusion can alter the DNA-binding specificity of NF-κB. It will thus be interesting to characterize the genomic binding sites of C11orf95-RELA and to compare them to the binding sites of the wildtype p65 protein in healthy tissue. The finding that C11orf95-RELA expression generates only one highly specific cancer type also illustrates the high plasticity and diversity of the NF-κB response.

6. Concluding Remarks

The NF-κB transcription factor system is of broad relevance for the formation and maintenance of many tumor entities. Its broad relevance for central physiological processes does not allow for its global inhibition and raises the need for the development of highly pathway-specific and cell-type specific components as exemplified by MALT1 inhibitors, which are currently tested in clinical trials for the treatment of lymphomas. Further work is needed in order to identify and characterize the auxiliary components that help to specify and regulate the transcriptional output programs of NF-κB, as these proteins could be additional druggable targets.

Acknowledgments: The work from M. Lienhard Schmitz and Michael Kracht is supported by grants from the Deutsche Forschungsgemeinschaft SFB/TRR81, SFB1021, SFB1213 and the Excellence Cluster Cardio-Pulmonary System (ECCPS, EXC 147/2). The work of M. Lienhard Schmitz is supported by grants from the Deutsche Forschungsgemeinschaft (SCHM 1417/8-3), the Deutsche Krebshilfe (111447) and the IMPRS-HLR program of the Max-Planck Society.

Author Contributions: Tabea Riedlinger has conceived the concept and prepared the figures, M. Lienhard Schmitz has written the first draft of the manuscript, Jana Haas, Julia Busch, Bart van de Sluis and Michael Kracht have provided substantial help for the correction of the manuscript.

Conflicts of Interest: The authors declare no conflict of interest.

References

1. Gilmore, T.D.; Wolenski, F.S. NF-κB: Where did it come from and why? *Immunol. Rev.* **2012**, *246*, 14–35. [CrossRef] [PubMed]
2. D'Ignazio, L.; Rocha, S. Hypoxia Induced NF-κB. *Cells* **2016**, *5*, 10. [CrossRef] [PubMed]
3. Miyamoto, S. Nuclear initiated NF-κB signaling: NEMO and ATM take center stage. *Cell Res.* **2011**, *21*, 116–130. [CrossRef] [PubMed]
4. Mitchell, S.; Vargas, J.; Hoffmann, A. Signaling via the NF-κB system. *Wiley. Interdiscip. Rev. Syst. Biol. Med.* **2016**, *8*, 227–241. [CrossRef] [PubMed]
5. Grimm, S.; Baeuerle, P.A. The inducible transcription factor NF-κB: Structure-function relationship of its protein subunits. *Biochem. J.* **1993**, *290*, 297–308. [CrossRef] [PubMed]
6. Yilmaz, Z.B.; Kofahl, B.; Beaudette, P.; Baum, K.; Ipenberg, I.; Weih, F.; Wolf, J.; Dittmar, G.; Scheidereit, C. Quantitative dissection and modeling of the NF-κB p100-p105 module reveals interdependent precursor proteolysis. *Cell Rep.* **2014**, *9*, 1756–1769. [CrossRef] [PubMed]
7. Schmitz, M.L.; Baeuerle, P.A. The p65 subunit is responsible for the strong transcription activating potential of NF-κB. *EMBO J.* **1991**, *10*, 3805–3817. [PubMed]
8. Hayden, M.S.; Ghosh, S. NF-κB, the first quarter-century: Remarkable progress and outstanding questions. *Genes Dev.* **2012**, *26*, 203–234. [CrossRef] [PubMed]
9. Schmitz, M.L.; Kracht, M.; Saul, V.V. The intricate interplay between RNA viruses and NF-κB. *Biochim. Biophys. Acta* **2014**, *1843*, 2754–2764. [CrossRef] [PubMed]
10. Ikeda, F. Linear ubiquitination signals in adaptive immune responses. *Immunol. Rev.* **2015**, *266*, 222–236. [CrossRef] [PubMed]
11. Rittinger, K.; Ikeda, F. Linear ubiquitin chains: Enzymes, mechanisms and biology. *Open. Biol.* **2017**, *7*, 170026. [CrossRef] [PubMed]
12. Senftleben, U.; Cao, Y.; Xiao, G.; Greten, F.R.; Krahn, G.; Bonizzi, G.; Chen, Y.; Hu, Y.; Fong, A.; Sun, S.C.; et al. Activation by IKKα of a second, evolutionary conserved, NF-κB signaling pathway. *Science* **2001**, *293*, 1495–1499. [CrossRef] [PubMed]
13. Xiao, G.; Harhaj, E.W.; Sun, S.C. NF-κB-inducing kinase regulates the processing of NF-κB2 p100. *Mol. Cell* **2001**, *7*, 401–409. [CrossRef]
14. Mabb, A.M.; Wuerzberger-Davis, S.M.; Miyamoto, S. PIASy mediates NEMO sumoylation and NF-κB activation in response to genotoxic stress. *Nat. Cell Biol.* **2006**, *8*, 986–993. [CrossRef] [PubMed]
15. Hinz, M.; Stilmann, M.; Arslan, S.C.; Khanna, K.K.; Dittmar, G.; Scheidereit, C. A cytoplasmic ATM-TRAF6-cIAP1 module links nuclear DNA damage signaling to ubiquitin-mediated NF-κB activation. *Mol. Cell* **2010**, *40*, 63–74. [CrossRef] [PubMed]
16. D'Ignazio, L.; Bandarra, D.; Rocha, S. NF-κB and HIF crosstalk in immune responses. *FEBS J.* **2016**, *283*, 413–424. [CrossRef] [PubMed]
17. Oeckinghaus, A.; Hayden, M.S.; Ghosh, S. Crosstalk in NF-κB signaling pathways. *Nat. Immunol.* **2011**, *12*, 695–708. [CrossRef] [PubMed]
18. Natoli, G. Tuning up inflammation: How DNA sequence and chromatin organization control the induction of inflammatory genes by NF-κB. *FEBS Lett.* **2006**, *580*, 2843–2849. [CrossRef] [PubMed]
19. Wong, D.; Teixeira, A.; Oikonomopoulos, S.; Humburg, P.; Lone, I.N.; Saliba, D.; Siggers, T.; Bulyk, M.; Angelov, D.; Dimitrov, S.; et al. Extensive characterization of NF-κB binding uncovers non-canonical motifs and advances the interpretation of genetic functional traits. *Genome Biol.* **2011**, *12*, R70. [CrossRef] [PubMed]
20. Stein, B.; Baldwin, A.S., Jr.; Ballard, D.W.; Greene, W.C.; Angel, P.; Herrlich, P. Cross-coupling of the NF-κB p65 and Fos/Jun transcription factors produces potentiated biological function. *EMBO J.* **1993**, *12*, 3879–3891. [PubMed]
21. Dryden, N.H.; Sperone, A.; Martin-Almedina, S.; Hannah, R.L.; Birdsey, G.M.; Khan, S.T.; Layhadi, J.A.; Mason, J.C.; Haskard, D.O.; Gottgens, B.; et al. The transcription factor Erg controls endothelial cell quiescence by repressing activity of nuclear factor (NF)-κB p65. *J. Biol. Chem.* **2012**, *287*, 12331–12342. [CrossRef] [PubMed]
22. Lim, C.A.; Yao, F.; Wong, J.J.; George, J.; Xu, H.; Chiu, K.P.; Sung, W.K.; Lipovich, L.; Vega, V.B.; Chen, J.; et al. Genome-wide mapping of RELA(p65) binding identifies E2F1 as a transcriptional activator recruited by NF-κB upon TLR4 activation. *Mol. Cell* **2007**, *27*, 622–635. [CrossRef] [PubMed]

23. Chen, L.F.; Mu, Y.; Greene, W.C. Acetylation of RelA at discrete sites regulates distinct nuclear functions of NF-κB. *EMBO J.* **2002**, *21*, 6539–6548. [CrossRef] [PubMed]
24. Liu, B.; Yang, R.; Wong, K.A.; Getman, C.; Stein, N.; Teitell, M.A.; Cheng, G.; Wu, H.; Shuai, K. Negative regulation of NF-κB signaling by PIAS1. *Mol. Cell Biol.* **2005**, *25*, 1113–1123. [CrossRef] [PubMed]
25. Park, S.W.; Huq, M.D.; Hu, X.; Wei, L.N. Tyrosine nitration on p65: A novel mechanism to rapidly inactivate nuclear factor-κB. *Mol. Cell Proteom.* **2005**, *4*, 300–309. [CrossRef] [PubMed]
26. Reintjes, A.; Fuchs, J.E.; Kremser, L.; Lindner, H.H.; Liedl, K.R.; Huber, L.A.; Valovka, T. Asymmetric arginine dimethylation of RelA provides a repressive mark to modulate TNFα/NF-κB response. *Proc. Natl. Acad. Sci. USA* **2016**, *113*, 4326–4331. [CrossRef] [PubMed]
27. Beg, A.A.; Sha, W.C.; Bronson, R.T.; Ghosh, S.; Baltimore, D. Embryonic lethality and liver degeneration in mice lacking the RelA component of NF-κB. *Nature* **1995**, *376*, 167–170. [CrossRef] [PubMed]
28. Dong, J.; Jimi, E.; Zhong, H.; Hayden, M.S.; Ghosh, S. Repression of gene expression by unphosphorylated NF-κB p65 through epigenetic mechanisms. *Genes Dev.* **2008**, *22*, 1159–1173. [CrossRef] [PubMed]
29. Dong, J.; Jimi, E.; Zeiss, C.; Hayden, M.S.; Ghosh, S. Constitutively active NF-κB triggers systemic TNFα-dependent inflammation and localized TNFα-independent inflammatory disease. *Genes Dev.* **2010**, *24*, 1709–1717. [CrossRef] [PubMed]
30. Rosenfeld, M.E.; Prichard, L.; Shiojiri, N.; Fausto, N. Prevention of hepatic apoptosis and embryonic lethality in RelA/TNFR-1 double knockout mice. *Am. J. Pathol.* **2000**, *156*, 997–1007. [CrossRef]
31. Poligone, B.; Hayden, M.S.; Chen, L.; Pentland, A.P.; Jimi, E.; Ghosh, S. A role for NF-κB activity in skin hyperplasia and the development of keratoacanthomata in mice. *PLoS ONE* **2013**, *8*, e71887. [CrossRef] [PubMed]
32. Moles, A.; Butterworth, J.A.; Sanchez, A.; Hunter, J.E.; Leslie, J.; Sellier, H.; Tiniakos, D.; Cockell, S.J.; Mann, D.A.; Oakley, F.; et al. A RelA(p65) Thr505 phospho-site mutation reveals an important mechanism regulating NF-κB-dependent liver regeneration and cancer. *Oncogene* **2016**, *35*, 4623. [CrossRef] [PubMed]
33. Msaki, A.; Sanchez, A.M.; Koh, L.F.; Barre, B.; Rocha, S.; Perkins, N.D.; Johnson, R.F. The role of RelA (p65) threonine 505 phosphorylation in the regulation of cell growth, survival, and migration. *Mol. Biol. Cell* **2011**, *22*, 3032–3040. [CrossRef] [PubMed]
34. Pradere, J.P.; Hernandez, C.; Koppe, C.; Friedman, R.A.; Luedde, T.; Schwabe, R.F. Negative regulation of NF-κB p65 activity by serine 536 phosphorylation. *Sci. Signal* **2016**, *9*, ra85. [CrossRef] [PubMed]
35. Geng, H.; Wittwer, T.; Dittrich-Breiholz, O.; Kracht, M.; Schmitz, M.L. Phosphorylation of NF-κB p65 at Ser468 controls its COMMD1-dependent ubiquitination and target gene-specific proteasomal elimination. *EMBO Rep.* **2009**, *10*, 381–386. [CrossRef] [PubMed]
36. Lawrence, T.; Bebien, M.; Liu, G.Y.; Nizet, V.; Karin, M. IKKα limits macrophage NF-κB activation and contributes to the resolution of inflammation. *Nature* **2005**, *434*, 1138–1143. [CrossRef] [PubMed]
37. Mao, X.; Gluck, N.; Li, D.; Maine, G.N.; Li, H.; Zaidi, I.W.; Repaka, A.; Mayo, M.W.; Burstein, E. GCN5 is a required cofactor for a ubiquitin ligase that targets NF-κB/RelA. *Genes Dev.* **2009**, *23*, 849–861. [CrossRef] [PubMed]
38. Riedlinger, T.; Dommerholt, M.B.; Wijshake, T.; Kruit, J.K.; Huijkman, N.; Dekker, D.; Koster, M.; Kloosterhuis, N.; Koonen, D.P.Y.; de Bruin, A.; et al. NF-κB p65 serine 467 phosphorylation sensitizes mice to weight gain and TNFα-or diet-induced inflammation. *Biochim. Biophys. Acta* **2017**, *1864*, 1785–1798. [CrossRef] [PubMed]
39. Kaltschmidt, B.; Kaltschmidt, C. NF-κB in Long-Term Memory and Structural Plasticity in the Adult Mammalian Brain. *Front. Mol. Neurosci.* **2015**, *8*, 69. [CrossRef] [PubMed]
40. Courtois, G.; Gilmore, T.D. Mutations in the NF-κB signaling pathway: Implications for human disease. *Oncogene* **2006**, *25*, 6831–6843. [CrossRef] [PubMed]
41. Hanahan, D.; Weinberg, R.A. Hallmarks of cancer: The next generation. *Cell* **2011**, *144*, 646–674. [CrossRef] [PubMed]
42. Beg, A.A.; Baltimore, D. An essential role for NF-κB in preventing TNF-alpha-induced cell death. *Science* **1996**, *274*, 782–784. [CrossRef] [PubMed]
43. Van Antwerp, D.J.; Martin, S.J.; Kafri, T.; Green, D.R.; Verma, I.M. Suppression of TNF-alpha-induced apoptosis by NF-κB. *Science* **1996**, *274*, 787–789. [CrossRef] [PubMed]
44. Wang, C.Y.; Mayo, M.W.; Baldwin, A.S., Jr. TNF- and cancer therapy-induced apoptosis: Potentiation by inhibition of NF-κB. *Science* **1996**, *274*, 784–787. [CrossRef] [PubMed]

45. Yan, H.Q.; Huang, X.B.; Ke, S.Z.; Jiang, Y.N.; Zhang, Y.H.; Wang, Y.N.; Li, J.; Gao, F.G. Interleukin 6 augments lung cancer chemotherapeutic resistance via ataxia-telangiectasia mutated/NF-κB pathway activation. *Cancer Sci.* **2014**, *105*, 1220–1227. [CrossRef] [PubMed]
46. Montagut, C.; Tusquets, I.; Ferrer, B.; Corominas, J.M.; Bellosillo, B.; Campas, C.; Suarez, M.; Fabregat, X.; Campo, E.; Gascon, P.; et al. Activation of nuclear factor-κB is linked to resistance to neoadjuvant chemotherapy in breast cancer patients. *Endocr. Relat Cancer* **2006**, *13*, 607–616. [CrossRef] [PubMed]
47. Godwin, P.; Baird, A.M.; Heavey, S.; Barr, M.P.; O'Byrne, K.J.; Gately, K. Targeting nuclear factor-κB to overcome resistance to chemotherapy. *Front Oncol.* **2013**, *3*, 120. [CrossRef] [PubMed]
48. Sohma, I.; Fujiwara, Y.; Sugita, Y.; Yoshioka, A.; Shirakawa, M.; Moon, J.H.; Takiguchi, S.; Miyata, H.; Yamasaki, M.; Mori, M.; et al. Parthenolide, an NF-κB inhibitor, suppresses tumor growth and enhances response to chemotherapy in gastric cancer. *Cancer Genom. Proteom.* **2011**, *8*, 39–47.
49. Enzler, T.; Sano, Y.; Choo, M.K.; Cottam, H.B.; Karin, M.; Tsao, H.; Park, J.M. Cell-selective inhibition of NF-κB signaling improves therapeutic index in a melanoma chemotherapy model. *Cancer Discov.* **2011**, *1*, 496–507. [CrossRef] [PubMed]
50. Kaltschmidt, B.; Kaltschmidt, C.; Hofmann, T.G.; Hehner, S.P.; Droge, W.; Schmitz, M.L. The pro- or anti-apoptotic function of NF-κB is determined by the nature of the apoptotic stimulus. *Eur. J. Biochem.* **2000**, *267*, 3828–3835. [CrossRef] [PubMed]
51. Jennewein, C.; Karl, S.; Baumann, B.; Micheau, O.; Debatin, K.M.; Fulda, S. Identification of a novel pro-apoptotic role of NF-κB in the regulation of T. *Oncogene* **2012**, *31*, 1468–1474. [CrossRef] [PubMed]
52. Akiyama, M.; Hideshima, T.; Hayashi, T.; Tai, Y.T.; Mitsiades, C.S.; Mitsiades, N.; Chauhan, D.; Richardson, P.; Munshi, N.C.; Anderson, K.C. Nuclear factor-κB p65 mediates tumor necrosis factor alpha-induced nuclear translocation of telomerase reverse transcriptase protein. *Cancer Res.* **2003**, *63*, 18–21. [PubMed]
53. Yin, L.; Hubbard, A.K.; Giardina, C. NF-κB regulates transcription of the mouse telomerase catalytic subunit. *J. Biol. Chem.* **2000**, *275*, 36671–36675. [CrossRef] [PubMed]
54. Ghosh, A.; Saginc, G.; Leow, S.C.; Khattar, E.; Shin, E.M.; Yan, T.D.; Wong, M.; Zhang, Z.; Li, G.; Sung, W.K.; et al. Telomerase directly regulates NF-κB-dependent transcription. *Nat. Cell Biol.* **2012**, *14*, 1270–1281. [CrossRef] [PubMed]
55. Yang, J.; Weinberg, R.A. Epithelial-mesenchymal transition: At the crossroads of development and tumor metastasis. *Dev. Cell* **2008**, *14*, 818–829. [CrossRef] [PubMed]
56. Huber, M.A.; Azoitei, N.; Baumann, B.; Grunert, S.; Sommer, A.; Pehamberger, H.; Kraut, N.; Beug, H.; Wirth, T. NF-κB is essential for epithelial-mesenchymal transition and metastasis in a model of breast cancer progression. *J. Clin. Investig.* **2004**, *114*, 569–581. [CrossRef] [PubMed]
57. Li, C.W.; Xia, W.; Huo, L.; Lim, S.O.; Wu, Y.; Hsu, J.L.; Chao, C.H.; Yamaguchi, H.; Yang, N.K.; Ding, Q.; et al. Epithelial-mesenchymal transition induced by TNF-alpha requires NF-κB-mediated transcriptional upregulation of Twist1. *Cancer Res.* **2012**, *72*, 1290–1300. [CrossRef] [PubMed]
58. Pires, B.R.; Mencalha, A.L.; Ferreira, G.M.; de Souza, W.F.; Morgado-Diaz, J.A.; Maia, A.M.; Correa, S.; Abdelhay, E.S. NF-κB Is Involved in the Regulation of EMT Genes in Breast Cancer Cells. *PLoS ONE* **2017**, *12*, e0169622. [CrossRef] [PubMed]
59. Collins, T.; Read, M.A.; Neish, A.S.; Whitley, M.Z.; Thanos, D.; Maniatis, T. Transcriptional regulation of endothelial cell adhesion molecules: NF-κB and cytokine-inducible enhancers. *FASEB J.* **1995**, *9*, 899–909. [CrossRef] [PubMed]
60. Maeda, S.; Hikiba, Y.; Sakamoto, K.; Nakagawa, H.; Hirata, Y.; Hayakawa, Y.; Yanai, A.; Ogura, K.; Karin, M.; Omata, M. Ikappa B kinasebeta/nuclear factor-κB activation controls the development of liver metastasis by way of interleukin-6 expression. *Hepatology* **2009**, *50*, 1851–1860. [CrossRef] [PubMed]
61. Warburg, O.; Wind, F.; Negelein, E. The metabolism of tumors in the body. *J. Gen. Physiol* **1927**, *8*, 519–530. [CrossRef] [PubMed]
62. Liberti, M.V.; Locasale, J.W. The Warburg Effect: How Does it Benefit Cancer Cells? *Trends Biochem. Sci.* **2016**, *41*, 211–218. [CrossRef] [PubMed]
63. Mauro, C.; Leow, S.C.; Anso, E.; Rocha, S.; Thotakura, A.K.; Tornatore, L.; Moretti, M.; De, S.E.; Beg, A.A.; Tergaonkar, V.; et al. NF-κB controls energy homeostasis and metabolic adaptation by upregulating mitochondrial respiration. *Nat. Cell Biol.* **2011**, *13*, 1272–1279. [CrossRef] [PubMed]
64. Kawauchi, K.; Araki, K.; Tobiume, K.; Tanaka, N. p53 regulates glucose metabolism through an IKK-NF-κB pathway and inhibits cell transformation. *Nat. Cell Biol.* **2008**, *10*, 611–618. [CrossRef] [PubMed]

65. Johnson, R.F.; Witzel, I.I.; Perkins, N.D. p53-dependent regulation of mitochondrial energy production by the RelA subunit of NF-κB. *Cancer Res.* **2011**, *71*, 5588–5597. [CrossRef] [PubMed]

66. Köntgen, F.; Grumont, R.J.; Strasser, A.; Metcalf, D.; Li, R.; Tarlinton, D.; Gerondakis, S. Mice lacking the c-rel proto-oncogene exhibit defects in lymphocyte proliferation, humoral immunity, and interleukin-2 expression. *Genes Dev.* **1995**, *9*, 1965–1977. [CrossRef] [PubMed]

67. Tumang, J.R.; Owyang, A.; Andjelic, S.; Jin, Z.; Hardy, R.R.; Liou, M.L.; Liou, H.C. c-Rel is essential for B lymphocyte survival and cell cycle progression. *Eur. J. Immunol.* **1998**, *28*, 4299–4312. [CrossRef]

68. Snapper, C.M.; Rosas, F.R.; Zelazowski, P.; Moorman, M.A.; Kehry, M.R.; Bravo, R.; Weih, F. B cells lacking RelB are defective in proliferative responses, but undergo normal B cell maturation to Ig secretion and Ig class switching. *J. Exp. Med.* **1996**, *184*, 1537–1541. [CrossRef] [PubMed]

69. Fullard, N.; Moles, A.; O'Reilly, S.; van Laar, J.M.; Faini, D.; Diboll, J.; Reynolds, N.J.; Mann, D.A.; Reichelt, J.; Oakley, F. The c-Rel subunit of NF-κB regulates epidermal homeostasis and promotes skin fibrosis in mice. *Am. J. Pathol.* **2013**, *182*, 2109–2120. [CrossRef] [PubMed]

70. Seitz, C.S.; Lin, Q.; Deng, H.; Khavari, P.A. Alterations in NF-κB function in transgenic epithelial tissue demonstrate a growth inhibitory role for NF-κB. *Proc. Natl. Acad. Sci. USA* **1998**, *95*, 2307–2312. [CrossRef] [PubMed]

71. Ge, Q.L.; Liu, S.H.; Ai, Z.H.; Tao, M.F.; Ma, L.; Wen, S.Y.; Dai, M.; Liu, F.; Liu, H.S.; Jiang, R.Z.; et al. RelB/NF-κB links cell cycle transition and apoptosis to endometrioid adenocarcinoma tumorigenesis. *Cell Death. Dis.* **2016**, *7*, e2402. [CrossRef] [PubMed]

72. Kaltschmidt, B.; Kaltschmidt, C.; Hehner, S.P.; Droge, W.; Schmitz, M.L. Repression of NF-κB impairs HeLa cell proliferation by functional interference with cell cycle checkpoint regulators. *Oncogene* **1999**, *18*, 3213–3225. [CrossRef] [PubMed]

73. Kisseleva, T.; Song, L.; Vorontchikhina, M.; Feirt, N.; Kitajewski, J.; Schindler, C. NF-κB regulation of endothelial cell function during LPS-induced toxemia and cancer. *J. Clin. Investig.* **2006**, *116*, 2955–2963. [CrossRef] [PubMed]

74. Grivennikov, S.I.; Greten, F.R.; Karin, M. Immunity, inflammation, and cancer. *Cell* **2010**, *140*, 883–899. [CrossRef] [PubMed]

75. Taniguchi, K.; Karin, M. NF-κB, inflammation, immunity and cancer: Coming of age. *Nat. Rev. Immunol.* **2018**. [CrossRef] [PubMed]

76. Volcic, M.; Karl, S.; Baumann, B.; Salles, D.; Daniel, P.; Fulda, S.; Wiesmuller, L. NF-κB regulates DNA double-strand break repair in conjunction with BRCA1-CtIP complexes. *Nucleic Acids Res.* **2012**, *40*, 181–195. [CrossRef] [PubMed]

77. Wang, J.; Jacob, N.K.; Ladner, K.J.; Beg, A.; Perko, J.D.; Tanner, S.M.; Liyanarachchi, S.; Fishel, R.; Guttridge, D.C. RelA/p65 functions to maintain cellular senescence by regulating genomic stability and DNA repair. *EMBO Rep.* **2009**, *10*, 1272–1278. [CrossRef] [PubMed]

78. Perkins, N.D. The diverse and complex roles of NF-κB subunits in cancer. *Nat. Rev. Cancer* **2012**, *12*, 121–132. [CrossRef] [PubMed]

79. Afonina, I.S.; Elton, L.; Carpentier, I.; Beyaert, R. MALT1—a universal soldier: Multiple strategies to ensure NF-κB activation and target gene expression. *FEBS J.* **2015**, *282*, 3286–3297. [CrossRef] [PubMed]

80. Coornaert, B.; Baens, M.; Heyninck, K.; Bekaert, T.; Haegman, M.; Staal, J.; Sun, L.; Chen, Z.J.; Marynen, P.; Beyaert, R. T cell antigen receptor stimulation induces MALT1 paracaspase-mediated cleavage of the NF-κB inhibitor A20. *Nat. Immunol.* **2008**, *9*, 263–271. [CrossRef] [PubMed]

81. Rebeaud, F.; Hailfinger, S.; Posevitz-Fejfar, A.; Tapernoux, M.; Moser, R.; Rueda, D.; Gaide, O.; Guzzardi, M.; Iancu, E.M.; Rufer, N.; et al. The proteolytic activity of the paracaspase MALT1 is key in T cell activation. *Nat. Immunol.* **2008**, *9*, 272–281. [CrossRef] [PubMed]

82. David, L.; Li, Y.; Ma, J.; Garner, E.; Zhang, X.; Wu, H. Assembly mechanism of the CARMA1-BCL10-MALT1-TRAF6 signalosome. *Proc. Natl. Acad. Sci. USA* **2018**, *115*, 1499–1504. [CrossRef] [PubMed]

83. Juilland, M.; Thome, M. Role of the CARMA1/BCL10/MALT1 complex in lymphoid malignancies. *Curr. Opin. Hematol.* **2016**, *23*, 402–409. [CrossRef] [PubMed]

84. Meininger, I.; Krappmann, D. Lymphocyte signaling and activation by the CARMA1-BCL10-MALT1 signalosome. *Biol. Chem.* **2016**, *397*, 1315–1333. [CrossRef] [PubMed]

85. Lim, S.M.; Jeong, Y.; Lee, S.; Im, H.; Tae, H.S.; Kim, B.G.; Park, H.D.; Park, J.; Hong, S. Identification of beta-Lapachone Analogs as Novel MALT1 Inhibitors To Treat an Aggressive Subtype of Diffuse Large B-Cell Lymphoma. *J. Med. Chem.* **2015**, *58*, 8491–8502. [CrossRef] [PubMed]

86. Nagel, D.; Spranger, S.; Vincendeau, M.; Grau, M.; Raffegerst, S.; Kloo, B.; Hlahla, D.; Neuenschwander, M.; Peter von Kries, J.; Hadian, K.; et al. Pharmacologic inhibition of MALT1 protease by phenothiazines as a therapeutic approach for the treatment of aggressive ABC-DLBCL. *Cancer Cell* **2012**, *22*, 825–837. [CrossRef] [PubMed]

87. Saba, N.S.; Wong, D.H.; Tanios, G.; Iyer, J.R.; Lobelle-Rich, P.; Dadashian, E.L.; Liu, D.; Fontan, L.; Flemington, E.K.; Nichols, C.M.; et al. MALT1 Inhibition Is Efficacious in Both Naive and Ibrutinib-Resistant Chronic Lymphocytic Leukemia. *Cancer Res.* **2017**, *77*, 7038–7048. [CrossRef] [PubMed]

88. Knies, N.; Alankus, B.; Weilemann, A.; Tzankov, A.; Brunner, K.; Ruff, T.; Kremer, M.; Keller, U.B.; Lenz, G.; Ruland, J. Lymphomagenic CARD11/BCL10/MALT1 signaling drives malignant B-cell proliferation via cooperative NF-κB and JNK activation. *Proc. Natl. Acad. Sci. USA* **2015**, *112*, E7230–E7238. [CrossRef] [PubMed]

89. Gilmore, T.D.; Temin, H.M. Different localization of the product of the v-rel oncogene in chicken fibroblasts and spleen cells correlates with transformation by REV-T. *Cell* **1986**, *44*, 791–800. [CrossRef]

90. Stephens, R.M.; Rice, N.R.; Hiebsch, R.R.; Bose, H.R., Jr.; Gilden, R.V. Nucleotide sequence of v-rel: The oncogene of reticuloendotheliosis virus. *Proc. Natl. Acad. Sci. USA* **1983**, *80*, 6229–6233. [CrossRef] [PubMed]

91. Carrasco, D.; Rizzo, C.A.; Dorfman, K.; Bravo, R. The v-rel oncogene promotes malignant T-cell leukemia/lymphoma in transgenic mice. *EMBO J.* **1996**, *15*, 3640–3650. [PubMed]

92. Walker, W.H.; Stein, B.; Ganchi, P.A.; Hoffman, J.A.; Kaufman, P.A.; Ballard, D.W.; Hannink, M.; Greene, W.C. The v-rel oncogene: Insights into the mechanism of transcriptional activation, repression, and transformation. *J. Virol.* **1992**, *66*, 5018–5029. [PubMed]

93. Ballard, D.W.; Dixon, E.P.; Peffer, N.J.; Bogerd, H.; Doerre, S.; Stein, B.; Greene, W.C. The 65-kDa subunit of human NF-κB functions as a potent transcriptional activator and a target for v-Rel-mediated repression. *Proc. Natl. Acad. Sci. USA* **1992**, *89*, 1875–1879. [CrossRef] [PubMed]

94. Gilmore, T.D.; Temin, H.M. v-rel oncoproteins in the nucleus and in the cytoplasm transform chicken spleen cells. *J. Virol.* **1988**, *62*, 703–714. [PubMed]

95. Sachdev, S.; Diehl, J.A.; McKinsey, T.A.; Hans, A.; Hannink, M. A threshold nuclear level of the v-Rel oncoprotein is required for transformation of avian lymphocytes. *Oncogene* **1997**, *14*, 2585–2594. [CrossRef] [PubMed]

96. Hrdlickova, R.; Nehyba, J.; Bose, H.R., Jr. Interferon regulatory factor 4 contributes to transformation of v-Rel-expressing fibroblasts. *Mol. Cell Biol.* **2001**, *21*, 6369–6386. [PubMed]

97. Liss, A.S.; Tiwari, R.; Kralova, J.; Bose, H.R., Jr. Cell transformation by v-Rel reveals distinct roles of AP-1 family members in Rel/NF-κB oncogenesis. *Oncogene* **2010**, *29*, 4925–4937. [CrossRef] [PubMed]

98. Sif, S.; Gilmore, T.D. Interaction of the v-Rel oncoprotein with cellular transcription factor Sp1. *J. Virol.* **1994**, *68*, 7131–7138. [PubMed]

99. Fan, Y.; Gelinas, C. An optimal range of transcription potency is necessary for efficient cell transformation by c-Rel to ensure optimal nuclear localization and gene-specific activation. *Oncogene* **2007**, *26*, 4038–4043. [CrossRef] [PubMed]

100. Starczynowski, D.T.; Reynolds, J.G.; Gilmore, T.D. Deletion of either C-terminal transactivation subdomain enhances the in vitro transforming activity of human transcription factor REL in chicken spleen cells. *Oncogene* **2003**, *22*, 6928–6936. [CrossRef] [PubMed]

101. Gilmore, T.D.; Gerondakis, S. The c-Rel Transcription Factor in Development and Disease. *Genes Cancer* **2011**, *2*, 695–711. [CrossRef] [PubMed]

102. Mao, X.; Orchard, G.; Lillington, D.M.; Russell-Jones, R.; Young, B.D.; Whittaker, S. Genetic alterations in primary cutaneous CD30+ anaplastic large cell lymphoma. *Genes Chromosomes. Cancer* **2003**, *37*, 176–185. [CrossRef] [PubMed]

103. Martin-Subero, J.I.; Klapper, W.; Sotnikova, A.; Callet-Bauchu, E.; Harder, L.; Bastard, C.; Schmitz, R.; Grohmann, S.; Hoppner, J.; Riemke, J.; et al. Chromosomal breakpoints affecting immunoglobulin loci are recurrent in Hodgkin and Reed-Sternberg cells of classical Hodgkin lymphoma. *Cancer Res.* **2006**, *66*, 10332–10338. [CrossRef] [PubMed]

104. Luo, W.J.; Takakuwa, T.; Ham, M.F.; Wada, N.; Liu, A.; Fujita, S.; Sakane-Ishikawa, E.; Aozasa, K. Epstein-Barr virus is integrated between REL and BCL-11A in American Burkitt lymphoma cell line (NAB-2). *Lab Investig.* **2004**, *84*, 1193–1199. [CrossRef] [PubMed]

105. Belguise, K.; Sonenshein, G.E. PKCtheta promotes c-Rel-driven mammary tumorigenesis in mice and humans by repressing estrogen receptor alpha synthesis. *J. Clin. Investig.* **2007**, *117*, 4009–4021. [PubMed]

106. Kalaitzidis, D.; Gilmore, T.D. Genomic organization and expression of the rearranged REL proto-oncogene in the human B-cell lymphoma cell line RC-K8. *Genes Chromosomes Cancer* **2002**, *34*, 129–135. [CrossRef] [PubMed]

107. Parker, M.; Mohankumar, K.M.; Punchihewa, C.; Weinlich, R.; Dalton, J.D.; Li, Y.; Lee, R.; Tatevossian, R.G.; Phoenix, T.N.; Thiruvenkatam, R.; et al. C11orf95-RELA fusions drive oncogenic NF-κB signalling in ependymoma. *Nature* **2014**, *506*, 451–455. [CrossRef] [PubMed]

biomedicines

MDPI

Review

NF-κB Activation in Lymphoid Malignancies: Genetics, Signaling, and Targeted Therapy

Paula Grondona, Philip Bucher, Klaus Schulze-Osthoff, Stephan Hailfinger * and Anja Schmitt

Interfaculty Institute for Biochemistry, Eberhard Karls University of Tuebingen, Hoppe-Seyler-Str. 4, 72076 Tuebingen, Germany; paula.grondona@student.uni-tuebingen.de (P.G.); philip.bucher@student.uni-tuebingen.de (P.B.); KSO@uni-tuebingen.de (K.S.-O.); anja.schmitt@ifib.uni-tuebingen.de (A.S.)
* Correspondence: Stephan.Hailfinger@uni-tuebingen.de; Tel.: +49-7071-2974175

Received: 5 March 2018; Accepted: 22 March 2018; Published: 26 March 2018

Abstract: The NF-κB transcription factor family plays a crucial role in lymphocyte proliferation and survival. Consequently, aberrant NF-κB activation has been described in a variety of lymphoid malignancies, including diffuse large B-cell lymphoma, Hodgkin lymphoma, and adult T-cell leukemia. Several factors, such as persistent infections (e.g., with *Helicobacter pylori*), the pro-inflammatory microenvironment of the cancer, self-reactive immune receptors as well as genetic lesions altering the function of key signaling effectors, contribute to constitutive NF-κB activity in these malignancies. In this review, we will discuss the molecular consequences of recurrent genetic lesions affecting key regulators of NF-κB signaling. We will particularly focus on the oncogenic mechanisms by which these alterations drive deregulated NF-κB activity and thus promote the growth and survival of the malignant cells. As the concept of a targeted therapy based on the mutational status of the malignancy has been supported by several recent preclinical and clinical studies, further insight in the function of NF-κB modulators and in the molecular mechanisms governing aberrant NF-κB activation observed in lymphoid malignancies might lead to the development of additional treatment strategies and thus improve lymphoma therapy.

Keywords: NF-κB; lymphoma; leukemia; CARMA1; CARD11; CD79; MyD88

1. NF-κB in Lymphocytes

The NF-κB transcription factors are involved in the regulation of a variety of biological processes, such as inflammation, survival, and proliferation. The NF-κB family comprises five structurally related members forming different homo- or heterodimers: RelA (also known as p65), c-Rel, RelB, NF-κB1 (p50 and its precursor p105), as well as NF-κB2 (p52 and its precursor p100). The NF-κB proteins share a conserved REL homology domain required for homo- or heterodimerization, the interaction with inhibitor of κB (IκB) proteins, nuclear localization, and DNA binding. In quiescent cells, the inactive transcription factors are retained in the cytoplasm either by binding to the classical IκB proteins IκBα, IκBβ, and IκBε or by interaction with the inactive precursors p105 and p100. NF-κB activation in response to extracellular cues is regulated by two distinct pathways: In the canonical NF-κB pathway, stimulus-dependent activation of the IκB kinase (IKK) complex, comprising the catalytic subunits IKKα and IKKβ as well as the regulatory subunit IKKγ (also known as NF-κB essential modulator; NEMO), results in the phosphorylation and subsequent proteasomal degradation of the IκB proteins [1,2]. This allows the nuclear translocation of the NF-κB transcription factors, preferentially heterodimers of p50 and RelA or c-Rel, as well as their subsequent DNA binding and target gene transcription. In normal lymphocytes, stimulation-induced NF-κB activation is only transient and rapidly terminated by feedback inhibition involving the NF-κB-dependent expression of negative regulators, such as IκBα, IκBε, and A20 [3]. In contrast, activation of the non-canonical pathway involves the NF-κB inducing kinase (NIK)-dependent activation of IKKα and the inducible proteolytic processing of the precursor protein p100 to generate p52, which preferentially forms transcriptionally active heterodimers with RelB [2,4]. In addition to the classical IκB proteins, the IκB family also includes the atypical or nuclear IκB proteins BCL3, IκBζ, IκBNS, and IκBη, which are normally expressed only in response to pro-inflammatory stimuli. Unlike the classical IκBs that function as cytoplasmic inhibitors, the atypical IκB proteins primarily act as transcriptional co-activators or co-repressors in the nucleus, where they modulate the expression of a subset of NF-κB target genes [5–7].

In lymphocytes, NF-κB signaling is transiently activated in response to engagement of various receptors, such as antigen receptors, TNF receptors as well as interleukin-1 (IL-1) and Toll-like receptors (TLRs), and plays a critical role in development, survival, and acquisition of effector functions [8]. A variety of lymphoid malignancies, however, exhibits pathological activation of NF-κB due to diverse genetic lesions which affect key components of the NF-κB signaling pathway [9,10]. The first evidence linking core components of the NF-κB signaling pathway to lymphomagenesis has been reported in studies on the viral oncogene product v-Rel which causes aggressive lymphomas in birds and other animals [11,12]. Several subsequent studies have identified genetic aberrations in the NF-κB protein family, such as amplifications in the *REL* locus, in different lymphoid cancers [13,14]. To date, several lymphoid malignancies, such as Hodgkin lymphoma (HL), diffuse large B-cell lymphoma (DLBCL) of the activated B cell-like (ABC) subtype, lymphomas of the mucosa-associated lymphoid tissue (MALT), primary mediastinal B-cell lymphoma (PMBL), mantle cell lymphoma (MCL), multiple myeloma, and chronic lymphocytic leukemia (CLL), have been associated with aberrant NF-κB signaling [15–24]. Whereas genetic aberrations affecting the NF-κB members themselves are relatively rare, deregulated NF-κB activation is frequently achieved by oncogenic events which trigger the constitutive activity of various upstream signaling pathways, culminating in enhanced transcriptional activity of NF-κB [9,25]. In this review, we will describe recurrent genetic lesions driving pathological NF-κB activation in lymphoid malignancies. We will particularly focus on the molecular mechanism of the affected, aberrantly expressed regulators as well as their impact on the composition and function of the signaling complexes involved in NF-κB regulation. The exact molecular characterization of the key oncogenic mechanisms of constitutive NF-κB activation either shared by several or unique to certain lymphoid malignancies might allow the rational design of therapeutic strategies tailored to the specific tumor entities and might thus significantly improve lymphoma therapy.

2. Oncogenic MyD88 Mutations

The aberrant activation of innate immune signaling cascades represents one mechanism to drive constitutive activation of NF-κB signaling in lymphoid malignancies [10]. B cells express TLRs which recognize a wide variety of pathogen-associated molecular patterns (PAMPs) derived from bacteria, viruses, or fungi independently of the B-cell receptor (BCR) [26]. Structurally, TLRs are characterized by a conserved Toll/IL-1 receptor (TIR) domain which undergoes a conformational change after receptor ligation, providing a platform for the interaction with cytoplasmic TIR domain-containing proteins, such as the adapter protein myeloid differentiation primary response protein 88 (MyD88) [27]. MyD88 comprises an N-terminal death domain which is connected to a C-terminal TIR domain by a linker region [28]. Ligand binding results in dimerization of the TLRs and subsequent recruitment of MyD88 homodimers via TIR–TIR interactions [29,30]. MyD88 forms a high-molecular weight signaling complex, the so-called Myddosome, through a series of sequential interactions (Figure 1): First, MyD88 oligomerizes and recruits the IL-1R-associated kinases 1, 2, and 4 (IRAK1, 2, and 4) via a homotypic interaction involving their death domain. In the Myddosome, IRAK4 is activated by auto-phosphorylation and in turn phosphorylates IRAK1 [28,31–34]. IRAK2 can functionally substitute IRAK1, implicating that IRAK1 and IRAK2 are redundant for downstream signaling [35]. Once IRAK1 is fully phosphorylated, it dissociates from the receptor complex and activates the E3 ubiquitin ligase TNF receptor-associated factor 6 (TRAF6). TRAF6-dependent lysine 63 (K63)-linked polyubiquitination of itself and several other proteins facilitates the recruitment of the IKK complex and TGFβ-activated kinase 1 (TAK1) via the ubiquitin binding domains of the regulatory subunit IKKγ and the adapter proteins TAK1-binding protein (TAB), respectively [36–38]. TAK1-dependent phosphorylation activates IKKβ which in turn induces the proteasomal degradation of the inhibitory protein IκBα, thus triggering canonical NF-κB activation [3,39].

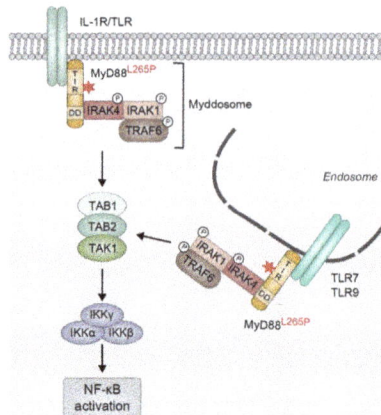

Figure 1. Oncogenic MyD88 mutations activate the canonical NF-κB pathway. Activation of canonical NF-κB signaling can be induced by the triggering of members of the IL-1 receptor or the Toll-like receptor family, which localize to the cell surface or, in the case of TLR7 and TLR9, to the endosomal compartment. Ligand binding induces the recruitment of MyD88 to the activated receptor via its TIR domain and triggers subsequent downstream signaling. MyD88 oligomers nucleate the formation of the Myddosome which ultimately results in the activation of IRAKs and the recruitment of the E3 ubiquitin ligase TRAF6. TRAF6 in turn recruits and activates TAK1 which mediates activation of the IKK complex resulting in canonical NF-κB activity. The oncogenic variant MyD88^{L265P} promotes spontaneous oligomerization and activation of downstream signaling independently of receptor stimulation. Mutant MyD88 is denoted with a red asterisk. DD, death domain; IL-1R, interleukin-1 receptor; IKK, IκB kinase; IRAK, IL-1R-associated kinase; MyD88, myeloid differentiation primary response protein 88; TAK1, TGFβ-activated kinase 1; TAB1/2, TAK1-binding protein 1/2; TIR; Toll/interleukin-1 receptor domain; TLR, Toll-like receptor; TRAF6, TNF receptor-associated factor 6.

Recurrent oncogenic mutations of the adapter protein MyD88 have been identified in a variety of B-cell malignancies. As approximately 40% of ABC DLCBL biopsies harbor MyD88 mutations, MyD88 represents the most frequently mutated oncogene in this tumor entity [40]. Whereas different somatic mutations of MyD88 have been reported, the most prevalent missense mutation encodes the amino acid substitution L265P within the TIR domain [40]. The L265P mutation of MyD88 occurs at a high frequency in ABC DLBCL (30% of cases) and in Waldenström's macroglobulinemia (WM; 90%) as well as less commonly in marginal-zone lymphoma (13%), gastric MALT lymphoma (9%), and CLL (3%) [40–44]. In contrast, gain-of-function mutations of MyD88 are rare or absent in other DLBCL subtypes, i.e., germinal center B cell-like (GCB) DLBCL and PMBL [40,45]. Ectopic expression of MyD88^{L265P}, but not of wild-type MyD88 in GCB DLBCL cell lines, which exhibit *per se* little to no NF-κB activity, potently induces NF-κB activation, demonstrating the oncogenic capacity of this MyD88 variant. This gain-of-function has been attributed to the ability of MyD88^{L265P} to spontaneously oligomerize and thus activate IRAK1 and IRAK4 independently of a TLR ligand (Figure 1) [40,46]. In mice, expression of MyD88^{L252P} (the orthologous position of the human L265) is sufficient to trigger the formation of lymphoma morphologically resembling the ABC DLBCL phenotype [47]. Interestingly, the L265P mutant TIR domain is able to recruit endogenous wild-type MyD88 to trigger downstream signaling in vitro [46,48,49]. Whereas the kinase activity of IRAK1 is dispensable for the capacity of mutant MyD88 to promote the survival of ABC DLBCL, NF-κB activation driven by oncogenic MyD88 mutations critically relies on the kinase activity of IRAK4, implicating this kinase as an interesting therapeutic target in lymphoid malignancies [40,50,51]. Indeed, the highly selective IRAK4 inhibitors ND-2158 and ND-2110 abrogate aberrant NF-κB activation induced by oncogenic MyD88^{L265P} and thus efficiently suppress the growth of ABC DLBCL cells in vitro and in vivo [50].

Similar to the requirement of chronically active BCR signaling in B-cell malignancies (discussed in Section 3), the importance of TLR-derived signals in lymphomagenesis is under debate. On the one hand, expression of a non-functional variant of Unc93b1, which is required for the endolysosomal localization of TLR3, 7, and 9, as well as TLR9 deficiency block the proliferation of primary B cells induced by the expression of ectopic MyD88^{L265P} in vitro, implicating a continued dependence on upstream TLR9 activation [52]. On the other hand, in vivo depletion of TLR9 in mice rather suggests an inhibitory role of TLRs in MyD88^{L265P}-transduced B cells [53]. The exact molecular and functional consequences of TLRs in MyD88^{L265P}-mutated tumor cells need to be addressed in future studies, especially since this could have implications for the use of TLR agonists/antagonists in lymphoma therapy.

3. Chronic B-Cell Receptor Signaling

The B-cell receptor complex comprises an immunoglobulin molecule (IgA, IgD, IgE, IgG, or IgM), which is anchored in the plasma membrane via a transmembrane domain, and a disulfide-linked CD79A/CD79B heterodimer essential for signal transmission. While recognition of the cognate antigen is achieved by the variable regions of the immunoglobulin chains (V_H and V_L, respectively), CD79A/B contain immunoreceptor tyrosine-based activation motifs (ITAMs) within their cytoplasmic domains which are essential for the initiation of an intracellular signaling cascade in response to receptor engagement (Figure 2a) [54,55]. Ligand binding induces BCR clustering and phosphorylation of two invariant tyrosine residues within the ITAMs of CD79A/B by the Src family tyrosine kinase LYN (Figure 2b) [56,57]. Subsequently, spleen tyrosine kinase (SYK) is recruited to the phosphorylated ITAMs via its SH2 domain, resulting in SYK auto-phosphorylation as well as phosphorylation of several downstream mediators, such as CD19, Bruton's tyrosine kinase (BTK) and B-cell linker protein (BLNK) [58]. Whereas CD19 phosphorylation leads to the recruitment of phosphoinositide 3-kinase (PI3K) culminating in activation of the AKT signaling axis, BLNK serves as a scaffold that binds both phospholipase Cγ2 (PLCγ2) and BTK, resulting in the BTK-dependent activation of PLCγ2 [59,60]. In turn, PLCγ2 converts phosphatidylinositol 4,5-bisphosphate (PIP$_2$) to generate the second messengers inositol 1,4,5-trisphosphate (IP$_3$) and diacylglycerol (DAG). Together, DAG and an increase in the intracellular Ca^{2+} levels induced by the action of IP$_3$ activate the protein kinase Cβ (PKCβ), which subsequently phosphorylates the scaffold protein caspase recruitment domain (CARD) membrane-associated guanylate kinase (MAGUK) protein 1 (CARMA1; discussed in Section 4), thus triggering downstream NF-κB activation [61].

Due to its capacity to induce NF-κB activation, BCR signaling plays an important role in the survival and proliferation of a subset of B-cell malignancies [21,62]. Accordingly, it has been reported that chronic infections with viral and bacterial pathogens are often associated with lymphoma development due to persistent antigen-driven activation and proliferation of the B cells (Figure 2b) [63]. Several foreign antigens, for instance derived from hepatitis C virus, have been reported to be associated with certain types of lymphoma and most likely govern lymphoma proliferation and survival in a BCR-dependent manner [63,64]. In contrast, the ABC subtype of DLBCL seems to rely on chronic BCR signaling driven by self-antigens, since the survival of ABC DLBCL cell lines is impaired upon substitution of the IgH variable region of their BCRs [65]. Interestingly, the BCRs of some ABC DLBCL cell lines are reactive towards self-antigens present in apoptotic debris or towards an invariant part of its own V region [65]. The toxicity caused by knockdown of the essential BCR subunits CD79A/B or of downstream signaling effectors, such as BTK, SYK, and PLCγ2, observed in most ABC DLBCL cells further corroborates the notion that these lymphoid tumors critically rely on chronic active BCR signaling [66,67]. Conversely, the GCB subtype of DLBCL is independent of chronic BCR activation but instead requires "tonic", antigen-independent BCR signals which promote survival by activating the PI3K/AKT pathway [66,68–70]. In line with chronic BCR triggering, cell lines as well as biopsies of ABC DLBCL typically exhibit BCR clustering on the cell surface, which correlates with increased levels of tyrosine phosphorylation and indicates sustained BCR signaling [67]. Chronic BCR signaling also plays a crucial role in CLL, since 30% of this cancer entity express a similarly rearranged BCR using a distinct subset of V_H, D_H, and J_H gene segments, which can also be found in ABC DLBCL [65,71]. These so-called "stereotyped" BCRs are thought to respond to similar antigens, most likely presented by the tumor itself, such as proteins of apoptotic cells or an epitope of the BCR [72–74]. Interestingly, expression of a CLL-derived IgH V region has been shown to sustain the survival of an ABC DLBCL cell line, suggesting that a similar (self-) antigen is driving chronic BCR signaling in a subset of CLL and ABC DLBCL [65].

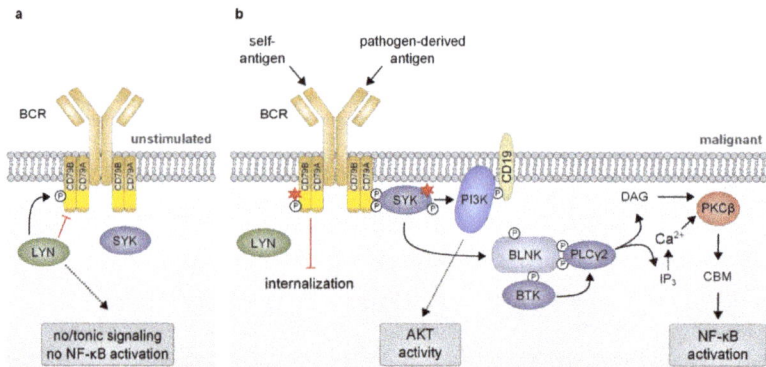

Figure 2. Mechanisms of chronic active BCR signaling in lymphoid malignancies. (**a**) In the absence of its cognate antigen, proximal BCR signaling remains inactive, and thus the protein tyrosine kinase SYK is not recruited to the co-receptors CD79A and CD79B. The Src kinase LYN prevents hyperactivation of BCR signaling by initiation of a negative feedback loop involving CD22 phosphorylation and subsequent activation of the phosphatase SHP-1. Tonic BCR signaling promotes B-cell survival via the PI3K/AKT pathway in an antigen-independent manner; (**b**) Engagement of the BCR, for instance by self-antigens or pathogen-derived antigens, results in receptor clustering and the induction of an intracellular signaling cascade. The subsequent phosphorylation of tyrosine residues in the ITAM regions of CD79A/B by Src kinases (e.g., LYN) allows the recruitment of SYK which in turn phosphorylates the adapter protein BLNK and thus promotes the formation of a proximal signaling complex involving BTK and PLCγ2. While activation of the AKT signaling axis is achieved by the SYK-mediated phosphorylation of CD19 and subsequent recruitment of PI3K, PLCγ2 activity generates the second messengers DAG and IP_3 the latter triggering the influx of Ca^{2+} into the cell. DAG and elevated Ca^{2+} levels activate PKCβ which induces activation of canonical NF-κB through the CBM signalosome. Whereas overexpression of SYK augments NF-κB activation in some lymphomas, mutations in the co-receptors CD79A/B prevent the internalization of activated BCRs and thus promote chronic BCR signaling. Proteins that are affected by recurrent genetic lesions in lymphoid malignancies are denoted with a red asterisk. BCR, B-cell receptor; BLNK, B-cell linker protein; BTK, Bruton's tyrosine kinase; CBM, CARMA1/BCL10/MALT1; DAG, diacylglycerol; IP_3, inositol 1,4,5-trisphosphate; ITAM, immunoreceptor tyrosine-based activation motif; PI3K, phosphoinositide 3-kinase; PKCβ, protein kinase Cβ; PLCγ2, phospholipase Cγ2; SHP-1, Src homology 2 domain phosphatase 1; SYK, spleen tyrosine kinase.

To maintain chronic BCR activation, approximately 20% of ABC DLBCL tumors exhibit somatic mutations in the co-receptors CD79B and less commonly CD79A (Figure 2b) [67]. A frequently occurring missense mutation present in 18% of ABC DLBCL biopsies involves the substitution of the membrane-proximal ITAM tyrosine (Y196) of CD79B. These CD79B mutations are associated with increased surface expression of the BCR in the context of chronic active BCR signaling and with a reduced activation of the tyrosine kinase LYN, which plays a dual role in BCR signaling [67,75]. Besides its positive regulatory role in the initial tyrosine phosphorylation cascade, LYN also exerts inhibitory effects on BCR-induced signaling. On the one hand, by phosphorylation of inhibitory motifs in CD22, LYN mediates the recruitment and activation of SHP-1, a phosphatase that quenches BCR signaling by the removal of ITAM phosphorylation [76–80]. On the other hand, LYN activity has been shown to promote BCR internalization, suggesting that reduced LYN activation in CD79-mutated ABC DLBCLs results in an increased surface expression of chronically activated BCRs [67]. A recent study has further highlighted the importance of *CD79B* mutations for surface expression of the BCR in ABC DLBCL cells and provided a rationale for the frequently observed co-occurrence of *CD79B* and *MYD88* mutations in B-cell malignancies [40]: While *CD79B* mutations alone are not sufficient to enhance NF-κB-mediated B-cell proliferation and *MYD88* mutations on their own decrease surface IgM/BCR expression reminiscent of anergic B cells, the combination of *CD79B* and *MYD88* mutations cooperates in plasmablastic differentiation [81]. Collectively, CD79 mutations sustain high BCR levels at the cell surface despite chronic BCR activation and thus prolong BCR-dependent signaling. It is tempting to speculate that the gene loss of LYN occurring in 60% of patients suffering from WM might, similar to the CD79A/B mutations, sustain surface BCR expression and thus chronic BCR signaling [82]. Additionally, overexpression of SYK as observed in MCL and peripheral T-cell lymphomas most likely also contributes to NF-κB activation, even though the molecular details have not been investigated so far [83,84].

From a clinical perspective, a multitude of new inhibitors targeting chronic BCR signaling are in the pipeline or already approved for the treatment of particular lymphoid malignancies [85]. Several of these inhibitors target BTK, such as ibrutinib or acalabrutinib, which are currently used or have been proposed for the treatment of CLL, MCL, WM, and DLBCL [86–90]. Other inhibitors targeting kinases involved in the proximal tyrosine phosphorylation cascade, such as SYK and LYN, may be able to potently reduce NF-κB activation caused by chronic BCR signaling [67]. While SYK can be targeted with fostamatinib [85,91,92], the small-molecule inhibitor dasatinib, which was initially utilized as an inhibitor of the oncogenic BCR-ABL fusion protein in chronic myelogenous leukemia, was shown to inhibit also LYN and BTK in CLL [93–95].

4. Genetic Alterations Driving Constitutive NF-κB Activation via the CBM Signalosome

In lymphocytes, the activation of canonical NF-κB signaling in response to antigen receptor ligation requires the formation of a multimeric signaling module, termed the CARMA1/BCL10/MALT1 (CBM) complex, which functionally links antigen receptor proximal signaling events with the activation of the IKK complex [96–98]. Antigen binding to the BCR triggers a signaling cascade involving the activation of Src kinases and ultimately leads to the activation of PKCβ (discussed in Section 3), which in turn phosphorylates the scaffold protein CARMA1 within the flexible linker region situated between its coiled-coil and C-terminal MAGUK domain (Figure 3) [99,100]. In quiescent lymphocytes, CARMA1 adopts an auto-inhibited, inactive conformation, which is stabilized by an intramolecular interaction between its inhibitory linker and the region spanning the CARD and coiled-coil domain. PKC-mediated phosphorylation activates CARMA1 by triggering a conformational change that facilitates the oligomerization of CARMA1 via the coiled-coil domain [99–102]. Subsequently, active CARMA1 nucleates the formation of long B-cell lymphoma 10 (BCL10) filaments through CARD–CARD-mediated homotypic interactions [103–106]. The mucosa-associated lymphoid tissue lymphoma translocation protein 1 (MALT1) is recruited to the fibrillary signaling complex due to its constitutive association with BCL10 [107–109]. The assembly of a high-molecular weight complex comprising CARMA1, oligomerized BCL10, and MALT1 is indispensable for NF-κB activation in response to antigen receptor ligation and represents a hallmark of lymphocyte activation [61,102,110]. Within the CBM complex, MALT1 recruits the ubiquitin ligase TRAF6, which in turn mediates polyubiquitination of MALT1, BCL10, and itself (Figure 3). These polyubiquitin chains serve as docking sites for the physical recruitment of the IKK complex via the ubiquitin binding motif of the regulatory subunit IKKγ [111–114]. Additionally, the linear ubiquitin chain assembly complex (LUBAC) is recruited to the CBM signalosome and in turn promotes activation of the IKK complex by mediating the linear ubiquitination of IKKγ [115–117]. Polyubiquitination also results in the recruitment of TAK1 via the ubiquitin binding domain of the adapter proteins TAK-binding protein 2/3 (TAB2/3). Collectively, the CBM complex serves as a signaling platform which facilitates the activation of the IKK complex through a series of ubiquitination-dependent interactions, culminating in TAK1-induced phosphorylation of the catalytic subunit IKKβ [39,112,113,118].

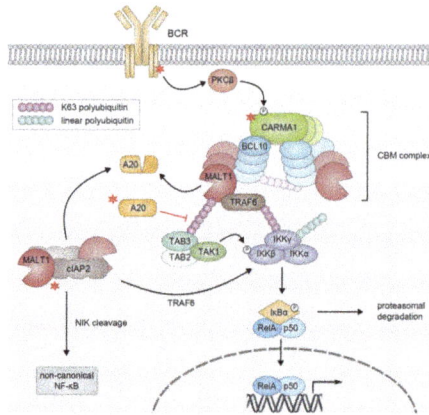

Figure 3. Increased activity of the CARMA1/BCL10/MALT1 signalosome drives constitutive NF-κB activation. Antigen-dependent triggering of the BCR results in the activation of PKCβ, which phosphorylates the scaffold protein CARMA1 at several residues within its inhibitory linker region. Activated CARMA1 in turn nucleates the formation of a fibrillary high-molecular weight signaling complex (CBM complex) comprising long BCL10 filaments and MALT1. Subsequently, the E3 ubiquitin ligase TRAF6 is recruited to the CBM complex via an interaction with MALT1 and catalyzes the K63-linked polyubiquitination of MALT1, BCL10, and itself. These K63-linked polyubiquitin chains as well as the linear polyubiquitin chains generated by the action of the LUBAC allow the recruitment and activation of the IKK complex. TAK1 which is recruited to the signaling complex via the ubiquitin binding domains of its accessory proteins TAB2 and TAB3 promotes IKK activation by phosphorylating IKKβ. IKK activity in turn results in the proteasomal degradation of IκBα and the subsequent nuclear translocation of transcriptionally active NF-κB heterodimers. In lymphoid malignancies, chronically active BCR signaling as well as gain-of-function mutations of the scaffold protein CARMA1 can lead to hyperactivation of the CBM signalosome and thus induce constitutive NF-κB activity. Genetic inactivation of A20, which negatively regulates IKK activity by deubiquitination of activated signaling effectors, augments NF-κB activation in several lymphoid cancers. Additionally, A20 can be proteolytically inactivated by the MALT1 protease, which is activated in the framework of the CBM complex. Expression of the oncogenic cIAP2-MALT1 fusion protein induces activation of canonical NF-κB in a TRAF6-dependent manner. The constitutive protease activity of cIAP2-MALT1 further promotes NF-κB activation by proteolytic inactivation of A20. Additionally, NIK cleavage by cIAP2-MALT1 also promotes non-canonical NF-κB activity. Proteins that are affected by recurrent genetic lesions in lymphoid malignancies are denoted with a red asterisk. BCL10, B-cell lymphoma 10; BCR, B-cell receptor; CARMA1, caspase recruitment domain membrane-associated guanylate kinase protein 1; cIAP2, cellular inhibitor of apoptosis protein 2; IκB, inhibitor of κB; IKK, IκB kinase; LUBAC, linear ubiquitin chain assembly complex; MALT1, mucosa-associated tissue lymphoma translocation protein 1; NIK, NF-κB inducing kinase; PKCβ, protein kinase Cβ; TAK1, TGFβ-activated kinase 1; TAB2/3, TAK1-binding protein 2/3; TRAF6, TNF receptor-associated factor 6.

In activated lymphocytes, the paracaspase MALT1 plays a dual role in promoting antigen-induced NF-κB activation: As a scaffold protein in the framework of the CBM signalosome, MALT1 on the one hand facilitates the physical recruitment and activation of the IKK complex by providing binding sites for the ubiquitin ligase TRAF6. On the other hand, the protease activity of MALT1 further potentiates pro-inflammatory signaling in response to antigen receptor stimulation [114]. The central protease domain of MALT1 shares homology with proteases of the caspase and metacaspase family and contains conserved cysteine and histidine residues essential for its catalytic activity [107,119]. In contrast to caspases, which catalyze substrate cleavage after the negatively charged amino acid aspartate, MALT1 cleaves its target proteins after positively charged arginine residues [61,120,121]. MALT1 dimerization via its protease domain is essential for the acquisition of a catalytically active conformation, occurs in the context of CBM complex assembly and is promoted by mono-ubiquitination of MALT1 [103,104,122–124]. Intriguingly, MALT1 protease activity potentiates and sustains NF-κB activation in an IKK-independent manner, presumably by the proteolytic inactivation of A20 and RelB [121,125]. MALT1-dependent cleavage and subsequent proteasomal degradation of RelB, which impedes classical NF-κB1 activation by the formation of transcriptionally inactive RelA/RelB heterodimers and/or by competing for DNA binding sites, results in enhanced DNA binding of RelA and c-Rel [125,126]. Additionally, proteolytic inactivation of the deubiquitinating enzyme A20, which removes K63-linked polyubiquitin chains from key signaling mediators, such as TRAF6, IKKγ, and MALT1, and thus negatively regulates NF-κB activation, sustains maximum NF-κB induction (Figure 3) [121,127]. Auto-processing of MALT1 is assumed to also be essential for NF-κB activation in lymphocytes, although the molecular mechanism of this contribution remains unclear thus far [128]. In contrast to the role of MALT1 catalytic activity in promoting NF-κB activation, MALT1-dependent cleavage of heme-oxidized iron-responsive element-binding 2 ubiquitin ligase-1 (HOIL-1), a component of LUBAC that promotes IKKγ ubiquitination and thus activation of the IKK complex, instead dampens NF-κB signaling and might be involved in the termination of CBM/IKK-mediated NF-κB activity [129–131].

While CBM-mediated NF-κB activation plays a critical role in lymphocyte proliferation and loss-of-function mutations result in immunodeficiency, aberrant constitutive NF-κB activation is not only associated with autoimmune diseases but also with the development of lymphoid malignancies [98,132,133]. Recurrent gain-of-function mutations in the genes encoding CBM proteins or their upstream regulators result in constitutive CBM-dependent NF-κB activation and have been detected in a wide range of lymphoid malignancies including ABC DLBCL, MCL, MALT lymphoma, acute T-cell leukemia/lymphoma (ATLL), and Sézary syndrome [16,23,134,135]. The toxicity of RNAi-mediated silencing of either CARMA1, BCL10 or MALT1 expression observed in ABC DLBCL cell lines further demonstrates the importance of CBM-mediated signaling in this tumor entity [66]. Hyperactivity of the CBM signalosome associated with gain-of-function mutations of the central scaffold protein CARMA1 or its upstream regulators (e.g., CD79A/B, discussed in Section 3) as well as with constitutive BCR signaling driven by self-antigens has emerged as a hallmark of lymphomagenesis [65,67,98,136,137].

4.1. Oncogenic CARMA1 Mutations

Oncogenic CARMA1 mutations driving constitutive signaling activity of the CBM complex have initially been discovered in approximately 10% of patients suffering from the aggressive ABC subtype of DLBCL which relies on constitutive NF-κB signaling for survival and proliferation [137]. At present, several tumor entities including ABC DLBCL, ATLL, and Sézary syndrome as well as a congenital B-cell lymphocytosis, a B cell proliferative syndrome associated with an increased risk of lymphoma development, have been found to harbor activating missense mutations of CARMA1 [135,138–140]. The majority of the identified somatic gain-of-function mutations of CARMA1 are located in the proximity of or within the region spanning the CARD and the coiled-coil domain. Mechanistically, these mutations are thought to disrupt auto-inhibition of CARMA1, thus favoring oligomerization and activation of downstream signaling (Figure 3) [137]. Indeed, ectopic expression of these CARMA1 mutants has been shown, one the one hand, to drive constitutive activation of CBM-mediated NF-κB signaling independently of upstream signals and, on the other hand, to potentiate NF-κB activation in response to antigen receptor stimulation [137].

4.2. Overexpression of BCL10/MALT1 and cIAP2-MALT1 Fusion Protein

Another tumor entity critically relying on constitutive CBM signaling is represented by lymphomas of the mucosa-associated lymphoid tissue, which occur most commonly in the stomach typically due to chronic infection, e.g., with *Helicobacter pylori*, and can be successfully treated at early stages by eradication of the source of inflammation [134,141,142]. At advanced stages, however, these MALT lymphomas are associated with distinctive chromosomal translocations, which either lead to the overexpression of BCL10 and MALT1 [143–145] or result in the generation of a constitutively active fusion protein comprising the N-terminal part of cellular inhibitor of apoptosis protein 2 (cIAP2, also known as API2) and the C-terminus of MALT1 [24,146,147]. The cIAP2-MALT1 fusion protein is able to auto-oligomerize independently of BCL10 and upstream signals via the baculovirus inhibitor of apoptosis repeat (BIR) domain of the cIAP2 moiety which binds heterotypically to the C-terminal region of MALT1 (Figure 3) [134,148,149]. Constitutive MALT1 protease activity and the capacity of cIAP2-MALT1 to potently activate both the canonical NF-κB1 (via the MALT1-dependent recruitment of TRAF6 and proteolytic inactivation of A20) and non-canonical NF-κB2 pathways (discussed in Section 5) drive the growth of these MALT lymphomas [24,107,121,148,149].

Aberrant induction of MALT1 protease activity is a major consequence of constitutive CBM signaling and has been reported to be essential for the survival of several lymphoid malignancies, such as ABC DLBCL, MCL, and CLL, making MALT1 an attractive therapeutic target in lymphoma treatment [150–153]. Even though at present no MALT1 inhibitors have entered the clinic, high-throughput screening revealed several small-molecule inhibitors targeting MALT1 protease activity [154–156]. Indeed, pharmacological inhibition of the MALT1 protease function has been reported to exert selective toxicity towards MALT1-dependent lymphomas both in vitro and in vivo using xenograft mouse models [154,155].

4.3. Inactivation of TNFAIP3/A20

In addition to constitutive signaling via the CBM complex, aberrant NF-κB activity in lymphoid malignancies can also be promoted by the genetic inactivation of A20, which negatively regulates IKK activation most likely by removing the K63-linked polyubiquitin chains from the activated CBM signalosome (Figure 3) [127,157–159]. Indeed, at least one allele of *TNFAIP3* encoding A20 is frequently targeted by mutations, deletions, or epigenetic silencing which result in a partial or complete loss of its negative regulatory function in several lymphoid malignancies, such as HL (approximately 45% of cases), PMBL (30%), ABC DLBCL (25%), and MALT lymphoma (20%) [160–163]. Recent reports suggest that loss of A20 function on its own might not result in sufficient NF-κB activation to support lymphomagenesis [164]. Instead, inactivation of A20 is often found to be associated with additional genetic aberrations, such as *MYD88* or *CD79A/B* mutations, which drive constitutive NF-κB activation [40,62]. The role of A20 as a tumor suppressor in B-cell lymphoma is further supported by the toxicity of ectopic expression of A20 in A20-deficient ABC DLBCL cell lines [160,161].

4.4. LUBAC Polymorphism

Genetic analyses of lymphomas have recently identified rare germ line polymorphisms which are enriched in ABC DLBCL patients and promote polyubiquitin-dependent NF-κB activation. These SNPs cause amino acid substitutions in HOIL-1 interacting protein (HOIP, encoded by *RNF31*), promote its interaction with HOIL-1 and result in an increased activity of the LUBAC [165]. Interestingly, RNAi-mediated silencing of LUBAC expression or inhibition of its activity have been reported to reduce constitutive NF-κB activity and to thus induce cell death in ABC DLBCL cells, suggesting an important role of linear ubiquitin in oncogenic NF-κB activation in these lymphomas [165,166].

5. Constitutive Activation of Non-Canonical NF-κB Signaling

In resting lymphocytes, the non-canonical, also termed alternative, NF-κB pathway is inactive, as NF-κB inducing kinase, a central player in this pathway, is constitutively targeted for proteasomal degradation [4,167]. Degradation of NIK relies on K48-linked polyubiquitination catalyzed by the E3 ubiquitin ligases cIAP1 and cIAP2 which are recruited to NIK via an interaction with TRAF3 (Figure 4a) [4]. Upon activation of certain members of the TNF receptor superfamily, such as CD40 or the BAFF receptor (BAFF-R), the complex comprising TRAF3, TRAF2, and cIAP1/2 is recruited to the receptor, resulting in the cIAP1/2-dependent K48-linked polyubiquitination and subsequent proteasomal degradation of TRAF3 (Figure 4b) [4]. Depletion of TRAF3 abrogates the interaction between NIK and cIAP1/2, thus stabilizing NIK which in turn phosphorylates and activates IKKα [168]. Subsequently, IKKα phosphorylates the precursor protein p100 (NF-κB2) and thus marks it for processing by the proteasome to generate p52 which forms a transcriptionally active heterodimer with RelB [168].

Several lymphoid cancers, particularly multiple myeloma but also Hodgkin lymphoma and cIAP2-MALT1 expressing MALT lymphoma (discussed in Section 4), harbor various genetic alterations which affect different regulators of the non-canonical NF-κB pathway and rely on constitutive nuclear activity of p52/RelB heterodimers [19,20,160,169,170]. Since activation of non-canonical NF-κB signaling is primarily regulated through the tight control of NIK protein levels, most of the genetic aberrations observed in lymphoid malignancies result in the increased expression or stabilization of NIK (Figure 4b). Indeed, increased NIK protein levels caused by copy number gains in the *MAP3K14* gene which encodes NIK or chromosomal translocations relocating *MAP3K14* into the proximity of immunoglobulin enhancer elements can be frequently observed in multiple myeloma and HL [20,170]. Similarly, a NIK fusion protein lacking the TRAF3-binding domain exhibits increased stability due to loss of TRAF3-dependent proteasomal degradation [19,20,171]. As an alternative oncogenic mechanism to augment NIK activity and thus non-canonical NF-κB activation, negative regulators of NIK protein stability are frequently inactivated by deletions, loss-of-function mutations or transcriptional silencing [9]. Whereas loss-of-function mutations of TRAF2 or cIAP2 have been described in MCL and DLBCL, inactivating mutations or homozygous deletions of the gene encoding TRAF3 have been reported in HL (15% of cases), DLBCL (15%), and in multiple myeloma (50%) [9,19,20,23,45,160,170,172].

Stabilization of NIK and consequently increased processing of p100 can also be achieved by overexpression or mutation of the receptors which induce non-canonical NF-κB activity in a stimulus-dependent manner in normal lymphocytes. Mechanistically, recruitment of TRAF3, TRAF2, and cIAP1/2 to activated or mutated TNF superfamily receptors, such as CD40, BAFF-R, RANK, and the lymphotoxin β receptor (LTβR), induces the cIAP1/2-dependent degradation of TRAF3 and promotes stabilization of NIK [19,20,160]. For instance, a missense mutation (H159Y) targeting the cytoplasmic tail of the BAFF-R identified in follicular lymphoma, DLBCL, and less commonly in MALT lymphoma results in the increased recruitment of TRAF2, TRAF3, and TRAF6 to the receptor [173]. Additionally, overexpression of the receptor CD40 resulting in enhanced p100 processing has been reported in rare cases of multiple myeloma [19,20]. Similarly, genetic alterations affecting LTβR and RANK have been reported in multiple myeloma and DLBCL [19,160].

Recently, oncogenic MyD88 mutations, such as L265P (discussed in Section 2), have been shown to induce the activation of NIK and thus increase processing of p100 and p105 in DLBCL (Figure 4b) [174]. Interestingly, p100 processing is required to maintain the ABC phenotype, since knockdown of p100 reduced the expression of genes, such as *IRF4* and *BCL6*, typically associated with the ABC subtype of DLBCL. Conversely, ectopic expression of MyD88[L265P] in GCB DLBCL cell lines has been shown to trigger p100 processing in a TAK1- and IKKα-dependent manner and to alter the B-cell differentiation status towards a phenotype resembling ABC DLBCL [174]. Besides genetic alterations targeting important regulators of alternative NF-κB activation, latent infection with the Epstein-Barr virus (EBV) induces non-canonical NF-κB signaling by introduction of the latent membrane protein 1 (LMP1) in approximately 40% of HL cases [175,176]. LMP1 is highly homologous to the cytoplasmic domain of the TNF receptor CD40 and induces IKKα-dependent p100 processing via the spontaneous formation of signaling aggregates [177–180]. Similarly, NF-κB2 activation mediated by the protein Tax of the human T-cell leukemia virus type 1 (HTLV-1) is often associated with ATLL [181].

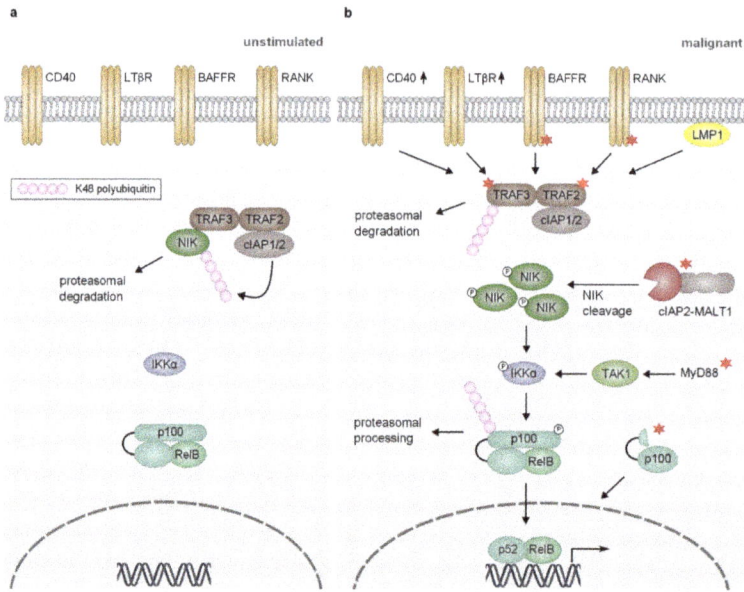

Figure 4. Genetic lesions driving non-canonical NF-κB activation in lymphoid malignancies. (a) In unstimulated cells, TRAF3 recruits NIK to an E3 ubiquitin ligase complex comprising TRAF2 and cIAP1/2 which constantly marks NIK for proteasomal degradation by K48-linked polyubiquitination. The heterodimer of RelB and the p52 precursor protein p100 is sequestered in the cytoplasm; (b) Activated or mutated members of the TNF receptor family, such as CD40, BAFF-R, LTβR, and RANK or the EBV-encoded CD40 mimic LMP1, recruit the complex consisting of TRAF3, TRAF2, and cIAP1/2 to the cell membrane and induce the proteasomal degradation of TRAF3. In the absence of TRAF3, NIK is released from the inhibitory E3 ubiquitin ligase complex and accumulates in the cytoplasm where it subsequently phosphorylates and activates IKKα. In turn, IKKα phosphorylates the precursor protein p100, thus targeting it for processing by the proteasome. Proteasomal processing of p100 generates p52 and results in the nuclear translocation of transcriptionally active p52/RelB heterodimers. Enhanced expression of NIK due to copy number gains and loss-of-function mutations of TRAF2, TRAF3, or cIAP2 result in increased NIK protein levels driving non-canonical activation of NF-κB in lymphoid malignancies. Similarly, proteolytic cleavage by the constitutively active cIAP2-MALT1 fusion protein stabilizes NIK by removal of the TRAF3-binding site. Additionally, oncogenic MyD88 mutations can promote the activation of IKKα in a TAK1-dependent manner. In certain lymphomas aberrant activation of non-canonical NF-κB is governed by the expression of a truncated, constitutively active version of the precursor p100 lacking the inhibitory ankyrin repeats. Proteins that are affected by recurrent genetic lesions in lymphoid malignancies are denoted with a red asterisk. BAFF-R, B cell-activating factor receptor; cIAP1/2, cellular inhibitor of apoptosis protein 1/2; EBV, Epstein-Barr virus; IKKα, IκB kinase α; LMP1, latent membrane protein 1; LTβR, lymphotoxin β receptor; MALT1, mucosa-associated tissue lymphoma translocation protein 1; MyD88, myeloid differentiation primary response protein 88; NIK, NF-κB inducing kinase; RANK, receptor activator of NF-κB; TAK1, TGFβ-activated kinase 1; TRAF2/3, TNF receptor-associated factor 2/3.

Furthermore, rearrangement or partial deletions within the *NFKB2* gene locus which disrupt the inhibitory ankyrin repeats at the C-terminus of the precursor p100 have been found to result in the generation of a truncated, constitutively active p100 protein (Figure 4b) [20,182,183]. Additionally, the deubiquitinase CYLD, which negatively regulates NF-κB activation by removing K63-linked polyubiquitin chains from IKKγ, TRAF2, and TRAF6, is frequently inactivated by deletion, mutation, or transcriptional silencing in multiple myeloma [19,20,184–186].

Approximately 25% of gastric MALT lymphomas harbor the chromosomal translocation t(11;18) which results in the expression of a cIAP2-MALT1 fusion protein retaining the proteolytic activity of MALT1 (discussed in Section 4.2) [146,147]. In addition to promoting canonical NF-κB signaling, the cIAP2-MALT1 fusion protein has also been found to potently induce activation of the non-canonical NF-κB pathway (Figures 3 and 4b) [24,107]. Oligomerization of the fusion protein via the cIAP2 moiety is assumed to stimulate constitutive protease activity of MALT1 [24,148]. Additionally, the cIAP2 portion mediates the recruitment of NIK which is subsequently cleaved by the MALT1 protease [24]. Proteolytic cleavage removes the TRAF3-binding domain while leaving the kinase domain of NIK intact and thus generates a truncated, constitutively active NIK which is resistant to negative regulation by proteasomal degradation and promotes constitutive p100 processing [24].

Collectively, loss of TRAF3 function, enhanced degradation of TRAF3 or increased expression of NIK augments processing of p100 and thus the nuclear accumulation of transcriptionally active p52/RelB heterodimers. Increased NIK activity can also promote canonical NF-κB activation, since NIK is also able to activate IKKβ [19,20,187]. As aberrant NIK activity driving constitutive activation of both canonical and non-canonical NF-κB signaling constitutes a common consequence of most of the genetic aberrations found in a large subset of multiple myeloma and Hodgkin lymphoma patients, pharmacological NIK inhibition represents an attractive therapeutic strategy for the treatment of these tumor entities. However, even though some NIK inhibitors have been developed, so far none has entered the clinics [188,189].

6. Aberrant Expression of IκB Proteins

6.1. Classical IκB Proteins

In non-activated cells, NF-κB dimers are sequestered in the cytoplasm by an interaction with classical IκB proteins (IκBα, IκBβ, and IκBε), which mask the nuclear localization signal (NLS) of the NF-κB subunits and thus prevent their nuclear translocation (Figure 5a). Stimulation-dependent proteasomal degradation of the IκB proteins allows the translocation of the NF-κB dimers to the nucleus, where they modulate the expression of a variety of genes [2,3]. The best characterized classical IκB protein, IκBα, is composed of a signal response domain, ankyrin repeats, a PEST domain as well as a nuclear export signal (NES) and binds preferentially RelA/p50 heterodimers. The presence of an NES suggests that besides its function as a cytoplasmic inhibitor, IκBα is also involved in the termination of NF-κB transcriptional activity by promoting both the dissociation of RelA/p50 complexes from the DNA and their subsequent nuclear export [3,190–192].

In lymphoid malignancies, such as HL or DLBCL, characteristic genetic aberrations targeting the classical IκB proteins can trigger NF-κB activation downstream of the IKK complex [193,194]. The malignant cellular entity in HL, the Hodgkin-Reed-Sternberg (HRS) cell, is present at a low frequency (<1% of the tumor), while the bulk of the tumor is formed by activated inflammatory cells [195]. Aberrant NF-κB activation in these malignant cells is not only driven by the inflammatory tumor microenvironment or by latent infection with EBV, but also by somatic mutations of key NF-κB regulators, such as classical IκB proteins [10,196–198]. In HL, various genetic lesions have been described that result in the generation of truncated IκBα isoforms which lack part of the ankyrin repeats and are thus unable to sequester the NF-κB dimers in the cytoplasm (Figure 5b) [193,194,199,200]. Interestingly, inactivating mutations of IκBα have been detected preferentially in EBV-negative cases of HL (approximately 20% of cases; discussed in Section 5), suggesting that inactivation of IκBα is selected for as an alternative strategy to sustain NF-κB activation [194,197,201]. Besides HL, mutations negatively affecting the function of IκBα have been reported, albeit at lower frequency, in MALT lymphoma and in DLBCL, similarly providing an alternative mechanism for NF-κB activation in these tumor entities [202–204].

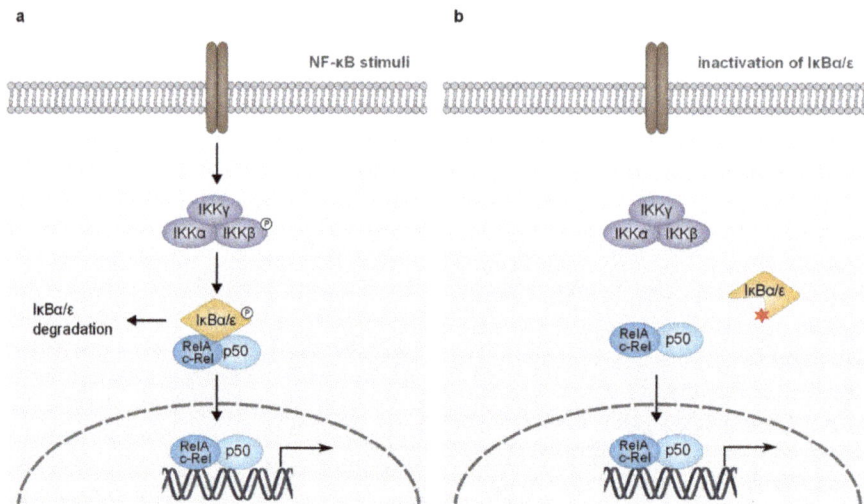

Figure 5. Inactivating mutations in classical IκB proteins promote canonical NF-κB activation. (a) The canonical NF-κB pathway is induced by the ligand-dependent activation of a variety of receptors, such as antigen receptors, the IL-1 receptor, Toll-like receptors and members of the TNF receptor family. Stimulus-dependent activation of the IKK complex results in the phosphorylation and subsequent proteasomal degradation of IκBα and IκBε. This allows the nuclear translocation of transcriptionally active RelA/p50 or c-Rel/p50 heterodimers; (b) Genetic alterations resulting in the expression of non-functional truncated versions of IκBα and IκBε, which lack part of the ankyrin repeat domain and are thus unable to bind to the NF-κB transcription factors, promote constitutive NF-κB activation in a subset of lymphoid malignancies. Proteins that are affected by recurrent genetic lesions in lymphoid malignancies are denoted with a red asterisk. IκB, inhibitor of κB; IKK, IκB kinase.

Analogous to inactivation of IκBα, loss-of-function mutations of IκBε, which result for instance in the expression of truncated versions lacking the ankyrin repeats essential for the interaction with NF-κB dimers, have been reported in HL, CLL, and PMBL as well as at a lower frequency in DLBCL and MCL (Figure 5b) [205–208]. Mechanistically, IκBε is assumed to limit the nuclear localization of Rel-containing dimers in a manner equivalent to IκBα [206,209,210]. While IκBα, however, predominantly regulates the cytoplasmic sequestration of RelA/p50 heterodimers, IκBε preferentially binds to c-Rel homodimers and c-Rel/p50 complexes [206,209]. Physiologically, stimulated B cells of IκBε-deficient mice exhibit increased proliferation and survival due to enhanced NF-κB activity [206,211]. Collectively, loss-of-function mutations targeting the classical IκB proteins IκBα and IκBε contribute to sustained NF-κB activation in lymphoid malignancies, indicating an important role of these negative regulators as tumor suppressors.

In lymphoid malignancies that rely on constitutive NF-κB activation but express functionally intact IκB proteins, preventing the proteasomal degradation of the IκB proteins constitutes an attractive therapeutic strategy. The proteasome inhibitor bortezomib is able to block the degradation of the classical IκB proteins and has been approved for multiple myeloma therapy, although it remains unclear if the beneficial effect of bortezomib can be attributed solely to NF-κB inhibition [212–214]. In clinical trials, bortezomib has also been shown to improve the efficacy of chemotherapy in ABC DLBCL [215].

6.2. Atypical IκB Proteins

Not only classical IκB proteins are targeted by genetic alterations in lymphoid malignancies, but also the expression of the so-called atypical IκBs including BCL3 and IκBζ can be affected in these cancers. Unlike the classical IκB proteins, atypical IκBs are not regulated by IKK phosphorylation and proteasomal degradation, but rather by their inducible expression. While atypical IκB proteins are generally not expressed in quiescent cells, they are strongly induced in the primary response upon NF-κB activation (Figure 6a,b) [5,7]. In contrast to their classical relatives, atypical IκBs interact with NF-κB proteins predominantly in the nucleus. Although the atypical IκB proteins were initially defined as NF-κB inhibitors, it is by now well established that they can act also as co-activators for a particular set of target genes [5,7]. Several studies have reported the importance of atypical IκB proteins in immune homeostasis and there is growing evidence for an involvement of these transcriptional regulators in the pathogenesis of lymphoid malignancies.

B-cell lymphoma 3 (BCL3) was first identified as a proto-oncogene in patients suffering from B-cell chronic lymphocytic leukemia [216–218]. Structurally, BCL3 is characterized by a conserved NLS and two transactivation domains (TAD) encompassing seven ankyrin repeats that mediate binding to NF-κB proteins [7,219–221]. Through an interaction with p50 or p52 homodimers, BCL3 can act both as an activator and as a repressor of NF-κB target gene transcription. How this dual function of BCL3 is realized on a molecular level remains unclear. It has been reported, on the one hand, that BCL3 is able to stabilize transcriptionally repressive p50 or p52 homodimers, and, on the other hand, that it binds to p50 or p52 homodimers and induces the transcription of target genes via its transactivation domains [221,222]. One explanation for the opposing BCL3 effects could lie in its capability to recruit both co-activator and co-repressor complexes comprising chromatin modifiers, such as p300 or HDAC1, respectively, to DNA-bound p50 and p52 homodimers in a context-dependent manner [223–225].

Different genetic aberrations affecting the NF-κB modulator BCL3 have been observed in lymphoid malignancies. The chromosomal translocation t(14;19)(q32;12) which juxtaposes *BCL3* with the *IGH* locus has been reported to result in an enhanced expression of BCL3 in a variety of lymphoid cancers, such as CLL and less commonly follicular lymphoma as well as marginal-zone lymphoma (Figure 6c) [226–228]. Additionally, amplification as well as alterations in the epigenetic modification status of the *BCL3* locus have been observed in HL and anaplastic large cell lymphomas [229–231]. The putative oncogenic role of BCL3 is supported by an *Eµ-BCL3* transgenic mouse model, which exhibits a lymphoproliferative disorder [232]. How exactly BCL3 exerts its oncogenic role in leukemia and lymphoma is unclear, but it has been proposed that BCL3 can promote cell proliferation and survival by transactivating a number of different target genes [233,234].

IκBζ (encoded by *NFKBIZ*) comprises a conserved NLS, a putative TAD as well as seven ankyrin repeats and is thus structurally highly homologous to the atypical IκB protein BCL3 [235,236]. Like BCL3, IκBζ interacts with NF-κB proteins, in particular with p50 and p52 homodimers, and is able to regulate the transcription of NF-κB target genes in a positive or negative manner [7]. Inhibition of target gene expression might be mediated by the stabilization of DNA-bound transcriptionally repressive p50 and p52 homodimers or by a competition between IκBζ and activating NF-κB members for DNA binding sites. Two molecular mechanisms have been proposed for its role as transcriptional activator: (I) Similar to BCL3, an intrinsic transactivation domain has been identified [237]. (II) IκBζ is capable to recruit the SWI/SNF complex, which mediates chromatin remodeling and thus allows transcription, to NF-κB consensus sites in the promoter of target genes [238].

In non-stimulated lymphocytes, IκBζ is not expressed but is rapidly induced upon engagement of receptors triggering NF-κB activity, such as TLRs and the antigen receptors [239,240]. Overexpression of IκBζ has been reported in various lymphoid malignancies, including ABC DLBCL, ATLL, and primary central nervous system lymphomas [241–243]. The expression of IκBζ is either promoted by genomic amplification of the *NFKBIZ* locus or by chronic NF-κB activation, which can be driven by deregulated BCR/IL-1R/TLR signaling or by viral proteins like Tax (HTLV-1) and LMP1 (EBV) (Figure 6d) [242–244].

The oncogenic function of IκBζ is best characterized in ABC DLBCL, since silencing of IκBζ reduced the growth of ABC DLBCL cell lines. Gene expression profiling revealed that IκBζ promotes the expression of several NF-κB target genes, including BCL-X_L, IL-6, and IL-10, which represent key regulators for ABC DLBCL survival [242].

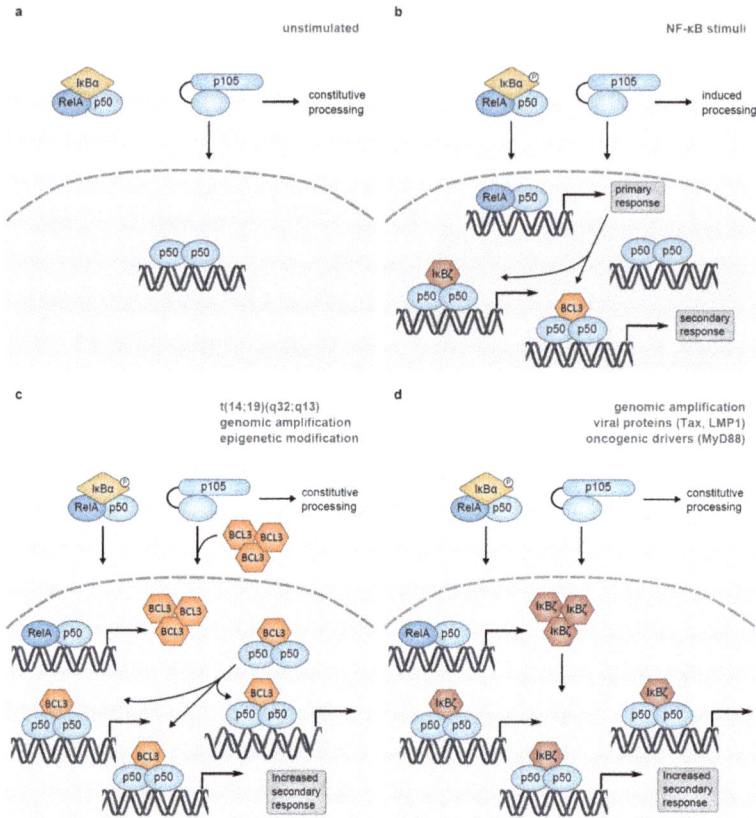

Figure 6. Abnormal expression of the atypical IκB proteins BCL3 and IκBζ promotes NF-κB activation in lymphoid malignancies. (**a**) In resting cells, DNA-bound homodimers of the NF-κB subunit p50 can be found in the nucleus due to constitutive processing of the p50 precursor p105; (**b**) Upon B-cell stimulation, the activated IKK complex induces both the generation of p50 by proteasomal processing of its precursor p105 and the proteasomal degradation of the inhibitor IκBα. This leads to the nuclear translocation of RelA/p50 heterodimers which drive the transcription of primary response genes, such as *BCL3* and *NFKBIZ*. In turn, BCL3 and IκBζ modulate the secondary NF-κB response by binding to DNA-bound p50 homodimers in the nucleus; (**c**) The chromosomal translocation t(14;18)(q32;q13) juxtaposing *BCL3* with the *IGH* locus, genomic amplifications, or epigenetic modifications result in enhanced expression of BCL3, which in turn increases the nuclear translocation of p50 and stabilizes DNA-bound p50 homodimers, thus augmenting the secondary NF-κB response; (**d**) Genomic amplification of the *NFKBIZ* locus as well as the expression of oncogenic MyD88^{L265P} or the viral proteins Tax and LMP1 leads to increased expression of IκBζ in a subset of lymphoid tumors. IκBζ promotes the secondary NF-κB response through binding to p50 homodimers in the nucleus. BCL3, B-cell lymphoma 3; *IGH*, immunoglobulin heavy chain (locus); IκB, inhibitor of κB; IKK, IκB kinase; LMP1, latent membrane protein 1; MyD88, myeloid differentiation primary response protein 88.

Collectively, modulation of the transcriptional activity of NF-κB in the nucleus by the atypical IκB proteins BCL3 and IκBζ potently affects the oncogenic potential of NF-κB in several lymphoid malignancies. As important regulators of cell proliferation and survival, BCL3 and IκBζ might emerge as attractive therapeutic targets to dampen excessive NF-κB activity in certain lymphoid cancers, possibly by pharmacologically preventing the interaction between p50 and BCL3 or IκBζ [245].

7. Conclusions and Implications for Lymphoma Therapy

To date, a large number of lymphoid malignancies has been found to harbor diverse genetic lesions that result in aberrant NF-κB activity. While some of these alterations are unique to specific lymphoma entities, other aberrations, such as inactivation of A20, commonly occur in a broad range of lymphoid tumors. Since several lymphoid cancers have been found to critically rely on constitutive NF-κB activity for their survival and since the anti-apoptotic effects of NF-κB activation can confer resistance towards cancer chemotherapy [246], inhibition of NF-κB activation represents an attractive therapeutic option in many lymphoid malignancies. As a master regulator of canonical NF-κB signaling, the IKK complex and in particular IKKβ, constitutes a promising therapeutic target. However, the crucial and pleiotropic role of NF-κB in many physiological processes is reflected, for instance, in the embryonic lethality associated with massive hepatocyte apoptosis in mice deficient for IKKγ and IKKβ, two major constituents of the IKK complex [247–249]. The expected systemic toxicity, immunosuppression and, paradoxically, increased IL-1β-mediated inflammation critically limit the therapeutic usefulness of general inhibition of canonical NF-κB, e.g., by pharmacological IKKβ inhibitors [250–252]. Thus, targeting of deregulated upstream pathways, such as chronic active BCR signaling, which drive constitutive NF-κB activation, potentially offers higher specificity for the malignant cells and represents an attractive alternative in the treatment of lymphoid malignancies. The validity of this concept has first been demonstrated by the therapeutic efficacy of the BTK inhibitor ibrutinib in ABC DLBCL and other lymphoid cancers relying on chronic BCR signaling [67,89,253,254]. Additional therapeutic targets in lymphoid tumors addicted to chronic BCR activation include SYK, LYN, and PKCβ [67,255,256]. The occurrence of primary resistance towards BTK inhibition due to oncogenic events targeting downstream effectors, such as CARMA1, as well as the acquisition of secondary resistance demonstrates the necessity of alternative therapeutic strategies and the rational stratification of patients regarding the mutational status of their lymphoid cancer [257]. While ABC DLBCL cells harboring CARMA1 mutations are insensitive towards inhibitors targeting upstream kinases involved in chronic BCR signaling, these cells still respond to treatment with inhibitors targeting MALT1 protease activity [256,258]. In addition to ABC DLBCL, MALT1 protease inhibition might also be beneficial for patients suffering from other lymphoid malignancies relying on constitutive CBM signaling or aberrant MALT1 protease activity, such as MCL and CLL as well as MALT lymphomas expressing the cIAP2-MALT1 fusion [150–153]. MALT1 protease inhibitors might prove especially valuable in the treatment of lymphomas that harbor mutations in signaling effectors downstream of BTK or that have acquired resistance towards BTK inhibition [98,136]. Besides aberrant activity of the CBM complex, constitutive NF-κB activation can also result from deregulation of the MyD88-IRAK4 signaling axis due to recurrent oncogenic MyD88 mutations. Recently, several small-molecule inhibitors selectively targeting IRAK4 have been shown to effectively abrogate aberrant NF-κB activation induced by MyD88[L265P] in ABC DLBCL, thus representing an attractive therapeutic strategy for the treatment of MyD88 mutant lymphoid malignancies [40,50]. In addition to the genetic lesions promoting the constitutive activation of the canonical NF-κB pathway, several lymphoid cancers rely on the aberrant activity of non-canonical NF-κB signaling. The majority of these genetic aberrations result in the stabilization or increased expression of the central kinase NIK, the activity of which is essential for the survival of these lymphomas [19,20]. Even though no NIK inhibitor has entered the clinic yet, NIK represents a promising therapeutic target that should be addressed more vigorously, particularly regarding the treatment of multiple myeloma and MALT lymphoma expressing the oncogenic cIAP2-MALT1 fusion protein [188,259].

Biomedicines **2018**, *6*, 38

In light of the variety of genetic lesions in lymphoid malignancies which can cause constitutive NF-κB activation through deregulation of distinct upstream signaling nodes, the ultimate goal in the treatment of cancer, i.e., the highly specific eradication of the tumor cells by a targeted therapy, requires the analysis of the relevant oncogenic lesions and deregulated signaling pathways in the respective lymphoid tumor. The knowledge about how a certain oncogenic lesion drives NF-κB activation as well as the identification and molecular characterization of novel oncogenic mechanisms governing lymphomagenesis will pave the way for the rational design of therapeutic strategies, for instance by simultaneously targeting complementary signaling pathways, and thus improve lymphoma therapy.

Acknowledgments: This work was supported by grants from the Emmy-Noether Program of the German Research Foundation, the SFB/TR 156 (to Stephan Hailfinger), and the SFB/TR 209 (to Klaus Schulze-Osthoff and Stephan Hailfinger). We also would like to acknowledge the support by the German Research Foundation and the Open Access Publishing Fund of the University of Tuebingen.

Author Contributions: Paula Grondona, Philip Bucher, Klaus Schulze-Osthoff, Stephan Hailfinger, and Anja Schmitt wrote paragraphs of the manuscript. Paula Grondona, Philip Bucher, and Anja Schmitt prepared the figures.

Conflicts of Interest: The authors declare no conflict of interest.

References

1. Liu, F.; Xia, Y.; Parker, A.S.; Verma, I.M. IKK biology. *Immunol. Rev.* **2012**, *246*, 239–253. [CrossRef] [PubMed]
2. Ghosh, S.; Hayden, M.S. New regulators of NF-κB in inflammation. *Nat. Rev. Immunol.* **2008**, *8*, 837–848. [CrossRef] [PubMed]
3. Oeckinghaus, A.; Ghosh, S. The NF-κB family of transcription factors and its regulation. *Cold Spring Harb. Perspect. Biol.* **2009**, *1*, a000034. [CrossRef] [PubMed]
4. Sun, S.C. The noncanonical NF-κB pathway. *Immunol. Rev.* **2012**, *246*, 125–140. [CrossRef] [PubMed]
5. Schuster, M.; Annemann, M.; Plaza-Sirvent, C.; Schmitz, I. Atypical IκB proteins–nuclear modulators of NF-κB signaling. *Cell Commun. Signal.* **2013**, *11*, 23. [CrossRef] [PubMed]
6. Hinz, M.; Arslan, S.C.; Scheidereit, C. It takes two to tango: IκBs, the multifunctional partners of NF-κB. *Immunol. Rev.* **2012**, *246*, 59–76. [CrossRef] [PubMed]
7. Annemann, M.; Plaza-Sirvent, C.; Schuster, M.; Katsoulis-Dimitriou, K.; Kliche, S.; Schraven, B.; Schmitz, I. Atypical IκB proteins in immune cell differentiation and function. *Immunol. Lett.* **2016**, *171*, 26–35. [CrossRef] [PubMed]
8. Kaileh, M.; Sen, R. NF-κB function in B lymphocytes. *Immunol. Rev.* **2012**, *246*, 254–271. [CrossRef] [PubMed]
9. Staudt, L.M. Oncogenic activation of NF-κB. *Cold Spring Harb. Perspect. Biol.* **2010**, *2*, a000109. [CrossRef] [PubMed]
10. Lim, K.H.; Yang, Y.; Staudt, L.M. Pathogenetic importance and therapeutic implications of NF-κB in lymphoid malignancies. *Immunol. Rev.* **2012**, *246*, 359–378. [CrossRef] [PubMed]
11. Beug, H.; Muller, H.; Grieser, S.; Doederlein, G.; Graf, T. Hematopoietic cells transformed in vitro by REVT avian reticuloendotheliosis virus express characteristics of very immature lymphoid cells. *Virology* **1981**, *115*, 295–309. [CrossRef]
12. Barth, C.F.; Ewert, D.L.; Olson, W.C.; Humphries, E.H. Reticuloendotheliosis virus REV-T(REV-A)-induced neoplasia: Development of tumors within the T-lymphoid and myeloid lineages. *J. Virol.* **1990**, *64*, 6054–6062. [PubMed]
13. Lu, D.; Thompson, J.D.; Gorski, G.K.; Rice, N.R.; Mayer, M.G.; Yunis, J.J. Alterations at the rel locus in human lymphoma. *Oncogene* **1991**, *6*, 1235–1241. [PubMed]
14. Gilmore, T.D.; Gerondakis, S. The c-Rel Transcription Factor in Development and Disease. *Genes Cancer* **2011**, *2*, 695–711. [CrossRef] [PubMed]
15. Bargou, R.C.; Leng, C.; Krappmann, D.; Emmerich, F.; Mapara, M.Y.; Bommert, K.; Royer, H.D.; Scheidereit, C.; Dorken, B. High-level nuclear NF-κB and Oct-2 is a common feature of cultured Hodgkin/Reed-Sternberg cells. *Blood* **1996**, *87*, 4340–4347. [PubMed]
16. Davis, R.E.; Brown, K.D.; Siebenlist, U.; Staudt, L.M. Constitutive nuclear factor κB activity is required for survival of activated B cell-like diffuse large B cell lymphoma cells. *J. Exp. Med.* **2001**, *194*, 1861–1874. [CrossRef] [PubMed]

17. Ni, H.; Ergin, M.; Huang, Q.; Qin, J.Z.; Amin, H.M.; Martinez, R.L.; Saeed, S.; Barton, K.; Alkan, S. Analysis of expression of nuclear factor kappa B (NF-κB) in multiple myeloma: Downregulation of NF-κB induces apoptosis. *Br. J. Haematol.* **2001**, *115*, 279–286. [CrossRef] [PubMed]

18. Savage, K.J.; Monti, S.; Kutok, J.L.; Cattoretti, G.; Neuberg, D.; De Leval, L.; Kurtin, P.; Dal Cin, P.; Ladd, C.; Feuerhake, F.; et al. The molecular signature of mediastinal large B-cell lymphoma differs from that of other diffuse large B-cell lymphomas and shares features with classical Hodgkin lymphoma. *Blood* **2003**, *102*, 3871–3879. [CrossRef] [PubMed]

19. Keats, J.J.; Fonseca, R.; Chesi, M.; Schop, R.; Baker, A.; Chng, W.J.; Van Wier, S.; Tiedemann, R.; Shi, C.X.; Sebag, M.; et al. Promiscuous mutations activate the noncanonical NF-κB pathway in multiple myeloma. *Cancer Cell* **2007**, *12*, 131–144. [CrossRef] [PubMed]

20. Annunziata, C.M.; Davis, R.E.; Demchenko, Y.; Bellamy, W.; Gabrea, A.; Zhan, F.; Lenz, G.; Hanamura, I.; Wright, G.; Xiao, W.; et al. Frequent engagement of the classical and alternative NF-κB pathways by diverse genetic abnormalities in multiple myeloma. *Cancer Cell* **2007**, *12*, 115–130. [CrossRef] [PubMed]

21. Shaffer, A.L., 3rd; Young, R.M.; Staudt, L.M. Pathogenesis of human B cell lymphomas. *Annu. Rev. Immunol.* **2012**, *30*, 565–610. [CrossRef] [PubMed]

22. Herishanu, Y.; Perez-Galan, P.; Liu, D.; Biancotto, A.; Pittaluga, S.; Vire, B.; Gibellini, F.; Njuguna, N.; Lee, E.; Stennett, L.; et al. The lymph node microenvironment promotes B-cell receptor signaling, NF-κB activation, and tumor proliferation in chronic lymphocytic leukemia. *Blood* **2011**, *117*, 563–574. [CrossRef] [PubMed]

23. Rahal, R.; Frick, M.; Romero, R.; Korn, J.M.; Kridel, R.; Chan, F.C.; Meissner, B.; Bhang, H.E.; Ruddy, D.; Kauffmann, A.; et al. Pharmacological and genomic profiling identifies NF-κB-targeted treatment strategies for mantle cell lymphoma. *Nat. Med.* **2014**, *20*, 87–92. [CrossRef] [PubMed]

24. Rosebeck, S.; Madden, L.; Jin, X.; Gu, S.; Apel, I.J.; Appert, A.; Hamoudi, R.A.; Noels, H.; Sagaert, X.; Van Loo, P.; et al. Cleavage of NIK by the API2-MALT1 fusion oncoprotein leads to noncanonical NF-κB activation. *Science* **2011**, *331*, 468–472. [CrossRef] [PubMed]

25. Young, R.M.; Shaffer, A.L., 3rd; Phelan, J.D.; Staudt, L.M. B-cell receptor signaling in diffuse large B-cell lymphoma. *Semin. Hematol.* **2015**, *52*, 77–85. [CrossRef] [PubMed]

26. Takeda, K.; Akira, S. Microbial recognition by Toll-like receptors. *J. Dermatol. Sci.* **2004**, *34*, 73–82. [CrossRef] [PubMed]

27. Medzhitov, R.; Preston-Hurlburt, P.; Kopp, E.; Stadlen, A.; Chen, C.; Ghosh, S.; Janeway, C.A., Jr. MyD88 is an adaptor protein in the hToll/IL-1 receptor family signaling pathways. *Mol. Cell* **1998**, *2*, 253–258. [CrossRef]

28. Lin, S.C.; Lo, Y.C.; Wu, H. Helical assembly in the MyD88-IRAK4-IRAK2 complex in TLR/IL-1R signalling. *Nature* **2010**, *465*, 885–890. [CrossRef] [PubMed]

29. Wesche, H.; Henzel, W.J.; Shillinglaw, W.; Li, S.; Cao, Z. MyD88: An adapter that recruits IRAK to the IL-1 receptor complex. *Immunity* **1997**, *7*, 837–847. [CrossRef]

30. Muzio, M.; Ni, J.; Feng, P.; Dixit, V.M. IRAK (Pelle) family member IRAK-2 and MyD88 as proximal mediators of IL-1 signaling. *Science* **1997**, *278*, 1612–1615. [CrossRef] [PubMed]

31. Vollmer, S.; Strickson, S.; Zhang, T.; Gray, N.; Lee, K.L.; Rao, V.R.; Cohen, P. The mechanism of activation of IRAK1 and IRAK4 by interleukin-1 and Toll-like receptor agonists. *Biochem. J.* **2017**, *474*, 2027–2038. [CrossRef] [PubMed]

32. Cheng, H.; Addona, T.; Keshishian, H.; Dahlstrand, E.; Lu, C.; Dorsch, M.; Li, Z.; Wang, A.; Ocain, T.D.; Li, P.; et al. Regulation of IRAK-4 kinase activity via autophosphorylation within its activation loop. *Biochem. Biophys. Res. Commun.* **2007**, *352*, 609–616. [CrossRef] [PubMed]

33. Kollewe, C.; Mackensen, A.C.; Neumann, D.; Knop, J.; Cao, P.; Li, S.; Wesche, H.; Martin, M.U. Sequential autophosphorylation steps in the interleukin-1 receptor-associated kinase-1 regulate its availability as an adapter in interleukin-1 signaling. *J. Biol. Chem.* **2004**, *279*, 5227–5236. [CrossRef] [PubMed]

34. Dossang, A.C.; Motshwene, P.G.; Yang, Y.; Symmons, M.F.; Bryant, C.E.; Borman, S.; George, J.; Weber, A.N.; Gay, N.J. The N-terminal loop of IRAK-4 death domain regulates ordered assembly of the Myddosome signalling scaffold. *Sci. Rep.* **2016**, *6*, 37267. [CrossRef] [PubMed]

35. Keating, S.E.; Maloney, G.M.; Moran, E.M.; Bowie, A.G. IRAK-2 participates in multiple toll-like receptor signaling pathways to NFκB via activation of TRAF6 ubiquitination. *J. Biol. Chem.* **2007**, *282*, 33435–33443. [CrossRef] [PubMed]

36. Ye, H.; Arron, J.R.; Lamothe, B.; Cirilli, M.; Kobayashi, T.; Shevde, N.K.; Segal, D.; Dzivenu, O.K.; Vologodskaia, M.; Yim, M.; et al. Distinct molecular mechanism for initiating TRAF6 signalling. *Nature* **2002**, *418*, 443–447. [CrossRef] [PubMed]

37. Deng, L.; Wang, C.; Spencer, E.; Yang, L.; Braun, A.; You, J.; Slaughter, C.; Pickart, C.; Chen, Z.J. Activation of the IκB kinase complex by TRAF6 requires a dimeric ubiquitin-conjugating enzyme complex and a unique polyubiquitin chain. *Cell* **2000**, *103*, 351–361. [CrossRef]

38. Takaesu, G.; Kishida, S.; Hiyama, A.; Yamaguchi, K.; Shibuya, H.; Irie, K.; Ninomiya-Tsuji, J.; Matsumoto, K. TAB2, a novel adaptor protein, mediates activation of TAK1 MAPKKK by linking TAK1 to TRAF6 in the IL-1 signal transduction pathway. *Mol. Cell* **2000**, *5*, 649–658. [CrossRef]

39. Wang, C.; Deng, L.; Hong, M.; Akkaraju, G.R.; Inoue, J.; Chen, Z.J. TAK1 is a ubiquitin-dependent kinase of MKK and IKK. *Nature* **2001**, *412*, 346–351. [CrossRef] [PubMed]

40. Ngo, V.N.; Young, R.M.; Schmitz, R.; Jhavar, S.; Xiao, W.; Lim, K.H.; Kohlhammer, H.; Xu, W.; Yang, Y.; Zhao, H.; et al. Oncogenically active MYD88 mutations in human lymphoma. *Nature* **2011**, *470*, 115–119. [CrossRef] [PubMed]

41. Puente, X.S.; Pinyol, M.; Quesada, V.; Conde, L.; Ordonez, G.R.; Villamor, N.; Escaramis, G.; Jares, P.; Bea, S.; Gonzalez-Diaz, M.; et al. Whole-genome sequencing identifies recurrent mutations in chronic lymphocytic leukaemia. *Nature* **2011**, *475*, 101–105. [CrossRef] [PubMed]

42. Yan, Q.; Huang, Y.; Watkins, A.J.; Kocialkowski, S.; Zeng, N.; Hamoudi, R.A.; Isaacson, P.G.; de Leval, L.; Wotherspoon, A.; Du, M.Q. BCR and TLR signaling pathways are recurrently targeted by genetic changes in splenic marginal zone lymphomas. *Haematologica* **2012**, *97*, 595–598. [CrossRef] [PubMed]

43. Treon, S.P.; Xu, L.; Yang, G.; Zhou, Y.; Liu, X.; Cao, Y.; Sheehy, P.; Manning, R.J.; Patterson, C.J.; Tripsas, C.; et al. MYD88 L265P somatic mutation in Waldenstrom's macroglobulinemia. *N. Engl. J. Med.* **2012**, *367*, 826–833. [CrossRef] [PubMed]

44. Troen, G.; Warsame, A.; Delabie, J. CD79B and MYD88 Mutations in Splenic Marginal Zone Lymphoma. *ISRN Oncol.* **2013**, *2013*, 252318. [CrossRef] [PubMed]

45. Pasqualucci, L.; Trifonov, V.; Fabbri, G.; Ma, J.; Rossi, D.; Chiarenza, A.; Wells, V.A.; Grunn, A.; Messina, M.; Elliot, O.; et al. Analysis of the coding genome of diffuse large B-cell lymphoma. *Nat. Genet.* **2011**, *43*, 830–837. [CrossRef] [PubMed]

46. Avbelj, M.; Wolz, O.O.; Fekonja, O.; Bencina, M.; Repic, M.; Mavri, J.; Kruger, J.; Scharfe, C.; Delmiro Garcia, M.; Panter, G.; et al. Activation of lymphoma-associated MyD88 mutations via allostery-induced TIR-domain oligomerization. *Blood* **2014**, *124*, 3896–3904. [CrossRef] [PubMed]

47. Knittel, G.; Liedgens, P.; Korovkina, D.; Seeger, J.M.; Al-Baldawi, Y.; Al-Maarri, M.; Fritz, C.; Vlantis, K.; Bezhanova, S.; Scheel, A.H.; et al. B-cell-specific conditional expression of Myd88p.L252P leads to the development of diffuse large B-cell lymphoma in mice. *Blood* **2016**, *127*, 2732–2741. [CrossRef] [PubMed]

48. Zhan, C.; Qi, R.; Wei, G.; Guven-Maiorov, E.; Nussinov, R.; Ma, B. Conformational dynamics of cancer-associated MyD88-TIR domain mutant L252P (L265P) allosterically tilts the landscape toward homo-dimerization. *Protein Eng. Des. Sel.* **2016**, *29*, 347–354. [CrossRef] [PubMed]

49. Loiarro, M.; Volpe, E.; Ruggiero, V.; Gallo, G.; Furlan, R.; Maiorino, C.; Battistini, L.; Sette, C. Mutational analysis identifies residues crucial for homodimerization of myeloid differentiation factor 88 (MyD88) and for its function in immune cells. *J. Biol. Chem.* **2013**, *288*, 30210–30222. [CrossRef] [PubMed]

50. Kelly, P.N.; Romero, D.L.; Yang, Y.; Shaffer, A.L., 3rd; Chaudhary, D.; Robinson, S.; Miao, W.; Rui, L.; Westlin, W.F.; Kapeller, R.; et al. Selective interleukin-1 receptor-associated kinase 4 inhibitors for the treatment of autoimmune disorders and lymphoid malignancy. *J. Exp. Med.* **2015**, *212*, 2189–2201. [CrossRef] [PubMed]

51. Yang, G.; Zhou, Y.; Liu, X.; Xu, L.; Cao, Y.; Manning, R.J.; Patterson, C.J.; Buhrlage, S.J.; Gray, N.; Tai, Y.T.; et al. A mutation in MYD88 (L265P) supports the survival of lymphoplasmacytic cells by activation of Bruton tyrosine kinase in Waldenstrom macroglobulinemia. *Blood* **2013**, *122*, 1222–1232. [CrossRef] [PubMed]

52. Wang, J.Q.; Jeelall, Y.S.; Beutler, B.; Horikawa, K.; Goodnow, C.C. Consequences of the recurrent MYD88(L265P) somatic mutation for B cell tolerance. *J. Exp. Med.* **2014**, *211*, 413–426. [CrossRef] [PubMed]

53. Wang, J.Q.; Beutler, B.; Goodnow, C.C.; Horikawa, K. Inhibiting TLR9 and other UNC93B1-dependent TLRs paradoxically increases accumulation of MYD88L265P plasmablasts in vivo. *Blood* **2016**, *128*, 1604–1608. [CrossRef] [PubMed]

54. Reth, M. Antigen receptors on B lymphocytes. *Annu. Rev. Immunol.* **1992**, *10*, 97–121. [CrossRef] [PubMed]

55. Wienands, J.; Engels, N. Multitasking of Ig-α and Ig-β to regulate B cell antigen receptor function. *Int. Rev. Immunol.* **2001**, *20*, 679–696. [CrossRef] [PubMed]

56. Johnson, S.A.; Pleiman, C.M.; Pao, L.; Schneringer, J.; Hippen, K.; Cambier, J.C. Phosphorylated immunoreceptor signaling motifs (ITAMs) exhibit unique abilities to bind and activate Lyn and Syk tyrosine kinases. *J. Immunol.* **1995**, *155*, 4596–4603. [PubMed]

57. Harwood, N.E.; Batista, F.D. Early events in B cell activation. *Annu. Rev. Immunol.* **2010**, *28*, 185–210. [CrossRef] [PubMed]

58. Kurosaki, T.; Hikida, M. Tyrosine kinases and their substrates in B lymphocytes. *Immunol. Rev.* **2009**, *228*, 132–148. [CrossRef] [PubMed]

59. Fu, C.; Turck, C.W.; Kurosaki, T.; Chan, A.C. BLNK: A central linker protein in B cell activation. *Immunity* **1998**, *9*, 93–103. [CrossRef]

60. Watanabe, D.; Hashimoto, S.; Ishiai, M.; Matsushita, M.; Baba, Y.; Kishimoto, T.; Kurosaki, T.; Tsukada, S. Four tyrosine residues in phospholipase C-γ2, identified as Btk-dependent phosphorylation sites, are required for B cell antigen receptor-coupled calcium signaling. *J. Biol. Chem.* **2001**, *276*, 38595–38601. [CrossRef] [PubMed]

61. Thome, M.; Charton, J.E.; Pelzer, C.; Hailfinger, S. Antigen receptor signaling to NF-κB via CARMA1, BCL10, and MALT1. *Cold Spring Harb. Perspect. Biol.* **2010**, *2*, a003004. [CrossRef] [PubMed]

62. Knittel, G.; Liedgens, P.; Korovkina, D.; Pallasch, C.P.; Reinhardt, H.C. Rewired NFκB signaling as a potentially actionable feature of activated B-cell-like diffuse large B-cell lymphoma. *Eur. J. Haematol.* **2016**, *97*, 499–510. [CrossRef] [PubMed]

63. Suarez, F.; Lortholary, O.; Hermine, O.; Lecuit, M. Infection-associated lymphomas derived from marginal zone B cells: A model of antigen-driven lymphoproliferation. *Blood* **2006**, *107*, 3034–3044. [CrossRef] [PubMed]

64. Quinn, E.R.; Chan, C.H.; Hadlock, K.G.; Foung, S.K.; Flint, M.; Levy, S. The B-cell receptor of a hepatitis C virus (HCV)-associated non-Hodgkin lymphoma binds the viral E2 envelope protein, implicating HCV in lymphomagenesis. *Blood* **2001**, *98*, 3745–3749. [CrossRef] [PubMed]

65. Young, R.M.; Wu, T.; Schmitz, R.; Dawood, M.; Xiao, W.; Phelan, J.D.; Xu, W.; Menard, L.; Meffre, E.; Chan, W.C.; et al. Survival of human lymphoma cells requires B-cell receptor engagement by self-antigens. *Proc. Natl. Acad. Sci. USA* **2015**, *112*, 13447–13454. [CrossRef] [PubMed]

66. Ngo, V.N.; Davis, R.E.; Lamy, L.; Yu, X.; Zhao, H.; Lenz, G.; Lam, L.T.; Dave, S.; Yang, L.; Powell, J.; et al. A loss-of-function RNA interference screen for molecular targets in cancer. *Nature* **2006**, *441*, 106–110. [CrossRef] [PubMed]

67. Davis, R.E.; Ngo, V.N.; Lenz, G.; Tolar, P.; Young, R.M.; Romesser, P.B.; Kohlhammer, H.; Lamy, L.; Zhao, H.; Yang, Y.; et al. Chronic active B-cell-receptor signalling in diffuse large B-cell lymphoma. *Nature* **2010**, *463*, 88–92. [CrossRef] [PubMed]

68. Havranek, O.; Xu, J.; Kohrer, S.; Wang, Z.; Becker, L.; Comer, J.M.; Henderson, J.; Ma, W.; Man Chun Ma, J.; Westin, J.R.; et al. Tonic B-cell receptor signaling in diffuse large B-cell lymphoma. *Blood* **2017**, *130*, 995–1006. [CrossRef] [PubMed]

69. Srinivasan, L.; Sasaki, Y.; Calado, D.P.; Zhang, B.; Paik, J.H.; DePinho, R.A.; Kutok, J.L.; Kearney, J.F.; Otipoby, K.L.; Rajewsky, K. PI3 kinase signals BCR-dependent mature B cell survival. *Cell* **2009**, *139*, 573–586. [CrossRef] [PubMed]

70. Kraus, M.; Alimzhanov, M.B.; Rajewsky, N.; Rajewsky, K. Survival of resting mature B lymphocytes depends on BCR signaling via the Igα/β heterodimer. *Cell* **2004**, *117*, 787–800. [CrossRef] [PubMed]

71. Agathangelidis, A.; Darzentas, N.; Hadzidimitriou, A.; Brochet, X.; Murray, F.; Yan, X.J.; Davis, Z.; van Gastel-Mol, E.J.; Tresoldi, C.; Chu, C.C.; et al. Stereotyped B-cell receptors in one-third of chronic lymphocytic leukemia: A molecular classification with implications for targeted therapies. *Blood* **2012**, *119*, 4467–4475. [CrossRef] [PubMed]

72. Catera, R.; Silverman, G.J.; Hatzi, K.; Seiler, T.; Didier, S.; Zhang, L.; Herve, M.; Meffre, E.; Oscier, D.G.; Vlassara, H.; et al. Chronic lymphocytic leukemia cells recognize conserved epitopes associated with apoptosis and oxidation. *Mol. Med.* **2008**, *14*, 665–674. [CrossRef] [PubMed]

73. Chu, C.C.; Catera, R.; Zhang, L.; Didier, S.; Agagnina, B.M.; Damle, R.N.; Kaufman, M.S.; Kolitz, J.E.; Allen, S.L.; Rai, K.R.; et al. Many chronic lymphocytic leukemia antibodies recognize apoptotic cells with exposed nonmuscle myosin heavy chain IIA: Implications for patient outcome and cell of origin. *Blood* **2010**, *115*, 3907–3915. [CrossRef] [PubMed]

74. Duhren-von Minden, M.; Ubelhart, R.; Schneider, D.; Wossning, T.; Bach, M.P.; Buchner, M.; Hofmann, D.; Surova, E.; Follo, M.; Kohler, F.; et al. Chronic lymphocytic leukaemia is driven by antigen-independent cell-autonomous signalling. *Nature* **2012**, *489*, 309–312. [CrossRef] [PubMed]

75. Gazumyan, A.; Reichlin, A.; Nussenzweig, M.C. Igβ tyrosine residues contribute to the control of B cell receptor signaling by regulating receptor internalization. *J. Exp. Med.* **2006**, *203*, 1785–1794. [CrossRef] [PubMed]

76. Chan, V.W.; Lowell, C.A.; DeFranco, A.L. Defective negative regulation of antigen receptor signaling in Lyn-deficient B lymphocytes. *Curr. Biol.* **1998**, *8*, 545–553. [CrossRef]

77. Doody, G.M.; Justement, L.B.; Delibrias, C.C.; Matthews, R.J.; Lin, J.; Thomas, M.L.; Fearon, D.T. A role in B cell activation for CD22 and the protein tyrosine phosphatase SHP. *Science* **1995**, *269*, 242–244. [CrossRef] [PubMed]

78. Nishizumi, H.; Horikawa, K.; Mlinaric-Rascan, I.; Yamamoto, T. A double-edged kinase Lyn: A positive and negative regulator for antigen receptor-mediated signals. *J. Exp. Med.* **1998**, *187*, 1343–1348. [CrossRef] [PubMed]

79. Chan, V.W.; Meng, F.; Soriano, P.; DeFranco, A.L.; Lowell, C.A. Characterization of the B lymphocyte populations in Lyn-deficient mice and the role of Lyn in signal initiation and down-regulation. *Immunity* **1997**, *7*, 69–81. [CrossRef]

80. Cornall, R.J.; Cyster, J.G.; Hibbs, M.L.; Dunn, A.R.; Otipoby, K.L.; Clark, E.A.; Goodnow, C.C. Polygenic autoimmune traits: Lyn, CD22, and SHP-1 are limiting elements of a biochemical pathway regulating BCR signaling and selection. *Immunity* **1998**, *8*, 497–508. [CrossRef]

81. Wang, J.Q.; Jeelall, Y.S.; Humburg, P.; Batchelor, E.L.; Kaya, S.M.; Yoo, H.M.; Goodnow, C.C.; Horikawa, K. Synergistic cooperation and crosstalk between MYD88(L265P) and mutations that dysregulate CD79B and surface IgM. *J. Exp. Med.* **2017**, *214*, 2759–2776. [CrossRef] [PubMed]

82. Hunter, Z.R.; Xu, L.; Yang, G.; Zhou, Y.; Liu, X.; Cao, Y.; Manning, R.J.; Tripsas, C.; Patterson, C.J.; Sheehy, P.; et al. The genomic landscape of Waldenstrom macroglobulinemia is characterized by highly recurring MYD88 and WHIM-like CXCR4 mutations, and small somatic deletions associated with B-cell lymphomagenesis. *Blood* **2014**, *123*, 1637–1646. [CrossRef] [PubMed]

83. Rinaldi, A.; Kwee, I.; Taborelli, M.; Largo, C.; Uccella, S.; Martin, V.; Poretti, G.; Gaidano, G.; Calabrese, G.; Martinelli, G.; et al. Genomic and expression profiling identifies the B-cell associated tyrosine kinase Syk as a possible therapeutic target in mantle cell lymphoma. *Br. J. Haematol.* **2006**, *132*, 303–316. [CrossRef] [PubMed]

84. Feldman, A.L.; Sun, D.X.; Law, M.E.; Novak, A.J.; Attygalle, A.D.; Thorland, E.C.; Fink, S.R.; Vrana, J.A.; Caron, B.L.; Morice, W.G.; et al. Overexpression of Syk tyrosine kinase in peripheral T-cell lymphomas. *Leukemia* **2008**, *22*, 1139–1143. [CrossRef] [PubMed]

85. Burger, J.A.; Wiestner, A. Targeting B cell receptor signalling in cancer: Preclinical and clinical advances. *Nat. Rev. Cancer* **2018**, *18*, 148–167. [CrossRef] [PubMed]

86. Burger, J.A.; Tedeschi, A.; Barr, P.M.; Robak, T.; Owen, C.; Ghia, P.; Bairey, O.; Hillmen, P.; Bartlett, N.L.; Li, J.; et al. Ibrutinib as Initial Therapy for Patients with Chronic Lymphocytic Leukemia. *N. Engl. J. Med.* **2015**, *373*, 2425–2437. [CrossRef] [PubMed]

87. Kim, E.S.; Dhillon, S. Ibrutinib: A review of its use in patients with mantle cell lymphoma or chronic lymphocytic leukaemia. *Drugs* **2015**, *75*, 769–776. [CrossRef] [PubMed]

88. Treon, S.P.; Tripsas, C.K.; Meid, K.; Warren, D.; Varma, G.; Green, R.; Argyropoulos, K.V.; Yang, G.; Cao, Y.; Xu, L.; et al. Ibrutinib in previously treated Waldenstrom's macroglobulinemia. *N. Engl. J. Med.* **2015**, *372*, 1430–1440. [CrossRef] [PubMed]

89. Wilson, W.H.; Young, R.M.; Schmitz, R.; Yang, Y.; Pittaluga, S.; Wright, G.; Lih, C.J.; Williams, P.M.; Shaffer, A.L.; Gerecitano, J.; et al. Targeting B cell receptor signaling with ibrutinib in diffuse large B cell lymphoma. *Nat. Med.* **2015**, *21*, 922–926. [CrossRef] [PubMed]

90. Byrd, J.C.; Harrington, B.; O'Brien, S.; Jones, J.A.; Schuh, A.; Devereux, S.; Chaves, J.; Wierda, W.G.; Awan, F.T.; Brown, J.R.; et al. Acalabrutinib (ACP-196) in Relapsed Chronic Lymphocytic Leukemia. *N. Engl. J. Med.* **2016**, *374*, 323–332. [CrossRef] [PubMed]

91. Kuiatse, I.; Baladandayuthapani, V.; Lin, H.Y.; Thomas, S.K.; Bjorklund, C.C.; Weber, D.M.; Wang, M.; Shah, J.J.; Zhang, X.D.; Jones, R.J.; et al. Targeting the Spleen Tyrosine Kinase with Fostamatinib as a Strategy against Waldenstrom Macroglobulinemia. *Clin. Cancer Res.* **2015**, *21*, 2538–2545. [CrossRef] [PubMed]

92. Flinn, I.W.; Bartlett, N.L.; Blum, K.A.; Ardeshna, K.M.; LaCasce, A.S.; Flowers, C.R.; Shustov, A.R.; Thress, K.S.; Mitchell, P.; Zheng, F.; et al. A phase II trial to evaluate the efficacy of fostamatinib in patients with relapsed or refractory diffuse large B-cell lymphoma (DLBCL). *Eur. J. Cancer* **2016**, *54*, 11–17. [CrossRef] [PubMed]

93. Hantschel, O.; Rix, U.; Schmidt, U.; Burckstummer, T.; Kneidinger, M.; Schutze, G.; Colinge, J.; Bennett, K.L.; Ellmeier, W.; Valent, P.; et al. The Btk tyrosine kinase is a major target of the Bcr-Abl inhibitor dasatinib. *Proc. Natl. Acad. Sci. USA* **2007**, *104*, 13283–13288. [CrossRef] [PubMed]

94. Amrein, P.C.; Attar, E.C.; Takvorian, T.; Hochberg, E.P.; Ballen, K.K.; Leahy, K.M.; Fisher, D.C.; Lacasce, A.S.; Jacobsen, E.D.; Armand, P.; et al. Phase II study of dasatinib in relapsed or refractory chronic lymphocytic leukemia. *Clin. Cancer Res.* **2011**, *17*, 2977–2986. [CrossRef] [PubMed]

95. Lindauer, M.; Hochhaus, A. Dasatinib. *Recent Results Cancer Res.* **2014**, *201*, 27–65. [PubMed]

96. Ruland, J.; Duncan, G.S.; Wakeham, A.; Mak, T.W. Differential requirement for Malt1 in T and B cell antigen receptor signaling. *Immunity* **2003**, *19*, 749–758. [CrossRef]

97. Thome, M. CARMA1, BCL-10 and MALT1 in lymphocyte development and activation. *Nat. Rev. Immunol.* **2004**, *4*, 348–359. [CrossRef] [PubMed]

98. Juilland, M.; Thome, M. Role of the CARMA1/BCL10/MALT1 complex in lymphoid malignancies. *Curr. Opin. Hematol.* **2016**, *23*, 402–409. [CrossRef] [PubMed]

99. Matsumoto, R.; Wang, D.; Blonska, M.; Li, H.; Kobayashi, M.; Pappu, B.; Chen, Y.; Wang, D.; Lin, X. Phosphorylation of CARMA1 plays a critical role in T Cell receptor-mediated NF-κB activation. *Immunity* **2005**, *23*, 575–585. [CrossRef] [PubMed]

100. Sommer, K.; Guo, B.; Pomerantz, J.L.; Bandaranayake, A.D.; Moreno-Garcia, M.E.; Ovechkina, Y.L.; Rawlings, D.J. Phosphorylation of the CARMA1 linker controls NF-κB activation. *Immunity* **2005**, *23*, 561–574. [CrossRef] [PubMed]

101. Tanner, M.J.; Hanel, W.; Gaffen, S.L.; Lin, X. CARMA1 coiled-coil domain is involved in the oligomerization and subcellular localization of CARMA1 and is required for T cell receptor-induced NF-κB activation. *J. Biol. Chem.* **2007**, *282*, 17141–17147. [CrossRef] [PubMed]

102. Blonska, M.; Lin, X. NF-κB signaling pathways regulated by CARMA family of scaffold proteins. *Cell Res.* **2011**, *21*, 55–70. [CrossRef] [PubMed]

103. Qiao, Q.; Yang, C.; Zheng, C.; Fontan, L.; David, L.; Yu, X.; Bracken, C.; Rosen, M.; Melnick, A.; Egelman, E.H.; et al. Structural architecture of the CARMA1/Bcl10/MALT1 signalosome: Nucleation-induced filamentous assembly. *Mol. Cell* **2013**, *51*, 766–779. [CrossRef] [PubMed]

104. David, L.; Li, Y.; Ma, J.; Garner, E.; Zhang, X.; Wu, H. Assembly mechanism of the CARMA1-BCL10-MALT1-TRAF6 signalosome. *Proc. Natl. Acad. Sci. USA* **2018**, *115*, 1499–1504. [CrossRef] [PubMed]

105. Bertin, J.; Wang, L.; Guo, Y.; Jacobson, M.D.; Poyet, J.L.; Srinivasula, S.M.; Merriam, S.; DiStefano, P.S.; Alnemri, E.S. CARD11 and CARD14 are novel caspase recruitment domain (CARD)/membrane-associated guanylate kinase (MAGUK) family members that interact with BCL10 and activate NF-κB. *J. Biol. Chem.* **2001**, *276*, 11877–11882. [CrossRef] [PubMed]

106. Gaide, O.; Martinon, F.; Micheau, O.; Bonnet, D.; Thome, M.; Tschopp, J. Carma1, a CARD-containing binding partner of Bcl10, induces Bcl10 phosphorylation and NF-κB activation. *FEBS Lett.* **2001**, *496*, 121–127. [CrossRef]

107. Uren, A.G.; O'Rourke, K.; Aravind, L.A.; Pisabarro, M.T.; Seshagiri, S.; Koonin, E.V.; Dixit, V.M. Identification of paracaspases and metacaspases: Two ancient families of caspase-like proteins, one of which plays a key role in MALT lymphoma. *Mol. Cell* **2000**, *6*, 961–967. [CrossRef]

108. Lucas, P.C.; Yonezumi, M.; Inohara, N.; McAllister-Lucas, L.M.; Abazeed, M.E.; Chen, F.F.; Yamaoka, S.; Seto, M.; Nunez, G. Bcl10 and MALT1, independent targets of chromosomal translocation in malt lymphoma, cooperate in a novel NF-κB signaling pathway. *J. Biol. Chem.* **2001**, *276*, 19012–19019. [CrossRef] [PubMed]

109. McAllister-Lucas, L.M.; Inohara, N.; Lucas, P.C.; Ruland, J.; Benito, A.; Li, Q.; Chen, S.; Chen, F.F.; Yamaoka, S.; Verma, I.M.; et al. Bimp1, a MAGUK family member linking protein kinase C activation to Bcl10-mediated NF-κB induction. *J. Biol. Chem.* **2001**, *276*, 30589–30597. [CrossRef] [PubMed]

110. Jaworski, M.; Thome, M. The paracaspase MALT1: Biological function and potential for therapeutic inhibition. *Cell. Mol. Life Sci.* **2016**, *73*, 459–473. [CrossRef] [PubMed]

111. Oeckinghaus, A.; Wegener, E.; Welteke, V.; Ferch, U.; Arslan, S.C.; Ruland, J.; Scheidereit, C.; Krappmann, D. Malt1 ubiquitination triggers NF-κB signaling upon T-cell activation. *EMBO J.* **2007**, *26*, 4634–4645. [CrossRef] [PubMed]

112. Sun, L.; Deng, L.; Ea, C.K.; Xia, Z.P.; Chen, Z.J. The TRAF6 ubiquitin ligase and TAK1 kinase mediate IKK activation by BCL10 and MALT1 in T lymphocytes. *Mol. Cell* **2004**, *14*, 289–301. [CrossRef]

113. Wu, C.J.; Ashwell, J.D. NEMO recognition of ubiquitinated Bcl10 is required for T cell receptor-mediated NF-κB activation. *Proc. Natl. Acad. Sci. USA* **2008**, *105*, 3023–3028. [CrossRef] [PubMed]

114. Thome, M. Multifunctional roles for MALT1 in T-cell activation. *Nat. Rev. Immunol.* **2008**, *8*, 495–500. [CrossRef] [PubMed]

115. Fujita, H.; Rahighi, S.; Akita, M.; Kato, R.; Sasaki, Y.; Wakatsuki, S.; Iwai, K. Mechanism underlying IκB kinase activation mediated by the linear ubiquitin chain assembly complex. *Mol. Cell Biol.* **2014**, *34*, 1322–1335. [CrossRef] [PubMed]

116. Tokunaga, F.; Sakata, S.; Saeki, Y.; Satomi, Y.; Kirisako, T.; Kamei, K.; Nakagawa, T.; Kato, M.; Murata, S.; Yamaoka, S.; et al. Involvement of linear polyubiquitylation of NEMO in NF-κB activation. *Nat. Cell Biol.* **2009**, *11*, 123–132. [CrossRef] [PubMed]

117. Rahighi, S.; Ikeda, F.; Kawasaki, M.; Akutsu, M.; Suzuki, N.; Kato, R.; Kensche, T.; Uejima, T.; Bloor, S.; Komander, D.; et al. Specific recognition of linear ubiquitin chains by NEMO is important for NF-κB activation. *Cell* **2009**, *136*, 1098–1109. [CrossRef] [PubMed]

118. Adhikari, A.; Xu, M.; Chen, Z.J. Ubiquitin-mediated activation of TAK1 and IKK. *Oncogene* **2007**, *26*, 3214–3226. [CrossRef] [PubMed]

119. Vercammen, D.; Declercq, W.; Vandenabeele, P.; Van Breusegem, F. Are metacaspases caspases? *J. Cell Biol.* **2007**, *179*, 375–380. [CrossRef] [PubMed]

120. Rebeaud, F.; Hailfinger, S.; Posevitz-Fejfar, A.; Tapernoux, M.; Moser, R.; Rueda, D.; Gaide, O.; Guzzardi, M.; Iancu, E.M.; Rufer, N.; et al. The proteolytic activity of the paracaspase MALT1 is key in T cell activation. *Nat. Immunol.* **2008**, *9*, 272–281. [CrossRef] [PubMed]

121. Coornaert, B.; Baens, M.; Heyninck, K.; Bekaert, T.; Haegman, M.; Staal, J.; Sun, L.; Chen, Z.J.; Marynen, P.; Beyaert, R. T cell antigen receptor stimulation induces MALT1 paracaspase-mediated cleavage of the NF-κB inhibitor A20. *Nat. Immunol.* **2008**, *9*, 263–271. [CrossRef] [PubMed]

122. Cabalzar, K.; Pelzer, C.; Wolf, A.; Lenz, G.; Iwaszkiewicz, J.; Zoete, V.; Hailfinger, S.; Thome, M. Monoubiquitination and activity of the paracaspase MALT1 requires glutamate 549 in the dimerization interface. *PLoS ONE* **2013**, *8*, e72051. [CrossRef] [PubMed]

123. Pelzer, C.; Cabalzar, K.; Wolf, A.; Gonzalez, M.; Lenz, G.; Thome, M. The protease activity of the paracaspase MALT1 is controlled by monoubiquitination. *Nat. Immunol.* **2013**, *14*, 337–345. [CrossRef] [PubMed]

124. Wiesmann, C.; Leder, L.; Blank, J.; Bernardi, A.; Melkko, S.; Decock, A.; D'Arcy, A.; Villard, F.; Erbel, P.; Hughes, N.; et al. Structural determinants of MALT1 protease activity. *J. Mol. Biol.* **2012**, *419*, 4–21. [CrossRef] [PubMed]

125. Hailfinger, S.; Nogai, H.; Pelzer, C.; Jaworski, M.; Cabalzar, K.; Charton, J.E.; Guzzardi, M.; Decaillet, C.; Grau, M.; Dorken, B.; et al. Malt1-dependent RelB cleavage promotes canonical NF-κB activation in lymphocytes and lymphoma cell lines. *Proc. Natl. Acad. Sci. USA* **2011**, *108*, 14596–14601. [CrossRef] [PubMed]

126. Marienfeld, R.; May, M.J.; Berberich, I.; Serfling, E.; Ghosh, S.; Neumann, M. RelB forms transcriptionally inactive complexes with RelA/p65. *J. Biol. Chem.* **2003**, *278*, 19852–19860. [CrossRef] [PubMed]

127. Duwel, M.; Welteke, V.; Oeckinghaus, A.; Baens, M.; Kloo, B.; Ferch, U.; Darnay, B.G.; Ruland, J.; Marynen, P.; Krappmann, D. A20 negatively regulates T cell receptor signaling to NF-κB by cleaving Malt1 ubiquitin chains. *J. Immunol.* **2009**, *182*, 7718–7728. [CrossRef] [PubMed]

128. Baens, M.; Bonsignore, L.; Somers, R.; Vanderheydt, C.; Weeks, S.D.; Gunnarsson, J.; Nilsson, E.; Roth, R.G.; Thome, M.; Marynen, P. MALT1 auto-proteolysis is essential for NF-κB-dependent gene transcription in activated lymphocytes. *PLoS ONE* **2014**, *9*, e103774. [CrossRef] [PubMed]

129. Klein, T.; Fung, S.Y.; Renner, F.; Blank, M.A.; Dufour, A.; Kang, S.; Bolger-Munro, M.; Scurll, J.M.; Priatel, J.J.; Schweigler, P.; et al. The paracaspase MALT1 cleaves HOIL1 reducing linear ubiquitination by LUBAC to dampen lymphocyte NF-κB signalling. *Nat. Commun.* **2015**, *6*, 8777. [CrossRef] [PubMed]

130. Elton, L.; Carpentier, I.; Staal, J.; Driege, Y.; Haegman, M.; Beyaert, R. MALT1 cleaves the E3 ubiquitin ligase HOIL-1 in activated T cells, generating a dominant negative inhibitor of LUBAC-induced NF-κB signaling. *FEBS J.* **2016**, *283*, 403–412. [CrossRef] [PubMed]

131. Hailfinger, S.; Schmitt, A.; Schulze-Osthoff, K. The paracaspase MALT1 dampens NF-κB signalling by cleaving the LUBAC subunit HOIL-1. *FEBS J.* **2016**, *283*, 400–402. [CrossRef] [PubMed]

132. Karin, M.; Cao, Y.; Greten, F.R.; Li, Z.W. NF-κB in cancer: From innocent bystander to major culprit. *Nat. Rev. Cancer* **2002**, *2*, 301–310. [CrossRef] [PubMed]

133. Li, Q.; Verma, I.M. NF-κB regulation in the immune system. *Nat. Rev. Immunol.* **2002**, *2*, 725–734. [CrossRef] [PubMed]

134. Rosebeck, S.; Rehman, A.O.; Lucas, P.C.; McAllister-Lucas, L.M. From MALT lymphoma to the CBM signalosome: Three decades of discovery. *Cell Cycle* **2011**, *10*, 2485–2496. [CrossRef] [PubMed]

135. Da Silva Almeida, A.C.; Abate, F.; Khiabanian, H.; Martinez-Escala, E.; Guitart, J.; Tensen, C.P.; Vermeer, M.H.; Rabadan, R.; Ferrando, A.; Palomero, T. The mutational landscape of cutaneous T cell lymphoma and Sezary syndrome. *Nat. Genet.* **2015**, *47*, 1465–1470. [CrossRef] [PubMed]

136. Young, R.M.; Staudt, L.M. Targeting pathological B cell receptor signalling in lymphoid malignancies. *Nat. Rev. Drug Discov.* **2013**, *12*, 229–243. [CrossRef] [PubMed]

137. Lenz, G.; Davis, R.E.; Ngo, V.N.; Lam, L.; George, T.C.; Wright, G.W.; Dave, S.S.; Zhao, H.; Xu, W.; Rosenwald, A.; et al. Oncogenic CARD11 mutations in human diffuse large B cell lymphoma. *Science* **2008**, *319*, 1676–1679. [CrossRef] [PubMed]

138. Brohl, A.S.; Stinson, J.R.; Su, H.C.; Badgett, T.; Jennings, C.D.; Sukumar, G.; Sindiri, S.; Wang, W.; Kardava, L.; Moir, S.; et al. Germline CARD11 Mutation in a Patient with Severe Congenital B Cell Lymphocytosis. *J. Clin. Immunol.* **2015**, *35*, 32–46. [CrossRef] [PubMed]

139. Snow, A.L.; Xiao, W.; Stinson, J.R.; Lu, W.; Chaigne-Delalande, B.; Zheng, L.; Pittaluga, S.; Matthews, H.F.; Schmitz, R.; Jhavar, S.; et al. Congenital B cell lymphocytosis explained by novel germline CARD11 mutations. *J. Exp. Med.* **2012**, *209*, 2247–2261. [CrossRef] [PubMed]

140. Wang, L.; Ni, X.; Covington, K.R.; Yang, B.Y.; Shiu, J.; Zhang, X.; Xi, L.; Meng, Q.; Langridge, T.; Drummond, J.; et al. Genomic profiling of Sezary syndrome identifies alterations of key T cell signaling and differentiation genes. *Nat. Genet.* **2015**, *47*, 1426–1434. [CrossRef] [PubMed]

141. Isaacson, P.G.; Du, M.Q. MALT lymphoma: From morphology to molecules. *Nat. Rev. Cancer* **2004**, *4*, 644–653. [CrossRef] [PubMed]

142. Wotherspoon, A.C.; Doglioni, C.; Diss, T.C.; Pan, L.; Moschini, A.; de Boni, M.; Isaacson, P.G. Regression of primary low-grade B-cell gastric lymphoma of mucosa-associated lymphoid tissue type after eradication of Helicobacter pylori. *Lancet* **1993**, *342*, 575–577. [CrossRef]

143. Streubel, B.; Lamprecht, A.; Dierlamm, J.; Cerroni, L.; Stolte, M.; Ott, G.; Raderer, M.; Chott, A. T(14;18)(q32;q21) involving IGH and MALT1 is a frequent chromosomal aberration in MALT lymphoma. *Blood* **2003**, *101*, 2335–2339. [CrossRef] [PubMed]

144. Willis, T.G.; Jadayel, D.M.; Du, M.Q.; Peng, H.; Perry, A.R.; Abdul-Rauf, M.; Price, H.; Karran, L.; Majekodunmi, O.; Wlodarska, I.; et al. Bcl10 is involved in t(1;14)(p22;q32) of MALT B cell lymphoma and mutated in multiple tumor types. *Cell* **1999**, *96*, 35–45. [CrossRef]

145. Zhang, Q.; Siebert, R.; Yan, M.; Hinzmann, B.; Cui, X.; Xue, L.; Rakestraw, K.M.; Naeve, C.W.; Beckmann, G.; Weisenburger, D.D.; et al. Inactivating mutations and overexpression of BCL10, a caspase recruitment domain-containing gene, in MALT lymphoma with t(1;14)(p22;q32). *Nat. Genet.* **1999**, *22*, 63–68. [CrossRef] [PubMed]

146. Dierlamm, J.; Baens, M.; Wlodarska, I.; Stefanova-Ouzounova, M.; Hernandez, J.M.; Hossfeld, D.K.; De Wolf-Peeters, C.; Hagemeijer, A.; Van den Berghe, H.; Marynen, P. The apoptosis inhibitor gene API2 and a novel 18q gene, MLT, are recurrently rearranged in the t(11;18)(q21;q21) associated with mucosa-associated lymphoid tissue lymphomas. *Blood* **1999**, *93*, 3601–3609. [PubMed]

147. Akagi, T.; Motegi, M.; Tamura, A.; Suzuki, R.; Hosokawa, Y.; Suzuki, H.; Ota, H.; Nakamura, S.; Morishima, Y.; Taniwaki, M.; et al. A novel gene, MALT1 at 18q21, is involved in t(11;18) (q21;q21) found in low-grade B-cell lymphoma of mucosa-associated lymphoid tissue. *Oncogene* **1999**, *18*, 5785–5794. [CrossRef] [PubMed]

148. Lucas, P.C.; Kuffa, P.; Gu, S.; Kohrt, D.; Kim, D.S.; Siu, K.; Jin, X.; Swenson, J.; McAllister-Lucas, L.M. A dual role for the API2 moiety in API2-MALT1-dependent NF-κB activation: Heterotypic oligomerization and TRAF2 recruitment. *Oncogene* **2007**, *26*, 5643–5654. [CrossRef] [PubMed]

149. Noels, H.; van Loo, G.; Hagens, S.; Broeckx, V.; Beyaert, R.; Marynen, P.; Baens, M. A Novel TRAF6 binding site in MALT1 defines distinct mechanisms of NF-κB activation by API2middle dotMALT1 fusions. *J. Biol. Chem.* **2007**, *282*, 10180–10189. [CrossRef] [PubMed]

150. Ferch, U.; Kloo, B.; Gewies, A.; Pfander, V.; Duwel, M.; Peschel, C.; Krappmann, D.; Ruland, J. Inhibition of MALT1 protease activity is selectively toxic for activated B cell-like diffuse large B cell lymphoma cells. *J. Exp. Med.* **2009**, *206*, 2313–2320. [CrossRef] [PubMed]

151. Hailfinger, S.; Lenz, G.; Ngo, V.; Posvitz-Fejfar, A.; Rebeaud, F.; Guzzardi, M.; Penas, E.M.; Dierlamm, J.; Chan, W.C.; Staudt, L.M.; et al. Essential role of MALT1 protease activity in activated B cell-like diffuse large B-cell lymphoma. *Proc. Natl. Acad. Sci. USA* **2009**, *106*, 19946–19951. [CrossRef] [PubMed]

152. Dai, B.; Grau, M.; Juilland, M.; Klener, P.; Horing, E.; Molinsky, J.; Schimmack, G.; Aukema, S.M.; Hoster, E.; Vogt, N.; et al. B-cell receptor-driven MALT1 activity regulates MYC signaling in mantle cell lymphoma. *Blood* **2017**, *129*, 333–346. [CrossRef] [PubMed]

153. Saba, N.S.; Wong, D.H.; Tanios, G.; Iyer, J.R.; Lobelle-Rich, P.; Dadashian, E.L.; Liu, D.; Fontan, L.; Flemington, E.K.; Nichols, C.M.; et al. MALT1 Inhibition Is Efficacious in Both Naive and Ibrutinib-Resistant Chronic Lymphocytic Leukemia. *Cancer Res.* **2017**, *77*, 7038–7048. [CrossRef] [PubMed]

154. Fontan, L.; Yang, C.; Kabaleeswaran, V.; Volpon, L.; Osborne, M.J.; Beltran, E.; Garcia, M.; Cerchietti, L.; Shaknovich, R.; Yang, S.N.; et al. MALT1 small molecule inhibitors specifically suppress ABC-DLBCL in vitro and in vivo. *Cancer Cell* **2012**, *22*, 812–824. [CrossRef] [PubMed]

155. Nagel, D.; Spranger, S.; Vincendeau, M.; Grau, M.; Raffegerst, S.; Kloo, B.; Hlahla, D.; Neuenschwander, M.; Peter von Kries, J.; Hadian, K.; et al. Pharmacologic inhibition of MALT1 protease by phenothiazines as a therapeutic approach for the treatment of aggressive ABC-DLBCL. *Cancer Cell* **2012**, *22*, 825–837. [CrossRef] [PubMed]

156. Bardet, M.; Unterreiner, A.; Malinverni, C.; Lafossas, F.; Vedrine, C.; Boesch, D.; Kolb, Y.; Kaiser, D.; Gluck, A.; Schneider, M.A.; et al. The T-cell fingerprint of MALT1 paracaspase revealed by selective inhibition. *Immunol. Cell Biol.* **2018**, *96*, 81–99. [CrossRef] [PubMed]

157. Wertz, I.E.; O'Rourke, K.M.; Zhou, H.; Eby, M.; Aravind, L.; Seshagiri, S.; Wu, P.; Wiesmann, C.; Baker, R.; Boone, D.L.; et al. De-ubiquitination and ubiquitin ligase domains of A20 downregulate NF-κB signalling. *Nature* **2004**, *430*, 694–699. [CrossRef] [PubMed]

158. Tavares, R.M.; Turer, E.E.; Liu, C.L.; Advincula, R.; Scapini, P.; Rhee, L.; Barrera, J.; Lowell, C.A.; Utz, P.J.; Malynn, B.A.; et al. The ubiquitin modifying enzyme A20 restricts B cell survival and prevents autoimmunity. *Immunity* **2010**, *33*, 181–191. [CrossRef] [PubMed]

159. Hymowitz, S.G.; Wertz, I.E. A20: From ubiquitin editing to tumour suppression. *Nat. Rev. Cancer* **2010**, *10*, 332–341. [CrossRef] [PubMed]

160. Compagno, M.; Lim, W.K.; Grunn, A.; Nandula, S.V.; Brahmachary, M.; Shen, Q.; Bertoni, F.; Ponzoni, M.; Scandurra, M.; Califano, A.; et al. Mutations of multiple genes cause deregulation of NF-κB in diffuse large B-cell lymphoma. *Nature* **2009**, *459*, 717–721. [CrossRef] [PubMed]

161. Honma, K.; Tsuzuki, S.; Nakagawa, M.; Tagawa, H.; Nakamura, S.; Morishima, Y.; Seto, M. TNFAIP3/A20 functions as a novel tumor suppressor gene in several subtypes of non-Hodgkin lymphomas. *Blood* **2009**, *114*, 2467–2475. [CrossRef] [PubMed]

162. Schmitz, R.; Hansmann, M.L.; Bohle, V.; Martin-Subero, J.I.; Hartmann, S.; Mechtersheimer, G.; Klapper, W.; Vater, I.; Giefing, M.; Gesk, S.; et al. TNFAIP3 (A20) is a tumor suppressor gene in Hodgkin lymphoma and primary mediastinal B cell lymphoma. *J. Exp. Med.* **2009**, *206*, 981–989. [CrossRef] [PubMed]

163. Kato, M.; Sanada, M.; Kato, I.; Sato, Y.; Takita, J.; Takeuchi, K.; Niwa, A.; Chen, Y.; Nakazaki, K.; Nomoto, J.; et al. Frequent inactivation of A20 in B-cell lymphomas. *Nature* **2009**, *459*, 712–716. [CrossRef] [PubMed]

164. Chu, Y.; Vahl, J.C.; Kumar, D.; Heger, K.; Bertossi, A.; Wojtowicz, E.; Soberon, V.; Schenten, D.; Mack, B.; Reutelshofer, M.; et al. B cells lacking the tumor suppressor TNFAIP3/A20 display impaired differentiation and hyperactivation and cause inflammation and autoimmunity in aged mice. *Blood* **2011**, *117*, 2227–2236. [CrossRef] [PubMed]

165. Yang, Y.; Schmitz, R.; Mitala, J.; Whiting, A.; Xiao, W.; Ceribelli, M.; Wright, G.W.; Zhao, H.; Yang, Y.; Xu, W.; et al. Essential role of the linear ubiquitin chain assembly complex in lymphoma revealed by rare germline polymorphisms. *Cancer Discov.* **2014**, *4*, 480–493. [CrossRef] [PubMed]

166. Dubois, S.M.; Alexia, C.; Wu, Y.; Leclair, H.M.; Leveau, C.; Schol, E.; Fest, T.; Tarte, K.; Chen, Z.J.; Gavard, J.; et al. A catalytic-independent role for the LUBAC in NF-κB activation upon antigen receptor engagement and in lymphoma cells. *Blood* **2014**, *123*, 2199–2203. [CrossRef] [PubMed]

167. Lin, X.; Mu, Y.; Cunningham, E.T., Jr.; Marcu, K.B.; Geleziunas, R.; Greene, W.C. Molecular determinants of NF-κB-inducing kinase action. *Mol. Cell Biol* **1998**, *18*, 5899–5907. [CrossRef] [PubMed]

168. Senftleben, U.; Cao, Y.; Xiao, G.; Greten, F.R.; Krahn, G.; Bonizzi, G.; Chen, Y.; Hu, Y.; Fong, A.; Sun, S.C.; et al. Activation by IKKα of a second, evolutionary conserved, NF-κB signaling pathway. *Science* **2001**, *293*, 1495–1499. [CrossRef] [PubMed]

169. Ranuncolo, S.M.; Pittaluga, S.; Evbuomwan, M.O.; Jaffe, E.S.; Lewis, B.A. Hodgkin lymphoma requires stabilized NIK and constitutive RelB expression for survival. *Blood* **2012**, *120*, 3756–3763. [CrossRef] [PubMed]

170. Otto, C.; Giefing, M.; Massow, A.; Vater, I.; Gesk, S.; Schlesner, M.; Richter, J.; Klapper, W.; Hansmann, M.L.; Siebert, R.; et al. Genetic lesions of the TRAF3 and MAP3K14 genes in classical Hodgkin lymphoma. *Br. J. Haematol.* **2012**, *157*, 702–708. [CrossRef] [PubMed]

171. Liao, G.; Zhang, M.; Harhaj, E.W.; Sun, S.C. Regulation of the NF-κB-inducing kinase by tumor necrosis factor receptor-associated factor 3-induced degradation. *J. Biol. Chem.* **2004**, *279*, 26243–26250. [CrossRef] [PubMed]

172. Zhang, B.; Calado, D.P.; Wang, Z.; Frohler, S.; Kochert, K.; Qian, Y.; Koralov, S.B.; Schmidt-Supprian, M.; Sasaki, Y.; Unitt, C.; et al. An oncogenic role for alternative NF-κB signaling in DLBCL revealed upon deregulated BCL6 expression. *Cell Rep.* **2015**, *11*, 715–726. [CrossRef] [PubMed]

173. Hildebrand, J.M.; Luo, Z.; Manske, M.K.; Price-Troska, T.; Ziesmer, S.C.; Lin, W.; Hostager, B.S.; Slager, S.L.; Witzig, T.E.; Ansell, S.M.; et al. A BAFF-R mutation associated with non-Hodgkin lymphoma alters TRAF recruitment and reveals new insights into BAFF-R signaling. *J. Exp. Med.* **2010**, *207*, 2569–2579. [CrossRef] [PubMed]

174. Guo, X.; Koff, J.L.; Moffitt, A.B.; Cinar, M.; Ramachandiran, S.; Chen, Z.; Switchenko, J.M.; Mosunjac, M.; Neill, S.G.; Mann, K.P.; et al. Molecular impact of selective NFKB1 and NFKB2 signaling on DLBCL phenotype. *Oncogene* **2017**, *36*, 4224–4232. [CrossRef] [PubMed]

175. Gruss, H.J.; Kadin, M.E. Pathophysiology of Hodgkin's disease: Functional and molecular aspects. *Baillieres Clin. Haematol.* **1996**, *9*, 417–446. [CrossRef]

176. Deacon, E.M.; Pallesen, G.; Niedobitek, G.; Crocker, J.; Brooks, L.; Rickinson, A.B.; Young, L.S. Epstein-Barr virus and Hodgkin's disease: Transcriptional analysis of virus latency in the malignant cells. *J. Exp. Med.* **1993**, *177*, 339–349. [CrossRef] [PubMed]

177. Kilger, E.; Kieser, A.; Baumann, M.; Hammerschmidt, W. Epstein-Barr virus-mediated B-cell proliferation is dependent upon latent membrane protein 1, which simulates an activated CD40 receptor. *EMBO J.* **1998**, *17*, 1700–1709. [CrossRef] [PubMed]

178. Eliopoulos, A.G.; Caamano, J.H.; Flavell, J.; Reynolds, G.M.; Murray, P.G.; Poyet, J.L.; Young, L.S. Epstein-Barr virus-encoded latent infection membrane protein 1 regulates the processing of p100 NF-κB2 to p52 via an IKKγ/NEMO-independent signalling pathway. *Oncogene* **2003**, *22*, 7557–7569. [CrossRef] [PubMed]

179. Graham, J.P.; Arcipowski, K.M.; Bishop, G.A. Differential B-lymphocyte regulation by CD40 and its viral mimic, latent membrane protein 1. *Immunol. Rev.* **2010**, *237*, 226–248. [CrossRef] [PubMed]

180. Luftig, M.; Prinarakis, E.; Yasui, T.; Tsichritzis, T.; Cahir-McFarland, E.; Inoue, J.; Nakano, H.; Mak, T.W.; Yeh, W.C.; Li, X.; et al. Epstein-Barr virus latent membrane protein 1 activation of NF-κB through IRAK1 and TRAF6. *Proc. Natl. Acad. Sci. USA* **2003**, *100*, 15595–15600. [CrossRef] [PubMed]

181. Xiao, G.; Cvijic, M.E.; Fong, A.; Harhaj, E.W.; Uhlik, M.T.; Waterfield, M.; Sun, S.C. Retroviral oncoprotein Tax induces processing of NF-κB2/p100 in T cells: Evidence for the involvement of IKKα. *EMBO J.* **2001**, *20*, 6805–6815. [CrossRef] [PubMed]

182. Migliazza, A.; Lombardi, L.; Rocchi, M.; Trecca, D.; Chang, C.C.; Antonacci, R.; Fracchiolla, N.S.; Ciana, P.; Maiolo, A.T.; Neri, A. Heterogeneous chromosomal aberrations generate 3′ truncations of the NFKB2/lyt-10 gene in lymphoid malignancies. *Blood* **1994**, *84*, 3850–3860. [PubMed]

183. Isogawa, M.; Higuchi, M.; Takahashi, M.; Oie, M.; Mori, N.; Tanaka, Y.; Aoyagi, Y.; Fujii, M. Rearranged NF-κB2 gene in an adult T-cell leukemia cell line. *Cancer Sci.* **2008**, *99*, 792–798. [CrossRef] [PubMed]

184. Brummelkamp, T.R.; Nijman, S.M.; Dirac, A.M.; Bernards, R. Loss of the cylindromatosis tumour suppressor inhibits apoptosis by activating NF-κB. *Nature* **2003**, *424*, 797–801. [CrossRef] [PubMed]

185. Kovalenko, A.; Chable-Bessia, C.; Cantarella, G.; Israel, A.; Wallach, D.; Courtois, G. The tumour suppressor CYLD negatively regulates NF-κB signalling by deubiquitination. *Nature* **2003**, *424*, 801–805. [CrossRef] [PubMed]

186. Trompouki, E.; Hatzivassiliou, E.; Tsichritzis, T.; Farmer, H.; Ashworth, A.; Mosialos, G. CYLD is a deubiquitinating enzyme that negatively regulates NF-κB activation by TNFR family members. *Nature* **2003**, *424*, 793–796. [CrossRef] [PubMed]

187. Ramakrishnan, P.; Wang, W.; Wallach, D. Receptor-specific signaling for both the alternative and the canonical NF-κB activation pathways by NF-κB-inducing kinase. *Immunity* **2004**, *21*, 477–489. [CrossRef] [PubMed]

188. Demchenko, Y.N.; Brents, L.A.; Li, Z.; Bergsagel, L.P.; McGee, L.R.; Kuehl, M.W. Novel inhibitors are cytotoxic for myeloma cells with NFkB inducing kinase-dependent activation of NFkB. *Oncotarget* **2014**, *5*, 4554–4566. [CrossRef] [PubMed]

189. Castanedo, G.M.; Blaquiere, N.; Beresini, M.; Bravo, B.; Brightbill, H.; Chen, J.; Cui, H.F.; Eigenbrot, C.; Everett, C.; Feng, J.; et al. Structure-Based Design of Tricyclic NF-κB Inducing Kinase (NIK) Inhibitors That Have High Selectivity over Phosphoinositide-3-kinase (PI3K). *J. Med. Chem.* **2017**, *60*, 627–640. [CrossRef] [PubMed]

190. Huxford, T.; Huang, D.B.; Malek, S.; Ghosh, G. The crystal structure of the IκBα/NF-κB complex reveals mechanisms of NF-κB inactivation. *Cell* **1998**, *95*, 759–770. [CrossRef]

191. Ito, C.Y.; Adey, N.; Bautch, V.L.; Baldwin, A.S., Jr. Structure and evolution of the human IKBA gene. *Genomics* **1995**, *29*, 490–495. [CrossRef] [PubMed]

192. Beg, A.A.; Sha, W.C.; Bronson, R.T.; Baltimore, D. Constitutive NF-κB activation, enhanced granulopoiesis, and neonatal lethality in IκBα-deficient mice. *Genes Dev.* **1995**, *9*, 2736–2746. [CrossRef] [PubMed]

193. Cabannes, E.; Khan, G.; Aillet, F.; Jarrett, R.F.; Hay, R.T. Mutations in the IkBa gene in Hodgkin's disease suggest a tumour suppressor role for IκBα. *Oncogene* **1999**, *18*, 3063–3070. [CrossRef] [PubMed]

194. Jungnickel, B.; Staratschek-Jox, A.; Brauninger, A.; Spieker, T.; Wolf, J.; Diehl, V.; Hansmann, M.L.; Rajewsky, K.; Kuppers, R. Clonal deleterious mutations in the IκBα gene in the malignant cells in Hodgkin's lymphoma. *J. Exp. Med.* **2000**, *191*, 395–402. [CrossRef] [PubMed]

195. Kanzler, H.; Kuppers, R.; Hansmann, M.L.; Rajewsky, K. Hodgkin and Reed-Sternberg cells in Hodgkin's disease represent the outgrowth of a dominant tumor clone derived from (crippled) germinal center B cells. *J. Exp. Med.* **1996**, *184*, 1495–1505. [CrossRef] [PubMed]

196. Krappmann, D.; Emmerich, F.; Kordes, U.; Scharschmidt, E.; Dorken, B.; Scheidereit, C. Molecular mechanisms of constitutive NF-κB/Rel activation in Hodgkin/Reed-Sternberg cells. *Oncogene* **1999**, *18*, 943–953. [CrossRef] [PubMed]

197. Staudt, L.M. The molecular and cellular origins of Hodgkin's disease. *J. Exp. Med.* **2000**, *191*, 207–212. [CrossRef] [PubMed]

198. Osborne, J.; Lake, A.; Alexander, F.E.; Taylor, G.M.; Jarrett, R.F. Germline mutations and polymorphisms in the NFKBIA gene in Hodgkin lymphoma. *Int. J. Cancer* **2005**, *116*, 646–651. [CrossRef] [PubMed]

199. Emmerich, F.; Meiser, M.; Hummel, M.; Demel, G.; Foss, H.D.; Jundt, F.; Mathas, S.; Krappmann, D.; Scheidereit, C.; Stein, H.; et al. Overexpression of IκBα without inhibition of NF-κB activity and mutations in the IκBα gene in Reed-Sternberg cells. *Blood* **1999**, *94*, 3129–3134. [PubMed]

200. Liu, X.; Yu, H.; Yang, W.; Zhou, X.; Lu, H.; Shi, D. Mutations of NFKBIA in biopsy specimens from Hodgkin lymphoma. *Cancer Genet. Cytogenet.* **2010**, *197*, 152–157. [CrossRef] [PubMed]

201. Lake, A.; Shield, L.A.; Cordano, P.; Chui, D.T.; Osborne, J.; Crae, S.; Wilson, K.S.; Tosi, S.; Knight, S.J.; Gesk, S.; et al. Mutations of NFKBIA, encoding IκBα, are a recurrent finding in classical Hodgkin lymphoma but are not a unifying feature of non-EBV-associated cases. *Int. J. Cancer* **2009**, *125*, 1334–1342. [CrossRef] [PubMed]

202. Thomas, R.K.; Wickenhauser, C.; Tawadros, S.; Diehl, V.; Kuppers, R.; Wolf, J.; Schmitz, R. Mutational analysis of the IκBα gene in activated B cell-like diffuse large B-cell lymphoma. *Br. J. Haematol.* **2004**, *126*, 50–54. [CrossRef] [PubMed]

203. Takahashi, H.; Feuerhake, F.; Monti, S.; Kutok, J.L.; Aster, J.C.; Shipp, M.A. Lack of IKBA coding region mutations in primary mediastinal large B-cell lymphoma and the host response subtype of diffuse large B-cell lymphoma. *Blood* **2006**, *107*, 844–845. [CrossRef] [PubMed]

204. Johansson, P.; Klein-Hitpass, L.; Grabellus, F.; Arnold, G.; Klapper, W.; Pfortner, R.; Duhrsen, U.; Eckstein, A.; Durig, J.; Kuppers, R. Recurrent mutations in NF-κB pathway components, KMT2D, and NOTCH1/2 in ocular adnexal MALT-type marginal zone lymphomas. *Oncotarget* **2016**, *7*, 62627–62639. [CrossRef] [PubMed]

205. Emmerich, F.; Theurich, S.; Hummel, M.; Haeffker, A.; Vry, M.S.; Dohner, K.; Bommert, K.; Stein, H.; Dorken, B. Inactivating IκBε mutations in Hodgkin/Reed-Sternberg cells. *J. Pathol.* **2003**, *201*, 413–420. [CrossRef] [PubMed]

206. Alves, B.N.; Tsui, R.; Almaden, J.; Shokhirev, M.N.; Davis-Turak, J.; Fujimoto, J.; Birnbaum, H.; Ponomarenko, J.; Hoffmann, A. IκBε is a key regulator of B cell expansion by providing negative feedback on cRel and RelA in a stimulus-specific manner. *J. Immunol.* **2014**, *192*, 3121–3132. [CrossRef] [PubMed]

207. Mansouri, L.; Sutton, L.A.; Ljungstrom, V.; Bondza, S.; Arngarden, L.; Bhoi, S.; Larsson, J.; Cortese, D.; Kalushkova, A.; Plevova, K.; et al. Functional loss of IκBε leads to NF-κB deregulation in aggressive chronic lymphocytic leukemia. *J. Exp. Med.* **2015**, *212*, 833–843. [CrossRef] [PubMed]

208. Mansouri, L.; Noerenberg, D.; Young, E.; Mylonas, E.; Abdulla, M.; Frick, M.; Asmar, F.; Ljungstrom, V.; Schneider, M.; Yoshida, K.; et al. Frequent NFKBIE deletions are associated with poor outcome in primary mediastinal B-cell lymphoma. *Blood* **2016**, *128*, 2666–2670. [CrossRef] [PubMed]

209. Whiteside, S.T.; Epinat, J.C.; Rice, N.R.; Israel, A. IκBε, a novel member of the IκB family, controls RelA and cRel NF-κB activity. *EMBO J.* **1997**, *16*, 1413–1426. [CrossRef] [PubMed]

210. Lee, S.H.; Hannink, M. Characterization of the nuclear import and export functions of IκB(ε). *J. Biol. Chem.* **2002**, *277*, 23358–23366. [CrossRef] [PubMed]

211. Memet, S.; Laouini, D.; Epinat, J.C.; Whiteside, S.T.; Goudeau, B.; Philpott, D.; Kayal, S.; Sansonetti, P.J.; Berche, P.; Kanellopoulos, J.; et al. IκBε-deficient mice: Reduction of one T cell precursor subspecies and enhanced Ig isotype switching and cytokine synthesis. *J. Immunol.* **1999**, *163*, 5994–6005. [PubMed]

212. Nencioni, A.; Grunebach, F.; Patrone, F.; Ballestrero, A.; Brossart, P. Proteasome inhibitors: Antitumor effects and beyond. *Leukemia* **2007**, *21*, 30–36. [CrossRef] [PubMed]

213. Jackson, G.; Einsele, H.; Moreau, P.; Miguel, J.S. Bortezomib, a novel proteasome inhibitor, in the treatment of hematologic malignancies. *Cancer Treat. Rev.* **2005**, *31*, 591–602. [CrossRef] [PubMed]

214. Richardson, P.G.; Sonneveld, P.; Schuster, M.W.; Irwin, D.; Stadtmauer, E.A.; Facon, T.; Harousseau, J.L.; Ben-Yehuda, D.; Lonial, S.; Goldschmidt, H.; et al. Bortezomib or high-dose dexamethasone for relapsed multiple myeloma. *N. Engl. J. Med.* **2005**, *352*, 2487–2498. [CrossRef] [PubMed]

215. Dunleavy, K.; Pittaluga, S.; Czuczman, M.S.; Dave, S.S.; Wright, G.; Grant, N.; Shovlin, M.; Jaffe, E.S.; Janik, J.E.; Staudt, L.M.; et al. Differential efficacy of bortezomib plus chemotherapy within molecular subtypes of diffuse large B-cell lymphoma. *Blood* **2009**, *113*, 6069–6076. [CrossRef] [PubMed]

216. Ohno, H.; Takimoto, G.; McKeithan, T.W. The candidate proto-oncogene bcl-3 is related to genes implicated in cell lineage determination and cell cycle control. *Cell* **1990**, *60*, 991–997. [CrossRef]

217. Kerr, L.D.; Duckett, C.S.; Wamsley, P.; Zhang, Q.; Chiao, P.; Nabel, G.; McKeithan, T.W.; Baeuerle, P.A.; Verma, I.M. The proto-oncogene bcl-3 encodes an IκB protein. *Genes Dev.* **1992**, *6*, 2352–2363. [CrossRef] [PubMed]

218. Wulczyn, F.G.; Naumann, M.; Scheidereit, C. Candidate proto-oncogene bcl-3 encodes a subunit-specific inhibitor of transcription factor NF-κB. *Nature* **1992**, *358*, 597–599. [CrossRef] [PubMed]

219. Hatada, E.N.; Nieters, A.; Wulczyn, F.G.; Naumann, M.; Meyer, R.; Nucifora, G.; McKeithan, T.W.; Scheidereit, C. The ankyrin repeat domains of the NF-κB precursor p105 and the protooncogene bcl-3 act as specific inhibitors of NF-κB DNA binding. *Proc. Natl. Acad. Sci. USA* **1992**, *89*, 2489–2493. [CrossRef] [PubMed]

220. Bours, V.; Franzoso, G.; Azarenko, V.; Park, S.; Kanno, T.; Brown, K.; Siebenlist, U. The oncoprotein Bcl-3 directly transactivates through kappa B motifs via association with DNA-binding p50B homodimers. *Cell* **1993**, *72*, 729–739. [CrossRef]

221. Fujita, T.; Nolan, G.P.; Liou, H.C.; Scott, M.L.; Baltimore, D. The candidate proto-oncogene bcl-3 encodes a transcriptional coactivator that activates through NF-κB p50 homodimers. *Genes Dev.* **1993**, *7*, 1354–1363. [CrossRef] [PubMed]

222. Carmody, R.J.; Ruan, Q.; Palmer, S.; Hilliard, B.; Chen, Y.H. Negative regulation of toll-like receptor signaling by NF-κB p50 ubiquitination blockade. *Science* **2007**, *317*, 675–678. [CrossRef] [PubMed]

223. Dechend, R.; Hirano, F.; Lehmann, K.; Heissmeyer, V.; Ansieau, S.; Wulczyn, F.G.; Scheidereit, C.; Leutz, A. The Bcl-3 oncoprotein acts as a bridging factor between NF-κB/Rel and nuclear co-regulators. *Oncogene* **1999**, *18*, 3316–3323. [CrossRef] [PubMed]

224. Viatour, P.; Dejardin, E.; Warnier, M.; Lair, F.; Claudio, E.; Bureau, F.; Marine, J.C.; Merville, M.P.; Maurer, U.; Green, D.; et al. GSK3-mediated BCL-3 phosphorylation modulates its degradation and its oncogenicity. *Mol. Cell* **2004**, *16*, 35–45. [CrossRef] [PubMed]

225. Na, S.Y.; Choi, J.E.; Kim, H.J.; Jhun, B.H.; Lee, Y.C.; Lee, J.W. Bcl3, an IκB protein, stimulates activating protein-1 transactivation and cellular proliferation. *J. Biol. Chem.* **1999**, *274*, 28491–28496. [CrossRef] [PubMed]

226. Canoz, O.; Rassidakis, G.Z.; Admirand, J.H.; Medeiros, L.J. Immunohistochemical detection of BCL-3 in lymphoid neoplasms: A survey of 353 cases. *Mod. Pathol.* **2004**, *17*, 911–917. [CrossRef] [PubMed]
227. McKeithan, T.W.; Takimoto, G.S.; Ohno, H.; Bjorling, V.S.; Morgan, R.; Hecht, B.K.; Dube, I.; Sandberg, A.A.; Rowley, J.D. BCL3 rearrangements and t(14;19) in chronic lymphocytic leukemia and other B-cell malignancies: A molecular and cytogenetic study. *Genes Chromosomes Cancer* **1997**, *20*, 64–72. [CrossRef]
228. Michaux, L.; Dierlamm, J.; Wlodarska, I.; Bours, V.; Van den Berghe, H.; Hagemeijer, A. t(14;19)/BCL3 rearrangements in lymphoproliferative disorders: A review of 23 cases. *Cancer Genet. Cytogenet.* **1997**, *94*, 36–43. [CrossRef]
229. Martin-Subero, J.I.; Wlodarska, I.; Bastard, C.; Picquenot, J.M.; Hoppner, J.; Giefing, M.; Klapper, W.; Siebert, R. Chromosomal rearrangements involving the BCL3 locus are recurrent in classical Hodgkin and peripheral T-cell lymphoma. *Blood* **2006**, *108*, 401–402; author reply 402–403. [CrossRef] [PubMed]
230. Rassidakis, G.Z.; Oyarzo, M.P.; Medeiros, L.J. BCL-3 overexpression in anaplastic lymphoma kinase-positive anaplastic large cell lymphoma. *Blood* **2003**, *102*, 1146–1147. [CrossRef] [PubMed]
231. Nishikori, M.; Maesako, Y.; Ueda, C.; Kurata, M.; Uchiyama, T.; Ohno, H. High-level expression of BCL3 differentiates t(2;5)(p23;q35)-positive anaplastic large cell lymphoma from Hodgkin disease. *Blood* **2003**, *101*, 2789–2796. [CrossRef] [PubMed]
232. Ong, S.T.; Hackbarth, M.L.; Degenstein, L.C.; Baunoch, D.A.; Anastasi, J.; McKeithan, T.W. Lymphadenopathy, splenomegaly, and altered immunoglobulin production in BCL3 transgenic mice. *Oncogene* **1998**, *16*, 2333–2343. [CrossRef] [PubMed]
233. Westerheide, S.D.; Mayo, M.W.; Anest, V.; Hanson, J.L.; Baldwin, A.S., Jr. The putative oncoprotein Bcl-3 induces cyclin D1 to stimulate G(1) transition. *Mol. Cell. Biol.* **2001**, *21*, 8428–8436. [CrossRef] [PubMed]
234. Mitchell, T.C.; Hildeman, D.; Kedl, R.M.; Teague, T.K.; Schaefer, B.C.; White, J.; Zhu, Y.; Kappler, J.; Marrack, P. Immunological adjuvants promote activated T cell survival via induction of Bcl-3. *Nat. Immunol.* **2001**, *2*, 397–402. [CrossRef] [PubMed]
235. Kitamura, H.; Kanehira, K.; Okita, K.; Morimatsu, M.; Saito, M. MAIL, a novel nuclear IκB protein that potentiates LPS-induced IL-6 production. *FEBS Lett.* **2000**, *485*, 53–56. [CrossRef]
236. Haruta, H.; Kato, A.; Todokoro, K. Isolation of a novel interleukin-1-inducible nuclear protein bearing ankyrin-repeat motifs. *J. Biol. Chem.* **2001**, *276*, 12485–12488. [CrossRef] [PubMed]
237. Motoyama, M.; Yamazaki, S.; Eto-Kimura, A.; Takeshige, K.; Muta, T. Positive and negative regulation of nuclear factor-kappaB-mediated transcription by IκB-ζ, an inducible nuclear protein. *J. Biol. Chem.* **2005**, *280*, 7444–7451. [CrossRef] [PubMed]
238. Tartey, S.; Matsushita, K.; Vandenbon, A.; Ori, D.; Imamura, T.; Mino, T.; Standley, D.M.; Hoffmann, J.A.; Reichhart, J.M.; Akira, S.; et al. Akirin2 is critical for inducing inflammatory genes by bridging IκB-ζ and the SWI/SNF complex. *EMBO J.* **2014**, *33*, 2332–2348. [CrossRef] [PubMed]
239. Yamamoto, M.; Yamazaki, S.; Uematsu, S.; Sato, S.; Hemmi, H.; Hoshino, K.; Kaisho, T.; Kuwata, H.; Takeuchi, O.; Takeshige, K.; et al. Regulation of Toll/IL-1-receptor-mediated gene expression by the inducible nuclear protein IκBζ. *Nature* **2004**, *430*, 218–222. [CrossRef] [PubMed]
240. Okamoto, K.; Iwai, Y.; Oh-Hora, M.; Yamamoto, M.; Morio, T.; Aoki, K.; Ohya, K.; Jetten, A.M.; Akira, S.; Muta, T.; et al. IκBζ regulates T(H)17 development by cooperating with ROR nuclear receptors. *Nature* **2010**, *464*, 1381–1385. [CrossRef] [PubMed]
241. Chapuy, B.; Roemer, M.G.; Stewart, C.; Tan, Y.; Abo, R.P.; Zhang, L.; Dunford, A.J.; Meredith, D.M.; Thorner, A.R.; Jordanova, E.S.; et al. Targetable genetic features of primary testicular and primary central nervous system lymphomas. *Blood* **2016**, *127*, 869–881. [CrossRef] [PubMed]
242. Nogai, H.; Wenzel, S.S.; Hailfinger, S.; Grau, M.; Kaergel, E.; Seitz, V.; Wollert-Wulf, B.; Pfeifer, M.; Wolf, A.; Frick, M.; et al. IκB-ζ controls the constitutive NF-κB target gene network and survival of ABC DLBCL. *Blood* **2013**, *122*, 2242–2250. [CrossRef] [PubMed]
243. Kimura, R.; Senba, M.; Cutler, S.J.; Ralph, S.J.; Xiao, G.; Mori, N. Human T cell leukemia virus type I tax-induced IκB-ζ modulates tax-dependent and tax-independent gene expression in T cells. *Neoplasia* **2013**, *15*, 1110–1124. [CrossRef] [PubMed]
244. Ishikawa, C.; Senba, M.; Mori, N. Induction of IκB-ζ by Epstein-Barr virus latent membrane protein-1 and CD30. *Int. J. Oncol.* **2015**, *47*, 2197–2207. [CrossRef] [PubMed]
245. Collins, P.E.; Kiely, P.A.; Carmody, R.J. Inhibition of transcription by B cell Leukemia 3 (Bcl-3) protein requires interaction with nuclear factor kappaB (NF-κB) p50. *J. Biol. Chem.* **2014**, *289*, 7059–7067. [CrossRef] [PubMed]

246. Nakanishi, C.; Toi, M. Nuclear factor-kappaB inhibitors as sensitizers to anticancer drugs. *Nat. Rev. Cancer* **2005**, *5*, 297–309. [CrossRef] [PubMed]

247. Li, Q.; Van Antwerp, D.; Mercurio, F.; Lee, K.F.; Verma, I.M. Severe liver degeneration in mice lacking the IκB kinase 2 gene. *Science* **1999**, *284*, 321–325. [CrossRef] [PubMed]

248. Tanaka, M.; Fuentes, M.E.; Yamaguchi, K.; Durnin, M.H.; Dalrymple, S.A.; Hardy, K.L.; Goeddel, D.V. Embryonic lethality, liver degeneration, and impaired NF-κB activation in IKK-β-deficient mice. *Immunity* **1999**, *10*, 421–429. [CrossRef]

249. Rudolph, D.; Yeh, W.C.; Wakeham, A.; Rudolph, B.; Nallainathan, D.; Potter, J.; Elia, A.J.; Mak, T.W. Severe liver degeneration and lack of NF-κB activation in NEMO/IKKγ-deficient mice. *Genes Dev.* **2000**, *14*, 854–862. [PubMed]

250. Gamble, C.; McIntosh, K.; Scott, R.; Ho, K.H.; Plevin, R.; Paul, A. Inhibitory kappa B Kinases as targets for pharmacological regulation. *Br. J. Pharmacol.* **2012**, *165*, 802–819. [CrossRef] [PubMed]

251. Wullaert, A.; Bonnet, M.C.; Pasparakis, M. NF-κB in the regulation of epithelial homeostasis and inflammation. *Cell Res.* **2011**, *21*, 146–158. [CrossRef] [PubMed]

252. Greten, F.R.; Arkan, M.C.; Bollrath, J.; Hsu, L.C.; Goode, J.; Miething, C.; Goktuna, S.I.; Neuenhahn, M.; Fierer, J.; Paxian, S.; et al. NF-κB is a negative regulator of IL-1β secretion as revealed by genetic and pharmacological inhibition of IKKβ. *Cell* **2007**, *130*, 918–931. [CrossRef] [PubMed]

253. Honigberg, L.A.; Smith, A.M.; Sirisawad, M.; Verner, E.; Loury, D.; Chang, B.; Li, S.; Pan, Z.; Thamm, D.H.; Miller, R.A.; et al. The Bruton tyrosine kinase inhibitor PCI-32765 blocks B-cell activation and is efficacious in models of autoimmune disease and B-cell malignancy. *Proc. Natl. Acad. Sci. USA* **2010**, *107*, 13075–13080. [CrossRef] [PubMed]

254. Herman, S.E.; Gordon, A.L.; Hertlein, E.; Ramanunni, A.; Zhang, X.; Jaglowski, S.; Flynn, J.; Jones, J.; Blum, K.A.; Buggy, J.J.; et al. Bruton tyrosine kinase represents a promising therapeutic target for treatment of chronic lymphocytic leukemia and is effectively targeted by PCI-32765. *Blood* **2011**, *117*, 6287–6296. [CrossRef] [PubMed]

255. Friedberg, J.W.; Sharman, J.; Sweetenham, J.; Johnston, P.B.; Vose, J.M.; Lacasce, A.; Schaefer-Cutillo, J.; De Vos, S.; Sinha, R.; Leonard, J.P.; et al. Inhibition of Syk with fostamatinib disodium has significant clinical activity in non-Hodgkin lymphoma and chronic lymphocytic leukemia. *Blood* **2010**, *115*, 2578–2585. [CrossRef] [PubMed]

256. Naylor, T.L.; Tang, H.; Ratsch, B.A.; Enns, A.; Loo, A.; Chen, L.; Lenz, P.; Waters, N.J.; Schuler, W.; Dorken, B.; et al. Protein kinase C inhibitor sotrastaurin selectively inhibits the growth of CD79 mutant diffuse large B-cell lymphomas. *Cancer Res.* **2011**, *71*, 2643–2653. [CrossRef] [PubMed]

257. Zhang, S.Q.; Smith, S.M.; Zhang, S.Y.; Lynn Wang, Y. Mechanisms of ibrutinib resistance in chronic lymphocytic leukaemia and non-Hodgkin lymphoma. *Br. J. Haematol.* **2015**, *170*, 445–456. [CrossRef] [PubMed]

258. Nagel, D.; Bognar, M.; Eitelhuber, A.C.; Kutzner, K.; Vincendeau, M.; Krappmann, D. Combinatorial BTK and MALT1 inhibition augments killing of CD79 mutant diffuse large B cell lymphoma. *Oncotarget* **2015**, *6*, 42232–42242. [CrossRef] [PubMed]

259. Gardam, S.; Beyaert, R. The kinase NIK as a therapeutic target in multiple myeloma. *Expert Opin. Ther. Targets* **2011**, *15*, 207–218. [CrossRef] [PubMed]

      MDPI

Review

NF-κB in Hematological Malignancies

Véronique Imbert * and Jean-François Peyron

Centre Méditerranéen de Médecine Moléculaire, INSERM U1065, Université Côte d'Azur, 06204 Nice, France;
Jean-Francois.PEYRON@unice.fr
* Correspondence: Veronique.IMBERT@unice.fr; Tel.: +33-489-064-315

Academic Editor: Véronique Baud
Received: 28 April 2017; Accepted: 26 May 2017; Published: 31 May 2017

Abstract: NF-κB (Nuclear Factor K-light-chain-enhancer of activated B cells) transcription factors are critical regulators of immunity, stress response, apoptosis, and differentiation. Molecular defects promoting the constitutive activation of canonical and non-canonical NF-κB signaling pathways contribute to many diseases, including cancer, diabetes, chronic inflammation, and autoimmunity. In the present review, we focus our attention on the mechanisms of NF-κB deregulation in hematological malignancies. Key positive regulators of NF-κB signaling can act as oncogenes that are often prone to chromosomal translocation, amplifications, or activating mutations. Negative regulators of NF-κB have tumor suppressor functions, and are frequently inactivated either by genomic deletions or point mutations. NF-κB activation in tumoral cells is also driven by the microenvironment or chronic signaling that does not rely on genetic alterations.

Keywords: NF-κB; leukemia; lymphoma

---

1. Introduction

The NF-κB family of transcription factors coordinates inflammatory responses, innate and adaptive immunity, cellular differentiation, proliferation, and survival in all multicellular organisms.

The NF-κB system is tightly controlled at various levels, and deregulations of NF-κB homeostasis have been implicated in a wide range of diseases, ranging from inflammatory and immune disorders to cancer [1,2]. In particular, NF-κB is a key link between chronic inflammation and cancer transformation [3].

The mammalian NF-κB family is composed of five members: RelA (p65), RelB, c-Rel, NFKB1 (p105/p50), and NFKB2 (p100/p52), which form various dimeric complexes that transactivate numerous target genes via binding to κB consensus DNA binding sites. They are tightly regulated by the IκBs protein family including typical IκBs (IκBα, IκBβ, IκBε), the precursor proteins p100 and p105, and the atypical IκBs (IκBζ, BCL-3, IκBη). The typical IκBs and the precursors are expressed in the cytoplasm to sequester NF-κB dimers by masking their nuclear localization sequence. On the contrary, the atypical IκBs are hardly expressed in resting cells, and are induced upon cell activation to interact with NF-κB dimers within the nucleus. They can act as activators or inhibitors of NF-κB, depending on which proteins of the transcriptional machinery they recruit.

NF-κB activation is mediated by two signaling pathways as detailed in [4–6].

The canonical or classical pathway—the most widely known route to NF-κB activation—is essentially mediated by the action of the RelA/p50 subunits. In resting conditions, RelA/p50 dimers are retained in the cytoplasm complexed with the IκBα inhibitor. Upon cell activation by pro-inflammatory cytokines, immune receptors engagement, and stress conditions, IκBα is first phosphorylated on two serine residues (Ser32 and 36 in human IκBα) by the IκB-kinase (IKK) complex, which is composed of two kinases—IKKα (or IKK1) and IKKβ (or IKK2)—associated with an essential scaffold protein IKKγ (or NEMO, NF-κB essential modulator). The activation of the IKK complex depends on extracellular signals: hence, T- and B-cell antigen receptors (TcR and BcR) are coupled to the CARMA/BCL10/MALT1 (CBM) complex to activate IKKs, whereas the receptors for IL1 (Interleukin 1), TNFα (Tumor Necrosis Factor-α), and toll-like receptors (TLRs) rather engage a signaling complex involving TRAFs (TNF Receptor Associated Factor), IRAKs (interleukin 1 receptor associated kinase 1), LUBAC (Linear UBiquitin Assembly Complex), and TAK1 (TGF-β Activated Kkinase 1) proteins. IκBα phosphorylation triggers the recruitment of the E3 ubiquitin ligase SCF/βTRCP (Skp, Cullin, F-box/transducing repeat containing protein) and the subsequent polyubiquitination of IκBα, marking it for degradation via the 26S proteasome. This degradation releases the RelA/p50 dimer that can translocate into the nucleus to activate the transcription of target genes.

The non-canonical or alternative pathway activates mainly RelB-p52 complexes through the inducible processing of p100. In contrast to the canonical pathway, this pathway is activated by a more restricted number of ligands, such as the B-cell-activating factor (BAFF) belonging to the TNF family, CD40L, lymphotoxin β (LTβ), receptor activator nuclear factor ligand (RANKL), or CD30L. The triggering of these cell surface molecules engages the assembly of a signaling complex that involves cellular inhibitor of apoptosis (cIAP1 and cIAP2), TRAF2, and TRAF3. Receptor engagement leads to the recruitment and activation of cIAP1/2 mediated by TRAF2, resulting in the degradation of TRAF3, stopping the continuous degradation of the NF-κB-inducing kinase (NIK) that is central to this pathway. NIK can then activate IKKα, which phosphorylates p100. This provides a signal for their recognition by the SCF/βTRCP ubiquitin ligase complex and the proteasome-mediated processing of p100 by in situ degradation of the IκB-like domain of the precursor proteins, producing mature p52 protein. The mature RelB/p52 dimers are then released, translocate into the nucleus to start the transcription of their target genes. NF-κB activation strongly relies on multiple connected biochemical reactions—in particular on three main types of ubiquitination events to build (K63, linear) or disassemble (K48) signaling complexes (see [7] for a review).

The two pathways are interconnected, as the canonical one regulates p100 and RelB levels [8]. Both canonical and non-canonical NF-κB activation pathways have been implicated in human hematological malignancies, mainly lymphoid leukemia and lymphoma. Because NF-κB regulates a large array of target genes, the constitutive activation of NF-κB can support most steps involved in cancer transformation: inhibition of cell differentiation and apoptosis, promotion of cell proliferation, angiogenesis, cancer-related inflammation and metastatic potential, and resistance to treatments. The constitutive activation may have different origins. It can result from rearrangements and mutations in genes encoding NF-κB or IκB members, or in genes encoding upstream components of the cascade, but it may also derive from persistent autocrine or paracrine signaling and/or hyper-activation of immune receptors (TcR, BcR, TLRs). In this review article, we present the role of the NF-κB pathway and subunits in human hematologic malignancies. Figure 1 summarizes the molecular partners involved in the two pathways to NF-κB activation, as well as the dysregulations that occur in hematologic malignancies. For a broader description of the role of NF-κB subunits in cancer, see [9].

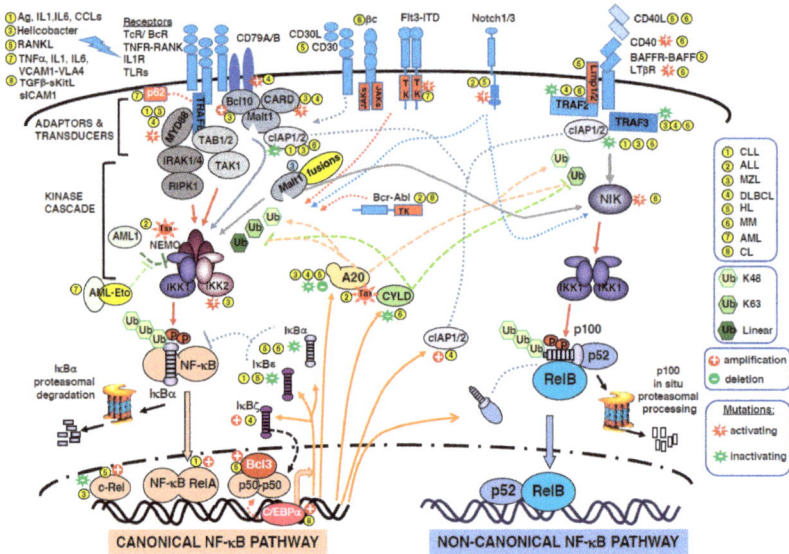

Figure 1. Deregulations of NF-κB in hematopoietic malignancies. The figure depicts the different actors in the canonical and non-canonical NF-κB activation pathways and their involvement in leukemia and lymphoma. The disease in which a given protein is involved is identified by a code number in a yellow filled circle ❶. 1- CLL: chronic lymphocytic leukemia; 2- ALL: acute lymphoblastic leukemia; 3-MZL: marginal zone lymphomas; 4- DLBCL: diffuse large B-cell lymphomas; 5- HL: Hodgkin lymphoma; 6- MM: multiple myeloma; 7- AML: acute myeloid leukemia; 8- CML: chronic myeloid leukemia. The participation of ubiquitination reactions in signaling is not detailed but highlighted with a green color code; K48 ubiquitination is responsible for degradation, K63 and linear are involved in the assembly of signaling complexes. Gene amplification/deletion are indicated by ⊕ and ⊖, respectively. Mutations affecting the genes for NF-κB subunits or actors of the NF-κB pathways are mentioned as activating 🟊 or inactivating ✳ mutations. → illustrates positive action whereas ⊣ illustrates inhibitory action. Color code: color of the arrow refers to the protein concerned by the depicted action. Plain orange arrow shows the proteins of interest after transcription via NF-κB. Plain red arrow shows phosphorylation events. Dotted red arrow shows tyrosine phosphorylation. Abbreviations used are: Ag: antigen; IL: interleukin; CCL: C-C motif chemokine ligand; RANKL: receptor activator of NF-κB ligand; TNF: Tumor Necrosis Factor; VCAM1: vascular cell adhesion molecule 1; VLA-4: very late antigen-4; TGF: transforming growth factor; KitL: kit ligand; ICAM1: intercellular adhesion molecule1; TcR/BcR: T-cell or B-cell antigen receptors; TNFR: TNF receptor; TLR: toll-like receptor; CD: cluster of differentiation; βc: common β chain of interleukin receptors; JAK: Janus kinase; TK: tyrosine kinase; Flt3-ITD: colony-stimulating factor receptor 1-like-3-internat tandem repeat; Notch: notch receptor; BAFF: B-cell activating factor; LTαR: lymphotoxin α receptor; LMP1: latent membrane protein; TRAF: TNF receptor associated factor; c-IAP: cellular-inhibitor of apoptosis; p62 = SQSTM1: sequestosome 1; Myd88: myeloid differentiation primary response gene 88; Bcl10: B-cell leukemia/lymphoma 10; CARD: protein with a caspase recruitment domain; Malt1: mucosa-associated lymphoid tissue lymphoma-1; TAB: TAK1 binding protein; TAK1: TGF-β activated kinase 1; IRAK: interleukin 1 receptor associated kinase 1; RIPK1: receptor (TNFR)-interacting serine-threonine kinase 1; Bcr-Abl: breakpoint cluster region-abelson kinase, product of the t(9; 22) translocation; NIK: NF-κB-inducing kinase; AML1: acute myeloid leukemia gene 1 = RUNX: Runt-related transcription factor 1; AML-Eto: product of the t(8; 21) translocation; Tax: HTLV-1 viral protein; IKK: inhibitor of κ-B kinase; NEMO: NF-κB essential modulator/inhibitor of NF-κB subunit γ; A20 = TNFAIP3: TNF α-induced protein 3; CYLD: cylindromatosis gene product lysine 63 deubiquitinase; IκB: inhibitor of NF-κB; Rel: reticuloendotheliosis proto-oncogene, NF-κB subunit; C/EBP: CCAAT/enhancer binding protein.

2. Lymphoid Malignancies

2.1. Leukemias

Chronic lymphocytic leukemia (CLL) is characterized by the progressive accumulation of mature monoclonal B lymphocytes in the peripheral blood (PB), bone marrow (BM), and secondary lymphoid organs such as lymph nodes (LN). NF-κB is constitutively activated in CLL patients [10,11]. The microenvironment exerts a critical role in the natural history of CLL. Indeed, signals from multiple receptors (BcR, TLR, CD40) result in the activation of downstream pathways, including NF-κB [12,13]. In addition, there is evidence that the NF-κB pathway and its upstream mediators can be targeted by recurrent genetic lesions in some minor cases. For example, 3% of CLL patients display the L265P mutation on MYD88 (MYeloid Differentiation primary response gene 88). This mutation changes the structure of MYD88 to allow spontaneous homodimerization and recruitment of the serine/threonine kinases IRAK1 and IRAK4 that are essential for NF-κB activation by the TLRs. This enhances the responses to TLR ligands, leading to a higher release of cytokines (IL6—Interleukin 6—, IL1RA—Interleukin 1 Receptor Agonist—) and chemokines (CCL2/CCL3/CCL4 —Chemokine Ligand 2/3/4—). These cytokines/chemokines have been reported to be important in attracting other cell types (e.g., T lymphocytes) by CLL cells to create an advantageous microenvironment supporting leukemic survival [14]. The most frequently mutated gene (7% of CLL cases) in CLL within the NF-κB pathway is *NFKBIE* that encodes IκBε—a negative NF-κB regulator. *NFKBIE* aberrations were highly enriched in poor-prognostic subgroups, demonstrating the supporting role of NF-κB for transformation. These aberrations lead to reduced IκBε protein levels, diminished interactions with RelA, as well as increased phosphorylated RelA and nuclear translocation [15]. *BIRC3—Baculoviral IAP repeat–containing protein 3*—(the gene encoding for the cIAP2 ubiquitin ligase) is also mutated in CLL. cIAP2, along with TRAF2 and TRAF3, cooperates in the same protein complex to negatively regulate NIK, the central activator of non-canonical NF-κB signaling. All inactivating *BIRC3* mutations detected in CLL are predicted to cause the elimination or truncation of the C-terminal RING domain, the E3 ubiquitin ligase activity which is essential for NIK proteasomal degradation. The mutations impair NIK ubiquitination, and thereby favor its stabilization, leading to the phosphorylation of NFKB2 and the processing of p100 to p52. This results in constitutive NF-κB activation, as evidenced by the detection of a higher p52/p100 ratio in *BIRC3*-mutated patients [16,17]. If *BIRC3* mutations are rare in early stages of CLL, they tend to accumulate as the disease progresses, suggesting a selective advantage for the transformed cells bearing this mutation. Finally, analysis of a cohort of 131 CLL patients revealed that DNA binding of RelA is constitutively elevated in patients with more aggressive disease, and is also further induced by conventional chemotherapy, which in turn seems to contribute to the depth of response to subsequent treatment cycles. That study identified RelA as a superior prognostic marker for the survival of CLL patients, and crucially demonstrates that RelA levels have the potential to predict the duration of the response to therapy [18].

Some epigenetic changes and aberrant microRNA expression have also been associated with NF-κB dysregulation in CLL. For example, the silencing of miR-9-3a tumor suppressor miRNA by methylation may account for the constitutive upregulation of *NFKB1*, and hence the constitutive activation of NF-κB in CLL patients [19]. In addition, the miR-708 enhancer is aberrantly methylated in CLL. miR708 directly targets IKKβ, and thereby leads to the repression of NF-κB signaling. CLL patients with high methylation of the miR708 enhancer present a poor prognosis [20].

Acute lymphoblastic leukemia (ALL) is mainly a disease of childhood that arises from recurrent genetic insults that block the differentiation of B- and T-cell precursors to drive their aberrant cell proliferation and survival. ALL subgroups with distinct biological characteristics are frequently characterized by dysregulated transcription factors or kinases—most prominently the fusion proteins TEL-AML1 (Translocation Ets Leukemia-Acute Myeloid Leukemia 1 protein) encoded by t(12;21)(p13;q22) in presumed good risk ALL, or Bcr-Abl (Breakpoint cluster region-Abelson murine leukemia virus) encoded by t(9;22)(q34;q11) and MLL-AF4 (Mixed Lineage Leukemia-ALL1 Fused gene on chromosome 4) by t(4;11)(q21;q23) in high risk ALL. ALL in adults is characterized by a higher frequency of high-risk cytogenetics and a lower incidence of favorable genetic abnormalities.

The vast majority of ALL patients present a constitutive activation of the canonical NF-κB pathway in the form of RelA/p50 complexes, which is an important switch to ensure the survival of ALL cells by blocking apoptosis or enhancing cell proliferation [21].

T-cell acute lymphoblastic leukemia (T-ALL) is an aggressive malignancy of transformed thymocytes that mainly affects children and adolescents. Although mutations in *NF-κB* genes have not been reported in T-ALL (unlike other lymphoid malignancies), constitutive activation of NF-κB frequently occurs in primary human T-ALL and T-ALL mouse models. Kordes and colleagues showed by electrophoretic mobility shift assays [21] that at diagnosis, childhood T-ALL cells (11 of 13 cases) displayed a constitutive activity of NF-κB consisting of RelA-p50 dimers associated with IκBα phosphorylation. Another study realized on human T-ALL cell lines depicted a constitutive NF-κB activity associated with constitutive IKK activity and nuclear localization of all the NF-κB members (p50, p105, RelA, RelB, and c-Rel), suggesting the activation of both canonical and non-canonical pathways. Somatic activating NOTCH1 mutations are found in more than 50% of human T-ALL cases, and they result in elevated levels of ICN1—the transcriptionally-active intracellular domain of NOTCH1. NOTCH1 can mobilize both the NF-κB canonical signaling by activating the IKKα/β/γ complex and the non-canonical pathway by inducing the expression of RelB and NFKB2 and activating IKKα homodimers [22]. The mechanism of Notch-induced NF-κB activation in T-ALL involves Hes1, which transcriptionally represses CYLD (cylindromatosis), a deubiquitinase which down-regulates NF-κB signaling by removing the activator K63 ubiquitin chains from different elements of the NF-κB signalosome [23]. In transgenic mice, NOTCH3 triggers both classical and alternative NF-κB activation pathways, depending on the expression of the pre-TcR. Indeed, in the absence of pre-Tα/pre-TcR, Notch 3 induces IKKα, inducing a higher p100 substrate processing and a release of RelB/p52 complexes [24]. Bcr-Abl expression in either T-ALL or B-ALL cells triggers an IKK-dependent activation of NF-κB that is crucial to the pathogenicity of Philadelphia positive (Ph+) leukemias [25].

Chronic inflammation directed by NF-κB is vital for the pathogenesis of adult T-cell leukemia (ATL)—an aggressive malignancy with a poor prognosis that is induced by the human T-cell leukemia virus type I. HTLV-1 encodes a 40 kDa oncoprotein (Tax) that regulates viral gene expression and plays vital roles in ATL leukemogenesis. Tax interactions with Ubc13 (E2-conjugating enzyme), NEMO, TAX_1BP_1 (Tax Binding Protein), and NRP (NEMO related Protein) are critical for activation of the IKK complex [26–28]. Tax also maintains persistent NF-κB activation by inactivating NF-κB negative regulators such as TNF α-induced protein 3 (TNFAIP3) and CYLD [29].

Finally, activation of NF-κB was identified as a mechanism for resistance to IFNβ in the poor-prognosis MLL-ALL subtype (t(4;11) translocation) [30].

2.2. Lymphomas

Marginal zone lymphomas (MZLs) are indolent small B-cell lymphomas classified in three subgroups, depending on the localization: mucosa-associated lymphoid tissue (MALT), splenic (sMZL) and nodal (nMZL) lymphomas. Both canonical and non-canonical NF-κB pathways can be activated through a variety of mechanisms, from molecular alterations to epigenetic modifications—especially in MALT and splenic MZL lymphomas.

In fact, MALT lymphomas represent an archetypal example of the link between chronic inflammation and tumor development. In particular, gastric MALT lymphomas develop from a background of chronic gastric infection with *Helicobater pylori*, while those from the skin and ocular adnexa or diseases of the small intestine are either associated with chronic infection with *Borrelia burgdorferi*, *Chlamydia psittaci*, or *Campylobacter jejuni*. The prolonged chronic microbial infection generates immune and inflammatory responses that transform the polyclonal B-cell population into a monoclonal B-cell lymphoma. At the early stage of MALT lymphoma, eradication of *H. pylori* infection by antibiotics causes tumor regression [31]. Thereafter, acquisition of chromosomal translocations constitutively activating NF-κB provides antigenic independence and antibiotic resistance. The t(11;18)(q21;q21) translocation represents the most frequent modification in 18% of MALT lymphomas, and can be enriched to 40% in gastric MALT lymphomas. It generates an cIAP2-MALT1 fusion which oligomerises, triggering TRAF6-dependent ubiquitination of NEMO, leading to the activation of the canonical NF-κB pathway [32]. In addition, the auto-oligomerization of cIAP2-MALT1 also induces the recruitment of NIK and its subsequent cleavage by the MALT1 protease domain, leading to degradation-resistant NIK kinase and deregulated non-canonical NF-κB signaling [33]. The t(14;18)(q32;q21) chromosomal translocation is very frequent in MALT lymphoma (10–20%), but is never found in the gastric forms. This translocation leads to the IGH–MALT1 fusion and causes MALT-1 overexpression, and thereby enhances the canonical NF-κB signaling [34]. The t(1;14) (p22;q32) translocation which juxtaposes the *BCL10* (B cell lymphoma/leukemia 10) gene under the regulatory control of the Ig heavy chain gene enhancer (*BCL10-IGH*) is a rare genetic aberration. Over-expression of BCL10 causes its constitutive activation through oligomerization via its N-terminal CARD/CARD (Caspase Activation and Recruitment Domain) interaction, and thus leads to enhanced NF-κB activity [35]. Intriguingly, BCL10 protein is aberrantly expressed in the nuclei of lymphoma cells, suggesting an as-of-yet unappreciated role of nuclear BCL10 in the pathogenesis of MALT lymphoma [35].

An aberrant NF-κB activation is found in 30–40% of splenic MZL patients. This involves dysregulation of the canonical pathways due to molecular lesions in a number of genes belonging to the NF-κB pathway. First, inactivation of the negative regulator *TNFAIP3* (A20) by non-sense or frame-shift mutations is found in 10–15% of patients [36]. TNFAIP3 is responsible for switching off signals converging from surface receptors on NF-κB as well as the inhibition of NF-κB proteins. Consequently, *TNFAIP3* disruption in MZL causes supraphysiological activation of NF-κB signaling. Two activating mutations (K171E and K171T) affecting IKKβ were found in 10% of splenic MZL [37,38]. The L265P mutation of MYD88—which affects 3% of MZL patients—changes the structure of MYD88 to allow spontaneous homodimerisation and recruitment of serine threonine kinases IRAK1 and IRAK4 essential for NF-κB activation by the toll-like receptors [39]. In addition, CARD11 coiled-coil domain mutations (5–10% of patients) promote spontaneous CARD11 multimerisation and association with other components of the CBM complex, thus leading to IKKβ kinase activation and NF-κB upregulation [39]. In some splenic MZL patients, activation of the non-canonical NF-κB pathway has also been described. BIRC3 RING domains mutations are found in 10–15% of patients. As detailed above, these mutations affect NF-κB activation by stabilizing NIK. Finally, in 5% of MZL patients, TRAF3 is inactivated by mutations causing elimination of the C-terminal domain of the protein. This domain is involved in the docking of NIK and its recruitment to BIRC3 degradation. So, TRAF3 mutations also stabilize NIK to upregulate non-canonical NF-κB signaling [39].

Rare mutations (0.3%) affecting an IKK phosphorylation site on c-Rel's transactivation domain have been described in two B-cell lymphomas [40]. Cells expressing this mutation appeared less sensitive to TNF-α-induced apoptosis, and consequently may be more prone to transformation by other oncogenic events affecting these cells.

Diffuse large B-cell lymphomas (DLBCL) are the most common types of non-Hodgkin lymphoma (40% all adult cases). They are divided into three molecular sub-types: ABC (activated B-cell) GCB (germinal center B-cell) and PMBL (primary mediastinal B-cell lymphoma). Initial evidence for the role of the canonical NF-κB pathway in DLBCL came from gene expression profiling studies that revealed a significant enrichment for NF-κB target genes specifically in the ABC subgroup, which display the worst prognosis [41]. This was then confirmed by the detection of RelA/p50 complexes in the nuclei of DLBCL tumors [42]. Constitutive NF-κB activation in ABC-DLBCL can result from mutations in components of the BcR signaling cascade, as ABC-DLBCL cells exhibit a chronic BCR activation. It is sustained by mutations in the two chains of the CD79 complex which forms the B-cell receptor with the membrane IgM molecules: CD79B chains are affected in 20% of the samples, whereas mutations in CD79A are less frequent. These mutations occur in the immunoreceptor tyrosine-based motif

(ITAM)—a unique module linking antigen and Fc receptors to downstream tyrosine kinase signaling cascades, in most cases replacing the first tyrosine residue (Y196) within the cytoplasmic tail [43]. In around 9% of ABC-DLBCL (and a smaller subset of GCB-DLBCL), activation of BCR and NF-κB can be attributed to gain of function mutations within the coiled-coil domain of the *CARD11/CARMA1* gene [44] component of the CBM complex. Finally, genetic gains/amplifications of the *BIRC2* and *BIRC3* loci, encoding the cIAP1 and cIAP2 E3 ubiquitin ligases, respectively, were detected in as many as 16% of ABC-DLBCL, but rarely in GCB-DLBCL [45]. In the canonical pathway, cIAP1/2 are now recognized as integral components of CBM signaling. In ABC-DLBCL, elevated cIAP1/2 leads to K63 auto-ubiquitination and thus controls the recruitment of LUBAC and IKK to the CBM complex, thereby inducing IKK2 activation and increasing NF-κB levels.

The *MYD88* gene coding for the adapter of TLRs is mutated in 30% of ABC-DLBC. The L265P MYD88 mutation induces a spontaneous activation of the downstream IRAK complex, leading to engagement of the NF-κB pathway [46].

Bi-allelic truncating mutations or deletions have been observed in the *TNFAIP3*/A20 gene in one third of ABC-DLBCL cases and fewer GBC-DLBCL cases [47]. TNFAIP3/A20 possesses dual ubiquitin-editing functions. In particular, the N-terminal domain of TNFAIP3 is a deubiquitinating enzyme for K63-linked polyubiquitinated signaling mediators such as TRAF2/6 and RIP1 (Receptor Interacting Protein 1), while its C-terminal domain is a E3 ubiquitin ligase for K48-linked degradative polyubiquitination of the same substrates. TNFAIP3 has a specificity for particular polyubiquitinated substrates to regulate NF-κB activation in the TNF, IL-1β, and TLR pathways. TNFAIP3 mutations likely contribute to lymphomagenesis by inducing unregulated prolonged NF-κB responses. Some DLBCL patient biopsies show a deregulation of the glycolytic enzyme GAPDH that resulted in the activation of a NF-κB/HIF-1α (Hypoxia inducible factor 1) axis that could support lymphoma growth and vascularization through the induction of *VEGFR* [48].

The non-canonical NF-κB pathway is also aberrantly dysregulated in DLBCL, as 10–15% of GCB- and ABC-DLBCL carry genetic lesions in TRAF2 and TRAF3 associated with an activation of this pathway [49]. Recently, a high expression of the atypical nuclear IκBζ was observed in ABC-DLCBL [50]. The underlying molecular mechanisms for this phenomenon remain unclear. However, the increase in IκBζ is important for lymphoma survival by inducing p50–p52 homodimer target genes.

Hodgkin lymphoma (HL) is an unusual type of lymphoid malignancy: less than 1% of the cells in the affected lymph node correspond to malignant tumor cells—so-called Hodgkin and Reed/Stenberg cells (HRS). Constitutive activity of NF-κB was first identified 20 years ago in Hodgkin's lymphoma-derived cell lines, as well as primary HRS cells [51,52]. Since then, numerous studies have been conducted and there is now evidence that both canonical and non-canonical pathways are enhanced in HRS cells. Indeed, RelB/p52 complexes can be found in HRS cell lines, indicating that the non-canonical pathway contributes to the neoplastic feature of HL [53]. This is further supported by the detection of NIK protein in HL cell lines and primary HRS cells [54].

HRS cells carry genetic lesions that lead to a gain or loss of function for several NF-κB pathway components. In about 50% of HL cases, the *REL* gene displays gains or amplifications. As a consequence, the levels of nuclear c-Rel are increased in HRS cells, likely contributing to constitutive NF-κB activation [55]. The two inhibitors IκBα and IκBε are also targets of mutations. Indeed, 10–20% of classical HL exhibit somatic inactivating mutations in the *IκBα* gene (*NFKBIA*), whereas mutations in the *IκBε* gene (*NFKBIE*) have been described in only one HL cell line and in one case of classical HL. This suggests that the genes encoding for NF-κB inhibitors may be considered as tumor suppressors in HL [56,57]. Some *BCL3* copy number gains leading to elevated expression of the proto-oncogene have been identified [58]. Bcl3 can enhance canonical NF-κB transcription and target gene expression after binding to p50 homodimers. Finally, recurrent deletions of the chromosomal region 6q23 involving *TNFAIP3/A20* have been found in HL cell lines and HRS cells, as well as inactivating *TNFAIP3* somatic mutations in several HL cell lines and in about 45% of classical Epstein-Barr virus (EBV)-negative HL cases [59].

Cell extrinsic mechanisms are also important contributors to NF-κB constitutive activation in HRS cells. Indeed, in a lymph node, the HRS malignant cells (less than 1% of the tumor mass) are surrounded by infiltrating T and B lymphocytes, monocytes, eosinophils, macrophages, and dendritic cells. These immune cells are attracted to the tumor site and activated by numerous cytokines/chemokines secreted by HRS cells, and in turn express various ligands that favor tumor survival. For example, CD40- and CD40L-expressing CD4$^+$ T-cells are abundant in the HL microenvironment. These T cells surround HRS cells to form rosettes, suggesting a direct engagement of CD40 and thereby its activation [60]. Stimulation of CD40 induces NF-κB signaling in HRS cells and promotes the expression of its target genes coding for CD40 itself, the tandem molecules CD80 and CD86 involved in the priming of T-cells by dendritic cells, and of a set of anti-apoptotic proteins (Bfl1/A1, Bcl-xL, cIAP1) [61,62]. Another important player in the HRS-microenvironment crosstalk is RANK (receptor activator of NF-κB)—a member of the TNFR family capable of activating NF-κB. RANK and its ligand RANKL are expressed by HRS cells, suggesting an autocrine stimulation. In addition, RANKL is primarily expressed by activated T cells and dendritic cells. In HRS cell lines the activity of NF-κB is positively correlated with the level of RANK expression, suggesting that RANK/RANKL may contribute to the constitutive NF-κB signaling [63].

HRS cells express both CD30 and its receptor CD30L (CD153), suggesting another autocrine mechanism. However, the overexpression of CD30 leads to the recruitment and self-aggregation of TRAF2 and TRAF5, resulting in constitutive NF-κB activation and cytokine expression, independently of CD30L [64]. In HRS cell lines, TRAF1 is induced by CD30 signaling. This leads to TRAF-dependent processing of p100, increased p52 levels, and hence activation of the non-canonical NF-κB pathway [65]. Aberrant Notch activity is an also essential upstream regulator of non-canonical NF-κB activation in HRS cells, as NOTCH1 induces the processing of the *NFKB2* gene product p100 into its p52 active form and leads to an enhanced DNA binding activity of RelB/p52 heterodimers [66].

In about 30–40% of classical Hodgkin lymphoma in the Western world, HRS cells are latently infected by Epstein-Barr virus (EBV). The EBV$^+$ HRS cells express three latent proteins: the EBV nuclear antigen (EBNA1) and the latent membrane proteins 1 and 2 (LMP1 and LMP2). The cytosolic domains of LMP1 carry two carboxyl terminal activating regions (CTAR) which bind TRAF proteins to mediate NF-κB activation, mimicking a constitutively-active CD40 receptor [67]. The important role of EBV in constitutive NF-κB activity is further supported by the observation that inactivating mutations in *TNFAIP3*/A20 gene and EBV positivity are largely mutually exclusive. In addition, *NFKBIA* mutations are mostly found in EBV-negative cells, suggesting that EBV infection can replace the role of some genetic lesions in HRS cells.

Multiple myeloma (MM) is an incurable plasma cell malignancy accounting for 13% of all hematological cancers. Disease progression involves clonal expansion of transformed plasma cells into tumors in the bone marrow. The heterogeneous tumor entity is characterized by long-lived plasmatic B-cells in the bone marrow. Elevated levels of NF-κB activity were found in relapsing MM, suggesting that NF-κB could be used as a prognostic marker as well as a target for therapy to prevent progression of the disease [68]. Constitutive activity of NF-κB has been found in multiple myeloma cell lines—especially in chemoresistant cell lines [69]. The same lab has also identified eight polymorphisms of the *IκBα* gene that were more frequently present in patients with MM [70]. The most frequent polymorphisms are located within exon 1, which encodes the N-terminal domain of IκBα containing the ser32/36, or within exon 6, encoding the PEST region. Gene expression analyses showed that 80% of myeloma biopsies display a high amount of *RelA* that correlates with enhanced expression of anti-apoptotic genes [71]. Keats et al. identified a promiscuous array of mutations that result in the constitutive activation of the non-canonical NF-κB pathway in approximately 20% of MM patients (inactivation of *TRAF2, TRAF3, CYLD, cIAP1/cIAP2*, and activation of *NFKB1, NFKB2, CD40*,

LTBR, TACI, and *NIK*) [72]. Another clue for the implication of the non-canonical pathway is the strong nuclear accumulation of RelB in primary MM samples [73].

In MM, the crosstalk between malignant cells and the bone marrow stromal cells plays an important role in the pathogenesis of the disease, and largely affects the status of NF-κB. Indeed, NF-κB can be activated in MM cells by diverse bone marrow-derived cytokines and growth factors (e.g., BAFF/BAFF-R [74], APRIL/BCMA [75], CD40L/CD40 [76]) and by direct physical contact between MM cells and stromal cells [77].

3. Myeloid Malignancies

Acute myeloid leukemia (AML) represents a heterogeneous group of clonal stem cell malignancies arising from so-called leukemic stem cells (LSC). LSCs give rise to leukemic myeloid blasts arrested at different maturation steps. High proliferation rates of AML blasts are responsible for the invasion of bone marrow, and are associated with a fatal outcome.

The first observation of a constitutive NF-κB activity in AML was reported 16 years ago [78]. Indeed, Craig Jordan's lab showed that primitive CD34$^+$/CD38$^-$/CD123$^+$ AML cells aberrantly express active NF-κB, and that in vitro cell treatment with proteasome inhibitors is sufficient to induce rapid cell death. One year later, the first evidence of a dysregulation of IKK signaling in AML was published [79]. An increased level of IKK activity was observed in AML blasts derived from both BM and PB, associated with NF-κB activation. A more extensive activation was found in FAB (French-American-British) subtypes M4/M5, representing myelomonocytic/monocytic blasts compared to M1/M2.

Studies using the proteasome inhibitor (MG132) or the IKKβ inhibitor (AS602868) demonstrated that the pharmacological targeting of NF-κB induced cell death in AML cells both in vitro and in vivo, without affecting normal bone marrow CD34$^+$CD38$^-$ cells [78,80,81]. This led to the conclusion that AML cells are highly dependent on NF-κB signaling for their survival compared to normal bone marrow cells—a dependency that constitutes a therapeutic window for NF-κB targeting strategies.

Genetic alterations of *NF-κB* genes have not frequently been found in myeloid leukemia, suggesting that the origin of sustained NF-κB activation is different in AML than in other lymphoid malignancies. NF-κB activation is therefore a direct consequence of the chromosomal translocations/mutations that are characteristic of the different AML subtypes. The t(8;21) translocation is one of the most frequent ones found in AML, and produces the chimeric protein AML1-ETO. In normal hematopoietic cells, AML1 represses NF-κB signaling by interacting with the IKK complex, preventing their activation. AML1/ETO is unable to interfere with the activation of the IKK complex, leading to increased IKK activity, and consequently to aberrant NF-κB signaling [82].

Mutations in the *C/EBPα* (CCAAT/enhancer binding protein α) gene are detected in 10–15% of AML patients. C/EBPα and C/EBPα leucine zipper mutants could bind to the *p50* gene promoter, thereby elevating the expression of p50 protein. Moreover, p50 regulates *C/EBPα* expression in a positive feedback loop. It has also been shown that C/EBPα oncoproteins physically interact with p50 homodimers at the promoter site of the anti-apoptotic genes *BCL-2* and *C-FLIP*. As a consequence, C/EBPα proteins have the capacity to displace histone deacetylases (HDAC) from p50 homodimers to activate NF-κB target genes whose expression would otherwise be repressed [83,84].

Deletions involving the long arm of chromosome 5—del(5q)—are a common cytogenetic defect in high-risk MDS/AML (Myelodysplastic Syndrome/Acute Myeloid Leukemia). del(5q) MDS/AML employs an intrachromosomal gene network involving the loss of miR-146a and haploid overexpression of p62 via NF-κB to sustain a TRAF6/NF-κB signaling for cell survival and proliferation [85]. p62 facilitates K63-linked polyubiquitination of TRAF6, and consequently initiates NF-κB signaling.

The receptor tyrosine kinase Fms-like tyrosine kinase 3 (FLT3) is highly expressed in most patients with AML, and nearly 30% of them possess an internal tandem duplication (ITD) within the juxtamembrane domain that is associated with a poor outcome. The constitutive activation of

FLT3-ITD is responsible for IKK activation through phosphorylation, thereby inducing the canonical NF-κB pathway [86]. However, FLT3-ITD also enhanced the phosphorylation and activation of TAK1 to activate the non-canonical NF-κB pathway [87]. TAK1-activated p52 binds to HDAC to repress transcription of DAPK1 (Death-Associated Protein Kinase 1)—an essential player in endoplasmic reticulum stress-induced apoptosis.

Since a proportional subset of AMLs express high levels of various cytokines, including the NF-κB activators TNFα, IL-1, and IL-6, autocrine signaling is also involved in constitutive NF-κB activation [88,89].

Disease progression is supported by cross-talk between AML cells and the bone marrow microenvironment. It has recently been shown that the engagement of VCAM-1 (vascular cell adhesion molecule 1) on stromal cells with VLA-4 (very late antigen 4) on AML cells induces reciprocal NF-κB activation in leukemia and stromal cells and triggers stroma-mediated chemoresistance [90].

Chronic myeloid leukemia (CML) is characterized by the chromosomal Bcr-Abl translocation, leading to the expression of the Bcr-Abl fusion oncoprotein that exhibits constitutive tyrosine kinase activity. Almost 20 years ago, NF-κB was described as an essential component of Bcr-Abl signaling [91,92]. Five years later, Kirchner et al. showed an NF-κB activation in primary CML samples from blast crisis patients for the first time [93]. Using a Bcr-Abl-transduced BaF3 cell line—an IL-3-dependent murine pro B-cell line—they showed that NF-κB activation is IKK-independent, but instead requires Ras activation. Nevertheless, other studies confirmed a contribution of IKKβ to the constitutive NF-κB activity observed in CML primary cells and cell lines. In these studies, two IKKβ inhibitors (PS-1145 [94] and AS602868 [95]) were able to promote apoptosis of CML primary cells. In particular, AS602868 was also efficient on CML cells expressing a T315I Bcr-Abl mutant that is resistant to all known tyrosine kinase inhibitors [95].

The microenvironment also plays a key role in the regulation of NF-κB signaling in CML. For instance, TNF-α supports the survival of CML stem/progenitor cells by promoting NF-κB/RelA pathway activity and expression of the interleukin 3 and granulocyte/macrophage-colony stimulating factor common β-chain receptor [96]. In CML, Bcr-Abl up-regulates TGF-β 1 expression, and released TGF-β 1 activates a PI3K/Akt/NF-κB/MMP9 signaling pathway from the stroma that subsequently results in the release of s-KitL and s-ICAM-1 (InterCellular Adhesion Molecule 1), ultimately enhancing the recruitment and mobilization of tumor stem cells to the peripheral circulation [97].

4. Conclusions

While a constitutive activation of NF-κB appears to be a frequent event in most (but not all) hematologic malignancies, it is amazing to observe that it results from a great variety of mechanisms and affects both the canonical and the non-canonical pathways. It is surprising that only rare mutations have been reported in the genes coding for the kinases that govern the two flavors of NF-κB activation—namely, IKKα and IKKβ for the classical pathway, and NIK and IKKα for the atypical one. For instance, it would have been expected that a mutated active IKKα that could activate both NF-κB pathways would have been found in many cases. Activating mutations have been reported for IKKβ in few (8/117) cases of MZL. It is possible that the active forms of these three kinases are too deleterious to be selected by cancer cells. In the opposite situation, inactivating mutations and deletions affecting the genes for IKKα, IKKβ, and NEMO have been described as generating several inherited diseases that display immune or developmental defects [98]. Constitutive NF-κB activation results from four main mechanisms: the presence of NF-κB activators in the leukemic environment, activating mutations in positive NF-κB regulators, invalidating mutations in NF-κB negative regulators, and finally, genetic defects resulting in amplifications of mutations of several NF-κB subunits. This diversity likely allows the stimulation of specific functions of NF-κB pathways without being too harmful. The specific consequences of activation of either the canonical or the non-canonical pathways on transformed cells are not presently known. One possibility to explore is that they may combine with different oncogenic events.

Interestingly, in several situations, it appears that the dysregulation of NF-κB follows two phases. In the first, the abnormal and sustained NF-κB activation results from influences from the leukemic microenvironment that secretes NF-κB activators or from chronic inflammation due to a bacterial infection and/or antigenic stimulation. Then, in a second phase, the occurrence of genetic events provides an autonomous activation of NF-κB that appears to lock on NF-κB activation to support further and complete transformation of the cell. A genetic activation of NF-κB is likely to be stronger and more sustained than upon chronic inflammation, which may result in a higher level of transcription of target genes, but could also modify the quality of the repertoire of activated genes.

The diversity in the defects that lead to abnormal NF-κB activation makes the possibility of finding a universal target difficult. Targeting IKKβ to block the canonical NF-κB pathway has been tempered by the fact that inhibition of IKKβ leads to enhanced IL1β expression and excessive inflammation [99]. Nevertheless, the different kinases [100] or proteases such as MALT1 [101] that participate in NF-κB activation are being evaluated as potential targets.

Author Contributions: Véronique Imbert and Jean-François Peyron wrote the paper.

Conflicts of Interest: The authors declare no conflict of interest.

References

1. Scheidereit, C. IκB kinase complexes: Gateways to NF-κB activation and transcription. *Oncogene* **2006**, *25*, 6685–6705. [CrossRef] [PubMed]
2. Zhang, Q.; Lenardo, M.J.; Baltimore, D. 30 years of NF-κB: A blossoming of relevance to human pathobiology. *Cell* **2017**, *168*, 37–57. [CrossRef] [PubMed]
3. Ben-Neriah, Y.; Karin, M. Inflammation meets cancer, with NF-κB as the matchmaker. *Nat. Immunol.* **2011**, *12*, 715–723. [CrossRef] [PubMed]
4. Ruland, J. Return to homeostasis: Downregulation of NF-κB responses. *Nat. Immunol.* **2011**, *12*, 709–714. [CrossRef] [PubMed]
5. Oeckinghaus, A.; Hayden, M.S.; Ghosh, S. Crosstalk in NF-κB signaling pathways. *Nat. Immunol.* **2011**, *12*, 695–708. [CrossRef] [PubMed]
6. Mitchell, S.; Vargas, J.; Hoffmann, A. Signaling via the NF-κB system. *Wiley Interdiscip. Rev. Syst. Biol. Med.* **2016**, *8*, 227–241. [CrossRef] [PubMed]
7. Hymowitz, S.G.; Wertz, I.E. A20: From ubiquitin editing to tumour suppression. *Nat. Rev. Cancer* **2010**, *10*, 332–341. [CrossRef] [PubMed]
8. Sun, S.C. The noncanonical NF-κB pathway. *Immunol. Rev.* **2012**, *246*, 125–140. [CrossRef] [PubMed]
9. Perkins, N.D. The diverse and complex roles of NF-κB subunits in cancer. *Nat. Rev. Cancer* **2012**, *12*, 121–132. [CrossRef] [PubMed]
10. Cuní, S.; Pérez-Aciego, P.; Pérez-Chacón, G.; Vargas, J.A.; Sánchez, A.; Martín-Saavedra, F.M.; Ballester, S.; García-Marco, J.; Jordá, J.; Durántez, A. A sustained activation of PI3K/NF-κB pathway is critical for the survival of chronic lymphocytic leukemia B cells. *Leukemia* **2004**, *18*, 1391–1400. [CrossRef] [PubMed]
11. Hewamana, S.; Alghazal, S.; Lin, T.T.; Clement, M.; Jenkins, C.; Guzman, M.L.; Jordan, C.T.; Neelakantan, S.; Crooks, P.A.; Burnett, A.K.; et al. The NF-κB subunit Rel A is associated with in vitro survival and clinical disease progression in chronic lymphocytic leukemia and represents a promising therapeutic target. *Blood* **2008**, *111*, 4681–4689. [CrossRef] [PubMed]
12. Herishanu, Y.; Pérez-Galán, P.; Liu, D.; Biancotto, A.; Pittaluga, S.; Vire, B.; Gibellini, F.; Njuguna, N.; Lee, E.; Stennett, L.; et al. The lymph node microenvironment promotes B-cell receptor signaling, NF-κB activation, and tumor proliferation in chronic lymphocytic leukemia. *Blood* **2011**, *117*, 563–574. [CrossRef] [PubMed]
13. Rosén, A.; Murray, F.; Evaldsson, C.; Rosenquist, R. Antigens in chronic lymphocytic leukemia—Implications for cell origin and leukemogenesis. *Semin. Cancer Biol.* **2010**, *20*, 400–409. [CrossRef] [PubMed]
14. Puente, X.S.; Pinyol, M.; Quesada, V.; Conde, L.; Ordóñez, G.R.; Villamor, N.; Escaramis, G.; Jares, P.; Beà, S.; González-Díaz, M.; et al. Whole-genome sequencing identifies recurrent mutations in chronic lymphocytic leukaemia. *Nature* **2011**, *475*, 101–105. [CrossRef] [PubMed]

15. Mansouri, L.; Sutton, L.A.; Ljungström, V.; Bondza, S.; Arngården, L.; Bhoi, S.; Larsson, J.; Cortese, D.; Kalushkova, A.; Plevova, K.; et al. Functional loss of IκBε leads to NF-κB deregulation in aggressive chronic lymphocytic leukemia. *J. Exp. Med.* **2015**, *212*, 833–843. [CrossRef] [PubMed]
16. Baliakas, P.; Hadzidimitriou, A.; Sutton, L.A.; Rossi, D.; Minga, E.; Villamor, N.; Larrayoz, M.; Kminkova, J.; Agathangelidis, A.; Davis, Z.; et al. Recurrent mutations refine prognosis in chronic lymphocytic leukemia. *Leukemia* **2015**, *29*, 329–336. [CrossRef] [PubMed]
17. Chiaretti, S.; Marinelli, M.; del Giudice, I.; Bonina, S.; Piciocchi, A.; Messina, M.; Vignetti, M.; Rossi, D.; di Maio, V.; Mauro, F.R.; et al. NOTCH1, SF3B1, BIRC3 and TP53 mutations in patients with chronic lymphocytic leukemia undergoing first-line treatment: Correlation with biological parameters and response to treatment. *Leuk. Lymphoma* **2014**, *55*, 2785–2792. [CrossRef] [PubMed]
18. Hewamana, S.; Lin, T.T.; Rowntree, C.; Karunanithi, K.; Pratt, G.; Hills, R.; Fegan, C.; Brennan, P.; Pepper, C. Rel a is an independent biomarker of clinical outcome in chronic lymphocytic leukemia. *J. Clin. Oncol.* **2009**, *27*, 763–769. [CrossRef] [PubMed]
19. Wang, L.Q.; Kwong, Y.L.; Kho, C.S.; Wong, K.F.; Wong, K.Y.; Ferracin, M.; Calin, G.A.; Chim, C.S. Epigenetic inactivation of miR-9 family microRNAs in chronic lymphocytic leukemia—Implications on constitutive activation of NF-κB pathway. *Mol. Cancer* **2013**, *12*, 173. [CrossRef] [PubMed]
20. Baer, C.; Oakes, C.C.; Ruppert, A.S.; Claus, R.; Kim-Wanner, S.Z.; Mertens, D.; Zenz, T.; Stilgenbauer, S.; Byrd, J.C.; Plass, C. Epigenetic silencing of miR-708 enhances NF-κB signaling in chronic lymphocytic leukemia. *Int. J. Cancer* **2015**, *137*, 1352–1361. [CrossRef] [PubMed]
21. Kordes, U.; Krappmann, D.; Heissmeyer, V.; Ludwig, W.D.; Scheidereit, C. Transcription factor NF-κB is constitutively activated in acute lymphoblastic leukemia cells. *Leukemia* **2000**, *14*, 399–402. [CrossRef] [PubMed]
22. Vilimas, T.; Mascarenhas, J.; Palomero, T.; Mandal, M.; Buonamici, S.; Meng, F.; Thompson, B.; Spaulding, C.; Macaroun, S.; Alegre, M.L.; et al. Targeting the NF-κB signaling pathway in Notch1-induced T-cell leukemia. *Nat. Med.* **2007**, *13*, 70–77. [CrossRef] [PubMed]
23. Espinosa, L.; Cathelin, S.; D'Altri, T.; Trimarchi, T.; Statnikov, A.; Guiu, J.; Rodilla, V.; Inglés-Esteve, J.; Nomdedeu, J.; Bellosillo, B.; et al. The Notch/Hes1 pathway sustains NF-κB activation through CYLD repression in T cell leukemia. *Cancer Cell* **2010**, *18*, 268–281. [CrossRef] [PubMed]
24. Vacca, A.; Felli, M.P.; Palermo, R.; di Mario, G.; Calce, A.; di Giovine, M.; Frati, L.; Gulino, A.; Screpanti, I. Notch3 and pre-TCR interaction unveils distinct NF-κB pathways in T-cell development and leukemia. *EMBO J.* **2006**, *25*, 1000–1008. [CrossRef] [PubMed]
25. Hsieh, M.Y.; van Etten, R.A. IKK-dependent activation of NF-κB contributes to myeloid and lymphoid leukemogenesis by BCR-ABL1. *Blood* **2014**, *123*, 2401–2411. [CrossRef] [PubMed]
26. Harhaj, E.W.; Good, L.; Xiao, G.; Uhlik, M.; Cvijic, M.E.; Rivera-Walsh, I.; Sun, S.C. Somatic mutagenesis studies of NF-κ B signaling in human T cells: Evidence for an essential role of IKK γ in NF-κB activation by T-cell costimulatory signals and HTLV-I Tax protein. *Oncogene* **2000**, *19*, 1448–1456. [CrossRef] [PubMed]
27. Shembade, N.; Harhaj, N.S.; Yamamoto, M.; Akira, S.; Harhaj, E.W. The human T-cell leukemia virus type 1 Tax oncoprotein requires the ubiquitin-conjugating enzyme Ubc13 for NF-κB activation. *J. Virol.* **2007**, *81*, 13735–13742. [CrossRef] [PubMed]
28. Journo, C.; Filipe, J.; About, F.; Chevalier, S.A.; Afonso, P.V.; Brady, J.N.; Flynn, D.; Tangy, F.; Israël, A.; Vidalain, P.O.; et al. NRP/Optineurin Cooperates with TAX1BP1 to potentiate the activation of NF-κB by human T-lymphotropic virus type 1 tax protein. *PLoS Pathog.* **2009**, *5*, e1000521. [CrossRef] [PubMed]
29. Wu, X.; Zhang, M.; Sun, S.C. Mutual regulation between deubiquitinase CYLD and retroviral oncoprotein Tax. *Cell Biosci.* **2011**, *1*, 27. [CrossRef] [PubMed]
30. Tracey, L.; Streck, C.J.; Du, Z.; Williams, R.F.; Pfeffer, L.M.; Nathwani, A.C.; Davidoff, A.M. NF-κB activation mediates resistance to IFN β in MLL-rearranged acute lymphoblastic leukemia. *Leukemia* **2010**, *24*, 806–812. [CrossRef] [PubMed]
31. Wotherspoon, A.C.; Doglioni, C.; Diss, T.C.; Pan, L.; Moschini, A.; de Boni, M.; Isaacson, P.G. Regression of primary low-grade B-cell gastric lymphoma of mucosa-associated lymphoid tissue type after eradication of Helicobacter pylori. *Lancet* **1993**, *342*, 575–577. [CrossRef]
32. Lucas, P.C.; Kuffa, P.; Gu, S.; Kohrt, D.; Kim, D.S.; Siu, K.; Jin, X.; Swenson, J.; McAllister-Lucas, L.M. A dual role for the API2 moiety in API2-MALT1-dependent NF-κB activation: Heterotypic oligomerization and TRAF2 recruitment. *Oncogene* **2007**, *26*, 5643–5654. [CrossRef] [PubMed]

33. Rosebeck, S.; Madden, L.; Jin, X.; Gu, S.; Apel, I.J.; Appert, A.; Hamoudi, R.A.; Noels, H.; Sagaert, X.; van Loo, P.; et al. Cleavage of NIK by the API2-MALT1 fusion oncoprotein leads to noncanonical NF-κB activation. *Science* **2011**, *331*, 468–472. [CrossRef] [PubMed]

34. Ye, H.; Gong, L.; Liu, H.; Hamoudi, R.A.; Shirali, S.; Ho, L.; Chott, A.; Streubel, B.; Siebert, R.; Gesk, S.; et al. MALT lymphoma with t(14;18)(q32;q21)/IGH-MALT1 is characterized by strong cytoplasmic MALT1 and BCL10 expression. *J. Pathol.* **2005**, *205*, 293–301. [CrossRef] [PubMed]

35. Liu, H.; Ye, H.; Dogan, A.; Ranaldi, R.; Hamoudi, R.A.; Bearzi, I.; Isaacson, P.G.; Du, M.Q. T(11;18)(q21;q21) is associated with advanced mucosa-associated lymphoid tissue lymphoma that expresses nuclear BCL10. *Blood* **2001**, *98*, 1182–1187. [CrossRef] [PubMed]

36. Parry, M.; Rose-Zerilli, M.J.; Ljungström, V.; Gibson, J.; Wang, J.; Walewska, R.; Parker, H.; Parker, A.; Davis, Z.; Gardiner, A.; et al. Genetics and prognostication in splenic marginal zone lymphoma: Revelations from deep sequencing. *Clin. Cancer Res.* **2015**, *21*, 4174–4183. [CrossRef] [PubMed]

37. Rossi, D.; Deaglio, S.; Dominguez-Sola, D.; Rasi, S.; Vaisitti, T.; Agostinelli, C.; Spina, V.; Bruscaggin, A.; Monti, S.; Cerri, M.; et al. Alteration of *BIRC3* and multiple other NF-κB pathway genes in splenic marginal zone lymphoma. *Blood* **2011**, *118*, 4930–4934. [CrossRef] [PubMed]

38. Kai, X.; Chellappa, V.; Donado, C.; Reyon, D.; Sekigami, Y.; Ataca, D.; Louissaint, A.; Mattoo, H.; Joung, J.K.; Pillai, S. IκB kinase β (IKBKB) mutations in lymphomas that constitutively activate canonical nuclear factor κB (NF-κB) signaling. *J. Biol. Chem.* **2014**, *289*, 26960–26972. [CrossRef] [PubMed]

39. Spina, V.; Rossi, D. NF-κB deregulation in splenic marginal zone lymphoma. *Semin. Cancer Biol.* **2016**, *39*, 61–67. [CrossRef] [PubMed]

40. Starczynowski, D.T.; Trautmann, H.; Pott, C.; Harder, L.; Arnold, N.; Africa, J.A.; Leeman, J.R.; Siebert, R.; Gilmore, T.D. Mutation of an IKK phosphorylation site within the transactivation domain of REL in two patients with B-cell lymphoma enhances REL's in vitro transforming activity. *Oncogene* **2007**, *26*, 2685–2694. [CrossRef] [PubMed]

41. Davis, R.E.; Brown, K.D.; Siebenlist, U.; Staudt, L.M. Constitutive nuclear factor κB activity is required for survival of activated B cell-like diffuse large B cell lymphoma cells. *J. Exp. Med.* **2001**, *194*, 1861–1874. [CrossRef] [PubMed]

42. Compagno, M.; Lim, W.K.; Grunn, A.; Nandula, S.V.; Brahmachary, M.; Shen, Q.; Bertoni, F.; Ponzoni, M.; Scandurra, M.; Califano, A.; et al. Mutations of multiple genes cause deregulation of NF-κB in diffuse large B-cell lymphoma. *Nature* **2009**, *459*, 717–721. [CrossRef] [PubMed]

43. Davis, R.E.; Ngo, V.N.; Lenz, G.; Tolar, P.; Young, R.M.; Romesser, P.B.; Kohlhammer, H.; Lamy, L.; Zhao, H.; Yang, Y.; et al. Chronic active B-cell-receptor signalling in diffuse large B-cell lymphoma. *Nature* **2010**, *463*, 88–92. [CrossRef] [PubMed]

44. Lenz, G.; Davis, R.E.; Ngo, V.N.; Lam, L.; George, T.C.; Wright, G.W.; Dave, S.S.; Zhao, H.; Xu, W.; Rosenwald, A.; et al. Oncogenic CARD11 mutations in human diffuse large B cell lymphoma. *Science* **2008**, *319*, 1676–1679. [CrossRef] [PubMed]

45. Yang, Y.; Kelly, P.; Shaffer, A.L.; Schmitz, R.; Yoo, H.M.; Liu, X.; Huang, D.W.; Webster, D.; Young, R.M.; Nakagawa, M.; et al. Targeting non-proteolytic protein ubiquitination for the treatment of diffuse large B cell lymphoma. *Cancer Cell* **2016**, *29*, 494–507. [CrossRef] [PubMed]

46. Ngo, V.N.; Young, R.M.; Schmitz, R.; Jhavar, S.; Xiao, W.; Lim, K.H.; Kohlhammer, H.; Xu, W.; Yang, Y.; Zhao, H.; et al. Oncogenically active MYD88 mutations in human lymphoma. *Nature* **2011**, *470*, 115–119. [CrossRef] [PubMed]

47. Kato, M.; Sanada, M.; Kato, I.; Sato, Y.; Takita, J.; Takeuchi, K.; Niwa, A.; Chen, Y.; Nakazaki, K.; Nomoto, J.; et al. Frequent inactivation of A20 in B-cell lymphomas. *Nature* **2009**, *459*, 712–716. [CrossRef] [PubMed]

48. Chiche, J; Pommier, S; Beneteau, M; Mondragón, L; Meynet, O; Zunino, B; Mouchotte, A; Verhoeyen, E; Guyot, M; Pagès, G.; et al. GAPDH enhances the aggressiveness and the vascularization of non-Hodgkin's B lymphomas via NF-κB-dependent induction of HIF-1α. *Leukemia* **2015**, *29*, 1163–1176. [CrossRef] [PubMed]

49. Zhang, B.; Calado, D.P.; Wang, Z.; Fröhler, S.; Köchert, K.; Qian, Y.; Koralov, S.B.; Schmidt-Supprian, M.; Sasaki, Y.; Unitt, C.; et al. An oncogenic role for alternative NF-κB signaling in DLBCL revealed upon deregulated BCL6 expression. *Cell Rep.* **2015**, *11*, 715–726. [CrossRef] [PubMed]

50. Nogai, H.; Wenzel, S.S.; Hailfinger, S.; Grau, M.; Kaergel, E.; Seitz, V.; Wollert-Wulf, B.; Pfeifer, M.; Wolf, A.; Frick, M.; et al. IκB-ζ controls the constitutive NF-κB target gene network and survival of ABC DLBCL. *Blood* **2013**, *122*, 2242–2250. [CrossRef] [PubMed]

51. Bargou, R.C.; Emmerich, F.; Krappmann, D.; Bommert, K.; Mapara, M.Y.; Arnold, W.; Royer, H.D.; Grinstein, E.; Greiner, A.; Scheidereit, C.; et al. Constitutive nuclear factor-κB-RelA activation is required for proliferation and survival of Hodgkin's disease tumor cells. *J. Clin. Investig.* **1997**, *100*, 2961–2969. [CrossRef] [PubMed]
52. Krappmann, D.; Emmerich, F.; Kordes, U.; Scharschmidt, E.; Dörken, B.; Scheidereit, C. Molecular mechanisms of constitutive NF-κB/Rel activation in Hodgkin/Reed-Sternberg cells. *Oncogene* **1999**, *18*, 943–953. [CrossRef] [PubMed]
53. Nonaka, M; Horie, R; Itoh, K; Watanabe, T; Yamamoto, N; Yamaoka, S. Aberrant NF-κB2/p52 expression in Hodgkin/Reed-Sternberg cells and CD30-transformed rat fibroblasts. *Oncogene* **2005**, *24*, 3976–3986.
54. Ranuncolo, S.M.; Pittaluga, S.; Evbuomwan, M.O.; Jaffe, E.S.; Lewis, B.A. Hodgkin lymphoma requires stabilized NIK and constitutive RelB expression for survival. *Blood* **2012**, *120*, 3756–3763. [CrossRef] [PubMed]
55. Barth, T.F.; Martin-Subero, J.I.; Joos, S.; Menz, C.K.; Hasel, C.; Mechtersheimer, G.; Parwaresch, R.M.; Lichter, P.; Siebert, R.; Mööller, P. Gains of 2p involving the REL locus correlate with nuclear c-Rel protein accumulation in neoplastic cells of classical Hodgkin lymphoma. *Blood* **2003**, *101*, 3681–3686. [CrossRef] [PubMed]
56. Cabannes, E.; Khan, G.; Aillet, F.; Jarrett, R.F.; Hay, R.T. Mutations in the *IkBa* gene in Hodgkin's disease suggest a tumour suppressor role for *IκBα*. *Oncogene* **1999**, *18*, 3063–3070. [CrossRef] [PubMed]
57. Emmerich, F.; Theurich, S.; Hummel, M.; Haeffker, A.; Vry, M.S.; Döhner, K.; Bommert, K.; Stein, H.; Dörken, B. Inactivating IκB epsilon mutations in Hodgkin/Reed-Sternberg cells. *J. Pathol.* **2003**, *201*, 413–420. [CrossRef] [PubMed]
58. Martín-Subero, J.I.; Kreuz, M.; Bibikova, M.; Bentink, S.; Ammerpohl, O.; Wickham-Garcia, E.; Rosolowski, M.; Richter, J.; Lopez-Serra, L.; Ballestar, E.; et al. New insights into the biology and origin of mature aggressive B-cell lymphomas by combined epigenomic, genomic, and transcriptional profiling. *Blood* **2009**, *113*, 2488–2497. [CrossRef] [PubMed]
59. Schmitz, R.; Hansmann, M.L.; Bohle, V.; Martin-Subero, J.I.; Hartmann, S.; Mechtersheimer, G.; Klapper, W.; Vater, I.; Giefing, M.; Gesk, S.; et al. *TNFAIP3* (A20) is a tumor suppressor gene in Hodgkin lymphoma and primary mediastinal B cell lymphoma. *J. Exp. Med.* **2009**, *206*, 981–989. [CrossRef] [PubMed]
60. Carbone, A.; Gloghini, A.; Gruss, H.J.; Pinto, A. CD40 ligand is constitutively expressed in a subset of T cell lymphomas and on the microenvironmental reactive T cells of follicular lymphomas and Hodgkin's disease. *Am. J. Pathol.* **1995**, *147*, 912–922. [PubMed]
61. Annunziata, C.M.; Safiran, Y.J.; Irving, S.G.; Kasid, U.N.; Cossman, J. Hodgkin disease: Pharmacologic intervention of the CD40-NF-κB pathway by a protease inhibitor. *Blood* **2000**, *96*, 2841–2848. [PubMed]
62. Hinz, M.; Löser, P.; Mathas, S.; Krappmann, D.; Dörken, B.; Scheidereit, C. Constitutive NF-κB maintains high expression of a characteristic gene network, including CD40, CD86, and a set of antiapoptotic genes in Hodgkin/Reed-Sternberg cells. *Blood* **2001**, *97*, 2798–2807. [CrossRef] [PubMed]
63. Fiumara, P.; Snell, V.; Li, Y.; Mukhopadhyay, A.; Younes, M.; Gillenwater, A.M.; Cabanillas, F.; Aggarwal, B.B.; Younes, A. Functional expression of receptor activator of nuclear factor κB in Hodgkin disease cell lines. *Blood* **2001**, *98*, 2784–2790. [CrossRef] [PubMed]
64. Horie, R; Watanabe, T; Ito, K; Morisita, Y; Watanabe, M; Ishida, T; Higashihara, M; Kadin, M.; Watanabe, T. Cytoplasmic aggregation of TRAF2 and TRAF5 proteins in the Hodgkin-Reed-Sternberg cells. *Am. J. Pathol.* **2002**, *160*, 1647–1654. [CrossRef]
65. Guo, F.; Sun, A.; Wang, W.; He, J.; Hou, J.; Zhou, P.; Chen, Z. TRAF1 is involved in the classical NF-κB activation and CD30-induced alternative activity in Hodgkin's lymphoma cells. *Mol. Immunol.* **2009**, *46*, 2441–2448. [CrossRef] [PubMed]
66. Schwarzer, R.; Dörken, B.; Jundt, F. Notch is an essential upstream regulator of NF-κB and is relevant for survival of Hodgkin and Reed-Sternberg cells. *Leukemia* **2012**, *26*, 806–813. [CrossRef] [PubMed]
67. Huen, D.S.; Henderson, S.A.; Croom-Carter, D.; Rowe, M. The Epstein-Barr virus latent membrane protein-1 (LMP1) mediates activation of NF-κB and cell surface phenotype via two effector regions in its carboxy-terminal cytoplasmic domain. *Oncogene* **1995**, *10*, 549–560. [PubMed]
68. Feinman, R.; Koury, J.; Thames, M.; Barlogie, B.; Epstein, J.; Siegel, D.S. Role of NF-κB in the rescue of multiple myeloma cells from glucocorticoid-induced apoptosis by bcl-2. *Blood* **1999**, *93*, 3044–3052. [PubMed]

69. Ma, M.H.; Yang, H.H.; Parker, K.; Manyak, S.; Friedman, J.M.; Altamirano, C.; Wu, Z.Q.; Borad, M.J.; Frantzen, M.; Roussos, E.; et al. The proteasome inhibitor PS-341 markedly enhances sensitivity of multiple myeloma tumor cells to chemotherapeutic agents. *Clin. Cancer Res.* **2003**, *9*, 1136–1144. [PubMed]

70. Parker, K.M.; Ma, M.H.; Manyak, S.; Altamirano, C.V.; Tang, Y.M.; Frantzen, M.; Mikail, A.; Roussos, E.; Sjak-Shie, N.; Vescio, R.A.; et al. Identification of polymorphisms of the *IκBα* gene associated with an increased risk of multiple myeloma. *Cancer Genet. Cytogenet.* **2002**, *137*, 43–48. [CrossRef]

71. Annunziata, C.M.; Davis, R.E.; Demchenko, Y.; Bellamy, W.; Gabrea, A.; Zhan, F.; Lenz, G.; Hanamura, I.; Wright, G.; Xiao, W.; et al. Frequent engagement of the classical and alternative NF-κB pathways by diverse genetic abnormalities in multiple myeloma. *Cancer Cell* **2007**, *12*, 115–130. [CrossRef] [PubMed]

72. Keats, J.J.; Fonseca, R.; Chesi, M.; Schop, R.; Baker, A.; Chng, W.J.; van Wier, S.; Tiedemann, R.; Shi, C.X.; Sebag, M.; et al. Promiscuous mutations activate the noncanonical NF-κB pathway in multiple myeloma. *Cancer Cell* **2007**, *12*, 131–144. [CrossRef] [PubMed]

73. Cormier, F.; Monjanel, H.; Fabre, C.; Billot, K.; Sapharikas, E.; Chereau, F.; Bordereaux, D.; Molina, T.J.; Avet-Loiseau, H.; Baud, V. Frequent engagement of RelB activation is critical for cell survival in multiple myeloma. *PLoS ONE* **2013**, *8*, e59127. [CrossRef] [PubMed]

74. Shen, X.; Zhu, W.; Zhang, X.; Xu, G.; Ju, S. A role of both NF-κB pathways in expression and transcription regulation of *BAFF-R* gene in multiple myeloma cells. *Mol. Cell. Biochem.* **2011**, *357*, 21–30. [CrossRef] [PubMed]

75. Tai, Y.T.; Acharya, C.; An, G.; Moschetta, M.; Zhong, M.Y.; Feng, X.; Cea, M.; Cagnetta, A.; Wen, K.; van Eenennaam, H.; et al. APRIL and BCMA promote human multiple myeloma growth and immunosuppression in the bone marrow microenvironment. *Blood* **2016**, *127*, 3225–3236. [CrossRef] [PubMed]

76. Tai, Y.T.; Li, X.; Tong, X.; Santos, D.; Otsuki, T.; Catley, L.; Tournilhac, O.; Podar, K.; Hideshima, T.; Schlossman, R.; et al. Human anti-CD40 antagonist antibody triggers significant antitumor activity against human multiple myeloma. *Cancer Res.* **2005**, *65*, 5898–5906. [CrossRef] [PubMed]

77. Landowski, T.H.; Olashaw, N.E.; Agrawal, D.; Dalton, W.S. Cell adhesion-mediated drug resistance (CAM-DR) is associated with activation of NF-κB (RelB/p50) in myeloma cells. *Oncogene* **2003**, *22*, 2417–2421. [CrossRef] [PubMed]

78. Guzman, M.L.; Neering, S.J.; Upchurch, D.; Grimes, B.; Howard, D.S.; Rizzieri, D.A.; Luger, S.M.; Jordan, C.T. Nuclear factor-κB is constitutively activated in primitive human acute myelogenous leukemia cells. *Blood* **2001**, *98*, 2301–2307. [CrossRef] [PubMed]

79. Baumgartner, B.; Weber, M.; Quirling, M.; Fischer, C.; Page, S.; Adam, M.; von Schilling, C.; Waterhouse, C.; Schmid, C.; Neumeier, D.; et al. Increased IκB kinase activity is associated with activated NF-κB in acute myeloid blasts. *Leukemia* **2002**, *16*, 2062–2071. [CrossRef] [PubMed]

80. Guzman, M.L.; Swiderski, C.F.; Howard, D.S.; Grimes, B.A.; Rossi, R.M.; Szilvassy, S.J.; Jordan, C.T. Preferential induction of apoptosis for primary human leukemic stem cells. *Proc. Natl. Acad. Sci. USA* **2002**, *99*, 16220–16225. [CrossRef] [PubMed]

81. Frelin, C.; Imbert, V.; Griessinger, E.; Peyron, A.C.; Rochet, N.; Philip, P.; Dageville, C.; Sirvent, A.; Hummelsberger, M.; Bérard, E.; et al. Targeting NF-κB activation via pharmacologic inhibition of IKK2-induced apoptosis of human acute myeloid leukemia cells. *Blood* **2005**, *105*, 804–811. [CrossRef] [PubMed]

82. Nakagawa, M.; Shimabe, M.; Watanabe-Okochi, N.; Arai, S.; Yoshimi, A.; Shinohara, A.; Nishimoto, N.; Kataoka, K.; Sato, T.; Kumano, K.; et al. AML1/RUNX1 functions as a cytoplasmic attenuator of NF-κB signaling in the repression of myeloid tumors. *Blood* **2011**, *118*, 6626–6637. [CrossRef] [PubMed]

83. Paz-Priel, I.; Ghosal, A.K.; Kowalski, J.; Friedman, A.D. C/EBPα or C/EBPα oncoproteins regulate the intrinsic and extrinsic apoptotic pathways by direct interaction with NF-κB p50 bound to the bcl-2 and FLIP gene promoters. *Leukemia* **2009**, *23*, 365–374. [CrossRef] [PubMed]

84. Paz-Priel, I.; Houng, S.; Dooher, J.; Friedman, A.D. C/EBPα and C/EBPα oncoproteins regulate *nfkb1* and displace histone deacetylases from NF-κB p50 homodimers to induce NF-κB target genes. *Blood* **2011**, *117*, 4085–4094. [CrossRef] [PubMed]

85. Fang, J.; Barker, B.; Bolanos, L.; Liu, X.; Jerez, A.; Makishima, H.; Christie, S.; Chen, X.; Rao, D.S.; Grimes, H.L.; et al. Myeloid malignancies with chromosome 5q deletions acquire a dependency on an intrachromosomal NF-κB gene network. *Cell Rep.* **2014**, *8*, 1328–1338. [CrossRef] [PubMed]

86. Grosjean-Raillard, J.; Adès, L.; Boehrer, S.; Tailler, M.; Fabre, C.; Braun, T.; de Botton, S.; Israel, A.; Fenaux, P.; Kroemer, G. Flt3 receptor inhibition reduces constitutive NF-κB activation in high-risk myelodysplastic syndrome and acute myeloid leukemia. *Apoptosis* **2008**, *13*, 1148–1161. [CrossRef] [PubMed]

87. Shanmugam, R.; Gade, P.; Wilson-Weekes, A.; Sayar, H.; Suvannasankha, A.; Goswami, C.; Li, L.; Gupta, S.; Cardoso, A.A.; Al Baghdadi, T.; et al. A noncanonical Flt3ITD/NF-κB signaling pathway represses DAPK1 in acute myeloid leukemia. *Clin. Cancer Res.* **2012**, *18*, 360–369. [CrossRef] [PubMed]

88. Dokter, W.H.; Tuyt, L.; Sierdsema, S.J.; Esselink, M.T.; Vellenga, E. The spontaneous expression of interleukin-1 β and interleukin-6 is associated with spontaneous expression of AP-1 and NF-κB transcription factor in acute myeloblastic leukemia cells. *Leukemia* **1995**, *9*, 425–432. [PubMed]

89. Kagoya, Y.; Yoshimi, A.; Kataoka, K.; Nakagawa, M.; Kumano, K.; Arai, S.; Kobayashi, H.; Saito, T.; Iwakura, Y.; Kurokawa, M. Positive feedback between NF-κB and TNF-α promotes leukemia-initiating cell capacity. *J. Clin. Investig.* **2014**, *124*, 528–542. [CrossRef] [PubMed]

90. Jacamo, R.; Chen, Y.; Wang, Z.; Ma, W.; Zhang, M.; Spaeth, E.L.; Wang, Y.; Battula, V.L.; Mak, P.Y.; Schallmoser, K.; et al. Reciprocal leukemia-stroma VCAM-1/VLA-4-dependent activation of NF-κB mediates chemoresistance. *Blood* **2014**, *123*, 2691–2702. [CrossRef] [PubMed]

91. Hamdane, M.; David-Cordonnier, M.H.; D'Halluin, J.C. Activation of p65 NF-κB protein by p210BCR-ABL in a myeloid cell line (P210BCR-ABL activates p65 NF-κB). *Oncogene* **1997**, *15*, 2267–2275. [CrossRef] [PubMed]

92. Reuther, J.Y.; Reuther, G.W.; Cortez, D.; Pendergast, A.M.; Baldwin, A.S. A requirement for NF-κB activation in Bcr-Abl-mediated transformation. *Genes Dev.* **1998**, *12*, 968–981. [CrossRef] [PubMed]

93. Kirchner, D.; Duyster, J.; Ottmann, O.; Schmid, R.M.; Bergmann, L.; Munzert, G. Mechanisms of Bcr-Abl-mediated NF-κB/Rel activation. *Exp. Hematol.* **2003**, *31*, 504–511. [CrossRef]

94. Cilloni, D.; Messa, F.; Arruga, F.; Defilippi, I.; Morotti, A.; Messa, E.; Carturan, S.; Giugliano, E.; Pautasso, M.; Bracco, E.; et al. The NF-κB pathway blockade by the IKK inhibitor PS1145 can overcome imatinib resistance. *Leukemia* **2006**, *20*, 61–67. [CrossRef] [PubMed]

95. Lounnas, N.; Frelin, C.; Gonthier, N.; Colosetti, P.; Sirvent, A.; Cassuto, J.P.; Berthier, F.; Sirvent, N.; Rousselot, P.; Dreano, M.; et al. NF-κB inhibition triggers death of imatinib-sensitive and imatinib-resistant chronic myeloid leukemia cells including T315I Bcr-Abl mutants. *Int. J. Cancer* **2009**, *125*, 308–317. [CrossRef] [PubMed]

96. Gallipoli, P.; Pellicano, F.; Morrison, H.; Laidlaw, K.; Allan, E.K.; Bhatia, R.; Copland, M.; Jørgensen, H.G.; Holyoake, T.L. Autocrine TNF-α production supports CML stem and progenitor cell survival and enhances their proliferation. *Blood* **2013**, *122*, 3335–3339. [CrossRef] [PubMed]

97. Zhu, X.; Wang, L.; Zhang, B.; Li, J.; Dou, X.; Zhao, R.C. TGF-β1-induced PI3K/Akt/NF-κB/MMP9 signalling pathway is activated in Philadelphia chromosome-positive chronic myeloid leukaemia hemangioblasts. *J. Biochem.* **2011**, *149*, 405–414. [CrossRef] [PubMed]

98. Senegas, A.; Gautheron, J.; Maurin, A.G.; Courtois, G. IKK-related genetic diseases: Probing NF-κB functions in humans and other matters. *Cell Mol. Life Sci.* **2015**, *72*, 1275–1287. [CrossRef] [PubMed]

99. Greten, F.R.; Arkan, M.C.; Bollrath, J.; Hsu, L.C.; Goode, J.; Miething, C.; Göktuna, S.I.; Neuenhahn, M.; Fierer, J.; Paxian, S.; et al. NF-κB is a negative regulator of IL-1β secretion as revealed by genetic and pharmacological inhibition of IKKβ. *Cell* **2007**, *130*, 918–931. [CrossRef] [PubMed]

100. Verstrepen, L.; Beyaert, R. Receptor proximal kinases in NF-κB signaling as potential therapeutic targets in cancer and inflammation. *Biochem. Pharmacol.* **2014**, *92*, 519–529. [CrossRef] [PubMed]

101. Demeyer, A.; Staal, J.; Beyaert, R. Targeting MALT1 Proteolytic Activity in Immunity, Inflammation and Disease: Good or Bad. *Trends Mol. Med.* **2016**, *22*, 135–150. [CrossRef] [PubMed]

MDPI

Review

The NF-κB Activating Pathways in Multiple Myeloma

Payel Roy †, Uday Aditya Sarkar and Soumen Basak *

Systems Immunology Laboratory, National Institute of Immunology, Aruna Asaf Ali Marg,
New Delhi 110067, India; payelroy87@nii.ac.in (P.R.); udayaditya2992@nii.ac.in (U.A.S.)
* Correspondence: sobasak@nii.ac.in
† Present Address: La Jolla Institute of Allergy and Immunology, La Jolla, CA 92037, USA.

Received: 12 March 2018; Accepted: 14 May 2018; Published: 16 May 2018

Abstract: Multiple myeloma(MM), an incurable plasma cell cancer, represents the second most prevalent hematological malignancy. Deregulated activity of the nuclear factor kappaB (NF-κB) family of transcription factors has been implicated in the pathogenesis of multiple myeloma. Tumor microenvironment-derived cytokines and cancer-associated genetic mutations signal through the canonical as well as the non-canonical arms to activate the NF-κB system in myeloma cells. In fact, frequent engagement of both the NF-κB pathways constitutes a distinguishing characteristic of myeloma. In turn, NF-κB signaling promotes proliferation, survival and drug-resistance of myeloma cells. In this review article, we catalog NF-κB activating genetic mutations and microenvironmental cues associated with multiple myeloma. We then describe how the individual canonical and non-canonical pathways transduce signals and contribute towards NF-κB -driven gene-expressions in healthy and malignant cells. Furthermore, we discuss signaling crosstalk between concomitantly triggered NF-κB pathways, and its plausible implication for anomalous NF-κB activation and NF-κB driven pro-survival gene-expressions in multiple myeloma. Finally, we propose that mechanistic understanding of NF-κB deregulations may provide for improved therapeutic and prognostic tools in multiple myeloma.

Keywords: NF-κB; multiple myeloma; canonical; non-canonical; mutations; microenvironment; cytokines; crosstalk; gene-expressions

---

1. General Introduction

Heterogeneous cancer-associated mutations often influence the interaction of malignant cells with their microenvironment that modifies therapeutic outcomes. Therefore, an understanding of the molecular mechanism underlying the coordinated functioning of genetic mutations and microenvironmental cues may have significance for the effective management of neoplastic diseases. Within the tumor microenvironment, immune cells as well as stromal cells secrete a diverse array of pro-inflammatory cytokines, which activate key pro-survival signaling pathways in malignant cells. Of particular importance is the NF-κB system, which forms a major link between cancer and inflammation. Interestingly, sequencing of cancer genomes revealed recurrent gain-of-function mutations in genes encoding key positive regulators of NF-κB signaling and inactivating genetic aberrations in negative regulators of this pathway. Multiple myeloma (MM), a plasma cell malignancy, provides one of the best examples where a number of mutations have been mapped onto the NF-κB pathway. In addition, microenvironment-derived signals were also shown to modulate NF-κB-dependent gene expressions in myeloma cells. Here, we briefly review multiple myeloma and the NF-κB signaling system. We then discuss the NF-κB-activating genetic lesions associated with MM and the role of the tumor microenvironment in reinforcing NF-κB signaling in cancerous cells. Finally, we elaborate interdependent regulations of NF-κB-activating pathways in MM.

2. Multiple Myeloma—Epidemiology and Aetiology

Multiple myeloma is the second most widespread hematologic malignancy after non-Hodgkin lymphoma, with a global estimate of 103,826 new cases and 72,453 mortalities annually [1]. The disease is more prevalent in men than in women, and the median age at diagnosis is 66 years [2]. The disease incidence also varies with ethnicity, being more prevalent among Caucasians than in Asians [1]. The American Cancer Society estimates that about 30,770 individuals (16,400 men and 14,370 women) will be diagnosed with this disease in the USA in 2018 (https://cancerstatisticscenter.cancer.org/#! /cancer-site/Myeloma; accessed on 10 March 2018). Indeed, it has been found that the incidence and the mortality rate are significantly higher in African Americans as compared to their Caucasian counterparts [3]. Various studies estimate that the incidence of MM in India is around 1.2–1.8 per 100,000 individuals.

MM is characterized by the clonal proliferation of cancerous plasma cells (PCs) in the bone marrow microenvironment and an associated increase in the level of monoclonal (M) protein in blood and serum [4]. Myeloma cells are characterized by high rates of somatic hypermutation of immunoglobulin (Ig) genes and isotype class switching, but differ from healthy PCs with respect to the abundance of certain cell surface molecules, including CD138 and CD38 [5,6]. In most cases, MM develops from monoclonal gammopathy of undetermined significance (MGUS) and smouldering MM (SMM) (Figure 1), conditions that involve high levels of M-protein and bone marrow (BM) plasmacytosis [7]. Clinical manifestations of MM include lytic bone lesions, hypercalcemia, cytopenia, renal dysfunction, hyperviscosity and peripheral neuropathy [8].

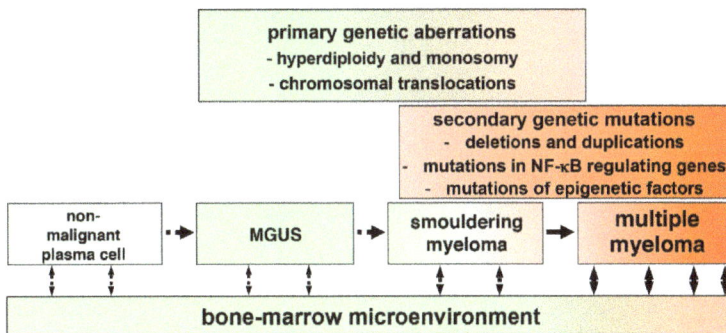

Figure 1. Development of multiple myeloma (MM) from plasma cells. Progression of MM from post-germinal center plasma cells to symptomatic myeloma occurs through intermediate MGUS (monoclonal gammopathy of undetermined significance), and SMM (smouldering MM) stages. Plasma cells associated with MGUS and SMM may display chromosomal abnormalities, which trigger cellular transformation. Cancerous plasma cells acquire additional, secondary genetic mutations, which support tumor growth by activating key signaling pathways in malignant cells. Finally, cell–cell communications, which involve physical interactions between cancerous cells and cancer-associated stromal cells, and secretion of cytokines and other soluble factors, promote survival and proliferation of myeloma cells within the bone marrow microenvironment.

Therapeutic intervention for MM currently involves six categories of medication—(1) immunomodulatory drugs (IMiDs) such as lenalidomide and pomalidomide; (2) proteasome inhibitors (PIs) such as bortezomib, carfilzomib, and ixazomib; (3) histone deacetylase inhibitors such as panobinostat; (4) monoclonal antibodies (MAbs) such as daratumumab and elotuzumab; (5) DNA alkylating agents; and (6) glucocorticosteroids [9]. Recent advancement in the management of the disease led to a substantial improvement in the 5-year survival rate in MM. The 5-year relative survival percent reported in the SEER database of the National Cancer Institute, USA steadily improved from 24.6% in between 1975–1977 to 52.4% in between 2008–2014 (https://seer.cancer.gov/csr/1975_ 2015/browse_csr.php?sectionSEL=18&pageSEL=sect_18_table.08.html; accessed on 14 May 2018). The disease, however, remains incurable because of widespread drug-resistance and relapse in most patients.

Although it is difficult to pinpoint the exact trigger of MM, karyotyping and DNA sequencing studies involving patient-derived cells or MMCLs identified several genomic translocations and mutations (Table 1 and Figure 1). The frequency and extent of these genetic aberrations were substantially augmented in individuals with advanced-stage disease or poor prognosis and in those with refractory MM unresponsive to therapy [10–13]. Nearly half of the MM tumors are hyperdiploid, characterized by trisomies of chromosomes 3, 5, 7, 9, 11, 15, 19 and 21 [14]. The non-hyperdiploid MM is associated with chromosomal translocations involving the immunoglobulin heavy-chain (IgH) locus and the loci encoding MMSET, FGFR3, CCND3 (cyclin D3), CCND1 (cyclin D1), MAF (c-Maf) or MAFB [15–20]. Moreover, duplications involving chromosome 1 and deletions affecting several other chromosomal arms have been associated with the onset of the disease [21–23]. Mutational events secondary to oncogenic transformation have also been reported in MM [24]. For example, frequent mutations have been observed in RAS-encoding *NRAS* and *KRAS* genes that triggered aberrant MAPK activity [25]. In addition, secondary mutations were mapped onto genes implicated in NF-κB regulations (discussed later), *CDKN2C* that encodes cyclin-dependent kinase inhibitor 2C [26], *TP53* expressing the tumor suppressor p53 [27] and *MYC* that codes for the proto-oncogene cMyc [28]. Disease progression is also influenced by physical or cytokine-mediated interactions between myeloma cells and the bone-marrow stromal cells. These cell-to-cell communications further inform the proliferative and anti-apoptotic program by modulating the NF-κB activity in myeloma cells.

The complex mutational landscape, which produces bioclinical heterogeneity, presents a significant challenge in the management of multiple myeloma. Risk-prediction based on traditional biomarkers measured using solely protein analysis and conventional cytologic assays often do not match with the actual patient outcomes and exhibit poor correlation with the minimal residual disease (MRD). Cutting-edge next-generation sequencing (NGS) technology has now made it possible to acquire a more comprehensive view of genomic alterations in MM. NGS not only offers genome-scale data but can also detect very low-frequency mutations [29]. Not surprisingly, NGS-based methods substantially improved the sensitivity of MRD detection in MM [30]. Furthermore, NGS provides for reliable measurement involving both bone-marrow samples and blood biopsies, enabling non-invasive diagnosis of MM. In fact, these high-throughput sequencing technologies helped to define the mutational landscape of bone-marrow resident as well as circulating myeloma cells at single-cell resolution [31]. As discussed later, NGS-based studies were also instrumental in charting NF-κB deregulating mutations in MM. In this article, we have focused on NF-κB deregulations in MM; for a more comprehensive description of MM and the underlying genetic as well as cell-signaling anomalies, please see [4,32].

Table 1. Genetic abnormalities in multiple myeloma (MM).

Genetic Abnormalities	Genes or Chromosomes Affected	Comments	References
Hyperdiploidy	chromosomes 3, 5, 7, 9, 11, 15, 19, 21	Functional role in the pathogenesis of MM remains elusive.	[12,14]
Monosomy	chromosome 13	Functional role of remains unclear.	[11]
Frequent IGH translocations	t(11;14)(q13;q32) *CCND1* t(4;14)(p16;q32) *FGFR3*	Upregulate the expression of oncogenes encoding cyclin D1 and fibroblast growth factor receptor 3.	[15,16]
	t(4;14)(p16;q32) *MMSET/WHSC1*	Upregulate the expression of *MMSET*, which methylates chromatin-associated proteins and modulates their functions.	[17]

Table 1. *Cont.*

Genetic Abnormalities	Genes or Chromosomes Affected	Comments	References
Relatively rare translocations	t(14;16)(q32;q23) *MAF* t(14;20)(q32;q11) *MAFB* t(6;14)(p21;q32) *CCND3*	These chromosomal translocations cause deregulated expressions of cell cycle regulators, including cyclin D2 and cyclin D3.	[18–20]
Duplication	chromosome 1 (1q)	Increased incidences in advanced MM, functional roles are unclear.	[23]
Deletions	1p, 6q, 8p, 12p, 14q, 16q, 17p, 20p	Functional role remains unclear.	[21,22]
	BIRC2 *BIRC3* *TRAF3* *CYLD*	Frequent homozygous deletions, which disrupt the function of various inhibitors of the NF-κB system.	[33,34]
	TNFAIP3	Heterozygous deletions, which inactivate the inhibitor of IKK, A20.	[35]
	CDKN2C	Abrogate the function of the tumor suppressor protein cyclin-dependent kinase inhibitor 2C.	[26]
Gene mutations	*NRAS* *KRAS* *BRAF*	Gain-of-function mutations in *NRAS* and *KRAS* trigger aberrant MAPK activity and are associated with the progression of MM, mutations in the *BRAF* oncogene promote myeloma growth.	[25]
	TP53	Abrogate the expression of p53 tumor suppressor in advanced MM.	[27]
8q24 locus rearrangements	*MYC*	Upregulate the expression of the *MYC* oncogene.	[28]

3. The NF-κB Signaling System

The nuclear factor kappa B (NF-κB) system functions in a wide variety of cells and coordinates innate and adaptive immune responses. Not surprisingly therefore, deregulated NF-κB activities have been implicated in several human ailments, including hematologic cancers [36,37]. The NF-κB family consists of five structurally related monomeric subunits: RelA (also called p65), RelB, c-Rel, p50 (encoded by *NFKB1* and produced as a precursor protein p105) and p52 (encoded by *NFKB2* and produced as a precursor p100). Combinatorial association of the mature subunits generate 15 possible homo- or heterodimeric transcription factors, with the most prevalent being the RelA:p50 and the RelB:p52 dimers. The NF-κB proteins possess a conserved Rel homology region (RHR) in their N-termini that contains the domains for dimerization and DNA binding as well as a nuclear localization sequence (NLS). NF-κB dimers recognize a broad consensus DNA sequence known as the κB motif, which is represented as: 5′-GGRN(W)YYCC-3′ (where R denotes A or G; N denotes A, C, G or T; W denotes an A or T; Y denotes T or C) [38]. In resting cells, NF-κB factors are held inactive in the cytoplasm by inhibitor proteins. Extracellular stimuli trigger the canonical (also known as classical) or the non-canonical (also known as alternative) NF-κB pathway to induce translocation of NF-κB dimers into the nucleus, where they mediate the expression of hundreds of immune and stress response genes as well as immune-differentiating and pro-survival factors.

4. The Canonical NF-κB Activation Pathway

Activation of canonical NF-κB signaling: inhibitory IκB proteins, including the major isoform IκBα as well as IκBβ and IκBε, sequester pre-existing NF-κB dimers in the cytoplasm of unstimulated cells [39]. Signals emanating from cytokine receptors such as tumor necrosis factor receptor-1 (TNFR1), pathogen-sensing receptors such as Toll-like receptors (TLRs), and B- or T-cell antigen receptor (BCR or TCR) engage the canonical NF-κB pathway to activate NF-κB dimers from the IκB-inhibited complexes (Figure 2A). Central to this pathway is the trimeric IκB kinase (IKK) complex, which is comprised of two catalytic subunits IKK1 (also known as IKKα) and IKK2 (also known as IKKβ) and a regulatory subunit NEMO (also known as IKKγ). However, the enzymatic activity of IKK2, and not IKK1, was found to be essential for triggering of the canonical pathway. Intracellular adaptor proteins belonging to the receptor-interacting serine/threonine-protein (RIP) and TNF receptor-associated factor (TRAF) families promote signal-induced activation of TGFβ activated kinase-1 (TAK1), which in turn phosphorylates and activates IKK2. In particular, K63-linked polyubiquitination of TRAF2 and TRAF6 as well as NEMO nucleates the assembly of a receptor-proximal complex, which coordinates the activation of upstream kinases of the canonical NF-κB pathway [40]. Indeed, deubiquitinating enzymes such as CYLD and A20 function as important negative regulators of NF-κB signaling. The activated IKK complex, in turn, phosphorylates IκBα, IκBβ and IκBε that induce K48-linked polyubiquitination and proteasomal degradation of these classical IκBs, and consequent release of the bound NF-κB dimers into the nucleus.

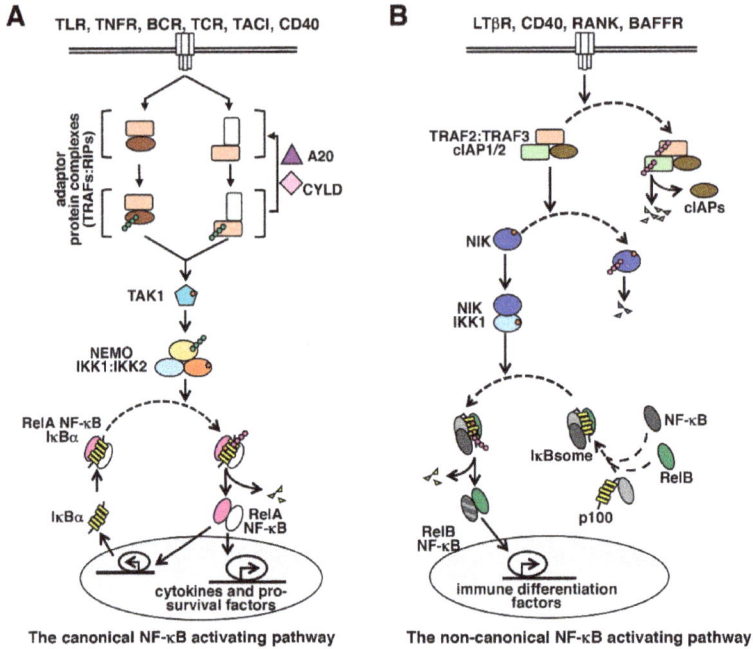

Figure 2. NF-κB activating pathways. The NF-κB system consists of (**A**) the canonical and (**B**) the non-canonical pathways. In general, immune-activating cues activate the canonical pathway, which stimulates the synthesis of pro-inflammatory and pro-survival factors involving the RelA NF-κB activity; and non-canonical signaling triggers RelB NF-κB activation during immune differentiation.

Target genes of the canonical NF-κB pathway: in most cell-types, canonical signaling activates RelA:p50 and c-Rel:p50 heterodimers. Once in the nucleus, these NF-κB heterodimers induce the transcription of an array of genes (see http://www.bu.edu/nf-kb/ for a complete list). Importantly, IκBα itself is encoded by an NF-κB target gene; NF-κB-dependent re-synthesis of IκBα constitutes a negative feedback loop that entails sequestration and nuclear export of NF-κB dimers by IκBα and post-induction attenuation of the canonical NF-κB response. Indeed, coordinated degradation and re-synthesis of classical IκBs allow a tight control over the amplitude and the duration of the NF-κB activity induced in the canonical pathway [41]. There are several other review articles, including that of Mitchell et al. [41], that provide further details of the regulatory mechanisms underlying the activation and post-induction attenuation of canonical NF-κB signaling.

Other NF-κB-target genes encode tumor-promoting, pro-inflammatory cytokines such as TNF, IL-1 and IL-6 (Interleukin 6); growth-stimulating cytokines and factors such as IL-2, GM-CSF (granulocyte-macrophage colony-stimulating factor), M-CSF and CD40L; cell-adhesion molecules such as ICAM-1 (intercellular adhesion molecule-1) and stress response mediators such as iNOS (inducible nitric oxide synthase). In addition, NF-κB factors also modulate cell-cycle progression by mediating the expression of cell-cycle regulators, such as cyclin D1. Notably, NF-κB proteins promote cell survival by triggering the expression of genes encoding anti-apoptotic proteins, such as cFLIP (cellular FLICE (FADD-like IL-1β-converting enzyme)-inhibitory protein), Bcl-2, Bcl-xL, XIAP, cIAPs (cellular inhibitors of apoptosis) and survivin. Indeed, RelA was shown to play a critical role in protecting cells from apoptotic death [39]. The canonical NF-κB signaling has been implicated in both innate and adaptive immune responses to a variety of microbial substances. In particular, B cell maturation antigen (BCMA), which activates the canonical NF-κB pathway, was shown to be required for the maintenance of long-lived plasma cells [42,43]. Previous studies, in fact, identified NF-κB transcriptional signatures in plasma cells isolated from the bone marrow of healthy individuals [34]. Moreover, canonical signaling supported the survival of plasma cells *in vitro* [44]. These observations indicated possible engagement of the canonical pathway in normal bone-resident plasma cells in physiological settings.

5. The Non-Canonical NF-κB Activation Pathway

Activation of non-canonical NF-κB signaling: the non-canonical NF-κB pathway is activated by cell-differentiating and organogenic cues, that engage B-cell maturating BAFFR (B cell activating factor receptor), CD40 (cluster of differentiation 40), osteoclast differentiating RANK (receptor activator of NF-κB) or lymph node-inducing lymphotoxin-β receptor (LTβR) [45]. The *Nfkb2*-encoded NF-κB molecule p100 regulates nuclear activation of RelB NF-κB dimers in the non-canonical pathway (Figure 2B). As a constituent of the multimeric IκBsome complex, p100 utilizes its C-terminal ankyrin repeat domain for sequestering primarily the RelB NF-κB factors in the cytoplasm of unstimulated cells [46,47]. Non-canonical signaling requires phosphorylation and activation of IKK1 by NF-κB inducing kinase (NIK) and does not involve IKK2 or NEMO [48]. NIK, in association with activated IKK1, phosphorylates p100 [48–50]. Subsequent K48-linked polyubiquitination and proteasomal action removes the C-terminal inhibitory domain of p100 that not only generates the mature p52 NF-κB subunit but also liberates the RelB:p52 and the RelB:p50 NF-κB heterodimers from the IκBsome into the nucleus [51]. Unlike canonical signaling, which triggers a strong yet transient RelA or c-Rel containing NF-κB response, non-canonical signaling elicits a sustained RelB NF-κB activity. In physiological settings, RelB activated by non-canonical signaling is thought to play a rather limited role in immune cell maturation and immune organogenesis, including B-cell maturation, dendritic cell activation, bone homeostasis and lymphoid organogenesis. Importantly, IKK1 and RelB as well as p100 have been implicated in the survival of post-germinal center plasma cells [52,53]. If the non-canonical NF-κB pathway modulates the survival of long-lived bone marrow plasma cells, the healthy counterpart of myeloma cells, remains unclear.

Interestingly, NIK is regulated at the level of protein stability. A complex composed of TRAF3, TRAF2 and cIAP1/2 targets NIK for K48-linked polyubiquitination that causes proteasomal degradation of NIK in resting cells. Non-canonical stimuli degrade TRAF3 and TRAF2 and rescue NIK from the constitutive destruction; NIK then marks p100 for proteasome processing [54,55]. In addition, it was proposed that IKK1-mediated phosphorylation triggers proteasomal degradation of NIK and consequent termination of non-canonical signaling [56].

Target genes of RelB NF-κB: while immune-activating substances trigger canonical signaling, immune-differentiating cues signal through the non-canonical pathway. RelA or c-creRel dimers activated by canonical signaling induce the expression of a wide spectrum of immune and stress response genes as well as pro-survival factors. It was originally suggested that RelB NF-κB dimers activated by the non-canonical pathway induce preferentially the expression of genes encoding organogenic chemokines and immune differentiating factors [45]. Indeed, the RelB:p52 dimer was shown to bind to the variant kappaB site present in the promoters of homeostatic chemokine genes, including stromal cell-derived factor 1α (SDF1α), and selectively activate their expressions [57–59]. Accordingly, it was postulated that stimulus-selective gene-expressions are achieved through the engagement of distinct NF-κB dimers, which possess non-redundant DNA binding specificity [38]. However, this notion has been contested by several other studies. In fact, protein-binding microarrays (PBMs) and surface plasmon resonance (SPR) analyses indicated that RelA:p50, c-Rel:p50 and RelB:p52 heterodimers not only bind to identical κB sites, but they also show similar affinities towards DNA [60]. Moreover, ChIP-seq analyses demonstrated that various NF-κB subunits occupy equivalent chromatin sites *in vivo* [61,62]. Finally, RelB was shown to activate the expression of at least a subset of RelA-target genes, including those encoding pro-inflammatory cytokines and pro-survival factors, in mouse-derived knockout cells, specialized immune cells, and cancerous cells [63–65]. Therefore, regulatory mechanisms separate from DNA binding appear to dictate stimulus-specific expressions of NF-κB-dependent genes [66]. It is likely that dynamical control of the activated NF-κB dimers, signal-induced chromatin modifying events, other co-regulated transcription factors and their potentially different interaction with NF-κB proteins impact stimulus specific gene-expressions. Clearly, additional studies are warranted for unraveling how these individual NF-κB-activating pathways direct specific gene-expression programs in physiological settings. For further reading on non-canonical NF-κB signaling, please see other review articles including that of Shao-Cong Sun [45].

6. NF-κB Deregulating Mutations in Multiple Myeloma

As early as in 2007, two research groups independently reported the presence of NF-κB-activating mutations in MM [33,34]. Annunziata et al. observed NF-κB-related genetic anomalies in 28% of the MM cell lines (MMCL) and 9% of the primary MM tumors [33]. Keats et al. examined 155 MM samples using high-resolution array-based comparative genomic hybridization (aCGH) and gene expression profiling (GEP) [34]. Their study mapped close to 20% of the mutational events onto the genes encoding the mediators and effectors of NF-κB signaling. These pioneering studies identified MM-associated mutations in both canonical and non-canonical arms of the NF-κB system (Table 1). The most common genetic abnormalities affecting the canonical pathway were: gain-of-function mutations in the gene encoding the TACI receptor (transmembrane activator and CAML interactor) that mediates canonical signaling; homozygous deletions of *CYLD*; and amplifications of *NFKB1*, which produces p50. Frequently occurring genetic aberrations pertinent to the non-canonical pathway included gain-of-function mutations and amplifications of genes encoding LTβR, CD40 and NIK; frameshift mutations and deletions associated with *NFKB2* that resulted in the production of a truncated p100 devoid of the C-terminal inhibitory domain; and mutational inactivation or deletion of genes encoding negative regulators, such as TRAF2, TRAF3, cIAP1 and cIAP2.

Subsequent studies substantiated the prevalence of NF-κB-activating mutations in MM. Parallel sequencing of 38 tumor genomes catalogued 10 point-mutations and four structural rearrangements associated with 11 distinct genes, which directly or indirectly control NF-κB functions [67]. The deregulated genes linked to canonical signaling included *TLR4*; *TNFRSF1A*, which encodes TNFR1, *IKBKB*, which expresses IKK2; *IKBIP*, which encodes an IKK2-interacting protein; *CARD11*, *MAP3K1* and *RIPK4* that code for various IKK2-activating proteins; *CYLD*; *BTRC*, which encodes a ubiquitin ligase involved in signal-induced degradation of IκBα. The mutated genes encoding non-canonical signal transducers included *MAP3K14*, which produces NIK, and *TRAF3*. In a separate study, samples derived from 177 and 26 myeloma patients were subjected to whole-exome and whole-genome sequencing, respectively [25]. This high-throughput analysis identified homozygous deletions in 32 genes, including *CYLD* and *TRAF3*. Similarly, 17% of the 463 patients enrolled in the National Cancer Research Institute (UK) Myeloma XI trial possessed NF-κB-activating mutations, which affected most frequently *CYLD* and *TRAF3* [68]. Finally, a recent study identified recurrent loss-of-function mutations in the gene encoding the negative regulator of IKK2, A20 [35].

These mutations are associated with a diverse set of genes, but result invariably in the pathological activation of a handful of key signaling pathways, particularly the NF-κB-activating pathways. Therefore, it is tempting to speculate that an "NF-κB-high phenotype," and not any particular genetic lesion, is enriched in MM and exacerbates the disease pathogenesis. MM-associated mutations potentially promote the nuclear activity of RelA and RelB heterodimers and trigger NF-κB-driven gene expressions. As such, heightened NF-κB activity has been implicated in the growth of myeloma cells, and in their resilience against apoptotic insults. Indeed, gene expression profiling of hyperdiploid multiple myeloma revealed a distinct patient cluster characterized by the overexpression of NF-κB-target genes, including anti-apoptotic genes [69]. Curiously, patients belonging to this NF-κB-signature cluster responded substantially better to bortezomib, a proteasome inhibitor that blocks degradation of IκBs, as compared to other patient groups (70% versus 29%; $p = 0.02$). This study underscored the importance of mutational activation of NF-κB signaling in the pathogenesis of MM.

7. NF-κB-Related Microenvironmental Cues in Multiple Myeloma

Unlike most other hematological malignancies, development of symptomatic myeloma obligatorily involves a tumor-promoting microenvironment that is provided by the bone marrow niche [70,71]. Within the bone marrow microenvironment, myeloma cells interact with accessory cells, including bone marrow stromal cells (BMSCs), osteoclasts, osteoblasts and endothelial cells. These physical contacts provoke tumor-associated non-cancerous cells to produce various soluble factors, including cytokines and chemokines. Cell–cell communications involving physical interactions as well as soluble mediators, which engage autocrine and paracrine loops, activate NF-κB and other key signaling pathways in both non-malignant and malignant cells. Heightened NF-κB signaling leads to further accumulation of tumor-promoting cytokines and growth factors, which support growth, survival and drug-resistance of myeloma cells [70].

For example, IL-6 acts as an important growth and survival factor in MM. Upon adhering to myeloma cells, BMSCs activate the canonical NF-κB pathway, which induces the expression of IL-6 [72]. Additionally, myeloma cell-derived IL-1β, also encoded by an NF-κB-target gene, promotes IL-6 production by inducing canonical NF-κB signaling in BMSCs [73]. In turn, IL-6 binds to the cognate receptor and induces pro-proliferative and pro-survival gene-expressions in myeloma cells. Furthermore, IL-6 induces the production of vascular endothelial growth factor (VEGF), which promotes angiogenesis and neovascularization in the tumor microenvironment [74]. Importantly, some of the VEGF isoforms are encoded by NF-κB-target genes [75]. Bone marrow stromal cells also secrete insulin-like growth factor 1 (IGF1), which in myeloma cells induces NF-κB-dependent expression of anti-apoptotic genes [76,77]. Pro-inflammatory cytokine TNF activates canonical NF-κB signaling in both BMSCs as well as myeloma cells, and is also produced by the canonical pathway. TNF not only supports the pro-inflammatory tumor microenvironment but also induces the expression of important NF-κB-target pro-survival factors in myeloma cells [78]. Indeed, disruption of the TNF autocrine loop was shown to sensitize myeloma cells to apoptosis-inducing anticancer drugs ex vivo [79]. More so, it has been suggested that therapeutic efficacy of thalidomide and lenalidomide in MM in part relies on their ability to inhibit pro-inflammatory cytokines such as TNF-α, IL-1β, and IL-6 [80,81].

MM is also associated with an increased abundance of B-cell activating factor (BAFF), which activates non-canonical NF-κB signaling via BAFFR in addition to TACI-mediated induction of the canonical pathway [82]. As such, BAFF is thought to support growth and survival of myeloma cells and contribute to poor disease prognosis [83]. Importantly, a proliferation-inducing ligand (APRIL) also signals through TACI as well as BCMA, and constitutes an important bone-marrow microenvironmental factor in MM [70]. Of note, targeting of BCMA in chimeric antigen receptor T cell (CAR-T cell) based therapy of MM produced promising results in initial clinical trials [84]. Moreover, myeloma cells produce RANKL, which promotes MM-associated bone loss by inducing osteoclast differentiation through the non-canonical NF-κB pathway [85]. It was shown that physical interaction of myeloma cells with the extracellular matrix (ECM) component fibronectin *per se* induces a RelB-dependent, pro-survival NF-κB signaling in cancerous cells [86]. Finally, non-canonical signaling in BMSCs produce the RelB-target homeostatic chemokine SDF1α, which directs homing of myeloma cells to the bone marrow niche [87]. In sum, autocrinal as well as paracrinal cytokine signals, cell–cell and cell–ECM adhesion mechanisms seem to activate both canonical and non-canonical NF-κB signaling in myeloma cells as well as in tumor-associated non-malignant cells.

8. Interactions between NF-κB Signaling Pathways and Multiple Myeloma

Concomitant activation of canonical and non-canonical NF-κB signaling in MM: a heightened activity of the canonical NF-κB pathway has been reported in several human malignancies. In MM, both genetic mutations and microenvironmental cues trigger this pathway. Canonical signaling induces the RelA NF-κB activity, which stimulates the transcription of genes encoding pro-inflammatory cytokines and pro-survival factors. However, canonical signaling generally induces a rather transient RelA activity ex vivo owing to negative feedback regulations. Genetic lesions and to some extent, tumor microenvironmental cues additionally activate the non-canonical NF-κB pathway in myeloma cells. In fact, frequent engagement of non-canonical signaling constitutes a prominent feature of MM [65]. However, the non-canonical pathway induces a sustained RelB NF-κB activity, which mediates the expression of mostly immune differentiating and immune organogenic factors in physiological settings. Considering relatively limited physiological functions of RelB in immune maturation, a preponderance of mutations in the non-canonical pathway in MM appears somewhat puzzling. Nevertheless, mutational activations of non-canonical signaling have been described in other hematological malignancies, including chronic lymphocytic leukemia (CLL), T-cell lymphoma and cutaneous B- and T-cell lymphomas [88]. Mechanistic analyses by several groups elucidated that the canonical and the non-canonical pathways are interdependently regulated [89]. As elaborated below (Figure 3), we propose that dysfunctions of an integrated NF-κB system, and not that of the individual canonical and non-canonical pathways, cause pathological NF-κB activity in MM. In other words, an integrated NF-κB system multiplies the effect of cancer-associated non-canonical signaling and causes deregulation of several NF-κB factors.

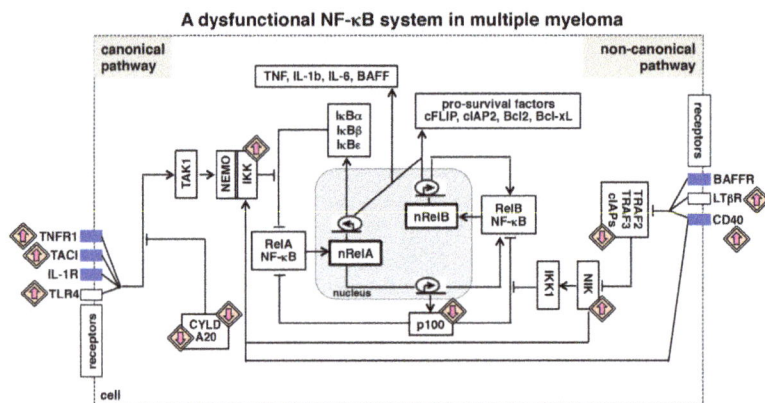

Figure 3. A dysfunctional NF-κB system in multiple myeloma. Various biochemical mechanisms interlink NF-κB-activating canonical and non-canonical pathways within an integrated NF-κB system. Also in myeloma cells, RelA and RelB containing dimers mediate expressions of overlapping set of anti-apoptotic genes. Multiple myeloma is associated with the gain-of-function mutation (upward directed purple arrow) of genes encoding NF-κB activators and the loss-of-function mutation (downward directed purple arrow) of genes encoding inhibitors of the NF-κB system. In addition, tumor microenvironment-derived factors trigger aberrant activation of cytokine receptor signaling (blue rectangle) in myeloma cells. As such, genetic and microenvironmental factors activate both the NF-κB activating pathways, and interdependent regulation of these pathways promotes pathological NF-κB activity and reinforces NF-κB-driven expressions of pro-survival genes in myeloma cells.

RelA NF-κB activation by non-canonical signals: Mechanistic studies involving mouse embryonic fibroblasts (MEFs) and B cells revealed that p100 retained not only RelB but also a sub-population of RelA and c-Rel subunits within the IκBsome complex [46,90–92]. Consistently, disruption of the IκBsome by LTβR-induced non-canonical signaling led to concomitant nuclear activation of both RelA and RelB NF-κB dimers [90]. Furthermore, NIK was shown to modulate directly the activity of the NEMO-IKK complex, which controls the canonical RelA NF-κB response [93]. Mutational activation of NIK indeed induced both canonical and non-canonical NF-κB signaling in MMCLs [94]. In comparison to targeting one or the other pathway, dual inhibition of mediators of both the pathways achieved significantly more anti-tumor activities in preclinical studies, which involved MMCLs, primary patient-derived cells, and xenograft murine models [95,96]. Therefore, it appears that NF-κB activating pathways are interlinked in MM and cooperate for promoting growth and survival of myeloma cells.

RelB:p50 NF-κB activation by canonical signals: interestingly, canonical TNF signaling induced an additional, long-lasting RelB:p50 activity in *Nfkb2*$^{-/-}$ MEFs, which lacked the expression of the non-canonical signal transducer p100 [63,97]. RelB is encoded by a NF-κB target gene [98]. Mechanistic studies clarified that RelA-dependent as well as autoregulatory mechanisms strengthened TNF-induced synthesis of RelB, which accumulated in the nucleus as an enduring NF-κB activity in the absence of the inhibitory p100 [63]. Curiously, it was demonstrated that Fbxw7 (F-box/WD40 repeat-containing protein 7), which supports the degradative K48-linked polyubiquitination of proteins, marked p100 for complete degradation in myeloma cells [99]. Likewise, mutational activation of the non-canonical pathway was associated with complete degradation of p100 in MMCLs [63]. Remarkably, NF-κB dependent RelB synthesis perpetuated a sustained RelB:p50 activity upon TNF stimulation of these myeloma cells. This TNF-induced RelB activity was implicated in the resistance of MMCLs harboring non-canonical pathway mutations to TRAIL-mediated apoptosis. Furthermore, RELB mRNA levels inversely correlated with the response of MM patients to therapeutic interventions [63]. In sum, genetic aberrations chronically activated the non-canonical pathway in myeloma cells that led to complete degradation of p100, instead of its partial processing into p52. p100-depleted malignant cells utilized the canonical TNF signal for eliciting a sustained, pro-survival NF-κB activity, which was composed of the RelB:p50 dimer and dependent on the NF-κB-driven RelB synthesis. Notably, RelA-dependent RelB synthesis was implicated in the activation of the RelB:p50 dimer in transformed B-cells [100] and invasive breast cancer cells [101]. Finally, MM-associated mutations in the non-canonical pathway seem to exacerbate the drug-resistance of malignant cells by altering the composition, and the dynamical control of the canonical NF-κB activity induced by microenvironmental cues, such as TNF.

Interlinked NF-κB pathways in the regulation of RelB NF-κB: canonical signaling also modulates the RelB:p52 activity induced by non-canonical inducers. RelA upregulates the transcription of *Nfkb2*, which encodes p100 [102]. It was demonstrated that BCR-induced canonical signaling abundantly produced p100 in mature B cells, and that apt conversion of this p100 into p52 by the BAFF-stimulated non-canonical pathway generated a robust RelB:p52 response [103]. On the other hand, an absence of RelA abrogated RelB synthesis and LTβR-induced RelB activity [97]. It was also proposed that p50 and p52 were generated in mouse-derived cells through an interdependent proteasomal processing mechanism involving the respective precursor proteins p105 and p100 [97,104]. Interestingly, CD40 simultaneously activates both the NF-κB pathways and is widely expressed on myeloma cells [105,106]. Engagement of CD40 enhances adhesion of myeloma cells to extracellular matrix proteins and BMSCs, and induces the production of IL-6 and VEGF by malignant cells. In fact, anti-CD40 monoclonal antibodies have been evaluated in clinical trials involving MM patients [107]. Despite the role of CD40 in the pathogenesis of the disease, CD40-mediated regulation of the RelB NF-κB activity has not been investigated in the context of multiple myeloma. It remains unclear if CD40-activated non-canonical signaling degrades p100 completely to promote canonical RelB:p50 activation, or crosstalk between CD40-induced NF-κB pathways contributes to aberrant RelB:p52 activation myeloma cells. It is likely that other microenvironmental and genetic factors, either alone or in combination, concurrently activate both the pathways in multiple myeloma. In this context, it will be important to investigate further if ongoing canonical signaling reinforces the RelB NF-κB activity induced by the non-canonical pathway in cancerous cells. Considering that dysfunctions associated with an integrated NF-κB system appears to cause myeloma-associated inappropriate NF-κB activation, further studies ought to delineate the molecular connectivity between NF-κB pathways in diseased cells.

9. NF-κB Driven Gene Expressions in Myeloma Cells

Aberrant activation of the NF-κB family of transcription factors in myeloma cells causes heightened expressions of a variety of tumor-promoting cytokines, which further act in an autocrine loop on malignant cells. Clinical studies involving serum and bone marrow samples derived from MM patients revealed a substantially augmented level of TNF, IL-1, IL-6 and BAFF [108–111]. Analysis of MMCLs confirmed that myeloma cells were indeed capable of inducing the expression of genes encoding tumor-promoting cytokines. In addition, enduring NF-κB signaling stimulated cancer cell-intrinsic expressions of pro-survival factors, including cFLIP, cIAP2, Bcl-xL, Bcl2 and Gadd45β [63,65,112]. These pro-survival factors promoted myeloma growth and imparted resilience in cancerous cells from apoptosis-inducing chemotherapeutic agents. In addition, NF-κB signaling has been also implicated in the elevated expression of proto-oncogenes encoding c-Myc as well as c-Myb [113,114]. Finally, cell adhesion molecules, such as ICAM1, were produced by tumor cells in a NF-κB-dependent manner.

It is generally thought that primarily RelA-containing heterodimers mediate the transcription of NF-κB target genes from the κB-driven promoters, while RelB heterodimers stimulate selectively the expression of immune differentiating factors. However, biophysical and biochemical analyses pointed out that RelA and RelB dimers possessed rather overlapping κB DNA binding specificities [60–62]. Furthermore, it was shown that a reduced p100 level in dendritic cells enabled RelB:p50 activation by canonical signaling and that the RelB:p50 heterodimer induced by the canonical pathway mediated the expression of certain RelA-target genes [64]. Subsequently, Roy et al. (2017) utilized a panel of mouse-derived knockout cells devoid of one or more NF-κB subunits to examine the gene-expression specificity of NF-κB heterodimers [63]. Their global-scale mRNA analyses established that RelB:p50 could largely substitute the RelA:p50 heterodimer for mediating TNF-induced expressions of NF-κB target genes, including those encoding pro-survival factors. Interestingly, over-expression of IKK, which selectively activates canonical signaling, or the non-canonical signal transducer NIK led to inductions of nearly overlapping set of genes in MMCLs [94]. These studies indicated that RelA and RelB heterodimers might coordinately control NF-κB driven gene-expressions in MM.

In contrast to its rather limited physiological functions, RelB was reported to play a critical role in promoting survival and drug resistance of myeloma cells. This pro-survival RelB function was attributed to RelB-dependent expressions of anti-apoptotic genes [63,65,86]. Cormier et al. (2013) demonstrated that ~40% of newly diagnosed MM patients possessed tumor cells with heightened nuclear RelB:p50 DNA binding activity. Their analysis further disclosed that RelB directly bound to the promoter of several NF-κB target pro-survival genes and activated their expressions in MMCLs [65]. In MMCLs harboring gain-of-function mutations in the non-canonical NF-κB module, TNF activated the RelB:p50 heterodimer, which similarly sustained the transcription of pro-survival genes, including *cFLIP* and *Bcl2* [63]. These studies identified RelB as a potent activator of pro-survival genes in myeloma cells. However, a recent study revealed that a HDAC4-RelB complex inhibited the transcription of pro-apoptotic gene *Bim* in MMCLs [54]. Further mechanistic investigations are warranted to unravel if RelB indeed functions as an activator and a repressor in different subsets of myeloma cells. Nevertheless, not only mutations onto the non-canonical NF-κB module are prevalent in MM, but RelB NF-κB dimers also appear to supplement pathological RelA functions in myeloma cells.

Taken together, intimately interlinked NF-κB pathways and a set of NF-κB transcription factors with overlapping gene-expression specificities perpetuate pro-survival signaling in myeloma cells.

10. Conclusions

Signals derived from diverse genetic and microenvironmental factors converge predominantly onto the NF-κB system in myeloma cells, highlighting the seminal role of NF-κB in the pathogenesis of MM. In myeloma cells, cell-intrinsic mutations activate mostly the non-canonical NF-κB pathway, whereas tumor microenvironment-derived cytokines trigger largely canonical signaling. Emerging evidence suggests that crosstalk between the NF-κB-activating pathways integrates mutational and microenvironmental signals to provoke anomalous NF-κB activation in MM. In particular, NIK-dependent non-canonical signaling amplifies RelA NF-κB responses and canonical signaling reinforces RelB NF-κB activity in myeloma cells. Importantly, RelB-containing heterodimers play a broader role in MM. They mediate the expression of pro-survival genes, which are traditionally thought to be RelA targets, and supplement the anti-apoptotic RelA function by protecting myeloma cells from apoptotic, chemotherapeutic drugs. We hypothesize that an integrated NF-κB system should be considered in sum, for understanding further NF-κB deregulations in MM.

The constituents of the non-canonical pathway contribute to the pro-survival and pro-proliferative program of myeloma cells *via* the integrated NF-κB system. In fact, an overarching role of non-canonical signaling in MM led to active consideration of NIK as a promising drug target [115]. Future studies should further characterize interdependent regulations of NF-κB pathways in MM, as these crosstalk connectivities may guide disease state-specific therapeutic interventions involving dual inhibition of both the NF-κB-activating pathways. Moreover, it remains unclear if the myeloma-associated RelB only activates the expression of RelA-target pro-survival genes or also mediates the expression of genes that are normally not induced by RelA. Global transcriptomic and chromatin-binding analyses involving mouse-derived knockout cells may offer valuable insights on the identity of RelA-target, RelB-target and generic NF-κB-dependent genes. These NF-κB gene signatures can be exploited for interrogating genome-scale data obtained from MM patients for assessing possible involvement of NF-κB pathways in the disease pathogenesis. Based on the outcome of such studies involving longitudinal cohorts, the NF-κB gene signatures can be further utilized as prognostic tools for predicting disease outcome, response to therapy, and the survival parameters of newly diagnosed MM patients. Furthermore, such analyses can potentially inform personalized treatment regimes for achieving efficacious and specific therapeutic interventions involving the NF-κB system. Finally, it is known that only a subset of MGUS progresses into MM. Recent high-throughput genome sequencing studies indicated that NF-κB-activating mutations are rather rare in the MGUS stage as compared to MM [116,117]. In this context, global transcriptomic analyses ought to address whether bone marrow-derived cues trigger uncontrolled NF-κB activation in MGUS and if such NF-κB deregulations precede the transition of MGUS into MM. An understanding of the mechanism that drives benign MGUS into malignant MM may enable the development of preventive measures. In conclusion, a multi-disciplinary approach, which involves mechanistic and clinical studies, will be indispensable for improving further therapeutic modalities in multiple myeloma.

Author Contributions: P.R. and U.A.S. carried out the literature survey under the supervision of S.B., S.B. wrote the review article with P.R.

Funding: Multiple Myeloma research and NF-κB signaling studies in the Systems Immunology Laboratory were funded by SERB, Department of Science and Technology, Govt. of India (EMR/2015/000658), the Wellcome Trust DBT India Alliance (500094/Z/09/Z) and NII-core.

Acknowledgments: We gratefully acknowledge critical comments and helpful suggestions received from the members of the Systems Immunology Laboratory on the manuscript.

Conflicts of Interest: The authors declare no conflict of interest.

References

1. Ferlay, J.; Shin, H.R.; Bray, F.; Forman, D.; Mathers, C.; Parkin, D.M. Estimates of worldwide burden of cancer in 2008: Globocan 2008. *Int. J. Cancer* **2010**, *127*, 2893–2917. [CrossRef] [PubMed]
2. Jemal, A.; Siegel, R.; Ward, E.; Hao, Y.; Xu, J.; Thun, M.J. Cancer statistics, 2009. *CA Cancer J. Clin.* **2009**, *59*, 225–249. [CrossRef] [PubMed]
3. Kazandjian, D. Multiple myeloma epidemiology and survival: A unique malignancy. *Semin. Oncol.* **2016**, *43*, 676–681. [CrossRef] [PubMed]
4. Palumbo, A.; Anderson, K. Multiple myeloma. *N. Engl. J. Med.* **2011**, *364*, 1046–1060. [CrossRef] [PubMed]
5. Flores-Montero, J.; de Tute, R.; Paiva, B.; Perez, J.J.; Böttcher, S.; Wind, H.; Sanoja, L.; Puig, N.; Lecrevisse, Q.; Vidriales, M.B.; et al. Immunophenotype of normal vs. Myeloma plasma cells: Toward antibody panel specifications for mrd detection in multiple myeloma. *Cytom. B Clin. Cytom.* **2016**, *90*, 61–72. [CrossRef] [PubMed]
6. Gonzalez, D.; van der Burg, M.; Garcia-Sanz, R.; Fenton, J.A.; Langerak, A.W.; Gonzalez, M.; van Dongen, J.J.; San Miguel, J.F.; Morgan, G.J. Immunoglobulin gene rearrangements and the pathogenesis of multiple myeloma. *Blood* **2007**, *110*, 3112–3121. [CrossRef] [PubMed]
7. Mahindra, A.; Hideshima, T.; Anderson, K.C. Multiple myeloma: Biology of the disease. *Blood Rev.* **2010**, *24* (Suppl. 1), S5–S11. [CrossRef]
8. Kyle, R.A.; Rajkumar, S.V. Multiple myeloma. *N. Engl. J. Med.* **2004**, *351*, 1860–1873. [CrossRef] [PubMed]
9. Chim, C.S.; Kumar, S.K.; Orlowski, R.Z.; Cook, G.; Richardson, P.G.; Gertz, M.A.; Giralt, S.; Mateos, M.V.; Leleu, X.; Anderson, K.C. Management of relapsed and refractory multiple myeloma: Novel agents, antibodies, immunotherapies and beyond. *Leukemia* **2018**, *32*, 252–262. [CrossRef] [PubMed]
10. Fonseca, R.; Bergsagel, P.L.; Drach, J.; Shaughnessy, J.; Gutierrez, N.; Stewart, A.K.; Morgan, G.; Van Ness, B.; Chesi, M.; Minvielle, S.; et al. International myeloma working group molecular classification of multiple myeloma: Spotlight review. *Leukemia* **2009**, *23*, 2210–2221. [CrossRef] [PubMed]
11. Avet-Loiseau, H.; Attal, M.; Moreau, P.; Charbonnel, C.; Garban, F.; Hulin, C.; Leyvraz, S.; Michallet, M.; Yakoub-Agha, I.; Garderet, L.; et al. Genetic abnormalities and survival in multiple myeloma: The experience of the intergroupe francophone du myelome. *Blood* **2007**, *109*, 3489–3495. [CrossRef] [PubMed]
12. Corre, J.; Munshi, N.; Avet-Loiseau, H. Genetics of multiple myeloma: Another heterogeneity level? *Blood* **2015**, *125*, 1870–1876. [CrossRef] [PubMed]
13. Robiou du Pont, S.; Cleynen, A.; Fontan, C.; Attal, M.; Munshi, N.; Corre, J.; Avet-Loiseau, H. Genomics of multiple myeloma. *J. Clin. Oncol.* **2017**, *35*, 963–967. [CrossRef] [PubMed]
14. Sawyer, J.R.; Waldron, J.A.; Jagannath, S.; Barlogie, B. Cytogenetic findings in 200 patients with multiple myeloma. *Cancer Genet. Cytogenet.* **1995**, *82*, 41–49. [CrossRef]
15. Chesi, M.; Bergsagel, P.L.; Brents, L.A.; Smith, C.M.; Gerhard, D.S.; Kuehl, W.M. Dysregulation of cyclin D1 by translocation into an IgH gamma switch region in two multiple myeloma cell lines. *Blood* **1996**, *88*, 674–681. [PubMed]
16. Chesi, M.; Nardini, E.; Brents, L.A.; Schrock, E.; Ried, T.; Kuehl, W.M.; Bergsagel, P.L. Frequent translocation t(4;14)(p16.3;q32.3) in multiple myeloma is associated with increased expression and activating mutations of fibroblast growth factor receptor 3. *Nat. Genet.* **1997**, *16*, 260–264. [CrossRef] [PubMed]
17. Chesi, M.; Nardini, E.; Lim, R.S.; Smith, K.D.; Kuehl, W.M.; Bergsagel, P.L. The t(4;14) translocation in myeloma dysregulates both FGFR3 and a novel gene, MMSET, resulting in igH/MMSET hybrid transcripts. *Blood* **1998**, *92*, 3025–3034. [PubMed]
18. Hurt, E.M.; Wiestner, A.; Rosenwald, A.; Shaffer, A.L.; Campo, E.; Grogan, T.; Bergsagel, P.L.; Kuehl, W.M.; Staudt, L.M. Overexpression of c-maf is a frequent oncogenic event in multiple myeloma that promotes proliferation and pathological interactions with bone marrow stroma. *Cancer Cell* **2004**, *5*, 191–199. [CrossRef]
19. Hanamura, I.; Iida, S.; Akano, Y.; Hayami, Y.; Kato, M.; Miura, K.; Harada, S.; Banno, S.; Wakita, A.; Kiyoi, H.; et al. Ectopic expression of MAFB gene in human myeloma cells carrying (14;20)(q32;q11) chromosomal translocations. *Jpn. J. Cancer Res.* **2001**, *92*, 638–644. [CrossRef] [PubMed]
20. Bergsagel, P.L.; Kuehl, W.M.; Zhan, F.; Sawyer, J.; Barlogie, B.; Shaughnessy, J., Jr. Cyclin d dysregulation: An early and unifying pathogenic event in multiple myeloma. *Blood* **2005**, *106*, 296–303. [CrossRef] [PubMed]

21. Avet-Loiseau, H.; Li, C.; Magrangeas, F.; Gouraud, W.; Charbonnel, C.; Harousseau, J.L.; Attal, M.; Marit, G.; Mathiot, C.; Facon, T.; et al. Prognostic significance of copy-number alterations in multiple myeloma. *J. Clin. Oncol.* **2009**, *27*, 4585–4590. [CrossRef] [PubMed]

22. Walker, B.A.; Leone, P.E.; Jenner, M.W.; Li, C.; Gonzalez, D.; Johnson, D.C.; Ross, F.M.; Davies, F.E.; Morgan, G.J. Integration of global snp-based mapping and expression arrays reveals key regions, mechanisms, and genes important in the pathogenesis of multiple myeloma. *Blood* **2006**, *108*, 1733–1743. [CrossRef] [PubMed]

23. Hanamura, I.; Stewart, J.P.; Huang, Y.; Zhan, F.; Santra, M.; Sawyer, J.R.; Hollmig, K.; Zangarri, M.; Pineda-Roman, M.; van Rhee, F.; et al. Frequent gain of chromosome band 1q21 in plasma-cell dyscrasias detected by fluorescence in situ hybridization: Incidence increases from MGUS to relapsed myeloma and is related to prognosis and disease progression following tandem stem-cell transplantation. *Blood* **2006**, *108*, 1724–1732. [CrossRef] [PubMed]

24. Chesi, M.; Bergsagel, P.L. Molecular pathogenesis of multiple myeloma: Basic and clinical updates. *Int. J. Hematol.* **2013**, *97*, 313–323. [CrossRef] [PubMed]

25. Lohr, J.G.; Stojanov, P.; Carter, S.L.; Cruz-Gordillo, P.; Lawrence, M.S.; Auclair, D.; Sougnez, C.; Knoechel, B.; Gould, J.; Saksena, G.; et al. Widespread genetic heterogeneity in multiple myeloma: Implications for targeted therapy. *Cancer Cell* **2014**, *25*, 91–101. [CrossRef] [PubMed]

26. Bolli, N.; Avet-Loiseau, H.; Wedge, D.C.; Van Loo, P.; Alexandrov, L.B.; Martincorena, I.; Dawson, K.J.; Iorio, F.; Nik-Zainal, S.; Bignell, G.R.; et al. Heterogeneity of genomic evolution and mutational profiles in multiple myeloma. *Nat. Commun.* **2014**, *5*, 2997. [CrossRef] [PubMed]

27. Neri, A.; Baldini, L.; Trecca, D.; Cro, L.; Polli, E.; Maiolo, A.T. p53 gene mutations in multiple myeloma are associated with advanced forms of malignancy. *Blood* **1993**, *81*, 128–135. [PubMed]

28. Affer, M.; Chesi, M.; Chen, W.G.; Keats, J.J.; Demchenko, Y.N.; Roschke, A.V.; Van Wier, S.; Fonseca, R.; Bergsagel, P.L.; Kuehl, W.M. Promiscuous MYC locus rearrangements hijack enhancers but mostly super-enhancers to dysregulate myc expression in multiple myeloma. *Leukemia* **2014**, *28*, 1725–1735. [CrossRef] [PubMed]

29. Lionetti, M.; Neri, A. Utilizing next-generation sequencing in the management of multiple myeloma. *Expert Rev. Mol. Diagn.* **2017**, *17*, 653–663. [CrossRef] [PubMed]

30. Bustoros, M.; Mouhieddine, T.H.; Detappe, A.; Ghobrial, I.M. Established and novel prognostic biomarkers in multiple myeloma. *Am. Soc. Clin. Oncol. Educ. Book* **2017**, *37*, 548–560. [CrossRef] [PubMed]

31. Lohr, J.G.; Kim, S.; Gould, J.; Knoechel, B.; Drier, Y.; Cotton, M.J.; Gray, D.; Birrer, N.; Wong, B.; Ha, G.; et al. Genetic interrogation of circulating multiple myeloma cells at single-cell resolution. *Sci. Transl. Med.* **2016**, *8*, 363ra147. [CrossRef] [PubMed]

32. Kumar, S.K.; Rajkumar, V.; Kyle, R.A.; van Duin, M.; Sonneveld, P.; Mateos, M.V.; Gay, F.; Anderson, K.C. Multiple myeloma. *Nat. Rev. Dis. Prim.* **2017**, *3*, 17046. [CrossRef] [PubMed]

33. Annunziata, C.M.; Davis, R.E.; Demchenko, Y.; Bellamy, W.; Gabrea, A.; Zhan, F.; Lenz, G.; Hanamura, I.; Wright, G.; Xiao, W.; et al. Frequent engagement of the classical and alternative NF-kappaB pathways by diverse genetic abnormalities in multiple myeloma. *Cancer Cell* **2007**, *12*, 115–130. [CrossRef] [PubMed]

34. Keats, J.J.; Fonseca, R.; Chesi, M.; Schop, R.; Baker, A.; Chng, W.J.; Van Wier, S.; Tiedemann, R.; Shi, C.X.; Sebag, M.; et al. Promiscuous mutations activate the noncanonical NF-kappaB pathway in multiple myeloma. *Cancer Cell* **2007**, *12*, 131–144. [CrossRef] [PubMed]

35. Troppan, K.; Hofer, S.; Wenzl, K.; Lassnig, M.; Pursche, B.; Steinbauer, E.; Wiltgen, M.; Zulus, B.; Renner, W.; Beham-Schmid, C.; et al. Frequent down regulation of the tumor suppressor gene a20 in multiple myeloma. *PLoS ONE* **2015**, *10*, e0123922. [CrossRef] [PubMed]

36. Baud, V.; Karin, M. Is NF-kappaB a good target for cancer therapy? Hopes and pitfalls. *Nat. Rev. Drug Discov.* **2009**, *8*, 33–40. [CrossRef] [PubMed]

37. DiDonato, J.A.; Mercurio, F.; Karin, M. NF-kappaB and the link between inflammation and cancer. *Immunol. Rev.* **2012**, *246*, 379–400. [CrossRef] [PubMed]

38. Hoffmann, A.; Natoli, G.; Ghosh, G. Transcriptional regulation via the NF-kappaB signaling module. *Oncogene* **2006**, *25*, 6706–6716. [CrossRef] [PubMed]

39. Oeckinghaus, A.; Ghosh, S. The NF-kappaB family of transcription factors and its regulation. *Cold Spring Harb. Perspect. Biol.* **2009**, *1*, a000034. [CrossRef] [PubMed]

40. Chen, Z.J. Ubiquitination in signaling to and activation of ikk. *Immunol. Rev.* **2012**, *246*, 95–106. [CrossRef] [PubMed]
41. Mitchell, S.; Vargas, J.; Hoffmann, A. Signaling via the nfkappab system. *Wiley Interdiscip. Rev. Syst. Biol. Med.* **2016**, *8*, 227–241. [CrossRef][PubMed]
42. O'Connor, B.P.; Raman, V.S.; Erickson, L.D.; Cook, W.J.; Weaver, L.K.; Ahonen, C.; Lin, L.L.; Mantchev, G.T.; Bram, R.J.; Noelle, R.J. Bcma is essential for the survival of long-lived bone marrow plasma cells. *J. Exp. Med.* **2004**, *199*, 91–98. [CrossRef] [PubMed]
43. Hatzoglou, A.; Roussel, J.; Bourgeade, M.F.; Rogier, E.; Madry, C.; Inoue, J.; Devergne, O.; Tsapis, A. TNF receptor family member BCMA (B cell maturation) associates with tnf receptor-associated factor (TRAF) 1, TRAF2, and TRAF3 and activates NF-kappa B, elk-1, c-jun N-terminal kinase, and p38 mitogen-activated protein kinase. *J. Immunol.* **2000**, *165*, 1322–1330. [CrossRef] [PubMed]
44. Jourdan, M.; Cren, M.; Robert, N.; Bollore, K.; Fest, T.; Duperray, C.; Guilloton, F.; Hose, D.; Tarte, K.; Klein, B. IL-6 supports the generation of human long-lived plasma cells in combination with either april or stromal cell-soluble factors. *Leukemia* **2014**, *28*, 1647–1656. [CrossRef] [PubMed]
45. Sun, S.C. The noncanonical NF-kappaB pathway. *Immunol. Rev.* **2012**, *246*, 125–140. [CrossRef] [PubMed]
46. Savinova, O.V.; Hoffmann, A.; Ghosh, G. The NFKB1 and NFKB2 proteins p105 and p100 function as the core of high-molecular-weight heterogeneous complexes. *Mol. Cell* **2009**, *34*, 591–602. [CrossRef] [PubMed]
47. Tao, Z.; Fusco, A.; Huang, D.B.; Gupta, K.; Young Kim, D.; Ware, C.F.; Van Duyne, G.D.; Ghosh, G. p100/ikappabdelta sequesters and inhibits NF-kappaB through kappaBsome formation. *Proc. Natl. Acad. Sci. USA* **2014**, *111*, 15946–15951. [CrossRef] [PubMed]
48. Xiao, G.; Harhaj, E.W.; Sun, S.C. NF-kappaB-inducing kinase regulates the processing of NF-kappaB2 p100. *Mol. Cell* **2001**, *7*, 401–409. [CrossRef]
49. Polley, S.; Passos, D.O.; Huang, D.B.; Mulero, M.C.; Mazumder, A.; Biswas, T.; Verma, I.M.; Lyumkis, D.; Ghosh, G. Structural basis for the activation of IKK1/alpha. *Cell Rep.* **2016**, *17*, 1907–1914. [CrossRef] [PubMed]
50. Senftleben, U.; Cao, Y.; Xiao, G.; Greten, F.R.; Krahn, G.; Bonizzi, G.; Chen, Y.; Hu, Y.; Fong, A.; Sun, S.C.; et al. Activation by IKKalpha of a second, evolutionary conserved, NF-kappa B signaling pathway. *Science* **2001**, *293*, 1495–1499. [CrossRef] [PubMed]
51. Fong, A.; Sun, S.C. Genetic evidence for the essential role of beta-transducin repeat-containing protein in the inducible processing of NF-kappa B2/p100. *J. Biol. Chem.* **2002**, *277*, 22111–22114. [CrossRef] [PubMed]
52. Mills, D.M.; Bonizzi, G.; Karin, M.; Rickert, R.C. Regulation of late B cell differentiation by intrinsic IKKalpha-dependent signals. *Proc. Natl. Acad. Sci. USA* **2007**, *104*, 6359–6364. [CrossRef] [PubMed]
53. De Silva, N.S.; Anderson, M.M.; Carette, A.; Silva, K.; Heise, N.; Bhagat, G.; Klein, U. Transcription factors of the alternative NF-kappaB pathway are required for germinal center B-cell development. *Proc. Natl. Acad. Sci. USA* **2016**, *113*, 9063–9068. [CrossRef] [PubMed]
54. Vallabhapurapu, S.; Matsuzawa, A.; Zhang, W.; Tseng, P.H.; Keats, J.J.; Wang, H.; Vignali, D.A.; Bergsagel, P.L.; Karin, M. Nonredundant and complementary functions of TRAF2 and TRAF3 in a ubiquitination cascade that activates NIK-dependent alternative NF-kappaB signaling. *Nat. Immunol.* **2008**, *9*, 1364–1370. [CrossRef] [PubMed]
55. Zarnegar, B.J.; Wang, Y.; Mahoney, D.J.; Dempsey, P.W.; Cheung, H.H.; He, J.; Shiba, T.; Yang, X.; Yeh, W.C.; Mak, T.W.; et al. Noncanonical NF-kappaB activation requires coordinated assembly of a regulatory complex of the adaptors CIAP1, CIAP2, TRAF2 and TRAF3 and the kinase nik. *Nat. Immunol.* **2008**, *9*, 1371–1378. [CrossRef] [PubMed]
56. Razani, B.; Zarnegar, B.; Ytterberg, A.J.; Shiba, T.; Dempsey, P.W.; Ware, C.F.; Loo, J.A.; Cheng, G. Negative feedback in noncanonical NF-kappaB signaling modulates NIK stability through IKKalpha-mediated phosphorylation. *Sci. Signal.* **2010**, *3*, ra41. [CrossRef] [PubMed]
57. Bonizzi, G.; Bebien, M.; Otero, D.C.; Johnson-Vroom, K.E.; Cao, Y.; Vu, D.; Jegga, A.G.; Aronow, B.J.; Ghosh, G.; Rickert, R.C.; et al. Activation of IKKalpha target genes depends on recognition of specific kappaB binding sites by RelB:P52 dimers. *EMBO J.* **2004**, *23*, 4202–4210. [CrossRef] [PubMed]
58. Fusco, A.J.; Huang, D.B.; Miller, D.; Wang, V.Y.; Vu, D.; Ghosh, G. NF-kappaB p52:ReLB heterodimer recognizes two classes of kappaB sites with two distinct modes. *EMBO Rep.* **2009**, *10*, 152–159. [CrossRef] [PubMed]

59. Mukherjee, T.; Chatterjee, B.; Dhar, A.; Bais, S.S.; Chawla, M.; Roy, P.; George, A.; Bal, V.; Rath, S.; Basak, S. A TNF-p100 pathway subverts noncanonical NF-kappaB signaling in inflamed secondary lymphoid organs. *EMBO J.* **2017**, *36*, 3501–3516. [CrossRef] [PubMed]

60. Siggers, T.; Chang, A.B.; Teixeira, A.; Wong, D.; Williams, K.J.; Ahmed, B.; Ragoussis, J.; Udalova, I.A.; Smale, S.T.; Bulyk, M.L. Principles of dimer-specific gene regulation revealed by a comprehensive characterization of NF-kappaB family DNA binding. *Nat. Immunol.* **2011**, *13*, 95–102. [CrossRef] [PubMed]

61. Zhao, B.; Barrera, L.A.; Ersing, I.; Willox, B.; Schmidt, S.C.; Greenfeld, H.; Zhou, H.; Mollo, S.B.; Shi, T.T.; Takasaki, K.; et al. The NF-kappaB genomic landscape in lymphoblastoid B cells. *Cell Rep.* **2014**, *8*, 1595–1606. [CrossRef] [PubMed]

62. De Oliveira, K.A.; Kaergel, E.; Heinig, M.; Fontaine, J.F.; Patone, G.; Muro, E.M.; Mathas, S.; Hummel, M.; Andrade-Navarro, M.A.; Hubner, N.; et al. A roadmap of constitutive NF-kappaB activity in hodgkin lymphoma: Dominant roles of p50 and p52 revealed by genome-wide analyses. *Genome Med.* **2016**, *8*, 28. [CrossRef] [PubMed]

63. Roy, P.; Mukherjee, T.; Chatterjee, B.; Vijayaragavan, B.; Banoth, B.; Basak, S. Non-canonical nfkappab mutations reinforce pro-survival tnf response in multiple myeloma through an autoregulatory RelB:P50 NFkappaB pathway. *Oncogene* **2017**, *36*, 1417–1429. [CrossRef] [PubMed]

64. Shih, V.F.; Davis-Turak, J.; Macal, M.; Huang, J.Q.; Ponomarenko, J.; Kearns, J.D.; Yu, T.; Fagerlund, R.; Asagiri, M.; Zuniga, E.I.; et al. Control of RelB during dendritic cell activation integrates canonical and noncanonical NF-kappaB pathways. *Nat. Immunol.* **2012**, *13*, 1162–1170. [CrossRef] [PubMed]

65. Cormier, F.; Monjanel, H.; Fabre, C.; Billot, K.; Sapharikas, E.; Chereau, F.; Bordereaux, D.; Molina, T.J.; Avet-Loiseau, H.; Baud, V. Frequent engagement of RelB activation is critical for cell survival in multiple myeloma. *PLoS ONE* **2013**, *8*, e59127. [CrossRef] [PubMed]

66. Smale, S.T. Dimer-specific regulatory mechanisms within the NF-kappaB family of transcription factors. *Immunol. Rev.* **2012**, *246*, 193–204. [CrossRef] [PubMed]

67. Chapman, M.A.; Lawrence, M.S.; Keats, J.J.; Cibulskis, K.; Sougnez, C.; Schinzel, A.C.; Harview, C.L.; Brunet, J.P.; Ahmann, G.J.; Adli, M.; et al. Initial genome sequencing and analysis of multiple myeloma. *Nature* **2011**, *471*, 467–472. [CrossRef] [PubMed]

68. Walker, B.A.; Boyle, E.M.; Wardell, C.P.; Murison, A.; Begum, D.B.; Dahir, N.M.; Proszek, P.Z.; Johnson, D.C.; Kaiser, M.F.; Melchor, L.; et al. Mutational spectrum, copy number changes, and outcome: Results of a sequencing study of patients with newly diagnosed myeloma. *J. Clin. Oncol.* **2015**, *33*, 3911–3920. [CrossRef] [PubMed]

69. Chng, W.J.; Kumar, S.; Vanwier, S.; Ahmann, G.; Price-Troska, T.; Henderson, K.; Chung, T.H.; Kim, S.; Mulligan, G.; Bryant, B.; et al. Molecular dissection of hyperdiploid multiple myeloma by gene expression profiling. *Cancer Res.* **2007**, *67*, 2982–2989. [CrossRef] [PubMed]

70. Hideshima, T.; Mitsiades, C.; Tonon, G.; Richardson, P.G.; Anderson, K.C. Understanding multiple myeloma pathogenesis in the bone marrow to identify new therapeutic targets. *Nat. Rev. Cancer* **2007**, *7*, 585–598. [CrossRef] [PubMed]

71. Fairfield, H.; Falank, C.; Avery, L.; Reagan, M.R. Multiple myeloma in the marrow: Pathogenesis and treatments. *Ann. N. Y. Acad. Sci.* **2016**, *1364*, 32–51. [CrossRef] [PubMed]

72. Chauhan, D.; Uchiyama, H.; Akbarali, Y.; Urashima, M.; Yamamoto, K.; Libermann, T.A.; Anderson, K.C. Multiple myeloma cell adhesion-induced interleukin-6 expression in bone marrow stromal cells involves activation of NF-kappa B. *Blood* **1996**, *87*, 1104–1112. [PubMed]

73. Costes, V.; Portier, M.; Lu, Z.Y.; Rossi, J.F.; Bataille, R.; Klein, B. Interleukin-1 in multiple myeloma: Producer cells and their role in the control of IL-6 production. *Br. J. Haematol.* **1998**, *103*, 1152–1160. [CrossRef] [PubMed]

74. Podar, K.; Anderson, K.C. Emerging therapies targeting tumor vasculature in multiple myeloma and other hematologic and solid malignancies. *Curr. Cancer Drug Targets* **2011**, *11*, 1005–1024. [CrossRef] [PubMed]

75. Chilov, D.; Kukk, E.; Taira, S.; Jeltsch, M.; Kaukonen, J.; Palotie, A.; Joukov, V.; Alitalo, K. Genomic organization of human and mouse genes for vascular endothelial growth factor C. *J. Biol. Chem.* **1997**, *272*, 25176–25183. [CrossRef] [PubMed]

76. Mitsiades, C.S.; Mitsiades, N.; Poulaki, V.; Schlossman, R.; Akiyama, M.; Chauhan, D.; Hideshima, T.; Treon, S.P.; Munshi, N.C.; Richardson, P.G.; et al. Activation of NF-kappaB and upregulation of intracellular

anti-apoptotic proteins via the IGF-1/AKT signaling in human multiple myeloma cells: Therapeutic implications. *Oncogene* **2002**, *21*, 5673–5683. [CrossRef] [PubMed]

77. Sprynski, A.C.; Hose, D.; Caillot, L.; Reme, T.; Shaughnessy, J.D., Jr.; Barlogie, B.; Seckinger, A.; Moreaux, J.; Hundemer, M.; Jourdan, M.; et al. The role of IGF-1 as a major growth factor for myeloma cell lines and the prognostic relevance of the expression of its receptor. *Blood* **2009**, *113*, 4614–4626. [CrossRef] [PubMed]

78. Musolino, C.; Allegra, A.; Innao, V.; Allegra, A.G.; Pioggia, G.; Gangemi, S. Inflammatory and anti-inflammatory equilibrium, proliferative and antiproliferative balance: The role of cytokines in multiple myeloma. *Med. Inflamm.* **2017**, *2017*, 1852517. [CrossRef] [PubMed]

79. Tsubaki, M.; Komai, M.; Itoh, T.; Imano, M.; Sakamoto, K.; Shimaoka, H.; Ogawa, N.; Mashimo, K.; Fujiwara, D.; Takeda, T.; et al. Inhibition of the tumour necrosis factor-alpha autocrine loop enhances the sensitivity of multiple myeloma cells to anticancer drugs. *Eur. J. Cancer* **2013**, *49*, 3708–3717. [CrossRef] [PubMed]

80. Corral, L.G.; Haslett, P.A.; Muller, G.W.; Chen, R.; Wong, L.M.; Ocampo, C.J.; Patterson, R.T.; Stirling, D.I.; Kaplan, G. Differential cytokine modulation and T cell activation by two distinct classes of thalidomide analogues that are potent inhibitors of TNF-alpha. *J. Immunol.* **1999**, *163*, 380–386. [PubMed]

81. Quach, H.; Ritchie, D.; Stewart, A.K.; Neeson, P.; Harrison, S.; Smyth, M.J.; Prince, H.M. Mechanism of action of immunomodulatory drugs (IMIDS) in multiple myeloma. *Leukemia* **2010**, *24*, 22–32. [CrossRef] [PubMed]

82. Moreaux, J.; Legouffe, E.; Jourdan, E.; Quittet, P.; Reme, T.; Lugagne, C.; Moine, P.; Rossi, J.F.; Klein, B.; Tarte, K. Baff and april protect myeloma cells from apoptosis induced by interleukin 6 deprivation and dexamethasone. *Blood* **2004**, *103*, 3148–3157. [CrossRef] [PubMed]

83. Hengeveld, P.J.; Kersten, M.J. B-cell activating factor in the pathophysiology of multiple myeloma: A target for therapy? *Blood Cancer J.* **2015**, *5*, e282. [CrossRef] [PubMed]

84. Carpenter, R.O.; Evbuomwan, M.O.; Pittaluga, S.; Rose, J.J.; Raffeld, M.; Yang, S.; Gress, R.E.; Hakim, F.T.; Kochenderfer, J.N. B-cell maturation antigen is a promising target for adoptive T-cell therapy of multiple myeloma. *Clin. Cancer Res.* **2013**, *19*, 2048–2060. [CrossRef] [PubMed]

85. Schmiedel, B.J.; Scheible, C.A.; Nuebling, T.; Kopp, H.G.; Wirths, S.; Azuma, M.; Schneider, P.; Jung, G.; Grosse-Hovest, L.; Salih, H.R. Rankl expression, function, and therapeutic targeting in multiple myeloma and chronic lymphocytic leukemia. *Cancer Res.* **2013**, *73*, 683–694. [CrossRef] [PubMed]

86. Landowski, T.H.; Olashaw, N.E.; Agrawal, D.; Dalton, W.S. Cell adhesion-mediated drug resistance (CAM-DR) is associated with activation of NF-kappa B (RelB/p50) in myeloma cells. *Oncogene* **2003**, *22*, 2417–2421. [CrossRef] [PubMed]

87. Roccaro, A.M.; Sacco, A.; Purschke, W.G.; Moschetta, M.; Buchner, K.; Maasch, C.; Zboralski, D.; Zollner, S.; Vonhoff, S.; Mishima, Y.; et al. SDF-1 inhibition targets the bone marrow niche for cancer therapy. *Cell Rep.* **2014**, *9*, 118–128. [CrossRef] [PubMed]

88. Thu, Y.M.; Richmond, A. NF-kappaB inducing kinase: A key regulator in the immune system and in cancer. *Cytokine Growth Factor Rev.* **2010**, *21*, 213–226. [CrossRef] [PubMed]

89. Shih, V.F.; Tsui, R.; Caldwell, A.; Hoffmann, A. A single nfkappab system for both canonical and non-canonical signaling. *Cell Res.* **2011**, *21*, 86–102. [CrossRef] [PubMed]

90. Basak, S.; Kim, H.; Kearns, J.D.; Tergaonkar, V.; O'Dea, E.; Werner, S.L.; Benedict, C.A.; Ware, C.F.; Ghosh, G.; Verma, I.M.; et al. A fourth IKappaB protein within the NF-kappaB signaling module. *Cell* **2007**, *128*, 369–381. [CrossRef] [PubMed]

91. Shih, V.F.; Kearns, J.D.; Basak, S.; Savinova, O.V.; Ghosh, G.; Hoffmann, A. Kinetic control of negative feedback regulators of NF-kappaB/RelA determines their pathogen- and cytokine-receptor signaling specificity. *Proc. Natl. Acad. Sci. USA* **2009**, *106*, 9619–9624. [CrossRef] [PubMed]

92. Almaden, J.V.; Tsui, R.; Liu, Y.C.; Birnbaum, H.; Shokhirev, M.N.; Ngo, K.A.; Davis-Turak, J.C.; Otero, D.; Basak, S.; Rickert, R.C.; et al. A pathway switch directs baff signaling to distinct NFkappaB transcription factors in maturing and proliferating B cells. *Cell Rep.* **2014**, *9*, 2098–2111. [CrossRef] [PubMed]

93. Zarnegar, B.; Yamazaki, S.; He, J.Q.; Cheng, G. Control of canonical NF-kappaB activation through the nik-ikk complex pathway. *Proc. Natl. Acad. Sci. USA* **2008**, *105*, 3503–3508. [CrossRef] [PubMed]

94. Demchenko, Y.N.; Glebov, O.K.; Zingone, A.; Keats, J.J.; Bergsagel, P.L.; Kuehl, W.M. Classical and/or alternative NF-kappaB pathway activation in multiple myeloma. *Blood* **2010**, *115*, 3541–3552. [CrossRef] [PubMed]

95. Hideshima, T.; Chauhan, D.; Kiziltepe, T.; Ikeda, H.; Okawa, Y.; Podar, K.; Raje, N.; Protopopov, A.; Munshi, N.C.; Richardson, P.G.; et al. Biologic sequelae of I{kappa}B kinase (IKK) inhibition in multiple myeloma: Therapeutic implications. *Blood* **2009**, *113*, 5228–5236. [CrossRef] [PubMed]

96. Fabre, C.; Mimura, N.; Bobb, K.; Kong, S.Y.; Gorgun, G.; Cirstea, D.; Hu, Y.; Minami, J.; Ohguchi, H.; Zhang, J.; et al. Dual inhibition of canonical and noncanonical NF-kappaB pathways demonstrates significant antitumor activities in multiple myeloma. *Clin. Cancer Res.* **2012**, *18*, 4669–4681. [CrossRef] [PubMed]

97. Derudder, E.; Dejardin, E.; Pritchard, L.L.; Green, D.R.; Korner, M.; Baud, V. RelB/p50 dimers are differentially regulated by tumor necrosis factor-alpha and lymphotoxin-beta receptor activation: Critical roles for p100. *J. Biol. Chem.* **2003**, *278*, 23278–23284. [CrossRef] [PubMed]

98. Basak, S.; Shih, V.F.; Hoffmann, A. Generation and activation of multiple dimeric transcription factors within the NF-kappaB signaling system. *Mol. Cell Biol.* **2008**, *28*, 3139–3150. [CrossRef] [PubMed]

99. Busino, L.; Millman, S.E.; Scotto, L.; Kyratsous, C.A.; Basrur, V.; O'Connor, O.; Hoffmann, A.; Elenitoba-Johnson, K.S.; Pagano, M. Fbxw7alpha- and GSK3-mediated degradation of p100 is a pro-survival mechanism in multiple myeloma. *Nat. Cell Biol.* **2012**, *14*, 375–385. [CrossRef] [PubMed]

100. Mineva, N.D.; Rothstein, T.L.; Meyers, J.A.; Lerner, A.; Sonenshein, G.E. CD40 ligand-mediated activation of the de novo RelB NF-kappaB synthesis pathway in transformed B cells promotes rescue from apoptosis. *J. Biol. Chem.* **2007**, *282*, 17475–17485. [CrossRef] [PubMed]

101. Wang, X.; Belguise, K.; Kersual, N.; Kirsch, K.H.; Mineva, N.D.; Galtier, F.; Chalbos, D.; Sonenshein, G.E. Oestrogen signalling inhibits invasive phenotype by repressing RelB and its target BCL2. *Nat. Cell Biol.* **2007**, *9*, 470–478. [CrossRef] [PubMed]

102. Dejardin, E.; Droin, N.M.; Delhase, M.; Haas, E.; Cao, Y.; Makris, C.; Li, Z.W.; Karin, M.; Ware, C.F.; Green, D.R. The lymphotoxin-beta receptor induces different patterns of gene expression via two NF-kappaB pathways. *Immunity* **2002**, *17*, 525–535. [CrossRef]

103. Stadanlick, J.E.; Kaileh, M.; Karnell, F.G.; Scholz, J.L.; Miller, J.P.; Quinn, W.J., 3rd; Brezski, R.J.; Treml, L.S.; Jordan, K.A.; Monroe, J.G.; et al. Tonic b cell antigen receptor signals supply an NF-kappaB substrate for prosurvival blys signaling. *Nat. Immunol.* **2008**, *9*, 1379–1387. [CrossRef] [PubMed]

104. Yilmaz, Z.B.; Kofahl, B.; Beaudette, P.; Baum, K.; Ipenberg, I.; Weih, F.; Wolf, J.; Dittmar, G.; Scheidereit, C. Quantitative dissection and modeling of the nf-kappab p100-p105 module reveals interdependent precursor proteolysis. *Cell Rep.* **2014**, *9*, 1756–1769. [CrossRef] [PubMed]

105. Tai, Y.T.; Li, X.; Tong, X.; Santos, D.; Otsuki, T.; Catley, L.; Tournilhac, O.; Podar, K.; Hideshima, T.; Schlossman, R.; et al. Human anti-CD40 antagonist antibody triggers significant antitumor activity against human multiple myeloma. *Cancer Res.* **2005**, *65*, 5898–5906. [CrossRef] [PubMed]

106. Zarnegar, B.; He, J.Q.; Oganesyan, G.; Hoffmann, A.; Baltimore, D.; Cheng, G. Unique CD40-mediated biological program in B cell activation requires both type 1 and type 2 NF-kappaB activation pathways. *Proc. Natl. Acad. Sci. USA* **2004**, *101*, 8108–8113. [CrossRef] [PubMed]

107. Bensinger, W.; Maziarz, R.T.; Jagannath, S.; Spencer, A.; Durrant, S.; Becker, P.S.; Ewald, B.; Bilic, S.; Rediske, J.; Baeck, J.; et al. A phase 1 study of lucatumumab, a fully human anti-CD40 antagonist monoclonal antibody administered intravenously to patients with relapsed or refractory multiple myeloma. *Br. J. Haematol.* **2012**, *159*, 58–66. [CrossRef] [PubMed]

108. Sati, H.I.; Apperley, J.F.; Greaves, M.; Lawry, J.; Gooding, R.; Russell, R.G.; Croucher, P.I. Interleukin-6 is expressed by plasma cells from patients with multiple myeloma and monoclonal gammopathy of undetermined significance. *Br. J. Haematol.* **1998**, *101*, 287–295. [CrossRef] [PubMed]

109. Cozzolino, F.; Torcia, M.; Aldinucci, D.; Rubartelli, A.; Miliani, A.; Shaw, A.R.; Lansdorp, P.M.; Di Guglielmo, R. Production of interleukin-1 by bone marrow myeloma cells. *Blood* **1989**, *74*, 380–387. [PubMed]

110. Frassanito, M.A.; Cusmai, A.; Iodice, G.; Dammacco, F. Autocrine interleukin-6 production and highly malignant multiple myeloma: Relation with resistance to drug-induced apoptosis. *Blood* **2001**, *97*, 483–489. [CrossRef] [PubMed]

111. Tai, Y.T.; Li, X.F.; Breitkreutz, I.; Song, W.; Neri, P.; Catley, L.; Podar, K.; Hideshima, T.; Chauhan, D.; Raje, N.; et al. Role of B-cell-activating factor in adhesion and growth of human multiple myeloma cells in the bone marrow microenvironment. *Cancer Res.* **2006**, *66*, 6675–6682. [CrossRef] [PubMed]

112. Tornatore, L.; Sandomenico, A.; Raimondo, D.; Low, C.; Rocci, A.; Tralau-Stewart, C.; Capece, D.; D'Andrea, D.; Bua, M.; Boyle, E.; et al. Cancer-selective targeting of the NF-kappaB survival pathway with GADD45beta/MKK7 inhibitors. *Cancer Cell* **2014**, *26*, 495–508. [CrossRef] [PubMed]

113. Romashkova, J.A.; Makarov, S.S. NF-kappaB is a target of akt in anti-apoptotic PDGF signalling. *Nature* **1999**, *401*, 86–90. [CrossRef] [PubMed]

114. Toth, C.R.; Hostutler, R.F.; Baldwin, A.S., Jr.; Bender, T.P. Members of the nuclear factor kappa B family transactivate the murine c-*myb* gene. *J. Biol. Chem.* **1995**, *270*, 7661–7671. [CrossRef] [PubMed]

115. Gardam, S.; Beyaert, R. The kinase NIK as a therapeutic target in multiple myeloma. *Expert Opin. Ther. Targets* **2011**, *15*, 207–218. [CrossRef] [PubMed]

116. Mailankody, S.; Kazandjian, D.; Korde, N.; Roschewski, M.; Manasanch, E.; Bhutani, M.; Tageja, N.; Kwok, M.; Zhang, Y.; Zingone, A.; et al. Baseline mutational patterns and sustained MRD negativity in patients with high-risk smoldering myeloma. *Blood Adv.* **2017**, *1*, 1911–1918. [CrossRef] [PubMed]

117. Mikulasova, A.; Wardell, C.P.; Murison, A.; Boyle, E.M.; Jackson, G.H.; Smetana, J.; Kufova, Z.; Pour, L.; Sandecka, V.; Almasi, M.; et al. The spectrum of somatic mutations in monoclonal gammopathy of undetermined significance indicates a less complex genomic landscape than that in multiple myeloma. *Haematologica* **2017**, *102*, 1617–1625. [CrossRef] [PubMed]

biomedicines

MDPI

Review

NF-κB Signalling in Glioblastoma

Vincent Soubannier and Stefano Stifani *

Department of Neurology and Neurosurgery, Montreal Neurological Institute, McGill University, Montreal, QC H3A2B4, Canada; vincent.soubannier@mcgill.ca
* Correspondence: stefano.stifani@mcgill.ca; Tel.: +1-514-398-3946

Received: 3 April 2017; Accepted: 7 June 2017; Published: 9 June 2017

Abstract: Nuclear factor-κB (NF-κB) is a transcription factor regulating a wide array of genes mediating numerous cellular processes such as proliferation, differentiation, motility and survival, to name a few. Aberrant activation of NF-κB is a frequent event in numerous cancers, including glioblastoma, the most common and lethal form of brain tumours of glial cell origin (collectively termed gliomas). Glioblastoma is characterized by high cellular heterogeneity, resistance to therapy and almost inevitable recurrence after surgery and treatment. NF-κB is aberrantly activated in response to a variety of stimuli in glioblastoma, where its activity has been implicated in processes ranging from maintenance of cancer stem-like cells, stimulation of cancer cell invasion, promotion of mesenchymal identity, and resistance to radiotherapy. This review examines the mechanisms of NF-κB activation in glioblastoma, the involvement of NF-κB in several mechanisms underlying glioblastoma propagation, and discusses some of the important questions of future research into the roles of NF-κB in glioblastoma.

Keywords: brain cancer; cancer stem-like cells; DNA damage repair; glioblastoma; mesenchymal glioblastoma subtype; NF-κB (nuclear factor-κB)

1. Glioblastoma

Glioblastoma (GBM), the most common and malignant of all gliomas, is a brain cancer characterized by remarkable cellular heterogeneity, aggressive growth, extensive invasion of brain tissue, and almost inevitable recurrence. It is virtually untreatable, even after combined surgery, radiation therapy and chemotherapy, and GBM patients have an average survival of less than two years [1–3].

GBM exhibits a build-up of a variety of poorly differentiated neural cells. At a minimum, GBM is hypothesized to contain three cell subpopulations. One pool is thought to include relatively rare cells with self-renewing and cancer (re)populating potential (hereafter referred to as "stem-like cells"). A separate pool, also expected to be infrequent, is proposed to harbour cells that have exited the cell cycle, have slow cellular turnover, but are not endowed with persistent self-renewal capacity (hereafter referred to as "quiescent cells"). The third pool is thought to be more prevalent and diverse and to include a collection of mitotic cells exhibiting various degrees of incomplete differentiation (hereafter operationally defined collectively as "differentiated cells"). The GBM stem-like cell pool is thought to be responsible for tumour (re)initiation. Both stem-like and quiescent cell pools have the potential to act as reservoirs of undifferentiated cells able to give rise to more rapidly dividing progeny cells. The differentiated GBM cell pool is considered as the main contributor to processes underlying GBM development, such as tumour growth and invasion [4–9].

The cellular heterogeneity of GBM is a major obstacle on the road to treatment because the diversity of cancer cells within the same tumour implies that they may respond differently to therapy. Current GBM treatments usually target the differentiated cell pools. Unfortunately, these approaches have failed to treat GBM. This situation suggests that stem-like and quiescent cell pools are not targeted by current therapies and may represent the population(s) responsible for GBM therapy resistance and recurrence after surgery and treatment. Unfortunately, our understanding of the pathobiology, and contribution to tumourigenesis, of different GBM cell pools remains limited.

To add to the complexity of GBM, emerging evidence suggests that at least some of the different GBM cell populations are in a fluent state and can switch from one phenotype to another. In this regard, recent studies suggest that exposure of glioma cells to therapeutic doses of temozolomide, the compound most commonly used in GBM chemotherapy, increases the stem-like cell pool as a result of a "differentiated-to-stem-like" phenotypic shift [9–11]. These observations suggest the existence of dynamic relationships between different GBM cell pools, a situation presenting a remarkable challenge to current and future GBM therapies. These findings underscore the necessity to improve our understanding of the molecular mechanisms underlying the pathobiology of different GBM cell populations and their contributions to gliomagenesis.

Numerous oncogenic pathways are active in GBM, and several previous reviews have addressed how they contribute to gliomagenesis by promoting processes ranging from proliferation to invasion to therapy resistance, e.g., [2,8,9,12,13]. This review will focus specifically on the involvement of nuclear factor-κB (NF-κB) signaling pathways in GBM and the implications of aberrant NF-κB activation in different GBM cell populations.

2. Nuclear Factor-κB Signalling

The term NF-κB signalling refers to several mechanisms, activated by a variety of stimuli, which ultimately converge on the NF-κB family of transcription factors. NF-κB is a dimeric DNA-binding complex composed of varying combinations of five family members including p50/NFKB1 (p50 hereafter for sake of clarity), p52/NFKB2 (p52), Rel-like domain containing protein A (RelA/RELA (RelA), RelB/RELB (RelB), and c-Rel/REL (c-Rel) [14–19]. These molecules act frequently as heterodimers (e.g., p50:RelA), but homodimers (e.g., RelA:RelA) have also been observed. They regulate the expression of a wide array of genes involved in important biological processes, such as cell proliferation, apoptosis, DNA repair, and immune and inflammatory responses, to name a few [14–19].

NF-κB dimers are expressed in many different cell types, where they are usually kept in an inactive (non-DNA bound) state by specific inhibitors of NF-κB (IκB) until the reception of activating signals occurs. IκB proteins such as NF-κB Inhibitor Alpha (NFKBIA)/IκBα (IκBα) bind NF-κB dimers and sequester them in the cytosol, thereby preventing DNA binding and transcriptional regulation by the latter.

In canonical NF-κB pathways, the activation of the IκB kinase (IKK/IKBK) complex (IKK) [composed of IKKα, IKKβ, and NF-κB essential modulator (NEMO)/IKKγ subunits] leads to IκBα phosphorylation by IKKβ, ubiquitination and proteosomal degradation of IκBα, and the ensuing dissociation of p50:RelA-containing NF-κB dimers from IκB and the translocation of NF-κB to the nucleus [14–19]. Canonical NF-κB activation is typically observed after stimulation of surface receptors such as the tumor necrosis factor alpha (TNFα) receptor 1 or the interleukin-1 receptor [14–19].

Activation of NF-κB can also occur through non-canonical mechanisms, involving NF-κB-inducing kinase (NIK; referred to as Mitogen-Activated Protein Kinase Kinase Kinase 14 (MAP3K14) in human cells) and p52-containing NF-κB dimers. In these pathways, acting downstream of receptors like the B cell activating factor receptor or the lymphotoxin beta receptor, phosphorylation of the p52 precursor form, p100 (which retains p100:RelB-containing NF-κB in the cytosol), leads to p100 proteolytic cleavage, generating p52:RelB NF-κB dimers that can translocate to the nucleus [14–19].

A third mode of NF-κB activation, termed atypical, has also been described, most characteristically in response to DNA double-strand breaks, replication stress or reactive oxygen species. Atypical NF-κB activation mechanisms can be divided into two main groups, depending on whether they are dependent or independent of the activity of the IKK complex [14,20]. In the case of IKK-dependent atypical pathways, stressful stimuli like genotoxic stress induce the NEMO protein to translocate to the nucleus, where it becomes sumoylated and ubiquitinated through a mechanism dependent on the ataxia and telangiectasia mutated (ATM) kinase. The sequential phosphorylation and ubiquitination of sumoylated nuclear NEMO results in the export of NEMO from the nucleus to the cytosol, where the latter activates IKKβ and induces NF-κB. IKK-independent atypical pathways involve p50-containing dimers but rely on other kinases, such as casein kinase-II, for the dissociation of NF-κB dimers from IκB and translocation to the nucleus [14,20].

Regulation of NF-κB signalling does not only involve nuclear translocation but also includes several other control mechanisms, ranging from post-translational modifications of specific NF-κB subunits, protein-protein interactions occurring at specific gene regulatory sites, as well as mechanisms of nuclear export [14–20]. Thus, different cellular responses to NF-κB usually result from a complex combination of specific mechanisms that may differ in their upstream stimuli and/or downstream targets.

3. Activation of NF-κB in Glioblastoma

Activation of NF-κB is a frequent event in several tumours. Deregulated NF-κB activation is often oncogenic through the promotion of tumor growth and invasion, the suppression of programmed cell death, as well as resistance to therapy [21,22]. Consistent with the complex nature of NF-κB activation in response to a variety of stimuli, aberrant NF-κB activity in cancer may ensue as a result of numerous events, including mutation or deregulated expression of genes encoding the NF-κB proteins or, more frequently, the perturbation of the mechanisms controlling the activation of NF-κB dimers [21–25].

Aberrant constitutive activation of NF-κB is a common event in GBM [26–30]. Numerous mechanisms have been proposed to contribute to deregulated NF-κB signalling in gliomas. For instance, receptor tyrosine kinases, most notably epidermal growth factor receptor (EGFR) and platelet derived growth factor receptor (PDGFR), which are often aberrantly activated in GBM, have been linked to NF-κB activation through a number of mechanisms, involving both protein kinase B/AKT (AKT)-dependent and -independent pathways. Oncogenic EGFR and PDGFR signalling mechanisms are important contributors to tumour growth and invasion in GBM, and NF-κB is implicated in at least some of the tumour promoting functions of these receptors [31–35].

Loss of tumour suppressors such as phosphatase and tensin homolog (PTEN) and neurofibromin 1 (NF1) has also been linked to aberrant NF-κB activation in GBM, at least in part as a result of increased PI3-kinase (PI3K) activity [32,35]. The depletion of the tumour suppressor, Krueppel-like factor 6 (KLF6), which acts as a negative NF-κB regulator, is also postulated to contribute to NF-κB activation in GBM [36], and so are perturbations of *TP53* biology, including both the loss and activation of the *TP53*-encoded protein, p53 [37–40].

Numerous other mechanisms are suggested to contribute to aberrant NF-κB signalling in GBM, including NF-κB activation by peptidyl-prolyl isomerase PIN1, mixed lineage kinase 4 (MLK4), heterozygous deletion of *NFKBIA*, the gene encoding IκBα, high levels of micro(mi)RNA-30e*, which targets IκBα, as well as DNA-damage [41–45].

These combined observations underscore the involvement of aberrant NF-κB pathways in multiple processes of GBM pathogenesis and identify several mechanisms upstream and downstream of NF-κB signalling.

4. Role of NF-κB in Glioblastoma Stem-Like Cells

4.1. Glioblastoma Patient-Derived Stem-Like Cell Cultures

Recent advances in GBM research have enabled the derivation of cultures of stem-like cells from surgically resected brain tumours from individual patients. These patient-derived cells exhibit stem-like cell behaviour in vitro and can serially propagate brain tumours when implanted into the crania of host mice under limiting dilution conditions [6–9]. GBM patient-derived stem-like cells (GSCs) share with normal neural stem cells the ability to generate progeny cultures comprised of a mixture of stem-like and more restricted non-stem-like descendants. This thereby offers the previously unavailable opportunity to study the behaviour of different GBM cell populations in mixed cultures initiated by single stem-like cells derived from individual patients. This situation provides a physiologically informative experimental model system to study mechanisms underlying the behaviour of different GBM cell pools contributing to tumour initiation, development and recurrence.

4.2. Involvement of NF-κB in Glioblastoma Stem-Like Cell Maintenance

Although the characterization of the precise pattern of NF-κB activation in different GBM cell populations remains elusive in surgically resected tissues, several studies have demonstrated the activation of NF-κB in patient-derived GSC cultures [42,46–49]. The involvement of NF-κB in GSCs has been examined using a number of biological assays, including studies based on the ability of cancer cells with stem-like properties to give rise to new populations of descendant cells (referred to as "tumourspheres") under limiting dilution conditions. In these assays, the number of cells/well required to generate at least one new tumoursphere is taken as an indication of the frequency of cells with repopulating ability, a typical feature of stem-like cells.

The inhibition of endogenous NF-κB activity in patient-derived GSC cultures, using the selective IKKβ antagonist, Compound A, or siRNA-mediated knockdown of IKKβ and/or RelA, was shown to markedly decrease tumoursphere formation frequency, suggesting that canonical NF-κB inhibition causes a reduction in the number of cells with self-renewal capacity. A similar effect was observed when the non-canonical NF-κB pathway was impaired through the attenuation of p52 or RelB [48]. A role for non-canonical NF-κB signalling in GSC self-renewal was also suggested by the independent observation that knockdown of RelB blocks the self-renewal activity of patient-derived glioma-initiating cells [49]. Together, these studies suggest that NF-κB signalling is involved in sustaining the GBM stem-like cell compartment and that both canonical and non-canonical branches of the NF-κB pathway are important for this function.

Rinkenbaugh and colleagues suggested that at least one source of NF-κB activation in GSCs is the transforming growth factor-β-activated kinase 1, a protein previously implicated in NF-κB signalling due to its ability to activate IKK [48]. Ohtsu et al. provided evidence suggesting that non-canonical NF-κB signalling is activated in GSCs downstream of epithelial V-like antigen 1, a protein originally identified as an immunoglobulin superfamily member expressed in developing thymus epithelial cells and involved in T-cell development in early mouse embryos. Epithelial V-like antigen 1 activates the non-canonical NF-κB signaling pathway through a TNF receptor-associated factor 2/cellular inhibitor of apoptosis-dependent accumulation of NIK [49].

Taken together, these observations suggest that a variety of mechanisms can lead to aberrant NF-κB activation in GSCs, consistent with the above-mentioned demonstration that a plethora of deregulated processes, including EGFR amplification, PTEN deletion, and monoallelic *NFKBIA* deletion, have been associated with deregulated NF-κB activation in GBM. It is therefore plausible that a variety of NF-κB-regulated mechanisms, activated in response to both canonical and non-canonical pathways, contribute to the regulation of the behaviour of GSCs.

The pathophysiological significance of these observations remains to be fully determined. NF-κB inhibition might result in depletion of the GBM stem-like cell pool because of impaired survival of these cells. Alternatively, the inhibition of NF-κB signalling in GSC cultures might enhance the pool of senescent cells at the expense of the stem-like compartment, a possibility that would be consistent with the demonstration that blockade of NF-κB signalling drives differentiating glioblastoma cells into replicative senescence [50]. It seems less likely that the stem-like cell compartment might be diminished as a result of an enhanced transition to a more proliferative, "transit-amplifying-like", state because the growth rate of NF-κB-attenuated GSCs was shown to be decreased, rather than increased, in ex vivo brain slice explants [48]. In the future, it will be important to characterize further the mechanisms upstream and downstream of NF-κB activation in the GSC compartment, as well as to perform in vivo brain tumour xenograft studies under limiting dilution conditions to better understand the contribution of NF-κB to stem-like cells and to GBM initiation and/or recurrence.

5. Involvement of NF-κB in Glioblastoma Invasion

5.1. Deregulated NF-κB Activation in Mesenchymal GBM Subtype

It has long been established that epithelial to mesenchymal transition (EMT) is a process associated with advanced malignancy in numerous cancers, and that NF-κB acts to promote EMT in several cell types [51]. Consistent with the latter finding, the examination of glioma patient-based mRNA expression databases first suggested that aberrant NF-κB activation was preferentially associated with GBM cases with a mesenchymal phenotype, rather than GBM subtypes with more neural features (many of which are defined as having a proneural phenotype) [52–54]. Mesenchymal features are a hallmark of glioma aggressiveness and are associated with poor patient outcome, due in part to the highly invasive nature of these tumours and increased radioresistance [55].

An EMT-like proneural-to-mesenchymal phenotypic shift occurs in GBM in response to factors in the microenvironment or cytotoxic treatments. This transition involves NF-κB, as demonstrated by the observation that canonical TNFα/NF-κB signalling can promote a proneural-to-mesenchymal transition in at least a subset of GBM patient-derived GSC cultures [42,47]. Moreover, RelB is highly expressed in the mesenchymal glioma subtype and the loss of RelB significantly attenuates glioma cell survival, motility and invasion. Importantly, RelB promotes the expression of mesenchymal genes in glioma cells [56]. Consistent with these findings, a mesenchymal signature, including NF-κB activation, is correlated with poor radiation response and shorter survival in GBM patients [47,56].

Further evidence for an important role for NF-κB in mesenchymal identity in GBM was provided by the demonstration that RelB-mediated NF-κB signalling is a critical mediator of GBM cell migration and invasion stimulated by the SMAC (second mitochondrial activator of caspases) mimetic, BV6, a molecule that antagonizes the inhibitor of apoptosis proteins and also triggers cell elongation, migration and invasion in GBM [57]. BV6-stimulated NF-κB activation leads to elevated *TNFα* mRNA levels, as well as increased levels of NF-κB target genes implicated in cell migration and invasion [57].

The mechanisms responsible for the activation of NF-κB in mesenchymal GBM tumours are beginning to be elucidated. The kinase MLK4 is overexpressed in mesenchymal, but not proneural, GSCs, where it is important for maintenance of the mesenchymal phenotype and for self-renewal, motility, tumourigenesis, and radioresistance. MLK4 binds to, and phosphorylates, IKKα in GSC cultures, thereby leading to activation of NF-κB signalling [42]. The peptidyl-prolyl isomerase PIN1, which recognizes phosphorylated Ser residues on RelA and promotes NF-κB activation, is up-regulated in GBM [41]. PIN1 attenuation decreases the amount of activated, phosphorylated RelA in the nucleus, with a concomitant decrease in the expression of the NF-κB target gene, *interleukin-8* (*IL-8*). These effects are associated with decreased glioma cell dissemination capacity. Conversely, two negative regulators of NF-κB signalling, inhibitor of growth family member 4 and protein inhibitor of activated STAT3, are decreased in GBM when compared to non-cancerous cells [58–60].

5.2. Involvement of NF-κB in Transcriptional Activation of Genes Promoting Epithelial-to-Mesenchymal Transition and Cell Motility

As discussed above, NF-κB signalling pathways play key roles in promoting and maintaining EMT in both healthy and cancer cells. This function is performed through regulation of the expression of numerous epithelial and mesenchymal genes [51,61]. For instance, studies in both invertebrate and vertebrate species identified NF-κB as a key player in the transactivation of mesenchymal genes such as *Snail*, *zinc finger E-Box binding homeobox 1* (*ZEB1*), *ZEB2*, *Twist*, and *matrix metalloproteinase* (*MMP*)-*2* and *MMP-9* [51,60,61].

There is evidence suggesting that NF-κB also activates the expression of mesenchymal genes in GBM. NF-κB binds to the *ZEB1* promoter in glioma cells in response to connective tissue growth factor, which is important for glioma invasion [62]. NF-κB is also involved in the activation of *MMP-2* and *MMP-9* in GBM cells, at least in part in response to protein kinase C (PKC) and mechanistic target of rapamycin (mTOR) signalling [63,64]. Moreover, RelB promotes the expression of the gene *YKL-40*, considered as a typical marker of the mesenchymal GBM subtype [56].

NF-κB has been further implicated in promoting GBM invasion by the demonstration of its involvement in the transactivation of the expression of several genes encoding molecules promoting cell motility and invasion. These include *fibroblast growth factor inducible 14 (FN14)*, a member of the TNF receptor superfamily. *FN14* is highly expressed in invading glioma cells in vivo. The *FN14* promoter region contains NF-κB binding sites important for sustained overexpression of *FN14* and enduring glioma cell invasion. Consistently, *FN14* gene expression levels increase with glioma grade and inversely correlate with patient survival [65]. Activated NF-κB in GBM also regulates other genes involved in cell migration, such as *IL-8, monocyte chemoattractant protein 1, cxc chemokine receptor 4*, to name a few [57].

In addition to controlling gene regulatory events promoting glioma invasion in a cell-autonomous manner, NF-κB also appears to promote GBM cell invasion in a non-cell autonomous ways. Specifically, high endogenous expression of receptor activator of NF-κB (RANKL) in GBM cells leads to the activation of neighbouring astrocytes in the tumour microenvironment through NF-κB signalling. Activated astrocytes in turn signal back to the glioma cells to promote glioma invasion [66]. These findings are consistent with the demonstration that intercellular communication between neighbouring astrocytes and GBM cells, possibly mediated by secreted extracellular vesicles, plays key roles in GBM growth and invasion [67,68].

These combined observations provide evidence for an important role of NF-κB in the promotion of more invasive and malignant mesenchymal features in GBM. The interpretation of these observations is complicated in part by the lack of information about the pattern of NF-κB activation in different GBM cell subpopulations. Presumably, enhanced dissemination potential should be a feature of more developmentally advanced GBM cells resembling the migratory neural cells in the healthy brain (e.g., glial precursor cells that leave their place of origin in the subventricular zone to reach their final destinations in the brain). If this were indeed the case, it would be reasonable to assume that NF-κB signalling is activated in, and promotes the migration of, more differentiated GBM cellular subtypes. A similar scenario could also result, however, by the activation of NF-κB in the GBM stem-like cell compartment, where its ability to promote proneural-to-mesenchymal transition could result in the generation of more restricted daughter cells with enhanced dissemination potential. It is also possible that NF-κB could enhance the migratory potential of most GBM cells in which it is activated, regardless of their stem-like or non-stem-like state. In the future, it will be important to better characterize the features of the cancer cells in which NF-κB is activated in GBM surgical specimens, as well as in the heterogeneous population of cells comprising typical GBM tumourspheres studied in vitro.

6. Other Roles of NF-κB in Glioblastoma

6.1. Resistance to Radiotherapy

Radiotherapy is a customary component of GBM treatment. Unfortunately, the majority of GBM patients exhibit radioresistance. Several mechanisms of cancer cell resistance to radiotherapy have been described, including enhanced ability to repair DNA damage, the presence of quiescent or slowly dividing cells that are less vulnerable to DNA damage, and resistance to apoptosis [69,70]. Numerous lines of studies suggest that aberrant NF-κB activation may contribute to radioresistance in GBM by modulating several of these processes.

In response to DNA damage, cells usually activate mechanisms aimed at restoring genomic stability or, in the case of more severe damage, leading to apoptosis. DNA damage is a well-known activator of NF-κB signalling; in turn, NF-κB plays important roles in DNA damage repair mechanisms. For instance, the DNA damage sensor poly-(ADP-ribose) polymerase-1 (PARP1) activates NF-κB through an ATM- and IKK-mediated pathway. NF-κB also participates in DNA repair through interaction with breast cancer-associated gene 1 (BRCA1), at least in part via interaction with

BRCA1-CtIP complexes, thereby promoting homologous recombination [44,71–73]. NF-κB activation also induces the expression of BRCA2, another important DNA repair proteins, as well as ATM, the key DNA double-strand-signalling kinase. Moreover, NF-κB activates the non-homologous end-joining recombination protein, ku70 [74,75]. It is worth mentioning that NF-κB also mediates chemoresistance to alkylating agents such as temozolomide. The mutagenic effects of these compounds are inhibited by the cellular DNA repair enzyme *O*-6-methylguanine-DNA-methyltransferase (MGMT), which removes alkyl/methyl adducts from DNA. Since MGMT becomes inactivated in the process, its de novo expression is a central mode of chemoresistance. NF-κB plays an important role in regulation of MGMT activity in glioma cells by activating *MGMT* gene expression through two NF-κB binding sites within the *MGMT* promoter [76].

As discussed above, NF-κB activation promotes maintenance of the GBM stem-like cell pool [48,49]. GBM stem-like cells are thought to promote radioresistance [77–79], suggesting that an additional mechanism through which NF-κB has an impact on GBM resistance to radiotherapy is by increasing the fraction of cancer cells with stem-like behaviour. The important role of NF-κB in promoting a mesenchymal GBM phenotype is also considered to be a contributor to radiation resistance because the mesenchymal subtype is associated with poor radiation response in GBM patients [47]. It is hypothesized that mesenchymal differentiation driven by NF-κB is associated with activation of checkpoint pathways, leading to enhanced DNA damage repair and unperturbed cell cycle progression in response to radiation [47].

Several lines of evidence suggest additional roles for NF-κB signalling in radiotherapy resistance in cancers by promoting the expression of genes implicated in evasion of apoptosis, the promotion of cell cycle progression, and the production of antioxidants [80,81]. NF-κB signalling regulates the expression of genes important for cell survival, including *Bcl2*, *Bcl-xL*, *survivin* and *inhibitor of apoptosis proteins*, as well as cell cycle progression genes like *Cyclin D1* [80–83]. Moreover, radiation-activated NF-κB signalling is associated with upregulated expression of NF-κB targets genes such as *IL-6* and *IL-8* in GBM. It was reported that IL-6 secreted by glioma cells enhances the invasive potential of these cells [84] and that IL-8 is important for glial tumour neovascularity and progression [85].

Together, these observations provide evidence for a role of NF-κB in numerous mechanisms underlying resistance to radiation therapy in several cancers, including GBM. These processes are likely to be particularly relevant in the context of the GBM stem-like cell pool, but may also impact on other GBM subpopulations in which NF-κB is activated.

6.2. Regulation of Cell Metabolism

As is the case in other cancers, aerobic glycolysis (production of lactate from glucose in the presence of oxygen) is a hallmark of GBM, underlying the need to synthesize molecules essential for tumour cell proliferation, including nucleotides, fatty acids, and proteins [86]. Consistent with the observation that NF-κB signalling plays an important role in regulating energy metabolism in several cancers [87,88], NF-κB contributes to the promotion of aerobic glycolysis in GBM through a number of mechanisms.

Pyruvate kinase M2 (PKM2), a member of the enzyme family regulating the rate-limiting step of glycolysis, is overexpressed in numerous cancers, including GBM [89,90]. Activation of EGFR in cancer cells results in increased glucose uptake and lactate production in a PKM2-dependent manner. The aberrant activation of *PKM2* expression occurs in response to EGFR-induced and PKCε monoubiquitylation-dependent activation of RelA, which interacts with transcription factor hypoxia-inducible factor 1-α to form a dimer that can bind to the *PKM2* promoter and activate *PKM2* transcription. Consistent with these findings, *PKM2* expression correlates with EGFR and IKKβ activity in human GBM specimens and with the grade of glioma malignancy [34].

NF-κB could also contribute to energy metabolism in GBM by upregulating mitochondrial respiration. In the presence of functional p53, RelA activates the expression of mitochondrial SCO2 (synthesis of cytochrome c oxidase assembly protein), a metallochaperone essential in the biogenesis of cytochrome c oxidase subunit II, enhancing oxidative metabolism [91]. Although this process may be altered in cancer cells without a functional p53, resulting in lower oxidative phosphorylation and increased glycolysis (possibly due to upregulation of metabolic genes such as high affinity glucose transporter 3), the possible significance of these mechanisms in the context of GBM remains to be defined.

In addition to glycolysis, glutamine metabolism is also important in cancer cells, both as another major energy source through α-ketoglutarate and as a nitrogen donor for nucleic acid synthesis. Several lines of evidence suggest an involvement of NF-κB in the control of glutamine metabolism. Glutaminase is likely the rate-limiting enzyme for glutamine consumption in cancer cells and increased glutaminase activity by Rho GTPase through a NF-κB-dependent mechanism was described as a process involved in meeting the elevated glutamine demand in cancer cells [92]. Additionally, NF-κB may exert a control on the expression of glutaminase in cancer cells through the regulation of miRNAs. In human leukemic Jurkat cells, RelA binds to the *miR-23a* promoter and inhibits *miR-23a* expression. In turn, *miR-23* targets *glutaminase* mRNA and inhibits expression of *glutaminase* [93].

Taken together, these findings point to NF-κB as an emerging important regulator of cell metabolism in numerous cancers, including GBM.

6.3. Involvement in Autophagy

Autophagy is an important cellular process that generally protects cells from stressful conditions by enabling continuous overhaul of cellular constituents through the degradation and recycling of damaged or non-essential macromolecular components, or even entire organelles such as mitochondria, ribosomes or endoplasmic reticulum [94–96]. Autophagy can play tumour-suppressing roles by maintaining cellular homeostasis through protection from accumulation of damaged proteins or reactive oxygen species. In cancer cells, however, autophagy plays a variety of tumour-promoting roles, including supporting cellular metabolism to promote cancer cell growth and survival and contributing to therapy resistance. Autophagy can also supports metastasis by protecting detached cells from anoikis, a form of programmed cell death [96–100].

The enhancement of pro-survival autophagy mechanisms is considered as an important contributor to cancer-promoting metabolism alterations in GBM. The remarkable resistance of GBM to chemo- and radiotherapy is also thought to result in part from the contribution of autophagy to the adaptation capabilities of this cancer [99–101]. On the basis of these considerations, the goal of inhibiting mechanisms of autophagy is emerging as an attractive therapeutic approach in GBM.

Recent studies suggest a complex interplay between NF-κB and autophagy [102]. NF-κB was initially proposed to have a negative effect on autophagy by activating the autophagy inhibitor mTOR [103,104]. However, more recent work has provided evidence that autophagy can activate NF-κB [105]. In agreement with this finding, there is evidence that autophagy and NF-κB mechanisms may cooperate in gliomas. Specifically, the mediator of autophagy, multifunctional scaffold protein p62, plays a role in the activation of IKKs and the resulting activation of NF-κB signaling [106–108]. A complex cross-talk between NF-κB and autophagy is further suggested by the observation that selenite-induced decrease in the expression of heat shock protein-90, which is associated with decreased autophagy and increased apoptosis, inhibits NF-κB signaling in human leukemia cells [109]. Consistently, selective targeting of heat shock protein-90 resulting in compensatory autophagy and unfolded protein response in mitochondria is correlated with the repression of NF-κB-dependent gene expression, enhanced tumor cell apoptosis, and reduced intracranial GBM growth in mice [110]. These observations suggest an important involvement of NF-κB in regulating the balance between autophagy and apoptosis in cancer.

In conclusion, these findings underscore important, but complex, interactions between autophagy and NF-κB during key cellular processes in cancer, including regulation of cellular metabolism, resistance to chemo- and radiotherapy, and/or responses to unfolded proteins. Understanding these mechanisms may enable more effective combinatorial therapeutic strategies.

Biomedicines **2017**, *5*, 29

6.4. Promotion of Angiogenesis

Angiogenesis is a common feature of the most aggressive gliomas and is correlated with poor patient prognosis. As discussed above, NF-κB activates the expression of *IL-8*, a pro-angiogenic gene, in glioma cells [85]. Moreover, NF-κB promotes the expression of vascular endothelial growth factor (VEGF), a major driver of angiogenesis [111]. Consistently, the impairment of NF-κB signaling significantly decreases GBM growth and angiogenesis in nude mice [111]. During the process of glioma angiogenesis, NF-κB is induced downstream of the transcription factor, BMI1 proto-oncogene, polycomb ring finger (BMI1), whose activity is required for expression of both NF-κB and VEGF [112]. Both the inhibition of NF-κB activity and knockdown of BMI1 result in decreased angiogenesis in orthotopically transplanted human gliomas [112]. These findings are in agreement with the demonstrated involvement of NF-κB in angiogenesis in other cancers [113] and further underscore the multiple roles of this transcription factor in gliomas.

Lastly, it is worth noting that several studies have shown that GSCs have the capacity to trans-differentiate into endothelial cells (ECs) that exhibit EC molecular profiles in vitro and can contribute to GBM vascularization in vivo [114–117]. It is possible that NF-κB may be involved in the EC trans-differentiation potential of GSCs. This is suggested by the recent demonstration that ovarian cancer stem-like cells can be induced to trans-differentiated into ECs by the activation of NF-κB through an autocrine loop mediated by the chemokine CCL5 [118]. CCL5 is synthesized and secreted by glioma-associated microglia [119], and thus a CCL5-mediated activation of NF-κB in GSCs may contribute to EC trans-differentiation in GBM, as is the case in other cells.

7. Concluding Comments

Aberrant NF-κB activation is a hallmark of numerous cancers, including GBM. This situation is consistent with the involvement of NF-κB-mediated signalling pathways in the control of a broad range of biological processes, including cell proliferation, survival, differentiation, motility, DNA repair, inflammation and angiogenesis. The fact that multiple stimuli and conditions can lead to NF-κB activation in GBM, often triggering multiple NF-κB pathways concurrently, implies that different modes of NF-κB activation may impact specifically on different mechanisms of cancer propagation. Thus, it will be of particular importance in the future to precisely understand not only the upstream stimuli but also the downstream targets of NF-κB signalling pathways during the various processes underlying gliomagenesis.

This information will also be essential if NF-κB-mediated pathways are to be considered as potentially attractive targets for GBM therapy. This possibility has been the subject of substantial interest, as addressed extensively in several previous reviews [21,23,30,120,121]. It should be emphasized, however, that although a number of approaches targeting NF-κB have shown promise in preclinical GBM models, those strategies that have been tested in clinical settings have thus far not shown satisfactory advantages over other approaches (reviewed in [30,120] and references therein). The promises and challenges of intervention strategies targeting NF-κB pathways will rely on the ability to design approaches directed at specific physiological mechanisms responsible for specific pathological outcomes in response to NF-κB activation. Moreover, given the role played by NF-κB in several vital processes, such as cellular homeostasis and immunity, prolonged NF-κB inhibition may possibly result in detrimental effects in tissues other than the cancerous ones.

In summary, understanding the range of contributions of NF-κB pathways to GBM both represents a remarkable scientific challenge and has the potential to provide important new insight into the management of this deadly cancer.

Acknowledgments: Grant support: Canadian Institutes of Health Research Operating Grants MOP-123270 and MOP-123500 to Stefano Stifani.

Conflicts of Interest: The authors declare no conflict of interest.

References

1. Omuro, A.; DeAngelis, L.M. Glioblastoma and other malignant gliomas: A clinical review. *JAMA* **2013**, *310*, 1842–1850. [CrossRef] [PubMed]
2. Aldape, K.; Zadeh, G.; Mansouri, S.; Reifenberger, G.; von Deimling, A. Glioblastoma: Pathology, molecular mechanisms and markers. *Acta Neuropathol.* **2015**, *129*, 829–848. [CrossRef] [PubMed]
3. Ellis, H.P.; Greenslade, M.; Powell, B.; Spiteri, I.; Sottoriva, A.; Kurian, K.M. Current challenges in glioblastoma: Intratumour heterogeneity, residual disease, and models to predict disease recurrence. *Front. Oncol.* **2015**, *5*, 251. [CrossRef] [PubMed]
4. Schneider, M.; Ströbele, S.; Nonnenmacher, L.; Siegelin, M.D.; Tepper, M.; Stroh, S.; Hasslacher, S.; Enzenmüller, S.; Strauss, G.; Baumann, B.; et al. A paired comparison between glioblastoma "stem cells" and differentiated cells. *Int. J. Cancer* **2016**, *138*, 1709–1718. [CrossRef] [PubMed]
5. Bonavia, R.; Inda, M.M.; Cavenee, W.K.; Furnari, F.B. Heterogeneity maintenance in glioblastoma: A social network. *Cancer Res.* **2011**, *71*, 4055–4060. [CrossRef] [PubMed]
6. Yan, K.; Yang, K.; Rich, J.N. The evolving landscape of glioblastoma stem cells. *Curr. Opin. Neurol.* **2013**, *26*, 701–770. [CrossRef] [PubMed]
7. Suvà, M.L.; Rheinbay, E.; Gillespie, S.M.; Patel, A.P.; Wakimoto, H.; Rabkin, S.D.; Riggi, N.; Chi, A.S.; Cahill, D.P.; Nahed, B.V.; et al. Reconstructing and reprogramming the tumor-propagating potential of glioblastoma stem-like cells. *Cell* **2014**, *157*, 580–594. [CrossRef] [PubMed]
8. Seymour, T.; Nowak, A.; Kakulas, F. Targeting aggressive cancer stem cells in glioblastoma. *Front. Oncol.* **2015**, *5*, 159. [CrossRef] [PubMed]
9. Lathia, J.D.; Mack, S.C.; Mulkearns-Hubert, E.E.; Valentim, C.L.; Rich, J.N. Cancer stem cells in glioblastoma. *Genes Dev.* **2015**, *29*, 1203–1217. [CrossRef] [PubMed]
10. Auffinger, B.; Tobias, A.L.; Han, Y.; Lee, G.; Guo, D.; Dey, M.; Lesniak, M.S.; Ahmed, A.U. Conversion of differentiated cancer cells into cancer stem-like cells in a glioblastoma model after primary chemotherapy. *Cell Death Differ.* **2014**, *21*, 1119–1131. [CrossRef] [PubMed]
11. Safa, A.R.; Saadatzadeh, M.R.; Cohen-Gadol, A.A.; Pollok, K.E.; Bijangi-Vishehsaraei, K. Glioblastoma stem cells (GSCs) epigenetic plasticity and interconversion between differentiated non-GSCs and GSCs. *Genes Dis.* **2015**, *2*, 152–163. [CrossRef] [PubMed]
12. Lasorella, A.; Sanson, M.; Iavarone, A. *FGFR–TACC* gene fusions in human glioma. *Neuro Oncol.* **2017**, *19*, 475–483. [CrossRef] [PubMed]
13. Li, X.; Wu, C.; Chen, N.; Gu, H.; Yen, A.; Cao, L.; Wang, E.; Wang, L. PI3K/Akt/mTOR signaling pathway and targeted therapy for glioblastoma. *Oncotarget* **2016**, *7*, 33440–33450. [CrossRef] [PubMed]
14. Perkins, N.D. Integrating cell-signalling pathways with NF-κB and IKK function. *Nat. Rev. Mol. Cell Biol.* **2007**, *8*, 49–62. [CrossRef] [PubMed]
15. Hayden, M.S.; Ghosh, S. NF-κB, the first quarter-century: Remarkable progress and outstanding questions. *Genes Dev.* **2012**, *26*, 203–234. [CrossRef] [PubMed]
16. Napetschnig, J.; Wu, H. Molecular basis of NF-κB signaling. *Annu. Rev. Biophys.* **2013**, *42*, 443–468. [CrossRef] [PubMed]
17. Mitchell, S.; Vargas, J.; Hoffmann, A. Signaling via the NF-κB system. *Wiley Interdiscip. Rev. Syst. Biol. Med.* **2016**, *8*, 227–241. [CrossRef] [PubMed]
18. Sun, S.C. Non-canonical NF-κB signaling pathway. *Cell Res.* **2011**, *21*, 71–85. [CrossRef] [PubMed]
19. Shih, V.F.; Tsui, R.; Caldwell, A.; Hoffmann, A. A single NF-κB system for both canonical and non-canonical signaling. *Cell Res.* **2011**, *21*, 86–102. [CrossRef] [PubMed]
20. Gloire, G.; Legrand-Poels, S.; Piette, J. NF-κB activation by reactive oxygen species: Fifteen years later. *Biochem. Pharmacol.* **2006**, *72*, 1493–1505. [CrossRef] [PubMed]
21. Xia, Y.; Shen, S.; Verma, I.M. NF-κB, an active player in human cancers. *Cancer Immunol. Res.* **2014**, *2*, 823–830. [CrossRef] [PubMed]
22. Rayet, B.; Gelinas, C. Aberrant *Rel/NF-κB* genes and activity in human cancer. *Oncogene* **1999**, *18*, 6938–6947. [CrossRef] [PubMed]
23. Bosman, M.C.; Schuringa, J.J.; Vellenga, E. Constitutive NF-κB activation in AML: Causes and treatment strategies. *Crit. Rev. Oncol. Hematol.* **2016**, *98*, 35–44. [CrossRef] [PubMed]

24. Shostak, K.; Chariot, A. EGFR and NF-κB: Partners in cancer. *Trends Mol. Med.* **2015**, *21*, 385–393. [CrossRef] [PubMed]
25. Prabhu, L.; Mundade, R.; Korc, M.; Loehrer, P.J.; Lu, T. Critical role of NF-κB in pancreatic cancer. *Oncotarget* **2014**, *5*, 10969–10975. [CrossRef] [PubMed]
26. Nagai, S.; Washiyama, K.; Kurimoto, M.; Takaku, A.; Endo, S.; Kumanishi, T. Aberrant NF-κB activity and its participation in the growth of human malignant astrocytoma. *J. Neurosurg.* **2002**, *96*, 909–917. [CrossRef] [PubMed]
27. Wang, H.; Wang, H.; Zhang, W.; Huang, H.J.; Liao, W.S.; Fuller, G.N. Analysis of the activation status of Akt, NF-κB, and STAT3 in human diffuse gliomas. *Lab. Investig.* **2004**, *84*, 941–951. [CrossRef] [PubMed]
28. Raychaudhuri, B.; Han, Y.; Lu, T.; Vogelbaum, M.A. Aberrant constitutive activation of NF-κB in glioblastoma multiforme drives invasive phenotype. *J. Neurooncol.* **2007**, *85*, 39–47. [CrossRef] [PubMed]
29. Smith, D.; Shimamura, T.; Barbera, S.; Bejcek, B.E. NF-κB controls growth of glioblastomas/astrocytomas. *Mol. Cell. Biochem.* **2008**, *307*, 141–147. [CrossRef] [PubMed]
30. Cahill, K.E.; Moshead, R.A.; Yamini, B. NF-κB in glioblastoma: Insights into regulators and targeted therapy. *Neuro Oncol.* **2016**, *18*, 329–339. [CrossRef] [PubMed]
31. Shih, A.H.; Holland, E.C. Platelet-derived growth factor (PDGF) and glial tumorigenesis. *Cancer Lett.* **2006**, *232*, 139–147. [CrossRef] [PubMed]
32. Romashkova, J.A.; Makarov, S.S. NF-κB is a target of AKT in anti-apoptotic PDGF signalling. *Nature* **1999**, *401*, 86–90. [PubMed]
33. Bonavia, R.; Inda, M.M.; Vandenberg, S.; Cheng, S.Y.; Nagane, M.; Hadwiger, P.; Tan, P.; Sah, D.W.; Cavenee, W.K.; Furnari, F.B. EGFRvIII promotes glioma angiogenesis and growth through the NF-κB, interleukin-8 pathway. *Oncogene* **2012**, *31*, 4054–4066. [CrossRef] [PubMed]
34. Yang, W.; Xia, Y.; Cao, Y.; Zheng, Y.; Bu, W.; Zhang, L.; You, M.J.; Koh, M.Y.; Cote, G.; Aldape, K.; et al. EGFR-induced and PKCε monoubiquitylation-dependent NF-κB activation upregulates PKM2 expression and promotes tumorigenesis. *Mol. Cell* **2012**, *48*, 771–784. [CrossRef] [PubMed]
35. Finco, T.S.; Westwick, J.K.; Norris, J.L.; Beg, A.A.; Der, C.J.; Baldwin, A.S., Jr. Oncogenic Ha-Ras-induced signaling activates NF-κB transcriptional activity, which is required for cellular transformation. *J. Biol. Chem.* **1997**, *72*, 24113–24116. [CrossRef]
36. Masilamani, A.P.; Ferrarese, R.; Kling, E.; Thudi, N.K.; Kim, H.; Scholtens, D.M.; Dai, F.; Hadler, M.; Unterkircher, T.; Platania, L.; et al. KLF6 depletion promotes NF-κB signaling in glioblastoma. *Oncogene* **2017**. [CrossRef] [PubMed]
37. Park, S.; Hatanpaa, K.J.; Xie, Y.; Mickey, B.E.; Madden, C.J.; Raisanen, J.M.; Ramnarain, D.B.; Xiao, G.; Saha, D.; Boothman, D.A.; et al. The receptor interacting protein 1 inhibits p53 induction through NF-κB activation and confers a worse prognosis in glioblastoma. *Cancer Res.* **2009**, *69*, 2809–2816. [CrossRef] [PubMed]
38. Weisz, L.; Damalas, A.; Liontos, M.; Karakaidos, P.; Fontemaggi, G.; Maor-Aloni, R.; Kalis, M.; Levrero, M.; Strano, S.; Gorgoulis, V.G.; et al. Mutant p53 enhances NF-κB activation by tumor necrosis factor α in cancer cells. *Cancer Res.* **2007**, *67*, 2396–2401. [CrossRef] [PubMed]
39. Rocha, S.; Perkins, N.D. ARF the integrator: Linking NF-κB, p53 and checkpoint kinases. *Cell Cycle* **2005**, *4*, 756–759. [CrossRef] [PubMed]
40. Schneider, G.; Kramer, O.H. NF-κB/p53 crosstalk-a promising new therapeutic target. *Biochim. Biophys. Acta* **2011**, *1815*, 90–103. [PubMed]
41. Atkinson, G.P.; Nozell, S.E.; Harrison, D.K.; Stonecypher, M.S.; Chen, D.; Benveniste, E.N. The prolyl isomerase Pin1 regulates the NF-κB signaling pathway and interleukin-8 expression in glioblastoma. *Oncogene* **2009**, *28*, 3735–3745. [CrossRef] [PubMed]
42. Kim, S.H.; Ezhilarasan, R.; Phillips, E.; Gallego-Perez, D.; Sparks, A.; Taylor, D.; Ladner, K.; Furuta, T.; Sabit, H.; Chhipa, R.; et al. Serine/threonine kinase MLK4 determines mesenchymal identity in glioma stem cells in an NF-κB-dependent manner. *Cancer Cell* **2016**, *29*, 201–213. [CrossRef] [PubMed]
43. Bredel, M.; Scholtens, D.M.; Yadav, A.K.; Alvarez, A.A.; Renfrow, J.J.; Chandler, J.P.; Yu, I.L.; Carro, M.S.; Dai, F.; Tagge, M.J.; et al. NFKBIA deletion in glioblastomas. *N. Engl. J. Med.* **2011**, *364*, 627–637. [CrossRef] [PubMed]
44. Xu, R.X.; Liu, R.Y.; Wu, C.M.; Zhao, Y.S.; Li, Y.; Yao, Y.Q.; Xu, Y.H. DNA damage-induced NF-κB activation in human glioblastoma cells promotes miR-181b expression and cell proliferation. *Cell. Physiol. Biochem.* **2015**, *35*, 913–925. [CrossRef] [PubMed]

45. Jiang, L.; Lin, C.; Song, L.; Wu, J.; Chen, B.; Ying, Z.; Fang, L.; Yan, X.; He, M.; Li, J.; et al. MicroRNA-30e* promotes human glioma cell invasiveness in an orthotopic xenotransplantation model by disrupting the NF-κB/IκBα negative feedback loop. *J. Clin. Investig.* **2012**, *122*, 33–47. [CrossRef] [PubMed]

46. Garner, J.M.; Fan, M.; Yang, C.H.; Du, Z.; Sims, M.; Davidoff, A.M.; Pfeffer, L.M. Constitutive activation of signal transducer and activator of transcription 3 (STAT3) and NF-κB signaling in glioblastoma cancer stem cells regulates the Notch pathway. *J. Biol. Chem.* **2013**, *288*, 26167–26176. [CrossRef] [PubMed]

47. Bhat, K.P.; Balasubramaniyan, V.; Vaillant, B.; Ezhilarasan, R.; Hummelink, K.; Hollingsworth, F.; Wani, K.; Heathcock, L.; James, J.D.; Goodman, L.D.; et al. Mesenchymal differentiation mediated by NF-κB promotes radiation resistance in glioblastoma. *Cancer Cell* **2013**, *24*, 331–346. [CrossRef] [PubMed]

48. Rinkenbaugh, A.L.; Cogswell, P.C.; Calamini, B.; Dunn, D.E.; Persson, A.I.; Weiss, W.A.; Lo, D.C.; Baldwin, A.S. IKK/NF-κB signaling contributes to glioblastoma stem cell maintenance. *Oncotarget* **2016**, *7*, 69173–69187. [CrossRef] [PubMed]

49. Ohtsu, N.; Nakatami, Y.; Yamashita, D.; Ohue, S.; Ohnishi, T.; Kondo, T. Eva1 maintains the stem-like character of glioblastoma-initiating cells by activating the noncanonical NF-κB signaling pathway. *Cancer Res.* **2016**, *76*, 171–181. [CrossRef] [PubMed]

50. Nogueira, L.; Ruiz-Ontañon, P.; Vazquez-Barquero, A.; Lafarga, M.; Berciano, M.T.; Aldaz, B.; Grande, L.; Casafont, I.; Segura, V.; Robles, E.F.; et al. Blockade of the NF-κB pathway drives differentiating glioblastoma-initiating cells into senescence both in vitro and in vivo. *Oncogene* **2011**, *30*, 3537–3548. [CrossRef] [PubMed]

51. Min, C.; Eddy, S.F.; Sherr, D.H.; Sonenshein, G.E. NF-κB and epithelial to mesenchymal transition of cancer. *J. Cell. Biochem.* **2008**, *104*, 733–744. [CrossRef] [PubMed]

52. Riddick, G.; Fine, H.A. Integration and analysis of genome-scale data from gliomas. *Nat. Rev. Neurol.* **2011**, *7*, 439–450. [CrossRef] [PubMed]

53. Verhaak, R.G.; Hoadley, K.A.; Purdom, E.; Wang, V.; Qi, Y.; Wilkerson, M.D.; Miller, C.R.; Ding, L.; Golub, T.; Mesirov, J.P.; et al. Integrated genomic analysis identifies clinically relevant subtypes of glioblastoma characterized by abnormalities in PDGFRA, IDH1, EGFR, and NF1. *Cancer Cell* **2010**, *17*, 98–110. [CrossRef] [PubMed]

54. Chandran, U.R.; Luthra, S.; Santana-Santos, L.; Mao, P.; Kim, S.H.; Minata, M.; Li, J.; Benos, P.V.; DeWang, M.; Hu, B.; et al. Gene expression profiling distinguishes proneural glioma stem cells from mesenchymal glioma stem cells. *Genom. Data* **2015**, *5*, 333–336. [CrossRef] [PubMed]

55. Carro, M.S.; Lim, W.K.; Alvarez, M.J.; Bollo, R.J.; Zhao, X.; Snyder, E.Y.; Sulman, E.P.; Anne, S.L.; Doetsch, F.; Colman, H.; et al. The transcriptional network for mesenchymal transformation of brain tumours. *Nature* **2010**, *463*, 318–325. [CrossRef] [PubMed]

56. Lee, D.W.; Ramakrishnan, D.; Valenta, J.; Parney, I.F.; Bayless, K.J.; Sitcheran, R. The NF-κB RelB protein is an oncogenic driver of mesenchymal glioma. *PLoS ONE* **2013**, *8*, e57489. [CrossRef] [PubMed]

57. Tchoghandjian, A.; Jennewein, C.; Eckhardt, I.; Rajalingam, K.; Fulda, S. Identification of non-canonical NF-κB signaling as a critical mediator of Smac mimetic-stimulated migration and invasion of glioblastoma cells. *Cell Death Dis.* **2013**, *4*, e564. [CrossRef] [PubMed]

58. Nozell, S.; Laver, T.; Moseley, D.; Nowoslawski, L.; de Vos, M.; Atkinson, G.P.; Harrison, K.; Nabors, L.B.; Benveniste, E.N. The ING4 tumor suppressor attenuates NF-κB activity at the promoter of target genes. *Mol. Cell. Biol.* **2008**, *28*, 6632–6645. [CrossRef] [PubMed]

59. Kesanakurti, D.; Chetty, C.; Rajasekhar, M.; Gujirati, M.; Rao, J.S. Essential role of cooperative NF-κB and Stat3 recruitment to ICAM-1 intronic consensus elements in the regulation of radiation-induced invasion and migration in glioma. *Oncogene* **2013**, *32*, 5144–5155. [CrossRef] [PubMed]

60. Brantley, E.C.; Nabors, L.B.; Gillespie, G.Y.; Choi, Y.H.; Palmer, C.A.; Harrison, K.; Roarty, K.; Benveniste, E.N. Loss of protein inhibitors of activated STAT-3 expression in glioblastoma multiforme tumors: Implications for STAT-3 activation and gene expression. *Clin. Cancer Res.* **2008**, *14*, 4694–4704. [CrossRef] [PubMed]

61. Bonavida, B.; Baritaki, S. The novel role of Yin Yang 1 in the regulation of epithelial to mesenchymal transition in cancer via the dysregulated NF-κB/Snail/YY1/RKIP/PTEN circuitry. *Crit. Rev. Oncog.* **2011**, *16*, 211–226. [CrossRef] [PubMed]

62. Edwards, L.A.; Woolard, K.; Son, M.J.; Li, A.; Lee, J.; Ene, C.; Mantey, S.A.; Maric, D.; Song, H.; Belova, G.; et al. Effect of brain-and tumor-derived connective tissue growth factor on glioma invasion. *J. Natl. Cancer Inst.* **2011**, *103*, 1162–1178. [CrossRef] [PubMed]

63. Park, M.H.; Ahn, B.H.; Hong, Y.K.; Min do, S. Overexpression of phospholipase D enhances matrix metalloproteinase-2 expression and glioma cell invasion via protein kinase C and protein kinase A/NF-κB/Sp1-mediated signaling pathways. *Carcinogenesis* **2009**, *30*, 356–365. [CrossRef] [PubMed]

64. Chandrika, G.; Natesh, K.; Ranade, D.; Chugh, A.; Shastry, P. Suppression of the invasive potential of glioblastoma cells by mTOR inhibitors involves modulation of NF-κB and PKC-α signaling. *Sci. Rep.* **2016**, *6*, 22455. [CrossRef] [PubMed]

65. Tran, N.L.; McDonough, W.S.; Savitch, B.A.; Fortin, S.P.; Winkles, J.A.; Symons, M.; Nakada, M.; Cunliffe, H.E.; Hostetter, G.; Hoelzinger, D.B.; et al. Increased fibroblast growth factor-inducible 14 expression levels promote glioma cell invasion via Rac1 and NF-κB and correlate with poor patient outcome. *Cancer Res.* **2006**, *66*, 9535–9542. [CrossRef] [PubMed]

66. Kim, J.K.; Jin, X.; Sohn, X.; Jeon, H.Y.; Kim, E.J.; Ham, S.W.; Jeon, H.M.; Chang, S.Y.; Oh, S.Y.; Yin, J.; et al. Tumoral RANKL activates astrocytes that promote glioma cell invasion through cytokine signaling. *Cancer Lett.* **2014**, *353*, 194–200. [CrossRef] [PubMed]

67. Okolie, O.; Bago, J.R.; Schmid, R.S.; Irvin, D.M.; Bash, R.E.; Miller, C.R.; Hingtgen, S.D. Reactive astrocytes potentiate tumor aggressiveness in a murine glioma resection and recurrence model. *Neuro Oncol.* **2016**, *18*, 1622–1633. [CrossRef] [PubMed]

68. Placone, L.; Quinones-Hinojosa, A.; Searson, P.C. The role of astrocytes in the progression of brain cancer: Complicating the picture of the tumor microenvironment. *Tumor Biol.* **2016**, *37*, 61–69. [CrossRef] [PubMed]

69. Safa, A.R. Resistance to cell death and its modulation in cancer stem cells. *Crit. Rev. Oncog.* **2016**, *21*, 203–219. [CrossRef] [PubMed]

70. Li, F.; Zhou, K.; Gao, L.; Zhang, B.; Li, W.; Yan, W.; Song, X.; Yu, H.; Wang, S.; Yu, N.; et al. Radiation induces the generation of cancer stem cells: A novel mechanism for cancer radioresistance. *Oncol. Lett.* **2016**, *12*, 3059–3065. [PubMed]

71. Huang, T.T.; Wuerzberger-Davis, S.M.; Seufzer, B.J.; Shumway, S.D.; Kurama, T.; Boothman, D.A.; Miyamoto, S. NF-κB activation by camptothecin. A linkage between nuclear DNA damage and cytoplasmic signaling events. *J. Biol. Chem.* **2000**, *275*, 9501–9509. [PubMed]

72. McCool, K.W.; Miyamoto, S. DNA damage-dependent NF-κB activation: NEMO turns nuclear signaling inside out. *Immunol. Rev.* **2012**, *246*, 311–326. [CrossRef] [PubMed]

73. Volcic, M.; Karl, S.; Baumann, B.; Salles, D.; Daniel, P.; Fulda, S.; Wiesmüller, L. NF-κB regulates DNA double-strand break repair in conjunction with BRCA1-CtIP complexes. *Nucleic Acids Res.* **2012**, *40*, 181–195. [CrossRef] [PubMed]

74. Wu, K.; Jiang, S.W.; Thangaraju, M.; Wu, G.; Couch, F.J. Induction of the BRCA2 promoter by NF-κB. *J. Biol. Chem.* **2000**, *275*, 35548–35556. [CrossRef] [PubMed]

75. Miyamoto, S. Nuclear initiated NF-κB signaling: NEMO and ATM take center stage. *Cell Res.* **2011**, *21*, 116–130. [CrossRef] [PubMed]

76. Lavon, I.; Fuchs, D.; Zrihan, D.; Efroni, G.; Zelikovitch, B.; Fellig, Y.; Siegal, T. Novel mechanism whereby NF-κB mediates DNA damage repair through regulation of O(6)-methylguanine-DNA-methyltransferase. *Cancer Res.* **2007**, *67*, 8952–8959. [CrossRef] [PubMed]

77. Erstad, D.J.; Cusack, J.C., Jr. Targeting the NF-κB pathway in cancer therapy. *Surg. Oncol. Clin. N. Am.* **2013**, *22*, 705–746. [CrossRef] [PubMed]

78. Bao, S.; Wu, Q.; McLendon, R.E.; Hao, Y.; Shi, Q.; Hjelmeland, A.B.; Dewhirst, M.W.; Bigner, D.D.; Rich, J.N. Glioma stem cells promote radioresistance by preferential activation of the DNA damage response. *Nature* **2006**, *444*, 756–760. [CrossRef] [PubMed]

79. Jamal, M.; Rath, B.H.; Tsang, P.S.; Camphausen, K.; Tofilon, P.J. The brain microenvironment preferentially enhances the radioresistance of CD133 (+) glioblastoma stem-like cells. *Neoplasia* **2012**, *14*, 150–158. [CrossRef] [PubMed]

80. Orlowski, R.Z.; Baldwin, A.S., Jr. NF-κB as a therapeutic target in cancer. *Trends Mol. Med.* **2002**, *8*, 385–389. [CrossRef]

81. Ahmed, K.M.; Li, J.J. NF-κB-mediated adaptive resistance to ionizing radiation. *Free Radic. Biol. Med.* **2008**, *44*, 1–13. [CrossRef] [PubMed]

82. Bai, M.; Ma, X.; Li, X.; Wang, X.; Mei, Q.; Li, X.; Wu, Z.; Han, W. The accomplices of NF-κB lead to radioresistance. *Curr. Protein Pept. Sci.* **2015**, *16*, 279–294. [CrossRef] [PubMed]

83. Koul, D.; Takada, Y.; Shen, R.; Aggarwal, B.B.; Yung, W.K. PTEN enhances TNF-induced apoptosis through modulation of NF-κB signaling pathway in human glioma cells. *Biochem. Biophys. Res. Commun.* **2006**, *350*, 463–471. [CrossRef] [PubMed]

84. Shan, Y.; He, X.; Song, W.; Han, D.; Niu, J.; Wang, J. Role of IL-6 in the invasiveness and prognosis of glioma. *Int. J. Clin. Exp. Med.* **2015**, *8*, 9114–9120. [PubMed]

85. Brat, D.J.; Bellail, A.C.; van Meir, E.G. The role of interleukin-8 and its receptors in gliomagenesis and tumoral angiogenesis. *Neuro Oncol.* **2005**, *7*, 122–133. [CrossRef] [PubMed]

86. Agnihotri, S.; Zadeh, G. Metabolic reprogramming in glioblastoma: The influence of cancer metabolism on epigenetics and unanswered questions. *Neuro Oncol.* **2016**, *18*, 160–172. [CrossRef] [PubMed]

87. Moretti, M.; Bennett, J.; Tornatore, L.; Thotakura, A.K.; Franzoso, G. Cancer: NF-κB regulates energy metabolism. *Int. J. Biochem. Cell Biol.* **2012**, *44*, 2238–2243. [CrossRef] [PubMed]

88. Johnson, R.F.; Perkins, N.D. NF-κB, p53, and mitochondria: Regulation of cellular metabolism and the Warburg effect. *Trends Biochem. Sci.* **2012**, *37*, 317–324. [CrossRef] [PubMed]

89. Yuen, C.A.; Asuthkar, S.; Guda, M.R.; Tsung, A.J.; Velpula, K.K. Cancer stem cell molecular reprogramming of the Warburg effect in glioblastomas: A new target gleaned from an old concept. *CNS Oncol.* **2016**, *5*, 101–108. [CrossRef] [PubMed]

90. Han, D.; Wei, W.; Chen, X.; Zhang, Y.; Wang, Y.; Zhang, J.; Wang, X.; Yu, T.; Hu, Q.; Liu, N.; et al. NF-κB/RelA-PKM2 mediates inhibition of glycolysis by fenofibrate in glioblastoma cells. *Oncotarget* **2015**, *6*, 26119–26128. [CrossRef] [PubMed]

91. Mauro, C.; Leow, S.C.; Anso, E.; Rocha, S.; Thotakura, A.K.; Tornatore, L.; Moretti, M.; de Smaele, E.; Beg, A.A.; Tergaonkar, V.; et al. NF-κB controls energy homeostasis and metabolic adaptation by upregulating mitochondrial respiration. *Nat. Cell Biol.* **2011**, *13*, 1272–1279. [CrossRef] [PubMed]

92. Wang, J.B.; Erickson, J.W.; Fuji, R.; Ramachandran, S.; Gao, P.; Dinavahi, R.; Wilson, K.F.; Ambrosio, A.L.B.; Dias, S.M.G.; Dang, C.V.; et al. Targeting mitochondrial glutaminase activity inhibits oncogenic transformation. *Cancer Cell* **2010**, *18*, 207–219. [CrossRef] [PubMed]

93. Rathore, M.G.; Saumet, A.; Rossi, J.F.; de Bettignies, C.; Tempé, D.; Lecellier, C.H.; Villalba, M. The NF-κB member p65 controls glutamine metabolism through miR-23a. *Int. J. Biochem. Cell Biol.* **2012**, *44*, 1448–1456. [CrossRef] [PubMed]

94. Viry, E.; Paggetti, J.; Baginska, J.; Mgrditchian, T.; Berchem, G.; Moussay, E.; Janji, B. Autophagy: An adaptive metabolic response to stress shaping the antitumor immunity. *Biochem. Pharmacol.* **2014**, *92*, 31–42. [CrossRef] [PubMed]

95. Galluzzi, L.; Bravo-San Pedro, J.M.; Levine, B.; Green, D.R.; Kroemer, G. Pharmacological modulation of autophagy: Therapeutic potential and persisting obstacles. *Nat. Rev. Drug Discov.* **2017**. [CrossRef] [PubMed]

96. Kimmelman, A.C.; White, E. Autophagy and tumor metabolism. *Cell Metab.* **2017**, *25*, 1037–1043. [CrossRef] [PubMed]

97. Martinez-Outschoorn, U.E.; Trimmer, C.; Lin, Z.; Whitaker-Menezes, D.; Chiavarina, B.; Zhou, J.; Wang, C.; Pavlides, S.; Martinez-Cantarin, M.P.; Capozza, F.; et al. Autophagy in cancer associated fibroblasts promotes tumor cell survival: Role of hypoxia, HIF1 induction and NF-κB activation in the tumor stromal microenvironment. *Cell Cycle* **2010**, *9*, 3515–3533. [CrossRef] [PubMed]

98. Ozpolat, B.; Benbrook, D.M. Targeting autophagy in cancer management—Strategies and developments. *Cancer Manag. Res.* **2015**, *7*, 291–299. [CrossRef] [PubMed]

99. Bischof, J.; Westhoff, M.A.; Wagner, J.E.; Halatsch, M.E.; Trentmann, S.; Knippschild, U.; Wirtz, C.R.; Burster, T. Cancer stem cells: The potential role of autophagy, proteolysis, and cathepsins in glioblastoma stem cells. *Tumour Biol.* **2017**, *39*. [CrossRef] [PubMed]

100. Strickland, M.; Stoll, E.A. Metabolic reprogramming in glioma. *Front. Cell Dev. Biol.* **2017**, *5*, 43. [CrossRef] [PubMed]

101. Wojton, J.; Meisen, W.H.; Kaur, B. How to train glioma cells to die: Molecular challenges in cell death. *J. Neuro Oncol.* **2016**, *126*, 377–384. [CrossRef] [PubMed]

102. Trocoli, A.; Djavaheri-Mergny, M. The complex interplay between autophagy and NF-κB signaling pathways in cancer cells. *Am. J. Cancer Res.* **2011**, *1*, 629–649. [PubMed]

103. Djavaheri-Mergny, M.; Amelotti, M.; Mathieu, J.; Besancon, F.; Bauvy, C.; Souquere, S.; Pierron, G.; Codogno, P. NF-κB activation represses tumor necrosis factor-α-induced autophagy. *J. Biol. Chem.* **2006**, *281*, 30373–30382. [CrossRef] [PubMed]

104. Schlottmann, S.; Buback, F.; Stahl, B.; Meierhenrich, R.; Walter, P.; Georgieff, M.; Senftleben, U. Prolonged classical NF-κB activation prevents autophagy upon *E. coli* stimulation in vitro: A potential resolving mechanism of inflammation. *Mediat. Inflamm.* **2008**. [CrossRef] [PubMed]

105. Criollo, A.; Chereau, F.; Malik, S.A.; Niso-Santano, M.; Mariño, G.; Galluzzi, L.; Maiuri, M.C.; Baud, V.; Kroemer, G. Autophagy is required for the activation of NF-κB. *Cell Cycle* **2012**, *11*, 194–199. [CrossRef] [PubMed]

106. Sanz, L.; Sanchez, P.; Lallena, M.J.; Diaz-Meco, M.T.; Moscat, J. The interaction of p62 with RIP links the atypical PKCs to NF-κB activation. *EMBO J.* **1999**, *18*, 3044–3053. [CrossRef] [PubMed]

107. Duran, A.; Linares, J.F.; Galvez, A.S.; Wikenheiser, K.; Flores, J.M.; Diaz-Meco, M.T.; Moscat, J. The signaling adaptor p62 is an important NF-κB mediator in tumorigenesis. *Cancer Cell* **2008**, *13*, 343–354. [CrossRef] [PubMed]

108. Su, J.; Liu, F.; Xia, M.; Xu, Y.; Li, X.; Kang, J.; Li, Y.; Sun, L. p62 participates in the inhibition of NF-κB signaling and apoptosis induced by sulfasalazine in human glioma U251 cells. *Oncol. Rep.* **2015**, *34*, 235–243. [CrossRef] [PubMed]

109. Jiang, Q.; Wang, Y.; Li, T.; Shi, K.; Li, Z.; Ma, Y.; Li, F.; Luo, H.; Yang, Y.; Xu, C.; et al. Heat shock protein 90-mediated inactivation of NF-κB switches autophagy to apoptosis through becn1 transcriptional inhibition in selenite-induced NB4 cells. *Mol. Biol. Cell* **2011**, *22*, 1167–1180. [CrossRef] [PubMed]

110. Siegelin, M.D.; Dohi, T.; Raskett, C.M.; Orlowski, G.M.; Powers, C.M.; Gilbert, C.A.; Ross, A.H.; Plescia, J.; Altieri, D.C. Exploiting the mitochondrial unfolded protein response for cancer therapy in mice and human cells. *J. Clin. Investig.* **2011**, *121*, 1349–1360. [CrossRef] [PubMed]

111. Xie, T.X.; Xia, Z.; Zhang, N.; Gong, W.; Huang, S. Constitutive NF-κB activity regulates the expression of VEGF and IL-8 and tumor angiogenesis of human glioblastoma. *Oncol. Rep.* **2010**, *23*, 725–732. [PubMed]

112. Jiang, L.; Song, L.; Wu, J.; Yang, Y.; Zhu, X.; Hu, B.; Cheng, S.Y.; Li, M. Bmi-1 promotes glioma angiogenesis by activating NF-κB signaling. *PLoS ONE* **2013**, *8*, e55527. [CrossRef] [PubMed]

113. El Hasasna, H.; Saleh, A.; Al Samri, H.; Athamneh, K.; Attoub, S.; Arafat, K.; Benhalilou, N.; Alyan, S.; Viallet, J.; Al Dhaheri, Y.; et al. Rhus coriaria suppresses angiogenesis, metastasis and tumor growth of breast cancer through inhibition of STAT3, NF-κB and nitric oxide pathways. *Sci. Rep.* **2016**, *6*, 21144. [CrossRef] [PubMed]

114. Ricci-Vitiani, L.; Pallini, R.; Biffoni, M.; Todaro, M.; Invernici, G.; Cenci, T.; Maira, G.; Parati, E.A.; Stassi, G.; Larocca, L.M.; et al. Tumour vascularization via endothelial differentiation of glioblastoma stem-like cells. *Nature* **2010**, *468*, 824–828. [CrossRef] [PubMed]

115. Wang, R.; Chadalavada, K.; Wilshire, J.; Kowalik, U.; Hovinga, K.E.; Geber, A.; Fligelman, B.; Leversha, M.; Brennan, C.; Tabar, V.; et al. Glioblastoma stem-like cells give rise to tumour endothelium. *Nature* **2010**, *468*, 829–833. [CrossRef] [PubMed]

116. Cheng, L.; Huang, Z.; Zhou, W.; Wu, Q.; Donnola, S.; Liu, J.K.; Fang, X.; Sloan, A.E.; Mao, Y.; Lathia, J.D.; et al. Glioblastoma stem cells generate vascular pericytes to support vessel function and tumor growth. *Cell* **2013**, *153*, 139–152. [CrossRef] [PubMed]

117. Hu, B.; Wang, Q.; Wang, Y.A.; Hua, S.; Sauvé, C.G.; Ong, D.; Lan, Z.D.; Chang, Q.; Ho, Y.W.; Monasterio, M.M.; et al. Epigenetic activation of WNT5A drives glioblastoma stem cell differentiation and invasive growth. *Cell* **2016**, *167*, 1281–1295.e18. [CrossRef] [PubMed]

118. Tang, S.; Xiang, T.; Huang, S.; Zhou, J.; Wang, Z.; Xie, R.; Long, H.; Zhu, B. Ovarian cancer stem-like cells differentiate into endothelial cells and participate in tumor angiogenesis through autocrine CCL5 signaling. *Cancer Lett.* **2016**, *376*, 137–147. [CrossRef] [PubMed]

119. Pan, Y.; Smithson, L.J.; Ma, Y.; Hambardzumyan, D.; Gutmann, D.H. Ccl5 establishes an autocrine high-grade glioma growth regulatory circuit critical for mesenchymal glioblastoma survival. *Oncotarget* **2017**, *8*, 32977–32989. [CrossRef] [PubMed]

120. Friedmann-Morvinski, D.; Narasimamurthy, R.; Xia, Y.; Myskiw, C.; Soda, Y.; Verma, I.M. Targeting NF-κB in glioblastoma: A therapeutic approach. *Sci. Adv.* **2016**, *2*, e1501292. [CrossRef] [PubMed]

121. Panday, A.; Inda, M.E.; Bagam, P.; Sahoo, M.K.; Osorio, D.; Batra, S. Transcription factor NF-κB: An update on intervention strategies. *Arch. Immunol. Ther. Exp.* **2016**, *64*, 463–483. [CrossRef] [PubMed]

biomedicines

MDPI

Review

Aspirin Prevention of Colorectal Cancer: Focus on NF-κB Signalling and the Nucleolus

Jingyu Chenand Lesley A. Stark *

Cancer Research UK Edinburgh Centre, Institute of Genetics and Molecular Medicine, University of Edinburgh, Crewe Rd., Edinburgh, Scotland EH4 2XU, UK; s1355550@sms.ed.ac.uk
* Correspondence: Lesley.Stark@IGMM.ed.ac.uk; Tel.: +44-131-651-8531

Received: 20 June 2017; Accepted: 13 July 2017; Published: 18 July 2017

Abstract: Overwhelming evidence indicates that aspirin and related non-steroidal anti-inflammatory drugs (NSAIDs) have anti-tumour activity and the potential to prevent cancer, particularly colorectal cancer. However, the mechanisms underlying this effect remain hypothetical. Dysregulation of the nuclear factor-kappaB (NF-κB) transcription factor is a common event in many cancer types which contributes to tumour initiation and progression by driving expression of pro-proliferative/anti-apoptotic genes. In this review, we will focus on the current knowledge regarding NSAID effects on the NF-κB signalling pathway in pre-cancerous and cancerous lesions, and the evidence that these effects contribute to the anti-tumour activity of the agents. The nuclear organelle, the nucleolus, is emerging as a central regulator of transcription factor activity and cell growth and death. Nucleolar function is dysregulated in the majority of cancers which promotes cancer growth through direct and indirect mechanisms. Hence, this organelle is emerging as a promising target for novel therapeutic agents. Here, we will also discuss evidence for crosstalk between the NF-κB pathway and nucleoli, the role that this cross-talk has in the anti-tumour effects of NSAIDs and ways forward to exploit this crosstalk for therapeutic purpose.

Keywords: Aspirin; non-steroidal anti-inflammatory drugs; nuclear factor kappaB; apoptosis; colon cancer; nucleolus; nucleolar; nucleoli; sequestration; stress; RelA; p65

1. Aspirin and Cancer

Incontrovertible evidence from laboratory, clinical and epidemiological studies indicates that aspirin and related non-steroidal anti-inflammatory drugs (NSAIDs) have anti-neoplastic properties and considerable potential as chemopreventative/therapeutic agents [1–4]. For example, at therapeutic concentrations, NSAIDs induce cell cycle arrest and atypical apoptosis in cancer cell lines [5–8]. In animal studies, NSAID administration significantly reduces tumour burden in the azoxymethane-induced rat model of colorectal cancer [9]. NSAIDs also reduce tumour burden and increase survival in the multiple intestinal neoplasia (*Min*/+) model of colorectal cancer [10–12]. However, in this model tumour burden is mostly affected when mice are exposed to NSAIDs in utero, suggesting the agents act at the early stages of tumour development [13,14]. Meta-analysis of randomised clinical trials (RCTs) for the prevention of vascular disease indicate daily aspirin (75 mg upwards) reduces cancer incidence and mortality. These effects are particularly evident for colorectal cancer where a 30% to 40% reduction in incidence and mortality are observed [15,16]. The risk of developing distant metastasis is also reduced in aspirin users, suggesting a potential benefit for patients with established disease [17,18]. RCTs for cancer prevention indicate aspirin limits recurrence of spontaneous and hereditary intestinal adenomas (the precursor lesion to most cancers). After long term followup, they also indicate aspirin prevents colorectal cancer in (1) women randomised to alternate day low dose (75 mg) aspirin; and (2) patients with Lynch syndrome (the most common type of hereditary colon cancer) [4,19–23]. The most compelling evidence for the chemopreventative effects of NSAIDs comes from epidemiological studies which have consistently demonstrated reduced cancer incidence and improved survival in persons who regularly take aspirin or other NSAIDs [16,24–26]. Again, this association is particularly strong for colorectal cancer, with other cancer types showing less consistent risk reduction.

The predominant anti-tumour activity of NSAIDs is recognized to be the selective induction of apoptosis in neoplastic cells [10,27]. However, the mechanisms underlying this pro-apoptotic activity are complex, interconnected, and remain controversial [4,28,29]. In 1982, John R Vane was awarded the Nobel Prize for discovering that aspirin irreversibly acetylates the cyclooxygenase enzymes, thereby blocking the conversion of arachidonic acid to prostaglandins [30]. Cyclo-oxygenase-2 (COX-2 (*PTGS2*)), the inducible form of the enzyme, is frequently upregulated in cancer and together with PGE_2, is implicated in several aspects of malignant growth including stem cell proliferation, migration, angiogenesis, apoptosis resistance, invasion, and metastasis [31–34]. Hence, inhibition of COX-2 activity was thought to be the main mechanism for the anti-tumour effects of NSAIDs. Indeed, a body of literature supports this suggestion [35–39]. More recently, it was proposed that aspirin acetylation of COX-1 in platelets, and the consequent inactivation of platelet function, is the only mechanism that can explain the anti-tumour properties of aspirin when taken at low dose [29,40,41]. However, NSAIDs induce cell cycle arrest and apoptosis in colon cancer cell lines that do not express COX-1 or COX-2 enzymes and in mouse embryo fibroblasts that are null for both COX-1 and COX-2 genes [42–44]. The growth inhibitory properties of NSAIDs cannot be reversed by addition of prostaglandins [4,9]. Furthermore, NSAID metabolites that do not appreciably affect the catalytic activity of COXs retain their anti-tumor properties in tissue culture [27] and animal models [45,46]. Hence, there is powerful evidence that inhibition of COX is not the only mechanism by which NSAIDs induce apoptosis and prevent the growth of neoplastic lesions [28,47]. A number of COX-independent targets have been identified including the WNT [10,48], AMPK [49,50] and MTOR [51] signalling pathways (reviewed in [4]). In the rest of this review we will focus on the role of nuclear factor-kappaB (NF-κB). In particular, we will examine the evidence for crosstalk between NF-κB signalling and nucleoli in the regulation of NF-κB transcriptional activity and NSAID-mediated apoptosis.

2. NF-κB, Cancer and Aspirin

NF-κB is the collective name for a family of ubiquitously expressed, inducible transcription factors that play a critical role in multiple processes including innate and adaptive immune response, inflammation, differentiation, proliferation and survival [52–54]. In mammalian cells there are five family members namely, RelA (p65), RelB, c-Rel, p105/p50 (NF-κB1), and p100/p52 (NF-κB2) [55]. These proteins homo- and hetero-dimerize through their Rel homology domain to create a variety of transcription factor complexes [56]. The most common form of NF-κB is p50/RelA heterodimers. In most cell types, this complex exists in the cytoplasm bound to a family of IκB inhibitory proteins (IκBα, IκBβ, IκBγ and Bcl-3). Following cellular stimulation by a plethora of stimuli including cytokines, pathogens, viruses and stresses, IκB proteins are phosphorylated by the IκB kinase (IKK) complex then degraded by the 26S proteasome [57]. Subsequently, NF-κB translocates to the nucleus where it regulates the transcription of target genes including those involved immune function, inflammation, cell adhesion, differentiation, cell growth, and apoptotic cell death.

In healthy cells, a number of feedback mechanisms ensure that activation of the NF-κB pathway is transient [53,56]. However, in chronic inflammatory conditions and cancer, NF-κB is aberrantly active which contributes to disease progression by promoting inflammation, blocking differentiation, driving stem cell proliferation and inhibiting apoptosis [53,54,58].

A substantial body of data supports a critical role for dysregulated NF-κB activity in intestinal tumorigenesis, the cancer type most responsive to aspirin treatment. For example, a recent meta-analysis of expression studies revealed that high expression of NF-κB is significantly associated with late stage colorectal cancer (TNM stage III–IV) and a worse overall 3 and 5-year survival [59]. Transgenic mice with constitutively active IKK in intestinal epithelial cells develop intestinal tumours and show accelerated adenoma development when crossed to *Min/+* mice [60]. Conversely, inactivation of IKK in intestinal epithelial or myeloid cells attenuates inflammation-associated tumour development [61]. Furthermore, deletion of *RelA* in intestinal epithelial cells prevents formation of adenomas in the *Min/+* model [62]. These data have identified inhibition of NF-κB activity as a promising therapeutic target for the treatment of this disease.

Targeting of the NF-κB pathway by NSAIDs was initially reported by Kopp and Ghosh in 1994, who demonstrated that the aspirin derivative, sodium salicylate, inhibits lipopolysaccharide (LPS) and phorbol 12-myristate 13-acetate (PMA)/phytohemagglutinin (PHA)-mediated degradation of IκB, nuclear translocation of NF-κB and NF-κB transcriptional activity [63]. Yin et al. subsequently demonstrated that salicylate specifically inhibits IKKβ activity in cell lines in vivo and when the agent is added to the kinase in vitro [64]. Since these early publications, NSAID modulation of the NF-κB pathway has been widely reported [28,65]. However, these studies have produced contrasting results dependent upon cell lines and experimental design. In most studies aimed at examining this relationship, cells are treated with NSAIDs for 1–2 h prior to activation of the NF-κB pathway by a potent stimulus (e.g., LPS, Interleukin-1 (IL-1), tumour necrosis factor (TNF)). Under these conditions, NSAIDs block activation of the NF-κB pathway and there is some evidence from in vitro and animal studies to suggest that inhibition of IκB degradation is responsible for the anti-tumour effect of the agents [66–68] (Figure 1). However, this experimental design it is entirely inconsistent with the protocol used to demonstrate NSAID-mediated apoptosis of cancer cells, where cells are exposed to the agents for prolonged periods in the absence of additional stimuli [5–8].

Examination of aspirin effects on NF-κB signalling using this alternative protocol revealed that prolonged treatment of colorectal cancer cells with pharmacologically relevant doses (0.5–5 mM) of aspirin alone actually stimulates the NF-κB pathway, as evidenced by phosphorylation/degradation of IκB and nuclear translocation of RelA [8] (Figure 1). Furthermore, using cells expressing degradation resistant IκB (super-repressor), Stark et al. demonstrated that this stimulation is absolutely required for the pro-apoptotic effects of the agent [8] (Figure 1). Interestingly, stimulation of the NF-κB pathway by aspirin, and the consequent induction of apoptosis, were particularly evident in colorectal cancer cells, which is in keeping with the increased sensitivity of this cancer type to the chemopreventative effects of the agent [7,8]. The NSAIDs diclofenac, sulindac, sulindac sulphone sulindac sulphide, Tolfenamic, indomethacin, celocoxib and ibuprofen, which are all known to protect against colorectal cancer, have also been shown to induce degradation of IκB and nuclear translocation of NF-κB in various cancer cell lines in the absence of additional NF-κB stimuli [69–75]. Furthermore, in the majority of these studies, NSAID-mediated activation of the NF-κB pathway was causally associated with the induction of apoptosis.

As the above data were generated using tissue culture systems, it was argued that the conditions are not representative of the tumour environment where inflammatory cytokines are abundant. To address this concern, our group examined the effects of aspirin on NF-κB signalling in colorectal neoplasia in vivo, using the HT-29 xenograft and *Min/+* mouse models. We found that aspirin (at doses resulting in serum salicylate levels relevant to humans (0.5–1.5 mM)) induces phosphorylation and degradation of IκBα, nuclear translocation of RelA and the induction of apoptosis in xenografted HT-29 tumours and in adenomas from *Min/+* mice [76]. Sulindac sulphide has also been shown to induce degradation of IκB and nuclear translocation of NF-κB in the proximal colons of mice [72]. Furthermore, exposure to low dose (100 μM) aspirin ex vivo was recently shown to stimulate the NF-κB pathway, as evidenced by increased phosphorylation of RelA at serine 536, in 5 of 6 freshly resected, human colorectal tumours [77]. These findings establish that aspirin and other NSAIDs activate the NF-κB pathway in neoplastic epithelial cells in the context of a whole tumour setting, and support the proposition that this effect is important for the anti-tumour activity of the agent.

In reality, NSAIDs likely both activate and suppress activation of the NF-κB pathway in cancer depending on the tumour type and microenvironment. Most solid malignancies require an intrinsic inflammatory response to promote a pro-tumorigenic microenvironment [78]. NSAIDs are thought to act against pre-malignant lesions, at least in part, by altering this response. That is, suppressing pro-tumorigenic immune cell populations while stimulating the adaptive immune system [79]. Notably, colorectal cancer response to NSAIDs is associated with a reduced number of tumour infiltrating lymphocytes [80]. Therefore, it is interesting to speculate that by blocking stimulation of the NF-κB pathway, NSAIDs modulate the tumour microenvironment to reduce the presence of inflammatory cells/cytokines, while stimulation of the pathway in a non-inflammatory environment mediates apoptosis of colorectal cancer cells.

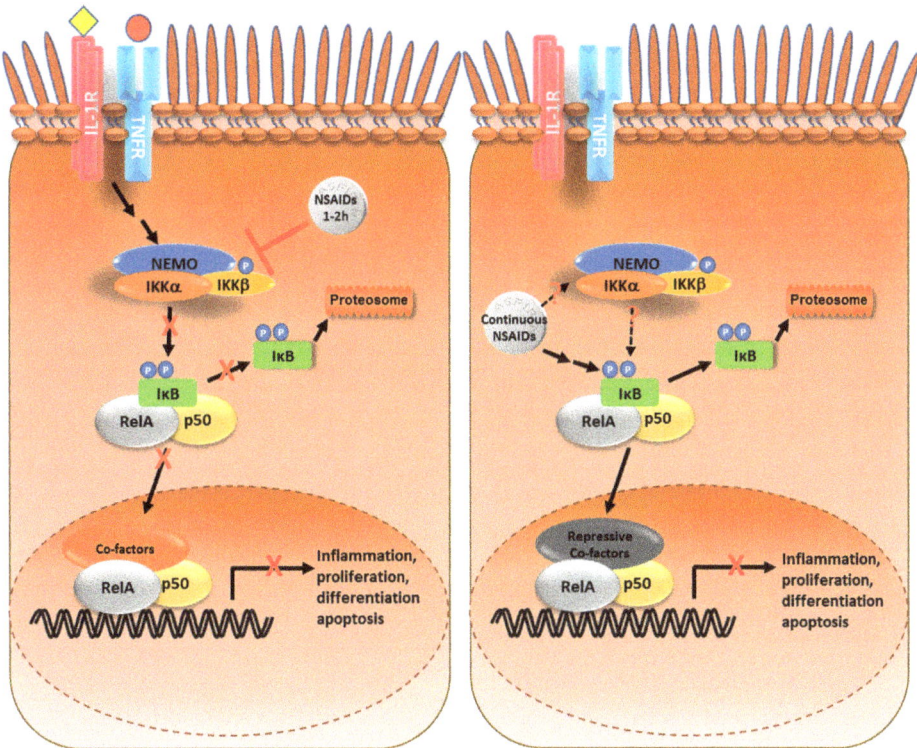

Figure 1. Aspirin modulation of the nuclear factor-kappaB (NF-κB) pathway. (**Left**) The NF-κB transcription factor, most commonly a hetero-dimer of the RelA (p65) and p50 polypeptides, is held in the cytoplasm by the inhibitory protein IκB. When the cell is stimulated by growth factors or cytokines (e.g., interleukin-1 (IL-1) or tumour necrosis factor (TNF)), IκB is phosphorylated by the IκB kinase (IKK) complex, which targets it for degradation by the proteasome. This allows NF-κB to translocate to the nucleus and regulate expression of target genes. In cancer cells, NF-κB is constitutively active which drives tumour progression. Short pre-treatment with aspirin or related non-steroidal anti-inflammatory drugs (NSAIDs) blocks cytokine-mediated activation of the pathway by inhibiting the IKK complex, particularly IKKβ; T bar: NSAIDs inhibit IKK kinase activity. IL-1R: IL-1 receptor; TNFR: TNF receptor; NEMO (IKKγ); (**Right**) In contrast, prolonged exposure to NSAIDs in the absence of additional NF-κB activators stimulates degradation of IκB and nuclear translocation of NF-κB. This NF-κB recruits specific complexes which lead to repression of NF-κB-driven transcription and the induction of apoptosis. Dotted lines: It remains unclear whether the IKK complex plays a role in the stimulatory pathway or whether NSAIDs target IκB by another pathway.

3. Crosstalk between the NF-κB Pathway and Nucleoli

As outlined above, several lines of data indicate that NSAIDs stimulate the NF-κB pathway in vitro and in vivo and that this is important for the anti-tumour activity of the agents. However, in most cases, stimulation of the NF-κB pathway by NSAIDs is associated with repression of NF-κB transcriptional activity and downregulation of NF-κB target genes [65,71,75] (Figure 1). In studies aimed at understanding the mechanisms responsible for this repression, a role for crosstalk between NF-κB signalling and the nuclear organelle, the nucleolus, has emerged.

The nucleolus is a highly dynamic, multifunctional organelle [81–84]. Its main role is in ribosome biogenesis which is the most energy consuming process in the cell and as such, is tightly linked to metabolic and proliferative activity. If cells are exposed to stresses or insults that threaten homeostasis (e.g., Ultraviolet-C (UV-C) radiation, nutrient deprivation, toxic agents), they respond by rapidly downregulating rDNA transcription. This triggers a cascade of nucleolar events that will either allow the cell to repair and regain homeostasis, or, if the damage is too great, undergo apoptosis. Over half of the 4500 proteins found within nucleoli are involved in processes out with ribosome biogenesis e.g., transcription, cell cycle regulation, ubiquitin modification, proliferation and apoptosis [85,86]. These regulatory proteins flux dynamically between this and other cellular compartments depending upon cellular environment [85,87]. While some are released from nucleoli under conditions of cell stress, others translocate to the organelle. For example, NF-κB repressing factor has recently been shown to accumulate in nucleoli in response to heat stress, causing repression of rDNA transcription [88]. P53 and a variety of ubiquitinated proteins accumulate in nucleoli in response to proteasome inhibition, while exposure of cells to heat shock, hypoxia and acidosis causes the accumulation of proteins with a specific nucleolar detention sequence (i.e., von Hippel-Lindau, DNA methyltransferase 1 (DNMT1) and the DNA polymerase subunit POLD1) in nucleolar foci [89–93]. Indeed, nucleolar sequestration of transcription factors and regulatory proteins is increasingly recognised as an important mechanism for controlling gene expression and maintaining cellular homeostasis under stress conditions.

Many proteins known to shuttle through nucleoli are regulators of the NF-κB pathway. For example, the nucleolar protein p14[ARF], which sequesters MDM2 in the nucleolus to regulate p53 stability, interacts with RelA and inhibits NF-κB-driven transcription [94]. In screens for NF-κB-interacting partners, the predominant proteins identified were the nucleolar proteins NFBP [95] and NPM [96]. The NF-κB regulators NIK (NF-κB-inducing kinase) [97] and NRF (NF-κB repressing factor) [98] also function through nucleolar shuttling. Disruption of nucleolar function is a common denominator for stresses that activate the NF-κB pathway [77]. Furthermore, proteins that have a role in stress-mediated activation of NF-κB reside within this organelle, such as CK2, which forms part of the PolI complex and phosphorylates IκB in response to UV-C [99,100] and EIF2α, that plays a role in NF-κB activation in response to multiple stresses [101,102].

When exploring repression of NF-κB-driven transcription associated with stimulation of the NF-κB pathway, it was found that in response to specific pro-apoptotic stress stimuli (e.g., aspirin, serum deprivation and UV-C radiation), the RelA component of NF-κB is sequestered in the nucleolus [65]. A nucleolar localization signal (NoLS) was identified at the N terminus of RelA and, using a dominant-negative mutant with a deletion of this motif, it was shown that nucleolar sequestration of RelA is causally involved in reduced basal NF-κB transcriptional activity and the induction of apoptosis [65] (Figure 2). Importantly, it was found that nucleolar translocation of RelA was absolutely required for the pro-apoptotic activity of aspirin [65,103]. Since this initial study, nucleolar localisation of RelA has been observed in response to the NSAIDs sulindac, sulindac sulphone and indomethacin [71], the naturally occurring derivative of estradiol and antitumor agent, 2-methoxyestradiol (2ME2) [104]; a potent Trk inhibitor and anti-tumour agent, K252a [105]; expression of the homeobox protein Hox-A5 (HOXA5) transcription factor [106], small molecule inhibitors of the CDK4 kinase [107] and the proteasome inhibitors MG132 and lactocystin [103]. In the majority of these studies, nucleolar sequestration of RelA is associated with a decrease in NF-κB-driven transcription. Furthermore, in all studies, it is associated with, or causally involved in, the induction of apoptosis.

Nucleolar sequestration of p50 has also been reported. Park et al. demonstrated that the anti-TNF therapy, infliximab, induces "massive" nucleolar localisation of NF-κB/p50 in the hippocampus of rats with a portacaval shunt (PCS). They also demonstrated that this nucleolar localisation is associated with a decrease in transcription of NF-κB target genes and a reduction in neuroinflammation [108].

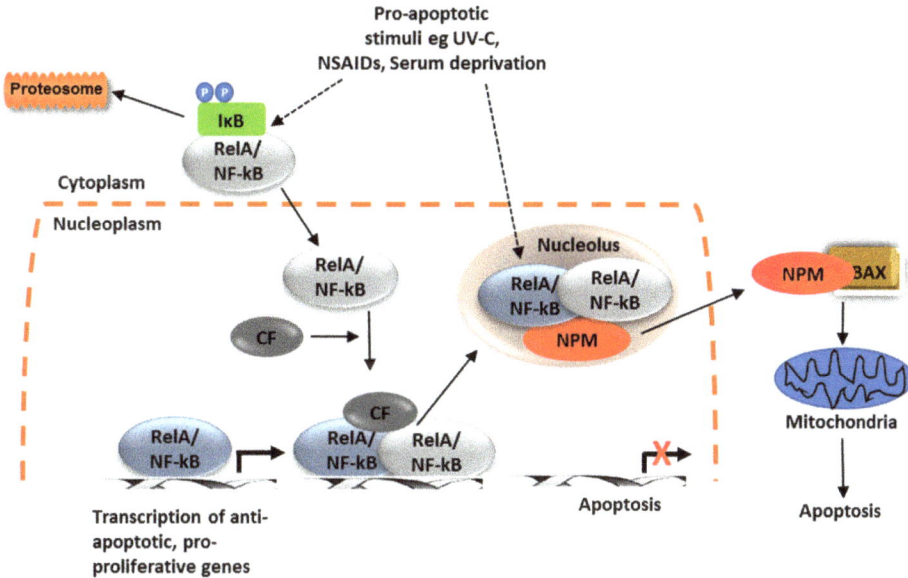

Figure 2. NF-κB-nucleolar crosstalk. Upon exposure of cells to specific pro-apoptotic stimuli, including NSAIDs and chemo toxic agents, IκB is degraded and RelA/NF-κB translocates to the nucleus. This induced NF-κB/RelA recruits specific co-factors (CF)/modifiers that target both constitutive and induced RelA to the nucleolus, reducing basal NF-κB transcriptional activity [62,101]. Once in the nucleolus, RelA induces the relocation of nucleophosmin (NPM) to the cytoplasm which in turn binds to BAX, then transports BAX to the mitochondria to mediate apoptosis [103]. An early response to stresses that induce nucleolar translocation of RelA is disruption of nucleolar morphology, which may "prime" this organelle for nucleolar residency of RelA. Dashed arrows: pathways still under exploration. Solid arrows: published pathways.

Given that nucleolar sequestration of RelA causes repression of constitutive NF-κB-driven transcription, it was assumed that the apoptotic effects were mediated through a reduction in transcription of NF-κB regulated, anti-apoptotic genes. However, it was found that once in the nucleolus, RelA triggers a cascade of events that actively promotes apoptosis [109] (Figure 2). That is, nucleolar RelA causes nucleophosmin (NPM)/B23 to relocate to the cytoplasm, bind BAX then transport BAX to the mitochondria to initiate apoptosis [110,111]. Indeed this, and a number of other studies have demonstrated a critical role for both BAX and NPM in the pro-apoptotic effects of NSAIDs [112]. Together, these data identify nucleolar-NF-κB crosstalk as an important regulator of NF-κB transcriptional activity and apoptosis and suggest that this is particular critical for the anti-tumour effects of NSAIDs and chemotherapeutic agents.

As mentioned above, aspirin irreversibly acetylates active site serines to inhibit the activity of cyclooxygenase enzymes. However, as an acetylating agent, it has the ability to acetylate other amino acid side chains [113,114]. Tatham et al. (2017) recently used isotopically labelled aspirin-d3, in combination with acetylated lysine purification and LC-MS/MS, to identify over 12,000 sites of aspirin-mediated lysine acetylation in cultured human cells [113]. Interestingly, gene ontology (GO) analysis indicated that acetylation of nucleolar proteins, including nucleophosmin, was one of the earliest responses to the agent. Immunocytochemical studies also suggest nucleolar morphology is altered as an early response to NSAIDs [65,71], suggesting the intriguing possibility that early effects of these agents on the organelle may enable cross-talk with the NF-κB pathway.

4. Conclusions

Despite the overwhelming proof that aspirin prevents colon and other cancers, these agents are still not recommended for cancer prevention in the general population due to their significant side effect profile. Identification of the precise pathway(s) by which daily aspirin inhibits the initiation/progression of cancer is now paramount so that patient populations who may benefit from exposure to the agent can be identified, and safer, more effective alternatives revealed. Inhibition of the cyclooxygenase enzymes, both in platelets and cancer cells, undoubtedly plays a role. However, given that aspirin acetylates many proteins [113,114], other pathways are more than likely involved. Indeed, multiple lines of evidence suggest the agents act in a COX-dependent and independent manner.

There is a consensus in the literature that NSAIDs induce repression of NF-κB-driven transcription, although the pathway to this repression appears to be cell type and context dependent. Nonetheless, given the critical role of de-regulated NF-κB activity in colorectal cancer initiation and progression, it is extremely likely that this repression contributes significantly to the anti-tumour effect of the agents in humans. There are several lines of cross-talk between the prostaglandin and NF-κB signaling pathways and so, it may be that NSAID inhibition of cyclooxygenases inhibits tumour growth through modulation of the NF-κB pathway. In this regard, Chan et al. suggested that NSAID inhibition of COX activity mediates apoptosis not by reducing prostaglandin levels, but by increasing the generation of ceramide, a potent cytotoxic agent that stimulates the NF-κB pathway to induce cell death [115,116]. As more large scale randomised clinical trials are initiated to examine the anti-tumour and chemopreventative effects of aspirin, it will be possible to definitively establish the effects of aspirin exposure on these signalling pathways in human pre-cancer and cancerous lesions, to understand their individual contributions and to determine how they may be interconnected.

Dysfunction of the nucleolus is now regarded a hallmark of cancer as it contributes to tumour growth not only by allowing the protein synthesis required for rapid cell proliferation, but also through de-regulation of critical nucleolar cell growth and death pathways. Hence, modulation of nucleolar function is emerging as an innovative therapeutic strategy. Recent work has uncovered an exciting new role for the nucleolus in the anti-tumour effects of NSAIDs and in particular, cross-talk between nucleoli and the NF-κB pathway. This evolving field is in its infancy and there are still a number of questions to be answered regarding the role of nucleolar sequestration of RelA in the regulation of NF-κB activity and apoptosis in vivo, and how this contributes to the chemopreventive effect of NSAIDs. Identification of the pathways responsible for nucleolar translocation of RelA would allow development of small molecules that act specifically on cancer cells by targeting chromatin bound RelA to nucleoli. Similarly, identification of the apoptotic pathways triggered by RelA within this organelle would allow the development of RelA mimetics that mediate apoptosis by targeting dysfunctional nucleoli. Indeed, further understanding in this area could reveal a whole new class of targets to be exploited for therapeutic purposes.

Acknowledgments: The work was supported by grants from the WWCR (formally AICR 10-0158 to LS), Rosetrees trust (A631 and JS16/M225 to LS) and MRC (MR/J001481/1).

Conflicts of Interest: The authors declare no conflict of interest.

References

1. Tougeron, D.; Sha, D.; Manthravadi, S.; Sinicrope, F.A. Aspirin and colorectal cancer: Back to the future. *Clin. Cancer Res.* **2014**, *20*, 1087–1094. [CrossRef] [PubMed]
2. Patrignani, P.; Patrono, C. Aspirin and Cancer. *J. Am. Coll. Cardiol.* **2016**, *68*, 967–976. [CrossRef] [PubMed]

3. Chan, A.T.; Arber, N.; Burn, J.; Chia, W.K.; Elwood, P.; Hull, M.A.; Logan, R.F.; Rothwell, P.M.; Schrör, K.; Baron, J.A. Aspirin in the chemoprevention of colorectal neoplasia: An overview. *Cancer Prev. Res.* **2012**, *5*, 164–178. [CrossRef] [PubMed]

4. Drew, D.A.; Cao, Y.; Chan, A.T. Aspirin and colorectal cancer: The promise of precision chemoprevention. *Nat. Rev. Cancer* **2016**, *16*, 173–186. [CrossRef] [PubMed]

5. Elder, D.J.; Hague, A.; Hicks, D.J.; Paraskeva, C. Differential growth inhibition by the aspirin metabolite salicylate in human colorectal tumor cell lines: Enhanced apoptosis in carcinoma and in vitro-transformed adenoma relative to adenoma relative to adenoma cell lines. *Cancer Res.* **1996**, *56*, 2273–2276. [PubMed]

6. Williams, C.S.; Smalley, W.; DuBois, R.N. Aspirin use and potential mechanisms for colorectal cancer prevention. *J. Clin. Invest.* **1997**, *100*, 1325–1329. [CrossRef] [PubMed]

7. Din, F.V.; Dunlop, M.G.; Stark, L.A. Evidence for colorectal cancer cell specificity of aspirin effects on NF-κB signalling and apoptosis. *Br. J. Cancer* **2004**, *91*, 381–388. [PubMed]

8. Stark, L.A.; Din, F.V.N.; Zwacka, R.M.; Dunlop, M.G. Aspirin-induced activation of the NF-κB signalling pathway: A novel mechanism for aspirin-mediated apoptosis in colon cancer cells. *FASEB J.* **2001**, *15*, 1273–1275. [PubMed]

9. Piazza, G.A.; Alberts, D.S.; Hixson, L.J.; Paranka, N.S.; Li, H.; Finn, T.; Bogert, C.; Guillen, J.M.; Brendel, K.; Gross, P.H.; et al. Sulindac sulfone inhibits azoxymethane-induced colon carcinogenesis in rats without reducing prostaglandin levels. *Cancer Res.* **1997**, *57*, 2909–2915. [PubMed]

10. Qiu, W.; Wang, X.; Leibowitz, B.; Liu, H.; Barker, N.; Okada, H.; Oue, N.; Yasui, W.; Clevers, H.; Scheen, R.E.; et al. Chemoprevention by nonsteroidal anti-inflammatory drugs eliminates oncogenic intestinal stem cells via SMAC-dependent apoptosis. *Proc. Natl. Acad. Sci. USA* **2010**, *107*, 20027–20032. [CrossRef] [PubMed]

11. Beazer-Barclay, Y.; Levy, D.B.; Moser, A.R.; Dove, W.F.; Hamilton, S.R.; Vogelstein, B.; Kinzler, K.M. Sulindac suppresses tumorigenesis in the Min mouse. *Carcinogenesis* **1996**, *17*, 1757–1760. [CrossRef] [PubMed]

12. Corpet, D.E.; Pierre, F. Point: From animal models to prevention of colon cancer. Systematic review of chemoprevention in min mice and choice of the model system. *Cancer Epidemiol. Biomark. Prev.* **2003**, *12*, 391–400.

13. Perkins, S.; Clarke, A.R.; Steward, W.; Gescher, A. Age-related difference in susceptibility of Apc$^{Min/+}$ mice towards the chemopreventive efficacy of dietary aspirin and curcumin. *Br. J. Cancer* **2003**, *88*, 1480–1483. [CrossRef] [PubMed]

14. Sansom, O.J.; Stark, L.A.; Dunlop, M.G.; Clarke, A.R. Suppression of intestinal and mammary neoplasia by lifetime administration of aspirin in *Apc*Min/+ and *Apc*Min/+, *Msh2*−/− mice. *Cancer Res.* **2001**, *61*, 7060–7064. [PubMed]

15. Rothwell, P.M.; Wilson, M.; Elwin, C.E.; Norrving, B.; Algra, A.; Warlow, C.P.; Meade, T.W. Long-term effect of aspirin on colorectal cancer incidence and mortality: 20-Year follow-up of five randomised trials. *Lancet* **2010**, *376*, 1741–1750. [CrossRef]

16. Rothwell, P.M.; Price, J.F.; Fowkes, F.G.; Zanchetti, A.; Roncaglioni, M.C.; Tognoni, G.; Lee, R.; Belch, J.F.F. Short-term effects of daily aspirin on cancer incidence, mortality, and non-vascular death: Analysis of the time course of risks and benefits in 51 randomised controlled trials. *Lancet* **2012**, *379*, 1602–1612. [CrossRef]

17. Rothwell, P.M.; Wilson, M.; Price, J.F.; Belch, J.F.; Meade, T.W.; Mehta, Z. Effect of daily aspirin on risk of cancer metastasis: A study of incident cancers during randomised controlled trials. *Lancet* **2012**, *379*, 1591–1601. [CrossRef]

18. Elwood, P.C.; Morgan, G.; Galante, J.; Chia, J.W.; Dolwani, S.; Graziano, J.M.; Kelson, M.; Lanas, A.; Longley, M.; Phillips, C.J.; et al. Systematic Review and Meta-Analysis of Randomised Trials to Ascertain Fatal Gastrointestinal Bleeding Events Attributable to Preventive Low-Dose Aspirin: No Evidence of Increased Risk. *PLoS ONE* **2016**, *11*, e0166166. [CrossRef] [PubMed]

19. Sandler, R.S.; Galanko, J.C.; Murray, S.C.; Helm, J.F.; Woosley, J.T. Aspirin and nonsteroidal anti-inflammatory agents and risk for colorectal adenomas. *Gastroenterology* **1998**, *114*, 441–447. [CrossRef]

20. Sandler, R.S.; Halabi, S.; Baron, J.A.; Budinger, S.; Paskett, E.; Keresztes, R.; Petrelli, N.; Pipas, J.M.; Karp, D.D.; Loprinzi, C.L.; et al. A randomized trial of aspirin to prevent colorectal adenomas in patients with previous colorectal cancer. *N. Engl. J. Med.* **2003**, *348*, 883–890. [CrossRef] [PubMed]

21. Cole, B.F.; Logan, R.F.; Halabi, S.; Benamouzig, R.; Sandler, R.S.; Grainge, M.J.; Chaussade, S.; Baron, J.A. Aspirin for the chemoprevention of colorectal adenomas: Meta-analysis of the randomized trials. *J. Natl. Cancer Inst.* **2009**, *101*, 256–266. [CrossRef] [PubMed]

22. Burn, J.; Gerdes, A.M.; Macrae, F.; Mecklin, J.P.; Moeslein, G.; Olschwang, S.; Ecclec, D.; Evans, D.G.; Maher, E.R.; Bertario, L.; et al. Long-term effect of aspirin on cancer risk in carriers of hereditary colorectal cancer: An analysis from the CAPP2 randomised controlled trial. *Lancet* **2011**, *378*, 2081–2087. [CrossRef]
23. Cook, N.R.; Lee, I.M.; Zhang, S.M.; Moorthy, M.V.; Buring, J.E. Alternate-day, low-dose aspirin and cancer risk: Long-term observational follow-up of a randomized trial. *Ann. Intern. Med.* **2013**, *159*, 77–85. [CrossRef] [PubMed]
24. Din, F.V.; Theodoratou, E.; Farrington, S.M.; Tenesa, A.; Barnetson, R.A.; Cetnarskyj, R.; Stark, L.; Porteous, M.; Campbell, H.; Dunlop, M.G. Effect of aspirin and NSAIDs on risk and survival from colorectal cancer. *Gut* **2010**, *59*, 1670–1679. [CrossRef] [PubMed]
25. Cuzick, J.; Thorat, M.A.; Bosetti, C.; Brown, P.H.; Burn, J.; Cook, N.R.; Ford, L.G.; Jacobs, E.J.; Jankowski, J.A.; Vecchia, C.L.; et al. Estimates of benefits and harms of prophylactic use of aspirin in the general population. *Ann. Oncol.* **2015**, *26*, 47–57. [CrossRef] [PubMed]
26. Cao, Y.; Nishihara, R.; Wu, K.; Wang, M.; Ogino, S.; Willett, W.C.; Spigelman, D.; Fuchs, C.S.; Giovannucci, E.L.; Chan, A.T. Population-wide Impact of Long-term Use of Aspirin and the Risk for Cancer. *JAMA Oncol.* **2016**, *1*, 762–769. [CrossRef] [PubMed]
27. Piazza, G.A.; Rahm, A.K.; Finn, T.S.; Fryer, B.H.; Li, H.; Stoumen, A.L.; Pamukcu, R.; Ahnen, D.J. Apoptosis primarily accounts for the growth-inhibitory properties of sulindac metabolites and involves a mechanism that is independent of cyclooxygenase inhibition, cell cycle arrest, and p53 induction. *Cancer Res.* **1997**, *57*, 2452–2459. [PubMed]
28. Alfonso, L.; Ai, G.; Spitale, R.C.; Bhat, G.J. Molecular targets of aspirin and cancer prevention. *Br. J. Cancer* **2014**, *111*, 61–67. [CrossRef] [PubMed]
29. Thun, M.J.; Jacobs, E.J.; Patrono, C. The role of aspirin in cancer prevention. *Nat. Rev. Clin. Oncol.* **2012**, *9*, 259–267. [CrossRef] [PubMed]
30. Vane, J.R. Inhibition of prostaglandin synthesis as a mechanism of action for aspirin-like drugs. *Nat. New. Biol.* **1971**, *231*, 232–235. [CrossRef] [PubMed]
31. Pang, L.Y.; Hurst, E.A.; Argyle, D.J. Cyclooxygenase-2: A Role in Cancer Stem Cell Survival and Repopulation of Cancer Cells during Therapy. *Stem Cells Int.* **2016**, *2016*, 2048731. [CrossRef] [PubMed]
32. Menter, D.G.; DuBois, R.N. Prostaglandins in cancer cell adhesion, migration, and invasion. *Int. J. Cell Biol.* **2012**, *2012*, 723419. [CrossRef] [PubMed]
33. Williams, C.S.; Mann, M.; DuBois, R.N. The role of cyclooxygenases in inflammation, cancer, and development. *Oncogene* **1999**, *18*, 7908–7916. [CrossRef] [PubMed]
34. Harris, R.E. Cyclooxygenase-2 (cox-2) and the inflammogenesis of cancer. *Subcell. Biochem.* **2007**, *42*, 93–126. [PubMed]
35. Chan, A.T.; Ogino, S.; Fuchs, C.S. Aspirin and the risk of colorectal cancer in relation to the expression of COX-2. *N. Engl. J. Med.* **2007**, *356*, 2131–2142. [CrossRef] [PubMed]
36. Chulada, P.C.; Thompson, M.B.; Mahler, J.F.; Doyle, C.M.; Gaul, B.W.; Lee, C.; Tiano, H.F.; Morham, S.G.; Smithies, O.; Langenbach, R. Genetic disruption of Ptgs-1, as well as Ptgs-2, reduces intestinal tumorigenesis in Min mice. *Cancer Res.* **2000**, *60*, 4705–4708. [PubMed]
37. Ruegg, C.; Zaric, J.; Stupp, R. Non steroidal anti-inflammatory drugs and COX-2 inhibitors as anti-cancer therapeutics: Hypes, hopes and reality. *Ann. Med.* **2003**, *35*, 476–487. [CrossRef] [PubMed]
38. Oshima, M.; Dinchuk, J.E.; Kargman, S.L.; Oshima, H.; Hancock, B.; Kwong, E.; Trzaskos, J.M.; Evans, J.F.; Taketo, M.M. Suppression of intestinal polyposis in Apc Δ716 knockout mice by inhibition of cyclooxygenase 2 (COX-2). *Cell* **1996**, *87*, 803–809. [CrossRef]
39. Gupta, R.A.; DuBois, R.N. Colorectal cancer prevention and treatment by inhibition of cyclooxygenase-2. *Nat. Rev. Cancer* **2001**, *1*, 11–21. [CrossRef] [PubMed]
40. Guillem-Llobat, P.; Dovizio, M.; Bruno, A.; Ricciotti, E.; Cufino, V.; Sacco, A.; Grande, R.; Alberti, S.; Arena, V.; Cirillo, M.; et al. Aspirin prevents colorectal cancer metastasis in mice by splitting the crosstalk between platelets and tumor cells. *Oncotarget* **2016**, *7*, 32462–32477. [CrossRef] [PubMed]
41. Guillem-Llobat, P.; Dovizio, M.; Alberti, S.; Bruno, A.; Patrignani, P. Platelets, cyclooxygenases, and colon cancer. *Semin. Oncol.* **2014**, *41*, 385–396. [CrossRef] [PubMed]
42. Hanif, R.; Pittas, A.; Feng, Y.; Koutsos, M.I.; Qiao, L.; Staiano-Coico, L.; Shiff, S.I.; Rigas, B. Effects of nonsteroidal anti-inflammatory drugs on proliferation and on induction of apoptosis in colon cancer cells by a prostaglandin-independent pathway. *Biochem. Pharmacol.* **1996**, *52*, 237–245. [CrossRef]

43. Rigas, B.; Shiff, S.J. Is inhibition of cyclooxygenase required for the chemopreventive effect of NSAIDs in colon cancer? A model reconciling the current contradiction. *Med. Hypotheses* **2000**, *54*, 210–215. [CrossRef] [PubMed]

44. Zhang, X.; Morham, S.G.; Langenbach, R.; Young, D.A. Malignant transformation and antineoplastic actions of nonsteroidal antiinflammatory drugs (NSAIDs) on cyclooxygenase-null embryo fibroblasts. *J. Exp. Med.* **1999**, *190*, 451–459. [CrossRef] [PubMed]

45. Mahmoud, N.N.; Boolbol, S.K.; Dannenberg, A.J.; Mestre, J.R.; Bilinski, R.T.; Martucci, C.; Newmark, H.L.; Chadburn, A.; Bertagnolli, M.M. The sulfide metabolite of sulindac prevents tumors and restores enterocyte apoptosis in a murine model of familial adenomatous polyposis. *Carcinogenesis* **1998**, *19*, 87–91. [CrossRef] [PubMed]

46. Reddy, B.S.; Kawamori, T.; Lubet, R.A.; Steele, V.E.; Kelloff, G.J.; Rao, C.V. Chemopreventive efficacy of sulindac sulfone against colon cancer depends on time of administration during carcinogenic process. *Cancer Res.* **1999**, *59*, 3387–3391. [PubMed]

47. Tegeder, I.; Pfeilschifter, J.; Geisslinger, G. Cyclooxygenase-independent actions of cyclooxygenase inhibitors. *FASEB J.* **2001**, *15*, 2057–2072. [CrossRef] [PubMed]

48. Bos, C.L.; Kodach, L.L.; van den Brink, G.R.; Diks, S.H.; van Santen, M.M.; Richel, D.J.; Peppelenbosch, M.P.; Hardwick, J.C.H. Effect of aspirin on the Wnt/β-catenin pathway is mediated via protein phosphatase 2A. *Oncogene* **2006**, *25*, 6447–6456. [CrossRef] [PubMed]

49. Hardie, D.G.; Ross, F.A.; Hawley, S.A. AMP-activated protein kinase: A target for drugs both ancient and modern. *Chem. Biol.* **2012**, *19*, 1222–1236. [CrossRef] [PubMed]

50. Hawley, S.A.; Fullerton, M.D.; Ross, F.A.; Schertzer, J.D.; Chevtzoff, C.; Walker, K.J.; Peggie, M.W.; Zibrova, D.; Green, K.A.; Mustard, K. The ancient drug salicylate directly activates AMP-activated protein kinase. *Science* **2012**, *336*, 918–922. [CrossRef] [PubMed]

51. Din, F.V.; Valanciute, A.; Houde, V.P.; Zibrova, D.; Green, K.A.; Sakamoto, K.; Alessi, D.R.; Dunlop, M.G. Aspirin inhibits mTOR signaling, activates AMP-activated protein kinase, and induces autophagy in colorectal cancer cells. *Gastroenterology* **2012**, *142*, 1504–1515. [CrossRef] [PubMed]

52. Basseres, D.S.; Baldwin, A.S. Nuclear factor-κB and inhibitor of κB kinase pathways in oncogenic initiation and progression. *Oncogene* **2006**, *25*, 6817–6830. [CrossRef] [PubMed]

53. DiDonato, J.A.; Mercurio, F.; Karin, M. NF-κB and the link between inflammation and cancer. *Immunol. Rev.* **2012**, *246*, 379–400. [CrossRef] [PubMed]

54. Perkins, N.D. NF-κB: Tumor promoter or suppressor? *Trends Cell Biol.* **2004**, *14*, 64–69. [CrossRef] [PubMed]

55. Gilmore, T.D. Introduction to NF-κB: Players, pathways, perspectives. *Oncogene* **2006**, *25*, 6680–6684. [CrossRef] [PubMed]

56. Hoesel, B.; Schmid, J.A. The complexity of NF-κB signaling in inflammation and cancer. *Mol. Cancer* **2013**, *12*, 86. [CrossRef] [PubMed]

57. Israel, A. The IKK complex: An integrator of all signals that activate NF-κB? *Trends Cell Biol.* **2000**, *10*, 129–133. [CrossRef]

58. Vlahopoulos, S.A.; Cen, O.; Hengen, N.; Agan, J.; Moschovi, M.; Critselis, E.; Adamaki, M.; Bacopoulou, F.; Copland, J.A.; Boldogh, I.; et al. Dynamic aberrant NF-κB spurs tumorigenesis: A new model encompassing the microenvironment. *Cytokine Growth Factor Rev.* **2015**, *26*, 389–403. [CrossRef] [PubMed]

59. Wu, D.; Wu, P.; Zhao, L.; Huang, L.; Zhang, Z.; Zhao, S.; Huang, J. NF-κB Expression and outcomes in solid tumors: A systematic review and meta-analysis. *Medicine* **2015**, *94*, e1687. [CrossRef] [PubMed]

60. Shaked, H.; Hofseth, L.J.; Chumanevich, A.; Chumanevich, A.A.; Wang, J.; Wang, Y.; Taniguchi, K.; Guma, M.; Shenouda, S.; Clevers, H.; et al. Chronic epithelial NF-κB activation accelerates APC loss and intestinal tumor initiation through iNOS up-regulation. *Proc. Natl. Acad. Sci. USA* **2012**, *109*, 14007–14012. [CrossRef] [PubMed]

61. Greten, F.R.; Eckmann, L.; Greten, T.F.; Park, J.M.; Li, Z.W.; Egan, L.J.; Kagnoff, M.F.; Karin, M. IKKβ links inflammation and tumorigenesis in a mouse model of colitis-associated cancer. *Cell* **2004**, *118*, 285–296. [CrossRef] [PubMed]

62. Myant, K.B.; Cammareri, P.; McGhee, E.J.; Ridgway, R.A.; Huels, D.J.; Cordero, J.B.; Schwitalla, S.; Kalna, G.; Ogg, E.-L.; Athineos, O.; et al. ROS production and NF-κB activation triggered by RAC1 facilitate WNT-driven intestinal stem cell proliferation and colorectal cancer initiation. *Cell Stem Cell* **2013**, *12*, 761–773. [CrossRef] [PubMed]

63. Kopp, E.; Ghosh, S. Inhibition of NF-κB by sodium salicylate and aspirin. *Science* **1994**, *265*, 956–959. [CrossRef] [PubMed]

64. Yin, M.J.; Yamamoto, Y.; Gaynor, R.B. The anti-inflammatory agents aspirin and salicylate inhibit the activity of IκB kinase-β. *Nature* **1998**, *396*, 77–80. [PubMed]

65. Stark, L.A.; Dunlop, M.G. Nucleolar sequestration of RelA (p65) regulates NF-κB-driven transcription and apoptosis. *Mol. Cell Biol.* **2005**, *25*, 5985–6004. [CrossRef] [PubMed]

66. Takada, Y.; Bhardwaj, A.; Potdar, P.; Aggarwal, B.B. Nonsteroidal anti-inflammatory agents differ in their ability to suppress NF-κB activation, inhibition of expression of cyclooxygenase-2 and cyclin D1, and abrogation of tumor cell proliferation. *Oncogene* **2004**, *23*, 9247–9258. [CrossRef] [PubMed]

67. Yamamoto, Y.; Yin, M.J.; Lin, K.M.; Gaynor, R.B. Sulindac inhibits activation of the NF-κB pathway. *J. Biol. Chem.* **1999**, *274*, 27307–27314. [CrossRef] [PubMed]

68. Liao, D.; Zhong, L.; Duan, T.; Zhang, R.H.; Wang, X.; Wang, G.; Hu, K.; Lv, X.; Kang, T. Aspirin suppresses the growth and metastasis of osteosarcoma through the NF-κB pathway. *Clin. Cancer Res.* **2015**, *21*, 5349–5359. [CrossRef] [PubMed]

69. Kim, S.H.; Song, S.H.; Kim, S.G.; Chun, K.S.; Lim, S.Y.; Na, H.K.; Kim, J.W.; Surh, Y.-J.; Bang, Y.-J.; Song, Y.-S. Celecoxib induces apoptosis in cervical cancer cells independent of cyclooxygenase using NF-κB as a possible target. *J. Cancer Res. Clin. Oncol.* **2004**, *130*, 551–560. [CrossRef] [PubMed]

70. Cho, M.; Gwak, J.; Park, S.; Won, J.; Kim, D.E.; Yea, S.S.; Cha, I.-J.; Kim, T.K.; Shin, J.-G.; Oh, S. Diclofenac attenuates Wnt/β-catenin signaling in colon cancer cells by activation of NF-κB. *FEBS Lett.* **2005**, *579*, 4213–4218. [CrossRef] [PubMed]

71. Loveridge, C.J.; Macdonald, A.D.; Thoms, H.C.; Dunlop, M.G.; Stark, L.A. The proapoptotic effects of sulindac, sulindac sulfone and indomethacin are mediated by nucleolar translocation of the RelA(p65) subunit of NF-κB. *Oncogene* **2008**, *27*, 2648–2655. [CrossRef] [PubMed]

72. Mladenova, D.; Pangon, L.; Currey, N.; Ng, I.; Musgrove, E.A.; Grey, S.T.; Kohonen-Corish, M.R.J. Sulindac activates NF-κB signaling in colon cancer cells. *Cell Commun. Signal.* **2013**, *11*, 73. [CrossRef] [PubMed]

73. Jeong, J.B.; Yang, X.; Clark, R.; Choi, J.; Baek, S.J.; Lee, S.H. A mechanistic study of the proapoptotic effect of tolfenamic acid: Involvement of NF-κB activation. *Carcinogenesis* **2013**, *34*, 2350–2360. [CrossRef] [PubMed]

74. Park, I.S.; Jo, J.R.; Hong, H.; Nam, K.Y.; Kim, J.B.; Hwang, S.H.; Choi, M.S.; Ryu, N.H.; Jang, H.J.; Lee, S.H.; et al. Aspirin induces apoptosis in YD-8 human oral squamous carcinoma cells through activation of caspases, down-regulation of Mcl-1, and inactivation of ERK-1/2 and AKT. *Toxicol. In Vitro* **2010**, *24*, 713–720. [CrossRef] [PubMed]

75. Greenspan, E.J.; Madigan, J.P.; Boardman, L.A.; Rosenberg, D.W. Ibuprofen inhibits activation of nuclear β-catenin in human colon adenomas and induces the phosphorylation of GSK-3β. *Cancer Prev. Res.* **2011**, *4*, 161–171. [CrossRef] [PubMed]

76. Stark, L.A.; Reid, K.; Sansom, O.J.; Din, F.V.; Guichard, S.; Mayer, I.; Jodrell, D.I.; Clarke, A.R.; Dunlop, M.G. Aspirin activates the NF-κB signalling pathway and induces apoptosis in intestinal neoplasia in two in vivo models of human colorectal cancer. *Carcinogenesis* **2007**, *28*, 968–976. [CrossRef] [PubMed]

77. Chen, J.; Lobb, I.; Pierre, M.; Sonia, M.N.; James, S.; Kathrin, K.; Kathrin, K.; Kathrin, K.; Oakley, F.; Stark, L.A. Disruption of the PolI complex links inhibition of CDK4 activity to activation of the NF-κB pathway. Available online: http://biorxiv.org/content/biorxiv/early/2017/01/13/100255.full.pdf (accessed on 13 January 2017).

78. Grivennikov, S.I.; Greten, F.R.; Karin, M. Immunity, inflammation, and cancer. *Cell* **2010**, *140*, 883–899. [CrossRef] [PubMed]

79. Marzbani, E.; Inatsuka, C.; Lu, H.; Disis, M.L. The invisible arm of immunity in common cancer chemoprevention agents. *Cancer Prev. Res.* **2013**, *6*, 764–773. [CrossRef] [PubMed]

80. Cao, Y.; Nishihara, R.; Qian, Z.R.; Song, M.; Mima, K.; Inamura, K.; Nowak, J.A.; Drew, D.A.; Lochhead, P.; Nosho, K.; et al. Regular aspirin use associates with lower risk of colorectal cancers with low numbers of tumor-infiltrating lymphocytes. *Gastroenterology* **2016**, *151*, 879–892. [CrossRef] [PubMed]

81. Boulon, S.; Westman, B.J.; Hutten, S.; Boisvert, F.M.; Lamond, A.I. The nucleolus under stress. *Mol. Cell* **2010**, *40*, 216–227. [CrossRef] [PubMed]

82. Grummt, I. The nucleolus-guardian of cellular homeostasis and genome integrity. *Chromosoma* **2013**, *122*, 487–497. [CrossRef] [PubMed]

83. James, A.; Wang, Y.; Raje, H.; Rosby, R.; DiMario, P. Nucleolar stress with and without p53. *Nucleus* **2014**, *5*, 402–426. [CrossRef] [PubMed]
84. Russo, A.; Russo, G. Ribosomal Proteins Control or Bypass p53 during Nucleolar Stress. *Int. J. Mol. Sci.* **2017**, *18*, 140. [CrossRef] [PubMed]
85. Thul, P.J.; Akesson, L.; Wiking, M.; Mahdessian, D.; Geladaki, A.; Ait, B.H.; Alm, T.; Asplund, A.; Björk, L.; Breckels, L.M.; et al. A subcellular map of the human proteome. *Science* **2017**, *356*. [CrossRef] [PubMed]
86. Andersen, J.S.; Lam, Y.W.; Leung, A.K.; Ong, S.E.; Lyon, C.E.; Lamond, A.I.; Mann, M. Nucleolar proteome dynamics. *Nature* **2005**, *433*, 77–83. [CrossRef] [PubMed]
87. Boisvert, F.M.; Lam, Y.W.; Lamont, D.; Lamond, A.I. A quantitative proteomics analysis of subcellular proteome localization and changes induced by DNA damage. *Mol. Cell. Proteom.* **2010**, *9*, 457–470. [CrossRef] [PubMed]
88. Coccia, M.; Rossi, A.; Riccio, A.; Trotta, E.; Santoro, M.G. Human NF-κB repressing factor acts as a stress-regulated switch for ribosomal RNA processing and nucleolar homeostasis surveillance. *Proc. Natl. Acad. Sci. USA* **2017**, *114*, 1045–1050. [CrossRef] [PubMed]
89. Rubbi, C.P.; Milner, J. Non-activated p53 co-localizes with sites of transcription within both the nucleoplasm and the nucleolus. *Oncogene* **2000**, *19*, 85–96. [CrossRef] [PubMed]
90. Audas, T.E.; Jacob, M.D.; Lee, S. Immobilization of proteins in the nucleolus by ribosomal intergenic spacer noncoding RNA. *Mol. Cell* **2012**, *45*, 147–157. [CrossRef] [PubMed]
91. Latonen, L. Nucleolar aggresomes as counterparts of cytoplasmic aggresomes in proteotoxic stress. *Bioessays* **2011**, *33*, 386–395. [CrossRef] [PubMed]
92. Latonen, L.; Moore, H.M.; Bai, B.; Jaamaa, S.; Laiho, M. Proteasome inhibitors induce nucleolar aggregation of proteasome target proteins and polyadenylated RNA by altering ubiquitin availability. *Oncogene* **2011**, *30*, 790–805. [CrossRef] [PubMed]
93. Ehm, P.; Nalaskowski, M.M.; Wundenberg, T.; Jucker, M. The tumor suppressor SHIP1 colocalizes in nucleolar cavities with p53 and components of PML nuclear bodies. *Nucleus* **2015**, *6*, 154–164. [CrossRef] [PubMed]
94. Rocha, S.; Campbell, K.J.; Perkins, N.D. p53- and Mdm2-independent repression of NF-κB transactivation by the ARF tumor suppressor. *Mol. Cell* **2003**, *12*, 15–25. [CrossRef]
95. Sweet, T.; Khalili, K.; Sawaya, B.E.; Amini, S. Identification of a novel protein from glial cells based on its ability to interact with NF-κB subunits. *J. Cell. Biochem.* **2003**, *90*, 884–891. [CrossRef] [PubMed]
96. Dhar, S.K.; Lynn, B.C.; Daosukho, C.; St Clair, D.K. Identification of nucleophosmin as an NF-κB co-activator for the induction of the human *SOD2* gene. *J. Biol. Chem.* **2004**. [CrossRef] [PubMed]
97. Birbach, A.; Bailey, S.T.; Ghosh, S.; Schmid, J.A. Cytosolic, nuclear and nucleolar localization signals determine subcellular distribution and activity of the NF-κB inducing kinase NIK. *J. Cell Sci.* **2004**, *117*, 3615–3624. [CrossRef] [PubMed]
98. Niedick, I.; Froese, N.; Oumard, A.; Mueller, P.P.; Nourbakhsh, M.; Hauser, H.; Köster, M. Nucleolar localization and mobility analysis of the NF-κB repressing factor NRF. *J. Cell Sci.* **2004**, *117*, 3447–3458. [CrossRef] [PubMed]
99. Bierhoff, H.; Dundr, M.; Michels, A.A.; Grummt, I. Phosphorylation by casein kinase 2 facilitates rRNA gene transcription by promoting dissociation of TIF-IA from elongating RNA polymerase I. *Mol. Cell. Biol.* **2008**, *28*, 4988–4998. [CrossRef] [PubMed]
100. Kato, T., Jr.; Delhase, M.; Hoffmann, A.; Karin, M. CK2 is a C-terminal IκB kinase responsible for NF-κB activation during the UV response. *Mol. Cell* **2003**, *12*, 829–839. [CrossRef]
101. Jiang, H.Y.; Wek, S.A.; McGrath, B.C.; Scheuner, D.; Kaufman, R.J.; Cavener, D.R.; Wek, R.C. Phosphorylation of the α subunit of eukaryotic initiation factor 2 is required for activation of NF-κB in response to diverse cellular stresses. *Mol. Cell. Biol.* **2003**, *23*, 5651–5663. [CrossRef] [PubMed]
102. Goldstein, E.N.; Owen, C.R.; White, B.C.; Rafols, J.A. Ultrastructural localization of phosphorylated eIF2α [eIF2α(P)] in rat dorsal hippocampus during reperfusion. *Acta Neuropathol.* **1999**, *98*, 493–505. [CrossRef] [PubMed]
103. Thoms, H.C.; Loveridge, C.J.; Simpson, J.; Clipson, A.; Reinhardt, K.; Dunlop, M.G.; Stark, L.A. Nucleolar targeting of RelA(p65) is regulated by COMMD1-dependent ubiquitination. *Cancer Res.* **2010**, *70*, 139–149. [CrossRef] [PubMed]

104. Parrondo, R.; de las, P.A.; Reiner, T.; Rai, P.; Perez-Stable, C. NF-κB activation enhances cell death by antimitotic drugs in human prostate cancer cells. *Mol. Cancer* **2010**, *9*, 182. [CrossRef] [PubMed]

105. Sniderhan, L.F.; Garcia-Bates, T.M.; Burgart, M.; Bernstein, S.H.; Phipps, R.P.; Maggirwar, S.B. Neurotrophin signaling through tropomyosin receptor kinases contributes to survival and proliferation of non-Hodgkin lymphoma. *Exp. Hematol.* **2009**, *37*, 1295–1309. [CrossRef] [PubMed]

106. Lee, D.H.; Forscher, C.; Di, V.D.; Koeffler, H.P. Induction of p53-independent apoptosis by ectopic expression of HOXA5 in human liposarcomas. *Sci. Rep.* **2015**, *5*, 12580. [CrossRef] [PubMed]

107. Thoms, H.C.; Dunlop, M.G.; Stark, L.A. p38-mediated inactivation of cyclin D1/cyclin-dependent kinase 4 stimulates nucleolar translocation of RelA and apoptosis in colorectal cancer cells. *Cancer Res.* **2007**, *67*, 1660–1669. [CrossRef] [PubMed]

108. Dadsetan, S.; Balzano, T.; Forteza, J.; Cabrera-Pastor, A.; Taoro-Gonzalez, L.; Hernandez-Rabaza, V.; Gil-Perotin, S.; Cubas-Nunez, L.; Garcia-Verdugo, J.M.; Agusti, A.; et al. Reducing peripheral inflammation with infliximab reduces neuroinflammation and improves cognition in rats with hepatic encephalopathy. *Front. Mol. Neurosci.* **2016**, *9*, 106. [CrossRef] [PubMed]

109. Khandelwal, N.; Simpson, J.; Taylor, G.; Rafique, S.; Whitehouse, A.; Hiscox, J.; Stark, L.A. Nucleolar NF-κB/RelA mediates apoptosis by causing cytoplasmic relocalization of nucleophosmin. *Cell Death Differ.* **2011**, *18*, 1889–1903. [CrossRef] [PubMed]

110. Kerr, L.E.; Birse-Archbold, J.L.; Short, D.M.; McGregor, A.L.; Heron, I.; Macdonald, D.C.; Thompson, J.; Carlson, G.J.; Kelly, J.S.; McCulloch, J.; et al. Nucleophosmin is a novel Bax chaperone that regulates apoptotic cell death. *Oncogene* **2007**, *26*, 2554–2562. [CrossRef] [PubMed]

111. Wang, Z.; Gall, J.M.; Bonegio, R.; Havasi, A.; Illanes, K.; Schwartz, J.H.; Borkan, S.C. Nucleophosmin, a critical Bax cofactor in ischemia-induced cell death. *Mol. Cell. Biol.* **2013**, *33*, 1916–1924. [CrossRef] [PubMed]

112. Zhang, L.; Yu, J.; Park, B.H.; Kinzler, K.W.; Vogelstein, B. Role of BAX in the apoptotic response to anticancer agents. *Science* **2000**, *290*, 989–992. [CrossRef] [PubMed]

113. Tatham, M.H.; Cole, C.; Scullion, P.; Wilkie, R.; Westwood, N.J.; Stark, L.A.; Hey, R.T. A proteomic approach to analyze the aspirin-mediated lysine acetylome. *Mol. Cell. Proteom.* **2017**, *16*, 310–326. [CrossRef] [PubMed]

114. Wang, J.; Zhang, C.J.; Zhang, J.; He, Y.; Lee, Y.M.; Chen, S.; Lim, T.K.; Ng, S.; Shen, H.M.; Lin, Q. Mapping sites of aspirin-induced acetylations in live cells by quantitative acid-cleavable activity-based protein profiling (QA-ABPP). *Sci. Rep.* **2015**, *5*, 7896. [CrossRef] [PubMed]

115. Chan, T.A.; Morin, P.J.; Vogelstein, B.; Kinzler, K.W. Mechanisms underlying nonsteroidal antiinflammatory drug-mediated apoptosis. *Proc. Natl. Acad. Sci. USA* **1998**, *95*, 681–686. [CrossRef] [PubMed]

116. Charruyer, A.; Grazide, S.; Bezombes, C.; Muller, S.; Laurent, G.; Jaffrezou, J.P. UV-C light induces raft-associated acid sphingomyelinase and JNK activation and translocation independently on a nuclear signal. *J. Biol. Chem.* **2005**, *280*, 19196–19204. [CrossRef] [PubMed]

MDPI

Review

Hypoxia and Inflammation in Cancer, Focus on HIF and NF-κB

Laura D'Ignazio, Michael Batie and Sonia Rocha *

Centre for Gene Regulation and Expression, School of Life Sciences, University of Dundee,
Dundee DD15EH, UK; l.dignazio@dundee.ac.uk (L.D.); m.t.batie@dundee.ac.uk (M.B.)
* Correspondence: s.rocha@dundee.ac.uk; Tel.: +44-013-823-85792

Academic Editor: Veronique Baud
Received: 31 March 2017; Accepted: 4 May 2017; Published: 9 May 2017

Abstract: Cancer is often characterised by the presence of hypoxia and inflammation. Paramount to the mechanisms controlling cellular responses under such stress stimuli, are the transcription factor families of Hypoxia Inducible Factor (HIF) and Nuclear Factor of κ-light-chain-enhancer of activated B cells (NF-κB). Although, a detailed understating of how these transcription factors respond to their cognate stimulus is well established, it is now appreciated that HIF and NF-κB undergo extensive crosstalk, in particular in pathological situations such as cancer. Here, we focus on the current knowledge on how HIF is activated by inflammation and how NF-κB is modulated by hypoxia. We summarise the evidence for the possible mechanism behind this activation and how HIF and NF-κB function impacts cancer, focusing on colorectal, breast and lung cancer. We discuss possible new points of therapeutic intervention aiming to harness the current understanding of the HIF-NF-κB crosstalk.

Keywords: NF-κB; hypoxia; inflammation; κB Kinase (IKK); Prolyl Hydroxylases (PHDs); cancer; Transforming Growth Factor-β-Activated Kinase 1 (TAK1); Factor Inhibiting HIF (FIH)

1. NF-κB Subunits and Signalling Pathways

NF-κB is the collective name of a family of transcription factors initially discovered in 1986 by Ranjan Sen and David Baltimore as a Nuclear Factor binding to the enhancer element of the immunoglobulin κ light-chain of activated B cells [1]. Over the years, a crucial role in controlling gene expression in response to inflammation, proliferation, differentiation, among other physiological processes, has been assigned to the five gene members of this protein family: RelA (p65), RelB, c-Rel, NF-κB1 (p105) and NF-κB2 (p100) [2,3]. All NF-κB subunits feature high structural similarity, with the N-terminal harbouring a Rel Homology Domain (RHD). This is essential to mediate DNA binding to κb sites in enhancers/promoters of target genes, as well as to dimerise with other subunits. In fact, formation of homo- and heterodimers determines the specificity of the transcriptional response in accordance with the cellular context [4]. Induction or repression of gene expression is achieved not only through combination of different NF-κB dimers, but also by physical association with co-activators, co-repressors, and other transcription factors, such as Signal Transducer and Activator of Transcription 3 (STAT3), p53 [5], or HIFs [6]. A further control of the transcriptional activity also derives from the C-terminal structure of the NF-κB subunits. In particular, RelA, RelB and c-Rel contain trans-activation domains (TAs), whereas p105 and p100, precursors of their active forms p50 and p52, contain IκB-like Ankyrin repeat (ANK) domains, acting as internal inhibitors. Consequently, p50 and p52 can only function as transcriptional activators in association with other subunits or proteins containing the trans-activation domain [5,7].

Activation of NF-κB can occur following canonical, non-canonical and atypical pathways, all triggered by different stimuli, as elegantly reviewed elsewhere [8,9]. Here, we will briefly focus only on the canonical and non-canonical mechanisms of activation (Figure 1), which are more likely to be involved in malignancies. In the canonical NF-κB signalling pathway, the binding of Tumour Necrosis Factor α (TNF-α), Lipopolysaccharides (LPS), or Interleukin-1 (IL-1) to their specific receptors on the cellular membrane, induces in the cytoplasm the recruitment of several adaptors and protein kinases, serving as activation platform, ultimately leading to the phosphorylation and activation of the Inhibition of κB Kinase (IKK) complex. The IKK complex is formed by two catalytic proteins (IKK1/IKKα and IKK2/IKKβ) and one regulatory protein (IKKγ/NEMO, NF-κB Essential Modulator). An activated IKK complex is then able to phosphorylate the IκB inhibitor molecules, which, in quiescent cells, hold the NF-κB subunits inactive in the cytoplasm. In humans, the most common NF-κB inhibitor protein is IκBα. Phosphorylation of IκBα on serines 32 and 36 is the key prerequisite for the lysine-48 polyubiquitin chain formation catalysed by SCF$^{\beta TrCP}$ E3 ubiquitin ligase, and subsequent degradation mediated by the proteasomal machinery. This event makes the NF-κB dimers free to translocate into the nucleus, binding the cognate DNA sequence to regulate gene transcription [10]. On the other hand, the non-canonical NF-κB signalling pathway depends on activation of different membrane receptors, such as Lymphotoxin β-Receptor (LTβR), B-cell Activation Factor Receptor (BAFFR), TNF Receptor 2 (TNFR2) and several others. The consequent signal transduction involves the activation of the NF-κB Inducing Kinase (NIK) that, in turns, phosphorylates and activates IKKα homodimers. This event is followed by specific phosphorylation of serines 866 and 870 at the C-terminal region of p100, resembling the phosphorylation site of IκBα. Upon binding of the SCF$^{\beta TrCP}$ ubiquitin ligase, the inhibitor ankyrin repeat domain of p100 is subject to proteasomal processing. Thus, the cleaved active form p52 originates and associates with RelB, serving as transcriptional activator heterodimer after translocation into the nucleus [7].

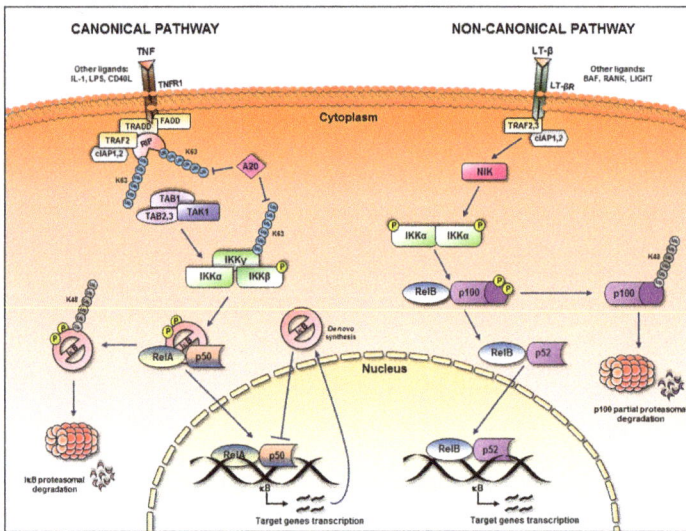

Figure 1. Activation pathways for canonical and non-canonical NF-κB signalling. Canonical NF-κB pathway is exemplified by TNF-α binding to its reception, while non-canonical NF-κB pathway is illustrated by binding of LT-β to its receptor.

In general, the fine tuning of the NF-κB response is controlled by different post-translational modification events, such as phosphorylation, as well as by an intricate series of protein-protein interactions and feedback loops. The fact that IκBα itself is one of the NF-κB target genes represents an example of negative feedback. The newly synthesized IκBα is able to relocate NF-κB subunits from the nucleus to the cytoplasm. Also, considering the important role of K63-linked or linear polyubiquitination as a platform for the activation of the NF-κB pathway, de-ubiquitinase (DUB) enzymes, such as Cyld or A20, can serve as negative feedback inhibitors of the pathways [5]. When the regulatory mechanisms described above become aberrant, NF-κB can become constitutively active or deregulated. For instance, NF-κB is chronically activated in several inflammatory diseases, such as arthritis, inflammatory bowel disease, asthma and many other pathological conditions, including cancers [11,12]. Here, we will discuss the role of NF-κB in cancer, focusing on how this transcription factor can be induced by hypoxia, and modulated by HIF, in a variety of tumour contexts.

2. NF-κB in Inflammation and Cancer

Activation of the NF-κB pathway is widely recognised as characteristic of inflammation. Inflammation is a defensive process used by the innate and adaptive immune systems to respond to bacterial and viral infections, facilitating wound healing or maintenance of tissue homeostasis. In the last fifteen years, chronically prolonged inflammatory response has been identified as a hallmark of cancer [13]. However, the role of NF-κB in malignancies remains quite controversial, acting as tumour promotor or tumour suppressor depending on the cellular context [14]. In cancers featuring a chronic inflammatory microenvironment, NF-κB is conspicuously activated. Thus, cancer development is further promoted, and tumour progression is reinforced. In fact, in these tissues, pro-inflammatory cytokines (i.e., TNF-α, IL-1, interleukin-6 (IL-6)), Reactive Oxygen and Nitrogen Species (RONS), prostaglandins, and microRNAs accumulate, contributing to create a pro-tumorigenic microenvironment. In particular, a constitutive activated NF-κB participates in carcinogenesis stimulating cell proliferation, inhibiting programmed cell death, regulating angiogenesis, promoting tumour metastasis and remodelling tumour metabolism [14]. NF-κB influences cell proliferation by controlling autocrine and paracrine production of cytokines, such as Interleukin-2 (IL-2) and Granulocyte-Macrophage Colony-Stimulating Factor (GM-CSF) [15]. Additionally, NF-κB controls gene expression of G1 cyclins, such as cyclin D1 [16], which is a crucial protein in the exit from the G1 phase of the cell cycle, frequently altered in numerous human cancers [17].

NF-κB plays a dual role in controlling apoptosis. Several reports have supported a pro-apoptotic activity of NF-κB [18,19]. However, NF-κB ability to prevent apoptosis seems to be more often studied, leading to the activation of target genes such as cellular Inhibitors of Apoptosis (cIAP1/2, XIAP), cellular FLICE (Caspase8/FADD-like IL-1β-Converting Enzyme)-Inhibitory Protein (c-FLIP), and members of the Bcl2 family, such as Bcl-xL [20]. Such anti-apoptotic roles of NF-κB might be determinant during cancer progression, when cells that have undergone DNA damage or chromosomal rearrangements can therefore escape from apoptosis, overcoming also other checkpoint controls normally operated by p53, a tumour suppressor mutual antagonist of NF-κB [21].

Expansion of vascular network, also termed angiogenesis, is one of the main hallmarks of tumour growth [13]. A number of NF-κB target genes are known to be involved in this process. Among them there are the chemokine Interleukin-8 (IL-8) [22], and the Vascular Endothelial Growth Factor (VEGF) [23]. Anti VEGF strategies have been successful in the clinic for treatment of patients with several cancers, including colorectal cancer (CRC) (reviewed in [24]), in the form of Bevacizumab, a monoclonal antibody against VEGFA, and Aflibercept, a recombinant fusion protein blocking VEGFA and VEGFB signalling. Interestingly, these genes can be also targets of HIF-1α [25,26], highlighting the existence of an intricate crosstalk between inflammation and hypoxia in cancer cells (see Section 4).

NF-κB also directly regulates expression of genes encoding for matrix metalloproteinases, such as MMP-9 [27]. Extracellular matrix remodelling not only facilitates the spread of endothelial cells (angiogenesis), but also of cancer cells (metastasis) in the surrounding areas. Notably, a role of NF-κB in tumour metastasis formation has been reported in human head and neck squamous cell carcinoma [28], and breast cancer [29].

As part of the immune response, NF-κB can exert also tumour suppressing functions. In particular, this occurs in some acute inflamed environments, when Cytotoxic T cells (CTL) are highly activated against malignant cells [30], or in chemically induced liver and skin cancers [31,32]. However, this response might be not able to eradicate all aberrant cells, which then escape the immune system. Chronic inflammation in the microenvironment, and, consequently, the increased cytokine release elevates the NF-κB activity, resulting in tumour promotion. However, the NF-κB activity can be enhanced also by mutation of NF-κB pathway components and/or oncogenes [33]. Mutation of NF-κB occurs in particular in lymphoid malignancies, such as human B cell- [34,35] or T cell-lymphomas [36]. Although less frequently, mutations of the NF-κB signalling pathway occur also in solid tumours. For instance, mutations in NF-κB1 have been detected in breast cancer [37]. In addition, most recently, high expression of IKKα has been associated with poor outcome in patients with Estrogen Receptor (ER)-positive invasive ductal breast cancer, although its expression appeared to be independent of NIK and RelB [38]. Overall, the NF-κB pathway has been found to be fundamental in development, maintenance, or invasiveness of multiple solid cancer types, including colon and lung cancers [39–41], hepatocellular carcinoma [42], and melanoma [43]. While not all cancers form in inflamed tissues, inflammation arises as a consequence of metabolic dysregulation associated with tumour development, facilitating the progression of the tumour itself. During tumour propagation, the high demand of oxygen and nutrients creates a pathologic hypoxic microenvironment in the tumour core, inducing the production of angiogenic growth factors and cytokines, to form new blood vessels and recruit more innate immune cells [44]. Indeed, hypoxia strongly impacts on tumour progression and metastasis, by activating specific transcriptional programmes, such as HIF and NF-κB [45].

3. Hypoxia and Hypoxia Inducible Factor (HIF) Pathway

Hypoxia, or diminished oxygen availability, is a common feature of the tumour microenvironment, where the oxygen level is often below 1%. This condition triggers a series of gene expression changes, affecting angiogenesis and metabolism, to enable tumour survival and progression [45]. Hypoxia in tumours occurs in different areas, from anoxic regions further away from the blood vessels to hypoxic areas, generated by highly metabolic cancer and immune cells [45,46], creating a non-uniform tumour microenvironment. The main molecular drivers of this response belong to a family of basic Helix-Loop-Helix-Per-ARNT-Sim (bHLH-PAS)-containing transcription factors, known as HIFs [47]. HIF is a heterodimeric complex formed by an oxygen-dependent α subunit and an oxygen-insensitive β subunit. In fact, the constitutively expressed HIF-1β (also called ARNT, Aryl Hydrocarbon Receptor Nuclear Translocator) is the obligatory binding partner for any HIF-α. So far, three α-forms (HIF-1α, -2α and -3α) have been identified in humans. HIF-1α, the most well-studied isoform, is ubiquitously expressed, while HIF-2α expression is restricted to endothelial cells, heart, lung, placenta and kidney [48]. Only recently the expression profile of the several HIF-3α spliced variants has been elucidated, with HIF-3α predominantly expressed in kidney and lung epithelial cells [49]. Interestingly, although HIF-1α and HIF-2α are structurally closely related, and despite some redundancy in their functions with the sharing of several common target genes, they also have distinct target gene cohorts [50]. Moreover, their roles can vary depending on tumour and cell types [51]. HIF-3α function remained unclear for long time, being mainly considered a dominant negative regulator of the other HIF-α isoforms, by competing for HIF-1β binding [50]. Only recently, a novel transactivation activity induced by hypoxia has been attributed to HIF-3α in zebrafish embryos, opening new scenarios in the regulation of transcriptional response following exposure to low oxygen [52].

The cellular response to hypoxia engages HIF primarily at the post-transcriptional level, where the HIF-α proteins stability is regulated by non-heme, Fe^{2+} and 2-oxoglutarate (2OG)-dependent dioxygenase enzymes called Prolyl Hydroxylases (PHDs). Under normal oxygen conditions, PHDs (PHD1, PHD2 and PHD3) hydroxylate specific proline residues within the Oxygen Dependent Domain (ODD) of the HIF-α subunit. Thus, the hydroxylated HIF-α is recognised by the von Hippel-Lindau (vHL) tumour suppressor protein, component of an E3-ubiquitin-ligase complex. This leads to HIF-α Lys48-linked poly-ubiquitination and subsequent proteasomal degradation. In addition, further control of transcriptional activity of HIFs escaping degradation is mediated by FIH (Factor Inhibiting HIF). This 2OG-dependent dioxygenase enzyme prevents HIF association with the co-activators p300/CREB

Binding Protein (CBP), by hydroxylation of a key asparagine in the transactivation domain of HIF-α. In hypoxia, PHDs and FIH are inactive, due to the absence of oxygen, essential cofactor for these enzymes [53]. This enables the hypoxic transcriptional programme to occur. The stabilised HIF-α subunit translocates into the nucleus, dimerizes with HIF-1β, and, upon binding to the consensus hypoxia response element (HRE), transactivates downstream target genes, involved in a large variety of processes, including glycolysis, angiogenesis, proliferation, migration, and apoptosis [54] (Figure 2). As crucial mediators of several biological and cellular processes, both HIF-1α and HIF-2α expression are elevated in numerous solid tumours, such as colon, breast and lung cancers [48]. For instance, high levels of HIF-1α correlate with poor clinical outcomes in human breast cancer, with HIF-1α being the master regulator of Epithelial-Mesenchymal Transition (EMT), invasion, extravasation, and metastasis in this type of tumor [55].

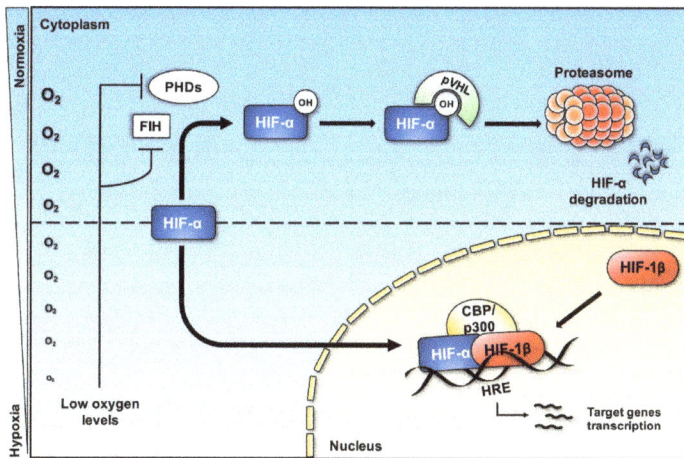

Figure 2. Activation of the Hypoxia Inducible Factor (HIF) pathway in response to hypoxia. HIF-α levels are controlled in normoxia by PHD-mediated hydroxylation (OH) and recognition by the E3-ligase complex containing the tumour suppressor pVHL. In hypoxia, Prolyl Hydroxylases (PHDs) and FIH are inhibited and HIF-α escapes VHL-dependent degradation. T-bars represent repression.

A relevant role as regulators of the hypoxic response has been recently attributed to the Jumonji C (JmjC) domain containing proteins [56], many of which are 2-OG dioxygenases functioning as protein demethylases. These enzymes, of which there are over 30 discovered in humans, mainly control histone methylation, and they are often deregulated in many cancers (reviewed in [57]). Structural work from the Schofield and Allshire laboratories has revealed that the domain responsible for demethylase activity of these enzymes (JmjC domain) has a fold that is remarkably similar to the catalytic core of FIH [58,59]. In addition, investigation of the oxygen dependency of two of these enzymes revealed a graded drop in activity over physiologically relevant ranges of oxygen [60,61]. These studies indicate the potential of JmjC enzymes to link chromatin structure to oxygen sensing and participate in the hypoxia mediated transcriptional response. Indeed, an increasing number of reports found elevated histone methylation marks in response to prolonged hypoxia [62–65], with the impaired JmjC enzyme activity being responsible for these changes [63,65]. Interestingly, many of these enzymes are hypoxia inducible (reviewed in [57]), with some, including Lysine (K)-specific demethylase 4B (KDM4B), KDM4C, KDM5B, KDM3A, KDM2A and KDM2B, being HIF targets [66–71]. This may point to negative feedback mechanism to help regulate histone methylation in a JmjC histone demethylase compromised environment, and possibly help the cell reset its oxygen sensing and response mechanisms after prolonged hypoxia and/or restoration of normoxia, as is seen with PHD regulation of HIF.

4. Crosstalk between Hypoxia and Inflammation in Cancer

Recently, an increasing number of studies supported a role of HIF beyond the hypoxia response [72]. HIF-1α activation has been detected following different bacterial infections under normal oxygen levels [73], whereas HIF-2α and HIF-1β have been found to regulate neutrophilic inflammation and myeloid cells function in wound healing, respectively [74,75]. In inflammation, mechanisms leading to HIF induction can be oxygen-independent, and mediated by other transcription factors, such as STAT3 [76] and NF-κB [77,78]. In recent years, an intimate crosstalk between HIF and NF-κB has been appreciated at different levels, as reviewed in [6]. Interestingly, this crosstalk is bi-directional. In fact, not only does NF-κB induce HIF (Figure 3A), but HIF regulates NF-κB (Figure 3B). Particularly, under inflammatory conditions, NF-κB transcriptional activity is restricted by HIF-1α in vivo and in vitro [25]. To date, the knowledge concerning HIF-2α and HIF-1β contribution to the NF-κB activity is still poor, despite the fact that these HIF subunits have been associated with NF-κB [79,80].

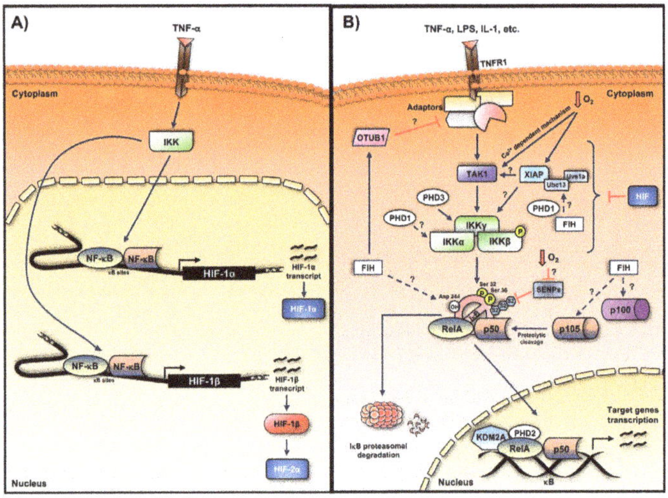

Figure 3. Levels of crosstalk between the HIF and the NF-κB pathways. (**A**) NF-κB control over the HIF pathway; (**B**) Reported points of interaction and control of the HIF pathway over the NF-κB signalling cascade. Red T-bars indicate inhibition points, whereas arrows with dashed lines indicate regulation events that have not been proved in an in vivo system yet. Question marks are used to highlight the unanswered questions into the HIF regulation of the NF-κB pathway.

HIF can directly contribute to the inflammatory response, inducing several pro-inflammatory chemokines and cytokines [12]. Importantly, numerous genes transcriptionally activated by HIF are also NF-κB target genes involved in tumorigenesis, such as IL-6, MMP9, cyclooxygenase 2 (COX2), as well as pro-survival genes, such as Bcl-2, among others [81]. Indeed, the cooperative relationship between HIF and NF-κB in the tumour-associated inflammation is evident. More pro-inflammatory mediators are produced in the hypoxic areas of the tumours, resulting in the recruitment of more immune cells at neoplastic sites. This determines a chronic inflammation in the tumour, with consequent high activation of NF-κB [12]. As previously stated, inflammation is a key player in tumour development and progression, therefore it is not surprising that chronic inflammatory diseases may predispose to cancer. One example is the Inflammatory Bowel Disease (IBD), a chronic intestinal disorder including Crohn's disease and ulcerative colitis [11]. Patients affected by this condition have a greater risk to develop colon cancer, particularly Colitis-Associated Colon cancer (CAC) [82].

In colon tumorigenesis, where hypoxic inflammation is significant, both HIF-1α and HIF-2α are expressed [48]. Notably, a recent study highlighted the importance of intestinal epithelial HIF-2α in the recruitment of neutrophils to colon tumour sites, supporting its prominent role in the inflammatory microenvironment [83]. Another example of hypoxia and inflammation crosstalk in cancer is observed in Hepatocellular Carcinoma (HCC), where TNF-α is one of the cytokines constantly activated by NF-κB, through the Tumour Associated Macrophages (TAMs). The expression of HIF-1α and HIF-2α in these immune cells seems to be particularly important in the HCC progression [12,84]. However, to better understand the precise mechanisms, by which hypoxia and immune cells contribute to the intra-tumoural inflamed microenvironment, further studies are needed.

5. Hypoxia-Induced NF-κB

As mentioned earlier, NF-κB is one of several transcription factors induced by hypoxia. Although mechanisms by which NF-κB is activated under low oxygen are still under investigation.

5.1. Role of the Oxygen Sensors in the Hypoxia Induction of NF-κB

5.1.1. Hydroxylases

2OG-dependent dioxygenases, such as PHDs and FIH, function as oxygen sensors stabilizing HIF in hypoxia, therefore these and other dioxygenases may confer oxygen sensitivity to additional pathways regulated by oxygen availability, including NF-κB. Indeed, the discovery of new and potential FIH and PHD targets supports this. Proteomics approaches have revealed NF-κB pathway components, including p105 and IκBα [85] as well as an upstream regulator of the NF-κB pathway, OTU dDe-ubiquitinase, Ubiquitin Aldehyde Binding 1 (OTUB1), as being hydroxylated by FIH [86,87]. Although these novel FIH targets provide promising links to NF-κB and oxygen sensing, with the exception of OTUB1 [87], mutational analysis has been unsuccessful in identifying any functional significance for these modifications [85,88]. Although OTUB1 hydroxylation by FIH regulates metabolic processes in the cell [87], a role of OTUB1 in activating the NF-κB pathway under hypoxic conditions has yet to be established. Interestingly, other players in the IL-1β pathway, the ubiquitin ligase enzymes Uve1a and Ubc13, are also targets for hydroxylation [86]. These enzymes have also been shown to be required for hypoxia induced NF-κB activity [89,90]. Thus, although speculative, oxygen regulated hydroxylation of these ubiquitin ligases could play a part in the NF-κB induction by hypoxia.

As with FIH, whilst there is evidence of PHD regulation of NF-κB induction following oxygen deprivation, a direct oxygen sensing mechanism mediated by PHD prolyl hydroxylase activity has not been discovered yet. PHDs have been found to antagonise NF-κB activity in various cell types [91–95]. Hypoxia induction of NF-κB via IKK activation has been shown to be regulated by PHD1 and, to a lesser extent, by PHD2 levels [91]. The authors of this study suggested a PHD prolyl hydroxylase dependent mechanism, providing evidence that IKKβ is a potential PHD target through the identification of conserved PHD prolyl hydroxylation motif, which is required for the hypoxia induction of IKKβ levels. Further supporting this finding, Zheng and colleagues identified IKKβ among the PHD1 substrates in a hydroxylation screening assay [96]. However, further studies are required to demonstrate IKK hydroxylation in cells. Conversely, another group has found that PHD3 inhibits NF-κB by a prolyl hydroxylase-independent inhibition of IKKγ ubiquitination [94]. In addition, a cooperative role of PHD2 with respect to NF-κB activity, functioning as coactivator of p65, has been shown [97]. Taken together, these studies exemplify the cell type and context specificity of hypoxia induced NF-κB regulation by PHDs.

5.1.2. JmjCs

The aforementioned JmjC enzymes also link NF-κB activities' in hypoxia to oxygen sensing. Despite methylated histone lysine residues being the prominent target for the demethylating activity of these enzymes, non-histone targets are emerging. KDM2A has been shown to demethylate p65, inhibiting expression of some of its target genes [98,99]. Various methylation sites have been identified on p65 [98,100–102]. Both K218 and K221 methylation sites are reversibly regulated by Nuclear Receptor Binding SET Domain Protein 1 (NSD1) and KDM2A, with NSD1 methylating them in response to IL-1β induction, aiding the activation of a subset of NF-κB target genes. Interestingly, KDM2A is NF-κB inducible [98] as well as hypoxia inducible in HIF-1 dependent manner, as recently shown by our group [71]. This represents another feedback loop of NF-κB regulating its own activity, and may confer another level of crosstalk between low oxygen availability and NF-κB function. KDM2B, the other member of KDM2 family member, is also hypoxia and NF-κB inducible [71,103]; however, it is currently unknown if this enzyme influences NF-κB activity. Another NF-κB induced JmjC enzyme, KDM6B, promotes activation of a subset of genes in LPS activated macrophages in a histone demethylase independent manner [104]. Furthermore, JMJD8 has been shown to positively regulate TNF induced NF-κB signalling, although the mechanism by which this occurs has not been elucidated [105]. These reports further highlight some known and potential crosstalk points between hypoxia and NF-κB. Emergence of new JmjC targets and functions may lead to new NF-κB regulatory links.

5.2. TAK and IKK in Hypoxia Induced NF-κB

An IKK independent mechanism of action was reported when hypoxic activation of NF-κB was initially discovered [106]. Since this seminal work, it has been shown that IKK dependent mechanisms of NF-κB activation in response to hypoxia do occur [72,107,108]. Our laboratory has shown that NF-κB responds rapidly to hypoxia in an IKK dependent manner in cancer and primary cell lines. Specifically, IKK mediates hypoxia induced phosphorylation of IκBα at serine 32 and 36 and also influences DNA binding of NF-κB [90,107]. This mechanism is dependent on the E2 ubiquitin conjugating enzyme Ubc13 [107], and XIAP may be one the E3 ligases interacting with Ubc13 in hypoxia-induced NF-κB activation [90]. We went on to show that hypoxia-induced IKK-mediated NF-κB activation is conserved in *Drosophila* [109]. This work also found that the MAPK family member TAK1 (Transforming Growth Factor-β-Activated Kinase 1) was part of the mechanism of hypoxia-induced NF-κB. Further investigation into other potential regulatory mechanisms of hypoxia induced NF-κB activation, particularly cell/tissue specific regulators, are required.

5.3. Role of IκBα in the Hypoxia Induction of NF-κB

As mentioned above, inactivation of the NF-κB inhibitor IκBα by TAK-IKK mediated serine phosphorylation can induce hypoxia-induced activation of NF-κB. Mutational analysis shows that this mechanism appears to be independent of IκBα tyrosine phosphorylation [107], for which a role was initially suggested [106]. Hypoxia induction of NF-κB is atypical since IκBα is not degraded as ubiquitination is inhibited and replaced with sumoylation in low oxygen environments [107]. There are currently various models of IκBα sumoylation influencing NF-κB activity in different cellular contexts [110,111]. Sumo 2/3 conjugation of IκBα may be important in hypoxia-induced NF-κB activation, whilst Sumo 1 conjugation has been shown to inhibit NF-κB. Inactivation of Sumo proteases in hypoxia is a potential mechanism whereby 2/3 conjugation of IκBα is present in hypoxia. Interestingly, polycomb complex regulated transcription has been shown to be influenced by nuclear IκBα phosphorylation and sumoylation [112]. Transcriptional control regulated by IκBα sumoylation in hypoxia represents a new area of research in the field of inflammation following hypoxia.

6. Hypoxia-Dependent NF-κB Activation in Cancer

6.1. Colorectal Cancer

Colorectal cancer (CRC) affects over 500,000 people each year, and is the fourth most common cause of cancer related mortalities [113]. The intestinal lumen of a CRC patient is characterised by both inflammatory and hypoxic regions (reviewed in [11]). NF-κB is activated in CRC in response to inflammation, promoting tumorigenesis and cancer progression [114]. Multiple pathways are implicated in NF-κB oncogenic role in CRC, including Reactive Oxygen Species (ROS) production, activation of pro-inflammatory cytokines, cell survival, EMT, cell proliferation, migration and angiogenesis (reviewed in [11]). Blocking NF-κB signalling has been shown to impair tumour growth in mouse models of CRC and CAC [40,115]. Moreover, anti-inflammatory drugs are used in the clinic to target chronic inflammation in CRC. These are mainly Non-Steroid Anti-Inflammatory Drugs (NSAIDs), which inhibit cyclooxygenase enzymes including COX2, upregulated by NF-κB. Like most solid tumours, hypoxia promotes tumorigenesis and progression in CRC. As mentioned earlier, effects of HIF-1α and HIF-2α on cancer is context specific. In CRC they have antagonising roles, with HIF-1α acting oncogenic and HIF-2α acting tumour suppressive [116]. High HIF-1α levels are associated with poor CRC prognosis. The role of HIF-1α in potentiating CRC through metastatic and angiogenic pathways has been characterised in several reports [117–121]. Conversely, transcript analysis on 120 CRC patient samples found that low HIF-2α mRNA is a prognostic factor, correlating with increased risk of mortality [122]. Another study performed immunohistochemically analysis on 63 primary tumour samples, finding an anti-correlation between HIF-2α levels and tumour stage [116]. The aforementioned study also used mouse xenografts showing siRNA depletion of HIF-1α reduces tumour growth whereas the opposite is seen in HIF-2α depletion. Additionally, HIF-1α knock-down in colon cancer cells increases cell proliferation, and, although HIF-2α has no effect on cell proliferation, colony formation was increased in a soft agar assay for anchorage independent growth [116]. Pharmacological inhibition of HIF has also been shown to result in tumour regression in a murine model of CAC, with a reduction in TAM infiltration [123]. As the above data demonstrates, NF-κB and HIF-1α are key components in driving CRC development and growth. Apoptosis, cell proliferation, angiogenesis and EMT are some overlapping pathways in the crosstalk between inflammatory and hypoxic signalling in CRC.

Around 70% of CRCs follow a distinct mutational sequence, starting with mutations in the tumour suppressor Adenomatous Polyposis Coli (APC) followed by V-Ki-ras2 Kirsten rat sarcoma viral oncogene homolog (KRAS), p53 and DCC (Deleted in Colorectal Cancer) mutations. The oncogene c-myc is also commonly overexpressed. The initial APC mutations trigger adenoma formation, which can develop into CRC [124]. APC supresses Wnt/β catenin signalling, which limits cell proliferation through the T-Cell Factor/Lymphoid Enhancer Factor (TCF/LEF) pathway [125,126]. APC, HIF-1α and β catenin are in cross regulatory network. APC is a HIF-1α target gene repressed in hypoxia, activating cell proliferation via increased Wnt/β catening signalling [127]. Moreover, APC can indirectly repress HIF-1α in a β catenin and NF-κB dependent fashion [127]. Furthermore, β catenin regulates NF-κB in a dose dependent manner. β catenin can activate NF-κB signalling through a positive feedback loop, however at higher levels β catenin can repress NF-κB (reviewed in [128]). β catenin levels have also been recently shown to be regulated by KDM2A and KDM2B [129]. Demethylation of non-phosphorylated β catenin by KDM2A and KDM2B induces nuclear degradation of β catenin and loss of Wnt/β catenin signalling. Given the importance of APC in CRC, gaining a better understanding of this complex crosstalk between HIF, NF-κB and APC pathways may give better insight into understanding molecular mechanisms behind CRC.

Another connection point between HIF and NF-κB pathways is c-myc. c-myc promotes cell proliferation and is typically overexpressed in transformed cells. HIF-1α inhibits c-myc activity and functioning via multiple mechanisms, including direct interaction, induction of Mx1 and activation of p21 [130–133]. Paradoxically, in an oncogenic environment where c-myc is overexpressed, HIF does not impair c-myc driven cell proliferation; instead, c-myc and HIF-1α collaborate to potentiate activation of metabolic proteins, such as Pyruvate Dehydrogenase Kinase 1 (PDK1) and Hexokinase 2 (HK2), driving the Warburg effect, and VEGF, driving angiogenesis (reviewed in [134]). c-myc is also transcriptionally upregulated by NF-κB, along with p21, conferring cell proliferative effects of NF-κB [16,135]. Furthermore, c-myc transcriptional activity on a subset of its targets driving cell growth is augmented by HIF-2α [136].

EMT is crucial for cancer progression, enabling invasion, and thus initiation of metastasis. The EMT driving transcription factors Snail and Twist are components co-regulated by NF-κB and HIF with potential clinical significance in CRC. Twist enhances EMT, Snail promotes lymph node metastasis in CRC [137]. Both are upregulated by hypoxia in a HIF dependent manner [138–140]. Also, HIF-1α upregulation of Twist in response to hypoxia or HIF-1α overexpression induce EMT and metastatic phenotypes [139]. TNF-α induction of NF-κB signalling, stabilising Snail and β catenin, also promotes EMT [141]. Moreover, high levels of Twist and NF-κB are associated with tumour metastasis to the lymph nodes [142]. Hypoxic and inflammatory stimuli in the tumour microenvironment can coordinate the infiltration TAMs to the tumour. TAMs can be tumour inhibitory or tumour promoting. Furthermore, low oxygen and chronic inflammation can subvert normal macrophage function from cancer killing (M1 classically activated) to cancer survival and growth (M2 alternatively activated). TAMs polarized towards the M2 state potentiate immunosuppressive, metastatic and angiogenic signals. NF-κB is activated by infiltrating macrophages and TAMs through release of growth factors and cytokines (reviewed in [143]), and is central to subverting TAM function. As mentioned earlier, pharmacological inhibition of HIF has also been shown to result in tumour regression in a murine model of CAC, with a reduction in TAMs infiltration [123]. The group identified a potential mechanism for the reduced TAM infiltration in this model through loss of Macrophage Colony Stimulating Factor Receptor (M-CSFR), a HIF target gene that is key signal for recruitment of macrophages to a tumour environment. The role played by TAMs in CRC progression is somewhat controversial. There is accumulating evidence for both tumour surviving and tumour killing phenotypes. It seems that which activity is dominant is dependent on where TAMs are located in CRC and the extent to which they have M1 or M2 like activities. Nonetheless the role of hypoxia and inflammation in tumour promoting functions of TAMs are well established. The development of therapies to polarize the TAM pool in CRC towards an M1 phenotype does seem attractive.

JmjC enzymes are emerging as new therapeutic targets and potential biomarkers in several cancers, including colon cancer. Given that KDM2A directly repressed NF-κB transcriptional activity via p65 demethylation [98], it would be anticipated that KDM2A activity may influence NF-κB driven cancer phenotypes. In the same study identifying KDM2A as demethylating p65, the authors showed that KDM2A impairs NF-κB dependent colon cancer cell growth [98]. Transcript analysis of colorectal liver metastases has identified the HIF-1α target KDM3A, as a biomarker for hypoxic tumour cells and potential prognostic marker and therapeutic target for CRC [69,144]. Another JmjC containing protein, KDM6B, is transcriptionally induced by vitamin D, and KDM6B mediates a subset of effects of vitamin D on colon cancer [145]. Specifically, KDM6B deletion induces the expression of pro EMT factors including Snail, and mesenchymal markers. KDM6B depletion was also shown to block vitamin D regulation of β catenin export. Moreover, KDM6B expression anti-correlates with Snail expression and correlates with vitamin D receptor expression [142].

PHD3 may function as a tumour suppressor in CRC through inhibition of IKKβ-mediated NF-κB signalling [146]. PHD3 has reduced expression in CRC compared to normal tissue and expression of PHD3 anti correlates with tumour severity [146]. Furthermore, PHD3 inhibits TNF-α induced NF-κB activity in colon cancer cell lines by blocking phosphorylation of IKKβ. This mechanism was shown to be independent of PHD3 hydroxylase activity, and via PHD3 blocking the interaction between IKKβ and Heat Shock Protein 90 (HSP90), an interaction required for IKKβ phosphorylation [146]. Control of p53 regulated apoptosis is another point of potential PHD-NF-κB crosstalk with clinical significance in CRC. PHD1 has been shown to aid the p53 mediated CRC resistance to genotoxic agents [147]. p53 phosphorylation, mediated by p38α kinase in response to chemotherapy, can inhibit chemotherapy-induced apoptosis through p53-induced DNA repair. Researchers found that silencing of PHD1, but not PHD2 or PHD3, prevented p53 activation in response to genotoxic treatment [147]. Moreover, PHD1 sensitized colorectal cancer to 5-FU treatment in mice [147]. Mechanistically, PHD1 was reported to enhance the p53-p38α kinase interaction, and subsequent p53 phosphorylation, in response to genotoxic damage in CRC in a hydroxylation-dependent and HIF-independent manner [147].

6.2. Breast and Lung Cancer

Breast and lung cancer are other examples of solid tumours where inflammation and hypoxia are fundamental drivers of disease progression. HIF and NF-κB signalling pathways have been established as playing oncogenic roles suitable for targeting. The usual players in inflammatory and hypoxic signalling crosstalk in cancer, such as TAMs, VEGF, and p53 have mechanistic importance in these diseases, as does the less well characterised crosstalk component, JmjC enzymes. In cell culture, IKK inhibition blocks Erb-B2 Receptor Tyrosine Kinase 2 (ERBB2) activation of NF-κB and induces apoptosis in ER negative ERBB2 positive breast cancer cells [148]. In vivo, inhibition of NF-κB impairs tumour progression in a murine model of breast cancer [148]. Furthermore, NF-κB dependent induction of EMT has been shown in breast cancer models [29,149]. There is also evidence of Epidermal Growth Factor (EGF) signalling enhancing NF-κB activity in breast cancer [150]. In this context, mechanism of activation of NF-κB by EGF may be similar to that described in lung cancer, where an IKK independent mechanism mediated by tyrosine kinase phosphorylation of IκBα is in place [151]. NF-κB activation is particularly prevalent in lung adenocarcinomas with constitutively active EGF-Receptor (EGFR) mutations. It can drive resistance of tumours to receptor tyrosine kinase inhibition, demonstrating the potential of targeting NF-κB to improve patient outcome in EGFR mutant lung cancer sufferers treated with receptor tyrosine kinase inhibitors [152]. An oncogenic role of HIF-1α in breast cancer has been well characterised (reviewed in [153]). Drugs inhibiting HIF, namely acriflavine, digoxin and topotecan impair tumour growth and metastasis in animal models of breast cancer [154–156]. Non-Small-Cell Lung Carcinoma (NSCLC) is highly metastatic and the most common form of lung cancer. HIF-1α and -2α subunits and frequently overexpressed in NSCLC, along with VEGF, which is targeted for treatment of NSCLC. HIF signalling has been shown to induce TAM mediated angiogenesis in human breast carcinoma models, and HIF-2α/TAM signalling may be a useful antiangiogenic breast cancer therapy [157,158]. Along with VEGF, Arginase 1 (Arg1) is transcriptionally induced by TAMs in response to HIF signalling. Arg1 also contributes to the tumour survival and growth activities of TAMs, and is elevated in mouse cancer model TAMs and in myeloid cells of breast cancer patients [159]. Several hypoxia-inducible HIF target JmjC enzymes are deregulated in breast and lung cancer. The KDM4 family members of JmjC enzymes, targeting di- and trimethylated H3K36 and H3K9, namely KDM4A, KDM4B and KDM4C are upregulated in breast cancer [160–162]. KDM4B has been found to mediate oestrogen stimulated cell proliferation of mammary cancer [163]. This study also found that KDM4B is transcriptionally induced by ERα in MCF7 cells and upregulates ERα target genes. Control of ERα target gene expression in mediated by fine tuning of H3K4 and H3K9 methylation at ER target promoters through a complex of KDM4B and the H3K4 methyltransferase MLL2 [163]. The H3K4 demethylase KDM5B has transcriptional repressor functions and is also a coactivator of the androgen receptor. KDM5B is overexpressed in both lung and breast cancer [164,165]. Furthermore, oncogenic functions of KDM5B have been demonstrated in vitro and in vivo. In MCF7 cells and in a mouse breast cancer model, depletion of KDM5B inhibits cell growth [166]. This correlates with suppression of the tumour suppressor gene BRCA1 [166]. More recently, high KDM5B expression was found to correlate with poor prognosis in breast cancer patients and enhanced breast cancer invasiveness in triple negative breast cancers [167]. Authors in this study also identified a mechanism for increased breast cancer invasion involving a KDM5B-metastasis-associated lung adenocarcinoma transcription (MALAT1)-hsa miR448 signalling axis [167]. A role of KDM5B is also present in lung cancer whereby KDM5B suppresses p53 expression [168]. This could have a functionally relevant link to NF-κB in lung cancer. Concomitant loss of p53 function and constitutively active KRAS control enhanced NF-κB activity in lung cancer cell lines and a mouse model of lung cancer [169]. Given the evidence for KMD5B driving cancer progression in the breast and lung cancers, it is perhaps no surprising that small molecule inhibitors against it are in development for use as potential cancer drugs. The small molecule inhibitor of KDM5B, EPT-103182 [170], has yielded promising results with anti-proliferative effects in various cancer cell lines and anti-tumour effects in mouse cancer models [171]. KDM2A along with KDM5B, have the highest frequency of gene amplifications and over expressions in breast cancer with respect to JmjC enzymes [172]. Interestingly the small isoform of KDM2A, which lacks the JmjC domain, is more highly expressed than the full isoform in a subset of breast cancer, suggesting an oncogenic role KDM2A independent of direct demethylase activity [172]. A potential mechanism for NSCLC progression in a subset of patients overexpressing KDM2A has been reported, with KDM2A stimulating cell proliferation through ERK1/2 signalling [173].

Studies on the importance of PHDs in breast and lung cancer are fairly limited. However, there is evidence of an important role of PHD1 in breast cancer through regulation of cell proliferation. Knockdown of PHD1 in breast cancer cell lines reduces cell proliferation and this correlates with loss of cyclin D1 [174]. Cyclin D1 stimulates cell cycle progression through its interaction with Cyclin Dependent Kinases (CDKs). Forkhead Box O3a (FOXO3a) was then identified as a new target for PHD1 [96]. Hydroxylation of FOXO3a by PHD1 has been shown to regulate cyclin D1 transcription, representing a potential mechanism by which PHD1 loss in breast cancer can impair cell growth. Conversely, treatment of breast cancer cell lines and a mouse xenograft model with docetaxel, a potential breast cancer drug, causes cell death in hypoxic conditions through c-Jun N-terminal Kinase 2 (JNK2)-PHD1 mediated HIF-1α degradation [175]. Interestingly, expression of individual PHD isoforms associates with good breast cancer patient outcome and PHD1 and PHD3 appear to be important in breast cancer in a HIF independent manner [176]. PHD3 is expressed highly in breast cancer patients with good prognosis and may be an important regulator of apoptosis in breast cancer [176]. PHD studies in the lung cancer are even more limited. Some groups have looked at the expression of PHD isoforms in lung cancer samples [177,178]. These studies show that PHDs are highly expressed in lung cancer compared to normal tissues, and both collective and individual PHD isoform expression are poor prognostic factors for NSCLC survival, independent of HIF levels [177,178]. As of yet, no clear mechanistic links between PHDs and NF-κB activity, independent of HIF, have been made in context of lung cancer and breast cancer.

7. Future Prospective in Cancer Therapeutics: Targeting HIF and the NF-κB Pathway

Due to the substantial contribution of HIF and NF-κB in carcinogenesis, over the years new therapeutic strategies have been developed to specifically target these two pathways. Moreover, considering their intimate crosstalk, it is possible that some of the therapeutics and modulators used might exert their function on both transcription factors. So far, HIF signalling has been altered mainly through PHD inhibitors, which have been beneficial in the treatment of several cancers, including those characterised by prominent inflammation. PHD inhibition can stabilise HIF-α, but also activate other pathways, such as NF-κB [72,91], promoting an inflammatory resolution. To date, five PHD inhibitors (BAY-853934, JTZ-951, FG-4592, AKB-6548 and GSK1278863) have entered clinical trials. However, some concerns about their side or off target effects exist, considering the different substrate specificity and cellular expression of the three enzymatic isoforms identified so far [179]. An alternative approach to target HIF stabilisation might be via the inhibition of the HIF-E3 ubiquitin ligase responsible for HIF degradation. Given that neddylation has been shown to modulate HIF via Cullin-2, MLN-4924, an adenosine monophosphate analog able to deneddylate cullin proteins, can act as a potent HIF stabilizer in vitro and in vivo, being also a viable tool in the treatment of cancer cells [180,181]. Additionally, VHL inhibitors represent an attractive alternative to PHD inhibitors. Recently, our group described VH298, a novel potent chemical probe blocking protein-protein interaction between VHL and HIF-α, downstream of HIF-α hydroxylation. Importantly, this small molecule highly selectively stabilises the hydroxylated form of HIF-1α and HIF-2α, in a concentration- and time-dependent manner, in both cancerous and non-cancerous primary cells, inducing a HIF-dependent hypoxic response [179]. On the other hand, several studies questioned the efficacy of a therapeutic activation of HIF, since this signaling pathway is directly involved in tumor promotion, as previously mentioned.

Therefore, several other compounds are currently in clinical trials as HIF-1α inhibitors, although, in many cases, they were originally intended to target different pathways (i.e., PI3K/mTOR inhibitors, microtubules targeting agents, cardiac glycosides, topoisomerase inhibitors, among others) [182].

HIF-1α and HIF-2α have often divergent roles in tumorigenesis, hence the necessity to selectively target one or the other. AKB-4924 is a PHD inhibitor tested in inflammation models, showing a relative selectivity for HIF-1α versus HIF-2α [183]. However, no HIF-1α specific inhibitors have been discovered so far, although great efforts have been made to identify inhibitors exerting their function by decreasing mRNA or protein level of HIF-1α, preventing its dimerization or DNA/co-activators binding. For long time, HIF-2α has been considered undruggable, till the revolutionary discovery of two small molecules, PT2399 and the closely related analogue PT2385. These compounds function as potent HIF-2α antagonists, being able to bind to a large cavity located in the PAS-B domain of HIF-2α, thus disrupting the hetero-dimerization between HIF-2α and HIF-1β [184,185]. These small molecules have a promising clinical potential, considering that they reduced tumour growth and decreased tumour vascular area in VHL$^{-/-}$ clear cell Renal Cell Carcinoma (ccRCC) patient derived xenographs, showing even greater efficacy than conventional treatments (i.e., sunitinib) [184–186].

As previously explained, not only HIF, but also NF-κB is a key player in many aspects of cancer development and progression. For this reason, the possibility to target directly NF-κB for cancer therapy has been an important subject of research. On one side, due to the deep link between inflammation and cancer, it would be ideal to prevent or treat tumour formation by blocking inflammation [44]. On the other side, in the light of the multiple functions of NF-κB in the innate and adaptive immune responses, the use of NF-κB inhibitors would not be recommended to treat malignancies, especially in the tumour-eliminating phase, when immune cells specifically target transformed cells. Nevertheless, using chemotherapeutics in combination with inhibitors of NF-κB seems to be currently the preferred approach, in particular when tumours feature chronic inflammation [5]. Many well-known NSAIDs, such as aspirin and ibuprofen, can be used to this scope. In fact, at low doses, aspirin has been suggested to prevent some types of cancer, including colorectal cancer [187], whereas at high doses, aspirin can inhibit the kinase activity of IKK or interfere with the degradation of IκBα, blocking NF-κB activity. Alongside NSAIDs, anti-inflammatory and anti-cancer activities have been recognised for a number of natural products, glucocorticoids, immunosuppressants, or inhibitors of other pathways, directly affecting the NF-κB induction or signalling cascade, the translocation of NF-κB to the nucleus, the DNA binding of the dimers or their interactions with the transcriptional machinery [3].

The use of NF-κB inhibitors has been shown to be important also to enhance the activity of immunotherapy. In a recent study, a combined use of IL-18 as immunotherapeutic agent alongside a targeted inhibition of the NF-κB pathway emerged as potentially effective against pancreatic cancer [188]. Recently, cancer immunotherapy effectiveness has been boosted by a better understanding of immune checkpoints. Programmed Death 1 (PD1)/Programmed Death Ligand 1 (PD-L1) and Cytotoxic T-Lymphocyte Associated Protein 4 (CTLA-4) are the most critical immune checkpoints. Both of them are used by tumours to escape host immune surveillance, although they have distinct expression and mechanisms of action in the regulation of T cell activity [189]. Given the importance of the hypoxic tumour microenvironment for TAMs and tumour-infiltrating myeloid cells, it is not surprising that PD-L1 was identified as a direct HIF-1α target gene in myeloid-derived suppressor cells (MDSCs) [190]. Several reports correlated a high expression of PD-L1 to a poor prognosis in a number of solid cancers. Thus, immune checkpoint blockade with monoclonal antibodies directed against PD-1 and PD-L1 has proven to be powerful compared to conventional chemotherapies, activating an anti-tumour immunity able to recognise specifically tumour derived antigens, even when they have undergone mutations [189]. Several PD1 and PD-L1 antibodies are currently in clinical trials (reviewed in [189]), representing surely the next generation of the immune modulators. However, whether HIF and NF-κB crosstalk extends to immunosuppression has yet to be formally investigated and surely is an area of extreme importance.

NF-κB signalling is upregulated in cells having a compromised expression of the tumour suppressor pVHL, such as ccRCC cells, where VHL is inactivated [191]. In this cellular context, Vascular Cell Adhesion Molecule 1 (VCAM-1) expression was found to be regulated by the non-canonical NF-κB pathway. Importantly, VCAM-1 decreased following VHL loss or after hypoxia exposure and PHD inactivation [192]. Recently, the mechanisms by which pVHL might directly impact on the NF-κB pathways have been proposed. Wang and colleagues demonstrated that pVHL mediates the K63-ubiquitination of IKKβ. Surprisingly, this modification does not lead to degradation, but prevents TAK1-IKKβ interaction, and consequent IKKβ phosphorylation and NF-κB activation [193]. Considering this novel function of pVHL regulating the NF-κB pathway, new therapeutic possibilities might be speculated, especially to inhibit the aberrant activation of the NF-κB pathway in some neoplastic contexts. Therefore, further studies are needed in this promising direction, as the main challenge for researchers in this field is still to directly target NF-κB, as well as HIF, only in transformed cells.

8. Conclusions

NF-κB and HIF crosstalk occurs at many levels, from shared activators to shared target genes. However, context specificity exists and this is an important determinant in whether this crosstalk can be used for future therapeutic intervention. Since, pharmacological interventions for both pathways are available, future studies investigating the role of HIF modulators on the NF-κB pathway are underway with the aim of broadening the use of these compounds in the clinic. Furthermore, the identification of novel compounds targeting specific transcription factor dimers such as HIF-2α-HIF-1β also opens the possibility of specifically targeting NF-κB dimers in the future.

Acknowledgments: Laura D'Ignazio is funded by a PhD studentship from the Wellcome Trust. Michael Batie is funded by a PhD studentship from the Medical Research Council (MRC). Sonia Rocha is funded by a Senior Research Fellowship from Cancer Research UK (C99667/A12918), and a Wellcome Trust Strategic Award (097945/B/11/Z).

Author Contributions: Laura D'Ignazio, Michael Batie and Sonia Rocha performed the literature review and contributed to the writing of the manuscript.

Conflicts of Interest: The authors declare no conflict of interest.

References

1. Sen, R.; Baltimore, D. Multiple nuclear factors interact with the immunoglobulin enhancer sequences. *Cell* **1986**, *46*, 705–716. [CrossRef]
2. Gupta, S.C.; Sundaram, C.; Reuter, S.; Aggarwal, B.B. Inhibiting NF-κB activation by small molecules as a therapeutic strategy. *Biochim. Biophys. Acta* **2010**, *1799*, 775–787. [CrossRef] [PubMed]
3. Park, M.H.; Hong, J.T. Roles of NF-κB in Cancer and inflammatory diseases and their therapeutic approaches. *Cells* **2016**, *5*, 15. [CrossRef] [PubMed]
4. Huxford, T.; Ghosh, G. A structural guide to proteins of the NF-κB signaling module. *Cold Spring Harb. Perspect. Biol.* **2009**, *1*, a000075. [CrossRef] [PubMed]
5. Hoesel, B.; Schmid, J.A. The complexity of NF-κB signaling in inflammation and cancer. *Mol. Cancer* **2013**, *12*, 86. [CrossRef] [PubMed]
6. Bandarra, D.; Rocha, S. Tale of two transcription factors: NF-κB and HIF crosstalk. *OA Mol. Cell Biol.* **2013**, *1*, 6. [CrossRef]
7. Sun, S.C. The noncanonical NF-κB pathway. *Immunol. Rev.* **2012**, *246*, 125–140. [CrossRef] [PubMed]
8. Oeckinghaus, A.; Hayden, M.S.; Ghosh, S. Crosstalk in NF-κB signaling pathways. *Nat. Immunol.* **2011**, *12*, 695–708. [CrossRef] [PubMed]
9. Gilmore, T.D. Introduction to NF-κB: Players, pathways, perspectives. *Oncogene* **2006**, *25*, 6680–6684. [CrossRef] [PubMed]
10. Israel, A. The IKK complex, a central regulator of NF-κB activation. *Cold Spring Harb. Perspect. Biol.* **2010**, *2*, a000158. [CrossRef] [PubMed]

11. Biddlestone, J.; Bandarra, D.; Rocha, S. The role of hypoxia in inflammatory disease (review). *Int. J. Mol. Med.* **2015**, *35*, 859–869. [CrossRef] [PubMed]

12. D'Ignazio, L.; Bandarra, D.; Rocha, S. NF-κB and HIF crosstalk in immune responses. *FEBS J.* **2016**, *283*, 413–424. [CrossRef] [PubMed]

13. Hanahan, D.; Robert, A. Weinberg, Hallmarks of cancer: The next generation. *Cell* **2011**, *144*, 646–674. [CrossRef] [PubMed]

14. Xia, Y.; Shen, S.; Verma, I.M. NF-κB, an active player in human cancers. *Cancer Immunol. Res.* **2014**, *2*, 823–830. [CrossRef] [PubMed]

15. Karin, M.; Cao, Y.; Greten, F.R.; Li, Z.W. NF-κB in cancer: From innocent bystander to major culprit. *Nat. Rev. Cancer* **2002**, *2*, 301–310. [CrossRef] [PubMed]

16. Guttridge, D.C.; Albanese, C.; Reuther, J.Y.; Pestell, R.G.; Baldwin, A.S., Jr. NF-κB controls cell growth and differentiation through transcriptional regulation of cyclin D1. *Mol. Cell. Biol.* **1999**, *19*, 5785–5799. [CrossRef] [PubMed]

17. Sherr, C.J. Cancer cell cycles. *Science* **1996**, *274*, 1672–1677. [CrossRef] [PubMed]

18. Pahl, H.L. Activators and target genes of Rel/NF-κB transcription factors. *Oncogene* **1999**, *18*, 6853–6866. [CrossRef] [PubMed]

19. Campbell, J.K.; Rocha, S.; Perkins, N.D. Active repression of antiapoptotic gene expression by RelA(p65) NF-κ B. *Mol. Cell* **2004**, *13*, 853–865. [CrossRef]

20. Karin, M.; Lin, A. NF-κB at the crossroads of life and death. *Nat. Immunol.* **2002**, *3*, 221–227. [CrossRef] [PubMed]

21. Webster, G.A.; Perkins, N.D. Transcriptional cross talk between NF-κB and p53. *Mol. Cell. Biol.* **1999**, *19*, 3485–3495. [CrossRef] [PubMed]

22. Koch, A.E.; Polverini, P.J.; Kunkel, S.L.; Harlow, L.A.; DiPietro, L.A.; Elner, V.M.; Elner, S.G.; Strieter, R.M. Interleukin-8 as a macrophage-derived mediator of angiogenesis. *Science* **1992**, *258*, 1798–1801. [CrossRef] [PubMed]

23. Huang, S.; Robinson, J.B.; Deguzman, A.; Bucana, C.D.; Fidler, I.J. Blockade of nuclear factor-κB signaling inhibits angiogenesis and tumorigenicity of human ovarian cancer cells by suppressing expression of vascular endothelial growth factor and interleukin 8. *Cancer Res.* **2000**, *60*, 5334–5339. [PubMed]

24. Marmol, I.; Sanchez-de-Diego, C.; Pradilla Dieste, A.; Cerrada, E.; Rodriguez Yoldi, M.J. Colorectal Carcinoma: A general overview and future perspectives in colorectal cancer. *Int. J. Mol. Sci.* **2017**, *18*. [CrossRef] [PubMed]

25. Bandarra, D.; Biddlestone, J.; Mudie, S.; Muller, H.A.; Rocha, S. HIF-1α restricts NF-κB-dependent gene expression to control innate immunity signals. *Dis. Models Mech.* **2015**, *8*, 169–181. [CrossRef] [PubMed]

26. Liu, W.; Shen, S.M.; Zhao, X.Y.; Chen, G.Q. Targeted genes and interacting proteins of hypoxia inducible factor-1. *Int. J. Biochem. Mol. Biol.* **2012**, *3*, 165–178. [PubMed]

27. Bond, M.; Fabunmi, R.P.; Baker, A.H.; Newby, A.C. Synergistic upregulation of metalloproteinase-9 by growth factors and inflammatory cytokines: An absolute requirement for transcription factor NF-κB. *FEBS Lett.* **1998**, *435*, 29–34. [CrossRef]

28. Yan, M.; Xu, Q.; Zhang, P.; Zhou, X.J.; Zhang, Z.Y.; Chen, W.T. Correlation of NF-κB signal pathway with tumor metastasis of human head and neck squamous cell carcinoma. *BMC Cancer* **2010**, *10*, 437. [CrossRef] [PubMed]

29. Huber, M.A.; Azoitei, N.; Baumann, B.; Grunert, S.; Sommer, A.; Pehamberger, H.; Kraut, N.; Beug, H.; Wirth, T. NF-κB is essential for epithelial-mesenchymal transition and metastasis in a model of breast cancer progression. *J. Clin. Investig.* **2004**, *114*, 569–581. [CrossRef] [PubMed]

30. Disis, M.L. Immune regulation of cancer. *J. Clin. Oncol.* **2010**, *28*, 4531–4538. [CrossRef] [PubMed]

31. Luedde, T.; Beraza, N.; Kotsikoris, V.; van Loo, G.; Nenci, A.; de Vos, R.; Roskams, T.; Trautwein, C.; Pasparakis, M. Deletion of NEMO/IKKγ in liver parenchymal cells causes steatohepatitis and hepatocellular carcinoma. *Cancer Cell* **2007**, *11*, 119–132. [CrossRef] [PubMed]

32. Van Hogerlinden, M.; Rozell, B.L.; Ahrlund-Richter, L.; Toftgard, R. Squamous cell carcinomas and increased apoptosis in skin with inhibited Rel/nuclear factor-κB signaling. *Cancer Res.* **1999**, *59*, 3299–3303. [PubMed]

33. Kim, H.J.; Hawke, N.; Baldwin, A.S. NF-κB and IKK as therapeutic targets in cancer. *Cell Death Differ.* **2006**, *13*, 738–747. [CrossRef] [PubMed]

34. Compagno, M.; Lim, W.K.; Grunn, A.; Nandula, S.V.; Brahmachary, M.; Shen, Q.; Bertoni, F.; Ponzoni, M.; Scandurra, M.; Califano, A.; et al. Mutations of multiple genes cause deregulation of NF-κB in diffuse large B-cell lymphoma. *Nature* **2009**, *459*, 717–721. [CrossRef] [PubMed]

35. Vaque, J.P.; Martinez, N.; Batlle-Lopez, A.; Perez, C.; Montes-Moreno, S.; Sanchez-Beato, M.; Piris, M.A. B-cell lymphoma mutations: Improving diagnostics and enabling targeted therapies. *Haematologica* **2014**, *99*, 222–231. [CrossRef] [PubMed]

36. Jost, P.J.; Ruland, J. Aberrant NF-κB signaling in lymphoma: Mechanisms, consequences, and therapeutic implications. *Blood* **2007**, *109*, 2700–2707. [PubMed]

37. Jiao, X.; Wood, L.D.; Lindman, M.; Jones, S.; Buckhaults, P.; Polyak, K.; Sukumar, S.; Carter, H.; Kim, D.; Karchin, R.; et al. Somatic mutations in the notch, NF-κB, PIK3CA, and Hedgehog pathways in human breast cancers. *Genes Chromosom. Cancer* **2012**, *51*, 480–489. [CrossRef] [PubMed]

38. Bennett, L.; Quinn, J.; McCall, P.; Mallon, E.A.; Horgan, P.G.; McMillan, D.C.; Paul, A.; Edwards, J. High IKKα expression is associated with reduced time to recurrence and cancer specific survival in oestrogen receptor (ER)-positive breast cancer. *Int. J. Cancer* **2017**, *140*, 1633–1644. [CrossRef] [PubMed]

39. Sakamoto, K.; Maeda, S.; Hikiba, Y.; Nakagawa, H.; Hayakawa, Y.; Shibata, W.; Yanai, A.; Ogura, K.; Omata, M. Constitutive NF-κB activation in colorectal carcinoma plays a key role in angiogenesis, promoting tumor growth. *Clin. Cancer Res.* **2009**, *15*, 2248–2258. [CrossRef] [PubMed]

40. Greten, F.R.; Eckmann, L.; Greten, T.F.; Park, J.M.; Li, Z.W.; Egan, L.J.; Kagnoff, M.F.; Karin, M. IKKβ links inflammation and tumorigenesis in a mouse model of colitis-associated cancer. *Cell* **2004**, *118*, 285–296. [CrossRef] [PubMed]

41. Chen, W.; Li, Z.; Bai, L.; Lin, Y. NF-κB in lung cancer, a carcinogenesis mediator and a prevention and therapy target. *Front. Biosci.* **2011**, *16*, 1172–1185. [CrossRef]

42. Luedde, T.; Schwabe, R.F. NF-κB in the liver—Linking injury, fibrosis and hepatocellular carcinoma. *Nat. Rev. Gastroenterol. Hepatol.* **2011**, *8*, 108–118. [CrossRef] [PubMed]

43. Ueda, Y.; Richmond, A. NF-κB activation in melanoma. *Pigment Cell Res.* **2006**, *19*, 112–124. [CrossRef] [PubMed]

44. Munn, L.L. Cancer and inflammation. *Wiley Interdiscip. Rev. Syst. Biol. Med.* **2017**, *9*. [CrossRef] [PubMed]

45. Muz, B.; de la Puente, P.; Azab, F.; Azab, A.K. The role of hypoxia in cancer progression, angiogenesis, metastasis, and resistance to therapy. *Hypoxia* **2015**, *3*, 83–92. [CrossRef] [PubMed]

46. Eales, K.L.; Hollinshead, K.E.; Tennant, D.A. Hypoxia and metabolic adaptation of cancer cells. *Oncogenesis* **2016**, *5*, e190. [CrossRef] [PubMed]

47. Rocha, S. Gene regulation under low oxygen: Holding your breath for transcription. *Trends Biochem. Sci.* **2007**, *32*, 389–397. [CrossRef] [PubMed]

48. Talks, K.L.; Turley, H.; Gatter, K.C.; Maxwell, P.H.; Pugh, C.W.; Ratcliffe, P.J.; Harris, A.L. The expression and distribution of the hypoxia-inducible factors HIF-1α and HIF-2α in normal human tissues, cancers, and tumor-associated macrophages. *Am. J. Pathol.* **2000**, *157*, 411–421. [CrossRef]

49. Yang, S.L.; Wu, C.; Xiong, Z.F.; Fang, X. Progress on hypoxia-inducible factor-3: Its structure, gene regulation and biological function (Review). *Mol. Med. Rep.* **2015**, *12*, 2411–2416. [CrossRef] [PubMed]

50. Kenneth, N.S.; Rocha, S. Regulation of gene expression by hypoxia. *Biochem. J.* **2008**, *414*, 19–29. [CrossRef] [PubMed]

51. Keith, B.; Johnson, R.S.; Simon, M.C. HIF1α and HIF2α: Sibling rivalry in hypoxic tumour growth and progression. *Nat. Rev. Cancer* **2011**, *12*, 9–22. [CrossRef] [PubMed]

52. Zhang, P.; Yao, Q.; Lu, L.; Li, Y.; Chen, P.J.; Duan, C. Hypoxia-inducible factor 3 is an oxygen-dependent transcription activator and regulates a distinct transcriptional response to hypoxia. *Cell Rep.* **2014**, *6*, 1110–1121. [CrossRef] [PubMed]

53. Ortmann, B.; Druker, J.; Rocha, S. Cell cycle progression in response to oxygen levels. *Cell. Mol. Life Sci.* **2014**, *71*, 3569–3582. [CrossRef] [PubMed]

54. Schofield, C.J.; Ratcliffe, P.J. Oxygen sensing by HIF hydroxylases. *Nat. Rev. Mol. Cell Biol.* **2004**, *5*, 343–354. [CrossRef] [PubMed]

55. Liu, Z.J.; Semenza, G.L.; Zhang, H.F. Hypoxia-inducible factor 1 and breast cancer metastasis. *J. Zhejiang Univ. Sci. B* **2015**, *16*, 32–43. [CrossRef] [PubMed]

56. Takeuchi, T.; Yamazaki, Y.; Katoh-Fukui, Y.; Tsuchiya, R.; Kondo, S.; Motoyama, J.; Higashinakagawa, T. Gene trap capture of a novel mouse gene, jumonji, required for neural tube formation. *Genes Dev.* **1995**, *9*, 1211–1222. [CrossRef] [PubMed]

57. Shmakova, A.; Batie, M.; Druker, J.; Rocha, S. Chromatin and oxygen sensing in the context of JmjC histone demethylases. *Biochem. J.* **2014**, *462*, 385–395. [CrossRef] [PubMed]

58. Elkins, J.M.; Hewitson, K.S.; McNeill, L.A.; Seibel, J.F.; Schlemminger, I.; Pugh, C.W.; Ratcliffe, P.J.; Schofield, C.J. Structure of factor-inhibiting hypoxia-inducible factor (HIF) reveals mechanism of oxidative modification of HIF-1 α. *J. Biol. Chem.* **2003**, *278*, 1802–1806. [CrossRef] [PubMed]

59. Trewick, S.C.; McLaughlin, P.J.; Allshire, R.C. Methylation: Lost in hydroxylation? *EMBO Rep.* **2005**, *6*, 315–320. [CrossRef] [PubMed]

60. Sanchez-Fernandez, E.M.; Tarhonskaya, H.; Al-Qahtani, K.; Hopkinson, R.J.; McCullagh, J.S.; Schofield, C.J.; Flashman, E. Investigations on the oxygen dependence of a 2-oxoglutarate histone demethylase. *Biochem. J.* **2013**, *449*, 491–496. [CrossRef] [PubMed]

61. Hancock, R.L.; Masson, N.; Dunne, K.; Flashman, E.; Kawamura, A. The Activity of JmjC Histone Lysine Demethylase KDM4A is Highly Sensitive to Oxygen Concentrations. *ACS Chem. Biol.* **2017**, *12*, 1011. [CrossRef] [PubMed]

62. Johnson, A.B.; Denko, N.; Barton, M.C. Hypoxia induces a novel signature of chromatin modifications and global repression of transcription. *Mutat. Res.* **2008**, *640*, 174–179. [CrossRef] [PubMed]

63. Zhou, X.; Sun, H.; Chen, H.; Zavadil, J.; Kluz, T.; Arita, A.; Costa, M. Hypoxia induces trimethylated H3 lysine 4 by inhibition of JARID1A demethylase. *Cancer Res.* **2010**, *70*, 4214–4221. [CrossRef] [PubMed]

64. Tausendschon, M.; Dehne, N.; Brune, B. Hypoxia causes epigenetic gene regulation in macrophages by attenuating jumonji histone demethylase activity. *Cytokine* **2011**, *53*, 256–262. [CrossRef] [PubMed]

65. Prickaerts, P.; Adriaens, M.E.; Beucken, T.V.; Koch, E.; Dubois, L.; Dahlmans, V.E.; Gits, C.; Evelo, C.T.; Chan-Seng-Yue, M.; Wouters, B.G.; et al. Hypoxia increases genome-wide bivalent epigenetic marking by specific gain of H3K27me3. *Epigenet. Chromatin* **2016**, *9*, 46. [CrossRef] [PubMed]

66. Beyer, S.; Kristensen, M.M.; Jensen, K.S.; Johansen, J.V.; Staller, P. The histone demethylases JMJD1A and JMJD2B are transcriptional targets of hypoxia-inducible factor HIF. *J. Biol. Chem.* **2008**, *283*, 36542–36552. [CrossRef] [PubMed]

67. Pollard, P.J.; Loenarz, C.; Mole, D.R.; McDonough, M.A.; Gleadle, J.M.; Schofield, C.J.; Ratcliffe, P.J. Regulation of Jumonji-domain-containing histone demethylases by hypoxia-inducible factor (HIF)-1α. *Biochem. J.* **2008**, *416*, 387–394. [CrossRef] [PubMed]

68. Wellmann, S.; Bettkober, M.; Zelmer, A.; Seeger, K.; Faigle, M.; Eltzschig, H.K.; Buhrer, C. Hypoxia upregulates the histone demethylase JMJD1A via HIF-1. *Biochem. Biophys. Res. Commun.* **2008**, *372*, 892–897. [CrossRef] [PubMed]

69. Krieg, A.J.; Rankin, E.B.; Chan, D.; Razorenova, O.; Fernandez, S.; Giaccia, A.J. Regulation of the histone demethylase JMJD1A by hypoxia-inducible factor 1 α enhances hypoxic gene expression and tumor growth. *Mol. Cell Biol.* **2010**, *30*, 344–353. [CrossRef] [PubMed]

70. Lee, H.Y.; Choi, K.; Oh, H.; Park, Y.K.; Park, H. HIF-1-dependent induction of Jumonji domain-containing protein (JMJD) 3 under hypoxic conditions. *Mol. Cells* **2014**, *37*, 43–50. [CrossRef] [PubMed]

71. Batie, M.; Druker, J.; D'Ignazio, L.; Rocha, S. KDM2 Family Members are Regulated by HIF-1 in Hypoxia. *Cells* **2017**, *6*, 8. [CrossRef] [PubMed]

72. D'Ignazio, L.; Rocha, S. Hypoxia Induced NF-κB. *Cells* **2016**, *5*, 10. [CrossRef] [PubMed]

73. Schaffer, K.; Taylor, C.T. The impact of hypoxia on bacterial infection. *FEBS J.* **2015**, *282*, 2260–2266. [CrossRef] [PubMed]

74. Scott, C.; Bonner, J.; Min, D.; Boughton, P.; Stokes, R.; Cha, K.M.; Walters, S.N.; Maslowski, K.; Sierro, F.; Grey, S.T.; et al. Reduction of ARNT in myeloid cells causes immune suppression and delayed wound healing. *Am. J. Physiol. Cell Physiol.* **2014**, *307*, C349–C357. [CrossRef] [PubMed]

75. Thompson, A.A.; Elks, P.M.; Marriott, H.M.; Eamsamarng, S.; Higgins, K.R.; Lewis, A.; Williams, L.; Parmar, S.; Shaw, G.; McGrath, E.E.; et al. Hypoxia-inducible factor 2α regulates key neutrophil functions in humans, mice, and zebrafish. *Blood* **2014**, *123*, 366–376. [CrossRef] [PubMed]

76. Pawlus, M.R.; Wang, L.; Hu, C.J. STAT3 and HIF1α cooperatively activate HIF1 target genes in MDA-MB-231 and RCC4 cells. *Oncogene* **2014**, *33*, 1670–1679. [CrossRef] [PubMed]

77. Van Uden, P.; Kenneth, N.S.; Rocha, S. Regulation of hypoxia-inducible factor-1α by NF-κB. *Biochem. J.* **2008**, *412*, 477–484. [CrossRef] [PubMed]

78. van Uden, P.; Kenneth, N.S.; Webster, R.; Muller, H.A.; Mudie, S.; Rocha, S. Evolutionary conserved regulation of HIF-1beta by NF-κB. *PLoS Genet.* **2011**, *7*, e1001285. [CrossRef] [PubMed]

79. Wright, C.W.; Duckett, C.S. The aryl hydrocarbon nuclear translocator alters CD30-mediated NF-κB-dependent transcription. *Science* **2009**, *323*, 251–255. [CrossRef] [PubMed]

80. Bracken, C.P.; Whitelaw, M.L.; Peet, D.J. Activity of hypoxia-inducible factor 2α is regulated by association with the NF-κB essential modulator. *J. Biol. Chem.* **2005**, *280*, 14240–14251. [CrossRef] [PubMed]

81. Balamurugan, K. HIF-1 at the crossroads of hypoxia, inflammation, and cancer. *Int. J. Cancer* **2016**, *138*, 1058–1066. [CrossRef] [PubMed]

82. Kim, E.R.; Chang, D.K. Colorectal cancer in inflammatory bowel disease: The risk, pathogenesis, prevention and diagnosis. *World J. Gastroenterol.* **2014**, *20*, 9872–9881. [CrossRef] [PubMed]

83. Triner, D.; Xue, X.; Schwartz, A.J.; Jung, I.; Colacino, J.A.; Shah, Y.M. Epithelial Hypoxia-Inducible Factor 2α Facilitates the Progression of Colon Tumors through Recruiting Neutrophils. *Mol. Cell. Biol.* **2017**, *37*, e00481-16. [CrossRef] [PubMed]

84. Triner, D.; Shah, Y.M. Hypoxia-inducible factors: A central link between inflammation and cancer. *J. Clin. Investig.* **2016**, *126*, 3689–3698. [CrossRef] [PubMed]

85. Cockman, M.E.; Lancaster, D.E.; Stolze, I.P.; Hewitson, K.S.; McDonough, M.A.; Coleman, M.L.; Coles, C.H.; Yu, X.; Hay, R.T.; Ley, S.C.; et al. Posttranslational hydroxylation of ankyrin repeats in IκB proteins by the hypoxia-inducible factor (HIF) asparaginyl hydroxylase, factor inhibiting HIF (FIH). *Proc. Natl. Acad. Sci. USA* **2006**, *103*, 14767–14772. [CrossRef] [PubMed]

86. Scholz, C.C.; Cavadas, M.A.; Tambuwala, M.M.; Hams, E.; Rodriguez, J.; von Kriegsheim, A.; Cotter, P.; Bruning, U.; Fallon, P.G.; Cheong, A.; et al. Regulation of IL-1beta-induced NF-κB by hydroxylases links key hypoxic and inflammatory signaling pathways. *Proc. Natl. Acad. Sci. USA* **2013**, *110*, 18490–18495. [CrossRef] [PubMed]

87. Scholz, C.C.; Rodriguez, J.; Pickel, C.; Burr, S.; Fabrizio, J.A.; Nolan, K.A.; Spielmann, P.; Cavadas, M.A.; Crifo, B.; Halligan, D.N.; et al. FIH Regulates Cellular Metabolism through Hydroxylation of the Deubiquitinase OTUB1. *PLoS Biol.* **2016**, *14*, e1002347. [CrossRef] [PubMed]

88. Devries, I.L.; Hampton-Smith, R.J.; Mulvihill, M.M.; Alverdi, V.; Peet, D.J.; Komives, E.A. Consequences of IκB α hydroxylation by the factor inhibiting HIF (FIH). *FEBS Lett.* **2010**, *584*, 4725–4730. [CrossRef] [PubMed]

89. Taylor, C.T.; Cummins, E.P. The role of NF-κB in hypoxia-induced gene expression. *Ann. N. Y. Acad. Sci.* **2009**, *1177*, 178–184. [CrossRef] [PubMed]

90. Melvin, A.; Mudie, S.; Rocha, S. Further insights into the mechanism of hypoxia-induced NFκB. *Cell Cycle* **2011**, *10*, 879–882. [CrossRef] [PubMed]

91. Cummins, E.P.; Berra, E.; Comerford, K.M.; Ginouves, A.; Fitzgerald, K.T.; Seeballuck, F.; Godson, C.; Nielsen, J.E.; Moynagh, P.; Pouyssegur, J.; et al. Prolyl hydroxylase-1 negatively regulates IκB kinase-β, giving insight into hypoxia-induced NFκB activity. *Proc. Natl. Acad. Sci. USA* **2006**, *103*, 18154–18159. [CrossRef] [PubMed]

92. Xie, X.; Xiao, H.; Ding, F.; Zhong, H.; Zhu, J.; Ma, N.; Mei, J. Over-expression of prolyl hydroxylase-1 blocks NF-κB-mediated cyclin D1 expression and proliferation in lung carcinoma cells. *Cancer Genet.* **2014**, *207*, 188–194. [CrossRef] [PubMed]

93. Fu, J.; Taubman, M.B. Prolyl hydroxylase EGLN3 regulates skeletal myoblast differentiation through an NF-κB-dependent pathway. *J. Biol. Chem.* **2010**, *285*, 8927–8935. [CrossRef] [PubMed]

94. Fu, J.; Taubman, M.B. EGLN3 inhibition of NF-κB is mediated by prolyl hydroxylase-independent inhibition of IκB kinase gamma ubiquitination. *Mol. Cell. Biol.* **2013**, *33*, 3050–3061. [CrossRef] [PubMed]

95. Takeda, Y.; Costa, S.; Delamarre, E.; Roncal, C.; Leite de Oliveira, R.; Squadrito, M.L.; Finisguerra, V.; Deschoemaeker, S.; Bruyere, F.; Wenes, M.; et al. Macrophage skewing by PHD2 haplodeficiency prevents ischaemia by inducing arteriogenesis. *Nature* **2011**, *479*, 122–126. [CrossRef] [PubMed]

96. Zheng, X.; Zhai, B.; Koivunen, P.; Shin, S.J.; Lu, G.; Liu, J.; Geisen, C.; Chakraborty, A.A.; Moslehi, J.J.; Smalley, D.M.; et al. Prolyl hydroxylation by EGLN2 destabilizes FOXO3a by blocking its interaction with the USP9x deubiquitinase. *Genes Dev.* **2014**, *28*, 1429–1444. [CrossRef] [PubMed]

97. Li, J.; Yuan, W.; Jiang, S.; Ye, W.; Yang, H.; Shapiro, I.M.; Risbud, M.V. Prolyl-4-hydroxylase domain protein 2 controls NF-κB/p65 transactivation and enhances the catabolic effects of inflammatory cytokines on cells of the nucleus pulposus. *J. Biol. Chem.* **2015**, *290*, 7195–7207. [CrossRef] [PubMed]

98. Lu, T.; Jackson, M.W.; Wang, B.; Yang, M.; Chance, M.R.; Miyagi, M.; Gudkov, A.V.; Stark, G.R. Regulation of NF-κB by NSD1/FBXL11-dependent reversible lysine methylation of p65. *Proc. Natl. Acad. Sci. USA* **2010**, *107*, 46–51. [CrossRef] [PubMed]

99. Lu, T.; Yang, M.; Huang, D.B.; Wei, H.; Ozer, G.H.; Ghosh, G.; Stark, G.R. Role of lysine methylation of NF-κB in differential gene regulation. *Proc. Natl. Acad. Sci. USA* **2013**, *110*, 13510–13515. [CrossRef] [PubMed]

100. Ea, C.K.; Baltimore, D. Regulation of NF-κB activity through lysine monomethylation of p65. *Proc. Natl. Acad. Sci. USA* **2009**, *106*, 18972–18977. [CrossRef]

101. Yang, X.D.; Huang, B.; Li, M.; Lamb, A.; Kelleher, N.L.; Chen, L.F. Negative regulation of NF-κB action by Set9-mediated lysine methylation of the RelA subunit. *EMBO J.* **2009**, *28*, 1055–1066. [CrossRef] [PubMed]

102. Levy, D.; Kuo, A.J.; Chang, Y.; Schaefer, U.; Kitson, C.; Cheung, P.; Espejo, A.; Zee, B.M.; Liu, C.L.; Tangsombatvisit, S.; et al. Lysine methylation of the NF-κB subunit RelA by SETD6 couples activity of the histone methyltransferase GLP at chromatin to tonic repression of NF-κB signaling. *Nat. Immunol.* **2011**, *12*, 29–36. [CrossRef] [PubMed]

103. Ge, R.; Wang, Z.; Zeng, Q.; Xu, X.; Olumi, A.F. F-box protein 10, an NF-κB-dependent anti-apoptotic protein, regulates TRAIL-induced apoptosis through modulating c-Fos/c-FLIP pathway. *Cell Death Differ.* **2011**, *18*, 1184–1195. [CrossRef] [PubMed]

104. De Santa, F.; Narang, V.; Yap, Z.H.; Tusi, B.K.; Burgold, T.; Austenaa, L.; Bucci, G.; Caganova, M.; Notarbartolo, S.; Casola, S.; et al. Jmjd3 contributes to the control of gene expression in LPS-activated macrophages. *EMBO J.* **2009**, *28*, 3341–3352. [CrossRef] [PubMed]

105. Yeo, K.S.; Tan, M.C.; Wong, W.Y.; Loh, S.W.; Lam, Y.L.; Tan, C.L.; Lim, Y.Y.; Ea, C.K. JMJD8 is a positive regulator of TNF-induced NF-κB signaling. *Sci. Rep.* **2016**, *6*, 34125. [CrossRef] [PubMed]

106. Koong, A.C.; Chen, E.Y.; Giaccia, A.J. Hypoxia causes the activation of nuclear factor κ B through the phosphorylation of IκB α on tyrosine residues. *Cancer Res.* **1994**, *54*, 1425–1430. [PubMed]

107. Culver, C.; Sundqvist, A.; Mudie, S.; Melvin, A.; Xirodimas, D.; Rocha, S. Mechanism of hypoxia-induced NF-κB. *Mol. Cell Biol.* **2010**, *30*, 4901–4921. [CrossRef] [PubMed]

108. Fitzpatrick, S.F.; Tambuwala, M.M.; Bruning, U.; Schaible, B.; Scholz, C.C.; Byrne, A.; O'Connor, A.; Gallagher, W.M.; Lenihan, C.R.; Garvey, J.F.; et al. An intact canonical NF-κB pathway is required for inflammatory gene expression in response to hypoxia. *J. Immunol.* **2011**, *186*, 1091–1096. [CrossRef] [PubMed]

109. Bandarra, D.; Biddlestone, J.; Mudie, S.; Muller, H.A.; Rocha, S. Hypoxia activates IKK- NF-κB and the immune response in *Drosophila melanogaster*. *Biosci. Rep.* **2014**, *34*. [CrossRef] [PubMed]

110. Liu, Q.; Li, J.; Khoury, J.; Colgan, S.P.; Ibla, J.C. Adenosine signaling mediates SUMO-1 modification of IκBα during hypoxia and reoxygenation. *J. Biol. Chem.* **2009**, *284*, 13686–13695. [CrossRef] [PubMed]

111. Aillet, F.; Lopitz-Otsoa, F.; Egana, I.; Hjerpe, R.; Fraser, P.; Hay, R.T.; Rodriguez, M.S.; Lang, V. Heterologous SUMO-2/3-ubiquitin chains optimize IκBα degradation and NF-κB activity. *PLoS ONE* **2012**, *7*, e51672. [CrossRef] [PubMed]

112. Mulero, M.C.; Ferres-Marco, D.; Islam, A.; Margalef, P.; Pecoraro, M.; Toll, A.; Drechsel, N.; Charneco, C.; Davis, S.; Bellora, N.; et al. Chromatin-bound IκBα regulates a subset of polycomb target genes in differentiation and cancer. *Cancer Cell* **2013**, *24*, 151–166. [CrossRef] [PubMed]

113. Arnold, M.; Sierra, M.S.; Laversanne, M.; Soerjomataram, I.; Jemal, A.; Bray, F. Global patterns and trends in colorectal cancer incidence and mortality. *Gut* **2017**, *66*, 683–691. [CrossRef] [PubMed]

114. Wang, S.; Liu, Z.; Wang, L.; Zhang, X. NF-κB signaling pathway, inflammation and colorectal cancer. *Cell. Mol. Immunol.* **2009**, *6*, 327–334. [CrossRef] [PubMed]

115. Fernandez-Majada, V.; Aguilera, C.; Villanueva, A.; Vilardell, F.; Robert-Moreno, A.; Aytes, A.; Real, F.X.; Capella, G.; Mayo, M.W.; Espinosa, L.; et al. Nuclear IKK activity leads to dysregulated notch-dependent gene expression in colorectal cancer. *Proc. Natl. Acad. Sci. USA* **2007**, *104*, 276–281. [CrossRef] [PubMed]

116. Imamura, T.; Kikuchi, H.; Herraiz, M.T.; Park, D.Y.; Mizukami, Y.; Mino-Kenduson, M.; Lynch, M.P.; Rueda, B.R.; Benita, Y.; Xavier, R.J.; et al. HIF-1α and HIF-2α have divergent roles in colon cancer. *Int. J. Cancer* **2009**, *124*, 763–771. [CrossRef] [PubMed]

117. Krishnamachary, B.; Berg-Dixon, S.; Kelly, B.; Agani, F.; Feldser, D.; Ferreira, G.; Iyer, N.; LaRusch, J.; Pak, B.; Taghavi, P.; et al. Regulation of colon carcinoma cell invasion by hypoxia-inducible factor 1. *Cancer Res.* **2003**, *63*, 1138–1143. [PubMed]

118. Jubb, A.M.; Pham, T.Q.; Hanby, A.M.; Frantz, G.D.; Peale, F.V.; Wu, T.D.; Koeppen, H.W.; Hillan, K.J. Expression of vascular endothelial growth factor, hypoxia inducible factor 1α, and carbonic anhydrase IX in human tumours. *J. Clin. Pathol.* **2004**, *57*, 504–512. [CrossRef] [PubMed]

119. Fukuda, R.; Hirota, K.; Fan, F.; Jung, Y.D.; Ellis, L.M.; Semenza, G.L. Insulin-like growth factor 1 induces hypoxia-inducible factor 1-mediated vascular endothelial growth factor expression, which is dependent on MAP kinase and phosphatidylinositol 3-kinase signaling in colon cancer cells. *J. Biol. Chem.* **2002**, *277*, 38205–38211. [CrossRef] [PubMed]

120. Dang, D.T.; Chen, F.; Gardner, L.B.; Cummins, J.M.; Rago, C.; Bunz, F.; Kantsevoy, S.V.; Dang, L.H. Hypoxia-inducible factor-1α promotes nonhypoxia-mediated proliferation in colon cancer cells and xenografts. *Cancer Res.* **2006**, *66*, 1684–1936. [CrossRef] [PubMed]
121. Giles, R.H.; Lolkema, M.P.; Snijckers, C.M.; Belderbos, M.; van der Groep, P.; Mans, D.A.; van Beest, M.; van Noort, M.; Goldschmeding, R.; van Diest, P.J.; et al. Interplay between VHL/HIF1α and Wnt/β-catenin pathways during colorectal tumorigenesis. *Oncogene* **2006**, *25*, 3065–3070. [CrossRef] [PubMed]
122. Rawluszko-Wieczorek, A.A.; Horbacka, K.; Krokowicz, P.; Misztal, M.; Jagodzinski, P.P. Prognostic potential of DNA methylation and transcript levels of HIF1A and EPAS1 in colorectal cancer. *Mol. Cancer Res.* **2014**, *12*, 1112–1127. [CrossRef] [PubMed]
123. Shay, J.E.; Imtiyaz, H.Z.; Sivanand, S.; Durham, A.C.; Skuli, N.; Hsu, S.; Mucaj, V.; Eisinger-Mathason, T.S.; Krock, B.L.; Giannoukos, D.N.; et al. Inhibition of hypoxia-inducible factors limits tumor progression in a mouse model of colorectal cancer. *Carcinogenesis* **2014**, *35*, 1067–1077. [CrossRef] [PubMed]
124. Fearon, E.R.; Vogelstein, B. A genetic model for colorectal tumorigenesis. *Cell* **1990**, *61*, 759–767. [CrossRef]
125. McCartney, B.M.; Nathke, I.S. Cell regulation by the Apc protein Apc as master regulator of epithelia. *Curr. Opin. Cell Biol.* **2008**, *20*, 186–193. [CrossRef] [PubMed]
126. Bienz, M.; Clevers, H. Linking colorectal cancer to Wnt signaling. *Cell* **2000**, *103*, 311–320. [CrossRef]
127. Newton, I.P.; Kenneth, N.S.; Appleton, P.L.; Nathke, I.; Rocha, S. Adenomatous polyposis coli and hypoxia-inducible factor-1α have an antagonistic connection. *Mol. Biol. Cell* **2010**, *21*, 3630–3638. [CrossRef] [PubMed]
128. Nathke, I.; Rocha, S. Antagonistic crosstalk between APC and HIF-1α. *Cell Cycle* **2011**, *10*, 1545–1547. [CrossRef] [PubMed]
129. Lu, L.; Gao, Y.; Zhang, Z.; Cao, Q.; Zhang, X.; Zou, J.; Cao, Y. Kdm2α/β Lysine demethylases regulate canonical wnt signaling by modulating the stability of nuclear β-catenin. *Dev. Cell* **2015**, *33*, 660–674. [CrossRef] [PubMed]
130. Zhang, H.; Gao, P.; Fukuda, R.; Kumar, G.; Krishnamachary, B.; Zeller, K.I.; Dang, C.V.; Semenza, G.L. HIF-1 inhibits mitochondrial biogenesis and cellular respiration in VHL-deficient renal cell carcinoma by repression of C-MYC activity. *Cancer Cell* **2007**, *11*, 407–420. [CrossRef] [PubMed]
131. Corn, P.G.; Ricci, M.S.; Scata, K.A.; Arsham, A.M.; Simon, M.C.; Dicker, D.T.; El-Deiry, W.S. Mxi1 is induced by hypoxia in a HIF-1-dependent manner and protects cells from c-Myc-induced apoptosis. *Cancer Biol. Ther.* **2005**, *4*, 1285–1294. [CrossRef] [PubMed]
132. Gordan, J.D.; Thompson, C.B.; Simon, M.C. HIF and c-Myc: Sibling rivals for control of cancer cell metabolism and proliferation. *Cancer Cell* **2007**, *12*, 108–113. [CrossRef] [PubMed]
133. Koshiji, M.; Kageyama, Y.; Pete, E.A.; Horikawa, I.; Barrett, J.C.; Huang, L.E. HIF-1α induces cell cycle arrest by functionally counteracting Myc. *EMBO J.* **2004**, *23*, 1949–1956. [CrossRef] [PubMed]
134. Podar, K.; Anderson, K.C. A therapeutic role for targeting c-Myc/HIF-1-dependent signaling pathways. *Cell. Cycle* **2010**, *9*, 1722–1728. [CrossRef] [PubMed]
135. Duyao, M.P.; Buckler, A.J.; Sonenshein, G.E. Interaction of an NF-κB-like factor with a site upstream of the c-myc promoter. *Proc. Natl. Acad. Sci. USA* **1990**, *87*, 4727–4731. [CrossRef] [PubMed]
136. Gordan, J.D.; Bertout, J.A.; Hu, C.J.; Diehl, J.A.; Simon, M.C. HIF-2α promotes hypoxic cell proliferation by enhancing c-myc transcriptional activity. *Cancer Cell* **2007**, *11*, 335–347. [CrossRef] [PubMed]
137. Fan, X.J.; Wan, X.B.; Yang, Z.L.; Fu, X.H.; Huang, Y.; Chen, D.K.; Song, S.X.; Liu, Q.; Xiao, H.Y.; Wang, L.; et al. Snail promotes lymph node metastasis and Twist enhances tumor deposit formation through epithelial-mesenchymal transition in colorectal cancer. *Hum. Pathol.* **2013**, *44*, 173–180. [CrossRef] [PubMed]
138. Evans, A.J.; Russell, R.C.; Roche, O.; Burry, T.N.; Fish, J.E.; Chow, V.W.; Kim, W.Y.; Saravanan, A.; Maynard, M.A.; Gervais, M.L.; et al. VHL promotes E2 box-dependent E-cadherin transcription by HIF-mediated regulation of SIP1 and snail. *Mol. Cell. Biol.* **2007**, *27*, 157–169. [CrossRef] [PubMed]
139. Yang, M.H.; Wu, M.Z.; Chiou, S.H.; Chen, P.M.; Chang, S.Y.; Liu, C.J.; Teng, S.C.; Wu, K.J. Direct regulation of TWIST by HIF-1α promotes metastasis. *Nat. Cell Biol.* **2008**, *10*, 295–305. [CrossRef] [PubMed]
140. Gort, E.H.; van Haaften, G.; Verlaan, I.; Groot, A.J.; Plasterk, R.H.; Shvarts, A.; Suijkerbuijk, K.P.; van Laar, T.; van der Wall, E.; Raman, V.; et al. The TWIST1 oncogene is a direct target of hypoxia-inducible factor-2α. *Oncogene* **2008**, *27*, 1501–1510. [CrossRef] [PubMed]
141. Wu, Y.; Zhou, B.P. TNF-α/ NF-κB/Snail pathway in cancer cell migration and invasion. *Br. J. Cancer* **2010**, *102*, 639–644. [CrossRef] [PubMed]

142. Schwitalla, S.; Ziegler, P.K.; Horst, D.; Becker, V.; Kerle, I.; Begus-Nahrmann, Y.; Lechel, A.; Rudolph, K.L.; Langer, R.; Slotta-Huspenina, J.; et al. Loss of p53 in enterocytes generates an inflammatory microenvironment enabling invasion and lymph node metastasis of carcinogen-induced colorectal tumors. *Cancer Cell* **2013**, *23*, 93–106. [CrossRef] [PubMed]

143. Perkins, N.D. The diverse and complex roles of NF-κB subunits in cancer. *Nat. Rev. Cancer* **2012**, *12*, 121–132. [CrossRef] [PubMed]

144. Uemura, M.; Yamamoto, H.; Takemasa, I.; Mimori, K.; Hemmi, H.; Mizushima, T.; Ikeda, M.; Sekimoto, M.; Matsuura, N.; Doki, Y.; et al. Jumonji domain containing 1A is a novel prognostic marker for colorectal cancer: In vivo identification from hypoxic tumor cells. *Clin. Cancer Res.* **2010**, *16*, 4636–4646. [CrossRef] [PubMed]

145. Pereira, F.; Barbachano, A.; Silva, J.; Bonilla, F.; Campbell, M.J.; Munoz, A.; Larriba, M.J. KDM6B/JMJD3 histone demethylase is induced by vitamin D and modulates its effects in colon cancer cells. *Hum. Mol. Genet.* **2011**, *20*, 4655–4665. [CrossRef] [PubMed]

146. Xue, J.; Li, X.; Jiao, S.; Wei, Y.; Wu, G.; Fang, J. Prolyl hydroxylase-3 is down-regulated in colorectal cancer cells and inhibits IKKβ independent of hydroxylase activity. *Gastroenterology* **2010**, *138*, 606–615. [CrossRef] [PubMed]

147. Deschoemaeker, S.; Di Conza, G.; Lilla, S.; Martin-Perez, R.; Mennerich, D.; Boon, L.; Hendrikx, S.; Maddocks, O.D.; Marx, C.; Radhakrishnan, P.; et al. PHD1 regulates p53-mediated colorectal cancer chemoresistance. *EMBO Mol. Med.* **2015**, *7*, 1350–1365. [CrossRef] [PubMed]

148. Biswas, D.K.; Shi, Q.; Baily, S.; Strickland, I.; Ghosh, S.; Pardee, A.B.; Iglehart, J.D. NF-κB activation in human breast cancer specimens and its role in cell proliferation and apoptosis. *Proc. Natl. Acad. Sci. USA* **2004**, *101*, 10137–10142. [CrossRef] [PubMed]

149. Chua, H.L.; Bhat-Nakshatri, P.; Clare, S.E.; Morimiya, A.; Badve, S.; Nakshatri, H. NF-κB represses E-cadherin expression and enhances epithelial to mesenchymal transition of mammary epithelial cells: Potential involvement of ZEB-1 and ZEB-2. *Oncogene* **2007**, *26*, 711–724. [CrossRef] [PubMed]

150. Biswas, D.K.; Cruz, A.P.; Gansberger, E.; Pardee, A.B. Epidermal growth factor-induced nuclear factor κB activation: A major pathway of cell-cycle progression in estrogen-receptor negative breast cancer cells. *Proc. Natl. Acad. Sci. USA* **2000**, *97*, 8542–8547. [CrossRef] [PubMed]

151. Sethi, G.; Ahn, K.S.; Chaturvedi, M.M.; Aggarwal, B.B. Epidermal growth factor (EGF) activates nuclear factor-κB through IκBα kinase-independent but EGF receptor-kinase dependent tyrosine 42 phosphorylation of IκBα. *Oncogene* **2015**, *34*, 5407. [CrossRef] [PubMed]

152. Bivona, T.G.; Hieronymus, H.; Parker, J.; Chang, K.; Taron, M.; Rosell, R.; Moonsamy, P.; Dahlman, K.; Miller, V.A.; Costa, C.; et al. FAS and NF-κB signalling modulate dependence of lung cancers on mutant EGFR. *Nature* **2011**, *471*, 523–526. [CrossRef] [PubMed]

153. Gilkes, D.M.; Semenza, G.L. Role of hypoxia-inducible factors in breast cancer metastasis. *Future Oncol.* **2013**, *9*, 1623–1636. [CrossRef] [PubMed]

154. Wong, C.C.; Zhang, H.; Gilkes, D.M.; Chen, J.; Wei, H.; Chaturvedi, P.; Hubbi, M.E.; Semenza, G.L. Inhibitors of hypoxia-inducible factor 1 block breast cancer metastatic niche formation and lung metastasis. *J. Mol. Med.* **2012**, *90*, 803–815. [CrossRef] [PubMed]

155. Lee, K.; Qian, D.Z.; Rey, S.; Wei, H.; Liu, J.O.; Semenza, G.L. Anthracycline chemotherapy inhibits HIF-1 transcriptional activity and tumor-induced mobilization of circulating angiogenic cells. *Proc. Natl. Acad. Sci. USA* **2009**, *106*, 2353–2358. [CrossRef] [PubMed]

156. Kummar, S.; Raffeld, M.; Juwara, L.; Horneffer, Y.; Strassberger, A.; Allen, D.; Steinberg, S.M.; Rapisarda, A.; Spencer, S.D.; Figg, W.D.; et al. Multihistology, target-driven pilot trial of oral topotecan as an inhibitor of hypoxia-inducible factor-1α in advanced solid tumors. *Clin. Cancer Res.* **2011**, *17*, 5123–5131. [CrossRef] [PubMed]

157. Leek, R.D.; Talks, K.L.; Pezzella, F.; Turley, H.; Campo, L.; Brown, N.S.; Bicknell, R.; Taylor, M.; Gatter, K.C.; Harris, A.L. Relation of hypoxia-inducible factor-2 α (HIF-2 α) expression in tumor-infiltrative macrophages to tumor angiogenesis and the oxidative thymidine phosphorylase pathway in human breast cancer. *Cancer Res.* **2002**, *62*, 1326–1329. [PubMed]

158. Giatromanolaki, A.; Sivridis, E.; Fiska, A.; Koukourakis, M.I. Hypoxia-inducible factor-2 α (HIF-2 α) induces angiogenesis in breast carcinomas. *Appl. Immunohistochem. Mol. Morphol.* **2006**, *14*, 78–82. [CrossRef] [PubMed]

159. De Boniface, J.; Mao, Y.; Schmidt-Mende, J.; Kiessling, R.; Poschke, I. Expression patterns of the immunomodulatory enzyme arginase 1 in blood, lymph nodes and tumor tissue of early-stage breast cancer patients. *Oncoimmunology* **2012**, *1*, 1305–1312. [CrossRef] [PubMed]

160. Pryor, J.G.; Brown-Kipphut, B.A.; Iqbal, A.; Scott, G.A. Microarray comparative genomic hybridization detection of copy number changes in desmoplastic melanoma and malignant peripheral nerve sheath tumor. *Am. J. Dermatopathol.* **2011**, *33*, 780–785. [CrossRef] [PubMed]

161. Yang, Z.Q.; Imoto, I.; Fukuda, Y.; Pimkhaokham, A.; Shimada, Y.; Imamura, M.; Sugano, S.; Nakamura, Y.; Inazawa, J. Identification of a novel gene, GASC1, within an amplicon at 9p23–24 frequently detected in esophageal cancer cell lines. *Cancer Res.* **2000**, *60*, 4735–4739. [PubMed]

162. Patani, N.; Jiang, W.G.; Newbold, R.F.; Mokbel, K. Histone-modifier gene expression profiles are associated with pathological and clinical outcomes in human breast cancer. *Anticancer Res.* **2011**, *31*, 4115–4125. [PubMed]

163. Shi, L.; Sun, L.; Li, Q.; Liang, J.; Yu, W.; Yi, X.; Yang, X.; Li, Y.; Han, X.; Zhang, Y.; et al. Histone demethylase JMJD2B coordinates H3K4/H3K9 methylation and promotes hormonally responsive breast carcinogenesis. *Proc. Natl. Acad. Sci. USA* **2011**, *108*, 7541–7546. [CrossRef] [PubMed]

164. Barrett, A.; Madsen, B.; Copier, J.; Lu, P.J.; Cooper, L.; Scibetta, A.G.; Burchell, J.; Taylor-Papadimitriou, J. PLU-1 nuclear protein, which is upregulated in breast cancer, shows restricted expression in normal human adult tissues: A new cancer/testis antigen? *Int. J. Cancer* **2002**, *101*, 581–588. [CrossRef] [PubMed]

165. Hayami, S.; Yoshimatsu, M.; Veerakumarasivam, A.; Unoki, M.; Iwai, Y.; Tsunoda, T.; Field, H.I.; Kelly, J.D.; Neal, D.E.; Yamaue, H.; et al. Overexpression of the JmjC histone demethylase KDM5B in human carcinogenesis: Involvement in the proliferation of cancer cells through the E2F/RB pathway. *Mol. Cancer* **2010**, *9*, 59. [CrossRef] [PubMed]

166. Yamane, K.; Tateishi, K.; Klose, R.J.; Fang, J.; Fabrizio, L.A.; Erdjument-Bromage, H.; Taylor-Papadimitriou, J.; Tempst, P.; Zhang, Y. PLU-1 is an H3K4 demethylase involved in transcriptional repression and breast cancer cell proliferation. *Mol. Cell* **2007**, *25*, 801–812. [CrossRef] [PubMed]

167. Bamodu, O.A.; Huang, W.C.; Lee, W.H.; Wu, A.; Wang, L.S.; Hsiao, M.; Yeh, C.T.; Chao, T.Y. Aberrant KDM5B expression promotes aggressive breast cancer through MALAT1 overexpression and downregulation of hsa-miR-448. *BMC Cancer* **2016**, *16*, 160. [CrossRef] [PubMed]

168. Shen, X.; Zhuang, Z.; Zhang, Y.; Chen, Z.; Shen, L.; Pu, W.; Chen, L.; Xu, Z. JARID1B modulates lung cancer cell proliferation and invasion by regulating p53 expression. *Tumour Biol.* **2015**, *36*, 7133–7142. [CrossRef] [PubMed]

169. Meylan, E.; Dooley, A.L.; Feldser, D.M.; Shen, L.; Turk, E.; Ouyang, C.; Jacks, T. Requirement for NF-κB signalling in a mouse model of lung adenocarcinoma. *Nature* **2009**, *462*, 104–107. [CrossRef] [PubMed]

170. Maes, T.; Carceller, E.; Salas, J.; Ortega, A.; Buesa, C. Advances in the development of histone lysine demethylase inhibitors. *Curr. Opin. Pharmacol.* **2015**, *23*, 52–60. [CrossRef] [PubMed]

171. Hancock, R.L.; Dunne, K.; Walport, L.J.; Flashman, E.; Kawamura, A. Epigenetic regulation by histone demethylases in hypoxia. *Epigenomics* **2015**, *7*, 791–811. [CrossRef] [PubMed]

172. Liu, H.; Liu, L.; Holowatyj, A.; Jiang, Y.; Yang, Z.Q. Integrated genomic and functional analyses of histone demethylases identify oncogenic KDM2A isoform in breast cancer. *Mol. Carcinog.* **2016**, *55*, 977–990. [CrossRef] [PubMed]

173. Wagner, K.W.; Alam, H.; Dhar, S.S.; Giri, U.; Li, N.; Wei, Y.; Giri, D.; Cascone, T.; Kim, J.H.; Ye, Y.; et al. KDM2A promotes lung tumorigenesis by epigenetically enhancing ERK1/2 signaling. *J. Clin. Investig.* **2013**, *123*, 5231–5246. [CrossRef] [PubMed]

174. Zhang, Q.; Gu, J.; Li, L.; Liu, J.; Luo, B.; Cheung, H.W.; Boehm, J.S.; Ni, M.; Geisen, C.; Root, D.E.; et al. Control of cyclin D1 and breast tumorigenesis by the EglN2 prolyl hydroxylase. *Cancer Cell* **2009**, *16*, 413–424. [CrossRef] [PubMed]

175. Oh, E.T.; Kim, C.W.; Kim, S.J.; Lee, J.S.; Hong, S.S.; Park, H.J. Docetaxel induced-JNK2/PHD1 signaling pathway increases degradation of HIF-1α and causes cancer cell death under hypoxia. *Sci. Rep.* **2016**, *6*, 27382. [CrossRef] [PubMed]

176. Peurala, E.; Koivunen, P.; Bloigu, R.; Haapasaari, K.M.; Jukkola-Vuorinen, A. Expressions of individual PHDs associate with good prognostic factors and increased proliferation in breast cancer patients. *Breast Cancer Res. Treat.* **2012**, *133*, 179–188. [CrossRef] [PubMed]

177. Chen, S.; Zhang, J.; Li, X.; Luo, X.; Fang, J.; Chen, H. The expression of prolyl hydroxylase domain enzymes are up-regulated and negatively correlated with BCL-2 in non-small cell lung cancer. *Mol. Cell. Biochem.* **2011**, *358*, 257–263. [CrossRef] [PubMed]

178. Andersen, S.; Donnem, T.; Stenvold, H.; Al-Saad, S.; Al-Shibli, K.; Busund, L.T.; Bremnes, R.M. Overexpression of the HIF hydroxylases PHD1, PHD2, PHD3 and FIH are individually and collectively unfavorable prognosticators for NSCLC survival. *PLoS ONE* **2011**, *6*, e23847. [CrossRef] [PubMed]

179. Frost, J.; Galdeano, C.; Soares, P.; Gadd, M.S.; Grzes, K.M.; Ellis, L.; Epemolu, O.; Shimamura, S.; Bantscheff, M.; Grandi, P.; et al. Potent and selective chemical probe of hypoxic signalling downstream of HIF-α hydroxylation via VHL inhibition. *Nat. Commun.* **2016**, *7*, 13312. [CrossRef] [PubMed]

180. Curtis, V.F.; Ehrentraut, S.F.; Campbell, E.L.; Glover, L.E.; Bayless, A.; Kelly, C.J.; Kominsky, D.J.; Colgan, S.P. Stabilization of HIF through inhibition of Cullin-2 neddylation is protective in mucosal inflammatory responses. *FASEB J.* **2015**, *29*, 208–215. [CrossRef] [PubMed]

181. Colgan, S.P. Targeting hypoxia in inflammatory bowel disease. *J. Investig. Med.* **2016**, *64*, 364–368. [CrossRef] [PubMed]

182. Tafani, M.; Pucci, B.; Russo, A.; Schito, L.; Pellegrini, L.; Perrone, G.A.; Villanova, L.; Salvatori, L.; Ravenna, L.; Petrangeli, E.; et al. Modulators of HIF1α and NFkB in cancer treatment: Is it a rational approach for controlling malignant progression? *Front. Pharmacol.* **2013**, *4*, 13. [CrossRef] [PubMed]

183. Okumura, C.Y.; Hollands, A.; Tran, D.N.; Olson, J.; Dahesh, S.; von Kockritz-Blickwede, M.; Thienphrapa, W.; Corle, C.; Jeung, S.N.; Kotsakis, A.; et al. A new pharmacological agent (AKB-4924) stabilizes hypoxia inducible factor-1 (HIF-1) and increases skin innate defenses against bacterial infection. *J. Mol. Med.* **2012**, *90*, 1079–1089. [CrossRef] [PubMed]

184. Cho, H.; Du, X.; Rizzi, J.P.; Liberzon, E.; Chakraborty, A.A.; Gao, W.; Carvo, I.; Signoretti, S.; Bruick, R.K.; Josey, J.A.; et al. On-target efficacy of a HIF-2α antagonist in preclinical kidney cancer models. *Nature* **2016**, *539*, 107–111. [CrossRef] [PubMed]

185. Wallace, E.M.; Rizzi, J.P.; Han, G.; Wehn, P.M.; Cao, Z.; Du, X.; Cheng, T.; Czerwinski, R.M.; Dixon, D.D.; Goggin, B.S.; et al. A small-molecule antagonist of HIF2α is efficacious in preclinical models of renal cell carcinoma. *Cancer Res.* **2016**, *76*, 5491–5500. [CrossRef] [PubMed]

186. Chen, W.; Hill, H.; Christie, A.; Kim, M.S.; Holloman, E.; Pavia-Jimenez, A.; Homayoun, F.; Ma, Y.; Patel, N.; Yell, P.; et al. Targeting renal cell carcinoma with a HIF-2 antagonist. *Nature* **2016**, *539*, 112–117. [CrossRef] [PubMed]

187. Drew, D.A.; Cao, Y.; Chan, A.T. Aspirin and colorectal cancer: The promise of precision chemoprevention. *Nat. Rev. Cancer* **2016**, *16*, 173–186. [CrossRef] [PubMed]

188. Guo, X.; Zheng, L.; Jiang, J.; Zhao, Y.; Wang, X.; Shen, M.; Zhu, F.; Tian, R.; Shi, C.; Xu, M.; et al. Blocking NF-κB is essential for the immunotherapeutic effect of recombinant IL18 in pancreatic cancer. *Clin. Cancer Res.* **2016**, *22*, 5939–5950. [CrossRef] [PubMed]

189. Iwai, Y.; Hamanishi, J.; Chamoto, K.; Honjo, T. Cancer immunotherapies targeting the PD-1 signaling pathway. *J. Biomed. Sci.* **2017**, *24*, 26. [CrossRef] [PubMed]

190. Noman, M.Z.; Desantis, G.; Janji, B.; Hasmim, M.; Karray, S.; Dessen, P.; Bronte, V.; Chouaib, S. PD-L1 is a novel direct target of HIF-1α, and its blockade under hypoxia enhanced MDSC-mediated T cell activation. *J. Exp. Med.* **2014**, *211*, 781–790. [CrossRef] [PubMed]

191. An, J.; Rettig, M.B. Mechanism of von Hippel-Lindau protein-mediated suppression of nuclear factor κB activity. *Mol. Cell. Biol.* **2005**, *25*, 7546–7556. [CrossRef] [PubMed]

192. Labrousse-Arias, D.; Martinez-Alonso, E.; Corral-Escariz, M.; Bienes-Martinez, R.; Berridy, J.; Serrano-Oviedo, L.; Conde, E.; Garcia-Bermejo, M.L.; Gimenez-Bachs, J.M.; Salinas-Sanchez, A.S.; et al. VHL promotes immune response against renal cell carcinoma via NF-κB-dependent regulation of VCAM-1. *J. Cell. Biol.* **2017**, *216*, 835–847. [CrossRef] [PubMed]

193. Wang, Y.; Zhao, W.; Gao, Q.; Fan, L.; Qin, Y.; Zhou, H.; Li, M.; Fang, J. pVHL mediates K63-linked ubiquitination of IKKβ, leading to IKKβ inactivation. *Cancer Lett.* **2016**, *383*, 1–8. [CrossRef] [PubMed]

biomedicines

MDPI

Review

BET Family Protein BRD4: An Emerging Actor in NFκB Signaling in Inflammation and Cancer

Azadeh Hajmirza [1], Anouk Emadali [1,2], Arnaud Gauthier [1], Olivier Casasnovas [3], Rémy Gressin [4] and Mary B. Callanan [1,5,*]

[1] INSERM U1209, CNRS UMR 5309, Institute for Advanced Biosciences, Université de Grenoble-Alpes, F-38042 Grenoble, France; azadeh.hajmirza@univ-grenoble-alpes.fr (A.H.); anouk.emadali@univ-grenoble-alpes.fr (A.E.); agauthier@chu-grenoble.fr (A.G.)

[2] Pôle Recherche, Grenoble-Alpes University Hospital, F-38043 Grenoble, France

[3] Département d'Hématologie Clinique, Dijon University Hospital, F-21000 Dijon, France; olivier.casasnovas@chu-dijon.fr

[4] Département d'Hématologie Clinique, Grenoble-Alpes University Hospital, F-38043 Grenoble, France; RGressin@chu-grenoble.fr

[5] Centre for Innovation in Cancer Genetics and Epigenetics, Dijon University Hospital, F-21000 Dijon, France

* Correspondance: mary.callanan@chu-dijon.fr

Received: 16 October 2017; Accepted: 1 February 2018; Published: 6 February 2018

Abstract: NFκB (Nuclear Factor-κ-light-chain-enhancer of activated B cells) signaling elicits global transcriptional changes by activating cognate promoters and through genome-wide remodeling of cognate regulatory elements called "super enhancers". BET (Bromodomain and Extra-Terminal domain) protein family inhibitor studies have implicated BET protein member BRD4 and possibly other BET proteins in NFκB-dependent promoter and super-enhancer modulation. Members of the BET protein family are known to bind acetylated chromatin to facilitate access by transcriptional regulators to chromatin, as well as to assist the activity of transcription elongation complexes via CDK9/pTEFb. BET family member BRD4 has been shown to bind non-histone proteins and modulate their activity. One such protein is RELA, the NFκB co-activator. Specifically, BRD4 binds acetylated RELA, which increases its transcriptional transactivation activity and stability in the nucleus. In aggregate, this establishes an intimate link between NFκB and BET signaling, at least via BRD4. The present review provides a brief overview of the structure and function of BET family proteins and then examines the connections between NFκB and BRD4 signaling, using the inflammatory response and cancer cell signaling as study models. We also discuss the potential of BET inhibitors for relief of aberrant NFκB signaling in cancer, focusing on non-histone, acetyl-lysine binding functions.

Keywords: NFκB; BET inhibition; transcription; chromatin looping; acetylation B cell non-Hodgkin lymphoma

---

1. Introduction

Epigenetic signaling refers to the chromatin-dependent mechanisms that directly or indirectly control genome activity. Essentially, this refers to the chemical modifications on DNA or chromatin (histone proteins) that, together with topological organization of the chromatin fiber in the nucleus, regulate chromatin compaction. This allows the formation of functionally distinct, dynamically reversible chromatin states called euchromatin and heterochromatin, respectively [1]. The former is characterized by loosely packed nucleosomes, while in heterochromatin nucleosomes are densely packed. Nucleosomes are the basic functional unit of chromatin and are comprised of an octamer of the core histones, H2A, H2B, H3, and H4 (two copies of each). DNA methylation and post-translational histone modifications allow regulation of genome function by the modulation of chromatin compaction- and thereby access to DNA by transcription, replication, and repair factors- by serving as docking platforms for specific chromatin-associated signaling complexes [1]. Enzymes that mediate chemical modifications of DNA or chromatin are referred to as "writers" of epigenetic information, while those proteins that dock or erase these chemical modifications are referred to as "readers" and "erasers" or epigenetic information, respectively [1].

Bromodomain and extra-terminal domain (BET) proteins constitute a novel class of epigenetic "readers" that is involved in the control of genome activity through the ability to bind acetylated lysine residues in both histone and non-histone proteins, including transcription factors. The mechanism by which BET proteins are recruited to acetylated lysine is discussed below. How this links to transcriptional control by NFκB signaling is reviewed.

BET proteins have emerged as key regulators of transcriptional control in development and cellular differentiation and have been identified as critical actors of disordered transcription in transformed cells, thereby rendering them hypersensitive to small molecule inhibition of BET protein activity [2,3]. Mechanisms of action are complex but rely on the capacity of these proteins to bind acetylated histone and non-histone proteins via their double bromodomains (Figure 1) [2].

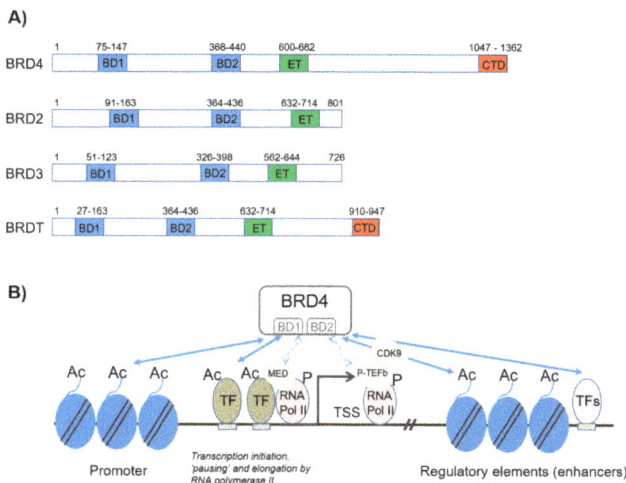

Figure 1. Schematic of domain organization of BET (bromodomain and extra-terminal domain) family proteins and the function of BRD4 in the regulation of promoter and enhancer activity. (**A**) Domain organization of human BET family members BRD4, BRD2, BRD3, and BRDT, as indicated. BET proteins contain two bromodomains (BD1 and BD2, respectively) and an extra-terminal domain (ET). BRD4 (long form) contains a carboxyterminal domain (CTD) that is not present in the other BET family members; (**B**) Schematic representation of the functions of BET family member BRD4 in the regulation of promoter and enhancer function (includes "super-enhancers"; see text for details). Through its BD1 and BD2 domains, BRD4 binds to acetylated lysines (Ac) in histones or transcription factors (TF). The binding of acetylated histones by BRD4, at transcription start sites (TSS), mediates transcriptional co-activation and elongation via RNA polymerase II (RNA pol II) and Mediator (Med) and pTEFb signaling complexes, respectively (see text for details). BRD4 can also bind acetylated lysines in histones or TF in enhancer elements, thereby contributing to long-range control of gene activity (see text for details). TF binding sites are depicted as horizontal rectangles.

2. Structure and Function of BET Proteins

The BET family of proteins comprises BRD2, BRD3, BRD4 (ubiquitously expressed), and BRDT (testis-specific expression). BET proteins are characterized by the presence of two conserved N-terminal bromodomains (BD1 and BD2) and a C-terminal "extra-terminal" domain (Figure 1). The bromodomain structure contains four alpha helices separated by a variable loop region that together allow the formation of a hydrophobic cavity that recognizes acetyl-lysine residues [4,5]. Structural data have established that acetylated lysine is recognized in this central hydrophobic pocket, by anchoring to a conserved asparagine residue. The BET bromodomain proteins can bind to two acetylated lysine histone marks that are simultaneously recognized by the same bromodomain module [5]. This property is shared by all members of the BET subclass of bromodomains. High-resolution co-crystal structures showed that the first acetylated lysine mark of histone H4 docks directly onto the conserved asparagine (Asn140 in the first bromodomain of BRD4). A network of hydrogen bonds within the acetyl-lysine binding cavity link to the second acetylated lysine mark thus stabilizing the peptide complex [4,5]. The largely hydrophobic nature of the central acetylated lysine binding pocket of the bromodomain, which is necessary to accommodate the charge-neutralized acetylated lysine, and the comparably weak interaction with its target sequences make these modules particularly attractive for the development of inhibitors targeting this protein–protein interaction. The interaction of BET proteins with mono-acetylated lysine seems to be of weak affinity compared to the interactions that occur at multiple closely spaced acetylated lysines [4,5].

As well as interacting with acetylated lysine in histones, BET proteins also interact with members of the transcription elongation complex and with other transcription factors. In the latter case, this can be through lysine acetylation-dependent or -independent mechanisms. As such, BET proteins are key "readers" of epigenetic information in both normal and transformed cells. The interaction of BDs at acetylated chromatin either at gene promoters or in long range *cis* regulatory elements called enhancers allows chromatin-dependent signaling to connect to transcription regulation [6].

2.1. BRD4 in Transcriptional Regulation by NFκB

BRD4 is a particularly well-studied member of the BET protein family. BRD4 contains two bromodomains associated to an extra-terminal domain (Figure 1A). As explained above, the BD domains allow BRD4 to interact with acetylated lysine in histone or non-histone proteins. How this property is involved in transcriptional regulation is discussed. New findings relating to BET proteins and pro-inflammatory- or cancer cell-specific NFκB signaling [7] are also reviewed.

2.1.1. Transcription Initiation and Elongation

BRD4 participates in the activation and elongation of transcription via interactions with transcription initiation and elongation complexes Mediator and pTEFb (positive transcription elongation factor B), respectively. The pTEFb complex is composed of the cyclin-dependent kinase, CDK9, and a regulatory subunit, Cyclin T1 or T2. The kinase activity of CDK9 inhibits negative regulators of RNA polymerase II activity while stimulating its elongation activity by phosphorylation [8]. At least two different regions of BRD4 interact directly with pTEFb. The C-terminal region interacts with Cyclin T1 and CDK9 and the BD2 region interacts with the acetylated region of Cyclin T1. The BRD4/pTEFb interaction plays a central role in the rapid initiation of transcription after the exit from mitosis [8].

The extra-terminal (ET) domain is involved in transcriptional regulation through interactions with histone modifiers such as JMJD6 (jmjC domain-containing protein 6), an arginine demethylase, and NSD3, a lysine methyltransferase [9,10]. Furthermore, the ET domain can associate with ATP-dependent chromatin remodelers such as the SWI-SNF and CHD2 [10]. These interactions are thought to allow BRD4 to remodel chromatin locally, although the regulatory significance of these events is not well understood. One possibility is that this allows the release of paused RNA pol II activity [9].

NFκB-dependent transcriptional control is regulated at multiple levels, including cytoplasmic signaling events leading to the nuclear translocation of NFκB, the binding of nuclear NFκB to various transcriptional factors or regulators, and the post-transcriptional modifications of histones and NFκB itself [7]. Within the nucleus, NFκB recognizes the cognate NFκB sites on the enhancer or promoter regions of its target genes and directs the binding of co-regulators to form the transcriptional machinery for target gene expression. In the setting of NFκB signaling, it has been found that pTEFb can be recruited by BRD4 to NFκB-dependent acetylated histones—a mechanism that is crucial for the transcription of primary response genes [6], and possibly pathological NFκB signaling in cancer cells, although the latter has not been investigated in any detail as yet.

2.1.2. Enhancer Regulation by BRD4 and Its Role in NFκB Signaling

The genome-wide distribution of BRD4 has been studied by chromatin immunoprecipitation and deep sequencing (ChIP-seq). These experiments have shown that BRD4 binds multiple promoters as well as intergenic regions, particularly enhancer sequences. The Mediator complex (MED), composed of 26 subunits in mammals, plays a key role in transcription initiation and elongation downstream of numerous signaling cascades as well as in the functional regulation of enhancer elements. BRD4 and MED have been found to co-occupy subsets of enhancers called "super-enhancers" [11], which are large enhancer regions that stimulate the transcription of growth-promoting and lineage-specific survival genes [6]. Super-enhancers are also co-enriched for histone H3 acetylated at lysine 27. Supporting a functional interaction of BRD4 and MED at super-enhancers, BET bromodomain inhibition releases the mediator complex from select *cis*-regulatory elements, at least in leukemia cells [12]. It is interesting to note that MED complex activity at super-enhancers involves reversible association with a subunit containing the cyclin dependent kinase CDK8 and the cofactors CCNC (CYCLIN C), MED12, and MED13. Mutations in the gene encoding MED12 have been described in chronic lymphocytic leukemia [13]. Furthermore, the MED complex contains both activating and inhibiting CDKs, the latter of which appear to constrain tumor suppressor and lineage identity gene-associated super-enhancers, which raises interest in combining BET and MED complex negative regulatory CDK inhibitors for anti-cancer treatment [14].

Numerous oncogenes have been shown to be under the control of super-enhancer elements in various cancer types. Remarkable examples are the deregulation of MYC in B cell non-Hodgkin lymphoma and multiple myeloma [15,16], EVI1 in acute myeloid leukemia with the inversion of chromosome 3q [17], and mutational processes that are predicted to alter super-enhancer activity in breast cancer [18]. The recruitment of BRD4 to enhancer regions seems to depend, at least in part, on the activity of specific transcription factors and on histone acetyl-transferases such as p300/CBP [19]. The recruitment of BRD4 is essential to the activity of numerous hematopoietic transcription factors such as PU.1, FLI1, ERG, C/EBPα, C/EBPβ, and MYB [19] and to the activity of NFκB at cognate enhancers, downstream of inflammatory responses in endothelial cells and macrophages [20]. Although the precise mechanism for BRD4 co-recruitment with the NFκB subunit RELA/p65 to pro-inflammatory genes is not yet deciphered, it may relate to both BRD4-dependent histone lysine acetylation docking as well as to the ability of BRD4 to bind acetylated p65/RELA, as discussed in detail below [21].

NFκB-dependent enhancer remodeling during pro-inflammatory responses has been shown to implicate the synthesis of enhancer RNA (eRNA) and local chromatin acetylation, followed by progressive H3 lysine 4 mono and di-methylation [22]. This process requires composite binding with other tissue-specific transcription factors. eRNAs belong to the non-coding RNAs and their transcripts are directed by enhancers, which is thought to favor the spatial repositioning ('chromatin looping') of distal enhancers close to their cognate promoters via RNA pol II. Thus, NFκB acts as a key modulator of de novo enhancer remodeling in the setting of pro-inflammatory signaling by modulating eRNA transcription [23].

In keeping with the above findings, BET inhibitors effectively suppress inflammatory responses mediated by NFκB, in particular at super-enhancers [20]. Likewise, the NFκB pathway is activated by LPS (lipopolysaccharide), and the pan-BET inhibitor I-BET762 has been shown to prevent or diminish the incidence of death in mice given lethal doses of lipopolysaccharide [24].

2.1.3. BRD4 Interaction with Acetylated RELA/p65; Impact on NFκB Signaling and Sensitivity to BET Inhibitors

BRD4/BET proteins have been shown to interact with acetylated lysines in non-histone proteins [6]. This facet of BRD4/BET activity has not been explored in detail in cancer, but is likely to underlie at least some of the clinical activity of BET inhibitors, in particular in relation to aberrant NFκB signaling. Indeed, BRD4 interacts with the NFκB subunit, p65/RELA, when the latter is acetylated at lysine 310 [21] (and references therein). Under NFκB activation conditions, this leads to the recruitment of pTEFb and stimulates the transcription of NFκB target genes [21], in a BRD4- and CDK9-dependent fashion. RELA/BRD4 co-recruitment to NFκB target promoters is observed under a variety of NFκB stimulation conditions and cell types, and is not observed at housekeeping genes, supporting a necessity for defined promoter features for this interaction. Mapping and amino acid exchange studies have identified RELA acetyl-lysine-310 as the key residue for this interaction. Indeed, lysine substitution by an arginine residue significantly reduces BRD4 co-recruitment to an NFκB target promoter, as measured by ChIP assays [21]. Importantly, this does not alter the recruitment of either CBP/p300 or non-acetylated RELA, while acetylated RELA recruitment is lost. Taken together, this points to a critical role for acetylated RELA lysine-310 in BRD4 recruitment at NFκB response promoters. The acetylation of RELA lysine 310 is mediated by p300/CBP (Figure 2).

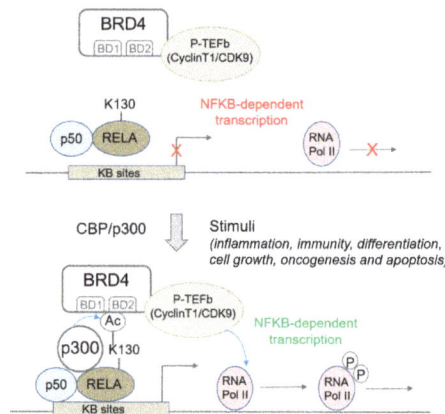

Figure 2. Schematic model for the binding of BRD4 to acetylated lysine-310 of RELA/p65 and the role of this interaction in the transcriptional activation of NFκB. Stimulus-dependent acetylation of RELA at lysine-310 by p300 induces the recruitment of BRD4 to the promoter via its bromodomains. BRD4 further activates CDK9 to phosphorylate the CTD (C-terminal domain) of RNA polymerase II and to facilitate the RNA pol II-mediated transcription of NFκB target genes. The p50 factor is a processed form of the REL family member p105 (NFκB1) that heterodimerizes with RELA/p65 and that is required for its transcriptional activation functions.

Of note, both bromodomains of BRD4 are necessary for the interaction with lysine-310-acetylated RELA, as evidenced by pull down assays using bromodomain deletion mutants, and in vivo immunoprecipitation assays. Moreover, loss of this interaction results in the inability of BRD4 to co-activate NFκB-dependent transcription. Structural studies have confirmed this by showing that both BD1 and BD2 bind to acetyl-lysine-310 peptide, and that this occurs through the conserved Asn140 in BD1 and Asn433 in BD2, respectively [25]. Consistent with this result, the dual bromodomain inhibitor JQ1 blocks the interaction between BRD4 and acetylated RELA, coincident with the suppression of NFκB-induced transcriptional responses [25]. Treatment with the BET inhibitor JQ1 releases RELA for ubiquitination and proteasome-mediated degradation. RELA transcript levels are not affected. In contrast, the levels of a second subunit of the NFκB signaling complex that binds RELA, p50, are unaltered. In this regard, it is worth noting that MS417, a derivative of the triazolo-thienodiazepine compound class, also prevents BRD4 binding to acetylated NFκB, thus reducing the transcriptional activity of this factor [26].

Previous studies have shown that the RELA/BRD4 interaction is favored by the phosphorylation of RELA/p65 at serines 276 and 536, which allows recruitment of histone acetyl-transferase p300/CBP [27]. Interestingly, enhanced STAT3-dependent RELA phosphorylation has been shown to increase RELA acetylation by p300/CBP and has been implicated in constitutive NFκB activity in human and murine tumors [25,28].

Of note, the mono-methylation of RELA lysine 310 by SETD6 has been identified and shown to couple NFκB activity to that of histone methyltransferase EHMT1/GLP (Euchromatic histone-lysine *N*-methyltransferase 1) at chromatin, thereby leading to the tonic repression of NFκB signaling at target promoters [29]. RELA/p65 lysine 310 mono-methylation is blocked by activation-dependent phosphorylation at Ser311, by protein kinase C-ζ(PKC-ζ). How this impacts subsequent lysine 310 acetylation is not known in detail.

In aggregate, the above data underscore the feasibility and potential clinical interest of targeting BRD4 interactions with non-histone acetylated lysines in anti-cancer therapy. Targeting the BRD4/ acetylated RELA axis should be of value in NFκB-dependent cancers, such as lymphoma, where aberrant NFκB signaling is frequent, including through genetic mechanisms [30].

3. BET Inhibition as an Anti-Cancer Therapy and Perspectives for Use in NFκB-Dependent Cancers

Numerous studies indicate that BET inhibition is an attractive therapeutic strategy in hematological and solid cancers. Single or dual agent therapy is showing promise pre-clinically [1,3,6,31,32]. Inhibitors are mostly small molecule inhibitors that are thought to mediate anti-cancer activity through the interruption of interactions between BET proteins and acetylated lysine in histones at promoters and enhancers. A number of these inhibitors are being tested in early phase clinical trials, in hematological and solid cancers, as either single agents or in combination with other treatments (Table 1; and https://clinicaltrials.gov/ for further details on BET inhibitor trials). Although published clinical data remain sparse, one BET inhibitor, OTX015, has shown efficacy in patients with refractory or relapsed hematological cancers [33,34].

Table 1. Summary of BET inhibitor clinical trials (open Dec. 2017).

Compound	Company	Indications	Phases	Completion Date	NCT Number
FT-1101	Forma Therapeutics, Inc. (Watertown, MA, USA)	Acute Myeloid Leukemia/Acute Myelogenous Leukemia/Myelodysplastic Syndrome	Phase 1	August 2018	NCT02543879
RO6870810 *	Hoffmann-La Roche (Basel, Switzerland)	Multiple Myeloma	Phase 1	15 January 2020	NCT03068351
CPI-0610	Constellation Pharmaceuticals (Cambridge, MA, USA)	Lymphoma	Phase 1	July 2018	NCT01949883
		Multiple Myeloma	Phase 1	March 2019	NCT02157636
		Acute Myeloid Leukemia/Myelodysplastic Syndrome (MDS)/ Myelodysplastic/Myeloproliferative Neoplasm, Unclassifiable/Myelofibrosis	Phase 1	January 2019	NCT02158858
		Peripheral Nerve Tumors	Phase 2	March 2020	NCT02986919
GSK525762	GSK (Brentford, UK)	Cancer	Phase 1	24 February 2020	NCT01943851
		Carcinoma, Midline	Phase 1	9 September 2019	NCT01587703
ZEN003694 **	Zenith Epigenetics (San Francisco, CA, USA)	Metastatic Castration-Resistant Prostate Cancer	Phase 1	April 2018	NCT02711956
BMS-986158	Bristol-Myers Squibb (New York, NY, USA)	Multiple Indications Cancer	Phase 1/Phase 2	17 December 2018	NCT02419417

* Alone or in combination with Daratumumab; ** in combination with Enzalutamide.

In this review, by focusing on BRD4-mediated regulation of NFκB signaling via acetylated RELA, we make a case for a second avenue of investigation for the development of therapeutic strategies utilizing BET inhibitors. Specifically, we propose that BRD4-dependent signaling via non-histone acetyl-lysine interactions will be of interest. Experimental data indicate that small molecule inhibition of the RELA-BRD4 interaction offers promise for the disruption of pathological NFκB signaling in cancer, a process which has been attributed to the inability of cancer cells to homeostatically control NFκB function [35]. Functional and chemical genomics approaches should allow progress in the identification and/or design of additional agents for pre-clinical testing in NFκB-dependent cancers such as lymphoma.

Acknowledgments: Mary B. Callanan acknowledges research support from the l'INCa 'épigénétique et cancer' program (2014–2017) and institutional support from Grenoble and Dijon University hospitals, Université Grenoble-Alpes, Université Bourgogne-Franche Comté, INSERM, and CNRS. Azadeh Hajmirza has been the recipient of doctoral funding from the Société Française d'Hématologie. Arnaud Gauthier was supported by the Auvergne-Rhone-Alpes region, under the program 'Année Recherche' (Masters 2 research training at Bart's Cancer Institute, Jude Fitzgibbon).

Author Contributions: Azadeh Hajmirza, Anouk Emadali, Arnaud Gauthier, Olivier Casasnovas, Rémy Gressin and Mary B. Callanan co-wrote the manuscript and approved the final version.

Conflicts of Interest: The authors declare no conflict of interest.

References

1. Dawson, M.A. The cancer epigenome: Concepts, challenges, and therapeutic opportunities. *Science* **2017**, *355*, 1147–1152. [CrossRef] [PubMed]
2. Basheer, F.; Huntly, B.J. BET bromodomain inhibitors in leukemia. *Exp. Hematol.* **2015**, *43*, 718–731. [CrossRef] [PubMed]
3. Shi, J.; Vakoc, C.R. The mechanisms behind the therapeutic activity of BET bromodomain inhibition. *Mol. Cell* **2014**, *54*, 728–736. [CrossRef] [PubMed]
4. Dhalluin, C.; Carlson, J.E.; Zeng, L.; He, C.; Aggarwal, A.K.; Zhou, M.M. Structure and ligand of a histone acetyltransferase bromodomain. *Nature* **1999**, *399*, 491–496. [PubMed]
5. Moriniere, J.; Rousseaux, S.; Steuerwald, U.; Soler-Lopez, M.; Curtet, S.; Vitte, A.L.; Govin, J.; Gaucher, J.; Sadoul, K.; Hart, D.J.; et al. Cooperative binding of two acetylation marks on a histone tail by a single bromodomain. *Nature* **2009**, *461*, 664–668. [CrossRef] [PubMed]
6. Filippakopoulos, P.; Knapp, S. Targeting bromodomains: Epigenetic readers of lysine acetylation. *Nat. Rev. Drug Discov.* **2014**, *13*, 337–356. [CrossRef] [PubMed]
7. Smale, S.T. Hierarchies of NF-kB target-gene regulation. *Nat. Immunol.* **2011**, *12*, 689–694. [CrossRef] [PubMed]
8. Zhou, Q.; Li, T.; Price, D.H. RNA polymerase II elongation control. *Annu. Rev. Biochem.* **2012**, *81*, 119–143. [CrossRef] [PubMed]
9. Liu, W.; Ma, Q.; Wong, K.; Li, W.; Ohgi, K.; Zhang, J.; Aggarwal, A.; Rosenfeld, M.G. Brd4 and JMJD6-associated anti-pause enhancers in regulation of transcriptional pause release. *Cell* **2013**, *155*, 1581–1595. [CrossRef] [PubMed]
10. Rahman, S.; Sowa, M.E.; Ottinger, M.; Smith, J.A.; Shi, Y.; Harper, J.W.; Howley, P.M. The Brd4 extraterminal domain confers transcription activation independent of pTEFb by recruiting multiple proteins, including NSD3. *Mol. Cell. Biol.* **2011**, *31*, 2641–2652. [CrossRef] [PubMed]
11. Loven, J.; Hoke, H.A.; Lin, C.Y.; Lau, A.; Orlando, D.A.; Vakoc, C.R.; Bradner, J.E.; Lee, T.I.; Young, R.A. Selective inhibition of tumor oncogenes by disruption of super-enhancers. *Cell* **2013**, *153*, 320–334. [CrossRef] [PubMed]
12. Bhagwat, A.S.; Roe, J.S.; Mok, B.Y.L.; Hohmann, A.F.; Shi, J.; Vakoc, C.R. BET Bromodomain Inhibition Releases the Mediator Complex from Select cis-Regulatory Elements. *Cell Rep.* **2016**, *15*, 519–530. [CrossRef] [PubMed]
13. Damm, F.; Mylonas, E.; Cosson, A.; Yoshida, K.; Della Valle, V.; Mouly, E.; Diop, M.; Scourzic, L.; Shiraishi, Y.; Chiba, K.; et al. Acquired initiating mutations in early hematopoietic cells of CLL patients. *Cancer Discov.* **2014**, *4*, 1088–1101. [CrossRef] [PubMed]

14. Pelish, H.E.; Liau, B.B.; Nitulescu, I.I.; Tangpeerachaikul, A.; Poss, Z.C.; Da Silva, D.H.; Caruso, B.T.; Arefolov, A.; Fadeyi, O.; Christie, A.L.; et al. Mediator kinase inhibition further activates super-enhancer-associated genes in AML. *Nature* **2015**, *526*, 273–276. [CrossRef] [PubMed]

15. Chapuy, B.; McKeown, M.R.; Lin, C.Y.; Monti, S.; Roemer, M.G.; Qi, J.; Rahl, P.B.; Sun, H.H.; Yeda, K.T.; Doench, J.G.; et al. Discovery and characterization of super-enhancer-associated dependencies in diffuse large B cell lymphoma. *Cancer Cell* **2013**, *24*, 777–790. [CrossRef] [PubMed]

16. Delmore, J.E.; Issa, G.C.; Lemieux, M.E.; Rahl, P.B.; Shi, J.; Jacobs, H.M.; Kastritis, E.; Gilpatrick, T.; Paranal, R.M.; Qi, J.; et al. BET bromodomain inhibition as a therapeutic strategy to target c-Myc. *Cell* **2011**, *146*, 904–917. [CrossRef] [PubMed]

17. Groschel, S.; Sanders, M.A.; Hoogenboezem, R.; de Wit, E.; Bouwman, B.A.M.; Erpelinck, C.; van der Velden, V.H.J.; Havermans, M.; Avellino, R.; van Lom, K.; et al. A single oncogenic enhancer rearrangement causes concomitant EVI1 and GATA2 deregulation in leukemia. *Cell* **2014**, *157*, 369–381. [CrossRef] [PubMed]

18. Glodzik, D.; Morganella, S.; Davies, H.; Simpson, P.T.; Li, Y.; Zou, X.; Diez-Perez, J.; Staaf, J.; Alexandrov, L.B.; Smid, M.; et al. A somatic-mutational process recurrently duplicates germline susceptibility loci and tissue-specific super-enhancers in breast cancers. *Nat. Genet.* **2017**, *49*, 341–348. [CrossRef] [PubMed]

19. Roe, J.S.; Mercan, F.; Rivera, K.; Pappin, D.J.; Vakoc, C.R. BET Bromodomain Inhibition Suppresses the Function of Hematopoietic Transcription Factors in Acute Myeloid Leukemia. *Mol. Cell* **2015**, *58*, 1028–1039. [CrossRef] [PubMed]

20. Brown, J.D.; Lin, C.Y.; Duan, Q.; Griffin, G.; Federation, A.; Paranal, R.M.; Bair, S.; Newton, G.; Lichtman, A.; Kung, A.; et al. NF-κB directs dynamic super enhancer formation in inflammation and atherogenesis. *Mol. Cell* **2014**, *56*, 219–231. [CrossRef] [PubMed]

21. Huang, B.; Yang, X.D.; Zhou, M.M.; Ozato, K.; Chen, L.F. Brd4 coactivates transcriptional activation of NF-kappaB via specific binding to acetylated RelA. *Mol. Cell. Biol.* **2009**, *29*, 1375–1387. [CrossRef] [PubMed]

22. Kaikkonen, M.U.; Spann, N.J.; Heinz, S.; Romanoski, C.E.; Allison, K.A.; Stender, J.D.; Chun, H.B.; Tough, D.F.; Prinjha, R.K.; Benner, C.; et al. Remodeling of the enhancer landscape during macrophage activation is coupled to enhancer transcription. *Mol. Cell* **2013**, *51*, 310–325. [CrossRef] [PubMed]

23. Hah, N.; Benner, C.; Chong, L.W.; Yu, R.T.; Downes, M.; Evans, R.M. Inflammation-sensitive super enhancers form domains of coordinately regulated enhancer RNAs. *Proc. Natl. Acad. Sci. USA* **2015**, *112*, E297–E302. [CrossRef] [PubMed]

24. Nicodeme, E.; Jeffrey, K.L.; Schaefer, U.; Beinke, S.; Dewell, S.; Chung, C.W.; Chandwani, R.; Marazzi, I.; Wilson, P.; Coste, H.; et al. Suppression of inflammation by a synthetic histone mimic. *Nature.* **2010**, *468*, 1119–1123. [CrossRef] [PubMed]

25. Zou, Z.; Huang, B.; Wu, X.; Zhang, H.; Qi, J.; Bradner, J.; Nair, S.; Chen, L.F. Brd4 maintains constitutively active NF-κB in cancer cells by binding to acetylated RelA. *Oncogene* **2014**, *33*, 2395–2404. [CrossRef] [PubMed]

26. Zhang, G.; Liu, R.; Zhong, Y.; Plotnikov, A.N.; Zhang, W.; Zeng, L.; Rusinova, E.; Gerona-Nevarro, G.; Moshkina, N.; Joshua, J.; et al. Down-regulation of NFκB transcriptional activity in HIV-associated kidney disease by BRD4 inhibition. *J. Biol. Chem.* **2012**, *287*, 28840–28851. [CrossRef] [PubMed]

27. Brasier, A.R.; Tian, B.; Jamaluddin, M.; Kalita, M.K.; Garofalo, R.P.; Lu, M. RelA Ser276 phosphorylation-coupled Lys310 acetylation controls transcriptional elongation of inflammatory cytokines in respiratory syncytial virus infection. *J. Virol.* **2011**, *85*, 11752–11769. [CrossRef] [PubMed]

28. Lee, H.; Herrmann, A.; Deng, J.H.; Kujawski, M.; Niu, G.; Li, Z.; Forman, S.; Jove, R.; Pardoll, D.M.; Yu, H. Persistently activated Stat3 maintains constitutive NF-kappaB activity in tumors. *Cancer Cell* **2009**, *15*, 283–293. [CrossRef] [PubMed]

29. Levy, D.; Kuo, A.J.; Chang, Y.; Schaefer, U.; Kitson, C.; Cheung, P.; Espejo, A.; Zee, B.M.; Liu, C.L.; Tangsombatvisit, S.; et al. Lysine methylation of the NF-kB subunit RelA by SETD6 couples activity of the histone methyltransferase GLP at chromatin to tonic repression of NF-kB signaling. *Nat. Immunol.* **2011**, *12*, 29–36. [CrossRef] [PubMed]

30. Nagel, D.; Vincendeau, M.; Eitelhuber, A.C.; Krappmann, D. Mechanisms and consequences of constitutive NF-κB activation in B-cell lymphoid malignancies. *Oncogene* **2014**, *33*, 5655–5665. [CrossRef] [PubMed]

31. Emadali, A.; Rousseaux, S.; Bruder-Costa, J.; Rome, C.; Duley, S.; Hamaidia, S.; Betton, P.; Debernardi, A.; Leroux, D.; Bernay, B.; et al. Identification of a novel BET bromodomain inhibitor-sensitive, gene regulatory circuit that controls Rituximab response and tumour growth in aggressive lymphoid cancers. *EMBO Mol. Med.* **2013**, *5*, 1180–1195. [CrossRef] [PubMed]

32. Emadali, A.; Hoghoughi, N.; Duley, S.; Hajmirza, A.; Verhoeyen, E.; Cosset, F.L.; Bertrand, P.; Roumier, C.; Roggy, A.; Suchaud-Martin, C.; et al. Haploinsufficiency for NR3C1, the gene encoding the glucocorticoid receptor, in blastic plasmacytoid dendritic cell neoplasms. *Blood* **2016**, *127*, 3040–3053. [CrossRef]

33. Berthon, C.; Raffoux, E.; Thomas, X.; Vey, N.; Gomez-Roca, C.; Yee, K.; Taussig, D.C.; Rezai, K.; Roumier, C.; Herait, P.; et al. Bromodomain inhibitor OTX015 in patients with acute leukaemia: A dose-escalation, phase 1 study. *Lancet Haematol.* **2016**, *3*, e186–e195. [CrossRef]

34. Amorim, S.; Stathis, A.; Gleeson, M.; Iyengar, S.; Magarotto, V.; Leleu, X.; Morschhauser, F.; Karlin, L.; Broussais, F.; Rezai, K.; et al. Bromodomain inhibitor OTX015 in patients with lymphoma or multiple myeloma: A dose-escalation, open-label, pharmacokinetic, phase 1 study. *Lancet Haematol.* **2016**, *3*, e196–e204. [CrossRef]

35. Baltimore, D. NF-κB is 25. *Nat. Immunol.* **2011**, *12*, 683–685. [CrossRef] [PubMed]

biomedicines

MDPI

Review

Roles of NF-κB Signaling in the Regulation of miRNAs Impacting on Inflammation in Cancer

Georgios S. Markopoulos [1,2,†], Eugenia Roupakia [1,2,†], Maria Tokamani [3], Georgia Alabasi [1,2], Raphael Sandaltzopoulos [3], Kenneth B. Marcu [2,4,5,6] and Evangelos Kolettas [1,2,*]

[1] Laboratory of Biology, School of Medicine, Faculty of Health Sciences, University of Ioannina, 45110 Ioannina, Greece; gmarkop@cc.uoi.gr (G.S.M.); ev.roupakia@gmail.com (E.R.); gwgw_al95@hotmail.com (G.A.)

[2] Biomedical Research Division, Institute of Molecular Biology and Biotechnology, Foundation for Research and Technology, 45110 Ioannina, Greece; kenneth.marcu@stonybrook.edu

[3] Department of Molecular Biology and Genetics, Democritus University of Thrace, 68100 Alexandroupolis, Greece; tokamanimaria@hotmail.com (M.T.); rmsandal@mbg.duth.gr (R.S.)

[4] Biomedical Research Foundation of the Academy of Athens, 4 Soranou Ephessiou Street, 115-27 Athens, Greece

[5] Departments of Biochemistry and Cell Biology, Microbiology and Pathology, Stony Brook University, Stony Brook, NY 11794-5215, USA

[6] Department of Biology, San Diego State University, San Diego, CA 92182-4614, USA

* Correspondence: ekoletas@cc.uoi.gr; Tel.: +30-26510-07578

† These authors contributed equally to this work.

Received: 5 March 2018; Accepted: 27 March 2018; Published: 30 March 2018

Abstract: The NF-κB family of transcription factors regulate the expression of genes encoding proteins and microRNAs (miRNA, miR) precursors that may either positively or negatively regulate a variety of biological processes such as cell cycle progression, cell survival, and cell differentiation. The NF-κB-miRNA transcriptional regulatory network has been implicated in the regulation of proinflammatory, immune, and stress-like responses. Gene regulation by miRNAs has emerged as an additional epigenetic mechanism at the post-transcriptional level. The expression of miRNAs can be regulated by specific transcription factors (TFs), including the NF-κB TF family, and vice versa. The interplay between TFs and miRNAs creates positive or negative feedback loops and also regulatory networks, which can control cell fate. In the current review, we discuss the impact of NF-κB-miRNA interplay and feedback loops and networks impacting on inflammation in cancer. We provide several paradigms of specific NF-κB-miRNA networks that can regulate inflammation linked to cancer. For example, the NF-κB-miR-146 and NF-κB-miR-155 networks fine-tune the activity, intensity, and duration of inflammation, while the NF-κB-miR-21 and NF-κB-miR-181b-1 amplifying loops link inflammation to cancer; and p53- or NF-κB-regulated miRNAs interconnect these pathways and may shift the balance to cancer development or tumor suppression. The availability of genomic data may be useful to verify and find novel interactions, and provide a catalogue of 162 miRNAs targeting and 40 miRNAs possibly regulated by NF-κB. We propose that studying active TF-miRNA transcriptional regulatory networks such as NF-κB-miRNA networks in specific cancer types can contribute to our further understanding of the regulatory interplay between inflammation and cancer, and also perhaps lead to the development of pharmacologically novel therapeutic approaches to combat cancer.

Keywords: miRNAs; NF-κB; transcriptional regulatory networks; oncogenic and tumor suppressor pathways; cancer; inflammation

1. Introduction

Carcinogenesis involves the accumulation of mutations in conjunction with epigenetic changes resulting in dominant alterations in gene expression and cellular physiology. NF-κB TFs and their signaling pathways play important roles in cellular growth and viability control and are often subject to deregulation in cancer. Oncogenic driver mutations and inactivating mutations in tumor suppressor genes along with epigenetic changes in normal cells, can lead to the growth of tumor containing cells with distinct phenotypic characteristics, known as the hallmarks of cancer [1]. Cancer cells are also characterized by extensive epigenetic alterations compared to their normal counterparts, as a result of deregulated tissue-specific gene regulatory mechanisms. Elucidating the interaction between genetic and epigenetic factors in cancer onset, development, and progression is considered as a main challenge in both our understanding of cancer biology and for the development of new therapeutic approaches [2,3].

Transcriptional control of gene expression involves binding of TFs to regulatory elements in gene promoters or enhancers. NF-κBs constitute a family of TFs that influence the expression of genes involved in many physiological processes, such as cell proliferation, cell survival, cell adhesion, inflammation, and immunity. The NF-κB signaling components are aberrantly expressed and/or activated in cancer [4–8]. NF-κBs play a central role in the regulation of inflammatory responses at the cellular and systemic levels, and can have tumor promoting effects [9]. However, the NF-κB biology is strikingly complex and NF-κB TFs and their upstream activating signaling components can have either tumorigenic or tumor suppressor roles in cell context-dependent manner and under certain conditions [6,10].

Genes encoding ~22 bp long, small non-coding RNAs, known as microRNAs (or miRNAs or miRs) are emerging as major epigenetic regulators of cell physiology and/or pathology [11–13]. MiRNAs regulate gene expression at the posttranscriptional level by acting as negative regulators of mRNA translation and/or stability resulting in the suppression of translation [13,14], and play an important role in inflammatory and immune responses [15–18] and cancer [11].

In cancer, miRNAs can act as oncogenes, targeting tumor suppressor mRNAs or as tumor suppressors, targeting oncogenic mRNAs. MiRNA genes can also be mutated or epigenetically altered, and suppressed or activated by transcription factors leading to changes in their expression [11]. Importantly, the balance between oncogenic and tumor suppressor miRNAs expressed in a cell, can be a major epigenetic factor that influences cancer onset, development, and progression [11]. Hence, the specific expression of miRNAs and their interplay may tip that balance towards cell proliferation, leading to tumor expansion, or cell cycle arrest, senescence or apoptosis leading to the impairment of tumor growth [11].

The epigenetic mechanisms of TF or miRNA regulation, act at different stages of gene expression, and have some unique features but also share some similarities [19]. TFs can target and regulate the expression of specific miRNAs and, vice versa, miRNAs can target TF mRNAs. This property of TF and miRNA regulation, offers the cells an opportunity to create genomic-scale regulatory networks in which positive or negative feedback loops can act in concert to influence the epigenomic landscape of cells [20]. In the current review, we discuss the specific roles of the regulatory networks between NF-κB TFs and miRNAs and their impact on the conditions of inflammation and cancer development, as well as their interactions.

2. NF-κB Signaling Pathway Activation and Its Multifaceted Functional Role in Cancer and Inflammation

The NF-κB TF family members are critical regulators of pro-inflammatory/stress-like responses. There are three protein subfamilies involved in NF-κB signaling: The NF-κB TF subunit subfamily (c-Rel, p65/RelA, RelB, p105/NF-κB1, and p100/NF-κB2), the regulatory family of NF-κB inhibitors, inhibitors of κB (IκBs), and the catalytic IKK (Inhibitor of NF-κB (IκB) kinase (IKK) complex) subfamily comprising the NF-κB upstream activating Ser/Thr kinases IKKα and IKKβ and a regulatory protein NEMO (NF-κB essential modulator)/IKKγ that together form a high molecular weight IKK signalosome complex, that activates NF-κB signaling in response to specific stimuli. Members of the NF-κB TF subfamily bind to DNA as hetero- or homodimers and can either activate or repress target gene transcription in different physiological contexts. Three of these NF-κB subunits (c-Rel, p65/RelA, and RelB) contain a transactivation domain (TAD), while the other two (p50 and p52) lack a TAD domain and are derived by proteolytic processing of the larger precursor proteins, p105/NF-κB1 and p100/NF-κB2, respectively. Activation of NF-κB signaling occurs by two major pathways: the canonical NF-κB pathway and the noncanonical or alternative NF-κB pathway. In unstimulated cells, heterodimers of p65/p50 subunits, involved in canonical NF-κB signaling, are retained in the cytoplasm by IκBs. Pro-inflammatory and stress stimuli lead to NEMO-dependent activation of IKKβ by phosphorylation of Ser177/181. Activated IKKβ then phosphorylates IκBα at Ser32/36 resulting in its proteasomal degradation and the release of p65/50 heterodimer that translocates to the nucleus where it binds and regulates target gene expression. In contrast, IKKα activation, through phosphorylation of Ser176/180, by adaptive immune response stimuli, is mediated by the NF-κB inducing kinase (NIK). IKKα kinase phosphorylates NF-κB2, inducing its proteasomal processing yielding the mature p52 subunit. Active p52/RelB heterodimers translocate to the nucleus and regulate distinct NF-κB target genes [4,6,7,21,22].

NF-κB target genes encode proteins and miRNAs that regulate a wide range of biological effects that together can be categorized as stress-like, pro-inflammatory reaction programming. The NF-κB signaling pathways have pleiotropic biological effects which may be context dependent. In cancer, NF-κB can exhibit tumor promoting and tumor suppressor activities in a cell context- and tissue-dependent manner [6]. Several mouse cancer models have shown a requirement for canonical NF-κB signaling in tumor onset, development and progression [6,23–37].

The tumor promoting effects of NF-κB are mediated by the activities of NF-κB-regulated genes that promote cancer cell survival, proliferation, metastasis, and angiogenesis, and modify the tumor microenvironment by inducing the secretion of proinflammatory cytokines. NF-κB also promotes a cancer cell metabolic switch from oxidative phosphorylation to glycolysis (Warburg effect) by inducing the expression of glycolytic enzymes while also directly repressing mitochondrial gene expression [38–41]. Thus, NF-κBs function as tumor promoters within transformed cells, but also influence the host's innate immune response to cancer cells by regulating functions of infiltrating lymphocytes and macrophages [22,42]. Although under physiological conditions NF-κB responses are self-limiting via the induction of negative feedback loops, such auto-regulatory loops often become deregulated in cancer cells. However, the regulatory circuitry that leads to dominant IKK/NF-κB-dependent effects in cancer is impressively complex [6,7,9,23].

2.1. Oncogenic Functions of NF-κB: A Link between Inflammation and Cancer

Epidemiological, clinical, genetic, and biochemical evidence obtained from cells, tissues, and mouse models indicate that NF-κB-dependent induction of pro-inflammatory cytokines are pivotal links between chronic inflammation and cancer development and progression [43–48].

Inflammation can either promote tumor growth, or it may be induced as a consequence of the tumor microenvironment leading to cancer progression [44]. Inflammation promotes cancer onset, development, and progression, and it also affects the immune surveillance and chemotherapy resistance of tumors. In addition, inflammation affects the crosstalk between infiltrating immune effector cells and tumor cells thereby linking immunity to tumor development [9,22,24,49–52].

NF-κB TFs have a central role in innate immunity, inflammation, and cancer [6–8,22,42,48–55]. NF-κBs induce inflammation and the secretion of inflammatory mediators enhances canonical NF-κB signaling [9], a feedback mechanism acting as tumor promoter [8,25,26,48,56], and a hallmark of cancer [1]. In chronic inflammation, canonical NF-κB that controls production of inflammatory mediators might prevent the elimination of genetically altered cells present in precancerous lesions by inhibiting their apoptosis [57]. Tumor-associated macrophages (TAMs) were shown to promote tumor growth in part by suppressing immune response to cancer cells but also by producing specific cytokines, most of which are dependent on IKKβ-mediated canonical NF-κB signaling (e.g., IL-6) that enhance tumor cell growth in vivo [9,56,58]. Canonical NF-κB also modifies the tumor microenvironment by inducing the secretion of proinflammatory cytokines such as IL-6, resulting in the activation of its responsive transcription factor STAT3 in *K-Ras*-mutant lung tumors [58]. IL-6 modifies the tumor microenvironment and promotes breast and lung cancer development and progression [58,59]. NF-κB functions in *K-Ras* oncogene transformation by suppressing immune surveillance of both innate and adaptive immune cells [60]. Moreover, canonical NF-κB pathway activation and the interplay with other signaling pathways such as those of STAT3 and p53, may affect tumor onset, development, and progression [44]. One of the critical contributing factors to the oncogenic functions of canonical NF-κB signaling is the induction of inflammation making NF-κB as the critical link between inflammation and cancer [27,28,44–46,48].

While the contribution of canonical NF-κB-activating IKKβ as a tumor promoter in oncogene and carcinogen-induced inflammation and non-small cell lung cancer (NSCLC) has been documented [25,30,31,34] functional studies on noncanonical NF-κB [61–64] and IKKα [65–67] suggest that they can act as tumor promoters or tumor suppressors and are involved in the resolution of inflammation [68–72], but an evolutionary conserved mechanism of action remains largely unknown. These different outcomes of canonical versus non-canonical NF-κB signaling pathways may be related to the preference of NF-κB dimers for binding to *κB* sites contained within the promoters or enhancers of target genes. Sensing the differences within *κB* sites, NF-κB dimers modulate physiological programs by activating, repressing, and altering the expression of effector genes [73–75].

A crosstalk between canonical and noncanonical NF-κB signaling pathways has also been shown. It was shown that NF-κB2 [76,77] and RelB [78] gene expression is induced by canonical NF-κB signaling. RelA/p65 suppresses RelB activity in response to TNFα and induces selective NF-κB target gene expression [79]. It was also shown that TNFα-induced canonical NF-κB signaling upregulates RelB expression that inhibits both basal and non-canonical NF-κB-dependent CXCL12 expression [80]. NIK which activates noncanonical NF-κB signaling may also contribute to the activation of canonical NF-κB [81]. While IKKα activates noncanonical NF-κB signaling, evidence show that it also inhibits the canonical NF-κB pathway [82–84]. It was also shown that nuclear IKKα is required for p65 DNA binding in a gene-specific manner [85].

NF-κB TFs are often deregulated and constitutively activated in many different types of cancer [4,6,53], leading to the development of different hallmarks of cancer [1]. NF-κB's function as a tumor promoter is also due to its role in driving cell proliferation and protecting cells from cell death under stress conditions by regulating the expression and activity of target genes involved in cell cycle progression and apoptosis [5–7,9,49,86–88]. Canonical NF-κB was shown to activate genes involved in cell cycle progression such as CcnD1 [5,86,89,90], E2F1 [5,86], and several E2F target genes [5] and the mitotic checkpoint Ser/Thr-protein kinase BUB1 [34]. It was also shown to suppress genes involved in apoptosis such as FOXO3a, leading to increased cell survival [4,21,91,92]. In keeping with this, miR-155, a canonical NF-κB regulated miRNA, was identified as a negative regulator of FOXO3a leading to increased gefitinib resistance and lung cancer stemness in vitro and in vivo [92]. NF-κB also suppresses the expression of c-Jun N-terminal kinase (JNK) via Gadd45β and blocks apoptosis [93,94]. Canonical NF-κB also contributes to chemoresistance of tumor cells such as leukemic cells, in part through its ability to induce p21$^{waf1/cip1}$ [95,96] and p27^{Kip1} [97].

NF-κB targets that play an important role in cancer progression are those involved in epithelial-to-mesenchymal cell transition (EMT), such as Snail, Twist, matrix metalloproteinases (MMPs) and cell adhesion molecules that promote metastasis, and pro-angiogenic genes such as Vascular Endothelial Growth Factor (VEGF), stimulating tumour neovascularization [8,48,98–100]. Canonical NF-κB also regulates the expression of matrix metalloproteinases involved in tissue remodeling, inflammatory diseases and cancer [101–105]. In addition, Timp1 (tissue inhibitor of metalloproteinase 1), was identified as a NF-κB target gene that contributes to mouse lung tumor growth [34], and it is highly expressed, and correlates with NF-κB activation in advanced lung-cancer patients with poor prognosis [106,107]. NF-κB is also a critical transcriptional regulator of HIF1α, and IKKβ-mediated canonical NF-κB activation is required for the hypoxia-induced accumulation of HIF1α and the expression of HIF1α target genes [108,109]. Several lines of evidence suggest a bi-directional crosstalk between NF-κB and HIF pathways, with the latter also contributing to inflammatory responses and cancer [109–113].

In physiological conditions, NF-κB activity is tightly regulated and inhibited after a short period of time through negative feedback loops [4]. Based on this concept, aberrant NF-κB signaling activation leading to chronic inflammation and increased cell proliferation and survival are additional factors contributing to the oncogenic function of NF-κB [6,23,47,48].

2.2. Tumor Suppressor Function of NF-κB

NF-κBs can also suppress tumor growth under certain conditions, a functional role dependent on the presence and crosstalk with tumor-suppressor-proteins, such as p53, which modulate NF-κB activity in cancer. These tumor suppressive functions of NF-κB are due to NF-κB-dependent activation of gene expression that can lead to the inhibition of cancer cell cycle progression and proliferation, apoptosis, suppression of cell invasion, and metastasis [6,55,114,115].

The tumor suppressive functions of canonical NF-κB may be due to the modulation of NF-κB activity by tumor suppressors such as p53 [6,61,116–118] or due to alterations in the phosphorylation status of NF-κB subunits [6,119–121] suppressing NF-κB's ability to induce the expression of genes that are associated with tumor growth and survival. Canonical NF-κB can also inhibit tumor growth by inducing the expression of tumor suppressors such as Bach2 induced in B-cells by c-Rel or RelA suggesting a tumor suppressive function of c-Rel in B-cell lymphoma [114]. *c-Myc* overexpression was shown to sensitize cells to NF-κB-induced apoptosis, and persistent inactivity of NF-κB signaling was shown to be a prerequisite for *c-myc*-mediated lymphomagenesis [122].

The tumor suppressive functions of canonical NF-κB may also be attributed to an attenuated inflammatory response.NF-κB p50 subunit functions as a transcriptional regulator either as a heterodimer with NF-κB subunits RelA, c-Rel, and RelB, or as a p50 homodimer. p50 heterodimers induce gene expression and are critical in inflammatory responses, while p50 homodimers generally act as transcriptional repressors [7,55,123]. The p50homodimer has an important function as suppressor of inflammation through repressing proinflammatory gene expression while enhancing the expression of anti-inflammatory genes [55,124,125]. *Nfkb1*(p105/p50)$^{-/-}$ mice display increased inflammation and susceptibility to DNA damaging agents, leading to cancer including lymphomas and liver cancer, and an ageing phenotype [55,126,127]. Reduced levels of p50 were observed in human tumor tissues from head and neck and glioblastoma cancers; and these results were further supported by xenograft models of human glioblastoma and breast cancer cell lines in mice [128].

The tumor suppressive functions of noncanonical NF-κB may be attributed to a reduced inflammatory response and oxidative stress [29,52,65,67,70,129,130]. For example, enforced expression of a kinase-dead IKKα mutant protein in mice led to spontaneous lung squamous cell cancer (SCC) development and the recruitment of TAMs, suggesting a tumor suppressor role for IKKα in lung SCC [65,130]. IKKα loss has also been reported to promote *K-Ras*-initiated NSCLC development through a redox regulatory pathway involving ROS accumulation [67].

Emerging evidence suggests that the tumor promoting or suppressive functions of NF-κB, in a cell- and tissue-dependent context may also be determined by miRNAs and their targets. Thus the IKK/NF-κB-miRNA transcriptional regulatory network may play a critical role in inflammation impacting on cancer [11].

3. MiRNAs: Epigenetic Regulators in Inflammation and Cancer

MiRNAs regulate gene expression at the post-transcriptional level acting as negative regulators of mRNA translation and/or stability by binding to complementary sequences in the 3′ untranslated region (3′ UTR) of their target mRNAs. Individual miRNAs may target several different mRNAs to inhibit their translation into polypeptides, partly because target sites on an mRNA require only partial base complementarity with their corresponding miRNAs. In cases of perfect complementarity, cleavage of the target mRNA is induced. Moreover, individual mRNAs may contain multiple binding sites for different miRNAs, resulting in complex regulatory networks. Conversely, binding sites for a specific miRNA may be limited to few mRNAs, while others may target a larger number of mRNAs. Hence, some miRNAs may regulate specific individual targets, while others can positively or negatively regulate a variety of cellular processes [11,131]. For example, the balance between oncogenic miRNAs (that target tumor suppressor genes) and tumor suppressive miRNAs (that target oncogenes) may influence tumor development. Sometimes, miRNAs act in concert with transcription factors, creating TF-miRNA transcriptional regulatory networks, such as the p53-miRNA and the NF-κB-miRNA networks that may also interconnect and influence each other [11].

4. General Concept: NF-κB Meets miRNAs

NF-κB TFs influence the expression of miRNAs, and importantly NF-κB signaling is also affected by miRNAs which target either the upstream NF-κB activating kinases or other NF-κB signaling components, in positive or negative feedback loops in several different cell types and under different conditions [6,16,52].

4.1. MiRNAs Regulated by NF-κB

Several miRNAs, including miR-9, miR-21, miR-30b, miR-143/miR-145, miR-146a, miR-155, miR-221/222, miR-224, miR-301a, and the miR-17-92 cluster have been validated as targets of the NF-κB transcription factors [11,16,52].

Most of these NF-κB-targeted miRNAs have been identified by low throughput methods or unbiased screens. Importantly, the availability of whole-genome data such as transcription factor binding sites based on Chip-Seq experiments, or whole-genome histone modification profiles and also RNA-Seq analyses makes it possible to objectively analyze and efficiently find transcription factors that regulate gene expression. By employing a bioinformatics tool that is used to characterize promoter regions of miRNAs (DIANA miRGen v3.0) [132], we additionally identified 40 miRNAs that contain experimentally verified NF-κB binding sites in their promoter regions (Table 1). Most of these miRNAs are novel potential targets and need further verification. Nevertheless, these data provide an additional, unbiased approach to verify known targets, and also to screen for possible novel targets of specific transcription factors under certain conditions.

Oncogenic miR-21 is an established NF-κB target [42]. NF-κB-dependent induction of miR-21 expression has been detected under different conditions, such as inflammation [16] or DNA damage responses [133] and can target multiple genes, such as *BCL2*, *MASPIN*, *PDCD4*, and *PTEN* [11]. For example, in breast cancer the NF-κB-dependent induction of miR-21 confers chemoresistance and induces cell invasion by repressing *PDCD4* expression which regulates apoptosis, and *PTEN* phosphatase, an inhibitor of Akt pathway that leads to cell survival [133].

In tumor-associated inflammation, the pro-inflammatory cytokine IL-1 leads to NF-κB activation and subsequent upregulation of miR-425 in gastric cancer cells. MiR-425 in turn acts as a tumor promoter by targeting *PTEN* to enhance cell survival [134].

In addition to oncogenic miRNAs, NF-κB can also upregulate tumor suppressive miRNAs, such as miR-143 and miR-145. The expression of these two miRNAs can lead to inhibition of cancer cell proliferation, and also metastasis and invasion by targeting oncogenes such as *MYC*, *ERK5*, and *KRAS*. Non-tumorigenic prostate cells secrete miR-143 to inhibit the growth exclusively of prostate cancer cells that bear activated oncogenes some of which have been mentioned above [11,18].

Table 1. MiRNAs containing experimentally verified NF-κB binding sites in their promoter (miRGen v3.0 tool).

miRNA Name	Chromosomal Location of Promoter (hg19)	Strand
hsa-let-7a-1	chr9:96929483–96929484	[+]
hsa-let-7d	chr9:96929483–96929484	[+]
hsa-let-7f-1	chr9:96929483–96929484	[+]
hsa-let-7i	chr12:62997400–62997401	[+]
hsa-mir-101-1	chr1:65532138–65532139	[−]
hsa-mir-1204	chr8:128806768–128806769	[+]
hsa-mir-1205	chr8:128806768–128806769	[+]
hsa-mir-1206	chr8:128806768–128806769	[+]
hsa-mir-1207	chr8:128806759–128806760	[+]
hsa-mir-1208	chr8:128806759–128806760	[+]
hsa-mir-124-1	chr8:9763203–9763204	[−]
hsa-mir-125b-1	chr11:121971206–121971207	[−]
hsa-mir-1289-1	chr20:34042503–34042504	[−]
hsa-mir-135b	chr1:205426509–205426510	[−]
hsa-mir-137	chr1:98520169–98520170	[−]
hsa-mir-146a	chr5:159894835–159894836	[+]
hsa-mir-148a	chr7:25990290–25990291	[−]
hsa-mir-193a	chr17:29886484–29886485	[+]
hsa-mir-22	chr17:1618561–1618562	[−]
hsa-mir-223	chrX:65219544–65219545	[+]
hsa-mir-23a	chr19:13953455–13953456	[−]
hsa-mir-24-2	chr19:13953455–13953456	[−]
hsa-mir-2682	chr1:98520169–98520170	[−]
hsa-mir-27a	chr19:13953455–13953456	[−]
hsa-mir-2861	chr9:130548069–130548070	[+]
hsa-mir-29a	chr7:130794752–130794753	[−]
hsa-mir-29b-1	chr7:130794752–130794753	[−]
hsa-mir-30a	chr6:72130555–72130556	[−]
hsa-mir-30c-2	chr6:72130555–72130556	[−]
hsa-mir-3142	chr5:159894835–159894836	[+]
hsa-mir-3199-2	chr22:28315414–28315415	[+]
hsa-mir-365b	chr17:29886484–29886485	[+]
hsa-mir-3667	chr22:50051180–50051181	[−]
hsa-mir-3672	chrX:120325891–120325892	[+]
hsa-mir-3679	chr2:134877461–134877462	[+]
hsa-mir-3960	chr9:130548069–130548070	[+]
hsa-mir-4725	chr17:29886484–29886485	[+]
hsa-mir-505	chrX:139015225–139015226	[−]
hsa-mir-5194	chr8:131028942–131028943	[−]
hsa-mir-612	chr11:65190256–65190257	[+]

NF-κB-miR-140 is another regulatory loop. MiR-140 acts as a liver tumor suppressor by negatively regulating NF-κB activity by directly targeting DNA methyltransferase 1 (Dnmt1) expression. In this cellular context, NF-κB suppresses miR-140 expression, resulting in the upregulation of *DNMT1* and increased NF-κB activity, forming a positive feedback loop that promotes liver cancer [135,136]. Aberrant miRNAs have been detected during inflammation and hepatocellular cancer (HCC). Many of these dysregulated miRNAs modulate the initiation and progression of inflammation-induced HCC, the majority of which are NF-κB-regulated miRNAs [137].

Finally, an interesting example of NF-κB-regulated miRNAs is that of miR-221/222, a miRNA family with a dual functional role, acting, in different cellular contexts, either as oncomiRs promoting cancer progression, or as tumor suppressors, promoting cellular senescence [11,12,138–140].

4.2. NF-κB-Regulating miRNAs

Multiple miRNAs have been shown to alter NF-κB activity. The current version of Tarbase v8 (http://carolina.imis.athena-innovation.gr/diana_tools/web/index.php?r=tarbasev8%2Findex, access date 20 January 2018), a database comprised of experimentally validated miRNA-gene interactions [141], contains a total of 163 miRNAs that target at least one of the main gene components of NF-κB signaling, either the NF-κB transcription subunits or the upstream NF-κB activating serine/threonine kinases, IKKα and IKKβ (Figure 1 and Supplementary Table S1).

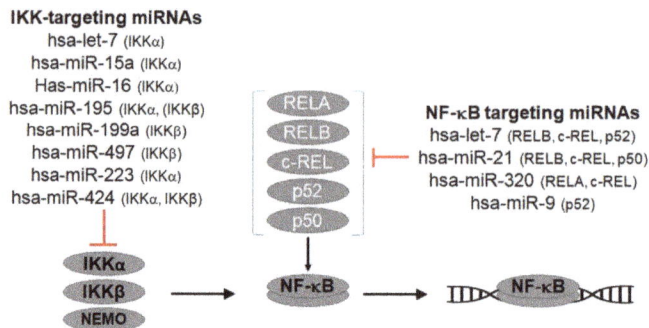

Figure 1. IKK and NF-κB targeting miRNAs. MiRNAs targeting at least one of the NF-κB signaling components such as one of the upstream NF-κB activating kinases, IKKα or IKKβ, or one of the NF-κB transcription factor subunits. For a comprehensive list of NF-κB targeting miRNAs, see Supplementary Table S1.

MiR-506 was shown to directly target and downregulate the expression of the NF-κB p65 subunit, leading to the generation of reactive oxygen species (ROS) and the induction of p53-dependent lung cancer cell apoptosis. Interestingly, the p53-dependent induction of miR-506, suggested that miR-506 in lung cancer cells is part of a regulatory network linking p53 and NF-κB signaling [142]. In prostate cancer, the tumor suppressive miR-497 regulates NF-κB signaling by targeting IKKβ, which activates canonical NF-κB signaling leading to inhibition of prostate cancer cell proliferation, migration, and invasion. Importantly, miR-497 expression is reduced in prostate cancer cells, leading to a more aggressive tumor phenotype [143].

The miR-520/373 family has also been shown to act as tumor suppressors in breast cancer, by targeting the RELA/p65 NF-κB subunit. The miR-520/373 family was identified in a genome-wide screen of miRNAs impacting on NF-κB signaling, using a luciferase-based reporter assay in HEK293T cells [144]. This screen identified 13 families of miRNAs, out of which let-7 and miR-181 are known to participate in NF-κB feedback loops [145,146] (discussed in the next section). MiR-520/373 was further analysed and was shown to inhibit NF-κB in estrogen-negative breast cancer cells, which further resulted in downregulation of NF-κB targets such as the pro-inflammatory cytokines IL-6, IL-8, CXCL1, and ICAM-1, leading to the inhibition of tumor-related inflammation, and suppression of tumor growth and metastasis [144]. In another functional screen for miRNAs regulating NF-κB, using a NF-κB reporter cell-line, miR-517a/c were found as potent activators of NF-κB signaling, upregulating the expression of the reporter more than 40-fold. In this case, the identified target of miR-517a/c leading to activation of NF-κB was TNIP1, an inhibitor of NF-κB signaling [147].

MiRNA-126a was shown to target the NF-κB inhibitor, IκBα, leading to canonical NF-κB activation thereby contributing to pathogenesis of ulcerative colitis [148], but paradoxically was shown to act as tumor suppressor for colon cancer [149].

MiRNA-223 was shown to suppress canonical NF-κB signaling in basal keratinocytes to dampen neutrophilic inflammation [150]. MiR-223 limits inflammation and prevents DNA damage and hematological and non-hematological malignancies [151]. MiR-223 is one of the most abundant miRNAs in macrophages and responds to stimuli to control the production of IL-6 and IL-1β [152]. MiR-223 was also associated with macrophage differentiation through targeting IKKα [153]. However, the role of miR-223 in cancer is cell-context dependent [150]. For example, miR-223 promotes the migration and invasion of gastric cancer cells, but has opposite effects in esophageal cancer cells and human cervical cancer [154–156].

Several miRNA sites were identified in IKKα including sites for let-7, miR-223, miR-16, and miRNA-142-5p and two target sites for miR15a, one of which overlapped the putative miR-16 site. Further experiments showed that miR15a, miR-16 and miR-223, which target IKKα and are downregulated during macrophage differentiation, they were responsible at least in part for the increase in IKKα protein expression observed during macrophage differentiation [153]. Regulation of IKKα by these miRNAs may contribute to cancer development [157].

MiR-199a negatively regulates the expression of IKKβ in ovarian cancer cells, and inhibits the secretion of pro-inflammatory cytokines, thereby causing suppression of tumor progression and chemoresistance [158]. IKKβ is also targeted by miR-497 in prostate cancer cells and inhibits their cell proliferation, migration, and invasion in vitro [143].

5. NF-κB-miRNA Feedback Loops and Transcriptional Regulatory Networks

Multiple feedback loops operating in a specific cell type can act in concert, creating functional networks that control cell fate. There are several NF-κB-miRNA feedback loops in the context of inflammation in normal cells and also during cancer development. These NF-κB-miRNA transcriptional regulatory loops may act in both physiological and pathological conditions, linking pro-inflammatory responses to oncogenic signals [11].

NF-κB signaling during inflammation is self-limiting. A novel feedback loop that has been identified recently involves miR-146a and miR-155, the combinatory action of which controls NF-κB activity during inflammation [18]. Their action is based on a two-step mechanism. First, miR-155 is rapidly upregulated by NF-κB only within the first 12 h of inflammatory response and, by targeting SHIP1, it activates the IKK signalosome complex in a PI3K/Akt-dependent manner, forming a positive feedback loop necessary for signal amplification. Secondly, miR-146a is rather gradually upregulated by NF-κB and forms a negative feedback loop by targeting IRAK1 and TRAF6, ultimately attenuating NF-κB activity in the late phase of inflammation. The combined action of these two positive (NF-κB-miR-155) and negative (NF-κB-miR-146a) NF-κB-miRNA regulatory loops provides optimal NF-κB activity during inflammatory stimuli, and eventually lead to the resolution of the inflammatory response [18].

Knockout of miR-146a in C57BL/6 mice leads to myeloid sarcomas and some lymphomas, and the animals exhibit chronic myeloproliferation in their bone marrow. The development of myeloid malignancies correlated with increased canonical NF-κB activity. Genetic ablation of NF-κB p50 suppressed myeloproliferation suggesting that NF-κB was required for myeloproliferative disease [159].

MiR-9 is induced by pro-inflammatory signals in a NF-κB-dependent manner in human monocytes [160]. MiR-9 targets the *NFKB1* gene, which encodes the p105/p50 precursor subunit and renders lung cancer cells sensitive to ionizing radiation [160]. In ovarian cancer, miR-9 also targets *NFKB1* and its downregulation in this cancer type, as compared to normal ovarian tissue is considered an additional tumor-promoting mechanism [161]. The fact that miR-9 is positively regulated by inflammation-induced canonical NF-κB (RelA/65-p50) signaling, taken together with the finding that miR-9 targets *NFKB1* (p105/p50), suggests a negative feedback loop mechanism fine tuning the inflammatory response with an impact in cancer.

Another negative feedback-loop in acute myeloid leukemia (AML), bearing *KIT* driver mutations, involves miR-29b and NF-κB. MiR-29b targets the Sp1 transcription factor. In *KIT*-driven AML, *KIT* upregulates Sp1, which in turn binds NF-κB and transactivates *KIT*. *Sp1* escapes from miR-29b downregulation through a negative feedback loop, in which Sp1-induced NF-κB recruits HDACs in the miR-29b promoter leading to its transcriptional repression [162].

A positive feedback loop that keeps NF-κB in an activated state operates in breast cancer cells after chemotherapy. In these cells, chemotherapy activates NF-κB which targets and downregulates miR-448 by binding to its promoter, leading to increased expression of the miR-448 target special AT-rich sequence-binding protein-1 (SATB1). SATB1 upregulation ultimately leads to Twist1 expression, a regulator of EMT; and it also further enhances NF-κB activity, forming a positive feedback loop that simultaneously promotes EMT [163].

One of the most well-defined regulatory networks that link inflammation and cancer has been extensively studied by Iliopoulos et al. and is formed by two distinct and complimentary feedback-loops involving either NF-κB, Lin28, let-7 miRNA and IL-6 or IL-6, miR-21, and miR-181b-1 miRNAs, *PTEN*, CYLD, and NF-κB [145,146]. During oncogenesis, proinflammatory signals that are mediated by NF-κB, upregulate Lin28, which downregulates the tumor suppressor let-7 miRNA [164] which targets IL-6. Let-7 downregulation results in increased IL-6 levels, further activating NF-κB, generating a feedback loop that sustains inflammation and promotes oncogenesis [145]. NF-κB can also remain active by a complimentary feedback-loop that involves miR-21 and miR-181b-1. IL-6 activates STAT3, an inducer of miR-21 and miR-181b-1 expression, which respectively target *PTEN* and *CYLD*. PTEN and CYLD inhibition further leads to NF-κB activation [146]. Therefore, the combined action of NF-κB and STAT3 leading to the induction of miR-21 and miR-181b-1 and let-7 downregulation, ultimately act as a feedback mechanism linking inflammation to cancer. In addition to NF-κB, STAT3 can also be further upregulated as a result of this feedback mechanism, since miR-181a/b induction by STAT3 can also activate the IL-6/STAT3 signaling pathway [146]. More recently, studies on the interplay between NF-κB and STAT3, two of the main transcription factors that regulate inflammation [44–46] have revealed that feedback mechanisms that involve these two factors also include several miRNAs [165]. Studies revealed the existence of a negative feedback loop mechanism between STAT3 and NF-κB involving miR-146b. In this mechanism, STAT3 targets miR-146b, which downregulates NF-κB, reducing IL-6 production. The reduction of IL-6 is the final step of a negative feedback loop, since IL-6 activates STAT3, contributing to chronic inflammation. This is also a mechanism linking inflammation and cancer in breast tissue, whereas in normal tissue miR-146b is upregulated, leading to resolution of inflammation, in breast cancer it is downregulated, leading to chronic inflammation, through deregulation of the above feedback loop and cancer development [166].

A constitutively activated feedforward circuit composed of IκBα/NF-κB(p65) and miR-196b-3p, was shown to drive castration-resistant prostate cancer (CRPC) development. Constitutive activation of IκBα/RelA(p65) in this circuit was independent of the activation of the canonical IKKβ/NF-κB pathway [167].

The availability of genomic data makes it possible to improve our knowledge of novel regulatory networks that exist in physiological or pathological conditions. Using bioinformatics tools and analysis we were able to identify candidate miRNAs regulated by NF-κB (Table 1) or targeting NF-κB pathway components (Figure 1 and Supplementary Table S1). Another such tool that offers a pathway-based approach, is the server of Diana miRpath for finding specific miRNAs involved in pathways or regulatory networks [168]. We believe that the exploitation of unbiased genomic data in conjunction with experimental validation may confirm biologically relevant findings and relate them to specific functions and (physiological or pathological) conditions.

6. Final Thoughts: Possible Therapeutic Approaches

In the current review, we focused on the interplay between NF-κB and miRNAs impacting on inflammation and cancer development. The functional role of miRNAs in these processes is due to their action as epigenetic switches that interconnect signaling pathways and cellular processes, integrating in larger regulatory networks. In this conceptual framework, the expression of miRNAs may offer the possibility to: (a) fine-tune the activity of a process in time, such as the expression of miR-155 and miR-146b regulating NF-κB expression and inflammation intensity and duration [18]; (b) amplify or attenuate the activity of a signaling pathway, by taking part in feedback-loops, such as the NF-κB-miRNA amplifying loops in inflammation linked to cancer [145,146]; and (c) interconnect TF-miRNA opposing regulatory pathways such as the p53-miRNA and NF-κB-miRNA networks. Certain NF-κB-regulated miRNAs can regulate p53, and vice versa, hence they can shift the balance towards apoptosis or cell survival and determine the fate of a cancer cell [11]. The complexity of epigenetic regulation requires taking into account aspects such as the expression of specific TFs and miRNAs and their possible interconnection.

Based on the dynamic nature of NF-κB signaling combined with the diverse actions and multiple targets of miRNAs, we believe that the NF-κB-miRNA feedback regulatory loop mechanisms discussed above or possibly novel ones yet to be discovered, should be considered when studying inflammatory responses linked to cancer initiation, progression, and development. Understanding of the NF-κB-miRNA transcription factor regulatory networks may offer opportunities for pharmacological exploitation and personalized treatments.

Supplementary Materials: The following are available online at http://www.mdpi.com/2227-9059/6/2/40/s1, Table S1: Validated miRNAs for targeting at least one of NF-κB signaling components. (# Denotes the number of NF-κB genes targeted).

Acknowledgments: We greatly acknowledge the financial support of the Fondation Santé, Stavros Niarchos Foundation (Archers, ref#SNF0031), and by the project "Advanced research activities in biomedical and agroalimentary technologies' which is implemented under the "Action for the Strategic Development on the Research and Technological Sector", funded by the Operational Program "Competitiveness, Entrepreneurship and Innovation" (NSRF 2014-2020) and co-financed by Greece and the European Union (European Regional Development Fund).

Conflicts of Interest: The authors declare no conflict of interest.

References

1. Hanahan, D.; Weinberg, R.A. Hallmarks of cancer: The next generation. *Cell* **2011**, *144*, 646–674. [CrossRef] [PubMed]
2. Dawson, M.A.; Kouzarides, T. Cancer epigenetics: From mechanism to therapy. *Cell* **2012**, *150*, 12–27. [CrossRef] [PubMed]
3. Rodríguez-Paredes, M.; Esteller, M. Cancer epigenetics reaches mainstream oncology. *Nat. Med.* **2011**, *17*, 330. [CrossRef] [PubMed]
4. Perkins, N.D. Integrating cell-signalling pathways with NF-κB and IKK function. *Nat. Rev. Mol. Cell Biol.* **2007**, *8*, 49–62. [CrossRef] [PubMed]
5. Penzo, M.; Massa, P.E.; Olivotto, E.; Bianchi, F.; Borzi, R.M.; Hanidu, A.; Li, X.; Li, J.; Marcu, K.B. Sustained NF-κB activation produces a short-term cell proliferation block in conjunction with repressing effectors of cell cycle progression controlled by E2F or FoxM1. *J. Cell. Physiol.* **2009**, *218*, 215–227. [CrossRef] [PubMed]
6. Perkins, N.D. The diverse and complex roles of NF-κB subunits in cancer. *Nat. Rev. Cancer* **2012**, *12*, 121–132. [CrossRef] [PubMed]
7. Hayden, M.S.; Ghosh, S. NF-κB, the first quarter-century: Remarkable progress and outstanding questions. *Genes Dev.* **2012**, *26*, 203–234. [CrossRef] [PubMed]
8. Chen, W.; Li, Z.; Bai, L.; Lin, Y. NF-κB in lung cancer, a carcinogenesis mediator and a prevention and therapy target. *Front. Biosci. (Landmark Ed.)* **2011**, *16*, 1172–1185. [CrossRef] [PubMed]
9. Taniguchi, K.; Karin, M. Nf-κB, inflammation, immunity and cancer: Coming of age. *Nat. Rev. Immunol.* **2018**. [CrossRef] [PubMed]
10. Perkins, N.D. Nf-κB: Tumor promoter or suppressor? *Trends Cell Biol.* **2004**, *14*, 64–69. [CrossRef] [PubMed]
11. Markopoulos, G.S.; Roupakia, E.; Tokamani, M.; Chavdoula, E.; Hatziapostolou, M.; Polytarchou, C.; Marcu, K.B.; Papavassiliou, A.G.; Sandaltzopoulos, R.; Kolettas, E. A step-by-step microRNA guide to cancer development and metastasis. *Cell. Oncol.* **2017**, *40*, 303–339. [CrossRef] [PubMed]
12. Markopoulos, G.S.; Roupakia, E.; Tokamani, M.; Vartholomatos, G.; Tzavaras, T.; Hatziapostolou, M.; Fackelmayer, F.O.; Sandaltzopoulos, R.; Polytarchou, C.; Kolettas, E. Senescence-associated microRNAs target cell cycle regulatory genes in normal human lung fibroblasts. *Exp. Gerontol.* **2017**, *96*, 110–122. [CrossRef] [PubMed]

13. Bartel, D.P. Metazoan microRNAs. *Cell* **2018**, *173*, 20–51. [CrossRef] [PubMed]
14. Bartel, D.P. MicroRNAs: Genomics, biogenesis, mechanism, and function. *Cell* **2004**, *116*, 281–297. [CrossRef]
15. Taganov, K.D.; Boldin, M.P.; Baltimore, D. MicroRNAs and immunity: Tiny players in a big field. *Immunity* **2007**, *26*, 133–137. [CrossRef] [PubMed]
16. Boldin, M.P.; Baltimore, D. MicroRNAs, new effectors and regulators of NF-κB. *Immunol. Rev.* **2012**, *246*, 205–220. [CrossRef] [PubMed]
17. Mehta, A.; Baltimore, D. MicroRNAs as regulatory elements in immune system logic. *Nat. Rev. Immunol.* **2016**, *16*, 279. [CrossRef] [PubMed]
18. Mann, M.; Mehta, A.; Zhao, J.L.; Lee, K.; Marinov, G.K.; Garcia-Flores, Y.; Baltimore, D. An NF-κB-microRNA regulatory network tunes macrophage inflammatory responses. *Nat. Commun.* **2017**, *8*, 851. [CrossRef] [PubMed]
19. Hobert, O. Gene regulation by transcription factors and microRNAs. *Science* **2008**, *319*, 1785–1786. [CrossRef] [PubMed]
20. Martinez, N.J.; Walhout, A.J. The interplay between transcription factors and microRNAs in genome-scale regulatory networks. *Bioessays* **2009**, *31*, 435–445. [CrossRef] [PubMed]
21. Chariot, A. The NF-κB-independent functions of IKK subunits in immunity and cancer. *Trends Cell Biol.* **2009**, *19*, 404–413. [CrossRef] [PubMed]
22. Karin, M.; Greten, F.R. NF-κB: Linking inflammation and immunity to cancer development and progression. *Nat. Rev. Immunol.* **2005**, *5*, 749–759. [CrossRef] [PubMed]
23. Bradford, J.W.; Baldwin, A.S. Chapter three-IKK/nuclear factor-κB and oncogenesis: Roles in tumor-initiating cells and in the tumor microenvironment. In *Advances in Cancer Research*; Tew, K.D., Fisher, P.B., Eds.; Academic Press: Cambridge, MA, USA, 2014; Volume 121, pp. 125–145.
24. Grivennikov, S.I.; Greten, F.R.; Karin, M. Immunity, inflammation, and cancer. *Cell* **2010**, *140*, 883–899. [CrossRef] [PubMed]
25. Takahashi, H.; Ogata, H.; Nishigaki, R.; Broide, D.H.; Karin, M. Tobacco smoke promotes lung tumorigenesis by triggering IKKβ-and JNK1-dependent inflammation. *Cancer Cell* **2010**, *17*, 89–97. [CrossRef] [PubMed]
26. Zaynagetdinov, R.; Stathopoulos, G.T.; Sherrill, T.P.; Cheng, D.-S.; McLoed, A.G.; Ausborn, J.A.; Polosukhin, V.V.; Connelly, L.; Zhou, W.; Fingleton, B. Epithelial nuclear factor-κB signaling promotes lung carcinogenesis via recruitment of regulatory T lymphocytes. *Oncogene* **2012**, *31*, 3164. [CrossRef] [PubMed]
27. Greten, F.R.; Eckmann, L.; Greten, T.F.; Park, J.M.; Li, Z.-W.; Egan, L.J.; Kagnoff, M.F.; Karin, M. IKKβ links inflammation and tumorigenesis in a mouse model of colitis-associated cancer. *Cell* **2004**, *118*, 285–296. [CrossRef] [PubMed]
28. Karin, M. NF-κB as a critical link between inflammation and cancer. *Cold Spring Harb. Perspect. Biol.* **2009**, *1*, a000141. [CrossRef] [PubMed]
29. Xia, Y.; Shen, S.; Verma, I.M. Nf-κB, an active player in human cancers. *Cancer Immunol. Res.* **2014**, *2*, 823–830. [CrossRef] [PubMed]
30. Stathopoulos, G.T.; Sherrill, T.P.; Cheng, D.S.; Scoggins, R.M.; Han, W.; Polosukhin, V.V.; Connelly, L.; Yull, F.E.; Fingleton, B.; Blackwell, T.S. Epithelial NF-κB activation promotes urethane-induced lung carcinogenesis. *Proc. Natl. Acad. Sci. USA* **2007**, *104*, 18514–18519. [CrossRef] [PubMed]
31. Basseres, D.S.; Ebbs, A.; Levantini, E.; Baldwin, A.S. Requirement of the NF-κB subunit p65/RelA for K-Ras-induced lung tumorigenesis. *Cancer Res.* **2010**, *70*, 3537–3546. [CrossRef] [PubMed]
32. Ben-Neriah, Y.; Karin, M. Inflammation meets cancer, with NF-κB as the matchmaker. *Nat. Immunol.* **2011**, *12*, 715–723. [CrossRef] [PubMed]
33. Vlantis, K.; Wullaert, A.; Sasaki, Y.; Schmidt-Supprian, M.; Rajewsky, K.; Roskams, T.; Pasparakis, M. Constitutive IKK2 activation in intestinal epithelial cells induces intestinal tumors in mice. *J. Clin. Investig.* **2011**, *121*, 2781–2793. [CrossRef] [PubMed]
34. Xia, Y.; Yeddula, N.; Leblanc, M.; Ke, E.; Zhang, Y.; Oldfield, E.; Shaw, R.J.; Verma, I.M. Reduced cell proliferation by IKK2 depletion in a mouse lung-cancer model. *Nat. Cell Biol.* **2012**, *14*, 257–265. [CrossRef] [PubMed]
35. Basseres, D.S.; Ebbs, A.; Cogswell, P.C.; Baldwin, A.S. IKK is a therapeutic target in Kras-induced lung cancer with disrupted p53 activity. *Genes Cancer* **2014**, *5*, 41–55. [PubMed]

36. Kim, C.; Pasparakis, M. Epidermal p65/NF-κB signalling is essential for skin carcinogenesis. *EMBO Mol. Med.* **2014**, *6*, 970–983. [CrossRef] [PubMed]

37. Koliaraki, V.; Pasparakis, M.; Kollias, G. IKKβ in intestinal mesenchymal cells promotes initiation of colitis-associated cancer. *J. Exp. Med.* **2015**, *212*, 2235–2251. [CrossRef] [PubMed]

38. Kawauchi, K.; Araki, K.; Tobiume, K.; Tanaka, N. P53 reulates glucose metabolism through an IKK-Nf-κB pathway and inhibits cell transformation. *Nat. Cell Biol.* **2008**, *10*, 611. [CrossRef] [PubMed]

39. Johnson, R.F.; Witzel, I.-I.; Perkins, N.D. P53-dependent regulation of mitochondrial energy production by the RelA subunit of NF-κB. *Cancer Res.* **2011**, *71*, 5588–5597. [CrossRef] [PubMed]

40. Mauro, C.; Leow, S.C.; Anso, E.; Rocha, S.; Thotakura, A.K.; Tornatore, L.; Moretti, M.; De Smaele, E.; Beg, A.A.; Tergaonkar, V.; et al. NF-κB controls energy homeostasis and metabolic adaptation by upregulating mitochondrial respiration. *Nat. Cell Biol.* **2011**, *13*, 1272–1279. [CrossRef] [PubMed]

41. Tornatore, L.; Thotakura, A.K.; Bennett, J.; Moretti, M.; Franzoso, G. The nuclear factor κB signaling pathway: Integrating metabolism with inflammation. *Trends Cell Biol.* **2012**, *22*, 557–566. [CrossRef] [PubMed]

42. Oeckinghaus, A.; Hayden, M.S.; Ghosh, S. Crosstalk in NF-κB signaling pathways. *Nat. Immunol.* **2011**, *12*, 695–708. [CrossRef] [PubMed]

43. Lin, W.-W.; Karin, M. A cytokine-mediated link between innate immunity, inflammation, and cancer. *J. Clin. Investig.* **2007**, *117*, 1175–1183. [CrossRef] [PubMed]

44. Mantovani, A.; Allavena, P.; Sica, A.; Balkwill, F. Cancer-related inflammation. *Nature* **2008**, *454*, 436. [CrossRef] [PubMed]

45. Colotta, F.; Allavena, P.; Sica, A.; Garlanda, C.; Mantovani, A. Cancer-related inflammation, the seventh hallmark of cancer: Links to genetic instability. *Carcinogenesis* **2009**, *30*, 1073–1081. [CrossRef] [PubMed]

46. Balkwill, F.R.; Mantovani, A. Cancer-Related Inflammation: Common Themes and Therapeutic Opportunities. *Semin. Cancer Biol.* **2012**, *22*, 33–40. [CrossRef] [PubMed]

47. Aggarwal, B.B.; Sung, B. Nf-κB in cancer: A matter of life and death. *Cancer Discov.* **2011**, *1*, 469–471. [CrossRef] [PubMed]

48. Allavena, P.; Garlanda, C.; Borrello, M.G.; Sica, A.; Mantovani, A. Pathways connecting inflammation and cancer. *Curr. Opin. Genet. Dev.* **2008**, *18*, 3–10. [CrossRef] [PubMed]

49. DiDonato, J.A.; Mercurio, F.; Karin, M. Nf-κB and the link between inflammation and cancer. *Immunol. Rev.* **2012**, *246*, 379–400. [CrossRef] [PubMed]

50. Liang, Y.; Zhou, Y.; Shen, P. NF-κB and its regulation on the immune system. *Cell Mol. Immunol.* **2004**, *1*, 343–350. [PubMed]

51. Disis, M.L. Immune regulation of cancer. *J. Clin. Oncol.* **2010**, *28*, 4531–4538. [CrossRef] [PubMed]

52. Hoesel, B.; Schmid, J.A. The complexity of Nf-κB signaling in inflammation and cancer. *Mol. Cancer* **2013**, *12*, 86. [CrossRef] [PubMed]

53. Baud, V.; Karin, M. Is Nf-κB a good target for cancer therapy? Hopes and pitfalls. *Nat. Rev. Drug Discov.* **2009**, *8*, 33. [CrossRef] [PubMed]

54. Baldwin, A.S. Regulation of cell death and autophagy by IKK and Nf-κB: Critical mechanisms in immune function and cancer. *Immunol. Rev.* **2012**, *246*, 327–345. [CrossRef] [PubMed]

55. Cartwright, T.; Perkins, N.D.; Wilson, C.L. NFKB1: A suppressor of inflammation, ageing and cancer. *FEBS J.* **2016**, *283*, 1812–1822. [CrossRef] [PubMed]

56. Moghaddam, S.J.; Li, H.; Cho, S.-N.; Dishop, M.K.; Wistuba, I.I.; Ji, L.; Kurie, J.M.; Dickey, B.F.; DeMayo, F.J. Promotion of lung carcinogenesis by chronic obstructive pulmonary disease—Like airway inflammation in a k-ras–induced mouse model. *Am. J. Respir. Cell Mol. Biol.* **2009**, *40*, 443–453. [CrossRef] [PubMed]

57. Meira, L.B.; Bugni, J.M.; Green, S.L.; Lee, C.-W.; Pang, B.; Borenshtein, D.; Rickman, B.H.; Rogers, A.B.; Moroski-Erkul, C.A.; McFaline, J.L. DNA damage induced by chronic inflammation contributes to colon carcinogenesis in mice. *J. Clin. Investig.* **2008**, *118*, 2516–2525. [CrossRef] [PubMed]

58. Caetano, M.S.; Zhang, H.; Cumpian, A.M.; Gong, L.; Unver, N.; Ostrin, E.J.; Daliri, S.; Chang, S.H.; Ochoa, C.E.; Hanash, S.; et al. Il6 blockade reprograms the lung tumor microenvironment to limit the development and progression of k-ras-mutant lung cancer. *Cancer Res.* **2016**, *76*, 3189–3199. [CrossRef] [PubMed]

59. Sansone, P.; Storci, G.; Tavolari, S.; Guarnieri, T.; Giovannini, C.; Taffurelli, M.; Ceccarelli, C.; Santini, D.; Paterini, P.; Marcu, K.B.; et al. Il-6 triggers malignant features in mammospheres from human ductal breast carcinoma and normal mammary gland. *J. Clin. Investig.* **2007**, *117*, 3988–4002. [CrossRef] [PubMed]

60. Wang, D.J.; Ratnam, N.M.; Byrd, J.C.; Guttridge, D.C. Nf-κB functions in tumor initiation by suppressing the surveillance of both innate and adaptive immune cells. *Cell Rep.* **2014**, *9*, 90–103. [CrossRef] [PubMed]
61. Wang, Y.; Cui, H.; Schroering, A.; Ding, J.L.; Lane, W.S.; McGill, G.; Fisher, D.E.; Ding, H.-F. Nf-κB2 p100 is a pro-apoptotic protein with anti-oncogenic function. *Nat. Cell Biol.* **2002**, *4*, 888. [CrossRef] [PubMed]
62. Jacque, E.; Billot, K.; Authier, H.; Bordereaux, D.; Baud, V. Relb inhibits cell proliferation and tumor growth through p53 transcriptional activation. *Oncogene* **2013**, *32*, 2661–2669. [CrossRef] [PubMed]
63. De Donatis, G.M.; Le Pape, E.; Pierron, A.; Cheli, Y.; Hofman, V.; Hofman, P.; Allegra, M.; Zahaf, K.; Bahadoran, P.; Rocchi, S.; et al. Nf-κB2 induces senescence bypass in melanoma via a direct transcriptional activation of EZH2. *Oncogene* **2016**, *35*, 2813. [CrossRef] [PubMed]
64. Wang, Y.; Xu, J.; Gao, G.; Li, J.; Huang, H.; Jin, H.; Zhu, J.; Che, X.; Huang, C. Tumor-suppressor NfκB2 p100 interacts with ERK2 and stabilizes *PTEN* mRNA via inhibition of mir-494. *Oncogene* **2016**, *35*, 4080–4090. [CrossRef] [PubMed]
65. Xiao, Z.; Jiang, Q.; Willette-Brown, J.; Xi, S.; Zhu, F.; Burkett, S.; Back, T.; Song, N.Y.; Datla, M.; Sun, Z.; et al. The pivotal role of IKKα in the development of spontaneous lung squamous cell carcinomas. *Cancer Cell* **2013**, *23*, 527–540. [CrossRef] [PubMed]
66. Xie, Y.; Xie, K.; Gou, Q.; Chen, N. IκB kinase α functions as a tumor suppressor in epithelial-derived tumors through an NF-κB-independent pathway (review). *Oncol. Rep.* **2015**, *34*, 2225–2232. [CrossRef] [PubMed]
67. Song, N.-Y.; Zhu, F.; Wang, Z.; Willette-Brown, J.; Xi, S.; Sun, Z.; Su, L.; Wu, X.; Ma, B.; Nussinov, R.; et al. IKKα inactivation promotes Kras-initiated lung adenocarcinoma development through disrupting major redox regulatory pathways. *Proc. Natl. Acad. Sci. USA* **2018**, *115*, E812–E821. [CrossRef] [PubMed]
68. Lawrence, T.; Bebien, M.; Liu, G.Y.; Nizet, V.; Karin, M. IKKα limits macrophage NF-κB activation and contributes to the resolution of inflammation. *Nature* **2005**, *434*, 1138–1143. [CrossRef]
69. Liu, B.; Willette-Brown, J.; Liu, S.; Chen, X.; Fischer, S.M.; Hu, Y. IKKα represses a network of inflammation and proliferation pathways and elevates c-myc antagonists and differentiation in a dose-dependent manner in the skin. *Cell Death Differ.* **2011**, *18*, 1854–1864. [CrossRef] [PubMed]
70. Li, N.; Wu, X.; Holzer, R.G.; Lee, J.H.; Todoric, J.; Park, E.J.; Ogata, H.; Gukovskaya, A.S.; Gukovsky, I.; Pizzo, D.P.; et al. Loss of acinar cell IKKα triggers spontaneous pancreatitis in mice. *J. Clin. Investig.* **2013**, *123*, 2231–2243. [CrossRef] [PubMed]
71. Liu, B.; Yang, Y.; Chernishof, V.; Loo, R.R.; Jang, H.; Tahk, S.; Yang, R.; Mink, S.; Shultz, D.; Bellone, C.J.; et al. Proinflammatory stimuli induce IKKα-mediated phosphorylation of pias1 to restrict inflammation and immunity. *Cell* **2007**, *129*, 903–914. [CrossRef] [PubMed]
72. Yang, L.; Cui, H.; Wang, Z.; Zhang, B.; Ding, J.; Liu, L.; Ding, H.F. Loss of negative feedback control of nuclear factor-κB2 activity in lymphocytes leads to fatal lung inflammation. *Am. J. Pathol.* **2010**, *176*, 2646–2657. [CrossRef] [PubMed]
73. Natoli, G.; De Santa, F. Shaping alternative NF-κB-dependent gene expression programs: New clues to specificity. *Cell Death Differ.* **2006**, *13*, 693–696. [CrossRef] [PubMed]
74. Wang, V.Y.; Huang, W.; Asagiri, M.; Spann, N.; Hoffmann, A.; Glass, C.; Ghosh, G. The transcriptional specificity of NF-κB dimers is coded within the κB DNA response elements. *Cell Rep.* **2012**, *2*, 824–839. [CrossRef] [PubMed]
75. Kolovos, P.; Georgomanolis, T.; Koeferle, A.; Larkin, J.D.; Brant, L.; Nikolicc, M.; Gusmao, E.G.; Zirkel, A.; Knoch, T.A.; van Ijcken, W.F.; et al. Binding of nuclear factor κB to noncanonical consensus sites reveals its multimodal role during the early inflammatory response. *Genome Res.* **2016**, *26*, 1478–1489. [CrossRef] [PubMed]
76. Dejardin, E.; Droin, N.M.; Delhase, M.; Haas, E.; Cao, Y.; Makris, C.; Li, Z.W.; Karin, M.; Ware, C.F.; Green, D.R. The lymphotoxin-β receptor induces different patterns of gene expression via two NF-κB pathways. *Immunity* **2002**, *17*, 525–535. [CrossRef]
77. Jin, J.; Xiao, Y.; Chang, J.H.; Yu, J.; Hu, H.; Starr, R.; Brittain, G.C.; Chang, M.; Cheng, X.; Sun, S.C. The kinase TBK1 controls IgA class switching by negatively regulating noncanonical NF-κB signaling. *Nat. Immunol.* **2012**, *13*, 1101–1109. [CrossRef] [PubMed]
78. Bren, G.D.; Solan, N.J.; Miyoshi, H.; Pennington, K.N.; Pobst, L.J.; Paya, C.V. Transcription of the relb gene is regulated by NF-κB. *Oncogene* **2001**, *20*, 7722–7733. [CrossRef] [PubMed]

79. Jacque, E.; Tchenio, T.; Piton, G.; Romeo, P.H.; Baud, V. Rela repression of relb activity induces selective gene activation downstream of tnf receptors. *Proc. Natl. Acad. Sci. USA* **2005**, *102*, 14635–14640. [CrossRef] [PubMed]

80. Madge, L.A.; May, M.J. Classical NF-κB activation negatively regulates noncanonical NF-κB-dependent cxcl12 expression. *J. Biol. Chem.* **2010**, *285*, 38069–38077. [CrossRef] [PubMed]

81. Ramakrishnan, P.; Wang, W.; Wallach, D. Receptor-specific signaling for both the alternative and the canonical NF-κB activation pathways by NF-κB-inducing kinase. *Immunity* **2004**, *21*, 477–489. [CrossRef] [PubMed]

82. Lam, L.T.; Davis, R.E.; Ngo, V.N.; Lenz, G.; Wright, G.; Xu, W.; Zhao, H.; Yu, X.; Dang, L.; Staudt, L.M. Compensatory IKKα activation of classical NF-κB signaling during IKKβ inhibition identified by an RNA interference sensitization screen. *Proc. Natl. Acad. Sci. USA* **2008**, *105*, 20798–20803. [CrossRef] [PubMed]

83. Shembade, N.; Pujari, R.; Harhaj, N.S.; Abbott, D.W.; Harhaj, E.W. The kinase IKKα inhibits activation of the transcription factor NF-κB by phosphorylating the regulatory molecule tax1bp1. *Nat. Immunol.* **2011**, *12*, 834–843. [CrossRef] [PubMed]

84. Pelzer, C.; Thome, M. IKKα takes control of canonical NF-κB activation. *Nat. Immunol.* **2011**, *12*, 815–816. [CrossRef] [PubMed]

85. Gloire, G.; Horion, J.; El Mjiyad, N.; Bex, F.; Chariot, A.; Dejardin, E.; Piette, J. Promoter-dependent effect of IKKα on NF-κB/p65 DNA binding. *J. Biol. Chem.* **2007**, *282*, 21308–21318. [CrossRef] [PubMed]

86. Araki, K.; Kawauchi, K.; Tanaka, N. IKK/Nf-κB signaling pathway inhibits cell-cycle progression by a novel rb-independent suppression system for E2F transcription factors. *Oncogene* **2008**, *27*, 5696. [CrossRef] [PubMed]

87. Sfikas, A.; Batsi, C.; Tselikou, E.; Vartholomatos, G.; Monokrousos, N.; Pappas, P.; Christoforidis, S.; Tzavaras, T.; Kanavaros, P.; Gorgoulis, V.G.; et al. The canonical NF-κB pathway differentially protects normal and human tumor cells from ros-induced DNA damage. *Cell. Signal.* **2012**, *24*, 2007–2023. [CrossRef] [PubMed]

88. Batsi, C.; Markopoulou, S.; Vartholomatos, G.; Georgiou, I.; Kanavaros, P.; Gorgoulis, V.G.; Marcu, K.B.; Kolettas, E. Chronic NF-κB activation delays rasv12-induced premature senescence of human fibroblasts by suppressing the DNA damage checkpoint response. *Mech. Ageing Dev.* **2009**, *130*, 409–419. [CrossRef] [PubMed]

89. Guttridge, D.C.; Albanese, C.; Reuther, J.Y.; Pestell, R.G.; Baldwin, A.S., Jr. NF-κB controls cell growth and differentiation through transcriptional regulation of cyclin d1. *Mol. Cell Biol.* **1999**, *19*, 5785–5799. [CrossRef] [PubMed]

90. Park, K.J.; Krishnan, V.; O'Malley, B.W.; Yamamoto, Y.; Gaynor, R.B. Formation of an IKKα-dependent transcription complex is required for estrogen receptor-mediated gene activation. *Mol. Cell* **2005**, *18*, 71–82. [CrossRef] [PubMed]

91. Hu, M.C.; Lee, D.F.; Xia, W.; Golfman, L.S.; Ou-Yang, F.; Yang, J.Y.; Zou, Y.; Bao, S.; Hanada, N.; Saso, H.; et al. IκB kinase promotes tumorigenesis through inhibition of forkhead foxo3a. *Cell* **2004**, *117*, 225–237. [CrossRef]

92. Chiu, C.F.; Chang, Y.W.; Kuo, K.T.; Shen, Y.S.; Liu, C.Y.; Yu, Y.H.; Cheng, C.C.; Lee, K.Y.; Chen, F.C.; Hsu, M.K.; et al. NF-κB-driven suppression of foxo3a contributes to egfr mutation-independent gefitinib resistance. *Proc. Natl. Acad. Sci. USA* **2016**, *113*, E2526–E2535. [CrossRef] [PubMed]

93. De Smaele, E.; Zazzeroni, F.; Papa, S.; Nguyen, D.U.; Jin, R.; Jones, J.; Cong, R.; Franzoso, G. Induction of gadd45β by NF-κB downregulates pro-apoptotic JNK signalling. *Nature* **2001**, *414*, 308–313. [CrossRef] [PubMed]

94. Papa, S.; Zazzeroni, F.; Bubici, C.; Jayawardena, S.; Alvarez, K.; Matsuda, S.; Nguyen, D.U.; Pham, C.G.; Nelsbach, A.H.; Melis, T.; et al. Gadd45β mediates the Nf-κB suppression of JNK signalling by targeting MKK7/JNKK2. *Nat. Cell Biol.* **2004**, *6*, 146–153. [CrossRef] [PubMed]

95. Wuerzberger-Davis, S.M.; Chang, P.Y.; Berchtold, C.; Miyamoto, S. Enhanced G_2-M arrest by nuclear factor-κB-dependent $p21^{waf1/cip1}$ induction. *Mol. Cancer Res.* **2005**, *3*, 345–353. [CrossRef] [PubMed]

96. Chang, P.Y.; Miyamoto, S. Nuclear factor-κB dimer exchange promotes a $p21^{waf1/cip1}$ superinduction response in human T leukemic cells. *Mol. Cancer Res.* **2006**, *4*, 101–112. [CrossRef] [PubMed]

97. Batsi, C.; Markopoulou, S.; Kontargiris, E.; Charalambous, C.; Thomas, C.; Christoforidis, S.; Kanavaros, P.; Constantinou, A.I.; Marcu, K.B.; Kolettas, E. Bcl-2 blocks 2-methoxyestradiol induced leukemia cell apoptosis by a p27(kip1)-dependent g1/s cell cycle arrest in conjunction with NF-κB activation. *Biochem. Pharmacol.* **2009**, *78*, 33–44. [CrossRef] [PubMed]

98. Julien, S.; Puig, I.; Caretti, E.; Bonaventure, J.; Nelles, L.; van Roy, F.; Dargemont, C.; de Herreros, A.G.; Bellacosa, A.; Larue, L. Activation of NF-κB by akt upregulates snail expression and induces epithelium mesenchyme transition. *Oncogene* **2007**, *26*, 7445–7456. [CrossRef] [PubMed]

99. Schmidt, D.; Textor, B.; Pein, O.T.; Licht, A.H.; Andrecht, S.; Sator-Schmitt, M.; Fusenig, N.E.; Angel, P.; Schorpp-Kistner, M. Critical role for nf-κB-induced junb in vegf regulation and tumor angiogenesis. *EMBO J.* **2007**, *26*, 710–719. [CrossRef] [PubMed]

100. Min, C.; Eddy, S.F.; Sherr, D.H.; Sonenshein, G.E. NF-κB and epithelial to mesenchymal transition of cancer. *J. Cell Biochem.* **2008**, *104*, 733–744. [CrossRef] [PubMed]

101. Yan, C.; Boyd, D.D. Regulation of matrix metalloproteinase gene expression. *J. Cell Physiol.* **2007**, *211*, 19–26. [CrossRef] [PubMed]

102. Clark, I.M.; Swingler, T.E.; Sampieri, C.L.; Edwards, D.R. The regulation of matrix metalloproteinases and their inhibitors. *Int. J. Biochem. Cell Biol.* **2008**, *40*, 1362–1378. [CrossRef] [PubMed]

103. Fanjul-Fernandez, M.; Folgueras, A.R.; Cabrera, S.; Lopez-Otin, C. Matrix metalloproteinases: Evolution, gene regulation and functional analysis in mouse models. *Biochim. Biophys. Acta* **2010**, *1803*, 3–19. [CrossRef] [PubMed]

104. Parks, W.C.; Wilson, C.L.; Lopez-Boado, Y.S. Matrix metalloproteinases as modulators of inflammation and innate immunity. *Nat. Rev. Immunol.* **2004**, *4*, 617–629. [CrossRef] [PubMed]

105. Fingleton, B. Matrix metalloproteinases as regulators of inflammatory processes. *Biochim. Biophys. Acta* **2017**, *1864*, 2036–2042. [CrossRef] [PubMed]

106. Fong, K.M.; Kida, Y.; Zimmerman, P.V.; Smith, P.J. Timp1 and adverse prognosis in non-small cell lung cancer. *Clin. Cancer Res.* **1996**, *2*, 1369–1372. [PubMed]

107. Aljada, I.S.; Ramnath, N.; Donohue, K.; Harvey, S.; Brooks, J.J.; Wiseman, S.M.; Khoury, T.; Loewen, G.; Slocum, H.K.; Anderson, T.M.; et al. Upregulation of the tissue inhibitor of metalloproteinase-1 protein is associated with progression of human non-small-cell lung cancer. *J. Clin. Oncol.* **2004**, *22*, 3218–3229. [CrossRef] [PubMed]

108. Rius, J.; Guma, M.; Schachtrup, C.; Akassoglou, K.; Zinkernagel, A.S.; Nizet, V.; Johnson, R.S.; Haddad, G.G.; Karin, M. NF-κB links innate immunity to the hypoxic response through transcriptional regulation of hif-1α. *Nature* **2008**, *453*, 807–811. [CrossRef] [PubMed]

109. D'Ignazio, L.; Batie, M.; Rocha, S. Hypoxia and inflammation in cancer, focus on hif and NF-κB. *Biomedicines* **2017**, *5*, E21. [CrossRef] [PubMed]

110. Bandarra, D.; Biddlestone, J.; Mudie, S.; Muller, H.A.; Rocha, S. Hif-1α restricts NF-κB-dependent gene expression to control innate immunity signals. *Dis. Models Mech.* **2015**, *8*, 169–181. [CrossRef] [PubMed]

111. Triner, D.; Shah, Y.M. Hypoxia-inducible factors: A central link between inflammation and cancer. *J. Clin. Investig.* **2016**, *126*, 3689–3698. [CrossRef] [PubMed]

112. D'Ignazio, L.; Bandarra, D.; Rocha, S. NF-κB and hif crosstalk in immune responses. *FEBS J.* **2016**, *283*, 413–424. [CrossRef] [PubMed]

113. Triner, D.; Xue, X.; Schwartz, A.J.; Jung, I.; Colacino, J.A.; Shah, Y.M. Epithelial hypoxia-inducible factor 2α facilitates the progression of colon tumors through recruiting neutrophils. *Mol. Cell Biol.* **2017**, *37*. [CrossRef] [PubMed]

114. Hunter, J.E.; Butterworth, J.A.; Zhao, B.; Sellier, H.; Campbell, K.J.; Thomas, H.D.; Bacon, C.M.; Cockell, S.J.; Gewurz, B.E.; Perkins, N.D. The Nf-κB subunit c-rel regulates bach2 tumour suppressor expression in b-cell lymphoma. *Oncogene* **2015**, *35*, 3476. [CrossRef] [PubMed]

115. Hunter, J.E.; Leslie, J.; Perkins, N.D. C-rel and its many roles in cancer: An old story with new twists. *Br. J. Cancer* **2016**, *114*, 1. [CrossRef] [PubMed]

116. Rocha, S.; Campbell, K.J.; Perkins, N.D. P53-and mdm2-independent repression of Nf-κB transactivation by the arf tumor suppressor. *Mol. Cell* **2003**, *12*, 15–25. [CrossRef]

117. Campbell, K.J.; Rocha, S.; Perkins, N.D. Active repression of antiapoptotic gene expression by RelA (p65) Nf-κB. *Mol. Cell* **2004**, *13*, 853–865. [CrossRef]

118. Campbell, K.J.; Witty, J.M.; Rocha, S.; Perkins, N.D. Cisplatin mimics arf tumor suppressor regulation of RelA (p65) nuclear factor-κB transactivation. *Cancer Res.* **2006**, *66*, 929–935. [CrossRef] [PubMed]
119. Msaki, A.; Sanchez, A.M.; Koh, L.F.; Barre, B.; Rocha, S.; Perkins, N.D.; Johnson, R.F. The role of RelA (p65) threonine 505 phosphorylation in the regulation of cell growth, survival, and migration. *Mol. Biol. Cell* **2011**, *22*, 3032–3040. [CrossRef] [PubMed]
120. Perkins, N.D. The importance of the p50 Nf-κB subunit. *Cell Cycle* **2015**, *14*, 2877. [CrossRef] [PubMed]
121. Moles, A.; Butterworth, J.A.; Sanchez, A.; Hunter, J.E.; Leslie, J.; Sellier, H.; Tiniakos, D.; Cockell, S.J.; Mann, D.A.; Oakley, F.; et al. A RelA(p65) thr505 phospho-site mutation reveals an important mechanism regulating Nf-κB-dependent liver regeneration and cancer. *Oncogene* **2016**, *35*, 4623. [CrossRef] [PubMed]
122. Klapproth, K.; Sander, S.; Marinkovic, D.; Baumann, B.; Wirth, T. The IKK2/Nf-κB pathway suppresses myc-induced lymphomagenesis. *Blood* **2009**, *114*, 2448–2458. [CrossRef] [PubMed]
123. Cao, S.; Zhang, X.; Edwards, J.P.; Mosser, D.M. Nf-κB1 (p50) homodimers differentially regulate pro-and anti-inflammatory cytokines in macrophages. *J. Biol. Chem.* **2006**, *281*, 26041–26050. [CrossRef] [PubMed]
124. Wang, D.; Paz-Priel, I.; Friedman, A.D. Nf-κB p50 regulates c/ebpα expression and inflammatory cytokine-induced neutrophil production. *J. Immunol.* **2009**, *182*, 5757–5762. [CrossRef] [PubMed]
125. Elsharkawy, A.M.; Oakley, F.; Lin, F.; Packham, G.; Mann, D.A.; Mann, J. The Nf-κB p50: P50: Hdac-1 repressor complex orchestrates transcriptional inhibition of multiple pro-inflammatory genes. *J. Hepatol.* **2010**, *53*, 519–527. [CrossRef] [PubMed]
126. Voce, D.J.; Schmitt, A.M.; Uppal, A.; McNerney, M.E.; Bernal, G.M.; Cahill, K.E.; Wahlstrom, J.S.; Nassiri, A.; Yu, X.; Crawley, C.D. NFKB1 is a haploinsufficient DNA damage-specific tumor suppressor. *Oncogene* **2015**, *34*, 2807. [CrossRef] [PubMed]
127. Wilson, C.; Jurk, D.; Fullard, N.; Banks, P.; Page, A.; Luli, S.; Elsharkawy, A.; Gieling, R.; Chakraborty, J.B.; Fox, C. NfκB1 is a suppressor of neutrophil-driven hepatocellular carcinoma. *Nat. Commun.* **2015**, *6*, 6818. [CrossRef] [PubMed]
128. Kravtsova-Ivantsiv, Y.; Shomer, I.; Cohen-Kaplan, V.; Snijder, B.; Superti-Furga, G.; Gonen, H.; Sommer, T.; Ziv, T.; Admon, A.; Naroditsky, I. Kpc1-mediated ubiquitination and proteasomal processing of Nf-κB1 p105 to p50 restricts tumor growth. *Cell* **2015**, *161*, 333–347. [CrossRef] [PubMed]
129. Martin, B.N.; Wang, C.; Willette-Brown, J.; Herjan, T.; Gulen, M.F.; Zhou, H.; Bulek, K.; Franchi, L.; Sato, T.; Alnemri, E.S. IKKα negatively regulates asc-dependent inflammasome activation. *Nat. Commun.* **2014**, *5*, 4977. [CrossRef] [PubMed]
130. Xiao, D.; Jia, J.; Shi, Y.; Fu, C.; Chen, L.; Jiang, Y.; Zhou, L.; Liu, S.; Tao, Y. Opposed expression of IKKα: Loss in keratinizing carcinomas and gain in non-keratinizing carcinomas. *Oncotarget* **2015**, *6*, 25499. [CrossRef] [PubMed]
131. Esteller, M. Non-coding RNAs in human disease. *Nat. Rev. Genet.* **2011**, *12*, 861–874. [CrossRef] [PubMed]
132. Georgakilas, G.; Vlachos, I.S.; Zagganas, K.; Vergoulis, T.; Paraskevopoulou, M.D.; Kanellos, I.; Tsanakas, P.; Dellis, D.; Fevgas, A.; Dalamagas, T. Diana-mirgen v3. 0: Accurate characterization of microRNA promoters and their regulators. *Nucleic Acids Res.* **2015**, *44*, D190–D195. [CrossRef] [PubMed]
133. Niu, J.; Shi, Y.; Tan, G.; Yang, C.H.; Fan, M.; Pfeffer, L.M.; Wu, Z.-H. DNA damage induces Nf-κB-dependent microRNA-21 up-regulation and promotes breast cancer cell invasion. *J. Biol. Chem.* **2012**, *287*, 21783–21795. [CrossRef] [PubMed]
134. Ma, J.; Liu, J.; Wang, Z.; Gu, X.; Fan, Y.; Zhang, W.; Xu, L.; Zhang, J.; Cai, D. NF-κB-dependent microRNA-425 upregulation promotes gastric cancer cell growth by targeting *PTEN* upon il-1β induction. *Mol. Cancer* **2014**, *13*, 40. [CrossRef] [PubMed]
135. Takata, A.; Otsuka, M.; Yoshikawa, T.; Kishikawa, T.; Hikiba, Y.; Obi, S.; Goto, T.; Kang, Y.J.; Maeda, S.; Yoshida, H.; et al. MicroRNA-140 acts as a liver tumor suppressor by controlling Nf-κB activity by directly targeting DNA methyltransferase 1 (dnmt1) expression. *Hepatology* **2013**, *57*, 162–170. [CrossRef] [PubMed]
136. Takata, A.; Otsuka, M.; Kojima, K.; Yoshikawa, T.; Kishikawa, T.; Yoshida, H.; Koike, K. MicroRNA-22 and microRNA-140 suppress Nf-κB activity by regulating the expression of Nf-κB coactivators. *Biochem. Biophys. Res. Commun.* **2011**, *411*, 826–831. [CrossRef] [PubMed]
137. Huan, L.; Liang, L.-H.; He, X.-H. Role of microRNAs in inflammation-associated liver cancer. *Cancer Biol Med.* **2016**, *13*, 407. [PubMed]

138. Le Sage, C.; Nagel, R.; Egan, D.A.; Schrier, M.; Mesman, E.; Mangiola, A.; Anile, C.; Maira, G.; Mercatelli, N.; Ciafre, S.A.; et al. Regulation of the p27(kip1) tumor suppressor by mir-221 and mir-222 promotes cancer cell proliferation. *EMBO J.* **2007**, *26*, 3699–3708. [CrossRef] [PubMed]

139. Voorhoeve, P.M.; Agami, R. Classifying microRNAs in cancer: The good, the bad and the ugly. *Biochim. Biophys. Acta* **2007**, *1775*, 274–282. [CrossRef] [PubMed]

140. Kedde, M.; van Kouwenhove, M.; Zwart, W.; Oude Vrielink, J.A.; Elkon, R.; Agami, R. A pumilio-induced RNA structure switch in p27-3′ utr controls mir-221 and mir-222 accessibility. *Nat. Cell Biol.* **2010**, *12*, 1014–1020. [CrossRef] [PubMed]

141. Karagkouni, D.; Paraskevopoulou, M.D.; Chatzopoulos, S.; Vlachos, I.S.; Tastsoglou, S.; Kanellos, I.; Papadimitriou, D.; Kavakiotis, I.; Maniou, S.; Skoufos, G. DIANA-TarBase v8: A decade-long collection of experimentally supported miRNA–gene interactions. *Nucleic Acids Res.* **2017**, *46*, D239–D245. [CrossRef] [PubMed]

142. Yin, M.; Ren, X.; Zhang, X.; Luo, Y.; Wang, G.; Huang, K.; Feng, S.; Bao, X.; He, X.; Liang, P. Selective killing of lung cancer cells by miRNA-506 molecule through inhibiting Nf-κB p65 to evoke reactive oxygen species generation and p53 activation. *Oncogene* **2015**, *34*, 691. [CrossRef]

143. Kong, X.-J.; Duan, L.-J.; Qian, X.-Q.; Xu, D.; Liu, H.-L.; Zhu, Y.-J.; Qi, J. Tumor-suppressive microRNA-497 targets IKKβ to regulate Nf-κB signaling pathway in human prostate cancer cells. *Am. J. Cancer Res.* **2015**, *5*, 1795–1804. [PubMed]

144. Keklikoglou, I.; Koerner, C.; Schmidt, C.; Zhang, J.; Heckmann, D.; Shavinskaya, A.; Allgayer, H.; Gückel, B.; Fehm, T.; Schneeweiss, A. MicroRNA-520/373 family functions as a tumor suppressor in estrogen receptor negative breast cancer by targeting Nf-κB and TGF-β signaling pathways. *Oncogene* **2012**, *31*, 4150. [CrossRef] [PubMed]

145. Iliopoulos, D.; Hirsch, H.A.; Struhl, K. An epigenetic switch involving Nf-κB, lin28, let-7 microRNA, and il6 links inflammation to cell transformation. *Cell* **2009**, *139*, 693–706. [CrossRef] [PubMed]

146. Iliopoulos, D.; Jaeger, S.A.; Hirsch, H.A.; Bulyk, M.L.; Struhl, K. Stat3 activation of mir-21 and mir-181b-1 via *PTEN* and cyld are part of the epigenetic switch linking inflammation to cancer. *Mol. Cell* **2010**, *39*, 493–506. [CrossRef] [PubMed]

147. Olarerin-George, A.O.; Anton, L.; Hwang, Y.-C.; Elovitz, M.A.; Hogenesch, J.B. A functional genomics screen for microRNA regulators of NF-κB signaling. *BMC Biol.* **2013**, *11*, 19. [CrossRef] [PubMed]

148. Feng, X.; Wang, H.; Ye, S.; Guan, J.; Tan, W.; Cheng, S.; Wei, G.; Wu, W.; Wu, F.; Zhou, Y. Up-regulation of microRNA-126 may contribute to pathogenesis of ulcerative colitis via regulating NF-κB inhibitor iκBα. *PLoS ONE* **2012**, *7*, e52782. [CrossRef] [PubMed]

149. Huang, W.; Lin, J.; Zhang, H. Mir-126: A novel regulator in colon cancer. *Biomed. Rep.* **2016**, *4*, 131–134. [CrossRef] [PubMed]

150. Zhou, W.; Pal, A.S.; Hsu, A.Y.; Gurol, T.; Zhu, X.; Wirbisky-Hershberger, S.E.; Freeman, J.L.; Kasinski, A.L.; Deng, Q. MicroRNA-223 suppresses the canonical NF-κB pathway in basal keratinocytes to dampen neutrophilic inflammation. *Cell Rep.* **2018**, *22*, 1810–1823. [CrossRef] [PubMed]

151. Haneklaus, M.; Gerlic, M.; O'Neill, L.A.; Masters, S.L. Mir-223: Infection, inflammation and cancer. *J. Intern. Med.* **2013**, *274*, 215–226. [CrossRef] [PubMed]

152. Chen, Q.; Wang, H.; Liu, Y.; Song, Y.; Lai, L.; Han, Q.; Cao, X.; Wang, Q. Inducible microRNA-223 down-regulation promotes tlr-triggered il-6 and il-1β production in macrophages by targeting stat3. *PLoS ONE* **2012**, *7*, e42971. [CrossRef] [PubMed]

153. Li, T.; Morgan, M.J.; Choksi, S.; Zhang, Y.; Kim, Y.S.; Liu, Z.G. MicroRNAs modulate the noncanonical transcription factor NF-κB pathway by regulating expression of the kinase IKKα during macrophage differentiation. *Nat. Immunol.* **2010**, *11*, 799–805. [CrossRef] [PubMed]

154. Li, S.; Li, Z.; Guo, F.; Qin, X.; Liu, B.; Lei, Z.; Song, Z.; Sun, L.; Zhang, H.T.; You, J.; et al. Mir-223 regulates migration and invasion by targeting artemin in human esophageal carcinoma. *J. Biomed. Sci.* **2011**, *18*, 24. [CrossRef] [PubMed]

155. Li, X.; Zhang, Y.; Zhang, H.; Liu, X.; Gong, T.; Li, M.; Sun, L.; Ji, G.; Shi, Y.; Han, Z.; et al. MiRNA-223 promotes gastric cancer invasion and metastasis by targeting tumor suppressor epb41l3. *Mol. Cancer Res.* **2011**, *9*, 824–833. [CrossRef] [PubMed]

156. Tang, Y.; Wang, Y.; Chen, Q.; Qiu, N.; Zhao, Y.; You, X. Mir-223 inhibited cell metastasis of human cervical cancer by modulating epithelial-mesenchymal transition. *Int. J. Clin. Exp. Pathol.* **2015**, *8*, 11224–11229. [PubMed]

157. Aqeilan, R.I.; Calin, G.A.; Croce, C.M. Mir-15a and mir-16-1 in cancer: Discovery, function and future perspectives. *Cell Death Differ.* **2009**, *17*, 215. [CrossRef] [PubMed]

158. Chen, R.; Alvero, A.B.; Silasi, D.A.; Kelly, M.G.; Fest, S.; Visintin, I.; Leiser, A.; Schwartz, P.E.; Rutherford, T.; Mor, G. Regulation of IKKβ by mir-199a affects Nf-κB activity in ovarian cancer cells. *Oncogene* **2008**, *27*, 4712. [CrossRef] [PubMed]

159. Zhao, J.L.; Rao, D.S.; Boldin, M.P.; Taganov, K.D.; O'Connell, R.M.; Baltimore, D. Nf-κB dysregulation in microRNA-146a—Deficient mice drives the development of myeloid malignancies. *Proc. Natl. Acad. Sci. USA* **2011**, *108*, 9184–9189. [CrossRef] [PubMed]

160. Bazzoni, F.; Rossato, M.; Fabbri, M.; Gaudiosi, D.; Mirolo, M.; Mori, L.; Tamassia, N.; Mantovani, A.; Cassatella, M.A.; Locati, M. Induction and regulatory function of mir-9 in human monocytes and neutrophils exposed to proinflammatory signals. *Proc. Natl. Acad. Sci. USA* **2009**, *106*, 5282–5287. [CrossRef] [PubMed]

161. Guo, L.M.; Pu, Y.; Han, Z.; Liu, T.; Li, Y.X.; Liu, M.; Li, X.; Tang, H. MicroRNA-9 inhibits ovarian cancer cell growth through regulation of Nf-κB1. *FEBS J.* **2009**, *276*, 5537–5546. [CrossRef] [PubMed]

162. Liu, S.; Wu, L.-C.; Pang, J.; Santhanam, R.; Schwind, S.; Wu, Y.-Z.; Hickey, C.J.; Yu, J.; Becker, H.; Maharry, K.; et al. Sp1/NFκB/HDAC/miR-29b regulatory network in KIT-driven myeloid leukemia. *Cancer Cell* **2010**, *17*, 333–347. [CrossRef] [PubMed]

163. Li, Q.Q.; Chen, Z.Q.; Cao, X.X.; Xu, J.D.; Xu, J.W.; Chen, Y.Y.; Wang, W.J.; Chen, Q.; Tang, F.; Liu, X.P.; et al. Involvement of Nf-κB/mir-448 regulatory feedback loop in chemotherapy-induced epithelial–mesenchymal transition of breast cancer cells. *Cell Death Differ.* **2010**, *18*, 16. [CrossRef] [PubMed]

164. Kumar, M.S.; Erkeland, S.J.; Pester, R.E.; Chen, C.Y.; Ebert, M.S.; Sharp, P.A.; Jacks, T. Suppression of non-small cell lung tumor development by the let-7 microRNA family. *Proc. Natl. Acad. Sci. USA* **2008**, *105*, 3903–3908. [CrossRef] [PubMed]

165. Iliopoulos, D. MicroRNA circuits regulate the cancer-inflammation link. *Sci. Signal.* **2014**, *7*, pe8. [CrossRef]

166. Xiang, M.; Birkbak, N.J.; Vafaizadeh, V.; Walker, S.R.; Yeh, J.E.; Liu, S.; Kroll, Y.; Boldin, M.; Taganov, K.; Groner, B. Stat3 induction of mir-146b forms a feedback loop to inhibit the Nf-κB to il-6 signaling axis and stat3-driven cancer phenotypes. *Sci. Signal.* **2014**, *7*, ra11. [CrossRef] [PubMed]

167. Jeong, J.-H.; Park, S.-J.; Dickinson, S.I.; Luo, J.-L. A constitutive intrinsic inflammatory signaling circuit composed of mir-196b, Meis2, PPP3CC, and p65 drives prostate cancer castration resistance. *Mol. Cell* **2017**, *65*, 154–167. [CrossRef] [PubMed]

168. Papadopoulos, G.L.; Alexiou, P.; Maragkakis, M.; Reczko, M.; Hatzigeorgiou, A.G. Diana-mirpath: Integrating human and mouse microRNAs in pathways. *Bioinformatics* **2009**, *25*, 1991–1993. [CrossRef] [PubMed]

MDPI

Review

The Many Roles of Ubiquitin in NF-κB Signaling

Gilles Courtois * and Marie-Odile Fauvarque

INSERM U1038/BGE/BIG, CEA Grenoble, 38054 Grenoble, France; marie-odile.fauvarque@cea.fr
* Correspondence: gilles.courtois@inserm.fr; Tel.: +33-438-786-107

Received: 15 March 2018; Accepted: 2 April 2018; Published: 10 April 2018

Abstract: The nuclear factor κB (NF-κB) signaling pathway ubiquitously controls cell growth and survival in basic conditions as well as rapid resetting of cellular functions following environment changes or pathogenic insults. Moreover, its deregulation is frequently observed during cell transformation, chronic inflammation or autoimmunity. Understanding how it is properly regulated therefore is a prerequisite to managing these adverse situations. Over the last years evidence has accumulated showing that ubiquitination is a key process in NF-κB activation and its resolution. Here, we examine the various functions of ubiquitin in NF-κB signaling and more specifically, how it controls signal transduction at the molecular level and impacts in vivo on NF-κB regulated cellular processes.

Keywords: nuclear factor κB; signal transduction; ubiquitin; ubiquitination/deubiquitination

1. Introduction

Regulation of gene expression in eukaryotic cells represents an essential process for the timely control of the production of proteins, while the fine-tuning of their final activities and/or fate often relies upon post-translational modifications (PTM). For decades, phosphorylation and dephosphorylation has been considered as the dominant switch controlling the fate and activity of proteins. In the eighties, ubiquitination was identified as an additional type of PTM, although primarily in the restricted context of protein degradation. Later, the ubiquitination/deubiquitination process was shown to expand to the regulation of almost any protein, with its own intricate modulation allowing the emergence of new layers of regulation.

Among the many cellular processes that are regulated by ubiquitination is signal transduction which needs to be tightly controlled for operating properly at the right place and time. One of the first identified and most extensively dissected signaling pathways regulated by ubiquitination is the NF-κB pathway that plays a critical role in inflammation, immunity and control of cell death and proliferation. Not surprisingly, this pathway represents a useful paradigm to illustrate how ubiquitination controls protein activity. Here, we will present the main participants in the NF-κB activation process, how ubiquitination generally operates and which specific steps of this process it regulates. Moreover, we will describe how perturbations in ubiquitination and ubiquitin recognition mechanisms in the NF-κB pathway impact on human health.

2. The NF-κB Signaling Pathway

NF-κB is a generic name for a collection of inducible transcription factors formed by the dimeric combination of members of the avian reticuloendotheliosis (Rel)/NF-κB family of proteins. The five members of this family (RelA, RelB, c-Rel, p50 and p52) share a conserved Rel homology domain at the N-terminus, which contains sequences involved in dimerization, nuclear localization and interaction with NF-κB inhibitors [1] (Figure 1A). In addition, RelA, RelB and c-Rel exhibit a transcriptional activator domain (TAD) at the C-terminus. Such a domain is absent in p50 and p52, which are both synthesized from precursors p105 and p100, respectively.

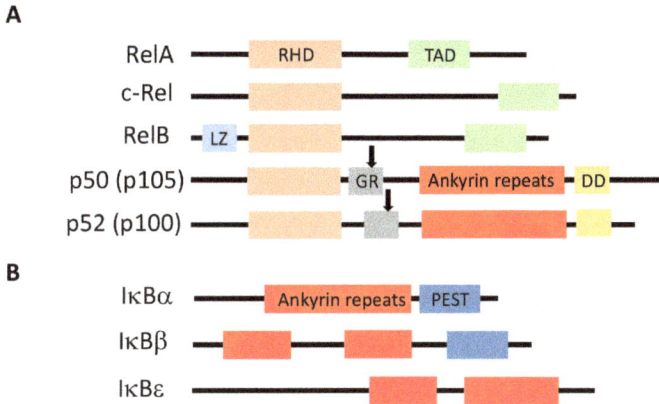

Figure 1. The NF-κB proteins and their inhibitors. (**A**) Members of the Rel/NF-κB family. The five NF-κB subunits are presented with their functional domains. RHD = Rel homology domain; TAD = transcription activation domain; LZ = leucine zipper; GR = glycine-rich domain; DD = death domain. The bold arrows indicate the C-terminus of p50 and p52 after processing of p105 and p100, respectively. (**B**) Members of the IκB family. The three IκB inhibitors are presented with their functional domains. PEST = proline/glutamic acid/serine/threonine-rich sequence.

NF-κB regulates the transcription of hundreds of genes participating in immunity, inflammation, cell proliferation and cell death and its activity is itself controlled by a plethora of stimuli such as pro-inflammatory cytokines, pathogen-associated molecular patterns (PAMPs), and oxidative stress [2]. Fast activation of NF-κB is achieved through its ubiquitous presence in the cytoplasm of resting cells as a latent form associated with inhibitors of the inhibitory κB (IκB) or IκB-like families, and its quick release from them in response to stimuli to become transcriptionally competent.

Members of the IκB family of NF-κB inhibitors are IκBα, IκBβ and IκBε (Figure 1B). They all share a similar structure with a conserved sequence at the N-terminus containing a DSGXXS motif and a series of Ankyrin repeats at the C-terminus. These repeats are responsible for the interaction with the Rel domain of NF-κB proteins and the masking of their nuclear localization sequence (NLS). Precursor proteins p105 and p100 also contain Ankyrin repeats at the C-terminus and can play the role, before processing to generate p50 and p52, of IκB-like proteins.

Activation of NF-κB can be achieved through two distinct modes designated as "canonical" and "non-canonical".

2.1. The Canonical Pathway of NF-κB Activation

The canonical pathway, which is induced by a large variety of external or internal cell stimuli (see below), involves a cytoplasmic kinase complex called IκB kinase (IKK) (Figure 2). This complex is composed of three main subunits: two catalytic ones with related structures, IKK1 (also called IKKα) and IKK2 (also called IKKβ), and a regulatory subunit, NF-κB essential modulator (NEMO) (also called IKKγ) [3] (Figure 3A). Upon cell activation, IKK phosphorylates the two Serine residues located in the DSGXXS motif of IκB inhibitors and this induces their degradation by the proteasome (see details below). Free NF-κB (usually dimers such as RelA/p50, RelA/RelA or c-Rel/p50) translocate in the nucleus and positively or negatively regulate the transcription of numerous target genes, encoding proteins mostly involved in immunity, inflammation, cell growth and cell survival.

In this pathway, IKK activation is often triggered by another kinase complex, the tumor growth factor β-activated kinase 1 (TAK1) complex, that contains, in addition to the kinase TAK1, three regulatory subunits: TAK1 binding protein 1 (TAB1), which regulates the catalytic activity of

TAK1, and TAB2 and TAB3, which participate in the activation of TAK1 by binding to ubiquitin (see below) [4] (Figure 3B).

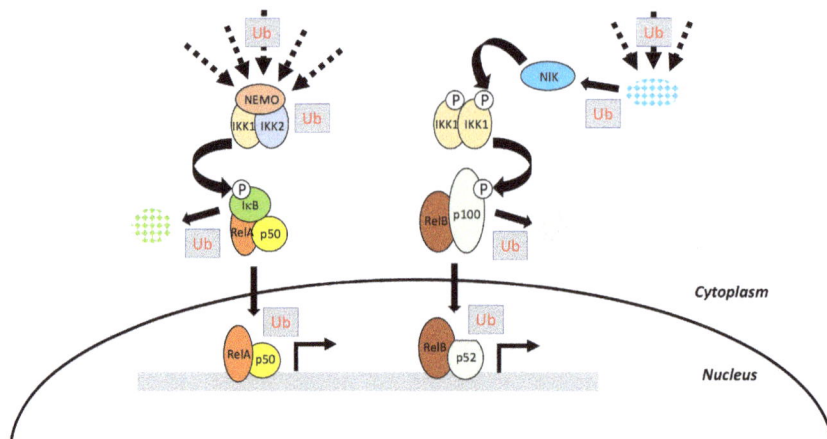

Figure 2. The canonical and non-canonical pathways of NF-κB activation. On the left is presented the canonical pathway which involves phosphorylation of IκBs by IKK to induce their degradation. On the right is presented the non-canonical pathway which is dependent on NIK stabilization and IKK1 activation. In each case specific NF-κB dimers are induced that regulates different classes of genes participating in various biological processes. Steps that are controlled by ubiquitination processes are indicated by "Ub". See text for details.

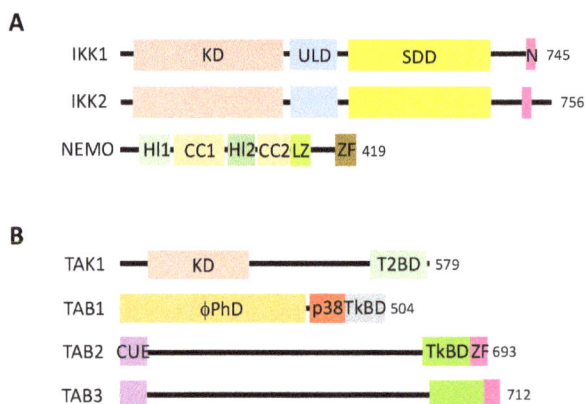

Figure 3. The subunits of IKK and TAK1 complexes. (**A**) IKK complex. The three subunits of this complex are presented with their functional domains. KD = kinase domain; ULD = ubiquitin-like domain; SDD = scaffold/dimerization domain; N = NEMO binding domain; Hl1/Hl2 = Helix 1 and 2; CC1/CC2 = coiled coil 1 and 2; LZ = leucine zipper; ZF = zinc finger; (**B**) TAK1 complex. The three subunits of this complex are presented with their functional domains. KD = kinase domain; T2BD = TAB2/TAB3-binding domain; φPhD = pseudo-phosphatase domain; p38 = p38-interacting domain; TkBD = TAK1-binding domain; CUE = coupling of ubiquitin conjugation to ER degradation domain; ZF = novel zinc finger (Npl4 class).

2.2. The Non-Canonical Pathway of NF-κB Activation

The non-canonical pathway is an alternative pathway which is activated by a limited set of stimuli (see below) and requires a cytoplasmic kinase called NF-κB inducing kinase (NIK) [5]. Following cell stimulation, NIK, which is normally constitutively degraded, starts to accumulate and further phosphorylates an IKK1 dimer, which in this setting, works independently from NEMO and IKK2 (Figure 2). Activated IKK1 then acts on p100 precursor, commonly bound to RelB, to induce its processing to p52. Active p52/RelB dimer translocate in the nucleus and regulate transcription of a set of specific genes distinct from those regulated by the canonical pathway.

Initially, these two modules of NF-κB activation may appear quite simple. Nevertheless, they require, in addition to phosphorylation, the linkage of various kinds of ubiquitin moieties on a number of the main actors of these pathways at various steps (Figure 2). This feature substantially complicates the picture and provides further layers of specificity and regulation that play a major role in proper cell physiology. Before discussing in depth how ubiquitination impacts on NF-κB signaling we will briefly present the basics of ubiquitin molecular machinery.

3. Ubiquitination: Players and Mechanisms

3.1. The Ubiquitination Process

Ubiquitin is an evolutionary conserved polypeptide of 76 amino acids that can be covalently attached through its terminal Gly residue to either the ε-amino group of a Lys residue (K) or to the amino group of the first Met (M1) of a protein target [6–8]. The linkage of ubiquitin depends on the successive action of E1 activating enzymes, E2 conjugating enzymes and E3 ligases (Figure 4A). The E1 catalyzes the ATP-dependent formation of a thioester bound between the C-terminus of ubiquitin and the active cysteine residue of E1. The ubiquitin is then trans-thiolated to the active cysteine of an E2 [9] and eventually transferred from the E2 to the target protein through the additional contribution of an E3 ligase. In this process, the E2–E3 complex is the main factor determining the specificity of the substrate. E3 ligases are quite diverse [10,11] but most of them possess a RING-finger domain and act as a bridging factor between the E2 and the substrate allowing the direct transfer of ubiquitin from the E2 to the target protein. Alternatively, E3 ligases possessing a homologous to the E6-AP carboxyl terminus (HECT) domain combine E2 and E3 activities and form a thioester intermediate with the active-site cysteine of the E3 prior to ubiquitin transfer to the substrate protein.

Ubiquitin itself contains seven Lys residues allowing for the formation of different ubiquitin polymerized chains by extension of the E2/E3 reaction or through the action of E4 enzymes [12] (Figure 4B). While the linkage of lysine-48-linked ubiquitin chains (Ub^{K48}) drives proteins for degradation by the proteasome, other ubiquitin moieties, such as linear (M1) linked chains, Ub^{K63} linked chains or ubiquitin monomers, regulate protein activity, protein subcellular localization or protein-protein interaction in a plethora of cellular processes, including endocytosis, signal transduction or DNA repair [13].

At the structural level, ubiquitin displays a β-grasp fold and possesses a hydrophobic surface patch which mediates the interaction with ubiquitin-binding domains (UBDs) containing proteins (see below). Ubiquitin is itself subjected to other kinds of PTMs including the linkage of ubiquitin-like proteins (ULPs) (see below), phosphorylation and acetylation, adding further structural complexity to the so-called "ubiquitin code" [14].

Ubiquitin proteases, also known as deubiquitinases (DUBs) catalyse the reverse reaction by hydrolysing ubiquitin from protein or polyubiquitin chains [15]. When acting at the level of the proteasome, they favour protein entry into the proteasome machinery and ubiquitin recycling. Alternatively, DUBs can act at earlier steps and save protein from degradation or interfere with any cell process. Mammalian genomes contain almost one hundred DUBs belonging to two main classes of proteases: the metalloproteases (JAB1/MPN/Mv34 metalloenzymes (JAMMs)) and the cysteine proteases (ubiquitin specific protease (USP), ubiquitin C-terminal hydrolases (UCHs), MIU-containing novel DUB family (Mindy), Machado-Joseph disease proteases (MJDs), ovarian tumor proteases (OTUs)], among which the USP subfamily represents the largest class [16]. DUBs contain a catalytic domain that has sequence similarity within subfamilies and unrelated flanking sequences that typically mediate protein-protein interaction and/or regulate the catalytic activity of the enzyme as shown for a number of DUBs containing ubiquitin-like domains (ULDs) that share a structure similar to ubiquitin. These flanking sequences, along with the catalytic core, can also contribute to the specific binding and cleavage of different polyubiquitin chains (see below). Actually, DUBs activity and substrate specificity are governed by many factors that are not fully elucidated and certainly depend on subcellular localization and target recognition through the integration of DUBs into large protein complexes [17]. Notably, DUBs and E3 ligases targeting a common substrate sometimes act within the same protein complex, possibly finely regulating protein activity through a ubiquitination/deubiquitination cycle [18].

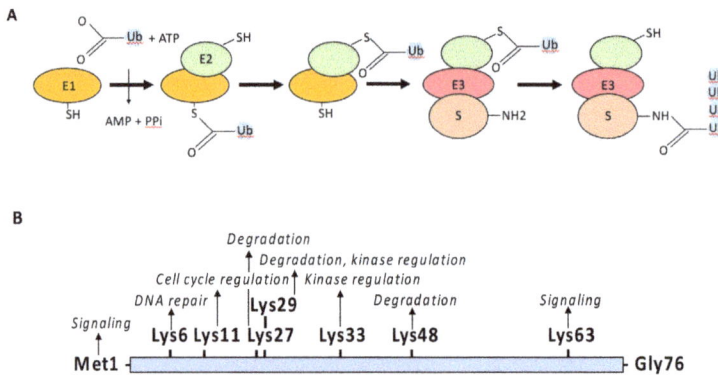

Figure 4. The ubiquitination process. (**A**) The enzymatic machinery. The three components (E1/E2/E3) involved in substrate (S) polyubiquitination and major steps of the ubiquitination process are shown; (**B**) key amino acids of ubiquitin. Indicated are Met1 and the seven internal Lys that can be used to form polyubiquitin chains through peptide (Met) or isopeptide (Lys) bonds involving Gly76. Main biological functions of these chains are indicated. See text for details.

A number of ubiquitin-like proteins (ULPs) have been reported [19], sharing a similar structure and conjugation mechanisms with ubiquitin, including the closely-related small ubiquitin-like modifier (SUMO), which plays an essential role in DNA repair, cell cycle, and signal transduction [20], and the neural precursor cell expressed, developmentally down-regulated 8 (NEDD8) [21], which modulates the function of the cell cycle and embryogenesis proteins.

Importantly, the number of E3 Ligases or DUBs mutations found to be associated with human pathologies such as inflammatory diseases, rare diseases, cancers and neurodegenerative disorders is rapidly increasing [22–24]. There is now clear evidence that many E3s and DUBs play critical roles in NF-κB signaling, as will be discussed in the next sections, and therefore represent attractive pharmacological targets in the field of cancers and inflammation or rare diseases.

3.2. E3 Ligase Families in NF-κB Signaling

Specific classes of E3 ligases participating in the NF-κB signaling pathways described in the following sections deserve a short introduction.

3.2.1. TRAFs

E3 ligases of the small TNF receptor associated factor (TRAF) family (seven members) [25] present a conserved organization with a really interesting new gene (RING) finger domain at the N-terminus, followed by a variable number of zinc fingers (ZFs) and in the second half a so-called TRAF domain, which is divided into two parts: the TRAF-N domain, which is a coiled-coil and the TRAF-C domain, with both of them participating in oligomerization and substrate recognition. Only two TRAFs lack one of these conserved domains: TRAF1, which does not contain any ring domain and behaves only as an inhibitor or adaptor and TRAF7, which does not contain a TRAF-C domain and therefore cannot necessarily be considered as a bona fide TRAF. The specific interaction of TRAFs with their targets often involves specific motifs recognized by the TRAF-C domains. PXQXT/S represents the consensus binding sequence for TRAF2, TRAF3 or TRAF5 whereas PXEXX acidic/aromatic is recognized by TRAF6.

3.2.2. TRIMs

Tripartite motif proteins (TRIMs) belong to a large family of E3 ligases of more than 70 members which display a conserved ring, B-box, coiled-coil (RBCC) domain at the N-terminus [26,27]. This domain includes a RING domain with E3 ligase activity, one or two B-box domains, and a coiled-coil, with these two latter domains participating in dimerization and higher-order oligomerization. The variable C-terminal regions of TRIMs, classified into 11 distinct classes, regulate target recognition and subcellular localization.

3.2.3. LUBAC

Linear ubiquitin chain assembly complex (LUBAC) is the only E3 ligase complex described so far that is able to synthesize M1-linked ubiquitin chains in mammalian cells [28,29]. It is composed of three subunits, haem-oxydized IRP2 ubiquitin Ligase 1L (HOIL-1L), HOIL-interacting protein (HOIP) and SHANK-associated RH domain interacting protein (SHARPIN), each of which exhibit specific functions. HOIP is the subunit that contains all the catalytic machinery to synthesize M1-linked chains of ubiquitin. This involves a ring between ring fingers (RBR) domain and the C-terminal linear ubiquitin chain determining domain (LDD), allowing HOIP to work as a RING/HECT hybrid E3.

HOIL-IL and SHARPIN are non-catalytic subunits that associate to the ubiquitin-associated (UBA) domain of HOIP through their ubiquitin-like (UBL) domains. The SHARPIN UBL domain may also recognize the Npl4 Zinc Finger 2 (NZF2) domain of HOIP. Other important sequences of LUBAC subunits are (1) the PNGase/UBA or UBX-containing proteins (PUB) domain of HOIP which interacts with DUBs acting in NF-κB signaling, OTU deubiquitinase with linear linkage specificity (OTULIN) and cylindromatosis (CYLD), directly in the former case and through SPATA2 in the latter, (2) the NZFs of HOIL-IL and SHARPIN, which recognize ubiquitin (see below), and (3) the Pleckstrin homology (PH) domain of SHARPIN, that acts as a dimerization module. Like HOIP, HOIL-IL contains an RBR domain at the C-terminus, but this domain is dispensable for catalysis in the context of LUBAC.

All these E3 ligases work with specific E2 conjugating enzymes to synthesize different kinds of ubiquitin chains during the NF-κB activation process. We will not describe them exhaustively in the following sections, mostly because they are much less characterized than the E3 ligases. Nevertheless, two of them should be mentioned. In many situations, if not all, the specific E2 conjugating K63-linked chains is ubiquitin-conjugating 13 (Ubc13) which forms a dimer with co-factor ubiquitin-conjugating enzyme variant 1A (Uev1A). Ubc13/Uev1A was originally identified as the E2 associated with TRAF6 in cell extracts used to study the activation of NF-κB [30] in vitro. Its broad involvement in NF-κB signaling has been confirmed in vivo [31]. In addition, the main E2 conjugating M1-linked chains to specific components of the NF-κB pathways (see below) with LUBAC is the ubiquitin-conjugating enzyme E2 L3 (UBE2L3) [32,33].

3.3. Ubiquitin Binding Domains in NF-κB Signaling

Interpretation of the "ubiquitin code" is achieved through the recognition of different kinds of ubiquitin moieties by specific UBD-containing proteins [34]. UBDs are quite diverse, belonging to more than twenty families, and their main characteristics can be summarized as follows: (1) They vary widely in size, amino acid sequences and three-dimensional structure; (2) The majority of them recognize the same hydrophobic patch on the β-sheet surface of ubiquitin, that includes Ile44, Leu8 and Val70; (3) Their affinity for ubiquitin is low (in the higher μM to lower mM range) but can be increased following polyubiquitination or through their repeated occurrence within a protein; (4) Using the topology of the ubiquitin chains, they discriminate between modified substrates to allow specific interactions or enzymatic processes. For instance, K11- and K48-linked chains adopt a rather closed conformation, whereas K63- or M1-linked chains are more elongated.

In the NF-κB signaling pathway, several key players such as TAB2/3, NEMO and LUBAC are UBD-containing proteins whose ability to recognize ubiquitin chains is at the heart of their functions.

Within the TAK1 complex, TAB2 and TAB3 are the UBD-containing subunits. They present a similar secondary structure to an N-terminal coupling of ubiquitin conjugation to endoplasmic reticulum-associated degradation (CUE) UBD, a coiled-coil, a TAK1-binding domain and, at the C-terminus, a NZF (Figure 3B). The NZF is responsible for the interaction with ubiquitin. Recognition of K63-linked ubiquitin chains requires binding of adjacent ubiquitin moieties by two binding sites, both of them involving the Ile44-containing hydrophobic patch [35,36]. The distal ubiquitin occupies the canonical NZF at residues Thr674 and Phe675 of TAB2, whereas the proximal one contacts residues Leu681, His678 and Glu685. As a consequence, the TAB2 NZF is surrounded by three ubiquitin molecules, two of them interacting with one NZF. This two-sided mode of interaction may be shared by other NZF-containing proteins but, very importantly, excludes recognition of M1-linked chains of ubiquitin in the case of TAB2/3.

Within the IKK complex, NEMO is the specialized subunit allowing protein scaffolding through the recognition of specific partners modified by ubiquitin. The interaction of NEMO with polyubiquitin involves two separate domains. First, the NEMO ubiquitin binding (NUB) domain, encompassing the CC2 and leucine zipper (LZ) domains, recognizes both M1- and K63-linked chains through distinct modes. Two M1-linked ubiquitin dimers can be recognized by the dimeric NUB domain through the interaction of the proximal ubiquitin (residues extending from Gln2 to Glu16 plus Glu64 and Thr66) with NEMO Arg and Glu residues located from 309 to 320 and interaction of the distal ubiquitin through its hydrophobic patch and NEMO residues centered around Asp304 [37]. This extended interaction interface, which also includes the linker region, explains the much higher affinity of the NUB domain for M1-linked chains than for K63-linked chains. Indeed, in the case of K63-linked chains, only the distal ubiquitin can be recognized because of a slight shift of the proximal ubiquitin caused by the K63 linkage [38]. Second, the ZF located at the very end of NEMO also displays affinity for ubiquitin, favoring recognition of K63-linked chains. Again, a two-sided interaction occurs, with the hydrophobic patch of distal ubiquitin binding to ZF residues Val414/Met415 and residue Phe395-centered patch connecting to the proximal ubiquitin [39]. It has been proposed that combined recognition of polyubiquitin chains by the NUB domain and the ZF occurs in the full-length protein [40], resulting in an affinity for both M1-linked and K63-linked chains, with a stronger affinity for the former ones [41]. This ability to recognize both kinds of chains may prove important considering the synthesis of mixed chains that can occur during cell stimulation (see below).

All LUBAC subunits contain similar NZFs of the Npl4 subtype but they appear to fulfill different functions. The NZF of HOIP displays a weak affinity for K63-linked chains compared to the ZF of SHARPIN; so SHARPIN is the driving force for recruitment to K63-linked partners [42]. Molecular characteristics of these NZF/ubiquitin interactions are not known. In contrast, HOIL-IL specifically recognizes M1-linked ubiquitin chains through a unique mode [43]. Indeed, its NZF binds both the canonical Ile44-centered hydrophobic surface on the distal ubiquitin and a Phe4-centered hydrophobic patch on the proximal ubiquitin. These distinct specificities help explaining how LUBAC operates within the NF-κB signaling pathways, as will be further detailed in Section 5.

4. Regulated Ubiquitination of IκBs and NF-κB Precursors

4.1. Regulated Ubiquitination of IκBs

As discussed above, the critical step in NF-κB activation is the phosphorylation-induced ubiquitination and degradation of IκBs, allowing NF-κB dimers to translocate into the nucleus. In the case of the three classic IκB proteins, IκBα, IκBβ and IκBε, a similar amino-acid sequence is targeted for phosphorylation by IKK. It includes two serine residues located within a DSGXXS motif [44] also found in other proteins whose activity is controlled by proteasome degradation such as β-catenin and mouse double minute 2 homolog (Mdm2). This "degron" is recognized by the E3 ligase Skp, Cullin, F-box (SCF) containing complex [45,46]. More specifically, phosphorylation of IκBs Serine residues (Ser32/36 for IκBα, Ser19/23 for IκBβ and Ser18/22 for IκBε) induces the recruitment of SCF through β-transducing repeat-containing protein (β-TrCP), an F-box/TrpAsp 40 aa (WD40)-repeat protein. Within the SCF complex β-TrCP interacts with S-phase kinase-associated protein 1 (Skp1), bringing other components such as Cullin, RING-box protein 1 (Rbx1) and the E2 conjugating enzymes UBCH5b, UBCH5c or cell division cycle 34 (CDC34)/UBC3 [47] in proximity to the end-terminus of IκB (Figure 5). This allows K48-linked chain addition to Lys21/22 of IκBα, Lys9 of IκBβ and Lys6 of IκBε for degradation by the 26S subunit of the proteasome. Differences exist between the degradation kinetics of IκBα, IκBβ and IκBε, and the generating waves of NF-κB dimers. This might be caused by differences in degrons environment influencing their phosphorylation by IKK [48].

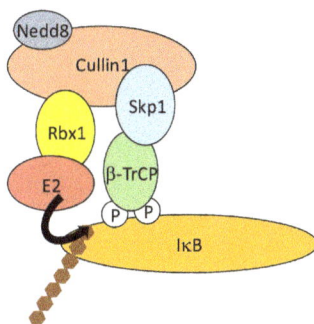

Figure 5. The degradation machinery of IκBs. The SCF E3 ligase complex that induces degradative ubiquitination of IκBs is depicted. K48-linked polyubiquitination is indicated with brown hexagons. See text for details.

The SCF activity is regulated by neddylation, a post-translational modification sharing similarities with ubiquitination (see above). The NEDD8-conjugated subunit is Cullin [49] and neddylation involves the E2 conjugating enzyme Ubc12, which helps recruit E3 ligases to SCF [50]. SCF activity can also be shut-off by E3 ligase TRIM9, at least in the brain [51]. E3 activity of TRIM9 is not required in this case. Instead, a phosphorylated degron within TRIM9 competes for β-TrCP binding to phosphorylated IκB. Finally, SCF components have been shown to be targeted by viral proteins to block NF-κB function in the anti-viral response. For instance, Non-Structural Protein 1 (NSP1), a rotavirus-derived protein induces the ubiquitination-dependent proteasomal degradation of β-TrCP [52].

Ubiquitination of IκBs is also attenuated by several DUBs, resulting in inhibition of the NF-κB activation process. First, USP11 has been shown to associate with IκBα and to catalyze its deubiquitination in the TNF-α signaling pathway [53]. Second, the constitutive photomorphogenesis 9 (COP9) signalosome, that regulates the assembly and activity of Cullin-E3 ligases, may inhibit sustained IκBα degradation by inducing its deubiquitination by associated USP15 [54]. Another ubiquitin moiety that may regulate IκBs degradation is monoubiquitination. Indeed, a pool of monoubiquitinated IκBα has been identified as insensitive to TNF-α-induced degradation through impaired phosphorylation [55]. Previously, it had been proposed that sumoylation of IκBα may also influence its stability or fate [56]. These processes remain poorly characterized, especially concerning the modified residues and the proteins involved.

4.2. Regulated Ubiquitination of p105

Processing of p105 to generate NF-κB subunit p50 occurs constitutively in resting cells, at low level, and can be amplified to a little extent upon stimulation. In addition, p105 associates with other NF-κB subunits, such as dimeric p50 or RelA, and can be either fully degraded or processed to p50 upon cell activation (Figure 6). Until now, how these various processes are coordinately regulated is still poorly understood, considering their dependency on the enzymatic machinery of the proteasome which is supposed to proteolyze its substrates to completion.

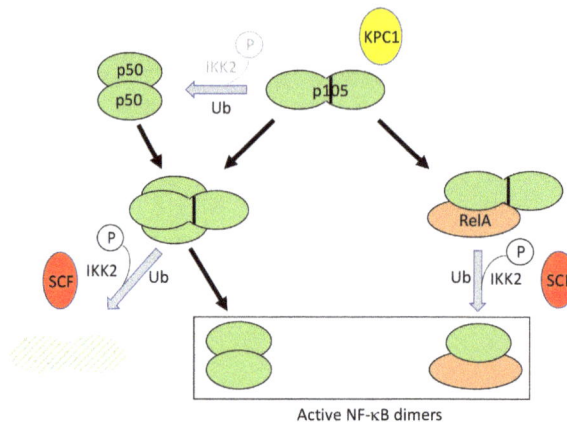

Figure 6. Regulated processing/degradation of p105. KPC1-dependent constitutive processing of p105 to generate p50, which can be slightly augmented upon IKK2 activation, is shown at the top. Complete proteolysis or limited processing to release active NF-κB dimers (p50/p50 or p50/RelA) upon cell activation is shown at the bottom. See text for details.

Originally, it was demonstrated that basal p105 processing to p50 required K48-linked ubiquitination and involved a long Gly-Ala repeat acting as a stop signal for degradation in the middle of the molecule [57]. Only recently the E3 ligase involved in p105 ubiquitination, Kip1 ubiquitylation-promoting complex 1 (KPC1), has been identified upon chromatographic purification [58]. It has been shown to interact with p105 through its Ankyrin repeats. Interestingly, this interaction is increased upon phosphorylation of p105 at Ser927 by IKK2.

In addition, p105 can also be totally degraded upon cell stimulation, releasing its associated NF-κB dimers [59]. In this case, an IKK2-dependent phosphorylation first occurs at residues 927 and 932, located in sequences exhibiting similarities to the IκB degrons (see above), inducing recognition and ubiquitination by the SCF/β-TrCP complex [60,61].

It still remains to be understood how the processing versus degradation choice is made. It has been proposed that the Gly-Ala repeats preceding, during nibbling by the proteasome, a tightly folded domain at a precise distance would be sufficient to halt proteolysis of p105 [62], but how this can be bypassed remains unknown.

Finally, processing and degradation of p105 can be negatively regulated by the DUB A20. During processing, A20 can interact with p105 through KPC1 and inhibit its ubiquitination [63].

4.3. Regulated Ubiquitination of p100

In contrast to p105, the processing of p100 for generating p52 is exclusively inducible and dependent on IKK1 activation by NIK [64–66]. At the C-terminus of p100, a ^{865}DSAYGS^{870} sequence is located, similar to the one found in IκB proteins. Upon its phosphorylation by IKK1, this sequence is recognized by SCF^{TrCP}, inducing ubiquitination of p100 with K48-linked chains and processing to generate p52 [67,68]. As for p105, it is unclear how a limited proteolysis of p100 by the proteasome is achieved.

The steady-state level of p100 itself is controlled by another SCF complex: SCF^{fbw7} [69,70]. Precursor p100 contains two conserved sequences which exhibit similarities to the TPPLSP degron recognized by F-box/WD repeat-containing protein 7 (Fbw7), a member of the F-box family of proteins. Ser and Thr residues within these sequences are constitutively phosphorylated by glycogen synthase kinase-3 (GSK3). This induces p100 recognition by SCF^{fbw7} and K48-linked polyubiquitination, triggering p100 degradation by the proteasome. How this control of the p100 amount influences the non-canonical NF-κB pathway remains unclear. It may limit the amount of p52 produced in the cytoplasm. Alternatively, as proposed by Busino et al. [70], p100 elimination through SCF^{fwb7} may occur in the nucleus, decreasing the level of an inhibitory molecule and resulting in more efficient NF-κB activation.

In the following sections we will describe the major upstream signaling pathways activating NF-κB, focusing on ubiquitin-related events permitting and regulating signal transduction. The protein/protein interfaces that are involved will not be described in detail although they provide the primary level of specificity. Instead, excellent reviews dealing with molecular organization of the mentioned proteins will be referred to. Ubiquitin modifications affecting components of the TAK1 and IKK complexes represent shared features of most, if not all, of the signaling pathways described and will be presented in Section 6.

5. Regulated Ubiquitination during Intracellular Signal Transduction

5.1. The TNF-R1 Signaling Pathway

TNF-α is a pleiotropic inflammatory cytokine that binds to two distinct receptors, TNF-R1 and TNF-R2, with TNF-R1 exhibiting the broader cellular distribution. The TNF-R1 signaling pathway is by far the most extensively analyzed NF-κB activation pathway and provides the best integrated example to illustrate how various ubiquitination processes contribute to the formation of multiprotein complexes and triggers signal transduction. So far, at least a dozen of distinct proteins has been involved in the building and activity of the so-called TNF-R1 complex 1 that initiates NF-κB signaling (Figure 7). They are all heavily regulated by PTMs, many of them involving ubiquitin. These modifications, which require several E3 ligases operating at distinct levels, generate numerous active interfaces through UBDs. Equally important is the role of negative regulators, mostly of the deubiquitinase family, that ensure the fine-tuning of signal transduction, controlling both the level and the duration of the activation process.

Upon TNF-α exposure TNF-R1 trimerizes and recruits the adaptor tumor necrosis factor receptor type 1-associated death domain protein (TRADD) at its intracytoplasmic domain [71,72]. This allows TRADD to attract both kinase receptor-interacting serine/threonine-protein kinase 1 (RIPK1) through its death domain and E3 ligase TRAF2 through its N-terminus, with the possible participation of adaptor Src-associated in mitosis 68 kDa (Sam68) [73]. Subsequent polyubiquitination of RIPK1 is a key node event in signal transmission [74]. Unexpectedly, TRAF2 does not behave as the RIPK1 E3 ligase [75] in this setting. Instead, it plays a role as a scaffold protein recruiting two other E3 ligases, cellular inhibitor of apoptosis protein-1 (c-IAP1) and c-IAP2, that add K63- and K11-linked chains to RIPK1. This TRAF2/c-IAP1/2 interaction and the resulting polyubiquitination of RIPK1 triggers the recruitment of the E3 LUBAC complex that targets several components of the complex 1, including RIPK1 and c-IAPs themselves, through the addition of M1-linked ubiquitin chains [76].

At this stage, polyubiquitinated RIPK1 attracts kinase complexes TAK1 and IKK, through their UBD-containing subunits TAB2/TAB3 and NEMO, respectively. This eventually triggers IKK phosphorylation by TAK1 and NF-κB activation. Importantly, although a model describing K63-linked ubiquitin chains attracting TAK1 through TAB2/TAB3 on one side and M1-linked chains attracting IKK through NEMO on the other was originally proposed, the recent discovery of synthesized mixed polyubiquitinated chains [77] suggests instead an even more promiscuous mechanism with optimal proximity between IKK and its kinase TAK1 [78], further amplified by LUBAC-induced M1-linked ubiquitination of NEMO itself inducing IKK auto-aggregation.

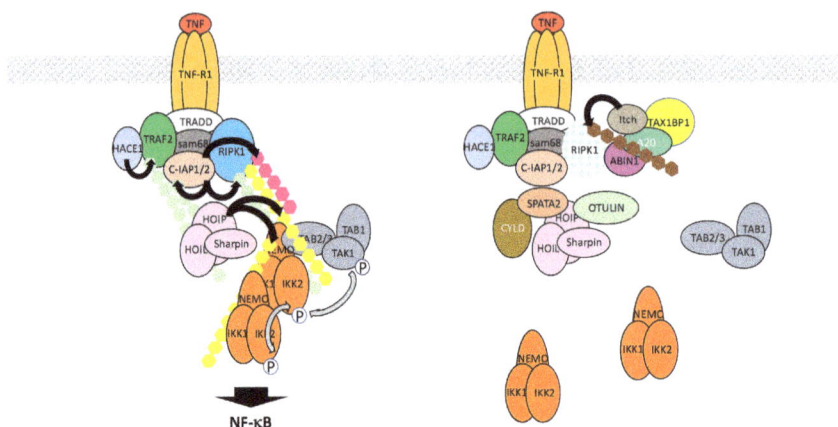

Figure 7. TNF-R1 signaling pathway. Components and mechanisms ensuring signal transduction in this pathway are depicted on the left, with black arrows indicating ubiquitination processes and grey arrows phosphorylation. M1-, K11- and K63-linked polyubiquitination is indicated with yellow, pink and green hexagons, respectively. Components and mechanisms participating in signal shut-off are presented on the right. An induced proteolysis of RIPK1, in addition to its deubiquitination, is indicated although its relevance in NF-κB signaling is uncertain. K48-linked polyubiquitination is indicated with brown hexagons. See text for details.

This set of events represents a consensual basic model of IKK/NF-κB activation upon TNF-α exposure (see below for its limitations and several controversial issues). In addition, several other layers of regulation are likely to operate, involving a collection of components. At this stage, it is still difficult to unequivocally distinguish their real contribution to the NF-κB activation process from their role in regulating the so-called RIPK1 «switch» that control the cell survival versus death decision [79]. Indeed, upon TNF-α exposure a cell is subject to a distinct fate depending on the post-translational status of RIPK1 and its interaction with specific partners. As discussed above, ubiquitination of RIPK1 in complex 1 at the cell membrane contributes to NF-κB activation by allowing recruitment of IKK and its activating kinase TAK1. Nevertheless, RIPK1 is also a critical component of the cytoplasmic complex 2 that forms with pro-caspase 8 and Fas-associated protein with death domain (FADD) upon TNF-α exposure and triggers apoptotic cell death. In the case of NF-κB activation, the complex 2 is kept inactive through the neutralization of pro-caspase 8, while in contrast in the absence of NF-κB activation, caspase 8 is activated. An additional NF-κB-related brake on complex 2 activation is the phosphorylation of RIPK1 by IKK [80]. Another RIPK1-containing complex (complex 3) can also form, containing the related protein RIPK3, when caspase 8 or FADD activity is abolished. It induces death through necroptosis and involves mixed lineage kinase domain-like pseudokinase (MLKL), a substrate of RIPK3 inducing membrane pores and cell lysis [81].

Among the regulators that affect ubiquitin-related events in NF-κB activation but also control the RIPK1 switch are several DUBs [82]. The first to be identified was A20, also known as TNF alpha induced protein 3 (TNFAIP3), which is encoded by an NF-κB regulated gene and is a member of the OTU family of deubiquitinases. It has been claimed that it also exhibits E3 ligase activity [83] but it remains unclear whether this is an intrinsic activity within cells or an activity due to its interaction with E3 ligases such as Itch and RING finger protein 11 (RNF11) [84]. As a consequence, A20 can deubiquitinate the K63-linked chains of RIPK1 but may also regulate its stability through direct or indirect induction of K48-linked polyubiquitination. Its own ubiquitin protease activity appears regulated by IKK2-induced phosphorylation [85] providing another regulatory link between this enzyme and the NF-κB signaling pathway. Finally, A20 interacts with A20-binding inhibitor of NF-κB activation 1 (ABIN1), a protein with an affinity for ubiquitin, and this interaction plays a critical function in its recruitment to TNF-R1 complex 1 and RIPK1 deubiquitination [86]. To what extent A20 impacts on NF-κB signaling remains uncertain. Wertz et al. [85] have shown that due to its specific affinity for K63-linked chains and the absence of M1-linked chains recognition by A20, it modestly affects the NF-κB activation process and acts mostly on TAK1 dependent signaling which also connects to MAPK pathways.

The second DUB regulating ubiquitination process upon TNF-α stimulation is CYLD. CYLD is a divergent member of the USP family that specifically hydrolyzes K63-linked or M1-linked chains [87]. To some extent its expression can be modulated by NF-κB but it seems to be already present in resting cells and could limit the activation process of NF-κB by deubiquitinating RIPK1 [88]. Other CYLD substrates may include its two direct partners NEMO or TRAF2 [89]. Recently, Spermatogenesis-associated protein 2 (SPATA2) has been identified as a critical protein for recruiting CYLD to LUBAC within complex 1 through HOIP interaction [90]. In several instances, often linked to cell transformation and cancer, an up-regulation of NF-κB activity has been associated with the decreased expression or activity of CYLD [91].

OTULIN is another DUB whose function has been extensively characterized in the context of TNF-R1 signaling. It is a member of the OTU family of DUBs that exhibits a very specific and strong affinity for M1-linked chains [92,93]. Consequently, it regulates LUBAC-induced ubiquitination after interacting with HOIP through a PUB/PIM interface [94,95]. Upon over-expression OTULIN blocks NF-κB activation, whereas its down-regulation results in amplified NF-κB activation, most likely by controlling the level of NEMO M1-linked ubiquitination.

Other USPs may also participate in the NF-κB activation process in response to TNF-α but their exact mode of action and specific target(s) remain uncertain. USP2 has been shown to negatively regulate NF-κB activity in one case [96] and positively in another [97]. Again, the level of RIPK1 ubiquitination has been shown to depend on this enzyme. Two other USPs have been reported to control the activity of the same target. First, USP4 negatively regulates TNF-α-induced NF-κB activation by removing K63-linked chains on RIPK1 upon interaction [98]. Incidentally, this DUB has been reported to also target TRAF2 [99] and additional components of the NF-κB signaling pathway (see Sections 5.2 and 6.1). Second, USP21 also displays affinity for RIPK1 and acts as a RIPK1 deubiquitinase negatively regulating NF-κB activation [100]. Whether these DUBs cooperate and work in the same way in different cell types or cooperate remain unknown.

The activity/stability of other complex 1 components can also be regulated by ubiquitination processes. First, TRAF2 has been shown to be modified by K63-linked polyubiquitin upon TNF-α exposure and this requires its previous phosphorylation at Thr117 by protein kinase Cδ (PKCδ) and PKCε [101]. The E3 ligase involved may be HECT domain and ankyrin repeat containing E3 ubiquitin protein ligase 1 (HACE1) [102]. K63-linked polyubiquitination of TRAF2 would help recruit TAK1 and IKK. TRAF2 expression level can also be controlled by K48-linked phosphorylation for proteasome recognition. This may be achieved by the carboxy terminus of Hsc70 interacting protein (CHIP), a U-box-dependent E3 ligase that interacts with TRAF2 [103].

The regulation of c-IAPs amount and activity in the TNF-R1 signaling pathway is also controlled by ubiquitin-related events. These proteins exhibit potent E3 ligase activity and are able to ubiquinate a collection of partners (TRAF2 may be one of them) or to auto-ubiquitinate, for proteasomal degradation.

This latter property, degradation through self-ubiquitination, can be exploited for therapeutic purpose with Smac mimetics [104]. The deubiquitinase OTUB1 modulates this step by disassembling K48-linked chains from c-IAPs [105]. USP19 has also been shown to interact with c-IAPs and to prevent c-IAP2 degradative ubiquitination, leading to protein stabilization [106]. It would be interesting to study how USP19 impacts on NF-κB, given its influence on apoptosis.

So far, nothing has been reported regarding the control of LUBAC stability during the normal TNF-R1 signaling process. Nevertheless, it is worth mentioning that invasion-plasmid antigen-H proteins 1.4/2.5 (IpaH1.4/2.5), *Shigella* modulators of innate immune signaling, blunt the NF-κB pathway by interacting with HOIL-1L and conjugating K48-linked chains to HOIP for degradation [107].

Despite the wealth of data regarding the components and mechanisms involved in NF-κB activation by TNF-α, important issues concerning the exact role of the key participants still need to be clarified. For instance, RIPK1 has been presented above as critical in IKK activation (and it is also crucial in cell death induction, but this is not the focus of this review) but specific situations in which it is not required for NF-κB activation have been reported. Indeed, Wong et al. [108] have shown that several murine cell types are still able to degrade IκBα in response to TNF-α when *Ripk1* is invalidated. One may therefore wonder whether what initially appears as a very elaborated molecular system for recruiting kinases to activate NF-κB, can be fairly well compensated by extensive polyubiquitination events occurring on other components of complex 1 such as TRAF2, c-IAPs and LUBAC, with only "subtle" effects on the timely regulation of the process. In this regard, Xu et al. [109] have shown, using an elegant ubiquitin replacement strategy, that K63-linked poly-ubiquitination is dispensable for TNF-α-mediated NF-κB activation. Thus, other modifications involving M1-linked ubiquitin chains may be sufficient to induce NF-κB, at least in cell culture conditions. Something in accord with this flexibility has been reported by Blackwell et al. [110]. Of course, these particularities may not apply to the RIPK1 "switch" controlling cell death.

The other important question relates to the true function of TRAF2 within complex 1. As mentioned above, TRAF2 is considered to play a role as a scaffold allowing the recruitment of c-IAPs for RIPK1 polyubiquitination. This is based on the fact that a catalytically inactive version of this E3 ligase, mutated in its catalytic RING finger domain, does not affect the NF-κB activation process [75]. In contrast, it has been proposed that the E3 ligase activity of TRAF2 participates in TNF-R1 signaling with sphingosine kinase 1 (Sphk1) as a co-factor [111]. However, recent studies using *Sphk1* KO cells do not support this model [112]. Nevertheless, TRAF2 has also been shown to be backed up by TRAF5, at least in MEFs, and enzymatic compensation may operate [113]. Finally, a paradoxical negative role of TRAF2 in TNF-R1 signaling has been observed in several instances. In particular, it has been reported that *Traf2/Traf5* KO MEFS exhibit up-regulation of NF-κB before and after TNF-α exposure [114]. Basal NF-κB activation may result from up-regulation of the non-canonical NF-κB pathway (see Section 7) but increased response to TNF-α is more difficult to explain. In any case, this suggests that while all primary players in the TNF-R1 signaling pathway might have been identified, details of their molecular functions still require further investigation.

5.2. The IL-1β R/TLR Signaling Pathways

Interleukin-1β receptor (IL-1βR), the receptor for inflammatory cytokine IL-1β, and toll-like receptors (TLRs), the receptors involved in the recognition of PAMPs, share similarities with regard to their mechanism of NF-κB activation [115,116]. This is due to the presence in their intracytoplasmic domain of the so-called toll-interleukin receptor (TIR) domain which interacts with a collection of related TIR-containing adaptor molecules [117]. These adaptors allow recruitment of the same set of proteins to transmit the signal that activates IKK. Nevertheless, differences exist that are mostly related to the ability of a subset of TLRs to additionally induce NF-κB from intracellular locations.

In the case of IL-1βR, interaction of a dimer composed of IL-1βR itself and IL1 receptor accessory protein (IL1RAP) with IL-1β allows the recruitment of toll-interleukin receptor adaptor protein (TIRAP) and myeloid differentiation primary response 88 (Myd88) at the TIR domain of IL-1βR, both proteins displaying TIR domains (Figure 8). The death domain (DD) of Myd88 then attracts sequentially DD-containing interleukin-1 receptor-associated kinase 4 (IRAK4), IRAK1 and IRAK2 kinases to form an oligomeric complex called the myddosome [118,119]. Through their PXEXXE motifs IRAK1 and 2 induce the recruitment and multimerization-induced activation of E3 ligase TRAF6 [120]. Concomitantly, the other E3 ligases Pellino1 and 2 interact with IRAK1 or 4 through their forkhead-associated (FHA) domain [121–123] and are activated by phosphorylation. Together with TRAF6, they synthesize K63-linked chains that are recognized by the TAK and IKK complex, triggering IKK activation. In this context, Pellino proteins appear to play the main role in K63-linked chain synthesis, mostly targeting IRAK1, while TRAF6 fulfills an additional scaffolding function by recruiting LUBAC. This allows full activation of NF-κB through the synthesis of M1-linked chains of ubiquitin and recruitment of IKK through NEMO. Again, K63- and M1-linked polyubiquitin chains may not be synthesized independently. Indeed, it has been shown that hybrid chains are formed and attached to IRAK1 [124]. In addition, unanchored chains are also synthesized [125] but their exact function remains uncertain.

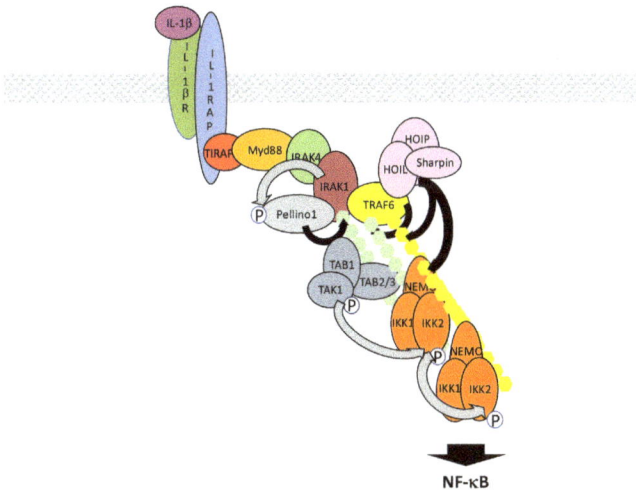

Figure 8. IL-1βR1 signaling pathway. Components and mechanisms ensuring signal transduction in this pathway are depicted, with black arrows indicating ubiquitination processes and grey arrows phosphorylation. M1- and K63-linked polyubiquitination is indicated with yellow and green hexagons, respectively. See text for details.

The whole process described here is tightly time-regulated and may rely on sub-complexes with distinct intracellular locations. It has been proposed that after initiation of the activation process at the membrane a TRAF6/TAK1-containing complex translocates to the cytosol [126], with this event requiring TRAF6/c-IAP-dependent degradation of Myd88-associated TRAF3 [127].

This dependency on intracellular localization is further illustrated with data concerning signal transmission by TLR family members. In the case of TLR4, which recognizes Gram-negative bacteria-derived LPS in combination with myeloid differentiation factor 2 (MD2) and/or cluster of differentiation 14 (CD14), the signaling pathway described above for IL-1βR operates and generates a first wave of NF-κB activation. Then, TLR4 is internalized and relocated to the endosome where it generates a second wave of NF-κB activation. At this step, the TIR domain of TLR4 attracts TRIF-related adaptor molecule (TRAM) which itself binds to protein TIR domain-containing adaptor-inducing interferon-β (TRIF) [128]. TRIF-dependent signaling to NF-κB involves members of the TRAF family such as TRAF2 and TRAF6 which both interact with TRIF through TRAF binding domains [129]. Interestingly, TRIF also interacts with RIPK1 through RIP homotypic interaction motif (RHIM)/RHIM homotypic binding [130,131]. K63-linked ubiquitination of TRIF and TRAF6 is supposed to allow recruitment of TAK1 and IKK for activating IKK but molecular details are missing, especially those concerning the exact function of RIP, maybe in relation with TRAF2. Another E3 ligase, Pellino1, has to be considered since it is necessary for TRIF-dependent NF-κB activation and would participate in RIPK1 recruitment and ubiquitination [132]. Intracellular members of the TLR family recognizing a diversity of ligands such as TLR3 (ligand: double strand RNA), TLR7 (ligand: single strand RNA) and TLR9 (ligand: CpG DNA), also activate NF-κB through this TRIF-dependent pathway.

Ubiquitin-related negative regulation of the IL-1β/TLR signaling pathways occurs at several levels. At the membrane level, membrane-associated E3 ligase membrane associated RING-CH 8 (MARCH8) down regulates IL1RAP amount through K48-linked ubiquitination [133]. This results in decreased recruitment of Myd88 and IRAK-1 after IL-1β stimulation. After its binding to Myd88, phospho-activated OTUD4 is the DUB that removes K63-linked ubiquitin chains from Myd88, IRAK1 and TRAF6, decreasing IL-1βR and TLR4 signaling [134]. Among these components ubiquitinated TRAF6 represents a regulatory target for a large collection of additional proteins. Among them are USPs such as USP2a, USP4 and USP20 that digest K63-linked ubiquitin chains and inhibit IL-1β/TLR signaling [135–137]. CYLD, A20, Itch, small heterodimer partner (SHP) and TRAF family member-associated NF-κB activator (TANK) also negatively regulate K63-linked ubiquitination of TRAF6 through distinct mechanisms. CYLD and A20 are DUBs, the latter one working with Itch [138,139], whereas SHP and TANK are TRAF6-binding proteins indirectly affecting the TRAF6 ubiquitination status [140,141]. In the case of TANK, a TANK-monocyte chemotactic protein-induced protein-1 (MCPIP1)-USP10 complex operates, with TANK playing an adaptor role for TRAF6 deubiquitination by USP10. Finally, RNF19A, through nod-like receptor protein 11 (NLRP11), WW domain-containing protein ligase 1 (WWP1) and TRIM38 are E3 ligases that conjugate TRAF6 with K48-linked chains, causing its degradation [142–144].

Another component of the TLR4-dependent NF-κB activation process whose function is fine-tuned by a ubiquitin-mediated process is TRAF3 (see above). Its ubiquitination-dependent degradation is down-regulated by USP25 [145].

5.3. The Nod1/Nod2 Signaling Pathway

Nucleotide-binding oligomerization domain 1 (Nod1) and Nod2 are two cytoplasmic members of the NOD-LRR (leucine-rich repeat) family with CARD (caspase recruitment domain) (NLRC) that recognize bacterial peptidoglycans [146–148]. Nod1 detects γ-D-glutamyl-mesodiaminopimelic acid (iE-DAP) from Gram-negative bacteria and a subset of Gram-positive bacteria. Nod2 detects muramyl dipeptide (MDP) structures from both Gram-positive and Gram-negative bacteria. They both exhibit one (Nod1) or two (Nod2) CARD domains at the N-terminus, followed by a NOD domain and a series of LRRs. These repeats are responsible for ligand binding.

In resting cells, Nod proteins present an autoinhibited monomeric conformation. Upon interaction with their ligands they self-oligomerize and recruit RIPK2 through homotypic CARD-CARD interactions. This triggers a complex set of phosphorylation and ubiquitination events whose ultimate function is to allow the recruitment of IKK and its kinase TAK1 (Figure 9). Surprisingly, the tyrosine kinase activity of RIPK2, but not the serine/threonine one, operates, inducing RIPK2 autophosphorylation at residue Tyr474 [149]. Ubiquitination of RIPK2, which involves residue Lys209, also occurs upon its interaction with Nod2 [150,151]. Several distinct E3 ligases may regulate this process. Among them are c-IAPs, X-linked inhibitor of apoptosis protein (XIAP), Pellino3 and TRAFs such as TRAF2, TRAF5 and TRAF6. C-IAPS have been shown to interact with RIPK2 and to induce its K63- and K48-linked ubiquination [152]. XIAP has also been reported to interact with the kinase domain of RIPK2 through its baculovirus inhibitor of apoptosis protein repeat 2 (Bir2) domain but it is not clear which kind of polyubiquitin chains it adds to RIPK2 [153]. Pellino is a third E3 ligase that interacts with RIPK2, through its FHA domain, and triggers its K63-linked polyubiquitination [154]. All these ligases appear necessary for inducing NF-κB in response to Nod2 activation, so it is likely that they perform distinct functions. C-IAP/RIPK2 and Pellino3/RIPK2 interactions are correlated to K63-linked ubiquitination of RIPK2 and this may induce the recruitment of the TAK complex. Moreover, it has been shown that c-IAP2 promotes RIPK2 tyrosine phosphorylation. In contrast, XIAP may play a scaffold role to attract both the TAK complex though direct interaction with its Bir1 domain and, more importantly, the E3 LUBAC complex [155,156]. As a consequence, LUBAC would synthesize M1-linked chains, allowing the recruitment of IKK through NEMO. As in other signaling pathways presented above, induced proximity of TAK and IKK complexes would result in IKK activation. Again, hybrid ubiquitin chains may represent the genuine anchor during this process [77].

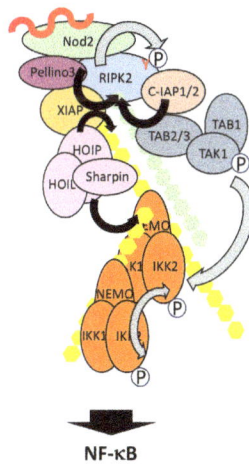

Figure 9. Nod1/2 signaling pathway. Components and mechanisms ensuring signal transduction in this pathway are depicted, with black arrows indicating ubiquitination processes and grey arrows phosphorylation. M1- and K63-linked polyubiquitination is indicated with yellow and green hexagons, respectively. Auto-activating Tyrosine phosphorylation of RIPK2 is indicated with a red Y. See text for details.

The role of TRAFs in Nod2 signaling is more controversial. It was originally reported that TRAF6 was essential for NF-κB activation [157] but this has not been confirmed by other studies. In particular, Hasegawa et al. [151] reported that instead of TRAF6, the other two TRAFs which often work jointly, TRAF2 and TRAF5 (see Section 5.1), were the active ligases for RIPK2. Considering what is said above about the other E3 targeting RIPK2, such observation requires further investigation.

Fine-tuning of Nod2 signaling and shut-off is achieved through several mechanisms. First, like Nod2, RIPK2 is kept in check in the cytoplasm before cell stimulation though its interaction with mitogen-activated protein/ extracellular signal-regulated kinase (ERK) kinase 4 (MEKK4) [158]. Upon Nod2 activation RIPK2 dissociates from MEKK4 to interact with Nod2. Second, modulation of activity and stability of several components are regulated by complex ubiquitination processes involving E3 ligases and DUBs. As described in other signaling pathways, OTULIN, through its recruitment by LUBAC, can hydrolyze M1-linked chains of ubiquitin to stop signal transmission [93]. LUBAC also provides a platform for recruiting SPATA2/CYLD with the same outcome [159,160]. Moreover, E3 ligase Itch acts as an inhibitor of Nod2 signaling though ubiquitination and inactivation of RIPK2 [161]. Interestingly, effective function of Itch requires RIPK2 tyrosine-phosphorylation. This suggests a dual positive and negative function of tyrosinated RIPK2: after allowing Nod2 signaling through a still unclear mechanism it may participate in its shut-off. Since A20 has been shown to negatively regulate Nod2/RIPK2 activity through RIPK2 deubiquitination [159,162] an A20/Itch complex may be at play, as in the TNF-R1 signaling pathway (see Section 5.1). Another E3 ligase, zinc and ring finger 4 (ZNRF4), also acts at the level of RIPK2 by controlling its level through K48-linked polyubiquination [163]. The stability of active Nod2 is itself regulated by TRIM27, which conjugates Nod2 with K48-linked polyubiquitinated chains for degradation by the proteasome [164]. Finally, the SH2-containing inositol phosphatase (SHIP) negatively regulates the Nod2/NF-κB signaling pathway by impairing interaction between XIAP and RIPK2 [165].

5.4. The MAVS Pathway

Intracellular retinoic-inducible gene-I (RIG-I)-like receptors (RLRs) regulate the synthesis of type-I interferons (IFNs) following virus-derived RNA recognition [166–168]. The RLR family includes RIG-I, melanoma differentiation-associated gene 5 (MDA5) and Laboratory of Genetics and Physiology 2 (LGP2), which all bind dsRNA through their death effector domain (DEAD)/H-box RNA helicase domain. Through their CARD, RIG-I and MDA5 then activate mitochondrial antiviral-signaling protein (MAVS), which in turn activates both the NF-κB and IRF signaling pathways to trigger IFNs production. LGP2, which is devoid of CARD, was originally thought to be a negative regulator of this process but may instead facilitate viral RNA recognition by RIG-I and MDA5 through its ATPase domain [169].

Not surprisingly, several ubiquitination events regulate MAVS-dependent signal transduction [170,171] (Figure 10). First, upon recognition of viral RNA exposure of CARD domains induces K63-linked polyubiquitination of RIG-I at Lys172 by TRIM25 [172], with the help of cyclophilin A (CypA) [173]. This induces formation of high order RIG-I oligomers through the CARDs of RIG-I that exhibit affinity for K63-linked chains [174]. Unanchored K63-linked chains may also participate in this process [175]. Other E3 ligases have been shown to also induce K63-linked polyubiquitination of RIG-I. Riplet could help opening RIG-I and facilitate its ubiquitination by TRIM25 [176]. Following the additional identification of TRIM4 and MEX-3 homolog C (C. elegans) (MEX3C) as E3 ligase for RIG-I [177,178] Sun et al. [179] have proposed a hierarchical model of RIG-I activation by K63-linked ubiquitination. The same process of induced activation by K63-linked polyubiquitination occurs with MDA5. In this case, TRIM65 represent the active E3 ligase [180]. Then, oligomerized RIG-I and MDA5 trigger extensive polymerization of MAVS at the mitochondria surface through a prion-like mechanism [181], allowing recruitment of proteins participating in NF-κB and/or IRF activation.

We will focus here on the proteins involved in NF-κB activation, neglecting both their additional function in IRF activation and the proteins specifically regulating this parallel process. Among these participants are primarily members of the TRAF family. Indeed, MAVS harbors binding motifs for TRAF2, TRAF3, TRAF5 and TRAF6. Mutations of all these motifs is required to completely abolish NF-κB activation, with TRAF6, TRAF2 and TRAF5 performing redundant functions [182–184] whereas the role of TRAF3 is still controversial. The E3 activity of TRAFs is required and K63-linked polyubiquitination is necessary for the recruitment of IKK through NEMO but the exact modified partner(s) is(are) unknown. Indeed, a mutated TRAF6 that cannot be ubiquitinated remains active in this pathway. Moreover, although TRAF2 is found associated with NEMO, inhibiting its ubiquitination does not affect the NF-κB activation process either. Nevertheless, when the same inhibition is combined with a loss of HOIP expression, NF-κB activity is not induced [182] suggesting another layer of redundancy involving M1-linked chains. The participation of TAK1, which often plays the role of IKK activating kinase, is still uncertain, leaving the question open for the final step in IKK activation and its putative similarities to those described for other signaling pathways. It is worth noting that a MAVS/TRAF2/TAK1 pathway controlling MAPK p38 activation has been reported but its role in NF-κB activation has not been analyzed [185].

Modulation of MAVS activation by ubiquitin-related PTMs is complex [170]. As outlined above, TRIM25 induces K63-linked polyubiquitination of RIG-I for initiating signaling. The amount of TRIM25 is itself controlled by the DUB USP15, which removes K48-linked chains [186]. At this level, LUBAC has been proposed to also act through an unorthodox dual mechanism. It could induce TRIM25 degradation through proteasome degradation and compete with TRIM25 for RIG-I recognition [187]. Activating K63-linked chains of RIG-I are proteolyzed by USP3, USP21 and CYLD [188–191], whereas RNF122 and RNF125 can induce RIG-1 degradation through K48-linked polyubiquitination [192–194]. This process is inhibited by USP4 [195].

RNF125 has a broad impact on the MAVS signaling pathway since it can also act on MDA5 and MAVS to induce their degradation [193,194]. It is not the only E3 ligase inducing MAVS degradation by the proteasome. The HEC domain containing E3 ligases Smad ubiquitin regulatory factor 1 (Smurf1), through Nedd4 family interacting protein 1 (Ndfip1), [196] and Smurf2 [197] inhibit interferon induction by promoting K48-linked polyubiquitination of MAVS. MARCH5 binds to MAVS, only after its polymerization, to induce its degradation [198]. RNF5 also acts after signal induction [199]. Activity of these two ligases may be controlled by inactive Rhomboid 2 (iRhom2) [200]. Finally, after its induced synthesis following viral infection, poly(RC) binding protein 2 (PCBP2) negatively regulates the RIG-I/MAVS signaling pathway by interacting with MAVS and recruiting E3 ligase atrophin-1-interacting protein 4 (AIP4) for MAVS degradation by the proteasome [201].

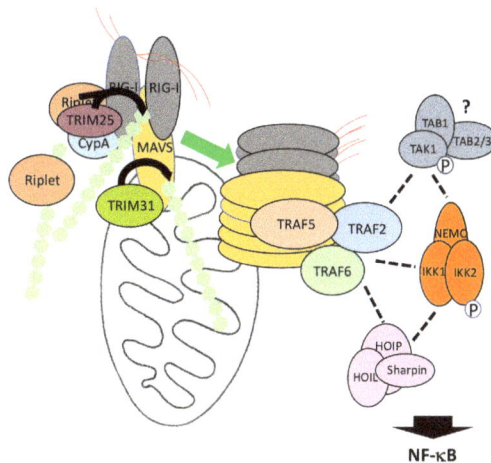

Figure 10. RIG-I/MAVS signaling pathway. Components and mechanisms ensuring signal transduction in this pathway are depicted, with black arrows indicating ubiquitination processes. K63-linked polyubiquitination is indicated with green hexagons. Components shown to be (formally or putatively (question mark)) required in this pathway but whose exact relationship is not defined are shown connected by broken lines. Red filaments represent activating double strand RNA. Activation occurs at the surface of the mitochondria. E3 ligases TRIM4 and MEX3C may also participate in RIG-I activating ubiquitination. See text for details.

5.5. The cGAS/STING Pathway

B-form DNA originating from invading/replicating pathogens or self-DNA causing auto-immune response is sensed in the cytoplasm by cyclic guanosine monophosphate (GMP)-adenosine monophosphate (AMP) synthase (cGAS) [202–204]. This induces the production of 2′-5′-cyclic guanosine adenosine monophosphate (2′-5′-cGAMP), an atypical cyclic dinucleotide second messenger that is recognized by endoplasmic reticulum (ER) membrane-associated protein stimulator of interferon genes (STING) [205] (Figure 11). Then, STING activates the NF-κB and IRF3 pathways to induce interferon production.

Details of NF-κB activation by STING are lacking but, as for other NF-κB pathways, this process is heavily regulated by ubiquitination. First, E3 ligases TRIM32 and TRIM56 promote K63-linked ubiquitination of STING to induce NF-κB signaling [206,207], most likely by recruitment of IKK through NEMO [206]. This step is negatively controlled by the DUB USP13 [208]. Second, the stability of STING is controlled by E3 ligases RNF5 and TRIM30α (murine specific) that target STING for degradation through K48-linked polyubiquitination [209,210]. This degradation is counteracted by E3 ligases USP18 and USP20 that deconjugate K48-linked chains from STING [211]. Through a distinct mechanism, involving SUMOylation, TRIM38 also interfere with K48 ubiquitination of STING [212].

Upstream of STING, cGAS activity is also regulated by ubiquitination. ER-associated E3 ligase RNF185 has been shown to induce K27-linked polyubiquitination of cGAS and to increase its activity [213]. Moreover, degradation of cGAS by K48-linked chains, carried out by an unknown E3 ligase, is inhibited by TRIM14 working in concert with USP14 [214].

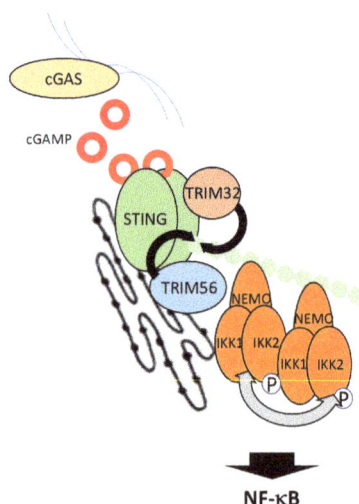

Figure 11. cGAS/STING signaling pathway. Components and mechanisms ensuring signal transduction in this pathway are depicted, with black arrows indicating ubiquitination processes and grey arrows phosphorylation. K63-linked polyubiquitination is indicated with green hexagons. Blue filaments represent activating double strand DNA. Activation occurs at the surface of the endoplasmic reticulum. See text for details.

5.6. The TCR/BCR Pathway

The T cell receptor (TCR) and B cell receptor (BCR) expressed on lymphocytes recognize foreign antigens and represent the pillars of adaptive immunity by allowing the mounting of the memory response. Both trigger NF-κB activation through similar components and processes. Proximal signaling at the TCR/BCR [215] will not be detailed here. We will only concentrate on the shared events following PKCθ and PKCβ activation in T cells and B cells, respectively [216,217]

(Figure 12). These kinases both targets CARD-containing a membrane-associated guanylate kinase (MAGUK) protein 1 (CARMA1), a member of a small family of adaptors displaying a MAGUK domain. In resting lymphocytes CARMA1 adopts a closed inactive conformation associated with the plasma membrane. Upon phosphorylation by PKCθ/β, CARMA1 opens and oligomerizes through its CC domains [218]. This provides a raft-associated platform for attracting through CARD/CARD interaction B cell lymphoma 10 (Bcl10), which is bound to mucosa-associated lymphoid tissue (MALT) lymphoma associated translocation protein 1 (MALT1). This ternary complex is called the CARMA1/Bcl10/MALT1 (CBM) complex. Through its TRAF6 binding motifs MALT1 then recruits TRAF6, which adds K63-linked polyubiquitin chains on several components of the CBM complex, among them MALT1 and Bcl10 [219]. MALT1/TRAF6 interaction may be reinforced by co-interaction with USP2a [220]. The resulting cluster of ubiquitinated proteins finally attracts the TAK and IKK complexes for NF-κB activation. This last step may occur in the cytosol [221,222].

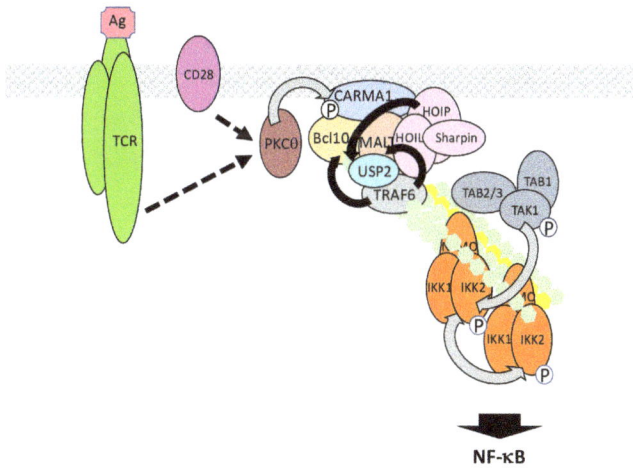

Figure 12. TCR signaling pathway. Components and mechanisms ensuring signal transduction in this pathway are depicted, with black arrows indicating ubiquitination processes and grey arrows phosphorylation. M1- and K63-linked polyubiquitination is indicated with yellow and green hexagons, respectively. Events occurring upstream of PKC activation are not shown and indicated by broken arrows. See text for details.

This basic model can be further refined taking into account the following observations. First, the E3 ligase involved in K63-linked CBM ubiquitination/IKK activation may not be uniquely TRAF6 since NF-κB activation still occurs in *Traf6* KO T lymphocytes [223]. TRAF2 could play redundant functions with TRAF6 to ensure CBM ubiquitination [219]. Alternatively, the E3 ligase mind bomb 2 (MIB2) may also be a candidate since it interacts with Bcl10 and induces K63-linked polyubiquitination of CBM components [224], Second, M1-linked chains of ubiquitin may also regulate TCR/BCR signaling. Nevertheless, the published data is controversial, even considering putative different requirements during TCR versus BCR signaling. Indeed, whereas LUBAC is necessary for TCR signaling, the catalytic activity of HOIP subunit is not required [225], suggesting a scaffold function of LUBAC maybe in relation with TRAF6. However, M1-linked polyubiquitination of Bcl10 has been observed following TCR stimulation [226] and during chronic B cell activation [227] suggesting the specific requirement of LUBAC-dependent linkage of M1 chains. A possible explanation for the observations of Dubois et al. [225] would be that quantitative evaluation of TCR/BCR signal transduction upon abolished E3 ligase activity of LUBAC is complicated by the remaining K63-linked dependent activation. Finally, the identity of the kinase activating IKK is not firmly defined in this particular pathway. TAK1 represents an obvious candidate, given its broad involvement in NF-κB-inducing pathways, but it is dispensable in some situations [228]. Another MEKK, MEKK3, may fulfill the same function in parallel or redundantly [229].

Several negative regulators of ubiquitin-related processes in TCR/BCR signaling have been identified. A20 hydrolyzes K63-linked chains on MALT1 to shut-off NF-κB activation [230]. This function is itself counteracted by MALT1, a member of the paracaspase family able to cleave and inactivate A20 [231]. MALT1 also acts the same way on CYLD, which deubiquitinates TAK1 [232], and possibly TRAF6 and NEMO during TCR signaling. In this case, the cleavage of CYLD does not affect the amplitude of NF-κB signaling but participates in JNK activation [233]. Finally, MALT1 cleaves HOIL-1 to generate a dominant-negative version of LUBAC [234]. Therefore, a key function of MALT1 is to control the amplitude of TCR/BCR signaling, by regulating LUBAC and A20 levels, after the activating signal has been transmitted.

Another DUB that down-regulates NF-κB activation upon TCR stimulation is USP34 which acts at a step downstream of IKK activation [235]. It may regulate IκBα degradation. How, and to what extent, USP34 is specifically activated by TCR stimulation still requires further investigation.

Finally, the ubiquitin-associated and SH3 domain containing 3A (UBASH3A) is an intriguing negative regulator of TCR signaling to NF-κB. UBASH3A down-regulates IKK activation and Ge et al. [236] have proposed that it may work by binding to TAK1 and NEMO, therefore hindering their recognition of K63-linked chains. Very interestingly, the gene coding for this ubiquitin-binding protein is located at a type 1 diabetes risk locus. In addition, genetic variants of *UBASH3A* are associated with several other autoimmune diseases.

5.7. The Genotoxic Stress Pathway

DNA damaging agents, such as ionizing radiation or chemotherapeutic drugs, induce a genotoxic stress that triggers NF-κB activation, resulting in cell protection from death [237,238]. Understanding the molecular details of this process is therefore of the utmost importance for improving the efficiency of several cancer treatments.

In this specific situation, unusual PTMs of NEMO have been discovered to play critical functions in NF-κB activation through IKK. More specifically, a pool of free NEMO exists in resting cells and can shuttle between the cytoplasm and the nucleus (Figure 13). Miyamoto et al. [239] were the first to show that NEMO sumoylation modulates this shuttling upon DNA damage, causing nuclear accumulation of NEMO. SUMO conjugation occurs at Lys277/Lys 309 and protein inhibitor of activated STAT (signal transducer and activator of transcription) y (PIASy) is the SUMO E3 ligase involved in this process [240]. Interestingly, the binding site of PIASy on NEMO overlaps with its IKK binding site confirming that sumoylation acts on a free pool of NEMO.

In the nucleus, sumoylated NEMO has been proposed to interact with P53-induced protein with a death domain (PIDD) and RIPK1 [241] but the requirement for PIDD in the sumoylation process is controversial [242]. Alternatively, PolyADP-ribose polymerase 1 (PARP-1) may provide a link between DNA damage sensing and NEMO sumoylation. It has been shown that PolyADP-ribose-modified PARP-1 triggers the formation of a complex containing NEMO, PIASy and ataxia telangiectasia mutated (ATM), a kinase responding to DNA double strand breaks [243]. Nevertheless, *Parp* KO mouse embryonic fibroblasts have been shown to properly activate NF-κB upon exposure to DNA damaging agents. Therefore, the identity of the proteins regulating sumoylation of NEMO in the nucleus remains uncertain. In contrast, the role of ATM in signal transmission is firmly established. Indeed, it regulates a NEMO-dependent activation process of NF-κB following DNA damage [244]. Moreover, the same DNA damaging treatment induces NEMO phosphorylation by ATM at Ser85 [245]. Phosphorylation of NEMO is not required for its sumoylation but instead, controls its monoubiquitination at Lys277 and Lys309. The identity of the E3 ligase involved in this monoubiquitination and how sumoylated sites are converted to monoubiquitinated sites remains unclear. In any case, this PTM results in the nuclear export of NEMO accompanied with a fraction of ATM.

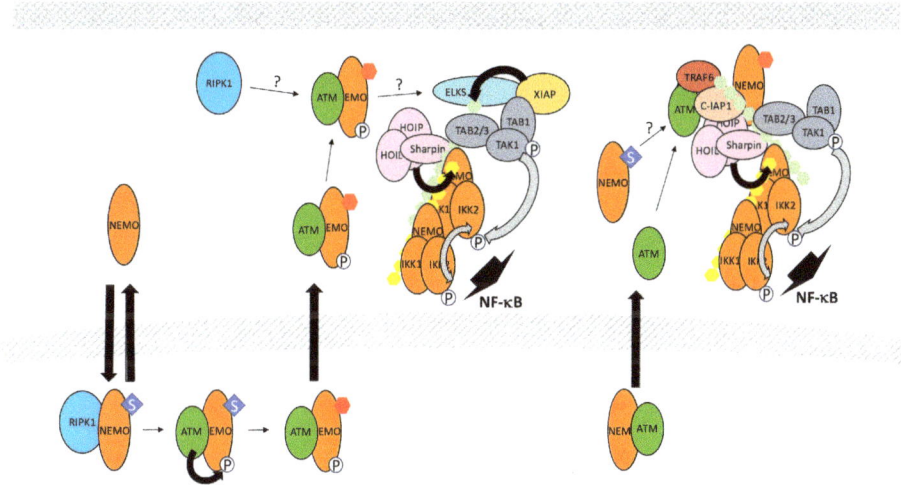

Figure 13. Genotoxic stress signaling pathway. Components and mechanisms ensuring signal transduction in the two proposed pathways are depicted, with black arrows indicating ubiquitination processes and grey arrows phosphorylation. M1- and K63-linked polyubiquitination is indicated with yellow and green hexagons, respectively. Blue squares labeled S indicate sumoylation whereas red hexagons indicate monoubiquitination. The participation of RIPK1 in the cytoplasmic events is indicated but with a question mark since how it relates to these two pathways, or another one, is unclear. See text for details.

How NEMO/ATM then participates in IKK activation in the cytoplasm is still a matter of debate. Clearly, the TAK1 complex is involved in this process since *Tak1* KO cells cannot activate NF-κB upon DNA damage. The complex NEMO/ATM may induce K63-linked ubiquitination of protein rich in amino acids E, L, K and S (ELKS) by E3 ligase XIAP through a mechanism that is still undefined [246]. This would then result in the recruitment and activation of TAK1. At the same time, the IKK complex would also be recruited and activated by TAK1, similarly to what has been described in other signaling pathways (see above). Additional proteins participate in this process. First, RIPK1, which plays a function in the nucleus events leading to NEMO modification, has been shown to also translocate in the cytoplasm, associated with NEMO and ATM, and is required for TAK complex recruitment [247]. Second, LUBAC also participates in the genotoxic stress pathway and modifies NEMO with M1-linked chains [248]. Surprisingly, in the Tergaonkar/Miyamoto study (246) the NUB domain of NEMO was shown to be dispensable for NF-κB activation upon DNA damage, something quite difficult to reconcile with a critical involvement of M1-linked and K63-linked polyubiquitin chains in TAK1-dependent NF-κB activation.

Alternatively, Hinz et al. [249] proposed that upon cytoplasmic release ATM interacts with TRAF6 through a TRAF6-binding domain and form a complex with c-IAP1 and NEMO to activate IKK through a "standard" TAK1-dependent mode. Importantly, monoubiquitination of NEMO was also observed in this study. It occurred after sumoylation, as reported by Wu et al. [245], although, in this case, exclusively in the cytoplasm. Moreover, monoubiquitination was observed only after recruitment of NEMO to ATM/TRAF6/c-IAP1 and required the NEMO NUB domain. The discrepancies between these studies and the relationship between these different complexes, if indeed they represent distinct entities, require further characterization. Interestingly, Jin et al. [250] demonstrated a non-redundant function of cIAPs and XIAP in the genotoxic stress pathway and proposed that c-IAP1 may be the E3 ligase responsible for NEMO monoubiquitination. This would explain the sequence of events reported by Hinz et al. [249] and further demonstrates the high versatility of c-IAP1, also able to induce 48-, K63-and, possibly, K11-linked polyubiquitination.

Negative regulation of the genotoxic stress pathway at the level of NF-κB induction is incompletely characterized. The NEMO sumoylation step is down-regulated by desumoylase SUMO-specific protease 2 (SENP2) [251]. Regarding the ubiquitin-related modifications, USP10 has been shown to decrease M1-linked ubiquitination of NEMO during DNA damage induction [252]. USP10 requires NF-κB-induced MCPIP1 for binding to NEMO, suggesting that MCPIP1 regulates a negative feedback mechanism. Interestingly, the same authors have shown that USP10 may also target ubiquitinated TRAF6 upon genotoxic stress through the TANK-MCPIP1-USP10 complex described above (see Section 5.2) [141].

6. Regulated Ubiquitination of TAK1 and IKK Complexes

In the above section we examined various signaling pathways and described the ubiquitin-related post-translational modifications of their specific components. Activation of these pathways eventually results in the recruitment of identical proteins that channel the signal towards NF-κB. Among them are the subunits of the TAK1 and IKK complexes, members of the TRAF family and core components of the NF-κB system (NF-κB subunits and IκBs). Here, we will describe how these core actors of NF-κB activation are modified through ubiquitination, affecting their function to impact on signaling. The broad versus specific involvement of these modifications is often not analyzed so we will only mention the reported associated pathways without further extrapolations to others.

6.1. Regulated Ubiquitination of TAK1 Complex Components

Not surprisingly, ubiquitination of the various components of the TAK1 complex regulates not only their activity but also their stability, impacting on NF-κB induction [253]. Following TNF-α and IL-1β exposure, K63-linked polyubiquitination of TAK1 occurs mainly at Lys158 but also possibly at Lys34, Lys209 and Lys562 [254,255]. TRAF2 and TRAF6 may be the respective E3 ligases involved in TAK1 ubiquitination but an alternative candidate is the TAK1-binding protein TRIM8 [256]. Ubiquitination of TAK1 could help to consolidate its interaction with the IKK complex through NEMO for full activation. As mentioned above, CYLD may negatively control this process and work with Itch as a partner to secondarily induce TAK1 degradation through K48-linked ubiquitination. Other deubiquitinases targeting TAK1 are USP4 and USP18. USP4, in particular, downregulates NF-κB activation by TNF-α and IL-1β [257] while USP18 was originally described as targeting TAK1 in the TCR pathway [258], it is also involved in TLR signaling [259]. Finally, Pellino3b negatively regulates TAK1-dependent NF-κB activation by IL-1β [260].

In a specific situation, during maternal-to-zygotic transition, E3 ligase RNF114 induces TAB1 degradation through K48-linked polyubiquitination [261]. Surprisingly, this results in NF-κB activation through a poorly defined mechanism. K63-linked polyubiquitination of TAB1 at several Lys residues has also been reported [262] but its relevance in NF-κB signaling has not been demonstrated. The same applies to Itch-dependent TAB1 degradation [263].

TAB2/3 stability has been shown to be controlled by at least three distinct E3 ligases. The first one, TRIM38, interacts with TAB2/3 in the TNF-α and IL-1β signaling pathways and induces their degradation by the lysosome through an E3 ligase-independent process [264]. TLR-induced TRIM30α appears to act the same way [265]. Finally, RNF4, also targets TAB2 for degradation by the lysosome pathways but, in this specific case, its E3 ligase activity is required [266]. This peculiar mode of disposal and its exact association with ubiquitination processes deserves additional investigation.

6.2. Regulated Ubiquitination of IKK Complex Components

As discussed above, the main function of NEMO is to recognize polyubiquitin chains, but it can also be modified by ubiquitination. Zhou et al. [267] were the first to identify a Bcl10-dependent site of ubiquitination on NEMO, located in the zinc finger (Lys399) and participating in NF-κB activation by the TCR. It is unlikely that Bcl10 is the E3 ligase involved in this pathway as originally proposed. Subsequently, Lys285 was identified as modified during Nod2 signaling [268]. In this situation, TRAF6 was proposed to be the required E3 ligase although its participation in Nod1/Nod2 signaling remains controversial (see Section 5.3). Finally, several residues targeted by TRAF6 have been reported to be affected by a *NEMO* mutation causing incontinentia pigmenti pathology (see below). In all these cases, the identified lysine residues are believed to be specifically modified by K63-linked chains.

More recently, M1-linked polyubiquitination has been shown to play a critical function in NF-κB activation, as explained above. This process involves LUBAC and its first identified target was NEMO. Lys285 and Lys309, which are located within the NUB domain represent preferentially modified residues [269].

The effective participation of these different ubiquitination sites of NEMO in general or pathway-specific NF-κB activation has been investigated in vivo. Mutating only Lys399 in mice does not generate a severe phenotype, notably at the T cell level, but a reduced response of macrophages to LPS is observed, which is associated with resistance to endotoxic shock [270]. Mutating both Lys 285 and 399 residues results in strong impairment of NF-κB activation, and early TNF-dependent male lethality, similar to the one seen in *Nemo* KO mice [271,272]. In addition, the same mice rescued with a *Tnfr1* KO display an impaired response of macrophages to Nod2, LPS or IL-1β [271]. Thus, NEMO ubiquitination plays an important and broad role in NF-κB signaling. Although originally identified individually, the various ubiquitinated Lys residues may fulfil a similar function, alone or in combination, i.e., to help reinforce, through K63- and M1-linked chains, interactions between TAK1 and IKK complexes for optimal IKK activation.

Other modifications of NEMO through ubiquitination have been reported. They can positively or negatively affect its function. First, as described in Section 5.7, NEMO can be monoubiquitinated in the nucleus following DNA damage and this is a key activating event. Second, TRIM23 has been shown to ubiquitinate NEMO, through K27-linked chains, in the RIG-I/MAVS pathway [273]. The precise function of this peculiar kind of modification remains unknown but it is required for NF-κB activation. Regarding the negative regulation of NEMO function by ubiquitination, Zotti et al. [274] have shown that atypical TRAF protein TRAF7 is a NEMO interactor able to induce its Lys29-linked polyubiquitination for degradation by the lysosome. This results in impaired NF-κB activation by TNF-α. In another setting, myogenic differentiation, TRAF7/NEMO interaction and NEMO ubiquitination would instead positively regulate NF-κB [275].

Other modes of NEMO ubiquitination modulation have been shown to control its activity and down-regulate NF-κB activation. For instance, USP18, already known to inhibit TLR signaling by deubiquitinating TAK1, as mentioned in Section 6.1, also acts on NEMO [259]. EGL nine homolog 3 (EGLN3), a member of a family of prolyl hydroxylase, may negatively regulate NF-κB signaling by inhibiting NEMO ubiquitination by c-IAP1 [276]. TRIM13, an ER resident E3 ligase, interacts with NEMO and can also induce its deubiquitination through an unknown process [277]. Finally, HSCARG has been proposed to negatively regulate TNF-α-induced NF-κB activation by interacting with NEMO and inducing its deubiquitination through the recruitment of USP7 [278].

In contrast to the NEMO regulatory subunit, little is known regarding ubiquitin-regulated modification of the IKK catalytic subunits. An intriguing observation made by Niida et al. [279] suggests that TRIM21 may monoubiquitinate activated IKK2 to induce its disposal by autophagosomes.

7. Regulated Ubiquitination in the Non-Canonical Pathway of NF-κB Activation

As mentioned above, NIK is constitutively degraded in resting cells and stabilized upon stimulation, inducing IKK1 activity. This critical switch is regulated by intricate ubiquitin-dependent events. First, basal degradation of NIK is dependent on E3 ligase TRAF3: newly synthesized NIK associates with TRAF3 and is ubiquitinated with K48-linked chains, inducing its degradation by the proteasome [280]. Second, upon cell stimulation, specific receptors such as the CD40 or B-cell activating factor receptor (BAFF-R) recruit TRAF3 for degradation. This results in NIK accumulation and activation of the non-canonical pathway.

This basic model has been substantially refined. Although TRAF3 is essential in controlling the amount of NIK, it does not work alone but within a multimolecular E3 complex also including TRAF2 and cIAP1/2 [281–283]. In this complex TRAF3 is not the genuine NIK E3 ligase but plays an adaptor role for recruiting TRAF2, which itself contains a cIAP binding site. As a consequence, cIAP1/2 bound to TRAF2 acts on NIK to induce its degradation in resting cells. In this situation, cIAP1 and c-IAP2 appear functionally redundant.

How this degradative process is interrupted following stimulation is not fully understood. It has been shown that when CD40 or BAFFR bind their ligands TRAF2 and TRAF3 are recruited to the plasma membrane lipid-raft compartment [284,285]. This recruitment could initiate TRAF2-mediated polyubiquitination of both TRAFs and their subsequent proteasome-mediated degradation, causing NIK accumulation. Alternatively, c-IAP1/2 may be the active player at this level also. In this case, c-IAP1/2 catalytic activity may be augmented by TRAF2-dependent K63 ubiquitination to fulfil this specific function [281,286]. TRAF3 ubiquitination following stimulation is also supposed to help recruiting OTUD7B, an OTU domain-containing DUB, which limits TRAF3 degradation [287]. Consequently, lack of OTUD7B results in hyperactivation of the non-canonical NF-κB pathway.

Notably, STING also has the ability to activate the non-canonical NF-κB pathway through TRAF3, but the molecular mechanism involved is unknown. This deserves closer examination given the recently identified importance of the non-canonical activation of NF-κB via STING in chromosomal instability-driven metastasis [288].

8. Regulated Ubiquitination of NF-κB Proteins

DNA transcription involves complex machinery involving core components for RNA synthesis, transcription factors and co-regulators. Ubiquitination regulates the activity of all these elements, including those participating in NF-κB-dependent gene transcription. Sacani et al. [289] in particular were the first to demonstrate that the promoter bound p50/RelA can be degraded by the proteasome for terminating the NF-κB signaling. Thus, in situ degradation of NF-κB appears as an important down-regulation mechanism in addition to the re-synthesis of IκB and its cytoplasm/nucleus shuttling which participates in NF-κB-dependent transcription shut-off by dissociating NF-κB dimers from DNA.

Over the years, several E3 ligases have been identified as targeting the RelA subunit and ensuring the proper level and timing of gene expression. They all conjugate K48-linked chains of ubiquitin to RelA, inducing its degradation by the nuclear proteasome. Among them are copper metabolism MURR1 domain-containing 1 (COMMD1), PDZ and LIM domain 2 (PDLIM2), peroxisome proliferator-activated receptor (PPARγ) and CHIP. COMMD1 induces RelA degradation by recruiting an E3 complex including Elongins B/C, Cul2 and the suppressor of cytokine signaling 1 (SOCS1) [290]. PDLIM2 binds RelA in the nucleus and promotes its ubiquitination at discrete intranuclear compartments [291]. Interestingly, another member of the LIM family, PDLIM1, inhibits NF-κB signaling by a different mechanism, i.e., by the sequestration of RelA in the cytoplasm [292]. Moreover, PPARγ shuts off NF-κB signaling by binding to RelA and inducing through Lys28 its K48-linked ubiquitination and degradation [293]. Finally, CHIP can induce the ubiquitination and degradation of several tumour related proteins, including RelA [294]. So far, the only identified proteins limiting the RelA degradative processes are the nuclear DUB USP48, which works in concert with the COP9 signalosome to stabilize RelA through removal of K48-linked chains [295], and USP7, which interacts with DNA-bound RelA and increases its residency time at promoters by antagonizing degradative ubiquitination [296].

Monoubiquitination of RelA also occurs in the nucleus and negatively impacts on its transcriptional activity, in particular by interfering with its ability to interact with its co-activator CBP [297]. Since monoubiquitination of RelA has been observed on a mutated form of the protein, i.e., upon mutations of its phospho-acceptor sites, or following proteasome inhibition, the physiological relevance of this observation has yet to be firmly established.

Nuclear p50 can also be modified by ubiquitination. As mentioned above, p50 does not contain any TAD, but upon association with co-transcription activator Bcl3, participates in the positive regulation of transcription either with RelA, as a p50/RelA heterodimer, or as a p50/p50 homodimer. In contrast, formation of p50 dimers in the absence of Bcl3 results in negative regulation of transcription. This situation also induces p50 destabilization by K48-linked polyubiquitination [298]. How Bcl3 protects p50 from degradation remains undefined.

Transcriptional activity of c-Rel subunit is also regulated by degradative ubiquitination. In T cells, c-Rel is ubiquitinated upon TCR stimulation by Pelino1, which in this setting, promotes K48-linked polyubiquitination of c-Rel and its proteasome-dependent degradation [299]. Since Pellino1 is mostly known to act through K63-linked polyubiquitination (see Section 5.2) how it regulates K48-linked polyubiquitin chain formation remains unclear. This degradation process impacts mostly on late-phase NF-κB activation, suggesting a specific effect on c-Rel-only containing NF-κB dimers. It is unknown if this mode of c-Rel regulation is stimulus specific. This might be the case since in another cellular setting, i.e., macrophages stimulated by LPS, c-Rel stability appears to be controlled by a distinct mechanism involving TRAF2 in combination with TRAF3 and c-IAP1 [300,301]. Molecular details are lacking but this TRAF2-driven degradation process of c-Rel limits the expression of proinflammatory cytokines.

Finally, RelB can be modified by ubiquitination and this affects not only its stability but also its activity. Indeed, nuclear ubiquitination of RelB, other than K63- or 48-linked polyubiquitination, is required for transcription [302]. Stimulus-dependent degradation of RelB can also occur [303]. Finally, sumoylation has also been reported to down-regulate RelB transcriptional activity [304].

9. In Vivo Relevance of Ubiquitin-Dependent NF-κB Processes

NF-κB-related ubiquitination/ubiquitin recognition processes described above at the protein level, regulate many important cellular/organismal functions impacting on human health. Indeed, several inherited pathologies recently identified are due to mutations on proteins involved in NF-κB signaling that impair ubiquitin-related processes [305]. Not surprisingly, given the close relationship existing between NF-κB and receptors participating in innate and acquired immunity, these diseases are associated with immunodeficiency and/or deregulated inflammation.

9.1. NEMO Mutations

In humans, *NEMO* mutations can cause two distinct pathologies [306]. Loss-of-function mutations of *NEMO* induce male lethality and are responsible in females for incontinentia pigmenti (IP), an X-linked disease mostly characterized by a severe skin inflammation starting at birth. In contrast, hypomorphic mutations of *NEMO*, causing anhidrotic ectodermal dysplasia with immunodeficiency (EDA-ID), affects surviving hemizygous males and is associated with life-threatening impaired immune responses. In both cases, *NEMO* mutations have been identified as affecting either NEMO ubiquitination or NEMO interaction with polyubiquitin chains. For instance, IP-related A323P mutation causes impaired TRAF6-induced polyubiquitination [307] whereas an IP-related internal deletion of NEMO disrupts its interaction with LUBAC subunit SHARPIN [308]. In addition, a large percentage of missense *NEMO* mutations causing EDA-ID affects one of the residues located either in the NUB domain or the zinc finger, producing a NEMO protein with reduced, but not completely abolished, affinity for polyubiquitin [309]. These mutations then provoke suboptimal NF-κB activation for a large set of signaling pathways regulating innate and acquired immunity.

9.2. LUBAC Mutations

The importance of M1-linked polyubiquitination in vivo is illustrated in patients bearing mutations in *HOIP* and *HOIL-IL*. In both cases, susceptibility to infection, due to T and B cell defects, and auto-inflammation resulting from complex and cell-specific deregulations of TNF-α and IL-1β/TLR signaling are observed [310,311]. These phenotypes result from impaired formation of M1-linked ubiquitin chains. Although the association of an immunodeficiency with auto-inflammation remains incompletely understood, the immune phenotype of LUBAC mutated patients confirms the essential modulatory function of LUBAC in immune/inflammatory processes.

9.3. OTULIN Mutations

Being identified as the only DUB able to remove M1-linked chains from LUBAC substrates, the ubiquitin protease OTULIN when overexpressed should attenuate NF-κB signaling in response to immune-specific receptors, whereas its reduced expression should result in up-regulation of NF-κB targets. This is actually what has been observed on cultured cells [92]. Remarkably, patients carrying *OTULIN* mutations also display such phenotypes. Indeed, the corresponding disease, called either OTULIN-related autoinflammatory syndrome (ORAS) [312] or otulipenia [313], is characterized by the over-production of inflammatory cytokines and autoimmunity associated with excessive M1-linked ubiquitination of LUBAC substrates.

10. Conclusions

Over the last fifteen years a wealth of studies has confirmed the critical function of ubiquitin in regulating essential processes such as signal transduction, DNA transcription, endocytosis or cell cycle. Focusing on the ubiquitin-dependent mechanisms of signal regulation and regulation of NF-κB pathways, as done here, illustrates the amazing versatility of ubiquitination in controlling the fate of protein, building of macromolecular protein complexes and fine-tuning regulation of signal transmission. All these molecular events are dependent on the existence of an intricate ubiquitin code that allows the scanning and proper translation of the various status of a given protein. Actually, this covalent addition of a polypeptide to a protein, a reaction that may seem to be a particularly energy consuming process, allows a crucial degree of flexibility and the occurrence of almost unlimited new layers of regulation. This latter point is particularly evident with ubiquitination/deubiquitination events regulating the fate and activity of primary targets often modulated themselves by ubiquitination/deubiquitination events regulating the fate and activity of ubiquitination effectors and so on.

Recurrent features emerge when comparing the various signaling pathways leading to NF-κB described here. In particular, the way in which IKK integrates so many signaling inputs appears to require a rather limited set of proteins (or protein families). In the canonical pathway of NF-κB activation, high variability is seen at distal initiation of signaling but ultimately results in polyubiquitin chain synthesis that attracts, in most situations the TAK1 complex and in all cases the IKK complex. These two elements require distinct kinds of poly-ubiquitin linkages that may be present within the same synthesized chains. The common requirement of TAK1 and IKK in most NF-κB signaling pathways disputes the notion that targeting the TAK1 and/or IKK complexes for therapeutic purpose would be an adequate choice both in term of pathway specificity and adverse side effects. Future research aimed at fully characterizing the specific components/features of each pathway is therefore a prerequisite to efficiently translate this knowledge into valuable clinical options.

Obviously, although our focus in this review has been on ubiquitination processes, low affinity recognition of ubiquitin on modified substrates is not enough to ensure specificity in signal transduction. It is the combined action of UBDs and protein/protein interfaces that ultimately dictates the efficiency of signal transduction. Such specific associations might represent encouraging targets with regard to the aforementioned therapeutic strategies.

Also worth mentioning, are several putative extra layers of complexity that were not discussed above. Their existence is suggested by a disparate collection of data that clearly requires further investigation. First, as already pointed in Section 5.1 concerning the sometime dispensable role of RIPK1, studies suggesting alternative/dual modes of NF-κB activation in major pathways have been published. For instance, it has been claimed that kinase MEKK3 is required for both TNF-α- and

IL-1β-dependent NF-κB activation [314,315]. How this enzyme fits into the picture is still unclear but two studies have proposed that it may participate in one of two parallel/sequential pathways of NF-κB activation following IL-1β stimulation [316,317]. All this is reminiscent of what has been observed during TLR4 signaling (see Section 5.2). Further strengthening these models of alternative/redundant modes of activation is the recent publication of Zhang et al. [318] showing TRAF6-dependent NF-κB activation without TAB2/3 subunits. Finally, the most provocative recent discovery, which expands the regulatory function played by polyubiquitination in NF-κB signaling, is the formation of branched chains of ubiquitin (with K48 and K63 links) that may favor TAK1 complex recruitment over CYLD enzymatic activity [319]. This needs to be integrated into signaling processes depending on mixed and unanchored chains of ubiquitin.

To the best of our knowledge the amazingly broad and intricate dependency of NF-κB signaling on ubiquitin has not been observed in any other major signaling pathways. It remains to be seen whether this is a unique property of the NF-κB signaling pathway or only due to a lack of exhaustive characterization of players involved in those other pathways.

Finally, supporting the crucial function of ubiquitin-related processes in NF-κB signaling is their strong evolutionary conservation. Indeed, the immune deficiency (imd) signaling pathway of Drosophila melanogaster, which participates in the fight against pathogens represents the equivalent of a mammalian NF-κB pathway [320]. This pathway shows many similarities to the TNF-R1 signaling pathway, both regarding the nature of the proteins involved and the regulation of their activity through ubiquitin-dependent processes.

Acknowledgments: Due to space constraints we apologize for not citing many excellent publications in the field. The work was funded by INSERM and CEA institutional grants. We thank Jérémie Gautheron for careful reading of the manuscript.

Conflicts of Interest: The authors declare no conflict of interest.

Abbreviations

ABIN	A20-Binding Inhibitor of NF-κB activation
AIP4	Atrophin-1-Interacting Protein 4
ATM	Ataxia Telangiectasia Mutated
BAFF-R	B-cell Activating Factor Receptor
Bcl	B cell lymphoma
BCR	B Cell Receptor
β-TrCP	β-Transducing repeat-Containing Protein
CARD	Caspase Recruitment Domain
CARMA1	CARD-containing a Membrane-Associated Guanylate Kinase (MAGUK) protein 1
CBM	CARMA1/Bcl10/MALT1
CD	Cluster of Differentiation
CDC	Cell Division Cycle
2′-5′-cGAMP	2′-5′-cyclic Guanosine Adenosine MonoPhosphate
cGAS	cyclic Guanosine MonoPhosphate (GMP)-Adenosine MonoPhosphate (AMP) Synthase
CHIP	Carboxy terminus of Hsc70 Interacting Protein
c-IAP	Cellular Inhibitor of Apoptosis Protein
COMMD1	COpper Metabolism MURR1 Domain-containing 1
COP	COnstitutive Photomorphogenesis
CUE	Coupling of Ubiquitin conjugation to Endoplasmic reticulum-associated degradation
CYLD	CYLinDromatosis
DD	Death Domain
DEAD	DEAth effector Domain
DUB	DeUBiquitinase
EDA-ID	anhidrotic ectodermal dysplasia with immunodeficiency
EGLN3	EGL Nine homolog 3

ER	Endoplasmic Reticulum
ERK	Extracellular signal-Regulated Kinase
FADD	Fas-Associated protein with Death Domain
Fbw	F-box/WD repeat-containing protein
FHA	ForkHead Associated
GSK3	Glycogen Synthase Kinase-3
HACE1	HECT domain and Ankyrin repeat Containing E3 ubiquitin protein ligase 1
HECT	Homologous to the E6-AP Carboxyl Terminus
HOIL-1L	Haem-Oxydized IRP2 ubiquitin Ligase 1L
HOIP	HOIL-Interacting Protein
iE-DAP	γ-D-glutamyl-mesoDiAminoPimelic acid
IFN	Interferon
IKK	IκB Kinase
IκB	Inhibitory κB
IL-1βR	Interleukin-1β Receptor
IL1RAP	IL1 Receptor Accessory Protein
Imd	immune deficiency
IP	incontinentia pigmenti
IpaH	Invasion-plasmid antigen-H protein
IRAK	Interleukin-1 Receptor Associated Kinase
iRhom2	inactive Rhomboid 2
JAMM	JAB1/MPN/Mv34 metalloenzyme
KPC1	Kipl ubiquitylation-Promoting Complex 1
LDD	Linear ubiquitin chain Determining Domain
LGP2	Laboratory of Genetics and Physiology 2
LUBAC	Linear UBiquitin chain Assembly Complex
LZ	Leucine Zipper
MAGUK	Membrane-Associated Guanylate Kinase
MALT	Mucosa-Associated Lymphoid Tissue
MALT1	MALT lymphoma associated translocation protein 1
MARCH	Membrane Associated RING-CH
MAVS	Mitochondrial AntiViral-Signaling Protein
MCPIP1	Monocyte Chemotactic Protein-Induced Protein-1
MDA5	Melanoma Differentiation Associated gene 5
Mdm2	Mouse double minute 2 homolog
MDP	Muramyl DiPeptide
MD2	Myeloid Differentiation factor 2
MEKK	Mitogen-activated protein/ERK Kinase Kinase
MEX3C	MEX-3 homolog C (C. elegans)
MJD	Machado-Joseph Diseases protease
MIB2	MIndBomb 2
Mindy	MIU-containing Novel DUB familY
MLKL	Mixed Lineage Kinase domain Like pseudokinase
Myd88	Myeloid differentiation primary response 88
Ndfip1	Nedd4 family interacting protein 1
NEDD8	Neural precursor cell Expressed, Developmentally Down-Regulated 8
NEMO	NF-κB Essential Modulator
NF-κB	Nuclear Factor κB
NIK	NF-κB Inducing Kinase
NLRC	NOD-LRR (Leucine-Rich Repeat) family with CARD
NLRP	Nod-Like Receptor Protein
Nod	Nucleotide-binding oligomerization domain
NSP	Non Structural Protein
NZF	Npl4 Zinc Finger

ORAS	OTULIN-Related Autoinflammatory Syndrome
OUT	Ovarian TUmor proteases
OTULIN	OTU deubiquitinase with LINear linkage specificity
PAMPs	Pathogen Associated Molecular Patterns
PARP-1	Poly[ADP-Ribose (PAR)] Polymerase 1
PPCBP2	Poly(RC) Binding Protein 2
PDLIM2	PDZ and LIM domain 2
PIASy.	Protein Inhibitor of Activated STAT (Signal Transducer and Activator of Transcription) y
PIDD	P53-Induced Protein with a Death Domain
PKC	Protein Kinase C
PPAR	Peroxisome Proliferator-Activated Receptor
PTM	Post-Translational Modifications
PUB	PNGase/UBA or UBX-containing proteins
RBCC	Ring, B-box, Coiled-Coil
Rbx1	RING-box protein 1
Rel	avian Reticuloendotheliosis
RBR	Ring Between Ring fingers
RHIM	RIP Homotypic Interaction Motif
RING	Really Interesting New Gene
RIPK1	Receptor-Interacting serine/threonine-Protein Kinase 1
RLRs	Retinoic-Inducible Gene-I (RIG-I) Like Receptors
RNF	RiNg Finger protein
Sam68	Src-Associated in Mitosis 68 kDa
SCF	Skp, Cullin, F-box
SENP2	SUMO-specific Protease 2
SHARPIN	SHANK-associated RH domain interacting ProteIN
SHIP	SH2-containing Inositol phosphatase
Skp1	S-phase kinase-associated protein 1
Smurf1	Smad ubiquitin regulatory factor 1
SOCS1	Suppressor Of Cytokine Signaling 1
SPATA2	SPermATogenesis-Associated protein 2
Sphk1	Sphingosine kinase 1
STAT	Signal Transducer and Activator of Transcription
STING	STimulator of INterferon Genes
SUMO	Small Ubiquitin-like Modifier
TAB	TAK1 Binding protein
TAD	transcriptional activator domain
TAK1	Tumour growth factor-Activated Kinase 1
TANK	TRAF family member-Associated NF-κB activator
TCR	T cell receptor
TIR	Toll-Interleukin Receptor
TIRAP	Toll-Interleukin Receptor Adaptor Protein
TLR	Toll-Like Receptor
TNFAIP3	TNF Alpha Induced Protein 3
TRADD	Tumor necrosis factor Receptor type 1-Associated Death Domain protein
TRAF	TNF Receptor Associated Factor
TRAM	TRIF-Related Adaptor Molecule
TRIF	TIR domain-containing adaptor-inducing Interferon-β
TRIM	TRIpartite Motif protein
UBA	UBiquitin-Associated
UBASH3A	UBiquitin Associated and SH3 domain containing 3A
Ubc13	Ubiquitin-conjugating 13
UBD	Ubiquitin-Binding Domain
UBE2L3	UBiquitin-conjugating Enzyme E2 L3

UBL	Ubiquitin-Like
UCH	Ubiquitin C-terminal Hydrolases
Uev1A	Ubiquitin-conjugating enzyme variant 1A
ULP	Ubiquitin-Like Proteins
USP	Ubiquitin Specific Protease
WD40	TrpAsp 40 amino acids
WWP	WW domain-containing Protein ligase
XIAP	X-linked Inhibitor of Apoptosis Protein
ZF	Zinc Finger
ZNRF	Zinc aNd Ring Finger

References

1. Karin, M.; Ben-Neriah, Y. Phosphorylation meets ubiquitination: The control of NF-κB activity. *Annu. Rev. Immunol.* **2000**, *18*, 621–663. [CrossRef] [PubMed]
2. Aggarwal, B.B.; Takada, Y.; Shishodia, S.; Gutierrez, A.M.; Oommen, O.V.; Ichikawa, H.; Baba, Y.; Kumar, A. Nuclear transcription factor NF-κB: Role in biology and medicine. *Indian J. Exp. Biol.* **2004**, *42*, 341–353. [PubMed]
3. Hinz, M.; Scheidereit, C. The IκB kinase complex in NF-κB regulation and beyond. *EMBO Rep.* **2014**, *15*, 46–61. [CrossRef] [PubMed]
4. Dai, L.; Aye Thu, C.; Liu, X.Y.; Xi, J.; Cheung, P.C. TAK1, more than just innate immunity. *IUBMB Life* **2012**, *64*, 825–834. [CrossRef] [PubMed]
5. Sun, S.C. The noncanonical NF-κB pathway. *Immunol. Rev.* **2012**, *246*, 125–140. [CrossRef] [PubMed]
6. Pickart, C.M.; Eddins, M.J. Ubiquitin: Structures, functions, mechanisms. *Biochim. Biophys. Acta* **2004**, *1695*, 55–72. [CrossRef] [PubMed]
7. Komander, D. The emerging complexity of protein ubiquitination. *Biochem. Soc. Trans.* **2009**, *37*, 937–953. [CrossRef] [PubMed]
8. Clague, M.J.; Heride, C.; Urbé, S. The demographics of the ubiquitin system. *Trends Cell Biol.* **2015**, *25*, 417–426. [CrossRef] [PubMed]
9. Stewart, M.D.; Ritterhoff, T.; Klevit, R.E.; Brzovic, P.S. E2 enzymes: More than just middle men. *Cell Res.* **2016**, *26*, 423–440. [CrossRef] [PubMed]
10. Metzger, M.B.; Hristova, V.A.; Weissman, A.M. HECT and RING finger families of E3 ubiquitin ligases at a glance. *J. Cell Sci.* **2012**, *125*, 531–537. [CrossRef] [PubMed]
11. Zheng, N.; Shabek, N. Ubiquitin Ligases: Structure, Function, and Regulation. *Annu. Rev. Biochem.* **2017**, *86*, 129–157. [CrossRef] [PubMed]
12. Hoppe, T. Multiubiquitinylation by E4 enzymes: "one size" doesn't fit all. *Trends Biochem. Sci.* **2005**, *30*, 183–187. [CrossRef] [PubMed]
13. Ohtake, F.; Tsuchiya, H. The emerging complexity of ubiquitin architecture. *J. Biochem.* **2017**, *161*, 125–133. [CrossRef] [PubMed]
14. Kwon, Y.T.; Ciechanover, A. The Ubiquitin Code in the Ubiquitin-Proteasome System and Autophagy. *Trends Biochem. Sci.* **2017**, *42*, 873–886. [CrossRef] [PubMed]
15. Mevissen, T.E.T.; Komander, D. Mechanisms of Deubiquitinase Specificity and Regulation. *Annu. Rev. Biochem.* **2017**, *86*, 159–192. [CrossRef] [PubMed]
16. Nijman, S.M.; Luna-Vargas, M.P.; Velds, A.; Brummelkamp, T.R.; Dirac, A.M.; Sixma, T.K.; Bernards, R. A genomic and functional inventory of deubiquitinating enzymes. *Cell* **2005**, *123*, 773–786. [CrossRef] [PubMed]
17. Leznicki, P.; Kulathu, Y. Mechanisms of regulation and diversification of deubiquitylating enzyme function. *J. Cell Sci.* **2017**, *130*, 1997–2006. [CrossRef] [PubMed]
18. Sowa, M.E.; Bennett, E.J.; Gygi, S.P.; Wade Harper, J. Defining the human deubiquitinating enzyme interaction landscape. *Cell* **2009**, *138*, 389–403. [CrossRef] [PubMed]
19. Cappadocia, L.; Lima, C.D. Ubiquitin-like Protein Conjugation: Structures, Chemistry, and Mechanism. *Chem. Rev.* **2018**, *118*, 889–918. [CrossRef] [PubMed]
20. Dohmen, R.J. SUMO protein modification. *Biochim. Biophys. Acta* **2004**, *1695*, 113–131. [CrossRef] [PubMed]

21. Boase, N.A.; Kumar, S. NEDD4: The founding member of a family of ubiquitin-protein ligases. *Gene* **2015**, *557*, 113–122. [CrossRef] [PubMed]

22. Ciechanover, A.; Schwartz, A.L. The ubiquitin system: Pathogenesis of human diseases and drug targeting. *Biochim. Biophys. Acta* **2004**, *1695*, 3–17. [CrossRef] [PubMed]

23. Pal, A.; Young, M.A.; Donato, N.J. Emerging potential of therapeutic targeting of ubiquitin-specific proteases in the treatment of cancer. *Cancer Res.* **2014**, *74*, 4955–4966. [CrossRef] [PubMed]

24. Liu, J.; Shaik, S.; Dai, X.; Wu, Q.; Zhou, X.; Wang, Z.; Wei, W. Targeting the ubiquitin pathway for cancer treatment. *Biochim. Biophys. Acta* **2015**, *1855*, 50–60. [CrossRef] [PubMed]

25. Xie, P. TRAF molecules in cell signaling and in human diseases. *J. Mol. Signal.* **2013**, *8*, 7. [CrossRef] [PubMed]

26. Tomar, D.; Singh, R. TRIM family proteins: Emerging class of RING E3 ligases as regulator of NF-κB pathway. *Biol. Cell* **2015**, *107*, 22–40. [CrossRef] [PubMed]

27. Van Tol, S.; Hage, A.; Giraldo, M.I.; Bharaj, P.; Rajsbaum, R. The TRIMendous Role of TRIMs in Virus-Host Interactions. *Vaccines* **2017**, *5*, 23. [CrossRef] [PubMed]

28. Tokunaga, F.; Iwai, K. LUBAC, a novel ubiquitin ligase for linear ubiquitination, is crucial for inflammation and immune responses. *Microbes Infect.* **2012**, *14*, 563–572. [CrossRef] [PubMed]

29. Rittinger, K.; Ikeda, F. Linear ubiquitin chains: Enzymes, mechanisms and biology. *Open Biol.* **2017**, *7*, 170026. [CrossRef] [PubMed]

30. Deng, L.; Wang, C.; Spencer, E.; Yang, L.; Braun, A.; You, J.; Slaughter, C.; Pickart, C.; Chen, Z.J. Activation of the IκB kinase complex by TRAF6 requires a dimeric ubiquitin-conjugating enzyme complex and a unique polyubiquitin chain. *Cell* **2000**, *103*, 351–361. [CrossRef]

31. Fukushima, T.; Matsuzawa, S.; Kress, C.L.; Bruey, J.M.; Krajewska, M.; Lefebvre, S.; Zapata, J.M.; Ronai, Z.; Reed, J.C. Ubiquitin-conjugating enzyme Ubc13 is a critical component of TNF receptor-associated factor (TRAF)-mediated inflammatory responses. *Proc. Natl. Acad. Sci. USA* **2007**, *104*, 6371–6376. [CrossRef] [PubMed]

32. Fu, B.; Li, S.; Wang, L.; Berman, M.A.; Dorf, M.E. The ubiquitin conjugating enzyme UBE2L3 regulates TNFα-induced linear ubiquitination. *Cell Res.* **2014**, *24*, 376–379. [CrossRef] [PubMed]

33. Lewis, M.J.; Vyse, S.; Shields, A.M.; Boeltz, S.; Gordon, P.A.; Spector, T.D.; Lehner, P.J.; Walczak, H.; Vyse, T.J. UBE2L3 polymorphism amplifies NF-κB activation and promotes plasma cell development, linking linear ubiquitination to multiple autoimmune diseases. *Am. J. Hum. Genet.* **2015**, *96*, 221–234. [CrossRef] [PubMed]

34. Scott, D.; Oldham, N.J.; Strachan, J.; Searle, M.S.; Layfield, R. Ubiquitin-binding domains: Mechanisms of ubiquitin recognition and use as tools to investigate ubiquitin-modified proteomes. *Proteomics* **2015**, *15*, 844–861. [CrossRef] [PubMed]

35. Kulathu, Y.; Akutsu, M.; Bremm, A.; Hofmann, K.; Komander, D. Two-sided ubiquitin binding explains specificity of the TAB2 NZF domain. *Nat. Struct. Mol. Biol.* **2009**, *16*, 1328–1330. [CrossRef] [PubMed]

36. Sato, Y.; Yoshikawa, A.; Yamashita, M.; Yamagata, A.; Fukai, S. Structural basis for specific recognition of Lys 63-linked polyubiquitin chains by NZF domains of TAB2 and TAB3. *EMBO J.* **2009**, *28*, 3903–3909. [CrossRef] [PubMed]

37. Lo, Y.C.; Lin, S.C.; Rospigliosi, C.C.; Conze, D.B.; Wu, C.J.; Ashwell, J.D.; Eliezer, D.; Wu, H. Structural basis for recognition of diubiquitins by NEMO. *Mol. Cell* **2009**, *33*, 602–615. [CrossRef] [PubMed]

38. Rahighi, S.; Ikeda, F.; Kawasaki, M.; Akutsu, M.; Suzuki, N.; Kato, R.; Kensche, T.; Uejima, T.; Bloor, S.; Komander, D.; et al. Specific recognition of linear ubiquitin chains by NEMO is important for NF-κB activation. *Cell* **2009**, *136*, 1098–1109. [CrossRef] [PubMed]

39. Ngadjeua, F.; Chiaravalli, J.; Traincard, F.; Raynal, B.; Fontan, E.; Agou, F. Two-sided ubiquitin binding of NF-κB essential modulator (NEMO) zinc finger unveiled by a mutation associated with anhidrotic ectodermal dysplasia with immunodeficiency syndrome. *J. Biol. Chem.* **2013**, *288*, 33722–33737. [CrossRef] [PubMed]

40. Laplantine, E.; Fontan, E.; Chiaravalli, J.; Lopez, T.; Lakisic, G.; Véron, M.; Agou, F.; Israël, A. NEMO specifically recognizes K63-linked poly-ubiquitin chains through a new bipartite ubiquitin-binding domain. *EMBO J.* **2009**, *28*, 2885–2895. [CrossRef] [PubMed]

41. Kensche, T.; Tokunaga, F.; Ikeda, F.; Goto, E.; Iwai, K.; Dikic, I. Analysis of nuclear factor-κB (NF-κB) essential modulator (NEMO) binding to linear and lysine-linked ubiquitin chains and its role in the activation of NF-κB. *J. Biol. Chem.* **2012**, *287*, 23626–23634. [CrossRef] [PubMed]

42. Shimizu, S.; Fujita, H.; Sasaki, Y.; Tsuruyama, T.; Fukuda, K.; Iwai, K. Differential Involvement of the Npl4 Zinc Finger Domains of SHARPIN and HOIL-1L in Linear Ubiquitin Chain Assembly Complex-Mediated Cell Death Protection. *Mol. Cell. Biol.* **2016**, *36*, 1569–1583. [CrossRef] [PubMed]

43. Sato, Y.; Fujita, H.; Yoshikawa, A.; Yamashita, M.; Yamagata, A.; Kaiser, S.E.; Iwai, K.; Fukai, S. Specific recognition of linear ubiquitin chains by the Npl4 zinc finger (NZF) domain of the HOIL-1L subunit of the linear ubiquitin chain assembly complex. *Proc. Natl. Acad. Sci. USA* **2011**, *108*, 20520–20525. [CrossRef] [PubMed]

44. Kanarek, N.; London, N.; Schueler-Furman, O.; Ben-Neriah, Y. Ubiquitination and degradation of the inhibitors of NF-κB. *Cold Spring Harb. Perspect. Biol.* **2010**, *2*, a000166. [CrossRef] [PubMed]

45. Spencer, E.; Jiang, J.; Chen, Z.J. Signal-induced ubiquitination of IκBα by the F-box protein Slimb/β-TrCP. *Genes Dev.* **1999**, *13*, 284–294. [CrossRef] [PubMed]

46. Winston, J.T.; Strack, P.; Beer-Romero, P.; Chu, C.Y.; Elledge, S.J.; Harper, J.W. The SCFβ-TRCP-ubiquitin ligase complex associates specifically with phosphorylated destruction motifs in IκBα and β-catenin and stimulates IκBα ubiquitination in vitro. *Genes Dev.* **1999**, *13*, 270–283. [CrossRef] [PubMed]

47. Gonen, H.; Bercovich, B.; Orian, A.; Carrano, A.; Takizawa, C.; Yamanaka, K.; Pagano, M.; Iwai, K.; Ciechanover, A. Identification of the ubiquitin carrier proteins, E2s, involved in signal-induced conjugation and subsequent degradation of IκBα. *J. Biol. Chem.* **1999**, *274*, 14823–14830. [CrossRef] [PubMed]

48. Wu, C.; Ghosh, S. Differential phosphorylation of the signal-responsive domain of IκBα and IκBβ by IκB kinases. *J. Biol. Chem.* **2003**, *278*, 31980–31987. [CrossRef] [PubMed]

49. Read, M.A.; Brownell, J.E.; Gladysheva, T.B.; Hottelet, M.; Parent, L.A.; Coggins, M.B.; Pierce, J.W.; Podust, V.N.; Luo, R.S.; Chau, V.; et al. Nedd8 modification of cul-1 activates SCF(β(TrCP))-dependent ubiquitination of IκBα. *Mol. Cell. Biol.* **2000**, *20*, 2326–2333. [CrossRef] [PubMed]

50. Kawakami, T.; Chiba, T.; Suzuki, T.; Iwai, K.; Yamanaka, K.; Minato, N.; Suzuki, H.; Shimbara, N.; Hidaka, Y.; Osaka, F.; et al. NEDD8 recruits E2-ubiquitin to SCF E3 ligase. *EMBO J.* **2001**, *20*, 4003–4012. [CrossRef] [PubMed]

51. Shi, M.; Cho, H.; Inn, K.S.; Yang, A.; Zhao, Z.; Liang, Q.; Versteeg, G.A.; Amini-Bavil-Olyaee, S.; Wong, L.Y.; et al. Negative regulation of NF-κB activity by brain-specific TRIpartite Motif protein 9. *Nat. Commun.* **2014**, *5*, 4820. [CrossRef] [PubMed]

52. Di Fiore, I.J.; Pane, J.A.; Holloway, G.; Coulson, B.S. NSP1 of human rotaviruses commonly inhibits NF-κB signalling by inducing β-TrCP degradation. *J. Gen. Virol.* **2015**, *96*, 1768–1776. [CrossRef] [PubMed]

53. Sun, W.; Tan, X.; Shi, Y.; Xu, G.; Mao, R.; Gu, X.; Fan, Y.; Yu, Y.; Burlingame, S.; Zhang, H.; et al. USP11 negatively regulates TNFα-induced NF-κB activation by targeting on IκBα. *Cell Signal.* **2010**, *22*, 386–394. [CrossRef] [PubMed]

54. Schweitzer, K.; Bozko, P.M.; Dubiel, W.; Naumann, M. CSN controls NF-kB by deubiquitinylation of IκBα. *EMBO J.* **2007**, *26*, 1532–1541. [CrossRef] [PubMed]

55. Da Silva-Ferrada, E.; Torres-Ramos, M.; Aillet, F.; Campagna, M.; Matute, C.; Rivas, C.; Rodríguez, M.S.; Lang, V. Role of monoubiquitylation on the control of IκBα degradation and NF-κB activity. *PLoS ONE* **2011**, *6*, e25397. [CrossRef] [PubMed]

56. Desterro, J.M.; Rodriguez, M.S.; Hay, R.T. SUMO-1 modification of IκBα inhibits NF-κB activation. *Mol. Cell* **1998**, *2*, 233–239. [CrossRef]

57. Orian, A.; Schwartz, A.L.; Israël, A.; Whiteside, S.; Kahana, C.; Ciechanover, A. Structural motifs involved in ubiquitin-mediated processing of the NF-κB precursor p105: Roles of the glycine-rich region and a downstream ubiquitination domain. *Mol. Cell. Biol.* **1999**, *19*, 3664–3673. [CrossRef] [PubMed]

58. Kravtsova-Ivantsiv, Y.; Shomer, I.; Cohen-Kaplan, V.; Snijder, B.; Superti-Furga, G.; Gonen, H.; Sommer, T.; Ziv, T.; Admon, A.; Naroditsky, I.; et al. KPC1-mediated ubiquitination and proteasomal processing of NF-κB1 p105 to p50 restricts tumor growth. *Cell* **2015**, *161*, 333–347. [CrossRef] [PubMed]

59. Heissmeyer, V.; Krappmann, D.; Hatada, E.N.; Scheidereit, C. Shared pathways of Iκkinase-induced SCF(βTrCP)-mediated ubiquitination and degradation for the NF-κB precursor p105 and IκBα. *Mol. Cell. Biol.* **2001**, *21*, 1024–1035. [CrossRef] [PubMed]

60. Orian, A.; Gonen, H.; Bercovich, B.; Fajerman, I.; Eytan, E.; Israël, A.; Mercurio, F.; Iwai, K.; Schwartz, A.L.; Ciechanover, A. SCF(β-TrCP) ubiquitin ligase-mediated processing of NF-κB p105 requires phosphorylation of its C-terminus by IκB kinase. *EMBO J.* **2000**, *19*, 2580–2591. [CrossRef] [PubMed]

61.   Amir, R.E.; Iwai, K.; Ciechanover, A. The NEDD8 pathway is essential for SCF(β-TrCP)-mediated ubiquitination and processing of the NF-κB precursor p105. *J. Biol. Chem.* **2002**, *277*, 23253–23259. [CrossRef] [PubMed]

62.   Tian, L.; Holmgren, R.A.; Matouschek, A. A conserved processing mechanism regulates the activity of transcription factors Cubitus interruptus and NF-κB. *Nat. Struct. Mol. Biol.* **2005**, *12*, 1045–1053. [CrossRef] [PubMed]

63.   Lapid, D.; Lahav-Baratz, S.; Cohen, S. A20 inhibits both the degradation and limited processing of the NF-κB p105 precursor: A novel additional layer to its regulator role. *Biochem. Biophys. Res. Commun.* **2017**, *493*, 52–57. [CrossRef] [PubMed]

64.   Heusch, M.; Lin, L.; Geleziunas, R.; Greene, W.C. The generation of nfκb2 p52: Mechanism and efficiency. *Oncogene* **1999**, *18*, 6201–6208. [CrossRef] [PubMed]

65.   Xiao, G.; Harhaj, E.W.; Sun, S.C. NF-κB-inducing kinase regulates the processing of NF-κB2 p100. *Mol. Cell* **2001**, *7*, 401–409. [CrossRef]

66.   Xiao, G.; Fong, A.; Sun, S.C. Induction of p100 processing by NF-κB-inducing kinase involves docking IκB kinase α (IKKα) to p100 and IKKα-mediated phosphorylation. *J. Biol. Chem.* **2004**, *279*, 30099–30105. [CrossRef] [PubMed]

67.   Liang, C.; Zhang, M.; Sun, S.C. β-TrCP binding and processing of NF-κB2/p100 involve its phosphorylation at serines 866 and 870. *Cell Signal.* **2006**, *18*, 1309–1317. [CrossRef] [PubMed]

68.   Amir, R.E.; Haecker, H.; Karin, M.; Ciechanover, A. Mechanism of processing of the NF-κB2 p100 precursor: Identification of the specific polyubiquitin chain-anchoring lysine residue and analysis of the role of NEDD8-modification on the SCF(β-TrCP) ubiquitin ligase. *Oncogene* **2004**, *23*, 2540–2547. [CrossRef] [PubMed]

69.   Fukushima, H.; Matsumoto, A.; Inuzuka, H.; Zhai, B.; Lau, A.W.; Wan, L.; Gao, D.; Shaik, S.; Yuan, M.; Gygi, S.P.; et al. SCF(Fbw7) modulates the NFκB signaling pathway by targeting NFκB2 for ubiquitination and destruction. *Cell Rep.* **2012**, *1*, 434–443. [CrossRef] [PubMed]

70.   Busino, L.; Millman, S.E.; Scotto, L.; Kyratsous, C.A.; Basrur, V.; O'Connor, O.; Hoffmann, A.; Elenitoba-Johnson, K.S.; Pagano, M. Fbxw7α- and GSK3-mediated degradation of p100 is a pro-survival mechanism in multiple myeloma. *Nat. Cell Biol.* **2012**, *14*, 375–385. [CrossRef] [PubMed]

71.   Silke, J. The regulation of TNF signalling: What a tangled web we weave. *Curr. Opin. Immunol.* **2011**, *23*, 620–626. [CrossRef] [PubMed]

72.   Kupka, S.; Reichert, M.; Draber, P.; Walczak, H. Formation and removal of poly-ubiquitin chains in the regulation of tumor necrosis factor-induced gene activation and cell death. *FEBS. J.* **2016**, *283*, 2626–2639. [CrossRef] [PubMed]

73.   Ramakrishnan, P.; Baltimore, D. Sam68 is required for both NF-κB activation and apoptosis signaling by the TNF receptor. *Mol. Cell* **2011**, *43*, 167–179. [CrossRef] [PubMed]

74.   Witt, A.; Vucic, D. Diverse ubiquitin linkages regulate RIP kinases-mediated inflammatory and cell death signaling. *Cell Death Differ.* **2017**, *24*, 1160–1171. [CrossRef] [PubMed]

75.   Vince, J.E.; Pantaki, D.; Feltham, R.; Mace, P.D.; Cordier, S.M.; Schmukle, A.C.; Davidson, A.J.; Callus, B.A.; Wong, W.W.; et al. TRAF2 must bind to cellular inhibitors of apoptosis for tumor necrosis factor (TNF) to efficiently activate NF-κB and to prevent TNF-induced apoptosis. *J. Biol. Chem.* **2009**, *284*, 35906–35915. [CrossRef] [PubMed]

76.   Haas, T.L.; Emmerich, C.H.; Gerlach, B.; Schmukle, A.C.; Cordier, S.M.; Rieser, E.; Feltham, R.; Vince, J.; Warnken, U.; Wenger, T.; et al. Recruitment of the linear ubiquitin chain assembly complex stabilizes the TNF-R1 signaling complex and is required for TNF-mediated gene induction. *Mol. Cell* **2009**, *36*, 831–844. [CrossRef] [PubMed]

77.   Emmerich, C.H.; Bakshi, S.; Kelsall, I.R.; Ortiz-Guerrero, J.; Shpiro, N.; Cohen, P. Lys63/Met1-hybrid ubiquitin chains are commonly formed during the activation of innate immune signaling. *Biochem. Biophys. Res. Commun.* **2016**, *474*, 452–461. [CrossRef] [PubMed]

78.   Zhang, J.; Clark, K.; Lawrence, T.; Peggie, M.W.; Cohen, P. An unexpected twist to the activation of IKKβ: TAK1 primes IKKβ for activation by autophosphorylation. *Biochem. J.* **2014**, *461*, 531–537. [CrossRef] [PubMed]

79.   Peltzer, N.; Darding, M.; Walczak, H. Holding RIPK1 on the Ubiquitin Leash in TNFR1 Signaling. *Trends Cell Biol.* **2016**, *26*, 445–461. [CrossRef] [PubMed]

80. Dondelinger, Y.; Jouan-Lanhouet, S.; Divert, T.; Theatre, E.; Bertin, J.; Gough, P.J.; Giansanti, P.; Heck, A.J.; Dejardin, E.; Vandenabeele, P.; et al. NF-κB-Independent Role of IKKα/IKKβ in Preventing RIPK1 Kinase-Dependent Apoptotic and Necroptotic Cell Death during TNF Signaling. *Mol. Cell* **2015**, *60*, 63–76. [CrossRef] [PubMed]

81. Grootjans, S.; Vanden Berghe, T.; Vandenabeele, P. Initiation and execution mechanisms of necroptosis: An overview. *Cell Death Differ.* **2017**, *24*, 1184–1195. [CrossRef] [PubMed]

82. Lork, M.; Verhelst, K.; Beyaert, R. CYLD, A20 and OTULIN deubiquitinases in NF-κB signaling and cell death: So similar, yet so different. *Cell Death Differ.* **2017**, *24*, 1172–1183. [CrossRef] [PubMed]

83. Wertz, I.E.; O'Rourke, K.M.; Zhou, H.; Eby, M.; Aravind, L.; Seshagiri, S.; Wu, P.; Wiesmann, C.; Baker, R.; Boone, D.L.; et al. De-ubiquitination and ubiquitin ligase domains of A20 downregulate NF-κB signalling. *Nature* **2004**, *430*, 694–699. [CrossRef] [PubMed]

84. Shembade, N.; Parvatiyar, K.; Harhaj, N.S.; Harhaj, E.W. The ubiquitin-editing enzyme A20 requires RNF11 to downregulate NF-κB signalling. *EMBO J.* **2009**, *28*, 513–522. [CrossRef] [PubMed]

85. Wertz, I.E.; Newton, K.; Seshasayee, D.; Kusam, S.; Lam, C.; Zhang, J.; Popovych, N.; Helgason, E.; Schoeffler, A.; Jeet, S.; et al. Phosphorylation and linear ubiquitin direct A20 inhibition of inflammation. *Nature* **2015**, *528*, 370–375. [CrossRef] [PubMed]

86. Dziedzic, S.A.; Su, Z.; Jean Barrett, V.; Najafov, A.; Mookhtiar, A.K.; Amin, P.; Pan, H.; Sun, L.; Zhu, H.; Ma, A.; et al. ABIN-1 regulates RIPK1 activation by linking Met1 ubiquitylation with Lys63 deubiquitylation in TNF-RSC. *Nat. Cell Biol.* **2018**, *20*, 58–68. [CrossRef] [PubMed]

87. Sun, S.C. CYLD: A tumor suppressor deubiquitinase regulating NF-κB activation and diverse biological processes. *Cell Death Differ.* **2010**, *17*, 25–34. [CrossRef] [PubMed]

88. Moquin, D.M.; McQuade, T.; Chan, F.K. CYLD deubiquitinates RIP1 in the TNFα-induced necrosome to facilitate kinase activation and programmed necrosis. *PLoS ONE* **2013**, *8*, e76841. [CrossRef] [PubMed]

89. Kovalenko, A.; Chable-Bessia, C.; Cantarella, G.; Israël, A.; Wallach, D.; Courtois, G. The tumour suppressor CYLD negatively regulates NF-κB signalling by deubiquitination. *Nature* **2003**, *424*, 801–805. [CrossRef] [PubMed]

90. Schlicher, L.; Maurer, U. SPATA2: New insights into the assembly of the TNFR signaling complex. *Cell Cycle* **2017**, *16*, 11–12. [CrossRef] [PubMed]

91. Massoumi, R. CYLD: A deubiquitination enzyme with multiple roles in cancer. *Future Oncol.* **2011**, *7*, 285–297. [CrossRef] [PubMed]

92. Keusekotten, K.; Elliott, P.R.; Glockner, L.; Fiil, B.K.; Damgaard, R.B.; Kulathu, Y.; Wauer, T.; Hospenthal, M.K.; Gyrd-Hansen, M.; Krappmann, D.; et al. OTULIN antagonizes LUBAC signaling by specifically hydrolyzing Met1-linked polyubiquitin. *Cell* **2013**, *153*, 1312–1326. [CrossRef] [PubMed]

93. Fiil, B.K.; Damgaard, R.B.; Wagner, S.A.; Keusekotten, K.; Fritsch, M.; Bekker-Jensen, S.; Mailand, N.; Choudhary, C.; Komander, D.; Gyrd-Hansen, M. OTULIN restricts Met1-linked ubiquitination to control innate immune signaling. *Mol. Cell* **2013**, *50*, 818–830. [CrossRef] [PubMed]

94. Schaeffer, V.; Akutsu, M.; Olma, M.H.; Gomes, L.C.; Kawasaki, M.; Dikic, I. Binding of OTULIN to the PUB domain of HOIP controls NF-κB signaling. *Mol. Cell* **2014**, *54*, 349–361. [CrossRef] [PubMed]

95. Elliott, P.R.; Nielsen, S.V.; Marco-Casanova, P.; Fiil, B.K.; Keusekotten, K.; Mailand, N.; Freund, S.M.; Gyrd-Hansen, M.; Komander, D. Molecular basis and regulation of OTULIN-LUBAC interaction. *Mol. Cell* **2014**, *54*, 335–348. [CrossRef] [PubMed]

96. Mahul-Mellier, A.L.; Pazarentzos, E.; Datler, C.; Iwasawa, R.; AbuAli, G.; Lin, B.; Grimm, S. De-ubiquitinating protease USP2a targets RIP1 and TRAF2 to mediate cell death by TNF. *Cell Death Differ.* **2012**, *19*, 891–899. [CrossRef] [PubMed]

97. Metzig, M.; Nickles, D.; Falschlehner, C.; Lehmann-Koch, J.; Straub, B.K.; Roth, W.; Boutros, M. An RNAi screen identifies USP2 as a factor required for TNF-α-induced NF-κB signaling. *Int. J. Cancer* **2011**, *129*, 607–618. [CrossRef] [PubMed]

98. Hou, X.; Wang, L.; Zhang, L.; Pan, X.; Zhao, W. Ubiquitin-specific protease 4 promotes TNF-α-induced apoptosis by deubiquitination of RIP1 in head and neck squamous cell carcinoma. *FEBS Lett.* **2013**, *587*, 311–316. [CrossRef] [PubMed]

99. Xiao, N.; Li, H.; Luo, J.; Wang, R.; Chen, H.; Chen, J.; Wang, P. Ubiquitin-specific protease 4 (USP4) targets TRAF2 and TRAF6 for deubiquitination and inhibits TNFα-induced cancer cell migration. *Biochem. J.* **2012**, *441*, 979–986. [CrossRef] [PubMed]

100. Xu, G.; Tan, X.; Wang, H.; Sun, W.; Shi, Y.; Burlingame, S.; Gu, X.; Cao, G.; Zhang, T.; Qin, J.; et al. Ubiquitin-specific peptidase 21 inhibits tumor necrosis factor α-induced nuclear factor κB activation via binding to and deubiquitinating receptor-interacting protein 1. *J. Biol. Chem.* **2010**, *285*, 969–978. [CrossRef] [PubMed]

101. Li, S.; Wang, L.; Dorf, M.E. PKC phosphorylation of TRAF2 mediates IKKα/β recruitment and K63-linked polyubiquitination. *Mol. Cell* **2009**, *33*, 30–42. [CrossRef] [PubMed]

102. Tortola, L.; Nitsch, R.; Bertrand, M.J.M.; Kogler, M.; Redouane, Y.; Kozieradzki, I.; Uribesalgo, I.; Fennell, L.M.; Daugaard, M.; Klug, H.; et al. The Tumor Suppressor Hace1 Is a Critical Regulator of TNFR1-Mediated Cell Fate. *Cell Rep.* **2016**, *15*, 1481–1492. [CrossRef] [PubMed]

103. Jang, K.W.; Lee, K.H.; Kim, S.H.; Jin, T.; Choi, E.Y.; Jeon, H.J.; Kim, E.; Han, Y.S.; Chung, J.H. Ubiquitin ligase CHIP induces TRAF2 proteasomal degradation and NF-κB inactivation to regulate breast cancer cell invasion. *J. Cell Biochem.* **2011**, *112*, 3612–3620. [CrossRef] [PubMed]

104. Fulda, S. Smac Mimetics to Therapeutically Target IAP Proteins in Cancer. *Int. Rev. Cell. Mol. Biol.* **2017**, *330*, 157–169. [PubMed]

105. Goncharov, T.; Niessen, K.; de Almagro, M.C.; Izrael-Tomasevic, A.; Fedorova, A.V.; Varfolomeev, E.; Arnott, D.; Deshayes, K.; Kirkpatrick, D.S.; Vucic, D. OTUB1 modulates c-IAP1 stability to regulate signalling pathways. *EMBO J.* **2013**, *32*, 1103–1114. [CrossRef] [PubMed]

106. Mei, Y.; Hahn, A.A.; Hu, S.; Yang, X. The USP19 deubiquitinase regulates the stability of c-IAP1 and c-IAP2. *J. Biol. Chem.* **2011**, *286*, 35380–35387. [CrossRef] [PubMed]

107. De Jong, M.F.; Liu, Z.; Chen, D.; Alto, N.M. Shigella flexneri suppresses NF-κB activation by inhibiting linear ubiquitin chain ligation. *Nat. Microbiol.* **2016**, *1*, 16084. [CrossRef] [PubMed]

108. Wong, W.W.; Gentle, I.E.; Nachbur, U.; Anderton, H.; Vaux, D.L.; Silke, J. RIPK1 is not essential for TNFR1-induced activation of NF-κB. *Cell Death Differ.* **2010**, *17*, 482–487. [CrossRef] [PubMed]

109. Xu, M.; Skaug, B.; Zeng, W.; Chen, Z.J. A ubiquitin replacement strategy in human cells reveals distinct mechanisms of IKK activation by TNFα and IL-1β. *Mol. Cell* **2009**, *36*, 302–314. [CrossRef] [PubMed]

110. Blackwell, K.; Zhang, L.; Workman, L.M.; Ting, A.T.; Iwai, K.; Habelhah, H. Two coordinated mechanisms underlie tumor necrosis factor a-induced immediate and delayed IκB kinase activation. *Mol. Cell. Biol.* **2013**, *33*, 1901–1915. [CrossRef] [PubMed]

111. Alvarez, S.E.; Harikumar, K.B.; Hait, N.C.; Allegood, J.; Strub, G.M.; Kim, E.Y.; Maceyka, M.; Jiang, H.; Luo, C.; Kordula, T.; et al. Sphingosine-1-phosphate is a missing cofactor for the E3 ubiquitin ligase TRAF2. *Nature* **2010**, *465*, 1084–1088. [CrossRef] [PubMed]

112. Etemadi, N.; Chopin, M.; Anderton, H.; Tanzer, M.C.; Rickard, J.A.; Abeysekera, W.; Hall, C.; Spall, S.K.; Wang, B.; Xiong, Y.; et al. TRAF2 regulates TNF and NF-κB signalling to suppress apoptosis and skin inflammation independently of Sphingosine kinase 1. *Elife* **2015**, *4*, E10592. [CrossRef] [PubMed]

113. Tada, K.; Okazaki, T.; Sakon, S.; Kobarai, T.; Kurosawa, K.; Yamaoka, S.; Hashimoto, H.; Mak, T.W.; Yagita, H.; Okumura, K.; et al. Critical roles of TRAF2 and TRAF5 in tumor necrosis factor-induced NF-κB activation and protection from cell death. *J. Biol. Chem.* **2001**, *276*, 36530–36534. [CrossRef] [PubMed]

114. Zhang, L.; Blackwell, K.; Thomas, G.S.; Sun, S.; Yeh, W.C.; Habelhah, H. TRAF2 suppresses basal IKK activity in resting cells and TNFα can activate IKK in TRAF2 and TRAF5 double knockout cells. *J. Mol. Biol.* **2009**, *389*, 495–510. [CrossRef] [PubMed]

115. Martin, M.U.; Wesche, H. Summary and comparison of the signaling mechanisms of the Toll/interleukin-1 receptor family. *Biochim. Biophys. Acta* **2002**, *1592*, 265–280. [CrossRef]

116. Kawasaki, T.; Kawai, T. Toll-like receptor signaling pathways. *Front. Immunol.* **2014**, *5*, 461. [CrossRef] [PubMed]

117. O'Neill, L.A.; Bowie, A.G. The family of five: TIR-domain-containing adaptors in Toll-like receptor signalling. *Nat. Rev. Immunol.* **2007**, *7*, 353–364. [CrossRef] [PubMed]

118. Motshwene, P.G.; Moncrieffe, M.C.; Grossmann, J.G.; Kao, C.; Ayaluru, M.; Sandercock, A.M.; Robinson, C.V.; Latz, E.; Gay, N.J. An oligomeric signaling platform formed by the Toll-like receptor signal transducers MyD88 and IRAK-4. *J. Biol. Chem.* **2009**, *284*, 25404–25411. [CrossRef] [PubMed]

119. Ferrao, R.; Zhou, H.; Shan, Y.; Liu, Q.; Li, Q.; Shaw, D.E.; Li, X.; Wu, H. IRAK4 dimerization and trans-autophosphorylation are induced by Myddosome assembly. *Mol. Cell* **2014**, *55*, 891–903. [CrossRef] [PubMed]

120. Ye, H.; Arron, J.R.; Lamothe, B.; Cirilli, M.; Kobayashi, T.; Shevde, N.K.; Segal, D.; Dzivenu, O.K.; Vologodskaia, M.; Yim, M.; et al. Distinct molecular mechanism for initiating TRAF6 signalling. *Nature* **2002**, *418*, 443–447. [CrossRef] [PubMed]

121. Moynagh, P.N. The roles of Pellino E3 ubiquitin ligases in immunity. *Nat. Rev. Immunol.* **2014**, *14*, 122–131. [CrossRef] [PubMed]

122. Medvedev, A.E.; Murphy, M.; Zhou, H.; Li, X. E3 ubiquitin ligases Pellinos as regulators of pattern recognition receptor signaling and immune response. *Immunol. Rev.* **2015**, *266*, 109–122. [PubMed]

123. Lin, C.C.; Huoh, Y.S.; Schmitz, K.R.; Jensen, L.E.; Ferguson, K.M. Pellino proteins contain a cryptic FHA domain that mediates interaction with phosphorylated IRAK1. *Structure* **2008**, *16*, 1806–1816. [CrossRef] [PubMed]

124. Cohen, P.; Strickson, S. The role of hybrid ubiquitin chains in the Myd88 and other innate immune signaling pathways. *Cell Death Differ.* **2017**, *24*, 1153–1159. [PubMed]

125. Xia, Z.P.; Sun, L.; Chen, X.; Pineda, G.; Jiang, X.; Adhikari, A.; Zeng, W.; Chen, Z.J. Direct activation of protein kinases by unanchored polyubiquitin chains. *Nature* **2009**, *461*, 114–119. [CrossRef] [PubMed]

126. Cui, W.; Xiao, N.; Xiao, H.; Zhou, H.; Yu, M.; Gu, J.; Li, X. β-TrCP-mediated IRAK1 degradation releases TAK1-TRAF6 from the membrane to the cytosol for TAK1-dependent NF-κB activation. *Mol. Cell. Biol.* **2012**, *32*, 3990–4000. [CrossRef] [PubMed]

127. Matsuzawa, A.; Tseng, P.H.; Vallabhapurapu, S.; Luo, J.L.; Zhang, W.; Wang, H.; Vignali, D.A.; Gallagher, E.; Karin, M. Essential cytoplasmic translocation of a cytokine receptor-assembled signaling complex. *Science* **2008**, *321*, 663–668. [CrossRef] [PubMed]

128. Kagan, J.C.; Su, T.; Horng, T.; Chow, A.; Akira, S.; Medzhitov, R. TRAM couples endocytosis of Toll-like receptor 4 to the induction of interferon-β. *Nat. Immunol.* **2008**, *9*, 361–368. [CrossRef] [PubMed]

129. Sasai, M.; Tatematsu, M.; Oshiumi, H.; Funami, K.; Matsumoto, M.; Hatakeyama, S.; Seya, T. Direct binding of TRAF2 and TRAF6 to TICAM-1/TRIF adaptor participates in activation of the Toll-like receptor 3/4 pathway. *Mol. Immunol.* **2010**, *47*, 1283–1291. [CrossRef] [PubMed]

130. Cusson-Hermance, N.; Khurana, S.; Lee, T.H.; Fitzgerald, K.A.; Kelliher, M.A. Rip1 mediates the Trif-dependent toll-like receptor 3- and 4-induced NF-κB activation but does not contribute to interferon regulatory factor 3 activation. *J. Biol. Chem.* **2005**, *280*, 36560–36566. [CrossRef] [PubMed]

131. Kaiser, W.J.; Offermann, M.K. Apoptosis induced by the toll-like receptor adaptor TRIF is dependent on its receptor interacting protein homotypic interaction motif. *J. Immunol.* **2005**, *174*, 4942–4952. [CrossRef] [PubMed]

132. Chang, M.; Jin, W.; Sun, S.C. Peli1 facilitates TRIF-dependent Toll-like receptor signaling and proinflammatory cytokine production. *Nat. Immunol.* **2009**, *10*, 1089–1095. [CrossRef] [PubMed]

133. Chen, R.; Li, M.; Zhang, Y.; Zhou, Q.; Shu, H.B. The E3 ubiquitin ligase MARCH8 negatively regulates IL-1β-induced NF-κB activation by targeting the IL1RAP coreceptor for ubiquitination and degradation. *Proc. Natl. Acad. Sci. USA* **2012**, *109*, 14128–14133. [CrossRef] [PubMed]

134. Zhao, Y.; Mudge, M.C.; Soll, J.M.; Rodrigues, R.B.; Byrum, A.K.; Schwarzkopf, E.A.; Bradstreet, T.R.; Gygi, S.P.; Edelson, B.T.; Mosammaparast, N. OTUD4 Is a Phospho-Activated K63 Deubiquitinase That Regulates MyD88-Dependent Signaling. *Mol. Cell* **2018**, *69*, 505–516. [CrossRef] [PubMed]

135. He, X.; Li, Y.; Li, C.; Liu, L.J.; Zhang, X.D.; Liu, Y.; Shu, H.B. USP2a negatively regulates IL-1β- and virus-induced NF-κB activation by deubiquitinating TRAF6. *J. Mol. Cell. Biol.* **2013**, *5*, 39–47. [CrossRef] [PubMed]

136. Zhou, F.; Zhang, X.; van Dam, H.; Ten Dijke, P.; Huang, H.; Zhang, L. Ubiquitin-specific protease 4 mitigates Toll-like/interleukin-1 receptor signaling and regulates innate immune activation. *J. Biol. Chem.* **2012**, *287*, 11002–11010. [CrossRef] [PubMed]

137. Yasunaga, J.; Lin, F.C.; Lu, X.; Jeang, K.T. Ubiquitin-specific peptidase 20 targets TRAF6 and human T cell leukemia virus type 1 tax to negatively regulate NF-κB signaling. *J. Virol.* **2011**, *85*, 6212–6219. [CrossRef] [PubMed]

138. Yoshida, H.; Jono, H.; Kai, H.; Li, J.D. The tumor suppressor cylindromatosis (CYLD) acts as a negative regulator for toll-like receptor 2 signaling via negative cross-talk with TRAF6 and TRAF7. *J. Biol. Chem.* **2005**, *280*, 41111–41121. [CrossRef] [PubMed]

139. Heyninck, K.; Beyaert, R. The cytokine-inducible zinc finger protein A20 inhibits IL-1-induced NF-κB activation at the level of TRAF6. *FEBS Lett.* **1999**, *442*, 147–150. [CrossRef]

140. Yuk, J.M.; Shin, D.M.; Lee, H.M.; Kim, J.J.; Kim, S.W.; Jin, H.S.; Yang, C.S.; Park, K.A.; Chanda, D.; Kim, D.K.; et al. The orphan nuclear receptor SHP acts as a negative regulator in inflammatory signaling triggered by Toll-like receptors. *Nat. Immunol.* **2011**, *12*, 742–751. [CrossRef] [PubMed]

141. Wang, W.; Huang, X.; Xin, H.B.; Fu, M.; Xue, A.; Wu, Z.H. TRAF Family Member-associated NF-κB Activator (TANK) Inhibits Genotoxic Nuclear Factor κB Activation by Facilitating Deubiquitinase USP10-dependent Deubiquitination of TRAF6 Ligase. *J. Biol. Chem.* **2015**, *290*, 13372–13385. [CrossRef] [PubMed]

142. Wu, C.; Su, Z.; Lin, M.; Ou, J.; Zhao, W.; Cui, J.; Wang, R.F. NLRP11 attenuates Toll-like receptor signalling by targeting TRAF6 for degradation via the ubiquitin ligase RNF19A. *Nat. Commun.* **2017**, *8*, 1977. [CrossRef] [PubMed]

143. Lin, X.W.; Xu, W.C.; Luo, J.G.; Guo, X.J.; Sun, T.; Zhao, X.L.; Fu, Z.J. WW domain containing E3 ubiquitin protein ligase 1 (WWP1) negatively regulates TLR4-mediated TNF-α and IL-6 production by proteasomal degradation of TNF receptor associated factor 6 (TRAF6). *PLoS ONE* **2013**, *8*, e67633. [CrossRef] [PubMed]

144. Zhao, W.; Wang, L.; Zhang, M.; Yuan, C.; Gao, C. E3 ubiquitin ligase tripartite motif 38 negatively regulates TLR-mediated immune responses by proteasomal degradation of TNF receptor-associated factor 6 in macrophages. *J. Immunol.* **2012**, *188*, 2567–2574. [CrossRef] [PubMed]

145. Zhong, B.; Liu, X.; Wang, X.; Liu, X.; Li, H.; Darnay, B.G.; Lin, X.; Sun, S.C.; Dong, C. Ubiquitin-specific protease 25 regulates TLR4-dependent innate immune responses through deubiquitination of the adaptor protein TRAF3. *Sci. Signal.* **2013**, *6*, ra35. [CrossRef] [PubMed]

146. Caruso, R.; Warner, N.; Inohara, N.; Núñez, G. NOD1 and NOD2: Signaling, host defense, and inflammatory disease. *Immunity* **2014**, *41*, 898–908. [CrossRef] [PubMed]

147. Boyle, J.P.; Parkhouse, R.; Monie, T.P. Insights into the molecular basis of the NOD2 signalling pathway. *Open Biol.* **2014**, *4*, 140178. [CrossRef] [PubMed]

148. Tigno-Aranjuez, J.T.; Abbott, D.W. Ubiquitination and phosphorylation in the regulation of NOD2 signaling and NOD2-mediated disease. *Biochim. Biophys. Acta* **2012**, *1823*, 2022–2028. [CrossRef] [PubMed]

149. Tigno-Aranjuez, J.T.; Asara, J.M.; Abbott, D.W. Inhibition of RIP2's tyrosine kinase activity limits NOD2-driven cytokine responses. *Genes Dev.* **2010**, *24*, 2666–2677. [CrossRef] [PubMed]

150. Yang, Y.; Yin, C.; Pandey, A.; Abbott, D.; Sassetti, C.; Kelliher, M.A. NOD2 pathway activation by MDP or Mycobacterium tuberculosis infection involves the stable polyubiquitination of Rip2. *J. Biol. Chem.* **2007**, *282*, 36223–36229. [CrossRef] [PubMed]

151. Hasegawa, M.; Fujimoto, Y.; Lucas, P.C.; Nakano, H.; Fukase, K.; Núñez, G.; Inohara, N. A critical role of RICK/RIP2 polyubiquitination in Nod-induced NF-κB activation. *EMBO J.* **2008**, *27*, 373–383. [CrossRef] [PubMed]

152. Bertrand, M.J.; Doiron, K.; Labbé, K.; Korneluk, R.G.; Barker, P.A.; Saleh, M. Cellular inhibitors of apoptosis cIAP1 and cIAP2 are required for innate immunity signaling by the pattern recognition receptors NOD1 and NOD2. *Immunity* **2009**, *30*, 789–801. [PubMed]

153. Krieg, A.; Correa, R.G.; Garrison, J.B.; Le Negrate, G.; Welsh, K.; Huang, Z.; Knoefel, W.T.; Reed, J.C. XIAP mediates NOD signaling via interaction with RIP2. *Proc. Natl. Acad. Sci. USA* **2009**, *106*, 14524–14529. [CrossRef] [PubMed]

154. Yang, S.; Wang, B.; Humphries, F.; Jackson, R.; Healy, M.E.; Bergin, R.; Aviello, G.; Hall, B.; McNamara, D.; Darby, T.; et al. Pellino3 ubiquitinates RIP2 and mediates Nod2-induced signaling and protective effects in colitis. *Nat. Immunol.* **2013**, *14*, 927–936. [CrossRef] [PubMed]

155. Damgaard, R.B.; Nachbur, U.; Yabal, M.; Wong, W.W.; Fiil, B.K.; Kastirr, M.; Rieser, E.; Rickard, J.A.; Bankovacki, A.; Peschel, C.; et al. The ubiquitin ligase XIAP recruits LUBAC for NOD2 signaling in inflammation and innate immunity. *Mol. Cell* **2012**, *46*, 746–758. [CrossRef] [PubMed]

156. Damgaard, R.B.; Fiil, B.K.; Speckmann, C.; Yabal, M.; zur Stadt, U.; Bekker-Jensen, S.; Jost, P.J.; Ehl, S.; Mailand, N.; Gyrd-Hansen, M. Disease-causing mutations in the XIAP BIR2 domain impair NOD2-dependent immune signalling. *EMBO Mol. Med.* **2013**, *5*, 1278–1295. [CrossRef] [PubMed]

157. Abbott, D.W.; Yang, Y.; Hutti, J.E.; Madhavarapu, S.; Kelliher, M.A.; Cantley, L.C. Coordinated regulation of Toll-like receptor and NOD2 signaling by K63-linked polyubiquitin chains. *Mol. Cell. Biol.* **2007**, *27*, 6012–6025. [CrossRef] [PubMed]

158. Clark, N.M.; Marinis, J.M.; Cobb, B.A.; Abbott, D.W. MEKK4 sequesters RIP2 to dictate NOD2 signal specificity. *Curr. Biol.* **2008**, *18*, 1402–1408. [CrossRef] [PubMed]

159. Draber, P.; Kupka, S.; Reichert, M.; Draberova, H.; Lafont, E.; de Miguel, D.; Spilgies, L.; Surinova, S.; Taraborrelli, L.; Hartwig, T.; et al. LUBAC-Recruited CYLD and A20 Regulate Gene Activation and Cell Death by Exerting Opposing Effects on Linear Ubiquitin in Signaling Complexes. *Cell Rep.* **2015**, *13*, 2258–2272. [CrossRef] [PubMed]

160. Hrdinka, M.; Fiil, B.K.; Zucca, M.; Leske, D.; Bagola, K.; Yabal, M.; Elliott, P.R.; Damgaard, R.B.; Komander, D.; Jost, P.J.; et al. CYLD Limits Lys63- and Met1-Linked Ubiquitin at Receptor Complexes to Regulate Innate Immune Signaling. *Cell Rep.* **2016**, *14*, 2846–2858. [PubMed]

161. Tao, M.; Scacheri, P.C.; Marinis, J.M.; Harhaj, E.W.; Matesic, L.E.; Abbott, D.W. ITCH K63-ubiquitinates the NOD2 binding protein, RIP2, to influence inflammatory signaling pathways. *Curr. Biol.* **2009**, *19*, 1255–1263. [CrossRef] [PubMed]

162. Hitotsumatsu, O.; Ahmad, R.C.; Tavares, R.; Wang, M.; Philpott, D.; Turer, E.E.; Lee, B.L.; Shiffin, N.; Advincula, R.; Malynn, B.A.; et al. The ubiquitin-editing enzyme A20 restricts nucleotide-binding oligomerization domain containing 2-triggered signals. *Immunity* **2008**, *28*, 381–390. [CrossRef] [PubMed]

163. Bist, P.; Cheong, W.S.; Ng, A.; Dikshit, N.; Kim, B.H.; Pulloor, N.K.; Khameneh, H.J.; Hedl, M.; Shenoy, A.R.; Balamuralidhar, V.; et al. E3 Ubiquitin ligase ZNRF4 negatively regulates NOD2 signalling and induces tolerance to MDP. *Nat. Commun.* **2017**, *8*, 15865. [CrossRef] [PubMed]

164. Zurek, B.; Schoultz, I.; Neerincx, A.; Napolitano, L.M.; Birkner, K.; Bennek, E.; Sellge, G.; Lerm, M.; Meroni, G.; Söderholm, J.D. TRIM27 negatively regulates NOD2 by ubiquitination and proteasomal degradation. *PLoS ONE* **2012**, *7*, e41255. [CrossRef] [PubMed]

165. Condé, C.; Rambout, X.; Lebrun, M.; Lecat, A.; Di Valentin, E.; Dequiedt, F.; Piette, J.; Gloire, G.; Legrand, S. The inositol phosphatase SHIP-1 inhibits NOD2-induced NF-κB activation by disturbing the interaction of XIAP with RIP2. *PLoS ONE* **2012**, *7*, e41005. [CrossRef] [PubMed]

166. Yoneyama, M.; Fujita, T. RNA recognition and signal transduction by RIG-I-like receptors. *Immunol. Rev.* **2009**, *227*, 54–65. [CrossRef] [PubMed]

167. Wilkins, C.; Gale, M., Jr. Recognition of viruses by cytoplasmic sensors. *Curr. Opin. Immunol.* **2010**, *22*, 41–47. [CrossRef] [PubMed]

168. Goubau, D.; Deddouche, S.; Reis e Sousa, C. Cytosolic sensing of viruses. *Immunity* **2013**, *38*, 855–869. [CrossRef] [PubMed]

169. Satoh, T.; Kato, H.; Kumagai, Y.; Yoneyama, M.; Sato, S.; Matsushita, K.; Tsujimura, T.; Fujita, T.; Akira, S.; Takeuchi, O. LGP2 is a positive regulator of RIG-I- and MDA5-mediated antiviral responses. *Proc. Natl. Acad. Sci. USA* **2010**, *107*, 1512–1517. [CrossRef] [PubMed]

170. Chiang, C.; Gack, M.U. Post-translational Control of Intracellular Pathogen Sensing Pathways. *Trends Immunol.* **2017**, *38*, 39–52. [CrossRef] [PubMed]

171. Liu, B.; Gao, C. Regulation of MAVS activation through post-translational modifications. *Curr. Opin. Immunol.* **2017**, *50*, 75–81. [CrossRef] [PubMed]

172. Gack, M.U.; Shin, Y.C.; Joo, C.H.; Urano, T.; Liang, C.; Sun, L.; Takeuchi, O.; Akira, S.; Chen, Z.; Inoue, S.; et al. TRIM25 RING-finger E3 ubiquitin ligase is essential for RIG-I-mediated antiviral activity. *Nature* **2007**, *446*, 916–920. [CrossRef] [PubMed]

173. Liu, W.; Li, J.; Zheng, W.; Shang, Y.; Zhao, Z.; Wang, S.; Bi, Y.; Zhang, S.; Xu, C.; Duan, Z. Cyclophilin A-regulated ubiquitination is critical for RIG-I-mediated antiviral immune responses. *Elife* **2017**, *6*, E24425. [CrossRef] [PubMed]

174. Jiang, X.; Kinch, L.N.; Brautigam, C.A.; Chen, X.; Du, F.; Grishin, N.V.; Chen, Z.J. Ubiquitin-induced oligomerization of the RNA sensors RIG-I and MDA5 activates antiviral innate immune response. *Immunity* **2012**, *36*, 959–973. [CrossRef] [PubMed]

175. Zeng, W.; Sun, L.; Jiang, X.; Chen, X.; Hou, F.; Adhikari, A.; Xu, M.; Chen, Z.J. Reconstitution of the RIG-I pathway reveals a signaling role of unanchored polyubiquitin chains in innate immunity. *Cell* **2010**, *141*, 315–330. [CrossRef] [PubMed]

176. Oshiumi, H.; Miyashita, M.; Matsumoto, M.; Seya, T. A distinct role of Riplet-mediated K63-Linked polyubiquitination of the RIG-I repressor domain in human antiviral innate immune responses. *PLoS Pathog.* **2013**, *9*, e1003533. [CrossRef] [PubMed]

177. Yan, J.; Li, Q.; Mao, A.P.; Hu, M.M.; Shu, H.B. TRIM4 modulates type I interferon induction and cellular antiviral response by targeting RIG-I for K63-linked ubiquitination. *J. Mol. Cell. Biol.* **2014**, *6*, 154–163. [CrossRef] [PubMed]

178. Kuniyoshi, K.; Takeuchi, O.; Pandey, S.; Satoh, T.; Iwasaki, H.; Akira, S.; Kawai, T. Pivotal role of RNA-binding E3 ubiquitin ligase MEX3C in RIG-I-mediated antiviral innate immunity. *Proc. Natl. Acad. Sci. USA* **2014**, *111*, 5646–5651. [CrossRef] [PubMed]

179. Sun, X.; Xian, H.; Tian, S.; Sun, T.; Qin, Y.; Zhang, S.; Cui, J. A Hierarchical Mechanism of RIG-I Ubiquitination Provides Sensitivity, Robustness and Synergy in Antiviral Immune Responses. *Sci. Rep.* **2016**, *6*, 29263. [CrossRef] [PubMed]

180. Lang, X.; Tang, T.; Jin, T.; Ding, C.; Zhou, R.; Jiang, W. TRIM65-catalized ubiquitination is essential for MDA5-mediated antiviral innate immunity. *J. Exp. Med.* **2017**, *214*, 459–473. [CrossRef] [PubMed]

181. Hou, F.; Sun, L.; Zheng, H.; Skaug, B.; Jiang, Q.X.; Chen, Z.J. MAVS forms functional prion-like aggregates to activate and propagate antiviral innate immune response. *Cell* **2011**, *146*, 448–461. [CrossRef] [PubMed]

182. Liu, S.; Chen, J.; Cai, X.; Wu, J.; Chen, X.; Wu, Y.T.; Sun, L.; Chen, Z.J. MAVS recruits multiple ubiquitin E3 ligases to activate antiviral signaling cascades. *Elife* **2013**, *2*, e00785. [CrossRef] [PubMed]

183. Fang, R.; Jiang, Q.; Zhou, X.; Wang, C.; Guan, Y.; Tao, J.; Xi, J.; Feng, J.M.; Jiang, Z. MAVS activates TBK1 and IKKε through TRAFs in NEMO dependent and independent manner. *PLoS Pathog.* **2017**, *13*, e1006720. [CrossRef] [PubMed]

184. Tang, E.D.; Wang, C.Y. TRAF5 is a downstream target of MAVS in antiviral innate immune signaling. *PLoS ONE* **2010**, *5*, e9172. [CrossRef] [PubMed]

185. Mikkelsen, S.S.; Jensen, S.B.; Chiliveru, S.; Melchjorsen, J.; Julkunen, I.; Gaestel, M.; Arthur, J.S.; Flavell, R.A.; Ghosh, S.; Paludan, S.R. RIG-I-mediated activation of p38 MAPK is essential for viral induction of interferon and activation of dendritic cells: Dependence on TRAF2 and TAK1. *J. Biol. Chem.* **2009**, *284*, 10774–10782. [PubMed]

186. Pauli, E.K.; Chan, Y.K.; Davis, M.E.; Gableske, S.; Wang, M.K.; Feister, K.F.; Gack, M.U. The ubiquitin-specific protease USP15 promotes RIG-I-mediated antiviral signaling by deubiquitylating TRIM25. *Sci. Signal.* **2014**, *7*, ra3. [CrossRef] [PubMed]

187. Inn, K.S.; Gack, M.U.; Tokunaga, F.; Shi, M.; Wong, L.Y.; Iwai, K.; Jung, J.U. Linear ubiquitin assembly complex negatively regulates RIG-I- and TRIM25-mediated type I interferon induction. *Mol. Cell* **2011**, *41*, 354–365. [CrossRef] [PubMed]

188. Cui, J.; Song, Y.; Li, Y.; Zhu, Q.; Tan, P.; Qin, Y.; Wang, H.Y.; Wang, R.F. USP3 inhibits type I interferon signaling by deubiquitinating RIG-I-like receptors. *Cell Res.* **2014**, *24*, 400–416. [CrossRef] [PubMed]

189. Fan, Y.; Mao, R.; Yu, Y.; Liu, S.; Shi, Z.; Cheng, J.; Zhang, H.; An, L.; Zhao, Y.; Xu, X.; et al. USP21 negatively regulates antiviral response by acting as a RIG-I deubiquitinase. *J. Exp. Med.* **2014**, *211*, 313–328. [CrossRef] [PubMed]

190. Friedman, C.S.; O'Donnell, M.A.; Legarda-Addison, D.; Ng, A.; Cárdenas, W.B.; Yount, J.S.; Moran, T.M.; Basler, C.F.; Komuro, A.; Horvath, C.M.; et al. The tumour suppressor CYLD is a negative regulator of RIG-I-mediated antiviral response. *EMBO Rep.* **2008**, *9*, 930–936. [CrossRef] [PubMed]

191. Lin, W.; Zhang, J.; Lin, H.; Li, Z.; Sun, X.; Xin, D.; Yang, M.; Sun, L.; Li, L.; Wang, H.; et al. Syndecan-4 negatively regulates antiviral signalling by mediating RIG-I deubiquitination via CYLD. *Nat. Commun.* **2016**, *7*, 11848. [CrossRef] [PubMed]

192. Wang, W.; Jiang, M.; Liu, S.; Zhang, S.; Liu, W.; Ma, Y.; Zhang, L.; Zhang, J.; Cao, X. RNF122 suppresses antiviral type I interferon production by targeting RIG-I CARDs to mediate RIG-I degradation. *Proc. Natl. Acad. Sci. USA* **2016**, *113*, 9581–9586. [CrossRef] [PubMed]

193. Arimoto, K.; Takahashi, H.; Hishiki, T.; Konishi, H.; Fujita, T.; Shimotohno, K. Negative regulation of the RIG-I signaling by the ubiquitin ligase RNF125. *Proc. Natl. Acad. Sci. USA* **2007**, *104*, 7500–7505. [CrossRef] [PubMed]

194. Hao, Q.; Jiao, S.; Shi, Z.; Li, C.; Meng, X.; Zhang, Z.; Wang, Y.; Song, X.; Wang, W.; Zhang, R.; et al. A non-canonical role of the p97 complex in RIG-I antiviral signaling. *EMBO J.* **2015**, *34*, 2903–2920. [CrossRef] [PubMed]

195. Wang, L.; Zhao, W.; Zhang, M.; Wang, P.; Zhao, K.; Zhao, X.; Yang, S.; Gao, C. USP4 positively regulates RIG-I-mediated antiviral response through deubiquitination and stabilization of RIG-I. *J. Virol.* **2013**, *87*, 4507–4515. [CrossRef] [PubMed]

196. Wang, Y.; Tong, X.; Ye, X. Ndfip1 negatively regulates RIG-I-dependent immune signaling by enhancing E3 ligase Smurf1-mediated MAVS degradation. *J. Immunol.* **2012**, *189*, 5304–5313. [CrossRef] [PubMed]

197. Pan, Y.; Li, R.; Meng, J.L.; Mao, H.T.; Zhang, Y.; Zhang, J. Smurf2 negatively modulates RIG-I-dependent antiviral response by targeting VISA/MAVS for ubiquitination and degradation. *J. Immunol.* **2014**, *192*, 4758–4764. [CrossRef] [PubMed]

198. Yoo, Y.S.; Park, Y.Y.; Kim, J.H.; Cho, H.; Kim, S.H.; Lee, H.S.; Kim, T.H.; Sun Kim, Y.; Lee, Y.; Kim, C.J.; et al. The mitochondrial ubiquitin ligase MARCH5 resolves MAVS aggregates during antiviral signalling. *Nat. Commun.* **2015**, *6*, 7910. [CrossRef] [PubMed]

199. Zhong, B.; Zhang, Y.; Tan, B.; Liu, T.T.; Wang, Y.Y.; Shu, H.B. The E3 ubiquitin ligase RNF5 targets virus-induced signaling adaptor for ubiquitination and degradation. *J. Immunol.* **2010**, *184*, 6249–6255. [CrossRef] [PubMed]

200. Luo, W.W.; Li, S.; Li, C.; Zheng, Z.Q.; Cao, P.; Tong, Z.; Lian, H.; Wang, S.Y.; Shu, H.B.; Wang, Y.Y. iRhom2 is essential for innate immunity to RNA virus by antagonizing ER- and mitochondria-associated degradation of VISA. *PLoS Pathog.* **2017**, *13*, e1006693. [CrossRef] [PubMed]

201. You, F.; Sun, H.; Zhou, X.; Sun, W.; Liang, S.; Zhai, Z.; Jiang, Z. PCBP2 mediates degradation of the adaptor MAVS via the HECT ubiquitin ligase AIP4. *Nat. Immunol.* **2009**, *10*, 1300–1308. [CrossRef] [PubMed]

202. Chen, Q.; Sun, L.; Chen, Z.J. Regulation and function of the cGAS-STING pathway of cytosolic DNA sensing. *Nat. Immunol.* **2016**, *17*, 1142–1149.

203. Xia, P.; Wang, S.; Gao, P.; Gao, G.; Fan, Z. DNA sensor cGAS-mediated immune recognition. *Protein Cell* **2016**, *7*, 777–791. [CrossRef] [PubMed]

204. Ma, Z.; Damania, B. The cGAS-STING Defense Pathway and Its Counteraction by Viruses. *Cell Host Microbe* **2016**, *19*, 150–158. [CrossRef] [PubMed]

205. Abe, T.; Barber, G.N. Cytosolic-DNA-mediated, STING-dependent proinflammatory gene induction necessitates canonical NF-κB activation through TBK1. *J. Virol.* **2014**, *88*, 5328–5341. [CrossRef] [PubMed]

206. Zhang, J.; Hu, M.M.; Wang, Y.Y.; Shu, H.B. TRIM32 protein modulates type I interferon induction and cellular antiviral response by targeting MITA/STING protein for K63-linked ubiquitination. *J. Biol. Chem.* **2012**, *287*, 28646–28655. [CrossRef] [PubMed]

207. Fang, R.; Wang, C.; Jiang, Q.; Lv, M.; Gao, P.; Yu, X.; Mu, P.; Zhang, R.; Bi, S.; Feng, J.M.; et al. NEMO-IKKβ Are Essential for IRF3 and NF-κB Activation in the cGAS-STING Pathway. *J. Immunol.* **2017**, *199*, 3222–3233. [CrossRef] [PubMed]

208. Sun, H.; Zhang, Q.; Jing, Y.Y.; Zhang, M.; Wang, H.Y.; Cai, Z.; Liuyu, T.; Zhang, Z.D.; Xiong, T.C.; Wu, Y. USP13 negatively regulates antiviral responses by deubiquitinating STING. *Nat. Commun.* **2017**, *8*, 15534. [CrossRef] [PubMed]

209. Zhong, B.; Zhang, L.; Lei, C.; Li, Y.; Mao, A.P.; Yang, Y.; Wang, Y.Y.; Zhang, X.L.; Shu, H.B. The ubiquitin ligase RNF5 regulates antiviral responses by mediating degradation of the adaptor protein MITA. *Immunity* **2009**, *30*, 397–407. [CrossRef] [PubMed]

210. Wang, Y.; Lian, Q.; Yang, B.; Yan, S.; Zhou, H.; He, L.; Lin, G.; Lian, Z.; Jiang, Z.; Sun, B. TRIM30α Is a Negative-Feedback Regulator of the Intracellular DNA and DNA Virus-Triggered Response by Targeting STING. *PLoS Pathog.* **2015**, *11*, e1005012. [CrossRef] [PubMed]

211. Zhang, M.; Zhang, M.X.; Zhang, Q.; Zhu, G.F.; Yuan, L.; Zhang, D.E.; Zhu, Q.; Yao, J.; Shu, H.B.; Zhong, B. USP18 recruits USP20 to promote innate antiviral response through deubiquitinating STING/MITA. *Cell Res.* **2016**, *26*, 1302–1319. [CrossRef] [PubMed]

212. Hu, M.M.; Yang, Q.; Xie, X.Q.; Liao, C.Y.; Lin, H.; Liu, T.T.; Yin, L.; Shu, H.B. Sumoylation Promotes the Stability of the DNA Sensor cGAS and the Adaptor STING to Regulate the Kinetics of Response to DNA Virus. *Immunity* **2016**, *45*, 555–569. [CrossRef] [PubMed]

213. Wang, Q.; Huang, L.; Hong, Z.; Lv, Z.; Mao, Z.; Tang, Y.; Kong, X.; Li, S.; Cui, Y.; Liu, H.; et al. The E3 ubiquitin ligase RNF185 facilitates the cGAS-mediated innate immune response. *PLoS Pathog.* **2017**, *13*, e1006264. [CrossRef] [PubMed]

214. Chen, M.; Meng, Q.; Qin, Y.; Liang, P.; Tan, P.; He, L.; Zhou, Y.; Chen, Y.; Huang, J.; Wang, R.F.; et al. TRIM14 Inhibits cGAS Degradation Mediated by Selective Autophagy Receptor p62 to Promote Innate Immune Responses. *Mol. Cell* **2016**, *64*, 105–119. [CrossRef] [PubMed]

215. Courtney, A.H.; Lo, W.L.; Weiss, A. TCR Signaling: Mechanisms of Initiation and Propagation. *Trends Biochem. Sci.* **2018**, *43*, 108–123. [CrossRef] [PubMed]

216. Paul, S.; Schaefer, B.C. A new look at T cell receptor signaling to nuclear factor-κB. *Trends Immunol.* **2013**, *34*, 269–281. [CrossRef] [PubMed]

217. Meininger, I.; Krappmann, D. Lymphocyte signaling and activation by the CARMA1-BCL10-MALT1 signalosome. *Biol. Chem.* **2016**, *397*, 1315–1333. [CrossRef] [PubMed]

218. Sommer, K.; Guo, B.; Pomerantz, J.L.; Bandaranayake, A.D.; Moreno-García, M.E.; Ovechkina, Y.L.; Rawlings, D.J. Phosphorylation of the CARMA1 linker controls NF-κB activation. *Immunity* **2005**, *23*, 561–574. [CrossRef] [PubMed]

219. Sun, L.; Deng, L.; Ea, C.K.; Xia, Z.P.; Chen, Z.J. The TRAF6 ubiquitin ligase and TAK1 kinase mediate IKK activation by BCL10 and MALT1 in T lymphocytes. *Mol. Cell* **2004**, *14*, 289–301. [CrossRef]

220. Li, Y.; He, X.; Wang, S.; Shu, H.B.; Liu, Y. USP2a positively regulates TCR-induced NF-κB activation by bridging MALT1-TRAF6. *Protein Cell* **2013**, *4*, 62–70. [CrossRef] [PubMed]

221. Carvalho, G.; Le Guelte, A.; Demian, C.; Vazquez, A.; Gavard, J.; Bidère, N. Interplay between BCL10, MALT1 and IκBα during T-cell-receptor-mediated NFκB activation. *J. Cell Sci.* **2010**, *123*, 2375–2380. [CrossRef] [PubMed]

222. Paul, S.; Traver, M.K.; Kashyap, A.K.; Washington, M.A.; Latoche, J.R.; Schaefer, B.C. T cell receptor signals to NF-κB are transmitted by a cytosolic p62-Bcl10-Malt1-IKK signalosome. *Sci. Signal.* **2014**, *7*, ra45. [CrossRef] [PubMed]

223. King, C.G.; Kobayashi, T.; Cejas, P.J.; Kim, T.; Yoon, K.; Kim, G.K.; Chiffoleau, E.; Hickman, S.P.; Walsh, P.T.; et al. TRAF6 is a T cell-intrinsic negative regulator required for the maintenance of immune homeostasis. *Nat. Med.* **2006**, *12*, 1088–1092.

224. Stempin, C.C.; Chi, L.; Giraldo-Vela, J.P.; High, A.A.; Häcker, H.; Redecke, V. The E3 ubiquitin ligase mind bomb-2 (MIB2) protein controls B-cell CLL/lymphoma 10 (BCL10)-dependent NF-κB activation. *J. Biol. Chem.* **2011**, *286*, 37147–37157. [CrossRef] [PubMed]

225. Dubois, S.M.; Alexia, C.; Wu, Y.; Leclair, H.M.; Leveau, C.; Schol, E.; Fest, T.; Tarte, K.; Chen, Z.J.; Gavard, J.; et al. A catalytic-independent role for the LUBAC in NF-κB activation upon antigen receptor engagement and in lymphoma cells. *Blood* **2014**, *123*, 2199–2203. [CrossRef] [PubMed]

226. Yang, Y.K.; Yang, C.; Chan, W.; Wang, Z.; Deibel, K.E.; Pomerantz, J.L. Molecular Determinants of Scaffold-induced Linear Ubiquitinylation of B Cell Lymphoma/Leukemia 10 (Bcl10) during T Cell Receptor and Oncogenic Caspase Recruitment Domain-containing Protein 11 (CARD11) Signaling. *J. Biol. Chem.* **2016**, *291*, 25921–25936. [CrossRef] [PubMed]

227. Yang, Y.; Kelly, P.; Shaffer, A.L., III; Schmitz, R.; Yoo, H.M.; Liu, X.; Huang, D.W.; Webster, D.; Young, R.M.; Nakagawa, M.; et al. Targeting Non-proteolytic Protein Ubiquitination for the Treatment of Diffuse B Cell Lymphoma. *Cancer Cell* **2016**, *29*, 494–507. [CrossRef] [PubMed]

228. Wan, Y.Y.; Chi, H.; Xie, M.; Schneider, M.D.; Flavell, R.A. The kinase TAK1 integrates antigen and cytokine receptor signaling for T cell development, survival and function. *Nat. Immunol.* **2006**, *7*, 851–858. [CrossRef] [PubMed]

229. Shinohara, H.; Yamasaki, S.; Maeda, S.; Saito, T.; Kurosaki, T. Regulation of NF-κB-dependent T cell activation and development by MEKK3. *Int. Immunol.* **2009**, *21*, 393–401. [CrossRef] [PubMed]

230. Düwel, M.; Welteke, V.; Oeckinghaus, A.; Baens, M.; Kloo, B.; Ferch, U.; Darnay, B.G.; Ruland, J.; Marynen, P.; Krappmann, D. A20 negatively regulates T cell receptor signaling to NF-κB by cleaving Malt1 ubiquitin chains. *J. Immunol.* **2009**, *182*, 7718–7728. [CrossRef] [PubMed]

231. Coornaert, B.; Baens, M.; Heyninck, K.; Bekaert, T.; Haegman, M.; Staal, J.; Sun, L.; Chen, Z.J.; Marynen, P.; Beyaert, R. T cell antigen receptor stimulation induces MALT1 paracaspase-mediated cleavage of the NF-κB inhibitor A20. *Nat. Immunol.* **2008**, *9*, 263–271. [CrossRef] [PubMed]

232. Reiley, W.W.; Jin, W.; Lee, A.J.; Wright, A.; Wu, X.; Tewalt, E.F.; Leonard, T.O.; Norbury, C.C.; Fitzpatrick, L.; Zhang, M.; et al. Deubiquitinating enzyme CYLD negatively regulates the ubiquitin-dependent kinase Tak1 and prevents abnormal T cell responses. *J. Exp. Med.* **2007**, *204*, 1475–1485. [CrossRef] [PubMed]

233. Staal, J.; Driege, Y.; Bekaert, T.; Demeyer, A.; Muyllaert, D.; Van Damme, P.; Gevaert, K.; Beyaert, R. T-cell receptor-induced JNK activation requires proteolytic inactivation of CYLD by MALT1. *EMBO J.* **2011**, *30*, 1742–1752. [CrossRef] [PubMed]

234. Klein, T.; Fung, S.Y.; Renner, F.; Blank, M.A.; Dufour, A.; Kang, S.; Bolger-Munro, M.; Scurll, J.M.; Priatel, J.J.; Schweigler, P.; et al. The paracaspase MALT1 cleaves HOIL1 reducing linear ubiquitination by LUBAC to dampen lymphocyte NF-κB signalling. *Nat. Commun.* **2015**, *6*, 8777. [CrossRef] [PubMed]

235. Poalas, K.; Hatchi, E.M.; Cordeiro, N.; Dubois, S.M.; Leclair, H.M.; Leveau, C.; Alexia, C.; Gavard, J.; Vazquez, A.; Bidère, N. Negative regulation of NF-κB signaling in T lymphocytes by the ubiquitin-specific protease USP34. *Cell. Commun. Signal.* **2013**, *11*, 25. [CrossRef] [PubMed]

236. Ge, Y.; Paisie, T.K.; Newman, J.R.B.; McIntyre, L.M.; Concannon, P. UBASH3A Mediates Risk for Type 1 Diabetes Through Inhibition of T-Cell Receptor-Induced NF-κB Signaling. *Diabetes* **2017**, *66*, 2033–2043. [CrossRef] [PubMed]

237. Hadian, K.; Krappmann, D. Signals from the nucleus: Activation of NF-κB by cytosolic ATM in the DNA damage response. *Sci. Signal.* **2011**, *4*, pe2. [CrossRef] [PubMed]

238. McCool, K.W.; Miyamoto, S. DNA damage-dependent NF-κB activation: NEMO turns nuclear signaling inside out. *Immunol. Rev.* **2012**, *246*, 311–326. [CrossRef] [PubMed]

239. Huang, T.T.; Wuerzberger-Davis, S.M.; Wu, Z.H.; Miyamoto, S. Sequential modification of NEMO/IKKγ by SUMO-1 and ubiquitin mediates NF-κB activation by genotoxic stress. *Cell* **2003**, *115*, 565–576. [CrossRef]

240. Mabb, A.M.; Wuerzberger-Davis, S.M.; Miyamoto, S. PIASy mediates NEMO sumoylation and NF-κB activation in response to genotoxic stress. *Nat. Cell Biol.* **2006**, *8*, 986–993. [CrossRef] [PubMed]

241. Janssens, S.; Tinel, A.; Lippens, S.; Tschopp, J. PIDD mediates NF-κB activation in response to DNA damage. *Cell* **2005**, *123*, 1079–1092. [CrossRef] [PubMed]

242. Bock, F.J.; Krumschnabel, G.; Manzl, C.; Peintner, L.; Tanzer, M.C.; Hermann-Kleiter, N.; Baier, G.; Llacuna, L.; Yelamos, J.; Villunger, A. Loss of PIDD limits NF-κB activation and cytokine production but not cell survival or transformation after DNA damage. *Cell Death Differ.* **2013**, *20*, 546–557. [CrossRef] [PubMed]

243. Stilmann, M.; Hinz, M.; Arslan, S.C.; Zimmer, A.; Schreiber, V.; Scheidereit, C. A nuclear poly(ADP-ribose)-dependent signalosome confers DNA damage-induced IκB kinase activation. *Mol. Cell* **2009**, *36*, 365–378. [CrossRef] [PubMed]

244. Li, N.; Banin, S.; Ouyang, H.; Li, G.C.; Courtois, G.; Shiloh, Y.; Karin, M.; Rotman, G. ATM is required for IκB kinase (IKK) activation in response to DNA double strand breaks. *J. Biol. Chem.* **2001**, *276*, 8898–8903. [CrossRef] [PubMed]

245. Wu, Z.H.; Shi, Y.; Tibbetts, R.S.; Miyamoto, S. Molecular linkage between the kinase ATM and NF-κB signaling in response to genotoxic stimuli. *Science* **2006**, *311*, 1141–1146. [CrossRef] [PubMed]

246. Wu, Z.H.; Wong, E.T.; Shi, Y.; Niu, J.; Chen, Z.; Miyamoto, S.; Tergaonkar, V. ATM- and NEMO-dependent ELKS ubiquitination coordinates TAK1-mediated IKK activation in response to genotoxic stress. *Mol. Cell* **2010**, *40*, 75–86. [CrossRef] [PubMed]

247. Yang, Y.; Xia, F.; Hermance, N.; Mabb, A.; Simonson, S.; Morrissey, S.; Gandhi, P.; Munson, M.; Miyamoto, S.; Kelliher, M.A. A cytosolic ATM/NEMO/RIP1 complex recruits TAK1 to mediate the NF-κB and p38 mitogen-activated protein kinase (MAPK)/MAPK-activated protein 2 responses to DNA damage. *Mol. Cell. Biol.* **2011**, *31*, 2774–2786. [CrossRef] [PubMed]

248. Niu, J.; Shi, Y.; Iwai, K.; Wu, Z.H. LUBAC regulates NF-κB activation upon genotoxic stress by promoting linear ubiquitination of NEMO. *EMBO J.* **2011**, *30*, 3741–3753. [CrossRef] [PubMed]

249. Hinz, M.; Stilmann, M.; Arslan, S.Ç.; Khanna, K.K.; Dittmar, G.; Scheidereit, C. A cytoplasmic ATM-TRAF6-cIAP1 module links nuclear DNA damage signaling to ubiquitin-mediated NF-κB activation. *Mol. Cell* **2010**, *40*, 63–74. [CrossRef] [PubMed]

250. Jin, H.S.; Lee, D.H.; Kim, D.H.; Chung, J.H.; Lee, S.J.; Lee, T.H. cIAP1, cIAP2, and XIAP act cooperatively via nonredundant pathways to regulate genotoxic stress-induced nuclear factor-κB activation. *Cancer Res.* **2009**, *69*, 1782–1791. [CrossRef] [PubMed]

251. Lee, M.H.; Mabb, A.M.; Gill, G.B.; Yeh, E.T.; Miyamoto, S. NF-κB induction of the SUMO protease SENP2: A negative feedback loop to attenuate cell survival response to genotoxic stress. *Mol. Cell* **2011**, *43*, 180–191. [CrossRef] [PubMed]

252. Niu, J.; Shi, Y.; Xue, J.; Miao, R.; Huang, S.; Wang, T.; Wu, J.; Fu, M.; Wu, Z.H. USP10 inhibits genotoxic NF-κB activation by MCPIP1-facilitated deubiquitination of NEMO. *EMBO J.* **2013**, *32*, 3206–3219. [CrossRef] [PubMed]

253. Hirata, Y.; Takahashi, M.; Morishita, T.; Noguchi, T.; Matsuzawa, A. Post-Translational Modifications of the TAK1-TAB Complex. *Int. J. Mol. Sci.* **2017**, *18*, 205. [CrossRef] [PubMed]

254. Hamidi, A.; von Bulow, V.; Hamidi, R.; Winssinger, N.; Barluenga, S.; Heldin, C.H.; Landström, M. Polyubiquitination of transforming growth factor β (TGFβ)-associated kinase 1 mediates nuclear factor-κB activation in response to different inflammatory stimuli. *J. Biol. Chem.* **2012**, *287*, 123–133. [CrossRef] [PubMed]

255. Fan, Y.; Yu, Y.; Shi, Y.; Sun, W.; Xie, M.; Ge, N.; Mao, R.; Chang, A.; Xu, G.; Schneider, M.D.; et al. Lysine 63-linked polyubiquitination of TAK1 at lysine 158 is required for tumor necrosis factor α- and interleukin-1β-induced IKK/NF-κB and JNK/AP-1 activation. *J. Biol. Chem.* **2010**, *285*, 5347–5360. [CrossRef] [PubMed]

256. Li, Q.; Yan, J.; Mao, A.P.; Li, C.; Ran, Y.; Shu, H.B.; Wang, Y.Y. Tripartite motif 8 (TRIM8) modulates TNFα- and IL-1β-triggered NF-κB activation by targeting TAK1 for K63-linked polyubiquitination. *Proc. Natl. Acad. Sci. USA* **2011**, *108*, 19341–19346. [CrossRef] [PubMed]

257. Fan, Y.H.; Yu, Y.; Mao, R.F.; Tan, X.J.; Xu, G.F.; Zhang, H.; Lu, X.B.; Fu, S.B.; Yang, J. USP4 targets TAK1 to downregulate TNFα-induced NF-κB activation. *Cell Death Differ.* **2011**, *18*, 1547–1560. [CrossRef] [PubMed]

258. Liu, X.; Li, H.; Zhong, B.; Blonska, M.; Gorjestani, S.; Yan, M.; Tian, Q.; Zhang, D.E.; Lin, X.; Dong, C. USP18 inhibits NF-κB and NFAT activation during Th17 differentiation by deubiquitinating the TAK1-TAB1 complex. *J. Exp. Med.* **2013**, *210*, 1575–1590. [CrossRef] [PubMed]

259. Yang, Z.; Xian, H.; Hu, J.; Tian, S.; Qin, Y.; Wang, R.F.; Cui, J. USP18 negatively regulates NF-κB signaling by targeting TAK1 and NEMO for deubiquitination through distinct mechanisms. *Sci. Rep.* **2015**, *5*, 12738. [CrossRef] [PubMed]

260. Xiao, H.; Qian, W.; Staschke, K.; Qian, Y.; Cui, G.; Deng, L.; Ehsani, M.; Wang, X.; Qian, Y.W.; Chen, Z.J.; et al. Pellino 3b negatively regulates interleukin-1-induced TAK1-dependent NF κB activation. *J. Biol. Chem.* **2008**, *283*, 14654–14664. [CrossRef] [PubMed]

261. Yang, Y.; Zhou, C.; Wang, Y.; Liu, W.; Liu, C.; Wang, L.; Liu, Y.; Shang, Y.; Li, M.; Zhou, S.; et al. The E3 ubiquitin ligase RNF114 and TAB1 degradation are required for maternal-to-zygotic transition. *EMBO Rep.* **2017**, *18*, 205–216. [CrossRef] [PubMed]

262. Charlaftis, N.; Suddason, T.; Wu, X.; Anwar, S.; Karin, M.; Gallagher, E. The MEKK1 PHD ubiquitinates TAB1 to activate MAPKs in response to cytokines. *EMBO J.* **2014**, *33*, 2581–2596. [CrossRef] [PubMed]

263. Theivanthiran, B.; Kathania, M.; Zeng, M.; Anguiano, E.; Basrur, V.; Vandergriff, T.; Pascual, V.; Wei, W.Z.; Massoumi, R.; et al. The E3 ubiquitin ligase Itch inhibits p38α signaling and skin inflammation through the ubiquitylation of Tab1. *Sci. Signal.* **2015**, *8*, ra22. [CrossRef] [PubMed]

264. Hu, M.M.; Yang, Q.; Zhang, J.; Liu, S.M.; Zhang, Y.; Lin, H.; Huang, Z.F.; Wang, Y.Y.; Zhang, X.D.; Zhong, B. TRIM38 inhibits TNFα- and IL-1β-triggered NF-κB activation by mediating lysosome-dependent degradation of TAB2/3. *Proc. Natl. Acad. Sci. USA* **2014**, *111*, 1509–1514. [CrossRef] [PubMed]

265. Shi, M.; Deng, W.; Bi, E.; Mao, K.; Ji, Y.; Lin, G.; Wu, X.; Tao, Z.; Li, Z.; Cai, X.; et al. TRIM30α negatively regulates TLR-mediated NF-κB activation by targeting TAB2 and TAB3 for degradation. *Nat. Immunol.* **2008**, *9*, 369–377. [CrossRef] [PubMed]

266. Tan, B.; Mu, R.; Chang, Y.; Wang, Y.B.; Wu, M.; Tu, H.Q.; Zhang, Y.C.; Guo, S.S.; Qin, X.H.; Li, T.; et al. RNF4 negatively regulates NF-κB signaling by down-regulating TAB2. *FEBS Lett.* **2015**, *589*, 2850–2858. [CrossRef] [PubMed]

267. Zhou, H.; Wertz, I.; O'Rourke, K.; Ultsch, M.; Seshagiri, S.; Eby, M.; Xiao, W.; Dixit, V.M. Bcl10 activates the NF-κB pathway through ubiquitination of NEMO. *Nature* **2004**, *427*, 167–171. [CrossRef] [PubMed]

268. Abbott, D.W.; Wilkins, A.; Asara, J.M.; Cantley, L.C. The Crohn's disease protein, NOD2, requires RIP2 in order to induce ubiquitinylation of a novel site on NEMO. *Curr. Biol.* **2004**, *14*, 2217–2227. [CrossRef] [PubMed]

269. Tokunaga, F.; Sakata, S.; Saeki, Y.; Satomi, Y.; Kirisako, T.; Kamei, K.; Nakagawa, T.; Kato, M.; Murata, S.; Yamaoka, S.; et al. Involvement of linear polyubiquitylation of NEMO in NF-κB activation. *Nat. Cell. Biol.* **2009**, *11*, 123–132. [CrossRef] [PubMed]

270. Ni, C.Y.; Wu, Z.H.; Florence, W.C.; Parekh, V.V.; Arrate, M.P.; Pierce, S.; Schweitzer, B.; Van Kaer, L.; Joyce, S.; Miyamoto, S.; et al. K63-linked polyubiquitination of NEMO modulates TLR signaling and inflammation in vivo. *J. Immunol.* **2008**, *180*, 7107–7111. [CrossRef] [PubMed]

271. Jun, J.C.; Kertesy, S.; Jones, M.B.; Marinis, J.M.; Cobb, B.A.; Tigno-Aranjuez, J.T.; Abbott, D.W. Innate immune-directed NF-κB signaling requires site-specific NEMO ubiquitination. *Cell Rep.* **2013**, *4*, 352–361. [CrossRef] [PubMed]

272. Schmidt-Supprian, M.; Bloch, W.; Courtois, G.; Addicks, K.; Israël, A.; Rajewsky, K.; Pasparakis, M. NEMO/IKKγ-deficient mice model incontinentia pigmenti. *Mol. Cell* **2000**, *5*, 981–992. [CrossRef]

273. Arimoto, K.; Funami, K.; Saeki, Y.; Tanaka, K.; Okawa, K.; Takeuchi, O.; Akira, S.; Murakami, Y.; Shimotohno, K. Polyubiquitin conjugation to NEMO by triparite motif protein 23 (TRIM23) is critical in antiviral defense. *Proc. Natl. Acad. Sci. USA* **2010**, *107*, 15856–15861. [CrossRef] [PubMed]

274. Zotti, T.; Uva, A.; Ferravante, A.; Vessichelli, M.; Scudiero, I.; Ceccarelli, M.; Vito, P.; Stilo, R. TRAF7 protein promotes Lys-29-linked polyubiquitination of IκB Kinase (IKKγ)/NF-κB Essential Modulator (NEMO) and p65/RelA protein and represses NF-κB activation. *J. Biol. Chem.* **2011**, *286*, 22924–22933. [CrossRef] [PubMed]

275. Tsikitis, M.; Acosta-Alvear, D.; Blais, A.; Campos, E.I.; Lane, W.S.; Sánchez, I.; Dynlacht, B.D. Traf7, a MyoD1 transcriptional target regulates nuclear factor-κB activity during myogenesis. *EMBO Rep.* **2010**, *11*, 969–976. [CrossRef] [PubMed]

276. Fu, J.; Taubman, M.B. EGLN3 inhibition of NF-κB is mediated by prolyl hydroxylase-independent inhibition of IκB kinase γ ubiquitination. *Mol. Cell. Biol.* **2013**, *33*, 3050–3061. [CrossRef] [PubMed]

277. Tomar, D.; Singh, R. TRIM13 regulates ubiquitination and turnover of NEMO to suppress TNF induced NF-κB activation. *Cell Signal.* **2014**, *26*, 2606–2613. [CrossRef] [PubMed]

278. Li, T.; Guan, J.; Li, S.; Zhang, X.; Zheng, X. HSCARG downregulates NF-κB signaling by interacting with USP7 and inhibiting NEMO ubiquitination. *Cell Death Dis.* **2014**, *5*, e1229. [CrossRef] [PubMed]

279. Niida, M.; Tanaka, M.; Kamitani, T. Downregulation of active IKKβ by Ro52-mediated autophagy. *Mol. Immunol.* **2010**, *47*, 2378–2387. [CrossRef] [PubMed]

280. Liao, G.; Zhang, M.; Harhaj, E.W.; Sun, S.C. Regulation of the NF-κB-inducing kinase by tumor necrosis factor receptor-associated factor 3-induced degradation. *J. Biol. Chem.* **2004**, *279*, 26243–26250. [CrossRef] [PubMed]

281. Vallabhapurapu, S.; Matsuzawa, A.; Zhang, W.; Tseng, P.H.; Keats, J.J.; Wang, H.; Vignali, D.A.; Bergsagel, P.L.; Karin, M. Nonredundant and complementary functions of TRAF2 and TRAF3 in a ubiquitination cascade that activates NIK-dependent alternative NF-κB signaling. *Nat. Immunol.* **2008**, *9*, 1364–1370. [CrossRef] [PubMed]

282. Zarnegar, B.J.; Wang, Y.; Mahoney, D.J.; Dempsey, P.W.; Cheung, H.H.; He, J.; Shiba, T.; Yang, X.; Yeh, W.C.; Mak, T.W.; et al. Noncanonical NF-κB activation requires coordinated assembly of a regulatory complex of the adaptors cIAP1, cIAP2, TRAF2 and TRAF3 and the kinase NIK. *Nat. Immunol.* **2008**, *9*, 1371–1378. [CrossRef] [PubMed]

283. Lee, S.; Challa-Malladi, M.; Bratton, S.B.; Wright, C.W. Nuclear factor-κB-inducing kinase (NIK) contains an amino-terminal inhibitor of apoptosis (IAP)-binding motif (IBM) that potentiates NIK degradation by cellular IAP1 (c-IAP1). *J. Biol. Chem.* **2014**, *289*, 30680–30689. [CrossRef] [PubMed]

284. Lin, W.W.; Hostager, B.S.; Bishop, G.A. TRAF3, ubiquitination, and B-lymphocyte regulation. *Immunol. Rev.* **2015**, *266*, 46–55. [CrossRef] [PubMed]

285. Yang, X.D.; Sun, S.C. Targeting signaling factors for degradation, an emerging mechanism for TRAF functions. *Immunol. Rev.* **2015**, *266*, 56–71. [CrossRef] [PubMed]

286. Fotin-Mleczek, M.; Henkler, F.; Hausser, A.; Glauner, H.; Samel, D.; Graness, A.; Scheurich, P.; Mauri, D.; Wajant, H. Tumor necrosis factor receptor-associated factor (TRAF) 1 regulates CD40-induced TRAF2-mediated NF-κB activation. *J. Biol. Chem.* **2004**, *279*, 677–685. [CrossRef] [PubMed]

287. Hu, H.; Brittain, G.C.; Chang, J.H.; Puebla-Osorio, N.; Jin, J.; Zal, A.; Xiao, Y.; Cheng, X.; Chang, M.; Fu, Y.X.; et al. OTUD7B controls non-canonical NF-κB activation through deubiquitination of TRAF3. *Nature* **2013**, *494*, 371–374. [CrossRef] [PubMed]

288. Bakhoum, S.F.; Ngo, B.; Laughney, A.M.; Cavallo, J.A.; Murphy, C.J.; Ly, P.; Shah, P.; Sriram, R.K.; Watkins, T.B.K.; Taunk, N.K.; et al. Chromosomal instability drives metastasis through a cytosolic DNA response. *Nature* **2018**, *553*, 467–472. [CrossRef] [PubMed]

289. Saccani, S.; Marazzi, I.; Beg, A.A.; Natoli, G. Degradation of promoter-bound p65/RelA is essential for the prompt termination of the nuclear factor κB response. *J. Exp. Med.* **2004**, *200*, 107–113. [CrossRef] [PubMed]

290. Maine, G.N.; Burstein, E. COMMD proteins and the control of the NFκB pathway. *Cell Cycle* **2007**, *6*, 672–676. [CrossRef] [PubMed]

291. Tanaka, T.; Grusby, M.J.; Kaisho, T. PDLIM2-mediated termination of transcription factor NF-κB activation by intranuclear sequestration and degradation of the p65 subunit. *Nat. Immunol.* **2007**, *8*, 584–591. [CrossRef] [PubMed]

292. Ono, R.; Kaisho, T.; Tanaka, T. PDLIM1 inhibits NF-κB-mediated inflammatory signaling by sequestering the p65 subunit of NF-κB in the cytoplasm. *Sci. Rep.* **2015**, *5*, 18327. [CrossRef] [PubMed]

293. Hou, Y.; Moreau, F.; Chadee, K. PPARγ is an E3 ligase that induces the degradation of NFκB/p65. *Nat. Commun.* **2012**, *3*, 1300. [CrossRef] [PubMed]

294. Wang, S.; Wu, X.; Zhang, J.; Chen, Y.; Xu, J.; Xia, X.; He, S.; Qiang, F.; Li, A.; Shu, Y.; et al. CHIP functions as a novel suppressor of tumour angiogenesis with prognostic significance in human gastric cancer. *Gut* **2013**, *62*, 496–508. [CrossRef] [PubMed]

295. Schweitzer, K.; Naumann, M. CSN-associated USP48 confers stability to nuclear NF-κB/RelA by trimming K48-linked Ub-chains. *Biochim. Biophys. Acta* **2015**, *1853*, 453–469. [CrossRef] [PubMed]

296. Colleran, A.; Collins, P.E.; O'Carroll, C.; Ahmed, A.; Mao, X.; McManus, B.; Kiely, P.A.; Burstein, E.; Carmody, R.J. Deubiquitination of NF-κB by Ubiquitin-Specific Protease-7 promotes transcription. *Proc. Natl. Acad. Sci. USA* **2013**, *110*, 618–623. [CrossRef] [PubMed]

297. Hochrainer, K.; Racchumi, G.; Zhang, S.; Iadecola, C.; Anrather, J. Monoubiquitination of nuclear RelA negatively regulates NF-κB activity independent of proteasomal degradation. *Cell. Mol. Life Sci.* **2012**, *69*, 2057–2073. [CrossRef] [PubMed]

298. Carmody, R.J.; Ruan, Q.; Palmer, S.; Hilliard, B.; Chen, Y.H. Negative regulation of toll-like receptor signaling by NF-κB p50 ubiquitination blockade. *Science* **2007**, *317*, 675–678. [CrossRef] [PubMed]

299. Chang, M.; Jin, W.; Chang, J.H.; Xiao, Y.; Brittain, G.C.; Yu, J.; Zhou, X.; Wang, Y.H.; Cheng, X.; Li, P.; et al. The ubiquitin ligase Peli1 negatively regulates T cell activation and prevents autoimmunity. *Nat. Immunol.* **2011**, *12*, 1002–1009. [CrossRef] [PubMed]

300. Liu, W.H.; Kang, S.G.; Huang, Z.; Wu, C.J.; Jin, H.Y.; Maine, C.J.; Liu, Y.; Shepherd, J.; Sabouri-Ghomi, M.; Gonzalez-Martin, A. A miR-155-Peli1-c-Rel pathway controls the generation and function of T follicular helper cells. *J. Exp. Med.* **2016**, *213*, 1901–1919. [CrossRef] [PubMed]

301. Jin, J.; Xiao, Y.; Hu, H.; Zou, Q.; Li, Y.; Gao, Y.; Ge, W.; Cheng, X.; Sun, S.C. Proinflammatory TLR signalling is regulated by a TRAF2-dependent proteolysis mechanism in macrophages. *Nat. Commun.* **2015**, *6*, 5930. [CrossRef] [PubMed]

302. Leidner, J.; Palkowitsch, L.; Marienfeld, U.; Fischer, D.; Marienfeld, R. Identification of lysine residues critical for the transcriptional activity and polyubiquitination of the NF-κB family member RelB. *Biochem. J.* **2008**, *416*, 117–127. [CrossRef] [PubMed]

303. Marienfeld, R.; Berberich-Siebelt, F.; Berberich, I.; Denk, A.; Serfling, E.; Neumann, M. Signal-specific and phosphorylation-dependent RelB degradation: A potential mechanism of NF-κB control. *Oncogene* **2001**, *20*, 8142–8147. [CrossRef] [PubMed]

304. Leidner, J.; Voogdt, C.; Niedenthal, R.; Möller, P.; Marienfeld, U.; Marienfeld, R.B. SUMOylation attenuates the transcriptional activity of the NF-κB subunit RelB. *J. Cell. Biochem.* **2014**, *115*, 1430–1440. [CrossRef] [PubMed]

305. Zhang, Q.; Lenardo, M.J.; Baltimore, D. 30 Years of NF-κB: A Blossoming of Relevance to Human Pathobiology. *Cell* **2017**, *168*, 37–57. [CrossRef] [PubMed]

306. Maubach, G.; Naumann, M. NEMO Links Nuclear Factor-κB to Human Diseases. *Trends Mol. Med.* **2017**, *23*, 1138–1155. [CrossRef] [PubMed]

307. Sebban-Benin, H.; Pescatore, A.; Fusco, F.; Pascuale, V.; Gautheron, J.; Yamaoka, S.; Moncla, A.; Ursini, M.V.; Courtois, G. Identification of TRAF6-dependent NEMO polyubiquitination sites through analysis of a new NEMO mutation causing incontinentia pigmenti. *Hum. Mol. Genet.* **2007**, *16*, 2805–2815. [CrossRef] [PubMed]

308. Bal, E.; Laplantine, E.; Hamel, Y.; Dubosclard, V.; Boisson, B.; Pescatore, A.; Picard, C.; Hadj-Rabia, S.; Royer, G.; Steffann, J.; et al. Lack of interaction between NEMO and SHARPIN impairs linear ubiquitination and NF-κB activation and leads to incontinentia pigmenti. *J. Allergy Clin. Immunol.* **2017**, *140*, 1671–1682. [CrossRef] [PubMed]

309. Senegas, A.; Gautheron, J.; Gentil-Dit-Maurin, A.; Courtois, G. IKK-related genetic diseases: Probing NF-κB functions in humans and other matters. *Cell. Mol. Life Sci.* **2015**, *72*, 1275–1287. [CrossRef] [PubMed]

310. Boisson, B.; Laplantine, E.; Prando, C.; Giliani, S.; Israelsson, E.; Xu, Z.; Abhyankar, A.; Israël, L.; Trevejo-Nunez, G.; Bogunovic, D.; et al. Immunodeficiency, autoinflammation and amylopectinosis in humans with inherited HOIL-1 and LUBAC deficiency. *Nat. Immunol.* **2012**, *13*, 1178–1186. [CrossRef] [PubMed]

311. Boisson, B.; Laplantine, E.; Dobbs, K.; Cobat, A.; Tarantino, N.; Hazen, M.; Lidov, H.G.; Hopkins, G.; Du, L.; Belkadi, A.; et al. Human HOIP and LUBAC deficiency underlies autoinflammation, immunodeficiency, amylopectinosis, and lymphangiectasia. *J. Exp. Med.* **2015**, *212*, 939–951. [CrossRef] [PubMed]

312. Damgaard, R.B.; Walker, J.A.; Marco-Casanova, P.; Morgan, N.V.; Titheradge, H.L.; Elliott, P.R.; McHale, D.; Maher, E.R.; McKenzie, A.N.J.; Komander, D. The Deubiquitinase OTULIN Is an Essential Negative Regulator of Inflammation and Autoimmunity. *Cell.* **2016**, *166*, 1215–1230. [CrossRef] [PubMed]

313. Zhou, Q.; Yu, X.; Demirkaya, E.; Deuitch, N.; Stone, D.; Tsai, W.L.; Kuehn, H.S.; Wang, H.; Yang, D.; Park, Y.H.; et al. Biallelic hypomorphic mutations in a linear deubiquitinase define otulipenia, an early-onset autoinflammatory disease. *Proc. Natl. Acad. Sci. USA* **2016**, *113*, 10127–10132. [CrossRef] [PubMed]

314. Yang, J.; Lin, Y.; Guo, Z.; Cheng, J.; Huang, J.; Deng, L.; Liao, W.; Chen, Z.; Liu, Z.; Su, B. The essential role of MEKK3 in TNF-induced NF-κB activation. *Nat. Immunol.* **2001**, *2*, 620–624. [CrossRef] [PubMed]

315. Huang, Q.; Yang, J.; Lin, Y.; Walker, C.; Cheng, J.; Liu, Z.G.; Su, B. Differential regulation of interleukin 1 receptor and Toll-like receptor signaling by MEKK3. *Nat. Immunol.* **2004**, *5*, 98–103. [CrossRef] [PubMed]

316. Yao, J.; Kim, T.W.; Qin, J.; Jiang, Z.; Qian, Y.; Xiao, H.; Lu, Y.; Qian, W.; Gulen, M.F.; Sizemore, N. Interleukin-1 (IL-1)-induced TAK1-dependent versus MEKK3-dependent NFκB activation pathways bifurcate at IL-1 receptor-associated kinase modification. *J. Biol. Chem.* **2007**, *282*, 6075–6089. [CrossRef] [PubMed]

317. Yamazaki, K.; Gohda, J.; Kanayama, A.; Miyamoto, Y.; Sakurai, H.; Yamamoto, M.; Akira, S.; Hayashi, H.; Su, B.; Inoue, J. Two mechanistically and temporally distinct NF-κB activation pathways in IL-1 signaling. *Sci. Signal.* **2009**, *2*, ra66. [CrossRef] [PubMed]

318. Zhang, J.; Macartney, T.; Peggie, M.; Cohen, P. Interleukin-1 and TRAF6-dependent activation of TAK1 in the absence of TAB2 and TAB3. *Biochem. J.* **2017**, *474*, 2235–2248. [CrossRef] [PubMed]

319. Ohtake, F.; Saeki, Y.; Ishido, S.; Kanno, J.; Tanaka, K. The K48-K63 Branched Ubiquitin Chain Regulates NF-κB Signaling. *Mol. Cell* **2016**, *64*, 251–266. [CrossRef] [PubMed]

320. Myllymäki, H.; Valanne, S.; Rämet, M. The Drosophila imd signaling pathway. *J. Immunol.* **2014**, *192*, 3455–3462. [CrossRef] [PubMed]

MDPI

Review

NF-κB, the Importance of Being Dynamic: Role and Insights in Cancer

Federica Colombo [1,2], Samuel Zambrano [1,3,*] and Alessandra Agresti [1,*]

1   Division of Genetics and Cell Biology, San Raffaele Scientific Institute, 20132 Milan, Italy;
    colombo.federica@hsr.it
2   Department of Electronics, Information and Bioengineering, Politecnico di Milano, 20133 Milan, Italy
3   Vita-Salute San Raffaele University, 20132 Milan, Italy
*   Correspondence: zambrano.samuel@hsr.it (S.Z.); agresti.alessandra@hsr.it (A.A.);
    Tel.: +39-02-2643-4756 (S.Z.); +39-02-2643-3660 (A.A.)

Received: 14 March 2018; Accepted: 13 April 2018; Published: 17 April 2018

Abstract: In this review, we aim at describing the results obtained in the past years on dynamics features defining NF-κB regulatory functions, as we believe that these developments might have a transformative effect on the way in which NF-κB involvement in cancer is studied. We will also describe technical aspects of the studies performed in this context, including the use of different cellular models, culture conditions, microscopy approaches and quantification of the imaging data, balancing their strengths and limitations and pointing out to common features and to some open questions. Our emphasis in the methodology will allow a critical overview of literature and will show how these cutting-edge approaches can contribute to shed light on the involvement of NF-κB deregulation in tumour onset and progression. We hypothesize that this "dynamic point of view" can be fruitfully applied to untangle the complex relationship between NF-κB and cancer and to find new targets to restrain cancer growth.

Keywords: NF-κB; dynamics; live cell imaging; microfluidics; cancer; transcription

---

1. Introduction: NF-κB, the Importance of Being Dynamic

NF-κB is a family of transcription factor dimers formed by the combination of p50/NFKB1, p52/NFKB2, c-Rel, p65/RelA and RelB. These transcription factors are involved in development, inflammation and immune response and play a crucial role in chronic inflammation as well as cancer initiation and progression [1–4].

In physiological conditions, NF-κB homo- and heterodimers are constrained in the cytoplasm by the interactions with IκB inhibitors (existing in three isoforms IκBα, IκBβ, IκBε) [5]. Different extracellular signals, from inflammatory cytokines like the Tumour Necrosis Factor TNF-α, to the component of the bacterial wall (Lipopolysaccharide, LPS) activate mainly three NF-κB family members (RelA/p65, p50 and cRel) that translocate simultaneously into the nucleus as a result of repressors degradation modulated by IKK kinases activity. Once in the nucleus, NF-κB activates 200–500 target genes, including those coding for IκBs, which shape the delayed inhibitory feedback loop in the system. Newly translated IκBs enter the nucleus, bind to NF-κB and relocate the complex in the cytoplasm [6]. A second independent negative feedback is represented by the ubiquitin-editing enzyme TNFAIP3, historically known as A20 (Figure 1). Not to forget, the described canonical pathway is flanked by an alternative one that is activated through different receptors (BCR, BAFFR, CD40, RANK, LTβR [7]) and their ligands, recruits the other two NF-κB family members, p52 and cRel and activates transcription. However, the mentioned alternative players as well as the interactions between the canonical and the alternative pathways have rarely been considered in dynamics studies and this underscores our superficial knowledge on the NF-κB system in normal and cancer tissues.

The presence of at least two strong negative feedback loops provides both system flexibility and a tight dynamical control of the response to external stimuli. As a consequence of this sensitive wiring, a constant stimulation with inflammatory stimuli translates in oscillations of nuclear NF-κB concentrations that were first observed via microscopy almost 15 years ago [8] and then in a variety of studies (revised in [9]). Importantly, it has been demonstrated that such oscillations are fundamental to modulate gene expression [10–13] in a functionally relevant way [14], as we will discuss in detail below.

Research in the past recent years using single-cell dynamics approaches has shown that the NF-κB circuit is one example among several that have been found to have rich dynamics [15]. Such list includes also the tumour-suppressor pro-apoptotic p53 whose switch from oscillating to sustained dynamics is able to determine the cell fate [16]. These approaches have also shown that the dynamics of NF-κB in single cells are quite heterogeneous according to each cell's susceptibility and to the inherent stochasticity of the system [8,14,17,18]. This heterogeneity was averaged out and hence went unnoticed in the first biochemical studies, where NF-κB appeared almost as non-oscillating [5,19]. Further insights on the role of NF-κB dynamics have also been provided by microfluidic cell culture technologies [20], able to reproduce, at least in part, complex time-varying signals that single cells receive from the environment. Importantly in this context, data-driven mathematical modelling of the NF-κB system [8,14,19,21] has reached a considerable sophistication and has contributed to integrate in a quantitative way all the information. This is a topic of great interest, but we will not review for sake of space here, although we will further outline its importance later.

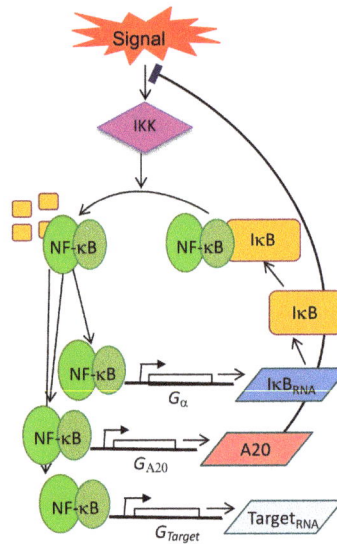

Figure 1. Schematic representation of NF-κB circuitry, with particular emphasis on the negative feedback loops controlled by IκBs and A20 proteins. Such circuitry is extended to the control of a set of target genes IκBα, A20 and "Target" genes, suggesting that NF-κB dynamics can directly operate the dynamics of the transcriptional output.

We will instead focus on the technological developments and the fundamental experimental insights on the role of NF-κB, since we believe that they might have a transformative effect on how NF-κB involvement in cancer is studied. The availability of cancer genomic and proteomic datasets has allowed the identification of many genes in both signalling pathways that are silenced, overexpressed or expressed below the normal threshold [22]. Since NF-κB inhibits apoptosis and fosters proliferation, the deregulated expression of these genes become central in cancer pathobiology [3,23,24]. However, this again provides a quite static view of the role of NF-κB in cancer, that might be enriched by new insights on its role as a dynamic player.

We aim at revising and highlighting recent findings on the output of NF-κB dynamics, with the cutting-edge approaches that have made them possible. We believe that the increasing importance of NF-κB dynamics can impact on our understanding of its involvement in cancer and would potentially help identify new and maybe more specific targets for cancer therapies. Before starting our description, in the next section we will briefly review the different levels at which NF-κB regulation can be affected in cancer.

2. NF-κB and Cancer

Many possible combinations of intrinsic and extrinsic factors account for NF-κB deregulation in a wide variety of cancers. "Intrinsic factors" are represented by mutations or epigenetic alterations either in the coding and regulatory sequences of genes belonging to the NF-κB pathway, or in their targets. The NF-κB subunits RelA/p65, p52 and RelB are highly expressed and chronically activated in almost 80% of the tumours [25–27] and thought to fosters NF-kB anti-apoptotic and pro-proliferative activities as well as contribute to all the remaining hallmarks of cancer [28,29]. Gain-of-function and loss-of-function mutations affecting regulatory genes like A20, IκBα and CYLD (TNFAIP3, Nfkbia, Cyld Lysine 63 Deubiquitinase, respectively) result in the activation of both the classical and alternative NF-κB pathways as evaluated taking into account the processing rate of p50 and IκBα, the phosphorylation of IκBα and p65/RelA, the nuclear localization of NF-kB dimers and, last and extremely relevant, their DNA-binding activity together with the expression of NF-κB target genes [26,30]

Even though the abundance of mutations would argue for a prominent role of non-canonical NF-κB signalling in cancer pathogenesis, a tight crosstalk between the two pathway leads to a diffuse activation involving also the canonical one [24,30–34]. Indeed, the detection of mutations in specific genes cannot predict the global outcome for NF-κB activation, nor whether the crosstalk between the canonical or alternative pathways might result blocked or fuelled [35] thus obliterating the relevance of Exome, DNA or RNA sequencing at population or tissue level.

The association of NF-κB activity and cancer has been exclusively provided at population level and inferred from static pictures of proteins localizations and dosage. Importantly, these data do not positively contribute to the prognosis since the overexpression of one of the NF-κB proteins can be either favourable or unfavourable in different tumours [36–40] as happens for p52/NF-κB2 in melanoma or renal and colorectal cancers, respectively (https://www.proteinatlas.org) or for the tumour suppressor TP53 in endometrial cancer (favourable) and prostate cancer (unfavourable) [41].

The term "extrinsic factors" collectively includes the autocrine/paracrine signalling provided by soluble molecules produced by the many cell types that add up to the signals originating from cell-to-cell interactions in the ever-changing inflammatory microenvironment [42] which of course modulates the degree and extent of NF-κB activation in all the cells. This complexity points to the need of a deeper understanding on how intrinsic and extrinsic factors interact to produce NF-κB deregulation in cancer [43].

As an example, regarding the complex tumour-environment interactions [44], two recent studies on NF-κB signalling in stromal fibroblasts in colitis-associated models of cancer showed that the inhibition IKKβ, a major activator of the canonical NF-κB pathway, results in either the promotion or the repression of cancer development. The comparison of the procedures to knock IKKβ out, identified the different time schedule of IKKβ genetic ablation [45] as the most probable culprit and highlighted the necessity of proper and time—defined dynamic interplay between intrinsic and extrinsic factors in tumour onset, progression and response to therapies.

A paradigmatic example for NF-κB dependent tumour-stroma interactions is represented by Multiple Myeloma (MM). In MM, constitutive NF-κB activity due to mutations in both pathways [23] critically contribute to tumour progression in the bone marrow (BM) microenvironment. Indeed,

NF-κB in plasma cells (PCs) is activated by SDF1 (Stromal Derived Factor-1 or CXCL12) released by stromal cells in the BM. Activated PCs, in turn, produce IL-6, an inflammatory cytokine that induces the stromal components to produce more SDF-1. This positive feedback loop, entirely based on NF-κB activity and crucial for physiologic PCs survival and proliferation, leads to NF-κB hyperactivation in both cell types and drives cells toward the malignant phenotypes independently from the stromal support [46].

An additional example comes from the establishment of bone metastasis as a result of the reciprocal engagement between tumour cells and normal host cells of the bone microenvironment (for example, osteoclasts, stromal cells, vascular cells, etc.). In particular, cancer cells activate the production of the NF-κB regulated paracrine factor RANKL (Receptor Activator of NF-κB Ligand) in stromal cells which in turn pathologically increases osteoclasts activity and bone resorption, both hallmarks of bone metastases and critical for metastatic establishment and progression [47,48]

Overall, this non-exhaustive list of examples highlights how NF-κB deregulation in cancer is extremely diversified, being the result of interactions of many players at many levels: mutations can occur in NF-κB proteins or in their target genes, either in the regulatory or in the coding sequences. Additionally, the resulting inflammatory or perturbed microenvironment might have an unpredictable contribution to progression.

As mentioned above, several dedicated groups [8,10–12,14,18] have shown that NF-κB signalling dynamics at single cell level are very rich and the cell populations have heterogeneous features that have been overlooked using the static approaches.

In our opinion, the recently developed techniques could be instrumental to untangle the complex interaction of intrinsic and extrinsic factors in cancer onset and progression. For these reasons in the next section we will highlight some of these techniques, summarize the main insights that have followed and try to draw connections to potential implications in the study of cancer.

3. Single-Cell Dynamics: Experimental Methodologies

3.1. Cell Systems and Fluorescent Tags

The availability of fluorescent labels for proteins triggered the investigation on single cell dynamics. The easiness of use made cancer-derived cell lines the preferred model for these studies. For example, in the pioneering work of Nelson et al. [8], SK-N-AS cells (human S type neuroblastoma cell line) and HeLa cells (human cervical carcinoma cell line) were stably transfected with a vector expressing RelA fused at the N terminus of a red fluorescent protein [8,10]. An N-term tagged pEGFP-RelA vector have also been used in a number of studies using HeLa cells [12,49]; liver hepatocellular carcinoma HepG2 cells [50] and in breast tumour cell lines [51].

A first note of caution is mandatory at this point. NF-κB dynamics and transcription profiles obtained from a cancer cell line will be informative only on that specific cell line, or on the specific cancer the cell line it has been derived from. In fact, phenotypic and genotypic changes during cancer development in vivo, or procedure for cell line isolation, in vitro, might have generated adaptive pressures on the sensitive NF-κB equilibrium that has further drifted away from the original one [52]. Therefore, it would be wise to dedicate attention at obtaining information on NF-κB dynamics also from the normal tissue/cell line counterpart, beside the tumour one. However, significant differences in dynamics might not emerge from the comparison until perturbations (activating stimuli) exacerbating hidden features are applied to cells.

The wide heterogeneity in NF-κB responses among cell systems encountered in NF-κB literature might just reflect this technical approximation. We are still far from having a clear picture of *bona fide* physiological NF-κB dynamics and probably many flavours exist in unrelated cells and tissues. We would just need to understand how NF-κB is regulated in normal tissues.

Other popular non-tumour cell lines like RAW264.7, a murine macrophage-like line, have been used in some studies [53–55]. NIH-3T3 cells in which the RelA gene has been knock-out and its expression replaced by a transduced fluorescent RelA–DsRed fusion protein under control of the endogenous RelA promoter [9,56,57] have been used as well.

The use of a fluorescently labelled p65 on a p65 knockout background would guarantee absence of interference between the endogenous and the exogenous versions of the protein—an issue that is seldom addressed properly. We must keep in mind, however, that p65/knock-out cells might have adapted their natural NF-κB signalling pathway and changed their phenotype during genetic manipulation procedures. Luckily enough, physiologic NF-κB expression level can be obtained by tagging the endogenous protein and for this specific reason a homozygous EGFP/RelA knock-in mouse was generated [58]. Fibroblasts (shown in Figure 2A) were derived from mouse embryos, immortalized and successfully employed in a number of works [11,14,18]. However, studies on primary cells from this mouse have not yet been published, presumably because they are more challenging to culture and the necessary procedures to isolate primary cells can activate them. As a reminder, the fluorescence intensity in homozygous knock-in cells is challenging low as expected from cells containing 25–30,000 tagged p65 proteins (unpublished data). We must be aware that the transient RelA expression from constitutive promoters produces a fluorescent signal that is at least 100-fold brighter than the fluorescence in p65 knock-in cells. Therefore, the sensitive NF-κB pathway might dramatically react to overexpression by reorganising itself.

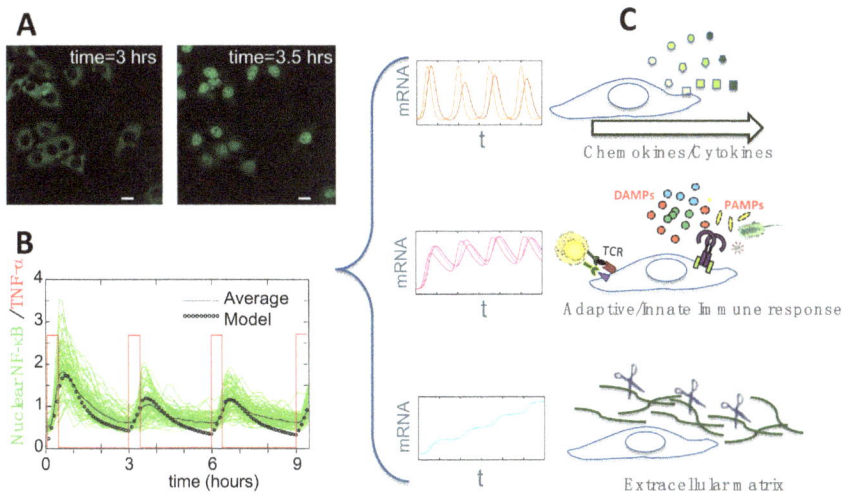

Figure 2. (**A**) Activation of GFP-p65 mouse embryonic fibroblasts upon stimulation. Untreated cells, left panel; cells stimulated with TNF-α for 30 min, right panel. Scale bar: 10μm. (**B**) Such activation can be modulated via a microfluidics device that delivers squared pulses of TNF-α (red profile). Synchronous oscillations from hundreds of cells can be measured (green lines) and compared with the averaged profile (black line) and with the dynamics predicted using a mathematical model (black dotted line). (**C**) Genome-wide gene expression profiling of the synchronized population that shows oscillations locked to the pulsed stimulus in B, revealed that genes can be clustered in three distinct dynamical patterns, each enriched in genes engaged in discrete cell functions.

The new CRISPR/Cas9-based gene-editing techniques [59] are expected to provide an important methodological leap forward in this context, allowing to tag NF-κB genes in their endogenous loci in cell types of choice, including cancer cells. This approach has been well validated to knock-out selected genes by introducing indel mutations in the coding frame. Actually, it still is in its infancy when a coding sequence must be inserted in frame with an endogenous gene. Unfortunately, although extremely useful to engineer tagged genes with physiological expression, this approach seems to be still in its infancy and the scientific community is looking forward to further and quick improvements [60]. In sum, different strategies have been utilized to fluorescently label NF-κB, each with its advantages and limitations that should be weighed when studying assorted aspects of NF-κB in tumour onset and progression.

Overall, the careful choice of the expression system for the tagged protein is necessarily a trade-off between sensitivity, specificity and physiological conditions that should also be weighed when studying the involvement of NF-κB in a given cancer type.

3.2. Experimental Observation and Quantification of Single-Cell NF-κB Dynamics

Once the cell line expressing NF-κB fused with a fluorescent protein is obtained, the next goal would be to extract high-quality data on NF-κB localization, that are able to represent well the expected heterogeneity as well as any features that would be blurred in population-level biochemical assays.

Concerning quantitative microscopy, no gold reference exists to quantify NF-κB fluorescence at single cell level. With a fluorescence widefield microscope or with a confocal microscope with a well open pinhole [18], it is possible to get a 2D fluorescent signal coming from the whole cell thickness. This implies though that the layers of cytosol above and below the nucleus will contribute to the "nuclear intensity" of the signal. This is not a problem for adherent cells as fibroblasts, where such layers are very thin compared to nuclear thickness. However, this approach can be misleading when round cell, such as monocytes, are observed. In this case, acquire a confocal z-stack and reconstruct the geometry of the cells to estimate the total amount of NF-κB in each compartment would be the solution. A note of caution in this case suggests that z-scanning requires repetitive and invasive imaging, so photobleaching and photodamage might become an issue for cells health and signalling.

A point of particular interest that is seldom discussed in detail regards the parameters that best describe NF-κB concentration in single cells. Being a transcriptional activator, the nuclear amount of NF-κB is probably the most informative one and different related magnitudes have been shown to be predictive of the transcriptional output [12,17,61]. However, in long time lapses, photobleaching plus variations of the focus and of laser intensity might alter the proportionality between the light emitted from a cell compartment and the protein content. That is why internally normalized measures are often used in the field and include ratios of nuclear-to-cytoplasm intensities [8,14,50,57], nuclear-to-total intensities [18] or average intensities [10,54]. As we reasoned elsewhere [18], the nuclear-to-total ratio is proportional to the total nuclear amount as long as the protein amount per cell is constant throughout the time lapse. This is taken as a given system parameter in mathematical models (see [62] and references therein) and its constancy has been shown at population levels for cells under TNF-α [14,19]. However, a recent work [54] showed the existence of a positive feedback loop that leads to an increase in the p65 total levels in macrophages stimulated with LPS, an increase that would be missed by using internally normalized measures. Hence intensity ratios, although able to eliminate distortions, should be used with awareness.

Finally, to extract information from these time-lapse experiments, most approaches require the segmentation of nuclei (based on either nuclear dyes such as Hoechst 33342 [14,18,55] or fluorescent histones [17,56]). Alternatively, tracks of nuclei position can be recorded without labelling before stimulation, when the nuclei are empty and hence easily visible but only in rather immobile cells [11,55]. Dedicated software propose the automatic segmentation of nuclei and cytosols to quantify single-cell NF-κB oscillations and can be found in a number of references [14,18,55]. However, in our experience a visual image inspection is always appropriate and common-sense to double-check software errors and technical flaws.

3.3. The Emerging Importance of Microfluidics Technologies

Microfluidic devices are becoming an important tool to gain further insight in the heterogeneity of NF-κB dynamics, potentially with single cell resolution under time-varying stimulation or multiple stimuli [63]. Both commercial or "custom-made" microfluidic devices are reported in the literature to study NF-κB dynamics [56,64–69]. A pioneering example of a microfluidic platform for mammalian cell culture is the so called "Cell Culture Chip," with 96 individually controlled chambers that can support the growth of 1–1,000 cells [64]. The device is made of two layers: the "flow layer", which contains cells and reagents and the automatically controlled "control layer" that creates microfluidic valves. This device was used to study NF-κB dynamics in 3T3 cells expressing p65-DsRed protein under different doses of TNF-α [17] and to analyse the dynamics under periodic (sawtooth-like) TNF-α pulses [56,57]. A commercial alternative is the CellASIC® ONIX Microfluidic Platform (EMD-Millipore, 2018 Merck KGaA, Darmstadt, Germany) used to study NF-κB dynamics in GFP-p65 knock-in MEFs under different periodic stimulations [14] (see Figure 2B). Time varying external stimuli can also be obtained using a simple "custom-made" microfluidic chip, where pulses can be generated by manipulating the relative heights of two reservoirs connected to the cell chambers and containing the medium plus the stimulus/drug and the medium alone [49,66].

Microfluidics can also be employed for other scopes, as for the real-time imaging of host-pathogen interactions using devices able to trap host cells and put them in contact with bacteria or LPS [53]. Interestingly, the CellASIC® ONIX system was used to generate transient interactions between pathogenic bacteria (*Salmonella typhimurium*) and mammalian cells. Unexpectedly, the infected cells showed impaired NF-κB response upon a second challenge with bacteria or inflammatory cytokines [67].

Microfluidic devices have also been used to perform other assays and complement the information obtained from single cell microscopy. In example, "Imstain" is a microfluidic system built for culturing, stimulating, fixing and immunostaining of cells [65]. Furthermore, the device described in [55] adapted a single-cell trapping for RNA-sequencing to make it also compatible with live cell imaging (before the sequencing).

Taken together, microfluidic technologies allow a more accurate and physiologic manipulation of cells and their environment for live cell imaging, as well as the combination with complementary biochemical assays. Microfluidic approaches then can become a fundamental tool to study in vitro the complex crosstalk between cancer and stroma, that we have just started to unravel.

4. Single-Cell Dynamics: Main Insights on NF-κB Regulation of Gene Expression

4.1. Single-Cell NF-κB Dynamics: Main Insights

We have seen so far that a variety of techniques have been used to quantify single-cell NF-κB dynamics. Such techniques have been applied in a myriad of papers appearing in the last 15 years and gave rise to important insights on NF-κB and its role. Here we will attempt to summarize some of the main features emerging from these analyses.

NF-κB oscillates: probably the most acknowledged feature of NF-κB dynamics is the presence of oscillations upon external stimuli. This was predicted by mathematical models of the NF-κB regulatory pathway [19] and later verified using microscopy approaches in single cells [8]. The oscillatory dynamics has later been observed for different cell types [11,14,18,50,56]. The period of the oscillations is widely accepted to be close to 90 min. The oscillatory pulses can be more or less spiky, as predicted by mathematical models but peak shapes might be due either to the nature of quantifiers used to asses NF-κB dynamics or to the cell type used [18].

Heterogeneity in the response: the fraction of cells responding depends on the dose, as well as the amplitude of the oscillatory peaks and in general the dynamics can be heterogeneous in the population of cells [8,17,18]. However, in some of the cells responsive to TNF-α, such as the widely used HeLa cells, sustained oscillations are infrequent upon inflammatory stimulation [8,12]. The type of dynamics observed can be determined by the stimulus applied; for example, NF-κB nuclear localization exhibits oscillatory dynamics when 3T3 cells are stimulated with TNF-α but a stable nuclear accumulation occurs when cells are stimulated with LPS [68]. Moreover, Lee et al. demonstrated that molecules of TNF-α produced during LPS stimulation lead to a secondary paracrine NF-κB activation in the cell neighbourhood [69]. Oscillations are not present either in LPS-stimulated murine primary pEGFP-p65 knock-in macrophages (unpublished data) and in the murine monocytic cell line RAW264.7 stimulated with LPS [54]. Unexplained discrepancies on different dynamics in similar cell systems suggest that our understanding of NF-κB regulation requires further improvements. In particular the debate is still open on whether the NF-κB oscillations are sustained, meaning that cycles would go on as long as the cells are under stimulus, as shown for 3T3 p65 knock-out fibroblasts expressing exogenous GFP-tagged p65 [56], or damped, meaning that the cycling oscillatory peaks tend to decrease their amplitude until complete disappearance, as shown for immortalized MEFs in [14]. We feel confident to exclude that dampening might be due to TNF-α degradation/internalization, since in microfluidics devices cells are grown in small chambers where the stimulus is continuously renewed.

Mathematical models, indeed, predict the presence of both sustained and damped oscillations upon the same stimulation that depend on how the negative feedbacks of the NF-κB regulatory system are tuned [14]. This was already suggested through biochemical approaches when the regulatory circuitry of NF-κB was unveiled [19]. Hoffmann and collaborators showed that the knockouts for the inhibitors, IκBβ and IκBε led to more pronounced, or more synchronized oscillations, than in wild type cells. Indeed, the relative intensity and timing of the feedback can impact on the overall dynamics and might play a role in maximizing population heterogeneity in the oscillations [21]. When oscillations were first described in single cells [8], mathematical modelling also suggested that higher expression level of NF-κB could be responsible for the dampening of the oscillations [70] but this feature was not confirmed experimentally [71]. Overall, we can anticipate that even subtle variations in players of the NF-κB regulatory circuitry can lead to diverse dynamics and such diversity represents the necessary adaptive flexibility to coordinate diverse biological functions, a hypothesis that has recently started to be explored systematically using synthetic NF-κB genetic circuits [72].

Microfluidics: On the other hand, microfluidics allows to explore more complex questions, such as how the NF-κB system responds to time-varying stimuli (mimicking the dynamic environment where the inflammatory response takes place). Pulsatile stimuli were indeed shown to produce a synchronous oscillatory response in the cell population [10]. Using cells showing sustained oscillations upon chronic stimulations, Kellog et al. recreated a sawtooth-like periodic profile of TNF-α stimulation with a home-made microfluidic chip and found that oscillations synchronised to some – not every – stimulation frequencies, a behaviour that could be referred to as "entrainment" [56]. Mathematical models suggest that noise can widen the range of acceptable synchronization modes (i.e. the acceptable ratio between oscillation frequency and the frequency of the external perturbation) and in turn leads to the hopping between them [57]. Damped NF-κB oscillations instead seem to adapt quickly to a wide range of frequencies without any preferred one and with no memory of the synchrony [14]. HeLa cells have also been found to respond quite synchronously as extremely damped oscillators to pulses of external stimuli [73].

Overall, the use of these techniques enlightens how the NF-κB system is able to adapt its activation dynamics to the time-varying external stimuli, although with subtle differences between cell types and stimuli. This is of critical importance in the study of cancer cells where the NF-κB circuitry is deranged. Furthermore, the use of microfluidic devices is now allowing to explore how NF-κB responds to complex environmental cues and this can also lead to important insights on the role of NF-κB in the interaction between the cancer cells and their microenvironment.

4.2. Connecting Dynamics to Transcription

The molecular control of transcription by NF-κB has been the subject of intensive research [61] but the combination with data from single-cell dynamics data has brought novel interesting insights. It was quickly identified that in oscillatory populations target genes were expressed with different dynamics, which included early, intermediate and late-responding genes [8,12]. By constraining RelA in the nucleus by means of LMB or CHX (Leptomycin or Cycloheximide, respectively) the transcriptional output of early and late gene expression was severely impaired or enhanced, respectively, indicating that the oscillatory dynamics is a fundamental ingredient to produce the timely transcription of target genes [11]. Other works showed that the amount of mRNA of NF-κB targets in the population would nicely scale with the area under the first peak of nuclear activation of the fraction of cells responding to the stimulus [17].

Time varying external signals also provide insights in this context. The use of a pulsatile stimulation showed that the dynamics of distinct genes could vary depending on the amplitude and the frequency of the pulsatile input. Thanks to microfluidics, this kind of studies have been extended. Microfluidics-generated sawtooth-like stimulation of cells, able to entrain sustained oscillatory NF-κB dynamics, would also lead to an increased transcriptional output [56]. A similar experimental setup based on squared pulses of TNF-α allowed us to show [14] that synchronous NF-κB oscillations translate into three transcriptional outputs: a fraction of genes would oscillate in sync with the nuclear-to-cytoplasm NF-κB translocations; a second group steadily accumulated, while intermediate dynamics were also possible (Figure 2C). Each pattern of gene expression was enriched in genes with similar biological functions, underlining the connection existing between dynamics and function of genes under the control of NF-κB [14] (Figure 2C). Interestingly, a recent microfluidics study employing single-cell culture chambers connected NF-κB dynamics to single cell transcription. Three main patterns of NF-κB dynamics under LPS activation were identified and each pattern correlated with a specific pattern of relative abundance of the common transcripts [55]. Finally, the use of a precise microfluidics device has shown that differences in the stimulus duration can lead to the activation of either pro or anti- apoptotic transcriptional programs, which impact directly on cell fate [49].

Considering the heterogeneity in gene expression even in a population of identical cells [15,74], it is important to characterize how single-cell NF-κB dynamics modulates (amplifies or tames) this stochasticity. The first works on single-cell NF-κB dynamics [8] indeed linked NF-κB dynamics to the translation of a fluorescently labelled IκBα reporter and confirmed what the model predicted: IκBα levels would oscillate synchronously with NF-κB nuclear localization but out of phase. A similar approach at single cell level showed the expression of a mCherry-reporter under the control of the TNF-α promoter (paradigmatic example of NF-κB controlled gene) whose accumulation appeared coordinates with the activation of NF-κB dynamics [54]. By using RNA-FISH a recent study has correlated the expression of different genes in single cells with dynamical features of NF-κB activation, finding that nuclear concentration fold change (and not the area under the curve of the activation peak) is the best predictor for the number of RNA copies produced [12]. Only through a more thorough characterization—including a wider variety of cell types—we will be able to understand how NF-κB modulates gene expression in single cells.

5. Conclusions

Cancer is a complex system in which intrinsic and extrinsic factors contribute to its growth, progression and disguise from the host immune defence. However, these factors are never at the equilibrium, rather they are constantly evolving both in number and intensity, over time.

In the past years NF-κB dynamics emerged as key regulators of cell life and death. This family of transcription factors, which in healthy tissues controls tissue homeostasis, responds to external stimuli and coordinates cell growth and differentiation, is often deregulated in cancer cells. A deranged signalling represents an important source of variability and heterogeneity in cancer growth that blunts therapeutic efforts.

In this review, we thus described the cutting-edge technological approaches that contributed to these novel insights.

In our opinion, the knowledge about the relationships between NF-κB regulation and cancer could be greatly enhanced by live cell imaging and microfluidics technologies described so far by following the workflow summarized in Figure 3. In fact, a dissection of intrinsic from extrinsic components causing NF-κB deregulation in cancer, should provide a clearer picture and would potentially help identify new and maybe more specific targets for cancer therapies.

This is actually a difficult task that can be accomplished only by the joint effort of scientists in the field taking advantage of tools from the System Biology field. Mathematical and computational models are increasingly used to help interpret dynamic data from cell biology and imaging experiments. Importantly, mathematical simulations of complex biological processes are best fit to generate hypotheses and suggest experiments to further challenge and validate the model.

Circular iterations of model implementation will allow more accurate predictions, (Figure 3) and can pave the way to fundamental new insights in NF-κB regulation in cancer cells.

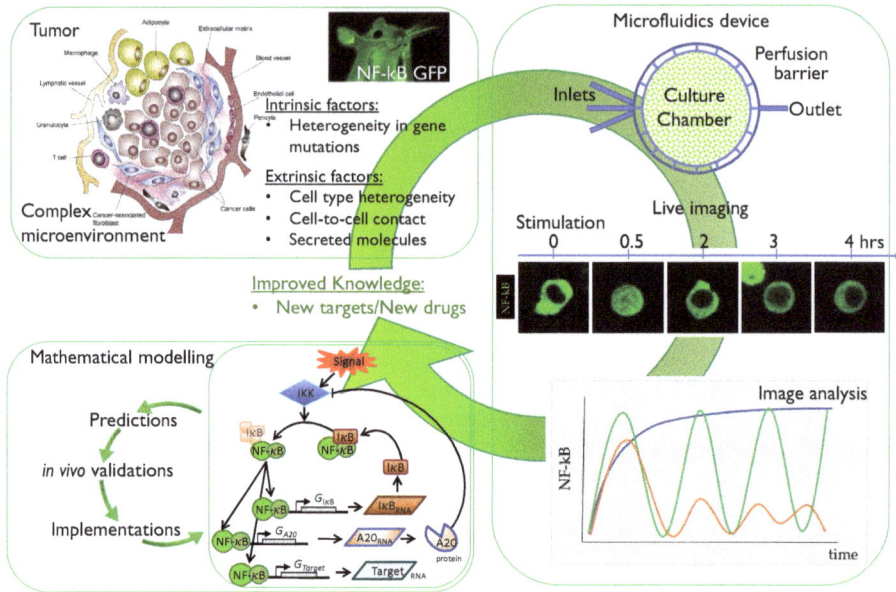

Figure 3. Workflow for the analysis of NF-κB dynamics to improve our knowledge in tumour biology. Starting from the tumour microenvironment [75], the intrinsic factors (mainly gene mutations in cancer cells) and extrinsic factors (cell to cell contact and secreted molecules) can be untangled and analysed by means of live imaging of cells expressing fluorescent NF-κB. In particular, microfluidics can help in recreating the external signals found in vivo. Finally, mathematical modelling represents a fundamental tool to fit live imaging data and to make predictions on NF-κB dynamics when either extrinsic, intrinsic or both components are modulated. The whole set-up may eventually improve our knowledge on tumour biology with the final aim of finding new molecular targets for pharmacological intervention.

Acknowledgments: Alessandra Agresti, Samuel Zambrano and Federica Colombo are funded by the 2017 AIRC IG Grant 18687 to Alessandra Agresti and by the Ospedale San Raffaele 2018 Seed Grant to Samuel Zambrano also covering the costs to publish in open access. Federica Colombo is a PhD student in the Bioengineering course at Politecnico di Milano, Italy. We are grateful to Lambert, Prajapati and Elsevier for the granted permission to use Figure 1 from the publication Journal of Bone Oncology, 5/3, Prajapati, Priyanka and Lambert, Daniel W, Cancer-associated fibroblasts—Not-so-innocent bystanders in metastasis to bone? 128–131, Copyright (2018), with permission from Elsevier to Alessandra Agresti (License Number 4325290449542, License date 10 April 2018).

Author Contributions: Alessandra Agresti, Federica Colombo and Samuel Zambrano wrote the paper and prepared figures.

Conflicts of Interest: The authors declare no conflict of interest. The funding sponsors had no role in the design of the study; in the collection, analyses, or interpretation of data; in the writing of the manuscript and in the decision to publish the results.

References

1. Karin, M. Nuclear factor-kappaB in cancer development and progression. *Nature* **2006**, *441*, 431–436. [CrossRef] [PubMed]
2. Hayden, M.S.; Ghosh, S. Shared principles in NF-kappaB signaling. *Cell* **2008**, *132*, 344–362. [CrossRef] [PubMed]
3. Hoesel, B.; Schmid, J.A. The complexity of NF-κB signaling in inflammation and cancer. *Mol. Cancer* **2013**, *12*, 86. [CrossRef] [PubMed]
4. Ledoux, A.C.; Perkins, N.D. NF-κB and the cell cycle. *Biochem. Soc. Trans.* **2014**, *42*, 76–81. [CrossRef] [PubMed]
5. Kearns, J.D.; Basak, S.; Werner, S.L.; Huang, C.S.; Hoffmann, A. IκBε provides negative feedback to control NF-κB oscillations, signaling dynamics, and inflammatory gene expression. *J. Cell Biol.* **2006**, *173*, 659–664. [CrossRef] [PubMed]
6. Gilmore, T.D. Introduction to NF-kappaB: Players, pathways, perspectives. *Oncogene* **2006**, *25*, 6680–6684. [CrossRef] [PubMed]
7. Sun, S.-C. Non-canonical NF-κB signaling pathway. *Cell Res.* **2010**, *21*, 71–85. [CrossRef] [PubMed]
8. Nelson, D.E.; Ihekwaba, A.E.C.; Elliott, M.; Johnson, J.R.; Gibney, C.A.; Foreman, B.E.; Nelson, G.; See, V.; Horton, C.A.; Spiller, D.G.; et al. Oscillations in NF-kappaB Signaling Control the Dynamics of Gene Expression. *Science* **2004**, *306*, 704–708. [CrossRef] [PubMed]
9. Lee, T.K.; Covert, M.W. High-throughput single-cell NF-κB dynamics. *Curr. Opin. Genet. Dev.* **2010**, *20*, 677–683. [CrossRef] [PubMed]
10. Ashall, L.; Horton, C.A.; Nelson, D.E.; Paszek, P.; Harper, C.V.; Sillitoe, K.; Ryan, S.; Spiller, D.G.; Unitt, J.F.; Broomhead, D.S.; et al. Pulsatile stimulation determines timing and specificity of NF-κB-dependent transcription. *Science* **2009**, *324*, 242–246. [CrossRef] [PubMed]
11. Sung, M.H.; Salvatore, L.; De Lorenzi, R.; Indrawan, A.; Pasparakis, M.; Hager, G.L.; Bianchi, M.E.; Agresti, A. Sustained oscillations of NF-κB produce distinct genome scanning and gene expression profiles. *PLoS ONE* **2009**, *4*, e7163. [CrossRef] [PubMed]
12. Lee, R.E.C.; Walker, S.R.; Savery, K.; Frank, D.A.; Gaudet, S. Fold change of nuclear NF-κB determines TNF-induced transcription in single cells. *Mol. Cell* **2014**, *53*, 867–879. [CrossRef] [PubMed]
13. Adamson, A.; Boddington, C.; Downton, P.; Rowe, W.; Bagnall, J.; Lam, C.; Maya-Mendoza, A.; Schmidt, L.; Harper, C.V.; Spiller, D.G.; et al. Signal transduction controls heterogeneous NF-κB dynamics and target gene expression through cytokine-specific refractory states. *Nat. Commun.* **2016**, *7*. [CrossRef] [PubMed]
14. Zambrano, S.; de Toma, I.; Piffer, A.; Bianchi, M.E.; Agresti, A. NF-κB oscillations translate into functionally related patterns of gene expression. *Elife* **2016**, *5*, 1–38. [CrossRef] [PubMed]
15. Levine, J.H.; Lin, Y.; Elowitz, M.B. Functional roles of pulsing in genetic circuits. *Science* **2013**, *342*, 1193–1200. [CrossRef] [PubMed]
16. Purvis, J.E.; Karhohs, K.W.; Mock, C.; Batchelor, E.; Loewer, A.; Lahav, G. p53 Dynamics Control Cell Fate. *Science* **2012**, *336*, 1440–1444. [CrossRef] [PubMed]
17. Tay, S.; Hughey, J.J.; Lee, T.K.; Lipniacki, T.; Quake, S.R.; Covert, M.W. Single-cell NF-κB dynamics reveal digital activation and analogue information processing. *Nature* **2010**, *466*, 267–271. [CrossRef] [PubMed]
18. Zambrano, S.; Bianchi, M.E.; Agresti, A. High-throughput analysis of NF-kB dynamics in single cells reveals basal nuclear localization of NF-kB and spontaneous activation of oscillations. *PLoS ONE* **2014**, *9*, e90104. [CrossRef] [PubMed]
19. Hoffmann, A. The IkappaB–NF-kappaB Signaling Module: Temporal Control and Selective Gene Activation. *Science* **2002**, *1241*, 1241–1245. [CrossRef] [PubMed]
20. Mehling, M.; Tay, S. Microfluidic cell culture. *Curr. Opin. Biotechnol.* **2014**, *25*, 95–102. [CrossRef] [PubMed]
21. Paszek, P.; Ryan, S.; Ashall, L.; Sillitoe, K.; Harper, C.V.; Spiller, D.G.; Rand, D.A.; White, M.R.H. Population robustness arising from cellular heterogeneity. *Proc. Natl. Acad. Sci. USA* **2010**, *107*, 11644–11649. [CrossRef] [PubMed]
22. Kim, E.J.; Lewis, D.J.; Duvic, M. Novel Mutations Involving NF-κB and B-Cell Signaling Pathways in Primary Cutaneous Large B-Cell Lymphoma, Leg-Type and Comparison with Sézary Syndrome. *J. Investig. Dermatol.* **2017**, *137*, 1831–1833. [CrossRef] [PubMed]

23. Baud, V.; Karin, M. Is NF-kappaB a good target for cancer therapy? Hopes and pitfalls. *Nat. Rev. Drug Discov.* **2009**, *8*, 33–40. [CrossRef] [PubMed]

24. Page, A.; Navarro, M.; Suárez-Cabrera, C.; Bravo, A.; Ramirez, A. Context-dependent role of IKKβ in cancer. *Genes* **2017**, *8*, 376. [CrossRef] [PubMed]

25. Basseres, D.S.; Baldwin, A.S. Nuclear factor-kappaB and inhibitor of kappaB kinase pathways in oncogenic initiation and progression. *Oncogene* **2006**, *25*, 6817–6830. [CrossRef] [PubMed]

26. Keats, J.J.; Fonseca, R.; Chesi, M.; Schop, R.; Baker, A.; Chng, W.J.; Van Wier, S.; Tiedemann, R.; Shi, C.X.; Sebag, M.; et al. Promiscuous Mutations Activate the Noncanonical NF-κB Pathway in Multiple Myeloma. *Cancer Cell* **2007**, *12*, 131–144. [CrossRef] [PubMed]

27. Wu, D.; Wu, P.; Zhao, L.; Huang, L.; Zhang, Z.; Zhao, S.; Huang, J. *NF-κB Expression and Outcomes in Solid Tumors: A Systematic Review and Meta-Analysis*; Wolters Kluwer Health: Philadelphia, PA, USA, 2015; Volume 94, p. e1687. [CrossRef] [PubMed]

28. Hanahan, D.; Weinberg, R.A. Hallmarks of cancer: The next generation. *Cell* **2011**, *144*, 646–674. [CrossRef] [PubMed]

29. Staudt, L.M. Oncogenic activation of NF-kappaB. *Cold Spring Harb. Perspect. Biol.* **2010**, *2*, a000109. [CrossRef] [PubMed]

30. Annunziata, C.M.; Davis, R.E.; Demchenko, Y.; Bellamy, W.; Gabrea, A.; Zhan, F.; Lenz, G.; Hanamura, I.; Wright, G.; Xiao, W.; et al. Frequent Engagement of the Classical and Alternative NF-κB Pathways by Diverse Genetic Abnormalities in Multiple Myeloma. *Cancer Cell* **2007**, *12*, 115–130. [CrossRef] [PubMed]

31. Hideshima, T.; Neri, P.; Tassone, P.; Yasui, H.; Ishitsuka, K.; Raje, N.; Chauhan, D.; Podar, K.; Mitsiades, C.; Dang, L.; et al. MLN120B, a novel IkappaB kinase beta inhibitor, blocks multiple myeloma cell growth in vitro and in vivo. *Clin. Cancer Res.* **2006**, *12*, 5887–5894. [CrossRef] [PubMed]

32. Jourdan, M.; Moreaux, J.; Vos, J.D.; Hose, D.; Mahtouk, K.; Abouladze, M.; Robert, N.; Baudard, M.; Rème, T.; Romanelli, A.; et al. Targeting NF-kappaB pathway with an IKK2 inhibitor induces inhibition of multiple myeloma cell growth. *Br. J. Haematol.* **2007**, *138*, 160–168. [CrossRef] [PubMed]

33. Cartwright, T.; Perkins, N.D.; Wilson, C.L. NFKB1: A suppressor of inflammation, ageing and cancer. *FEBS J.* **2016**, *283*, 1812–1822. [CrossRef] [PubMed]

34. Bennett, L.; Mallon, E.A.; Horgan, P.G.; Paul, A.; McMillan, D.C.; Edwards, J. The relationship between members of the canonical NF-κB pathway, components of tumour microenvironment and survival in patients with invasive ductal breast cancer. *Oncotarget* **2017**, *8*, 33002–33013. [CrossRef] [PubMed]

35. Demchenko, Y.N.; Brents, L.A.; Li, Z.; Bergsagel, L.P.; McGee, L.R.; Kuehl, M.W. Novel inhibitors are cytotoxic for myeloma cells with NFkB inducing kinase-dependent activation of NFkB. *Oncotarget* **2014**, *5*, 4554–4566. [CrossRef] [PubMed]

36. Yeh, H.-C.; Huang, C.-H.; Yang, S.-F.; Li, C.-C.; Chang, L.-L.; Lin, H.-H.; Ke, H.-L.; Wei, Y.-C.; Wu, W.-J. Nuclear factor-κB activation predicts an unfavourable outcome in human upper urinary tract urothelial carcinoma. *BJU Int.* **2010**, *106*, 1223–1229. [CrossRef] [PubMed]

37. Al-Saad, S.; Al-Shibli, K.; Donnem, T.; Persson, M.; Bremnes, R.M.; Busund, L.T. The prognostic impact of NF-κB p105, vimentin, E-cadherin and Par6 expression in epithelial and stromal compartment in non-small-cell lung cancer. *Br. J. Cancer* **2008**, *99*, 1476–1483. [CrossRef] [PubMed]

38. Ueda, Y.; Richmond, A. NF-κB activation in melanoma. *Pigment Cell Res.* **2006**, *19*, 112–124. [CrossRef] [PubMed]

39. Peri, S.; Devarajan, K.; Yang, D.H.; Knudson, A.G.; Balachandran, S. Meta-Analysis Identifies NF-κB as a Therapeutic Target in Renal Cancer. *PLoS ONE* **2013**, *8*, e76746. [CrossRef] [PubMed]

40. Hassanzadeh, P. Colorectal cancer and NF-κB signaling pathway. *Gastroenterol. Hepatol. Bed Bench* **2011**, *4*, 127–132. [PubMed]

41. Jin, R.; Yi, Y.; Yull, F.E.; Blackwell, T.S.; Clark, P.E.; Koyama, T.; Smith, J.A.; Matusik, R.J. NF-κB gene signature predicts prostate cancer progression. *Cancer Res.* **2014**, *74*, 2763–2772. [CrossRef] [PubMed]

42. Didonato, J.A.; Mercurio, F.; Karin, M. NF-κB and the link between inflammation and cancer. *Immunol. Rev.* **2012**, *246*, 379–400. [CrossRef] [PubMed]

43. Nagel, D.; Vincendeau, M.; Eitelhuber, A.C.; Krappmann, D. Mechanisms and consequences of constitutive NF-kappaB activation in B-cell lymphoid malignancies. *Oncogene* **2014**, *33*, 5655–5665. [CrossRef] [PubMed]

44. Nakasone, E.S.; Askautrud, H.A.; Kees, T.; Park, J.-H.; Plaks, V.; Ewald, A.J.; Fein, M.; Rasch, M.G.; Tan, Y.-X.; Qiu, J.; et al. Imaging tumor-stroma interactions during chemotherapy reveals contributions of the microenvironment to resistance. *Cancer Cell* **2012**, *21*, 488–503. [CrossRef] [PubMed]

45. Wagner, E.F. Cancer: Fibroblasts for all seasons. *Nature* **2016**, *530*, 42–43. [CrossRef] [PubMed]

46. Bouyssou, J.M.C.; Ghobrial, I.M.; Roccaro, A.M. Targeting SDF-1 in multiple myeloma tumor microenvironment. *Cancer Lett.* **2016**, *380*, 315–318. [CrossRef] [PubMed]

47. Lacey, D.L.; Boyle, W.J.; Simonet, W.S.; Kostenuik, P.J.; Dougall, W.C.; Sullivan, J.K.; Martin, J.S.; Dansey, R. Bench to bedside: Elucidation of the OPG–RANK–RANKL pathway and the development of denosumab. *Nat. Rev. Drug Discov.* **2012**, *11*, 401–419. [CrossRef] [PubMed]

48. Dougall, W.C.; Holen, I.; González Suárez, E. Targeting RANKL in metastasis. *Bonekey Rep.* **2014**, *3*, 519. [CrossRef] [PubMed]

49. Lee, R.E.C.; Qasaimeh, M.A.; Xia, X.; Juncker, D.; Gaudet, S. NF-κB signalling and cell fate decisions in response to a short pulse of tumour necrosis factor. *Sci. Rep.* **2016**, *6*, 1–12. [CrossRef] [PubMed]

50. Di, Z.; Herpers, B.; Fredriksson, L.; Yan, K.; van de Water, B.; Verbeek, F.J.; Meerman, J.H.N. Automated Analysis of NF-κB Nuclear Translocation Kinetics in High-Throughput Screening. *PLoS ONE* **2012**, *7*, e52337. [CrossRef] [PubMed]

51. Sero, J.E.; Sailem, H.Z.; Ardy, R.C.; Almuttaqi, H.; Zhang, T.; Bakal, C. Cell shape and the microenvironment regulate nuclear translocation of NF-κB in breast epithelial and tumor cells. *Mol. Syst. Biol.* **2015**, *11*, 790. [CrossRef] [PubMed]

52. Gapuzan, M.E.R.; Schmah, O.; Pollock, A.D.; Hoffmann, A.; Gilmore, T.D. Immortalized fibroblasts from NF-κB RelA knockout mice show phenotypic heterogeneity and maintain increased sensitivity to tumor necrosis factor α after transformation by v-Ras. *Oncogene* **2005**, *24*, 6574–6583. [CrossRef] [PubMed]

53. James, C.D.; Moorman, M.W.; Carson, B.D.; Branda, C.S.; Lantz, J.W.; Manginell, R.P.; Martino, A.; Singh, A.K. Nuclear translocation kinetics of NF-κB in macrophages challenged with pathogens in a microfluidic platform. *Biomed. Microdevices* **2009**, *11*, 693–700. [CrossRef] [PubMed]

54. Sung, M.; Li, N.; Lao, Q.; Gottschalk, R.A.; Hager, G.L.; Fraser, I.D.C. Switching of the Relative Dominance Between Feedback Mechanisms in Lipopolysaccharide-Induced NF-κB Signaling. *Sci. Signal.* **2014**, *7*, ra6. [CrossRef] [PubMed]

55. Lane, K.; Van Valen, D.; DeFelice, M.M.; Macklin, D.N.; Kudo, T.; Jaimovich, A.; Carr, A.; Meyer, T.; Pe'er, D.; Boutet, S.C.; et al. Measuring Signaling and RNA-Seq in the Same Cell Links Gene Expression to Dynamic Patterns of NF-κB Activation. *Cell Syst.* **2017**, *4*, 458–469. [CrossRef] [PubMed]

56. Kellogg, R.A.; Tay, S. Noise facilitates transcriptional control under dynamic inputs. *Cell* **2015**, *160*, 381–392. [CrossRef] [PubMed]

57. Heltberg, M.; Kellogg, R.A.; Krishna, S.; Tay, S.; Jensen, M.H. Noise Induces Hopping between NF-κB Entrainment Modes. *Cell Syst.* **2016**, *3*, 532–539. [CrossRef] [PubMed]

58. De Lorenzi, R.; Gareus, R.; Fengler, S.; Pasparakis, M. GFP-p65 knock-in mice as a tool to study NF-κB dynamics in vivo. *Genesis* **2009**, *47*, 323–329. [CrossRef] [PubMed]

59. Ran, F.A.; Hsu, P.D.; Wright, J.; Agarwala, V.; Scott, D.A.; Zhang, F. Genome engineering using the CRISPR-Cas9 system. *Nat. Protoc.* **2013**, *8*, 2281–2308. [CrossRef] [PubMed]

60. Peng, R.; Lin, G.; Li, J. Potential pitfalls of CRISPR/Cas9-mediated genome editing. *FEBS J.* **2016**, *283*, 1218–1231. [CrossRef] [PubMed]

61. Giorgetti, L.; Siggers, T.; Tiana, G.; Caprara, G.; Notarbartolo, S.; Corona, T.; Pasparakis, M.; Milani, P.; Bulyk, M.L.; Natoli, G. Noncooperative Interactions between Transcription Factors and Clustered DNA Binding Sites Enable Graded Transcriptional Responses to Environmental Inputs. *Mol. Cell* **2010**, *37*, 418–428. [CrossRef] [PubMed]

62. Zambrano, S.; Bianchi, M.E.; Agresti, A. A simple model of NF-κB dynamics reproduces experimental observations. *J. Theor. Biol.* **2014**, *347*, 44–53. [CrossRef] [PubMed]

63. Halldorsson, S.; Lucumi, E.; Gómez-Sjöberg, R.; Fleming, R.M.T. Advantages and challenges of microfluidic cell culture in polydimethylsiloxane devices. *Biosens. Bioelectron.* **2015**, *63*, 218–231. [CrossRef] [PubMed]

64. Gómez-Sjöberg, R.; Leyrat, A.A.; Pirone, D.M.; Chen, C.S.; Quake, S.R. Versatile, fully automated, microfluidic cell culture system. *Anal. Chem.* **2007**, *79*, 8557–8563. [CrossRef] [PubMed]

65. Cheong, R.; Wang, C.J.; Levchenko, A. High Content Cell Screening in a Microfluidic Device. *Mol. Cell. Proteom.* **2009**, *8*, 433–442. [CrossRef] [PubMed]

66. Menolascina, F.; Fiore, G.; Orabona, E.; De Stefano, L.; Ferry, M.; Hasty, J.; di Bernardo, M.; di Bernardo, D. In-Vivo Real-Time Control of Protein Expression from Endogenous and Synthetic Gene Networks. *PLoS Comput. Biol.* **2014**, *10*, e1003625. [CrossRef] [PubMed]

67. Ramos-Marquès, E.; Zambrano, S.; Tiérrez, A.; Bianchi, M.E.; Agresti, A.; García-del Portillo, F. Single-cell analyses reveal an attenuated NF-κB response in the Salmonella-infected fibroblast. *Virulence* **2017**, *8*, 719–740. [CrossRef] [PubMed]

68. Covert, M.W.; Leung, T.H.; Gaston, J.E.; Baltimore, D. Achieving stability of lipopolysaccharide-induced NF-kappaB activation. *Science* **2005**, *309*, 1854–1857. [CrossRef] [PubMed]

69. Lee, T.K.; Denny, E.M.; Sanghvi, J.C.; Gaston, J.E.; Maynard, N.D.; Hughey, J.J.; Covert, M.W. A Noisy Paracrine Signal Determines the Cellular NF-κB Response to Lipopolysaccharide. *Sci. Signal.* **2009**, *2*, ra65. [CrossRef] [PubMed]

70. Barken, D.; Wang, C.J.; Kearns, J.; Cheong, R.; Hoffmann, A.; Levchenko, A. Comment on "Oscillations in NF-kB Signaling Control the Dynamics of Gene Expression". *Science* **2005**, *308*, 52a. [CrossRef] [PubMed]

71. Nelson, D.E. Response to Comment on "Oscillations in NF-κB Signaling Control the Dynamics of Gene Expression. " *Science* **2005**, *308*, 52. [CrossRef]

72. Zhang, Z.B.; Wang, Q.Y.; Ke, Y.X.; Liu, S.Y.; Ju, J.Q.; Lim, W.A.; Tang, C.; Wei, P. Design of Tunable Oscillatory Dynamics in a Synthetic NF-κB Signaling Circuit. *Cell Syst.* **2017**, *5*, 460–470.e5. [CrossRef] [PubMed]

73. Zhang, Q.; Gupta, S.; Schipper, D.L.; Kowalczyk, G.J.; Mancini, A.E.; Faeder, J.R.; Lee, R.E.C.; Zhang, Q.; Gupta, S.; Schipper, D.L.; et al. NF-kB Dynamics Discriminate between TNF Doses in Single Cells. *Cell Syst.* **2017**, *5*, 638–645. [CrossRef] [PubMed]

74. Elowitz, M.B.; Levine, A.J.; Siggia, E.D.; Swain, P.S. Stochastic gene expression in a single cell. *Science* **2002**, *297*, 1183–1186. [CrossRef] [PubMed]

75. Prajapati, P.; Lambert, D.W. Cancer-associated fibroblasts—Not-so-innocent bystanders in metastasis to bone? *J. Bone Oncol.* **2016**, *5*, 128–131. [CrossRef] [PubMed]

biomedicines

MDPI

Review

Subunit-Specific Role of NF-κB in Cancer

Barbara Kaltschmidt [1], Johannes F. W. Greiner [2], Hussamadin M. Kadhim [2] and Christian Kaltschmidt [2,*]

[1] AG Molecular Neurobiology, University of Bielefeld, 33615 Bielefeld, Germany;
    Barbara.Kaltschmidt@uni-bielefeld.de
[2] Department of Cell Biology, University of Bielefeld, 33615 Bielefeld, Germany;
    johannes.greiner@uni-bielefeld.de (J.F.W.G.); kadhim.hussamadin@uni-bielefeld.de (H.M.K.)
*   Correspondence: c.kaltschmidt@uni-bielefeld.de; Tel.: +49-521-106-5797

Received: 1 March 2018; Accepted: 12 April 2018; Published: 17 April 2018

Abstract: The transcription factor NF-κB is a key player in inflammation, cancer development, and progression. NF-κB stimulates cell proliferation, prevents apoptosis, and could promote tumor angiogenesis as well as metastasis. Extending the commonly accepted role of NF-κB in cancer formation and progression, different NF-κB subunits have been shown to be active and of particular importance in distinct types of cancer. Here, we summarize overexpression data of the NF-κB subunits RELA, RELB, and c-REL (referring to the v-REL, which is the oncogene of Reticuloendotheliosis virus strain T) as well as of their upstream kinase inhibitor, namely inhibitor of κB kinases (IKK), in different human cancers, assessed by database mining. These data argue against a universal mechanism of cancer-mediated activation of NF-κB, and suggest a much more elaborated mode of NF-κB regulation, indicating a tumor type-specific upregulation of the NF-κB subunits. We further discuss recent findings showing the diverse roles of NF-κB signaling in cancer development and metastasis in a subunit-specific manner, emphasizing their specific transcriptional activity and the role of autoregulation. While non-canonical NF-κB RELB signaling is described to be mostly present in hematological cancers, solid cancers reveal constitutive canonical NF-κB RELA or c-REL activity. Providing a linkage to cancer therapy, we discuss the recently described pivotal role of NF-κB c-REL in regulating cancer-targeting immune responses. In addition, current strategies and ongoing clinical trials are summarized, which utilize genome editing or drugs to inhibit the NF-κB subunits for cancer treatment.

Keywords: NF-κB; RELA; cREL; RELB; tumor; cancer; transformation; inflammation; gene expression; tumor necrosis factor; Treg

---

1. The NF-κB Family—An Introduction

The transcription factor nuclear factor "kappa-light-chain-enhancer" of activated B-cells (NF-κB) [1,2] plays a key role in a broad range of cellular processes like cell growth, apoptosis, inflammation, learning, and memory as well as immunity [3,4]. The transcription factor is ubiquitously expressed and responds to diverse stimuli, particularly including infectious agents, cytokines, or growth factors [5,6]. According to its various cellular functions, deregulation of NF-κB signaling is strongly associated with cancer formation and progression [7,8].

The NF-κB family is composed of five subunits, namely, RELA (p65), RELB, c-REL, p50, and p52 (Figure 1A), all comprising a conserved REL homology domain (RHD) near the N-terminus. This domain is crucial for DNA binding (N-terminal part of RHD), dimerization of NF-κB family members, as well as interaction with the inhibitors of κB (IκBs) (C-terminal part of RHD). Via the RHD, NF-κB family members can form homo- or heterodimers, like p50/RELAp65, RELB/p50, p52/c-REL, or RELA/RELA. In addition, the subunits RELA, RELB, and c-REL contain a C-terminal transactivation domain (TAD) [9,10].

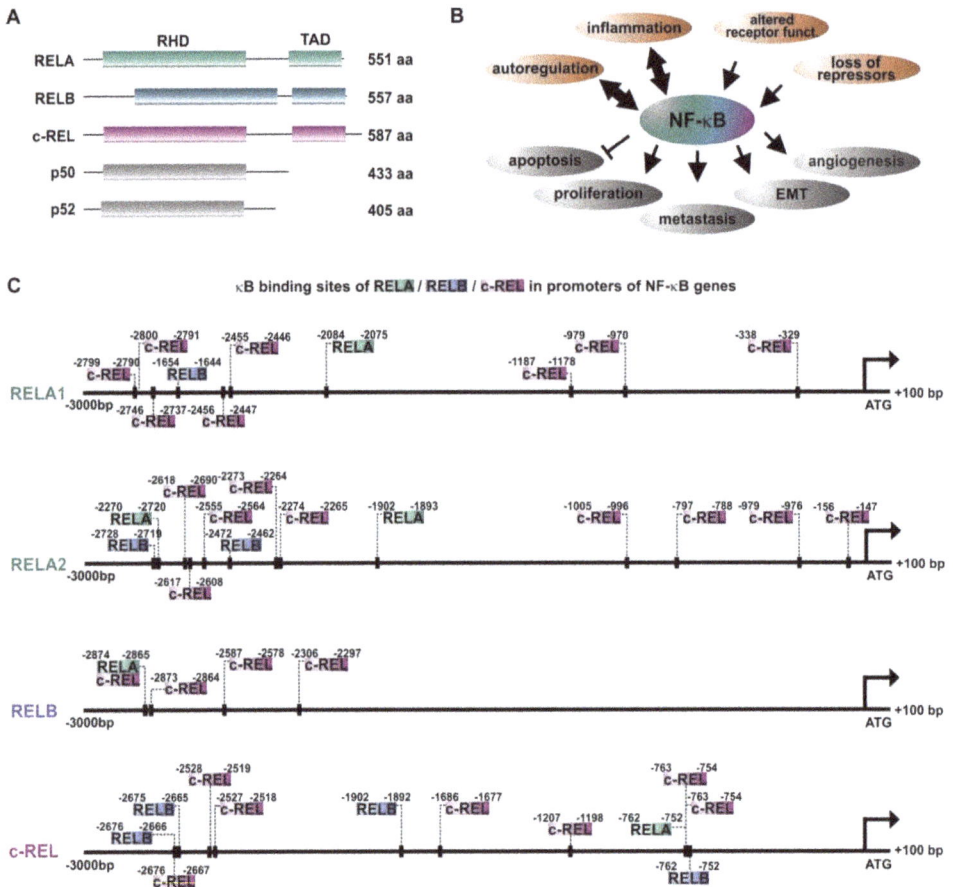

Figure 1. NF-κB and autoregulation of NF-κB subunits in cancer. (**A**) Schematic view of the NF-κB-family ([11]). (**B**) Principal mechanisms causing overexpression/activation of NF-κB as well as the cellular effects of NF-κB acitivity leading to cancer development and progression. RHD: REL homology domain, TAD: transactivation domain. (**C**) The promoters of NF-κB subunits RELA, RELB, and c-REL contain various κB sites enabling autoregulation of NF-κB in cancer. Promoter analysis was done as described in [12]. Briefly, sequences of promoter regions (3000 bp downstream and 100 bp upstream of the respective ATG) of interest were taken from Eukaryotic Promoter Database (epd.vital-ti.ch) for Homo sapiens. The binding sites for the gene of interest in the chosen promoter sequence were looked up using the JASPAR Tool [13]) with a relative score threshold of 85%.

Inactive NF-κB dimers are localized within the cytoplasm, since the NLS (nuclear localization signal) within the RHD is masked by IκBs. During canonical NF-κB signaling, binding of ligands such as cytokines, growth factors, or lipopolysaccharides to their respective receptors (see below, Section 2) leads to the phosphorylation of the IκB kinase (IKK) complex comprised of IKK1/IKK2 (IKKα/IKKβ) and NEMO (NF-κB essential modulator). Phosphorylated IKKs, in particular IKK2, in turn phosphorylate IκBα, which subsequently undergoes proteasome-mediated degradation via polyubiquitinylation. Degradation of IκBα leads to demasking of the nuclear translocation site of the NF-κB p50/RELA heterodimer. In turn, translocation into the nucleus occurs. This results in the expression of NF-κB-target genes via binding to the respective target sites [4,9]. On the contrary, non-canonical NF-κB signaling induced by distinct members of the tumor necrosis factor (TNF) family like lymphotoxin-β relies on the phosphorylation of IKK1 via NIK (NF-κB-inducing kinase). IKK1 mediates the phosphorylation of p100, associated to RELB, inducing the proteasomal processing of p100 to p52 [14]. The p52/RELB heterodimer is able to enter the nucleus and activate specific target genes via binding to selective κB sites. Both the canonical and the non-canonical pathway have been described to be closely linked to cancer formation and progression [15] (Figure 1B, see also Section 2). In addition, atypical NF-κB pathways, as in the case of epidermal growth factor receptor (EGFR) tyrosine kinase-dependent NF-κB activation, were likewise described to promote cancer [16].

2. NF-κB in Inflammation and Cancer

In response to physical or physiological stress, injury, or infection, inflammation takes place as a key defense process of innate immunity aiming to restore the physiological situation. NF-κB is broadly described to be one of the key transcription factors regarding pro-inflammatory signaling, particularly activated by the presence of pro-inflammatory cytokines (like TNFα or IL-1), lipopolysaccharides (LPS) of the bacterial cell wall [17], or viral and bacterial nucleic acids [18]. Recognition of cytokines or LPS species is mediated by the respective receptors, such as TNF receptors or Toll-like microbial pattern recognition receptors (TLRs). As described above, binding of such ligands to their respective receptors leads to canonical NF-κB signaling, ultimately resulting in the translocation of released NF-κB p50/RELA into the nucleus and binding onto κB elements located in distinct target genes. Among the broad range of target genes of NF-κB, the most prominent ones in terms of inflammation are also pro-inflammatory cytokines, such as TNFα [19,20], IL-1 [21], and T cell regulatory ones, such as IL-2 [22] (proliferation) or IL-8 [23] (recruitment). The resulting feed-forward loops of NF-κB-activation, particularly in the case of TNFα, make NF-κB a booster of pro-inflammatory signaling, which augments the inflammation. In the case of cancer, these signaling cascades and the resulting production of pro-inflammatory cytokines likewise recruit cytotoxic immune cells targeting and eliminating the transformed cells [24]. However, the presence of active NF-κB in cancer is a double-edged sword. Although being a mediator of immune responses eliminating cancer cells, NF-κB was observed to be constitutively active in many types of cancer arising from a prolonged chronic inflammatory microenvironment or induced by various oncogenic mutations [8,25]. In a seminal review, Baud and Karin listed 11 types of blood-born cancers (including frequent ones such as acute myeloid leukemia (AML)) and 23 solid tumors (including frequent ones such colon cancer), which showed activated NF-κB signaling [26]. By way of example, elevated NF-κB activity resulting in the accumulation of pro-inflammatory cytokines in the tumor was reported to directly contribute to a pro-tumorigenic microenvironment in colon cancer [27]. Despite this close relation between inflammatory NF-κB signaling and cancer, NF-κB directly mediates vital tumor-promoting mechanisms. NF-κB activity was shown to stimulate cell proliferation, prevent apoptosis, and promote tumor angiogenesis, epithelial-to-mesenchymal transition, invasiveness, as well as metastasis [8,28,29] (Figure 1B). For further details, see a recent review by Taniguchi and Karin [30]. Extending this commonly accepted role of NF-κB in cancer formation and progression, different NF-κB subunits have been shown to be active and of particular importance in distinct types of cancers [11]. In the following, we will discuss the current literature depicting the roles of different NF-κB subunits, their autoregulation, and specific transcriptional activity in cancer and outline how particular subunits and upstream kinases contribute to cancer progression.

3. Autoregulation of NF-κB—A Potential Driver on the Road to Cancer Development?

In addition to the canonical, non-canonical, and atypical activation of NF-κB (see also Section 1), NF-κB RELA, RELB, and c-REL have been described to be activated by autoregulation [31–34].

Accordingly, the promoter analysis of NF-κB-subunits performed in the present study depicted the presence of various κB-binding sites for RELA, RELB, and c-REL (Figure 1C). While RELA and c-REL promoters contain binding sites for all three transactivating subunits RELA, RELB, and c-REL, the promoter of RELB showed only binding sites for cREL and RELA/c-REL in its proximal region. Next to the transactivating subunits, p50 and p52 are likewise known to be autoregulated [35,36]. As depicted in Table 1, several tumor types show various levels of overexpression of the NF-κB-transactivating subunits. A mechanistic reason for this observation might be a feed-forward autoregulation. In this line, a broad range of different κB binding sites within the NF-κB promoters shown here suggest NF-κB feed-forward loops to act as boosters of vital tumor-promoting mechanisms, like cell proliferation, angiogenesis, invasiveness, and metastasis. In addition, these autoregulatory mechanisms may at least in part account for the constitutive activity of NF-κB observed in a broad range of cancers [25,26].

Table 1. Overexpression of NF-κB subunits in distinct human cancer tissues. COSMIC was used for database mining [37]. Parts of this table are published in part [12]). n.a: not assessed.

Cancer Tissue	RELA		RELB		c-REL	
	% Overexpressed	No. Tested	% Overexpressed	No. Tested	% Overexpressed	No. Tested
Ovary	11.65	266	3.38	266	7.52	266
Lung	2.36	1019	4.12	1019	7.26	1019
Urinary tract	2.45	408	4.41	408	7.11	408
Endometrium	1.99	602	8.8	602	6.81	602
Pancreas	2.79	179	6.7	179	6.7	179
Haematopoietic and lymphoid	4.07	221	1.36	221	6.33	221
Soft tissue	3.42	263	1.9	263	6.08	263
Cervix	1.3	307	7.17	307	5.86	307
Upper aerodigestive tract	2.49	522	4.02	522	5.75	522
Kidney	2.83	600	4.5	600	5.5	600
Thyroid	1.36	513	3.7	513	5.46	513
Large intestine	1.87	610	5.25	610	4.92	610
Stomach	7.02	285	7.37	285	4.91	285
Liver	3.75	373	6.97	373	4.83	373
Central nervous system(CNS)	4.45	697	3.73	697	4.73	697
Prostate	4.62	498	5.02	498	4.62	498
Breast	4.17	1104	4.26	1104	3.71	1104
Skin	6.34	473	4.23	473	3.59	473
Oesophagus	2.4	125	2.4	125	3.2	125
Adrenal gland	12.66	79	5.06	79	2.53	79
Nervous system (NS)	n.a.	n.a.	n.a.	n.a.	n.a.	n.a.
Bone	n.a.	n.a.	n.a.	n.a.	n.a.	n.a.

4. Activity of Distinct NF-κB Upstream Kinases in Cancer

To investigate the role of the upstream regulators of NF-κB-signaling IKK1 and IKK2 in human cancers, we applied database mining using COSMIC to determine their levels of overexpression (Table 2) [37,38].

Here, IKK1 and IKK2 showed distinct levels of overexpression in different types of cancer, with IKK2 being overexpressed in cancers arising in the large intestine, the oesophagus, and the lung (Table 2). Accordingly, data from a lung cancer mouse model indicated that tumor cell proliferation was significantly impaired after deletion of IKK2 [39]. Interestingly, IKK-mediated phosphorylation of IκB was shown to mainly depend on the IKK2 catalytic subunit of the IKK complex in mice [40], particularly in terms of prevention of apoptosis [41]. On the contrary, we recently observed TNF-α-mediated cell death only in human cells lacking IKK1 and IKK2 and not in single CRISPR/Cas-mediated IKK knockouts, suggesting that both IKK1 and IKK2 are required for functional TNF-signaling [38] (Figure 2). However, knockout of IKK2 was shown to be associated with about a one-third reduced number of tumors in a colitis-associated cancer model. Surprisingly, deletion of IKK2 in enterocytes led to an increased expression of COX-2, IL-6, and MIP-2, whereas TNF-α, IL-1, and ICAM were not affected. In the myeloid compartment, the number of tumors per mouse was reduced by about 50% after deletion of IKK2 [42]. Constitutive IKK2 activation in intestinal epithelial cells was further demonstrated to induce intestinal tumors in mice [43]. These findings are in accordance with the profound overexpression of IKK2 observed in cancers of the large intestine (Table 2) [38]. On the functional level, IKK2 was shown to directly promote the development of lung cancer in an inflammation-dependent manner triggered by tobacco smoke, which was abrogated by ablation of IKK2 in myeloid cells [44]. Applying a model of breast cancer progression, Huber and colleagues showed IKK2-dependent activation of NF-κB to be essential for epithelial-to-mesenchymal transition and metastasis [45]. Furthermore, the activation of NF-κB by overexpression of constitutively active IKK-2 in prostate cancer cell lines promoted the growth of prostate cancer cells in bone [46]. Accordingly, IKK1 activated by cytokines was shown to control prostate cancer metastasis, with the amount of active nuclear IKK1 correlating with metastatic progression of mouse and human prostate cancer [47].

Table 2. Overexpression of IκB kinases IKK1 and IKK2 in distinct human cancer tissues. COSMIC was used for database mining [37].

Cancer Tissue	IKK1		IKK2	
	% Overexpressed	No. Tested	% Overexpressed	No. Tested
Breast	7.07	1104	9.6	1104
Lung	5.1	1019	7.16	1019
Adrenal Gland	5.06	79	1.27	79
Endometrium	4.98	602	13.12	602
Oesophagus	4.8	125	24.8	125
Liver	4.56	373	5.36	373
Pancreas	4.47	179	4.47	179
Urinary tract	4.41	408	4.9	408
Stomach	4.21	285	7.72	285
Ovary	4.14	266	7.52	266
Thyroid	4.09	513	2.34	513
Prostate	3.21	498	5.02	498
Haematopoietic and lymphoid	3.17	221	5.43	221
Upper aerodigestive tract	2.87	522	6.13	522
Large intestine	2.46	610	18.52	610
Central nervous system(CNS)	2.44	697	3.59	697
Cervix	1.95	307	5.54	307
Soft tissue	1.9	263	6.08	263
Kidney	1.83	600	3.33	600
Skin	1.48	473	8.25	473
Biliary tract	n.a.	n.a.	n.a.	n.a.
Bone	n.a.	n.a.	n.a.	n.a.
Nervous system (NS)	n.a.	n.a.	n.a.	n.a.
Pituitary	n.a.	n.a.	n.a.	n.a.
Salivary gland	n.a.	n.a.	n.a.	n.a.
Testis	n.a.	n.a.	n.a.	n.a.

Next to IKKs, downstream signaling of IκBs is likewise associated to cancer development and progression. Pikarsky and colleagues reported a super-repressor of IκB in hepatocytes to act as a tumor promoter in inflammation-induced liver cancer [48]. Furthermore, in Hodgkin's disease, a hematologic malignancy, the overexpression of a truncated form of IκB is linked to constitutive NF-κB (p50/RELA) activity [49]. In addition, we observed reduced cell growth and a retarded G1/S transition in human cervical cancer cells, accompanied by an increase in cyclin D1-dependent kinase activity after overexpression of IκBα. We further demonstrated a crosstalk of IκBα overexpression with cell cycle checkpoints via a reduction of transcription factor p53 and elevation of p21WAF [50].

5. Differential Roles of NF-κB Subunits in Cancer

To provide an overview on the occurrences of distinct NF-κB subunits in cancer subtypes, we assessed the overexpression of the NF-κB subunits RELA, RELB, and c-REL in human cancers by database mining, using COSMIC (Table 1) [12,37].

In line with the concept of subunit-specific gene regulation in cancer [11], we found profound differences in the overexpression of particular NF-κB subunits in distinct types of cancer. On the contrary, gene amplification and/or mutations within the coding region were only found in neglectable amounts in the COSMIC database. For instance, RELA is most dominantly overexpressed in ovarian cancer and cancer of adrenal glands in comparison to RELB and c-REL, while the overexpression of c-REL is most abundantly found in lung cancers, compared to that of the other subunits (Table 1). In 2016, Scheidereit and coworkers reported the cell survival of Hodgkin lymphoma (HL) cells to be predominantly controlled by the non-canonical NF-κB pathway. In particular, knockdown of p52/RELB in HL cells resulted in 95% reduction of viability. Using combined ChIP-sequencing and microarray analyses, the authors further showed a low frequency of RELA bound to DNA, but a high frequency of DNA-bound p50- and p52-containing complexes, also including p50/p52 heterodimers [51]. Non-canonical NF-κB signaling was further reported to be active in 20% of the samples from 155 multiple myeloma patients. Here, constitutive activation of the non-canonical RELB/p52 pathway was associated with abnormalities like bi-allelic deletion events, mutations, and gene rearrangements in the genes *NFKB1 (p50/p105)* and *NFKB2 (p52/p100)* [52]. Furthermore ectopic expression of RELB can inhibit the growth of tumor xenografts in mice [53]. C-REL is frequently amplified in B cell lymphoma and could function as a tumor-promoting transcription factor, but c-rel-/-mice also could develop an earlier onset of B cell lymphoma [54]. In summary, non-canonical NF-κB-signaling seems to predominantly contribute to hematological cancers (Figure 2).

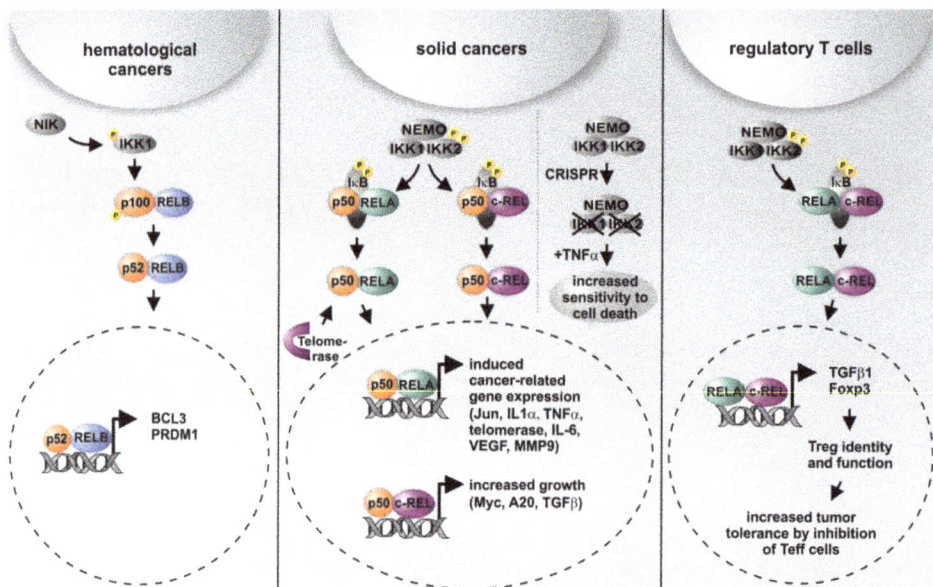

Figure 2. Graphical overview on the differential roles of NF-κB subunits and their transcriptional activity in distinct types of cancer and in regulatory T cells. While non-canonical signaling is mostly present in hematological cancer, solid cancer shows predominantly canonical signaling via p50/RELA or p50/cREL. In addition, CRISPR/Cas-mediated double knockout (KO) of IKK1/2 was recently shown to result in increased sensitivity to TNF-α-mediated cell death [38]. In regulatory T cells (Tregs), activation of RELA/cREL results in distinct target gene expression leading to active Tregs inhibiting effector T cells (Teff), which infiltrate the tumor [11,12,52,55–58].

In contrast to its non-canonical counterpart, canonical NF-κB signaling is described to be present in solid cancers (Figure 2). Shukla and colleagues reported an increased expression of RELA and p50 in human high-grade prostate adenocarcinomas, leading to constitutive NF-κB activity. Active NF-κB p50/RELA led to increased expression of NF-κB target genes *MMP9* and *VEGF*, commonly involved in cell migration and vascularization. Accordingly, NF-κB activity was related to tumor progression due to transcriptional regulation of these NF-κB target genes [57,59]. An increased NF-κB RELA signaling was likewise observed in tumor-initiating stem-like cells in human prostate cancer [60]. Applying a set of 1826 fully annotated prostate cancers, Gannon and colleagues showed a significant association between an increase in the nuclear frequency of NF-κB RELA and Gleason score, which is used to score prostate cancer grade, although the contribution of NF-κB RELA to a post-surgical predictive model appears modest [61]. In lung cancer, NF-κB RELA is known to be required for K-Ras-induced lung tumorigenesis, while lung tumors with RELA deletion show increased apoptosis accompanied by reduced spread and a lower grade [62]. In addition, Mukhopadhyay and colleagues showed highly increased levels of NF-κB p50 in nine of 11 non-small-cell lung carcinoma tissues [63]. KrasG12D-induced IKK2/NF-κB activation, resulting in increased expression of IL1-α and p62 and respective feed-forward loops, was demonstrated to be required for the development of pancreatic ductal adenocarcinoma [64]. In 2003, Nair and coworkers showed a constitutive activation of NF-κB RELA during human cervical cancer progression. Here, NF-κB RELA was demonstrated to be particularly activated in high-grade squamous intraepithelial lesions and squamous cell carcinomas of the human uterine cervix [65]. Interestingly, NF-κB-dependent transcription was recently shown to be directly regulated by telomerase. In particular, telomerase directly bound to the NF-κB RELA subunit, thus regulating NF-κB-dependent gene expression by binding κB sites in the promoter regions of IL-6 and TNF-α, both critical for inflammation and cancer progression [58] (Figure 2). The effect of telomerase on the strong activation of colony formation of tumor stem cells could be repressed by siRNA knockdown of RELA. Given the transcriptional regulation of telomerase by NF-κB RELA, Gosh and coworkers suggested a feed-forward regulation, linking chronic inflammation to increased activity of telomerase in human cancer [58]. Next to RELA, the NF-κB subunit c-REL was likewise shown to possess a key role in tumor formation. In 2000, Cogswell and colleagues revealed the induction of mammary tumors by c-REL expression in mouse models of breast cancer [66]. Shehata and coworkers further demonstrated a sixfold slower cell growth in cultivated cervical cancer cells after expression of the c-REL homolog Xrel3 from *Xenopus laevi* [67]. We recently investigated the role of c-REL in human cervical cancer cells using CRISPR/Cas9n-mediated gene editing. Knockout of c-REL resulted in significantly decreased basal expression levels of Myc, A20, and TGFβ, accompanied by a significantly reduced proliferative behavior and a significant delay in the prometaphase of mitosis (see also Figure 2 for overview). Compared to the wild type, an increased resistance against chemotherapeutic agents was observable in c-REL knockout cells [12]. Next, by directly promoting cancer cell growth and proliferation, c-REL was very recently shown to possess an important role in cancer-targeting immune responses with highly promising implications for therapeutic approaches [68]. Enabling tumor progression, activated CD4+Foxp3+ regulatory T cells (Tregs) are known to migrate to the tumor site and inhibit of CD8 effector T cells (Teffs), which are mainly responsible for anti-tumor immune responses [69]. In melanoma, large amounts of Tregs have been observed [70] and associated with impaired prognosis, while a lesser amount of Tregs was associated with increased survival in stage 4 melanoma patients [71]. In their groundbreaking study, Grindberg-Bleyer and colleagues demonstrated NF-κB cREL as the critical subunit for identity and function of activated CD4+Foxp3+ Tregs in melanoma (see also Figure 2). Notably, deletion or inhibition of c-REL, but not of RELA, in Tregs resulted in reduced melanoma growth and potentiated anti-PDL1 therapy, a ligand presented by cancer cells and dendritic cells to evade the immune system by binding to the immunosuppressive programmed death (PD) receptor on CD8+ Teff cells [68]. In the following, we will emphasize the therapeutic implications of these findings as wells current strategies to utilize genome editing or drugs for targeted deletion/inhibition of NF-κB subunits in cancer therapy.

6. Targeting NF-κB Subunits via Genome Editing or Drugs—Therapeutic Implications

Given the important roles of distinct NF-κB subunits in cancer development and progression, we aim to summarize currently used drugs targeting NF-κB subunits for cancer treatment. One drug utilized in the clinics is the NF-κB inhibitor Bortezomib [72] (developed by Millenium Pharmaceuticals (Cambridge, MA, USA) as Velcade, also known as Neomib (Getwell Pharmaceuticals, Gurgaon, Haryana, India) or Bortecad (Cadila Healthcare, Ahmedabad, Gujarat, India), a reversible 26S proteasome inhibitor of IκB-α degradation. This drug is certified in Europe as monotherapy for pre-treated adult patients with progressive multiple myeloma. Next to Bortezomib, the NF-κB inhibitor Thalidomide is also clinically applied. In 2002, Majumdar and colleagues showed Thalidomide to abrogate TNFα-dependent activation of IKKs and I-Bα [73]. First evaluated in patients with refractory multiple myeloma in the 1990s, Thalidomide is now known to cause responses in 30–50% of myeloma patients as a single agent and acts synergistically with corticosteroids and chemotherapy [74,75]. In addition, a phase III clinical trial is presently studying a combination of Aspirin (an IKK inhibitor) and Esomeprazole (a proton pump inhibitor) to prevent esophageal cancer in patients with Barrett's metaplasia (ClinicalTrials.gov Identifier: NCT00357682). Furthermore, a phase 3 clinical trial is using high-dose ascorbic acid, a well-known NF-κB inhibitor [76], as a pharmaceutical for a combination therapy for colorectal cancer (ClinicalTrials.gov Identifier: NCT02969681). Recently, a subunit-specific inhibitor for c-REL was discovered, which might be useful for inhibiting Tregs (patent filed, IPO: WO2017058881A1). Thus, the inhibition of c-REL might be a new way to treat tumors pharmacologically. In addition to the application of NF-κB-inhibiting drugs, a recent increase in clinical studies applying CRISPR/Cas-mediated knockout strategies suggest that gene therapy might be considered in future therapeutic approaches (e.g., five clinical trials with PD-1 knockout engineered T cells; information retrieved in February 2018 from ClinicalTrials.gov).

7. Conclusions

Although NF-κB might be considered as a major factor in cancer development and progression, distinct NF-κB subunits seem to be active in different kinds of cancer. While non-canonical NF-κB RELB signaling is described to be mostly present in hematological cancers, solid cancers reveal canonical NF-κB RELA (p65) and/or c-REL activity. These particular subunits contribute to cancer formation and invasiveness as a result of their specific transcriptional activity, inter alia via feed-forward loops as in the case of TNFα or telomerase. Currently ongoing clinical trials target NF-κB-dependent signaling by application of drugs or CRISPR/Cas-mediated genome editing impinging on potentially NF-κB-driven processes. Thus, although the here summarized data emphasize the importance to assure subunit specificity, NF-κB seems to be a highly promising target for cancer treatment. Michael Karin and coworkers suggested over the years a universal activation of NF-κB in cancer by inflammatory cytokines [30]. It might be important to note that our analysis of the COSMIC database argues against this universal mechanism of cancer-mediated activation of NF-κB. Here, we suggest a much more elaborated mode of NF-κB regulation in terms of a tumor type-specific upregulation of the NF-κB subunits.

Acknowledgments: Studies in own lab were funded in part by the University of Bielefeld, which provides funds for open access publishing. Further funding was provided by the Thyssen-Stiftung and the European Union's Horizon 2020 research and innovation program under the Marie Sklodowska-Curie Grant Agreement No. 766181, project "DeLIVER", which also provides funds covering the costs to publish in open access. We thank the anonymous reviewers for valuable advice.

Conflicts of Interest: The authors declare no conflict of interest.

References

1. Sen, R.; Baltimore, D. Inducibility of kappa immunoglobulin enhancer-binding protein NF-kappaB by a posttranslational mechanism. *Cell* **1986**, *47*, 921–928. [CrossRef]
2. Sen, R.; Baltimore, D. Multiple nuclear factors interact with the immunoglobulin enhancer sequences. *Cell* **1986**, *46*, 705–716. [CrossRef]
3. Perkins, N.D. Integrating cell-signalling pathways with NF-kappaB and IKK function. *Nat. Rev. Mol. Cell Biol.* **2007**, *8*, 49–62. [CrossRef] [PubMed]
4. Kaltschmidt, B.; Kaltschmidt, C. NF-kappaB in the nervous system. *Cold Spring Harb. Perspect. Biol.* **2009**, *1*, a001271. [CrossRef] [PubMed]
5. Kaltschmidt, B.; Kaltschmidt, C. NF-kappaB in long-term memory and structural plasticity in the adult mammalian brain. *Front. Mol. Neurosci.* **2015**, *8*, 69. [CrossRef] [PubMed]
6. Hayden, M.S.; Ghosh, S. NF-kappaB, the first quarter-century: Remarkable progress and outstanding questions. *Genes Dev.* **2012**, *26*, 203–234. [CrossRef] [PubMed]
7. Ben-Neriah, Y.; Karin, M. Inflammation meets cancer, with NF-kappaB as the matchmaker. *Nat. Immunol.* **2011**, *12*, 715–723. [CrossRef] [PubMed]
8. Xia, Y.; Shen, S.; Verma, I.M. NF-kappaB, an active player in human cancers. *Cancer Immunol. Res.* **2014**, *2*, 823–830. [CrossRef] [PubMed]
9. Oeckinghaus, A.; Ghosh, S. The NF-kappaB family of transcription factors and its regulation. *Cold Spring Harb. Perspect. Biol.* **2009**, *1*, a000034. [CrossRef] [PubMed]
10. Chen, F.E.; Huang, D.B.; Chen, Y.Q.; Ghosh, G. Crystal structure of p50/p65 heterodimer of transcription factor NF-kappaB bound to DNA. *Nature* **1998**, *391*, 410–413. [CrossRef] [PubMed]
11. Perkins, N.D. The diverse and complex roles of NF-kappaB subunits in cancer. *Nat. Rev. Cancer* **2012**, *12*, 121–132. [CrossRef] [PubMed]
12. Slotta, C.; Schluter, T.; Ruiz-Perera, L.M.; Kadhim, H.M.; Tertel, T.; Henkel, E.; Hubner, W.; Greiner, J.F.W.; Huser, T.; Kaltschmidt, B.; et al. Crispr/cas9-mediated knockout of c-rel in hela cells results in profound defects of the cell cycle. *PLoS ONE* **2017**, *12*, e0182373. [CrossRef] [PubMed]
13. Khan, A.; Fornes, O.; Stigliani, A.; Gheorghe, M.; Castro-Mondragon, J.A.; van der Lee, R.; Bessy, A.; Cheneby, J.; Kulkarni, S.R.; Tan, G.; et al. Jaspar 2018: Update of the open-access database of transcription factor binding profiles and its web framework. *Nucleic Acids Res.* **2018**, *46*, D260–D266. [CrossRef] [PubMed]
14. Senftleben, U.; Cao, Y.; Xiao, G.; Greten, F.R.; Krahn, G.; Bonizzi, G.; Chen, Y.; Hu, Y.; Fong, A.; Sun, S.C.; et al. Activation by ikkalpha of a second, evolutionary conserved, nf-kappa b signaling pathway. *Science* **2001**, *293*, 1495–1499. [CrossRef] [PubMed]
15. Hoesel, B.; Schmid, J.A. The complexity of NF-kappaB signaling in inflammation and cancer. *Mol. Cancer* **2013**, *12*, 86. [CrossRef] [PubMed]
16. Alberti, C.; Pinciroli, P.; Valeri, B.; Ferri, R.; Ditto, A.; Umezawa, K.; Sensi, M.; Canevari, S.; Tomassetti, A. Ligand-dependent egfr activation induces the co-expression of il-6 and pai-1 via the nfkb pathway in advanced-stage epithelial ovarian cancer. *Oncogene* **2012**, *31*, 4139–4149. [CrossRef] [PubMed]
17. Greiner, J.F.; Muller, J.; Zeuner, M.T.; Hauser, S.; Seidel, T.; Klenke, C.; Grunwald, L.M.; Schomann, T.; Widera, D.; Sudhoff, H.; et al. 1,8-cineol inhibits nuclear translocation of NF-kappaB p65 and NF-kappaB-dependent transcriptional activity. *Biochim. Biophys. Acta* **2013**, *1833*, 2866–2878. [CrossRef] [PubMed]
18. Müller, J.; Greiner, J.F.; Zeuner, M.; Brotzmann, V.; Schafermann, J.; Wieters, F.; Widera, D.; Sudhoff, H.; Kaltschmidt, B.; Kaltschmidt, C. 1,8-cineole potentiates irf3-mediated antiviral response in human stem cells and in an ex vivo model of rhinosinusitis. *Clin. Sci.* **2016**, *130*, 1339–1352. [CrossRef] [PubMed]
19. Shakhov, A.N.; Collart, M.A.; Vassalli, P.; Nedospasov, S.A.; Jongeneel, C.V. Kappa b-type enhancers are involved in lipopolysaccharide-mediated transcriptional activation of the tumor necrosis factor alpha gene in primary macrophages. *J. Exp. Med.* **1990**, *171*, 35–47. [CrossRef] [PubMed]
20. Collart, M.A.; Baeuerle, P.; Vassalli, P. Regulation of tumor necrosis factor alpha transcription in macrophages: Involvement of four kappa B-like motifs and of constitutive and inducible forms of NF-kappaB. *Mol. Cell. Biol.* **1990**, *10*, 1498–1506. [CrossRef] [PubMed]
21. Mori, N.; Prager, D. Transactivation of the interleukin-1alpha promoter by human t-cell leukemia virus type i and type ii tax proteins. *Blood* **1996**, *87*, 3410–3417. [PubMed]

22. Serfling, E.; Barthelmas, R.; Pfeuffer, I.; Schenk, B.; Zarius, S.; Swoboda, R.; Mercurio, F.; Karin, M. Ubiquitous and lymphocyte-specific factors are involved in the induction of the mouse interleukin 2 gene in t lymphocytes. *EMBO J.* **1989**, *8*, 465–473. [PubMed]

23. Kunsch, C.; Rosen, C.A. NF-kappaB subunit-specific regulation of the interleukin-8 promoter. *Mol. Cell. Biol.* **1993**, *13*, 6137–6146. [CrossRef] [PubMed]

24. Disis, M.L. Immune regulation of cancer. *J. Clin. Oncol.* **2010**, *28*, 4531–4538. [CrossRef] [PubMed]

25. Nakshatri, H.; Bhat-Nakshatri, P.; Martin, D.A.; Goulet, R.J., Jr.; Sledge, G.W., Jr. Constitutive activation of NF-kappaB during progression of breast cancer to hormone-independent growth. *Mol. Cell. Biol.* **1997**, *17*, 3629–3639. [CrossRef] [PubMed]

26. Baud, V.; Karin, M. Is NF-kappaB a good target for cancer therapy? Hopes and pitfalls. *Nat. Rev. Drug Discov.* **2009**, *8*, 33–40. [CrossRef] [PubMed]

27. Terzic, J.; Grivennikov, S.; Karin, E.; Karin, M. Inflammation and colon cancer. *Gastroenterology* **2010**, *138*, 2101–2114. [CrossRef] [PubMed]

28. Hanahan, D.; Weinberg, R.A. Hallmarks of cancer: The next generation. *Cell* **2011**, *144*, 646–674. [CrossRef] [PubMed]

29. Huang, S.; Pettaway, C.A.; Uehara, H.; Bucana, C.D.; Fidler, I.J. Blockade of NF-kappaB activity in human prostate cancer cells is associated with suppression of angiogenesis, invasion, and metastasis. *Oncogene* **2001**, *20*, 4188–4197. [CrossRef] [PubMed]

30. Taniguchi, K.; Karin, M. NF-kappaB, inflammation, immunity and cancer: Coming of age. *Nat. Rev. Immunol.* **2018**. [CrossRef]

31. Hannink, M.; Temin, H.M. Structure and autoregulation of the c-rel promoter. *Oncogene* **1990**, *5*, 1843–1850. [PubMed]

32. Bren, G.D.; Solan, N.J.; Miyoshi, H.; Pennington, K.N.; Pobst, L.J.; Paya, C.V. Transcription of the relb gene is regulated by NF-kappaB. *Oncogene* **2001**, *20*, 7722–7733. [CrossRef] [PubMed]

33. Linker, R.A.; Baeuerle, P.A.; Kaltschmidt, C. Cloning of the murine rela (p65 nf-kappa b) gene and comparison to the human gene reveals a distinct first intron. *Gene* **1996**, *176*, 119–124. [CrossRef]

34. Capobianco, A.J.; Gilmore, T.D. Repression of the chicken c-rel promoter by vrel in chicken embryo fibroblasts is not mediated through a consensus NF-kappaB binding site. *Oncogene* **1991**, *6*, 2203–2210. [PubMed]

35. Lombardi, L.; Ciana, P.; Cappellini, C.; Trecca, D.; Guerrini, L.; Migliazza, A.; Maiolo, A.T.; Neri, A. Structural and functional characterization of the promoter regions of the NFKB2 gene. *Nucleic Acids Res.* **1995**, *23*, 2328–2336. [CrossRef] [PubMed]

36. Ten, R.M.; Paya, C.V.; Israel, N.; Le Bail, O.; Mattei, M.G.; Virelizier, J.L.; Kourilsky, P.; Israel, A. The characterization of the promoter of the gene encoding the p50 subunit of NF-kappaB indicates that it participates in its own regulation. *EMBO J.* **1992**, *11*, 195–203. [PubMed]

37. Forbes, S.A.; Beare, D.; Gunasekaran, P.; Leung, K.; Bindal, N.; Boutselakis, H.; Ding, M.; Bamford, S.; Cole, C.; Ward, S.; et al. Cosmic: Exploring the world's knowledge of somatic mutations in human cancer. *Nucleic Acids Res.* **2015**, *43*, D805–D811.38. [CrossRef] [PubMed]

38. Slotta, C.; Storm, J.; Pfisterer, N.; Henkel, E.; Kleinwachter, S.; Pieper, M.; Ruiz-Perera, L.M.; Greiner, J.F.W.; Kaltschmidt, B.; Kaltschmidt, C. Ikk1/2 protect human cells from tnf-mediated ripk1-dependent apoptosis in an nf-kappab-independent manner. *Biochim. Biophys. Acta* **2018**, in press. [CrossRef] [PubMed]

39. Xia, Y.; Yeddula, N.; Leblanc, M.; Ke, E.; Zhang, Y.; Oldfield, E.; Shaw, R.J.; Verma, I.M. Reduced cell proliferation by ikk2 depletion in a mouse lung-cancer model. *Nat. Cell Biol.* **2012**, *14*, 257–265. [CrossRef] [PubMed]

40. Li, Q.; Van Antwerp, D.; Mercurio, F.; Lee, K.F.; Verma, I.M. Severe liver degeneration in mice lacking the ikappab kinase 2 gene. *Science* **1999**, *284*, 321–325. [CrossRef] [PubMed]

41. Li, Z.W.; Chu, W.; Hu, Y.; Delhase, M.; Deerinck, T.; Ellisman, M.; Johnson, R.; Karin, M. The ikkbeta subunit of ikappab kinase (ikk) is essential for nuclear factor kappab activation and prevention of apoptosis. *J. Exp. Med.* **1999**, *189*, 1839–1845. [CrossRef] [PubMed]

42. Greten, F.R.; Eckmann, L.; Greten, T.F.; Park, J.M.; Li, Z.W.; Egan, L.J.; Kagnoff, M.F.; Karin, M. Ikkbeta links inflammation and tumorigenesis in a mouse model of colitis-associated cancer. *Cell* **2004**, *118*, 285–296. [CrossRef] [PubMed]

43. Vlantis, K.; Wullaert, A.; Sasaki, Y.; Schmidt-Supprian, M.; Rajewsky, K.; Roskams, T.; Pasparakis, M. Constitutive ikk2 activation in intestinal epithelial cells induces intestinal tumors in mice. *J. Clin. Investig.* **2011**, *121*, 2781–2793. [CrossRef] [PubMed]

44. Takahashi, H.; Ogata, H.; Nishigaki, R.; Broide, D.H.; Karin, M. Tobacco smoke promotes lung tumorigenesis by triggering ikkbeta- and jnk1-dependent inflammation. *Cancer Cell* **2010**, *17*, 89–97. [CrossRef] [PubMed]

45. Huber, M.A.; Azoitei, N.; Baumann, B.; Grunert, S.; Sommer, A.; Pehamberger, H.; Kraut, N.; Beug, H.; Wirth, T. NF-kappaB is essential for epithelial-mesenchymal transition and metastasis in a model of breast cancer progression. *J. Clin. Investig.* **2004**, *114*, 569–581. [CrossRef] [PubMed]

46. Jin, R.; Sterling, J.A.; Edwards, J.R.; DeGraff, D.J.; Lee, C.; Park, S.I.; Matusik, R.J. Activation of nf-kappa b signaling promotes growth of prostate cancer cells in bone. *PLoS ONE* **2013**, *8*, e60983. [CrossRef] [PubMed]

47. Luo, J.L.; Tan, W.; Ricono, J.M.; Korchynskyi, O.; Zhang, M.; Gonias, S.L.; Cheresh, D.A.; Karin, M. Nuclear cytokine-activated ikkalpha controls prostate cancer metastasis by repressing maspin. *Nature* **2007**, *446*, 690–694. [CrossRef] [PubMed]

48. Pikarsky, E.; Porat, R.M.; Stein, I.; Abramovitch, R.; Amit, S.; Kasem, S.; Gutkovich-Pyest, E.; Urieli-Shoval, S.; Galun, E.; Ben-Neriah, Y. NF-kappaB functions as a tumour promoter in inflammation-associated cancer. *Nature* **2004**, *431*, 461–466. [CrossRef] [PubMed]

49. Brandl, C.; Florian, C.; Driemel, O.; Weber, B.H.; Morsczeck, C. Identification of neural crest-derived stem cell-like cells from the corneal limbus of juvenile mice. *Exp. Eye Res.* **2009**, *89*, 209–217. [CrossRef] [PubMed]

50. Kaltschmidt, B.; Kaltschmidt, C.; Hehner, S.P.; Droge, W.; Schmitz, M.L. Repression of NF-kappaB impairs hela cell proliferation by functional interference with cell cycle checkpoint regulators. *Oncogene* **1999**, *18*, 3213–3225. [CrossRef] [PubMed]

51. De Oliveira, K.A.; Kaergel, E.; Heinig, M.; Fontaine, J.F.; Patone, G.; Muro, E.M.; Mathas, S.; Hummel, M.; Andrade-Navarro, M.A.; Hubner, N.; et al. A roadmap of constitutive NF-kappaB activity in hodgkin lymphoma: Dominant roles of p50 and p52 revealed by genome-wide analyses. *Genome Med.* **2016**, *8*, 28. [CrossRef] [PubMed]

52. Keats, J.J.; Fonseca, R.; Chesi, M.; Schop, R.; Baker, A.; Chng, W.J.; Van Wier, S.; Tiedemann, R.; Shi, C.X.; Sebag, M.; et al. Promiscuous mutations activate the noncanonical NF-kappaB pathway in multiple myeloma. *Cancer Cell* **2007**, *12*, 131–144. [CrossRef] [PubMed]

53. Toma, J.G.; Akhavan, M.; Fernandes, K.J.; Barnabe-Heider, F.; Sadikot, A.; Kaplan, D.R.; Miller, F.D. Isolation of multipotent adult stem cells from the dermis of mammalian skin. *Nat. Cell Biol.* **2001**, *3*, 778–784. [CrossRef] [PubMed]

54. Techawattanawisal, W.; Nakahama, K.; Komaki, M.; Abe, M.; Takagi, Y.; Morita, I. Isolation of multipotent stem cells from adult rat periodontal ligament by neurosphere-forming culture system. *Biochem. Biophys. Res. Commun.* **2007**, *357*, 917–923. [CrossRef] [PubMed]

55. Ruan, Q.; Kameswaran, V.; Tone, Y.; Li, L.; Liou, H.C.; Greene, M.I.; Tone, M.; Chen, Y.H. Development of foxp3(+) regulatory T cells is driven by the c-Rel enhanceosome. *Immunity* **2009**, *31*, 932–940. [CrossRef] [PubMed]

56. Libermann, T.A.; Baltimore, D. Activation of interleukin-6 gene expression through the NF-kappaB transcription factor. *Mol. Cell. Biol.* **1990**, *10*, 2327–2334. [CrossRef] [PubMed]

57. Shukla, S.; MacLennan, G.T.; Fu, P.; Patel, J.; Marengo, S.R.; Resnick, M.I.; Gupta, S. Nuclear factor-kappaB/p65 (Rel A) is constitutively activated in human prostate adenocarcinoma and correlates with disease progression. *Neoplasia* **2004**, *6*, 390–400. [CrossRef] [PubMed]

58. Ghosh, A.; Saginc, G.; Leow, S.C.; Khattar, E.; Shin, E.M.; Yan, T.D.; Wong, M.; Zhang, Z.; Li, G.; Sung, W.K.; et al. Telomerase directly regulates NF-kappaB-dependent transcription. *Nat. Cell Biol.* **2012**, *14*, 1270–1281. [CrossRef] [PubMed]

59. Widera, D.; Zander, C.; Heidbreder, M.; Kasperek, Y.; Noll, T.; Seitz, O.; Saldamli, B.; Sudhoff, H.; Sader, R.; Kaltschmidt, C.; et al. Adult palatum as a novel source of neural crest-related stem cells. *Stem Cells* **2009**, *27*, 1899–1910. [CrossRef] [PubMed]

60. Rajasekhar, V.K.; Studer, L.; Gerald, W.; Socci, N.D.; Scher, H.I. Tumour-initiating stem-like cells in human prostate cancer exhibit increased NF-kappaB signalling. *Nat. Commun.* **2011**, *2*, 162. [CrossRef] [PubMed]

61. Gannon, P.O.; Lessard, L.; Stevens, L.M.; Forest, V.; Begin, L.R.; Minner, S.; Tennstedt, P.; Schlomm, T.; Mes-Masson, A.M.; Saad, F. Large-scale independent validation of the nuclear factor-kappaB p65 prognostic biomarker in prostate cancer. *Eur. J. Cancer* **2013**, *49*, 2441–2448. [CrossRef] [PubMed]

62. Basseres, D.S.; Ebbs, A.; Levantini, E.; Baldwin, A.S. Requirement of the NF-kappaB subunit p65/RelA for K-Ras-induced lung tumorigenesis. *Cancer Res.* **2010**, *70*, 3537–3546. [CrossRef] [PubMed]

63. Mukhopadhyay, T.; Roth, J.A.; Maxwell, S.A. Altered expression of the p50 subunit of the NF-kappaB transcription factor complex in non-small cell lung carcinoma. *Oncogene* **1995**, *11*, 999–1003. [PubMed]

64. Ling, J.; Kang, Y.; Zhao, R.; Xia, Q.; Lee, D.F.; Chang, Z.; Li, J.; Peng, B.; Fleming, J.B.; Wang, H.; et al. Krasg12d-induced ikk2/beta/NF-kappaB activation by il-1alpha and p62 feedforward loops is required for development of pancreatic ductal adenocarcinoma. *Cancer Cell* **2012**, *21*, 105–120. [CrossRef] [PubMed]

65. Nair, A.; Venkatraman, M.; Maliekal, T.T.; Nair, B.; Karunagaran, D. NF-kappaB is constitutively activated in high-grade squamous intraepithelial lesions and squamous cell carcinomas of the human uterine cervix. *Oncogene* **2003**, *22*, 50–58. [CrossRef] [PubMed]

66. Cogswell, P.C.; Guttridge, D.C.; Funkhouser, W.K.; Baldwin, A.S., Jr. Selective activation of nf-kappa b subunits in human breast cancer: Potential roles for NF-kappaB2/p52 and for Bcl-3. *Oncogene* **2000**, *19*, 1123–1131. [CrossRef] [PubMed]

67. Shehata, M.; Shehata, F.; Pater, A. Apoptosis effects of xrel3 c-Rel/Nuclear factor-kappa B homolog in human cervical cancer cells. *Cell Biol. Int.* **2005**, *29*, 429–440. [CrossRef] [PubMed]

68. Grinberg-Bleyer, Y.; Oh, H.; Desrichard, A.; Bhatt, D.M.; Caron, R.; Chan, T.A.; Schmid, R.M.; Klein, U.; Hayden, M.S.; Ghosh, S. NF-kappaB c-Rel is crucial for the regulatory T cell immune checkpoint in cancer. *Cell* **2017**, *170*, 1096–1108. [CrossRef] [PubMed]

69. Nishikawa, H.; Sakaguchi, S. Regulatory t cells in tumor immunity. *Int. J. Cancer* **2010**, *127*, 759–767. [CrossRef] [PubMed]

70. Jandus, C.; Bioley, G.; Speiser, D.E.; Romero, P. Selective accumulation of differentiated foxp3(+) cd4 (+) T cells in metastatic tumor lesions from melanoma patients compared to peripheral blood. *Cancer Immunol. Immunother.* **2008**, *57*, 1795–1805. [CrossRef] [PubMed]

71. Baumgartner, J.M.; Gonzalez, R.; Lewis, K.D.; Robinson, W.A.; Richter, D.A.; Palmer, B.E.; Wilson, C.C.; McCarter, M.D. Increased survival from stage IV melanoma associated with fewer regulatory T cells. *J. Surg. Res.* **2009**, *154*, 13–20. [CrossRef] [PubMed]

72. Adams, J.; Kauffman, M. Development of the proteasome inhibitor velcade (bortezomib). *Cancer Investig.* **2004**, *22*, 304–311. [CrossRef]

73. Majumdar, S.; Lamothe, B.; Aggarwal, B.B. Thalidomide suppresses NF-kappa B activation induced by TNF and H2O2, but not that activated by ceramide, lipopolysaccharides, or phorbol ester. *J. Immunol.* **2002**, *168*, 2644–2651. [CrossRef] [PubMed]

74. Singhal, S.; Mehta, J. Thalidomide in cancer. *Biomed. Pharmacother.* **2002**, *56*, 4–12. [CrossRef]

75. Strasser, K.; Ludwig, H. Thalidomide treatment in multiple myeloma. *Blood Rev.* **2002**, *16*, 207–215. [CrossRef]

76. Carcamo, J.M.; Pedraza, A.; Borquez-Ojeda, O.; Golde, D.W. Vitamin C suppresses TNF alpha-induced NF kappa B activation by inhibiting I kappa B alpha phosphorylation. *Biochemistry* **2002**, *41*, 12995–13002. [CrossRef] [PubMed]

biomedicines

MDPI

Review

NF-κB Members Left Home: NF-κB-Independent Roles in Cancer

Carlota Colomer [†], Laura Marruecos [†], Anna Vert [†], Anna Bigas and Lluis Espinosa *

Stem Cells and Cancer Research Laboratory, CIBERONC. Institut Hospital del Mar Investigacions Mèdiques (IMIM), 08003 Barcelona, Spain; ccolomer@imim.es (C.C.); lmarruecos@imim.es (L.M.); Avert@imim.es (A.V.); abigas@imim.es (A.B.)
* Correspondence: lespinosa@imim.es; Tel./Fax: +34 933 160 589
† These authors contributed equally to this work.

Academic Editor: Veronique Baud
Received: 31 March 2017; Accepted: 19 May 2017; Published: 25 May 2017

Abstract: Nuclear factor-κB (NF-κB) has been long considered a master regulator of inflammation and immune responses. Additionally, aberrant NF-κB signaling has been linked with carcinogenesis in many types of cancer. In recent years, the study of NF-κB members in NF-κB unrelated pathways provided novel attractive targets for cancer therapy, specifically linked to particular pathologic responses. Here we review specific functions of IκB kinase complexes (IKKs) and IκBs, which have distinctly tumor promoting or suppressing activities in cancer. Understanding how these proteins are regulated in a tumor-related context will provide new opportunities for drug development.

Keywords: Cancer; NF-κB; Non-conventional pathways; IKKs; IκBs

1. Introduction

Since the discovery of the nuclear factor κB (NF-κB) more than 30 years ago [1] the NF-κB pathway has been the focus of multiple studies owing to its role in the regulation of essential biological processes, such as immune and stress responses, cell survival, or cell maturation. Due to its functional relevance, alterations in NF-κB signaling tend to affect organism homeostasis, leading to tissue damage and, in some cases, to cancer [2]. Thus, gaining insight into the function and regulation of particular NF-κB components is crucial for the future development of effective therapies against a wide variety of diseases that involves NF-κB, including diabetes [3,4], allergies and rheumatoid arthritis [5], Crohn's disease [6], Alzheimer's disease [7], or cancer, among others.

The mammalian NF-κB family consists of five transcription factors: p65 (RelA), RelB, c-Rel, p105/p50 (NF-κB1), and p100/p52 (NF-κB2) [8–10]. Although RelA, RelB, and c-Rel are synthesized as final proteins, p50 and p52 derive from p105 and p100, respectively, upon proteasomal processing. All of the members can form homo- and heterodimers, and shuttle from the cytoplasm to the nucleus in response to cell stimulation. NF-κB transcription factors are characterized by the presence of a highly-conserved Rel homology domain (RHD) which is responsible for dimerization, DNA binding, and interaction with the inhibitor of κB (IκB) proteins [10]. The IκB proteins, including IκBα, IκBβ, IκBε, IκBγ, IκBζ, Bcl-3, and the precursor Rel proteins p100 and p105, are characterized by the presence of multiple ankyrin repeats, which are protein-protein interaction domains that interact with NF-κB via the RHD [10]. IκBs control the activation of the NF-κB dimers (except for p52-RelB) by masking the nuclear localization signal (NLS) of Rel proteins, thus preventing its nuclear translocation and the subsequent activation of target genes. Therefore, IκB degradation is a tightly-regulated event that is triggered upon a stimulus-response activation of the IκB kinase (IKK) complex. The IKK complex is formed by two catalytic subunits, IKKα and IKKβ, and a regulatory subunit called IKKγ or NF-κB essential modulator (NEMO) [11].

There are a variety of ligands that can trigger the signal transduction resulting in the activation of specific IKK-dependent cascades, being the two principal the classical (or canonical) and the alternative (or non-canonical) NF-κB pathways. In the classical pathway, activated IKKβ by transforming growth factor-β activated kinase 1 (TAK1) is necessary to induce phosphorylation of IκBs on two N-terminal residues (IκBα on Ser32 and Ser36 and IκBβ on Ser19 and Ser23). This event leads to its ubiquitination by the Skp-1/Cul/F box (SCF) family and its proteasomal degradation [11,12]. On the other hand, the alternative pathway depends on the activation of IKKα by the NF-κB inducing kinase (NIK). The IKKα subunit phosphorylates p100 which, under resting conditions, is associated with RelB in the cytoplasm, inducing its processing to p52 [13]. In both pathways, after this processing, the NF-κB transcriptional factors are able to translocate to the nucleus, where they bind to promoter and enhancer regions containing κB sites with the consensus sequence GGGRNNYYCC (N = any base, R = purine, Y = pyrimidine).

As mentioned, NF-κB pathway play an important task in the development and maintenance of cancer, mainly associated with its normal role in inflammation and immune response. However, it is also true that particular NF-κB-related elements can be deregulated in cancer cells, thus exerting less conventional pro- or anti-tumorigenic functions. Examples include the aberrant activity of members of the pathway, genetic aberrations of genes coding for NF-κB family members, autocrine and paracrine production of pro-inflammatory cytokines by the tumor cells, as well as oncogenic activation of upstream signaling molecules. All of these mechanisms lead to altered expression of specific target genes or whole transcriptional programs which, in turn, modify cellular proliferation or apoptosis, tumor-associated angiogenesis, metastasis, or resistance to chemo- and radiotherapy [14–19]. In addition, particular members of the NF-kB pathway have been found to exert non-conventional and NF-κB-independent functions that are physiologically relevant, but can also impact some cancer cell capabilities. The present review focuses on the non-conventional functions of the NF-κB pathway family of proteins IKK and IκB that negatively or positively contribute to cancer initiation and progression.

2. Breast Cancer

Both IKKα and IKKβ display oncogenic functions in breast cancer cells that are independent of their role in the NF-κB pathway. In response to estrogen, IKKα increases phosphorylation and recruitment of estrogen receptor alpha (ERα) and steroid receptor coactivator 3 (SRC-3) to estrogen-responsive promoters, including *cyclin D1* and *c-myc*, leading to enhanced gene transcription. Activation of these genes increases estrogen-dependent proliferation of breast cancer cells [20]. IKKα can also cooperate with Notch-1 to induce the transcriptional activation of ERα-dependent genes [21]. On the other hand, IKKα promotes the estrogen-induced transcription of E2F Transcription Factor 1 (E2F1) and facilitates the subsequent activation of several E2F1-responsive genes such as *thymidine kinase 1* (*TK1*), *proliferating cell nuclear antigen* (*PCNA*), *cyclin E*, and *cdc25A*, which are required for cell cycle progression of breast cancer cells [22]. IKKα is also an important contributor to ErbB2-induced oncogenesis, as it supports the expansion of tumor-initiating cells from premalignant ErbB2-expressing mammary glands. Upon activation, IKKα enters into the nucleus of these cells and phosphorylates p27/Kip1 inducing its nuclear export, which results in enhanced cell proliferation [23] (Figure 1).

IKKβ also promotes breast cancer through the phosphorylation of forkhead box O3 (FOXO3a), which triggers its cytoplasmic export and proteasomal degradation, resulting in increased proliferation and tumorigenesis (Figure 1). This mechanism was primarily found in tumors lacking Akt activity since Akt is usually responsible for FOXO3a phosphorylation and degradation [24].

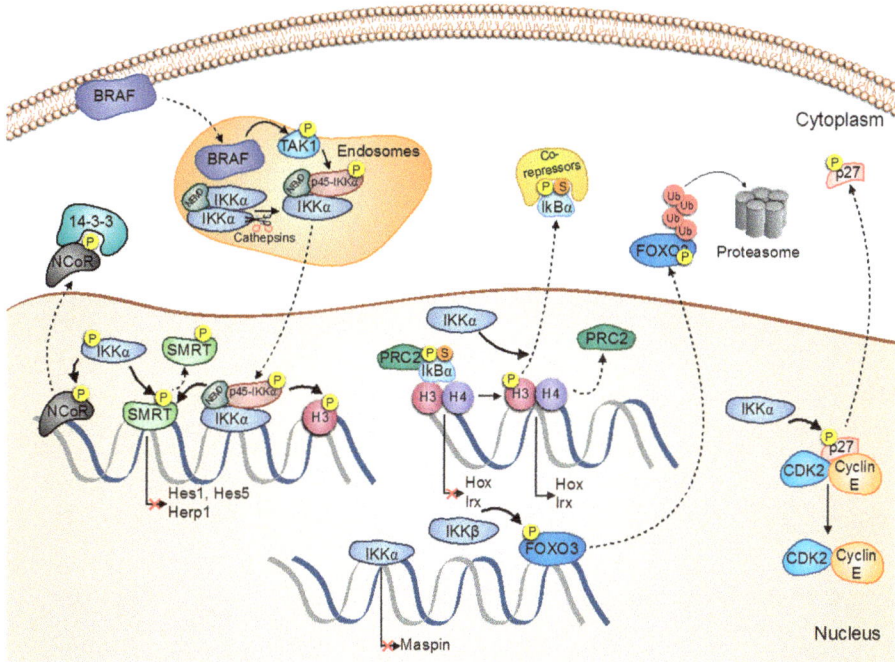

Figure 1. Pro-tumorigenic functions of the NF-κB members. In CRC, IKKα phosphorylates the nuclear co-repressors N-CoR and SMRT, inducing its dissociation from the chromatin. In prostate cancer cells IKKα regulates the gene transcription of the metastasis repressor Maspin. The proteolytic fragment p45-IKKα is activated by BRAF and TAK1 in the endosomal compartment, and upon activation can phosphorylate histone H3 and SMRT. Moreover, nuclear IKKα contributes to the chromatin release of IκBα, and stimulates the nuclear export of p27/Kip1, thereby supporting the proliferation and expansion of tumor cells. On the other hand, IKKβ phosphorylates FOXO3a, leading to its nuclear exclusion and protein degradation. Arrows: ⟶ Activation/Regulation/Phosphorylation; ⋯⟶ Migration; ⤬⟶ Inactivation.

3. Prostate Cancer

In prostate cancer, IKKα phosphorylates and activates the mammalian Target of Rapamycin Complex 1 (mTORC1) in phosphatase- and tensin homolog (PTEN)-null prostate cancer cells in a manner dependent on Akt, promoting cell proliferation [25,26]. Similarly, IKKα associates with, and enhances, mTORC2 kinase activity [27]. Of note, it is known that activated Akt promotes cell survival, cell growth and proliferation, and energy metabolism in prostate cancer [28]. IKKα can also phosphorylate the nuclear co-repressor silencing mediator for retinoid and thyroid receptors (SMRT), thus inducing its dissociation from the chromatin and its nuclear export mediated by 14-3-3. This event is a prerequisite for the recruitment of NF-κB to specific promoters such as the cellular inhibitor of apoptosis 2 (cIAP-2) and interleukin 8 (IL-8), leading to increased cell survival [29]. In castration-resistant tumors, nuclear active IKKα represses the transcription of the metastasis-suppressor gene *Maspin* (Figure 1). Accordingly, accumulation of nuclear active IKKα in human and mouse prostate tumors correlates with metastatic progression, reduced *Maspin* expression, and infiltration of receptor activator of nuclear factor κ-B ligand (RANKL)-expressing inflammatory cells [30]. A similar association between IKKα nuclear localization, *Maspin* levels, and cell migration or metastasis has been shown in squamous cell carcinoma cells (see details in Section 5).

4. Colorectal Cancer

For years, several groups, including our own, have investigated the role of IKKα in colorectal cancer (CRC). Initially, we found that IKKα was aberrantly activated and recruited to the promoter of different Notch target genes such as *hes1*, *hes5*, and *herp2*. Chromatin-bound IKKα constitutively phosphorylates SMRT, leading to its cytoplasmic export and the transcriptional activation of these genes (Figure 1). Conversely, IKKα inhibition, either pharmacologically or by expression of a dominant-negative form of the kinase, restores SMRT chromatin binding, inhibits Notch-dependent gene transcription, and reduces tumor size in a model of CRC xenografts [31]. Similarly, IKKα can phosphorylate the nuclear receptor co-repressor (N-CoR), a nuclear co-repressor homologous to SMRT, thus creating a functional 14-3-3-binding domain and promoting its nuclear export [32]. In a more recent study, we were able to identify the presence of a truncated form of IKKα with a predicted molecular weight of 45 KDa (p45-IKKα) that was specifically activated in the nucleus of CRC cells [33]. This truncated form of IKKα is generated by the proteolytic cleavage of full-length IKKα in the early endosomes by the action of cathepsins. The p45-IKKα form includes the kinase domain, but lacks some regulatory domains at the c-terminal [33]. Nuclear active p45-IKKα forms a complex with full length IKKα and NEMO, and regulates the phosphorylation of SMRT and histone H3. Activated p45-IKKα prevents apoptosis of CRC cells in vitro and it is required for the maintenance of tumor growth in vivo. Consistent with the fact that p45-IKKα is generated in the endosomes, inhibitors of endosome acidification abolish p45-IKKα activation and suppress CRC cell growth both in vitro and in vivo. Moreover, we demonstrated that BRAF activity is required and sufficient to induce p45-IKKα activation, which is TAK1-dependent [34] (Figure 1).

In a different set of experiments, mice deficient in the IKKα kinase activity were protected from intestinal tumor development, which was associated with an enhanced recruitment of interferon γ (IFNγ)-producing M1-like myeloid cells into the tumor. Polarization and accumulation of M1 macrophages in the mutant mice is not cell-autonomous, but depends on the interaction between IKKα-mutant epithelial cells and mutant stromal cells [35].

5. Skin Cancer

Nuclear IKKα is clearly involved in skin cancer progression, although some controversy exists about its contribution. Whereas different studies have definitively shown that nuclear IKKα in association with SMAD2/3 is required for physiologic skin differentiation [36–38], others also indicate that altered IKKα function can directly contribute to specific oncogenic functions. For example, IKKα can bind and repress the promoter of epidermal growth factor (EGF), among others, thus suppressing the EGF receptor/Ras/ERK pathway to prevent squamous cell carcinoma (SCC) [39]. Binding of IKKα to histone H3 at the 14-3-3 sigma locus prevents its hypermethylation by SUV39h1 and supports 14-3-3 sigma expression (Figure 2). Since 14-3-3 sigma controls the cytoplasmic export of the cell cycle-regulatory phosphatase CDC25, the absence of functional IKKα precludes G2/M cell cycle arrest in response to DNA damage, thus contributing to genomic instability and skin cancer [40].

Additional tumor suppressor activity for IKKα in SCC, which is again dependent on its nuclear localization and associated with the transforming growth factor β (TGFβ) pathway, is executed through Myc inhibition [41]. In the same direction, IKKα activates several anti-proliferative Myc antagonists, including Mad1, Mad2, and Ovol1, through Smad2/3, leading to enhanced keratinocyte differentiation [42] (Figure 2). In basal cell carcinoma, *LGR5* expression in also dependent on IKKα and STAT3, suggesting that increased IKKα activity can contribute to oncogenic transformation not only through inflammatory-related signals but also through the regulation of stemness-related genes [43]. In a different study, we found that IKKα induces the chromatin release of phospho-SUMO-IκBα (PS-IκBα), previously identified as a regulator of multiple developmental- and stemness-related genes, such as *HOX* and *IRX*, and its subsequent accumulation in the cytoplasm, which was linked to oncogenic keratinocyte transformation [44] (Figures 1 and 2). The mechanisms by which IKKα promote PS-IκBα inactivation are primarily unknown, but we speculate that nuclear IKKα might phosphorylate PS-IκBα and non-canonical, sites or regulate specific editing enzymes, phosphatases, SUMO-proteases or specific PS-IκBα-interacting proteins.

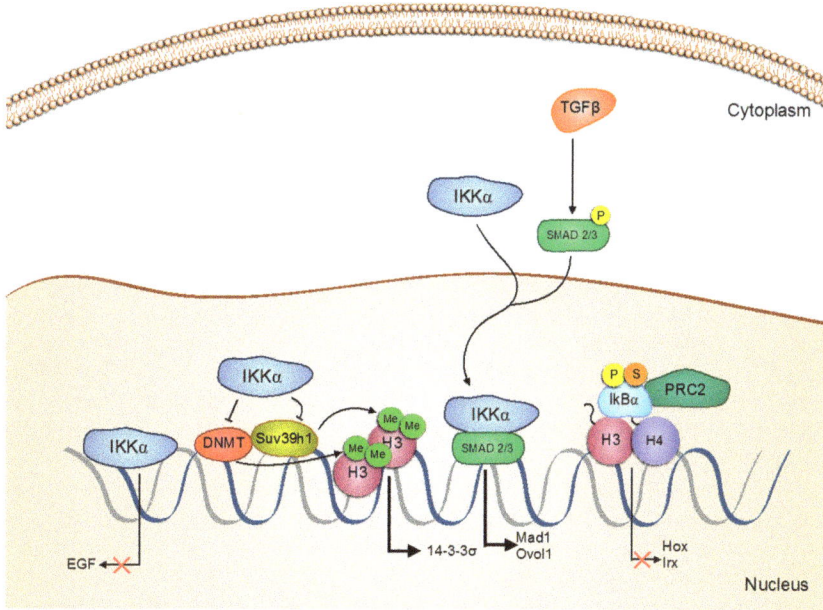

Figure 2. Tumor-suppressing functions of IKKα and IκBα. On one hand, IKKα increases SMAD transcriptional activity and decreases EGF transcription. It also promotes G2/M phase progression by de-repressing 14-3-3σ gene expression through preventing DNA and histone methylation on its promoter. On the other hand, IκBα is bound to histones and nuclear co-repressors, such as PRC2 regulating the expression of genes related to development and differentiation. Arrows: ⟶ Activation/Regulation/Phosphorylation; ⟶✕ Inactivation; ⊢ Inhibition

Recently, it was shown that mice carrying an IKKα variant that specifically localizes in the nucleus of the keratinocytes develop more aggressive tumors in response to chemical carcinogens than control mice. Nuclear IKKα seem to promote tumorigenesis by regulation of *c-myc*, *Maspin*, and *Integrin-α6*, and tumors with nuclear IKKα mimic the characteristics of human skin tumors with a high risk of metastasizing [45]. These results partially overlap our previous findings indicating that nuclear active IKKα plays oncogenic and pro-metastatic roles in SCC, being that its detection is predictive of higher metastatic capacity and worse patient outcome. We also found that nuclear active IKK levels inversely correlated with the levels of the metastasis suppressor Maspin (Figure 1), and tumors negative for this protein were exclusively found in the metastatic group [46].

As mentioned, PS-IκBα was previously detected in fibroblasts [47] and primary keratinocytes [44] as a protein capable of binding the chromatin through the N-terminal tail of histones H2A and H4 [44,47].

Importantly, PS-IκBα also binds histone deacetylases (HDACs) and the polycomb repressive complex 2 (PRC2) to regulate the expression of genes related to development and differentiation in a TNFα-dependent, but NF-κB-independent, manner [44]. Regulation of these genes might contribute to the maintenance of the skin homeostasis, as IκBα-deficient mice die five days after birth due to massive skin inflammation and defective skin differentiation [44,48–50]. Supporting a role for nuclear PS-IκBα in skin cancer, nuclear IκBα levels are significantly reduced, or totally lost, in aggressive human SCC and mouse transformed keratinocytes associated with an accumulation of cytoplasmic IκBα and altered *HOX* gene expression (Figure 2). In contrast, IκBα remains nuclear in the normal skin, and also in benign skin lesions, such as elastosis, psoriasis, actinic keratosis, and Bowen disease [44]. Our data might also help to understand previous and unexpected results obtained using a transgenic mouse carrying the non-degradable IκBα mutant, IκBα-SR (for IκBα super repressor) that showed increased and more aggressive tumorigenesis, even in the absence of NF-κB activity [51–54]. We propose that accumulation of IκBα-SR in the cytoplasm exerts pro-tumorigenic capacities by sequestering PRC2 and HDACs in the cytoplasm leading to inappropriate gene expression of PS-IκBα targets [29,31,32,47] (Figure 1).

6. Liver Cancer

Hepatocellular carcinoma (HCC) is one of the most common cancers worldwide and develops frequently in the context of chronic hepatitis, characterized by liver inflammation and hepatocyte apoptosis [55,56]. In this context, the NF-κB pathway can act as a tumor promoter or tumor suppressor [57]. Luedde and colleagues demonstrated that IKKα and IKKβ regulate biliary homeostasis and promote hepatocellular carcinoma by phosphorylating receptor-interacting protein kinase 1 (RIPK1), which is involved in both apoptosis and programmed necrotic cell death (necroptosis), independent of NF-κB. Specifically, loss of IKKα- and IKKβ-dependent RIPK1 phosphorylation in liver parenchymal cells inhibits compensatory proliferation and prevents the development of HCC, but promotes biliary cell paucity and cholestasis [58]. Moreover, IKKβ-depleted hepatocytes display sustained activation of the MKK4/7-JNK signaling cascade, previously identified as a mediator of hepatocellular carcinoma [59]. Deletion of the TAK1 kinase in these same cells induces hepatocyte dysplasia and early carcinogenesis in mice, and this tumor suppressor TAK1 activity is mediated by an NF-κB-independent, but NEMO-dependent, pathway [60].

On the contrary, other studies indicate that NEMO exerts a protective role against HCC through NF-κB-dependent and -independent pathways. In this sense, deletion of NEMO in the liver parenchymal cells (LPC) of 12-month-old mice results in spontaneous hepatocyte apoptosis, which triggers compensatory hepatocyte proliferation, inflammation, activation of liver progenitor cells and, finally, development of chronic hepatitis and HCC [61]. However, ablation of all three NF-κB proteins in LPC able of activating gene transcription (RelA, RelB, and c-Rel) has a limited effect on hepatocyte apoptosis at a young age, indicative of NF-κB-independent activity. Therefore, the canonical NF-κB pathway contributes to the survival of liver cells, but NEMO prevents liver tumorigenesis by NF-κB-independent functions. The mechanism by which NEMO prevents hepatocyte apoptosis is by inhibiting the formation of the death-inducing RIPK1/FADD/caspase-8 signaling complex. Thus, in the absence of NEMO, but high activity of the NF-κB pathway, which induces pro-survival genes, the RIPK1/FADD/caspase-8 complex imposes chronic liver damage, leading to HCC development [61–63]. All of these results are clinically relevant since NEMO expression is lost or low in a significant percentage of human HCC correlating with a poor five-year overall survival of patients [64].

7. Renal Cancer

Clear cell renal cell carcinomas (ccRCCs) are characterized by the loss of functional von Hippel-Lindau protein (pVHL), which leads to the stabilization of hypoxia-inducible factor alpha (HIFα) and activation of genes related to tumor development and progression, such as chemokine C-X-C motif (CXCR4) [65]. It was found that NEMO stabilizes HIFα via direct interaction and independently of NF-κB signaling. Moreover, NEMO inhibits apoptosis of tumor cells and activates the epithelial-to-mesenchymal transition, thus facilitating the metastatic process [66,67].

8. Lung Cancer

In lung cancer, it was shown that IKKα phosphorylates CBP to increase its affinity for NF-κB at the expense of CBP association to p53. Thus, IKKα activity causes increased NF-κB-mediated signaling, but decreased p53-dependent gene expression, leading to cell proliferation and tumor growth. In agreement with this finding, increased CBP phosphorylation and high levels of active IKKα are both detected in human lung tumor tissue compared to the adjacent normal tissue [68].

9. Conclusions

As mentioned, NF-κB is a complex and diverse pathway with a clear role in inflammation and immune response. However, there is now increasing evidence that specific elements of the pathway exert NF-κB-independent functions (Table 1), thus increasing the complexity of the NF-κB-related responses. This complexity is even higher in the context of cancer where particular elements could be mutated or aberrantly activated. Most of these functions are due to the accumulation of these members in the nucleus, regulating the expression of onco- or tumor suppressor genes. Here, we have examined some of the non-conventional functions for specific IKK and IκB members that are related to carcinogenesis, which might open new perspectives for future investigations with potential clinical applications.

Table 1. Table summarizing the published data on non-conventional functions of the NF-κB members in cancer. The red background shows pro-tumorigenic functions and the green background shows anti-tumorigenic activities. Abbreviations: ERα: estrogen receptor α; SRC-3: nuclear receptor coactivator-3; mTORC: mammalian target of rapamycin complex; SMRT: silencing mediator for retinoid and thyroid receptors; N-CoR: nuclear correpresor; IFNγ: interferon γ; EGF: epidermal growth factor; LGR5: leucine-rich repeat-containing G-protein coupled receptor 5; PS-IκBα: phospho-sumo inhibitor of κBα; EGFR: epidermal growth factor receptor; MMP-9: matric metallopeptidase 9; VEGF-A: vascular endothelial growth factor-A; RIPK1: receptor interacting serine/threonine kinase 1; FOXO3a: forkhead box O3; MKK4/7: mitogen-activated protein kinase kinases 4 and 7; JNK: c-Jun N-terminal kinase; HDAC: histone deacetylase; PRC2: polycomb Repressive Complex 2: NEMO: NFκB essential modulator; NFκB: nuclear factor κB; Casp8: caspase 8; HIFα: hypoxia-inducible factor α; CBP: CREB-binding protein; CRC: colorectal cancer; SCC: squamous cell carcinoma; BCC: basal cell carcinoma; NMSC: non-melanoma skin cancer; HCC: hepatocellular carcinoma; ccRCC: clear cell renal cell carcinoma.

Protein	Substrate	Effect	Cancer Type	References
IKKα	Phosphorylation of ERα and SRC-3	Estrogen-dependent gene transcription	Breast Cancer	[20]
	Cooperation with Notch1 to activate transcription of ERα-dependent genes	Cell proliferation	Breast Cancer	[21]
	E2F1 transcription	Cell cycle progression	Breast Cancer	[22]
	Phosphorylation of p27	Expansion of tumour-initiating cells	Breast Cancer	[23]
	Phosphorylation of mTORC	Cell proliferation	Prostate Cancer	[25,26]
	Activation of mTORC2	Akt activation	Prostate Cancer	[27]
	Phosphorylation of SMRT	Increased cell survival	Prostate Cancer	[29]
		Regulation of Notch-dependent gene transcription: Tumour growth	CRC	[31]
	Maspin gene repression	Metastasis induction	Prostate Cancer	[30]
			SCC	[46]
	Phosphorylation of NCoR	Increased gene transcription	CRC	[32]
	Regulation of IFNγ-expressing M1-like myeloid cells recruitment	Enhanced tumorigenesis	CRC	[35]
	Repression of EGF transcription	Prevention of SCC	SCC	[39]
	Prevents hypermethylation of 14-3-3sigma through Suv39h1	Maintenance of genomic stability in keratinocytes	Skin Cancer	[40]
	Myc inhibition	Tumour-suppressive activity	SCC	[41]
	Myc inhibition	Keratinocyte proliferation and differentiation	Skin Cancer	[42]
	LGR5 expression	Oncogenic transformation	BCC	[43]
	Chromatin release of PS-IκBα	Oncogenic transformation	Skin Cancer	[44]
	N: c-Myc, Maspin and Integrin-α6 expression: Cyt: Increases EGFR, MMP-9 and VEGF-A activity	Cancer progression	NMSC	[45]
	Phosphorylation of RIPK1	Regulation of cell viability	HCC	[58]

Table 1. *Cont.*

Protein	Substrate	Effect	Cancer Type	References
p45-IKKα	Phosphorylation of SMRT and Histone H3	Tumour maintenance and apoptosis inhibition	CRC	[33]
			CRC	[34]
	Regulation of anti-apoptotic and pro-metastatic genes	Tumour growth and metastasis		
IKKβ	Phosphorylation of FOXO3a	Increased proliferation	Breast Cancer	[24]
	Phosphorylation of RIPK1	Regulation of cell viability	HCC	[58]
	Repression of MKK4/7-JNK signalling cascade	Tumour suppressor	HCC	[59]
IκBα	Binding to HDACs and PRC2	Regulation of HOX and IRX: keratinocyte differentiation	SCC	[44]
TAK1	Suppression of specific NEMO function	Suppression of procarcinogenic and pronecrotic pathway	HCC	[60]
NEMO	NFκB activation	Tumour suppressor	HCC	[61] [62]
	Inhibition RIPK1 and Casp8	Suppression of hepatocyte apoptosis	HCC	[62]
	HIFα stabilization	Cell survival	ccRCC	[66]
	Phosphorylation of CBP	Cell proliferation	Lung Cancer	[68]

Among other elements of the pathway, IKKα seem to play a principal role in the regulation, both negatively and positively, of many types of cancer. However, IKKβ and NEMO that are essential components of the canonical IKK complex might also play a role, as it has already been shown in breast, liver, and renal cancer. The recent identification of chromatin-associated PS-IκBα, and its likely regulation by IKKα, add a novel layer of complexity and should lead to the re-evaluation of previous observations and conclusions about the role of IκBα inhibitors in cancer.

In conclusion, a better characterization of these non-canonical functions, how they are accumulated in the nucleus of cancer cells, and how they are integrated or not in the circuits involving NF-κB, should provide a clearer picture of the mechanisms controlling human cancer, thus providing novel elements for therapy assignment.

Conflicts of Interest: The authors declare no conflicts of interests.

References

1. Sen, R.; Baltimore, D. Inducibility of κ immunoglobulin enhancer-binding protein NF-κB by a posttranslational mechanism. *Cell* **1986**, *47*, 921–928. [CrossRef]
2. Karin, M.; Greten, F.R. NF-κB: Linking inflammation and immunity to cancer development and progression. *Nat. Rev. Immunol.* **2005**, *5*, 749–759. [CrossRef] [PubMed]
3. Cai, D.; Yuan, M.; Frantz, D.F.; Melendez, P.A.; Hansen, L.; Lee, J.; Shoelson, S.E. Local and systemic insulin resistance resulting from hepatic activation of IKK-β and NF-κB. *Nat. Med.* **2005**, *11*, 183–190. [CrossRef] [PubMed]
4. Arkan, M.C.; Hevener, A.L.; Greten, F.R.; Maeda, S.; Li, Z.W.; Long, J.M.; Wynshaw-Boris, A.; Poli, G.; Olefsky, J.; Karin, M. IKK-β links inflammation to obesity-induced insulin resistance. *Nat. Med.* **2005**, *11*, 191–198. [CrossRef] [PubMed]
5. Tak, P.P.; Firestein, G.S. NF-κB: A key role in inflammatory diseases. *J. Clin. Investig.* **2001**, *107*, 7–11. [CrossRef] [PubMed]
6. Ellis, R.D.; Goodlad, J.R.; Limb, G.A.; Powell, J.J.; Thompson, R.P.; Punchard, N.A. Activation of nuclear factor κB in Crohn's disease. *Inflamm. Res.* **1998**, *47*, 440–445. [CrossRef] [PubMed]
7. Tilstra, J.S.; Clauson, C.L.; Niedernhofer, L.J.; Robbins, P.D. NF-κB in aging and disease. *Aging Dis.* **2011**, *2*, 449–465. [PubMed]
8. Siebenlist, U.; Franzoso, G.; Brown, K. Structure, regulation and function of NF-κB. *Annu. Rev. Cell Biol.* **1994**, *10*, 405–455. [CrossRef] [PubMed]
9. Baldwin, A.S. The NF-κB and IκB proteins: New discoveries and insights. *Annu. Rev. Immunol.* **1996**, *14*, 649–683. [CrossRef] [PubMed]
10. Ghosh, S.; May, M.J.; Kopp, E.B. NF-κB and rel proteins: Evolutionarily conserved mediators of immune responses. *Annu. Rev. Immunol.* **1998**, *16*, 225–260. [CrossRef] [PubMed]
11. Karin, M.; Ben-Neriah, Y. Phosphorylation meets ubiquitination: The control of NF-κB activity. *Annu. Rev. Immunol.* **2000**, *18*, 621–663. [CrossRef] [PubMed]

12. Ben-Neriah, Y. Regulatory functions of ubiquitination in the immune system. *Nat. Immunol.* **2002**, *3*, 20–26. [CrossRef] [PubMed]

13. Senftleben, U.; Cao, Y.; Xiao, G.; Greten, F.R.; Krähn, G.; Bonizzi, G.; Chen, Y.; Hu, Y.; Fong, A.; Sun, S.C.; et al. Activation by IKKα of a second, evolutionary conserved, NF-κB signaling pathway. *Science* **2001**, *293*, 1495–1499. [CrossRef] [PubMed]

14. Rayet, B.; Gélinas, C. Aberrant *rel/nfkb* genes and activity in human cancer. *Oncogene* **1999**, *18*, 6938–6947. [CrossRef] [PubMed]

15. Karin, M.; Cao, Y.; Greten, F.R.; Li, Z.W. NF-κB in cancer: From innocent bystander to major culprit. *Nat. Rev. Cancer* **2002**, *2*, 301–310. [CrossRef] [PubMed]

16. Garg, A.; Aggarwal, B.B. Nuclear transcription factor-κB as a target for cancer drug development. *Leukemia* **2002**, *16*, 1053–1068. [CrossRef] [PubMed]

17. Lee, D.F.; Kuo, H.P.; Chen, C.T.; Hsu, J.M.; Chou, C.K.; Wei, Y.; Sun, H.L.; Li, L.Y.; Ping, B.; Huang, W.C.; et al. IKKβ suppression of tsc1 links inflammation and tumor angiogenesis via the mtor pathway. *Cell* **2007**, *130*, 440–455. [CrossRef] [PubMed]

18. Ben-Neriah, Y.; Karin, M. Inflammation meets cancer, with NF-κB as the matchmaker. *Nat. Immunol.* **2011**, *12*, 715–723. [CrossRef] [PubMed]

19. Sorriento, D.; Illario, M.; Finelli, R.; Iaccarino, G. To NFκB or not to NFκB: The dilemma on how to inhibit a cancer cell fate regulator. *Transl. Med. UniSa* **2012**, *4*, 73–85. [PubMed]

20. Park, K.J.; Krishnan, V.; O'Malley, B.W.; Yamamoto, Y.; Gaynor, R.B. Formation of an IKKα-dependent transcription complex is required for estrogen receptor-mediated gene activation. *Mol. Cell* **2005**, *18*, 71–82. [CrossRef] [PubMed]

21. Hao, L.; Rizzo, P.; Osipo, C.; Pannuti, A.; Wyatt, D.; Cheung, L.W.; Sonenshein, G.; Osborne, B.A.; Miele, L. Notch-1 activates estrogen receptor-α-dependent transcription via IKKα in breast cancer cells. *Oncogene* **2010**, *29*, 201–213. [CrossRef] [PubMed]

22. Tu, Z.; Prajapati, S.; Park, K.J.; Kelly, N.J.; Yamamoto, Y.; Gaynor, R.B. IKKα regulates estrogen-induced cell cycle progression by modulating E2F1 expression. *J. Biol. Chem.* **2006**, *281*, 6699–6706. [CrossRef] [PubMed]

23. Zhang, W.; Tan, W.; Wu, X.; Poustovoitov, M.; Strasner, A.; Li, W.; Borcherding, N.; Ghassemian, M.; Karin, M. A NIK-IKKα module expands ErbB2-induced tumor-initiating cells by stimulating nuclear export of p27/Kip1. *Cancer Cell* **2013**, *23*, 647–659. [CrossRef] [PubMed]

24. Hu, M.C.; Lee, D.F.; Xia, W.; Golfman, L.S.; Ou-Yang, F.; Yang, J.Y.; Zou, Y.; Bao, S.; Hanada, N.; Saso, H.; et al. IκB kinase promotes tumorigenesis through inhibition of forkhead foxo3a. *Cell* **2004**, *117*, 225–237. [CrossRef]

25. Dan, H.C.; Adli, M.; Baldwin, A.S. Regulation of mammalian target of rapamycin activity in PTEN-inactive prostate cancer cells by IκB kinase α. *Cancer Res.* **2007**, *67*, 6263–6269. [CrossRef] [PubMed]

26. Dan, H.C.; Ebbs, A.; Pasparakis, M.; Van Dyke, T.; Basseres, D.S.; Baldwin, A.S. Akt-dependent activation of mtorc1 complex involves phosphorylation of mtor (mammalian target of rapamycin) by IκB kinase α (IKKα). *J. Biol. Chem.* **2014**, *289*, 25227–25240. [CrossRef] [PubMed]

27. Dan, H.C.; Antonia, R.J.; Baldwin, A.S. PI3K/Akt promotes feedforward mTORC2 activation through IKKα. *Oncotarget* **2016**, *7*, 21064–21075. [CrossRef] [PubMed]

28. Majumder, P.K.; Sellers, W.R. Akt-regulated pathways in prostate cancer. *Oncogene* **2005**, *24*, 7465–7474. [CrossRef] [PubMed]

29. Hoberg, J.E.; Yeung, F.; Mayo, M.W. Smrt derepression by the IκB kinase α: A prerequisite to NF-κB transcription and survival. *Mol. Cell* **2004**, *16*, 245–255. [CrossRef] [PubMed]

30. Luo, J.L.; Tan, W.; Ricono, J.M.; Korchynskyi, O.; Zhang, M.; Gonias, S.L.; Cheresh, D.A.; Karin, M. Nuclear cytokine-activated IKKα controls prostate cancer metastasis by repressing maspin. *Nature* **2007**, *446*, 690–694. [CrossRef] [PubMed]

31. Fernández-Majada, V.; Aguilera, C.; Villanueva, A.; Vilardell, F.; Robert-Moreno, A.; Aytés, A.; Real, F.X.; Capella, G.; Mayo, M.W.; Espinosa, L.; et al. Nuclear ikk activity leads to dysregulated notch-dependent gene expression in colorectal cancer. *Proc. Natl. Acad. Sci. USA* **2007**, *104*, 276–281. [CrossRef] [PubMed]

32. Fernández-Majada, V.; Pujadas, J.; Vilardell, F.; Capella, G.; Mayo, M.W.; Bigas, A.; Espinosa, L. Aberrant cytoplasmic localization of N-CoR in colorectal tumors. *Cell Cycle* **2007**, *6*, 1748–1752. [CrossRef] [PubMed]

33. Margalef, P.; Fernández-Majada, V.; Villanueva, A.; Garcia-Carbonell, R.; Iglesias, M.; López, L.; Martínez-Iniesta, M.; Villà-Freixa, J.; Mulero, M.C.; Andreu, M.; et al. A truncated form of IKKα is responsible for specific nuclear IKK activity in colorectal cancer. *Cell Rep.* **2012**, *2*, 840–854. [CrossRef] [PubMed]
34. Margalef, P.; Colomer, C.; Villanueva, A.; Montagut, C.; Iglesias, M.; Bellosillo, B.; Salazar, R.; Martínez-Iniesta, M.; Bigas, A.; Espinosa, L. Braf-induced tumorigenesis is IKKα-dependent but NF-κB-independent. *Sci. Signal.* **2015**, *8*, ra38. [CrossRef] [PubMed]
35. Göktuna, S.I.; Canli, O.; Bollrath, J.; Fingerle, A.A.; Horst, D.; Diamanti, M.A.; Pallangyo, C.; Bennecke, M.; Nebelsiek, T.; Mankan, A.K.; et al. IKKα promotes intestinal tumorigenesis by limiting recruitment of M1-like polarized myeloid cells. *Cell Rep.* **2014**, *7*, 1914–1925.
36. Hu, Y.; Baud, V.; Delhase, M.; Zhang, P.; Deerinck, T.; Ellisman, M.; Johnson, R.; Karin, M. Abnormal morphogenesis but intact IKK activation in mice lacking the ikkalpha subunit of IκB kinase. *Science* **1999**, *284*, 316–320. [CrossRef] [PubMed]
37. Hu, Y.; Baud, V.; Oga, T.; Kim, K.I.; Yoshida, K.; Karin, M. IKKα controls formation of the epidermis independently of NF-κB. *Nature* **2001**, *410*, 710–714. [CrossRef] [PubMed]
38. Descargues, P.; Sil, A.K.; Karin, M. IKKα, a critical regulator of epidermal differentiation and a suppressor of skin cancer. *EMBO J.* **2008**, *27*, 2639–2647. [CrossRef] [PubMed]
39. Liu, B.; Xia, X.; Zhu, F.; Park, E.; Carbajal, S.; Kiguchi, K.; DiGiovanni, J.; Fischer, S.M.; Hu, Y. IKKα is required to maintain skin homeostasis and prevent skin cancer. *Cancer Cell* **2008**, *14*, 212–225. [CrossRef] [PubMed]
40. Zhu, F.; Xia, X.; Liu, B.; Shen, J.; Hu, Y.; Person, M. IKKα shields 14-3-3σ, a G_2/M cell cycle checkpoint gene, from hypermethylation, preventing its silencing. *Mol. Cell* **2007**, *27*, 214–227. [CrossRef] [PubMed]
41. Marinari, B.; Moretti, F.; Botti, E.; Giustizieri, M.L.; Descargues, P.; Giunta, A.; Stolfi, C.; Ballaro, C.; Papoutsaki, M.; Alemà, S.; et al. The tumor suppressor activity of IKKα in stratified epithelia is exerted in part via the TGF-β antiproliferative pathway. *Proc. Natl. Acad. Sci. USA* **2008**, *105*, 17091–17096. [CrossRef] [PubMed]
42. Descargues, P.; Sil, A.K.; Sano, Y.; Korchynskyi, O.; Han, G.; Owens, P.; Wang, X.J.; Karin, M. IKKα is a critical coregulator of a smad4-independent TGFβ-smad2/3 signaling pathway that controls keratinocyte differentiation. *Proc. Natl. Acad. Sci. USA* **2008**, *105*, 2487–2492. [CrossRef] [PubMed]
43. Jia, J.; Shi, Y.; Yan, B.; Xiao, D.; Lai, W.; Pan, Y.; Jiang, Y.; Chen, L.; Mao, C.; Zhou, J.; et al. Lgr5 expression is controled by IKKα in basal cell carcinoma through activating stat3 signaling pathway. *Oncotarget* **2016**, *7*, 27280–27294. [CrossRef] [PubMed]
44. Mulero, M.C.; Ferres-Marco, D.; Islam, A.; Margalef, P.; Pecoraro, M.; Toll, A.; Drechsel, N.; Charneco, C.; Davis, S.; Bellora, N.; et al. Chromatin-bound IκBα regulates a subset of polycomb target genes in differentiation and cancer. *Cancer Cell* **2013**, *24*, 151–166. [CrossRef] [PubMed]
45. Alameda, J.P.; Gaspar, M.; Ramírez, Á.; Navarro, M.; Page, A.; Suárez-Cabrera, C.; Fernández, M.G.; Mérida, J.R.; Paramio, J.M.; García-Fernández, R.A.; et al. Deciphering the role of nuclear and cytoplasmic IKKα in skin cancer. *Oncotarget* **2016**, *7*, 29531–29547. [PubMed]
46. Toll, A.; Margalef, P.; Masferrer, E.; Ferrándiz-Pulido, C.; Gimeno, J.; Pujol, R.M.; Bigas, A.; Espinosa, L. Active nuclear IKK correlates with metastatic risk in cutaneous squamous cell carcinoma. *Arch. Dermatol. Res.* **2015**, *307*, 721–729. [CrossRef] [PubMed]
47. Aguilera, C.; Hoya-Arias, R.; Haegeman, G.; Espinosa, L.; Bigas, A. Recruitment of IκBα to the hes1 promoter is associated with transcriptional repression. *Proc. Natl. Acad. Sci. USA* **2004**, *101*, 16537–16542. [CrossRef] [PubMed]
48. Beg, A.A.; Sha, W.C.; Bronson, R.T.; Ghosh, S.; Baltimore, D. Embryonic lethality and liver degeneration in mice lacking the rela component of NF-κB. *Nature* **1995**, *376*, 167–170. [CrossRef] [PubMed]
49. Klement, J.F.; Rice, N.R.; Car, B.D.; Abbondanzo, S.J.; Powers, G.D.; Bhatt, P.H.; Chen, C.H.; Rosen, C.A.; Stewart, C.L. IκBα deficiency results in a sustained NF-κB response and severe widespread dermatitis in mice. *Mol. Cell. Biol.* **1996**, *16*, 2341–2349. [CrossRef] [PubMed]
50. Rebholz, B.; Haase, I.; Eckelt, B.; Paxian, S.; Flaig, M.J.; Ghoreschi, K.; Nedospasov, S.A.; Mailhammer, R.; Debey-Pascher, S.; Schultze, J.L.; et al. Crosstalk between keratinocytes and adaptive immune cells in an IκBα protein-mediated inflammatory disease of the skin. *Immunity* **2007**, *27*, 296–307. [CrossRef] [PubMed]
51. Van Hogerlinden, M.; Rozell, B.L.; Ahrlund-Richter, L.; Toftgård, R. Squamous cell carcinomas and increased apoptosis in skin with inhibited rel/nuclear factor-κB signaling. *Cancer Res.* **1999**, *59*, 3299–3303. [PubMed]

52. Seitz, C.S.; Lin, Q.; Deng, H.; Khavari, P.A. Alterations in NF-κB function in transgenic epithelial tissue demonstrate a growth inhibitory role for NF-κB. *Proc. Natl. Acad. Sci. USA* **1998**, *95*, 2307–2312. [CrossRef] [PubMed]

53. Dajee, M.; Lazarov, M.; Zhang, J.Y.; Cai, T.; Green, C.L.; Russell, A.J.; Marinkovich, M.P.; Tao, S.; Lin, Q.; Kubo, Y.; et al. NF-κB blockade and oncogenic ras trigger invasive human epidermal neoplasia. *Nature* **2003**, *421*, 639–643. [CrossRef] [PubMed]

54. van Hogerlinden, M.; Rozell, B.L.; Toftgård, R.; Sundberg, J.P. Characterization of the progressive skin disease and inflammatory cell infiltrate in mice with inhibited NF-κB signaling. *J. Investig. Dermatol.* **2004**, *123*, 101–108. [CrossRef] [PubMed]

55. Motola-Kuba, D.; Zamora-Valdés, D.; Uribe, M.; Méndez-Sánchez, N. Hepatocellular carcinoma. An overview. *Ann. Hepatol.* **2006**, *5*, 16–24. [PubMed]

56. Okuda, K. Hepatocellular carcinoma. *J. Hepatol.* **2000**, *32*, 225–237. [CrossRef]

57. Vainer, G.W.; Pikarsky, E.; Ben-Neriah, Y. Contradictory functions of NF-κB in liver physiology and cancer. *Cancer Lett.* **2008**, *267*, 182–188. [CrossRef] [PubMed]

58. Koppe, C.; Verheugd, P.; Gautheron, J.; Reisinger, F.; Kreggenwinkel, K.; Roderburg, C.; Quagliata, L.; Terracciano, L.; Gassler, N.; Tolba, R.H.; et al. IκB kinaseα/β control biliary homeostasis and hepatocarcinogenesis in mice by phosphorylating the cell-death mediator receptor-interacting protein kinase 1. *Hepatology* **2016**, *64*, 1217–1231. [CrossRef] [PubMed]

59. Sakurai, T.; Maeda, S.; Chang, L.; Karin, M. Loss of hepatic NF-κB activity enhances chemical hepatocarcinogenesis through sustained c-jun n-terminal kinase 1 activation. *Proc. Natl. Acad. Sci. USA* **2006**, *103*, 10544–10551. [CrossRef] [PubMed]

60. Bettermann, K.; Vucur, M.; Haybaeck, J.; Koppe, C.; Janssen, J.; Heymann, F.; Weber, A.; Weiskirchen, R.; Liedtke, C.; Gassler, N.; et al. Tak1 suppresses a nemo-dependent but NF-κB-independent pathway to liver cancer. *Cancer Cell* **2010**, *17*, 481–496. [CrossRef] [PubMed]

61. Luedde, T.; Beraza, N.; Kotsikoris, V.; van Loo, G.; Nenci, A.; De Vos, R.; Roskams, T.; Trautwein, C.; Pasparakis, M. Deletion of nemo/IKKγ in liver parenchymal cells causes steatohepatitis and hepatocellular carcinoma. *Cancer Cell* **2007**, *11*, 119–132. [CrossRef] [PubMed]

62. Kondylis, V.; Polykratis, A.; Ehlken, H.; Ochoa-Callejero, L.; Straub, B.K.; Krishna-Subramanian, S.; Van, T.M.; Curth, H.M.; Heise, N.; Weih, F.; et al. Nemo prevents steatohepatitis and hepatocellular carcinoma by inhibiting ripk1 kinase activity-mediated hepatocyte apoptosis. *Cancer Cell* **2015**, *28*, 582–598. [CrossRef] [PubMed]

63. Ehlken, H.; Krishna-Subramanian, S.; Ochoa-Callejero, L.; Kondylis, V.; Nadi, N.E.; Straub, B.K.; Schirmacher, P.; Walczak, H.; Kollias, G.; Pasparakis, M. Death receptor-independent fadd signalling triggers hepatitis and hepatocellular carcinoma in mice with liver parenchymal cell-specific nemo knockout. *Cell Death Differ.* **2014**, *21*, 1721–1732. [CrossRef] [PubMed]

64. Aigelsreiter, A.; Haybaeck, J.; Schauer, S.; Kiesslich, T.; Bettermann, K.; Griessbacher, A.; Stojakovic, T.; Bauernhofer, T.; Samonigg, H.; Kornprat, P.; et al. Nemo expression in human hepatocellular carcinoma and its association with clinical outcome. *Hum. Pathol.* **2012**, *43*, 1012–1019. [CrossRef] [PubMed]

65. Shen, B.; Zheng, M.Q.; Lu, J.W.; Jiang, Q.; Wang, T.H.; Huang, X.E. Cxcl12-cxcr4 promotes proliferation and invasion of pancreatic cancer cells. *Asian Pac. J. Cancer Prev.* **2013**, *14*, 5403–5408. [CrossRef] [PubMed]

66. Nowicka, A.M.; Häuselmann, I.; Borsig, L.; Bolduan, S.; Schindler, M.; Schraml, P.; Heikenwalder, M.; Moch, H. A novel PVHl-independent but nemo-driven pathway in renal cancer promotes hif stabilization. *Oncogene* **2016**, *35*, 3125–3138. [CrossRef] [PubMed]

67. Bracken, C.P.; Whitelaw, M.L.; Peet, D.J. Activity of hypoxia-inducible factor 2α is regulated by association with the NF-κB essential modulator. *J. Biol. Chem.* **2005**, *280*, 14240–14251. [CrossRef] [PubMed]

68. Huang, W.C.; Ju, T.K.; Hung, M.C.; Chen, C.C. Phosphorylation of CBP by IKKα promotes cell growth by switching the binding preference of cbp from p53 to NF-κB. *Mol. Cell* **2007**, *26*, 75–87. [CrossRef] [PubMed]

biomedicines

MDPI

Review

The Crosstalk of Endoplasmic Reticulum (ER) Stress Pathways with NF-κB: Complex Mechanisms Relevant for Cancer, Inflammation and Infection

M. Lienhard Schmitz [1], M. Samer Shaban [2], B. Vincent Albert [2], Anke Gökçen [2] and Michael Kracht [2,3,*,†]

[1] Institute of Biochemistry, Justus Liebig University Giessen, D-35392 Giessen, Germany; lienhard.schmitz@biochemie.med.uni-giessen.de

[2] Rudolf-Buchheim-Institute of Pharmacology, Justus Liebig University Giessen, D-35392 Giessen, Germany; Mohammed.S.Shaban@pharma.med.uni-giessen.de (M.S.S.); Benadict.V.Albert@pharma.med.uni-giessen.de (B.V.A.); Anke.Goekcen@pharma.med.uni-giessen.de (A.G.)

[3] Rudolf-Buchheim-Institute of Pharmacology, Universities of Giessen and Marburg Lung Center (UGMLC), Schubertstrasse 81, D-35392 Giessen, Germany

* Correspondence: michael.kracht@pharma.med.uni-giessen.de; Tel.: +49-0641-99-47600; Fax: +49-0641-99-47619

† Member of the German Center for Lung Research.

Received: 21 April 2018; Accepted: 11 May 2018; Published: 16 May 2018

Abstract: Stressful conditions occuring during cancer, inflammation or infection activate adaptive responses that are controlled by the unfolded protein response (UPR) and the nuclear factor of kappa light polypeptide gene enhancer in B-cells (NF-κB) signaling pathway. These systems can be triggered by chemical compounds but also by cytokines, toll-like receptor ligands, nucleic acids, lipids, bacteria and viruses. Despite representing unique signaling cascades, new data indicate that the UPR and NF-κB pathways converge within the nucleus through ten major transcription factors (TFs), namely activating transcription factor (ATF)4, ATF3, CCAAT/enhancer-binding protein (CEBP) homologous protein (CHOP), X-box-binding protein (XBP)1, ATF6α and the five NF-κB subunits. The combinatorial occupancy of numerous genomic regions (enhancers and promoters) coordinates the transcriptional activation or repression of hundreds of genes that collectively determine the balance between metabolic and inflammatory phenotypes and the extent of apoptosis and autophagy or repair of cell damage and survival. Here, we also discuss results from genetic experiments and chemical activators of endoplasmic reticulum (ER) stress that suggest a link to the cytosolic inhibitor of NF-κB (IκB)α degradation pathway. These data show that the UPR affects this major control point of NF-κB activation through several mechanisms. Taken together, available evidence indicates that the UPR and NF-κB interact at multiple levels. This crosstalk provides ample opportunities to fine-tune cellular stress responses and could also be exploited therapeutically in the future.

Keywords: ER stress; cancer; infection; inflammation; unfolded protein response; NF-κB; IκBα; thapsigargin

---

1. Introduction

The NF-κB pathways regulate the activities of a family of five transcription factors (RELA (p65), RELB, c-REL, NFKB1 (p105/p50) and NFKB2 (p100/p52)) that play numerous roles in physiological, but also pathophysiological, conditions [1]. The pivotal role of NF-κB in promoting several of the ten hallmarks of cancer is well established and has been the subject of excellent reviews [2–4].

NF-κB is an inducible transcription factor that displays low constitutive background activity. It can be strongly induced by a variety of agents which increase the nuclear concentration of the

DNA-binding subunits and subsequently, promote the binding to specific cognate κB sites across the genome. The NF-κB system acts as a general stress sensor that is activated by a multitude of adverse conditions. These are extracellular activators, such as infection or inflammation, but also intracellular activators, such as DNA damage or the unfolded protein response (UPR). Here, we discuss recent progress in the analysis of (reciprocal) interactions between the NF-κB pathways and the UPR, a pathway that controls the folding capacities of the endoplasmic reticulum (reviewed in [5]).

2. The Unfolded Protein Response, ER Stress and Cancer

The ER is a specialized organelle that is responsible for the synthesis, assembly, folding, transport and degradation of a large number of membrane and secreted proteins [6]. The quality and fidelity of all of these steps is constantly monitored by the cell. The accumulation of unfolded or misfolded proteins in the ER lumen rapidly results in ER stress and activates the UPR process. The UPR combines several systems to slow down ongoing protein synthesis and to increase the folding capacity of the ER. If this reaction is successful, cellular protein synthesis resumes, and cellular homeostasis will be restored, facilitating survival of the ER stress condition. If ER stress persists, the UPR pathways will eventually induce oxidative stress and cell death [6]. Tumor cells have not only acquired the capacity to suppress death-inducing pathways, to induce angiogenesis and to reprogram their metabolism, as reviewed elsewhere [7], but also show an increased demand for protein synthesis and folding capacity. These demands favor an increased supply of nutrients and eventually, shape the tumor microenvironment, including the activities of invading immune cells. It is therefore not surprising that akin to NF-κB, ER stress has also been related to cancer, and this has been comprehensively reviewed recently [8–10]. NF-κB induction by the UPR does not only occur in highly proliferative tumor cells, but also in other pathophysiological situations, such as infection by viruses, as demonstrated for the hepatitis C virus or for human coronavirus 229E [11,12]. As discussed below, both the ER/UPR and NF-κB pathways operate through gene-regulatory mechanisms, ultimately inducing and fine-tuning the mRNA and protein expression of specific sets of genes. However, our knowledge on the interactions of the two systems and the levels and specificities of this type of crosstalk for certain types of cancer, for inflammatory and immune reactions or for infections is far from complete and represents an emerging area of investigation [13–17].

3. The ER Stress Sensors

ER stress is recognized by three sensors that are inserted into the ER membrane: protein kinase R (PKR)-like ER kinase (PERK), inositol-requiring protein 1α (IRE1α) and ATF6α (also called cyclic AMP-dependent transcription factor 6α). PERK and IRE1α share similar lumenal parts and possess cytosolic ser/thr kinase domains [18]. In non-stressed cells, the major ER chaperone binding-immunoglobulin protein (BIP), also called 78 kDa glucose-regulated protein (GPR78), binds to the ER-oriented parts of PERK and IRE1α and keeps them in a monomeric inactive state. Increased binding of BIP/GRP78 to misfolded proteins relieves both PERK and IRE1α and facilitates activation by dimerization (or oligomerization) followed by trans(auto)phosphorylation [19,20]. Active PERK then phosphorylates the eukaryotic translation initiation factor 2 (eIF2) subunit α to shut down translation but also activates the ATF4-dependent transcription program (see below). Phosphorylated IRE1α activates its own RNAase domain to catalyze the excision of 26 nucleotides of XBP1 mRNA, thereby generating spliced XBP1 mRNA. This transcript is translated into the active XBP1 protein, a multifunctional transcriptional regulator. ATF6α is a transcription factor that is processed into its mature form (ATF6f) by Golgi-associated proteases upon relief from BIP/GRP78 interactions. The released cytoplasmic part contains the basic-region leucine zipper (bZIP) transactivation domain. The three branches of the UPR often act in concert but can also be activated sequentially and with different strengths, allowing a multitude of outcomes spanning from the compensation of ER stress and the restoration of proteostasis to cell death as the ultimate effect [5]. As the mechanistic evidence for a connection of ATF6α/ATF6f and XBP1s to NF-κB is scarce, in this review, we concentrate on the crosstalk of PERK and IRE1α with the NF-κB system.

4. Regulation of eIF2-Dependent Translation Initiation by Phosphorylation

During de novo protein synthesis, a central event in translation initiation involves the assembly of a ternary complex composed of the multi subunit eIF2 complex loaded with both GTP and the initiator Met-tRNA with the ribosome at the start codon [21,22]. During translation initiation, eIF2-GTP is hydrolyzed, releasing eIF2-GDP from the ribosome. The recycling of eIF2-GDP to the GTP-bound form requires the guanine nucleotide exchange factor, eIF2β. Phosphorylation of the eIF2 subunit, eIF2α, by eIF2α kinases at ser51 converts the eIF2 initiation factor from a substrate to an inhibitor of eIF2β. The resulting drop in eIF2-GTP levels then suppresses general translation [21]. Phosphorylation of eIF2α is mediated by one of the four kinases: (i) heme-regulated inhibitor kinase (HRI, encoded by the gene *EIF2AK1*), (ii) RNA-activated protein kinase (PKR, encoded by the gene *EIF2AK2*), (iii) PERK (encoded by the gene *EIF2AK3*) and (iv) general control non-derepressible-2 (GCN2, encoded by the gene *EIF2AK4*). Phosphorylation of eIF2α is reversed by inducible growth arrest and DNA damage-inducible protein GADD34 (encoded by *PPP1R15a*) that targets protein phosphatase 1 (PP1) to dephosphorylate and inactivate eIF2α. PP1 activity is also under control of the constitutive repressor of eIF2α phosphorylation, CReP (encoded by *PPP1R15b*) [23–26]. The four eIF2α kinases are typically activated by infection (double stranded RNA, PKR), amino acid starvation (GCN2), heme-depletion (HRI) or by unfolded proteins in the ER (PERK). Despite their stimulus and cell type selectivity, they cooperate to mediate the phosphorylation of eIF2α, and mouse embryonic fibroblasts (Mefs) lacking PERK/GCN2/PKR have strongest reduction in phospho-eIF2α [27]. Notably, the eIF2 phosphorylation/dephosphorylation cycle is often disrupted in cancer to further promote or suppress translation, as comprehensively reviewed in [28].

5. Small Molecule Effectors as Prevailing Tools to Model ER Stress

A considerable number of reports addressing the mechanisms of ER stress have used chemical compounds to induce the UPR, in particular tunicamycin, thapsigargin, dithiotreitol (DTT), proteasome inhibitors and brefeldin A [6]. Tunicamycin inhibits ER-associated glycoprotein synthesis [29,30], while thapsigargin, a sesquiterpene lactone isolated from the plant, *Thapsia garganica* L., has long been known to potently inhibit the sarco-endoplasmic reticulum Ca^{2+}-ATPase (SERCA) [31]. Thereby, thapsigargin depletes Ca^{2+} from the ER and is highly cytotoxic [32]. This has led to the testing of thapsigargin as an anti-cancer agent [31]. Another UPR-inducing compound is DTT, a reducing agent that disrupts disulfide bonds and thus, results in the accumulation of unfolded proteins [33]. In this review, we will specifically refer to mechanistic conclusions derived from such approaches using chemical effectors that allow exact and reproducible control of the experimental conditions, as opposed to alternative, more physiological settings

6. Evidence for Activation of the Canonical NF-κB Pathway through Phosphorylation of eIF2α

In 1995, using electrophoretic mobility shift assays (EMSA) which assess the in vitro DNA-binding activity of NF-κB proteins in nuclear extracts, Pahl and Baeuerle discovered that thapsigargin activates p65/p50 NF-κB subunits [34,35]. In 2003, Jiang et al. reported that this effect was abolished in Mefs lacking PERK or expressing an eIF2α S51A mutant [36]. The same effect was seen in thapsigargin-treated Mefs subjected to leucine starvation or exposed to UV light. In this case, NF-κB activation was reduced in cells lacking GCN2, the eIF2α kinase that functions as a sensor for amino acid starvation [36,37]. PERK was also shown to be required for activation of a luciferase reporter gene driven by NF-κB binding sites, demonstrating that this pathway increases NF-κB-dependent transcriptional activity in intact cells [36]. These authors did not observe inducible phosphorylation of IκBα or its degradation in response to thapsigargin. Rather, they found evidence that thapsigargin favors the release of p65/p50 from cytosolic IκBα complexes, an effect that was absent in Mefs expressing the eIF2α S51A mutant. The phospho-eIF2α-dependent regulation of p65 by thapsigargin was confirmed by Deng and coworkers [38]. This group also designed experiments to investigate the role of eIF2α phosphorylation independent from stress signals. For this, they constructed a protein consisting of the cytoplasmic PERK kinase domain fused to a protein module that allowed conditional tethering (or oligomerization) of the PERK domains by a bivalent cell permeable compound. Indeed, the addition of the cross linker to stable cell lines induced phosphorylation of eIF2α, and nuclear translocation of p65 and induction of its transcriptional activity [38]. However, in marked contrast to Jiang et al., this was paralleled by partial degradation of IκBα. The authors concluded that the PERK–phospho-eIF2α pathway inhibits the synthesis of IκBα but does not affect the pre-existing protein, as revealed by pulse chase experiments that monitored de novo protein synthesis [38]. Thus, while both studies showed that the ER stress pathway activates NF-κB, they disagreed on the underlying cytosolic activation mechanism. In both of these studies, no evidence for the phosphorylation of IκBα or for activation of IκB kinases (IKKs) was found. Additionally, UV treatment reduces cellular IκBα protein levels, an effect that is absent in eIF2α S51A Mefs and thus, critically requires eIF2α phosphorylation [39]. UV treatment globally shuts down translation through the PERK–eIF2α pathway suggesting that this is the key mechanism for IκBα decay [40]. However, UV treatment was later shown to also induce degradation of IκBα through C-terminal phosphorylation of the PEST domain by casein kinase II (CK2) providing an alternative explanation for IκBα depletion [41]. To date, the paradigm of ER stress-mediated suppression of constitutive IκBα protein synthesis is still suggested to be the prevailing mechanism of ER–NF-κB crosstalk and this view has been emphasized in several reviews [13–15,42,43].

7. Regulation of IκBα Half-Life: The Major Control Point Affected by ER Stress?

The regulation of IκB protein levels is key to NF-κB activation, as also shown by the early observation that the inhibition of protein synthesis by cycloheximide (CHX) is sufficient for NF-κB activation [44]. This raises the question of what is known about the half-life of IκBα in diverse conditions and whether these observations match the conclusion that ER stress mainly activates NF-κB indirectly by suppressing constitutive IκBα steady state levels. Amongst others, the IκBα protein (and its related family members, IκBβ and IκBε) can be regarded as one of the most powerful and universal negative regulators of NF-κB [45]. A plethora of infectious and inflammatory conditions, including cytokines such as interleukin(IL)-1 or tumor necrosis factor(TNF)α trigger the active and rapid destruction of IκBα by the well-characterized phosphorylation-dependent proteasomal degradation pathway [46,47]. This allows NF-κB subunits to translocate to the nucleus and to induce transcription of numerous target genes, including the *NFKBIA* gene that encodes IκBα [48]. The newly synthesized IκBα sequesters NF-κB from promoters and enhancers, retaining it in the cytoplasm [49]. This major negative feedback loop shuts down NF-κB transcriptional activation in multiple cell types [45,50,51]. However, with respect to the regulation of basal IκBα protein levels, several studies have reported different half-lives in a number of cell types and suggested different ways in which this might be regulated [52–55]. Mathes et al. provided one of the most comprehensive studies on the regulation of basal IκBα levels. By studying genetically modified Mefs, they concluded that cells essentially contain two pools of IκBα, a small fraction of free IκBα (around 15% of total) and IκBα bound to NF-κB. The level of free IκBα is severely reduced in cells lacking the p105, c-REL and p65 (RELA) subunits, showing that the NF-κB subunits are required for protein stabilization of IκBα. They may also be required for basal transcription of the *NFKBIA* gene, an issue not addressed in this study. The half-life of free IκBα ranges from 10 min to 20 min, whereas that of the NF-κB-bound IκBα is more stable, ranging from 8 h to 10 h. Free IκBα is degraded by the proteasome through IKK- and ubiquitination-independent events involving the C-terminal PEST domain [56]. Mathes and coworkers also showed that the free IκBα degradation pathway allows for fast and maximal activation of NF-κB, as cells expressing a degradation-resistant C-terminal IκBα mutant (ΔC288) have a delayed and dampened TNFα-inducible NF-κB activation profile [56]. It follows that our understanding of the true role of the PERK–eIF2α pathway in the regulation of the IκBα–NF-κB complex is still hampered by a lack of data concerning a detailed analysis of the effects of ER stress on IκBα protein stability and the stoichiometry of free versus bound IκBα in physiological and pathophysiological situations, independent from highly toxic chemicals, such as thapsigargin.

8. Additional Levels of Cytosolic NF-κB Regulation by ER Stress

Emerging evidence has shown that ER stress also modulates (or requires) critical upstream regulators of the NF-κB pathway. Thapsigargin or tunicamycin can also induce NF-κB through the catalytic activity of the IRE1α kinase. In this case, IRE forms a complex with IκB kinase (IKK)β and tumor necrosis factor receptor (TNFR)-associated factor (TRAF)2. In this model, thapsigargin-induced cell death involves the strongly increased synthesis and secretion of TNFα which is blocked in cells expressing a non-degradable IκBα mutant or in cells lacking IRE1α [57,58]. TRAF2 is an ubiquitous adaptor protein of TNFα and toll-like receptor (TLR) pathways [59]. Another TLR adaptor protein called toll-interleukin-1 receptor domain-containing adapter protein inducing interferon β TRIF or TIR domain-containing adapter molecule 1 (TCAM-1) is essential for activation of the lipopolysaccharide (LPS) target gene, IL-1β. Pretreatment of macrophages with thapsigargin or tunicamycin strongly increases LPS-triggered synthesis, processing and release of mature IL-1β in a TRIF-dependent manner [60]. However, when macrophages are pretreated with low doses of LPS, TRIF specifically mediates the suppression of tunicamycin-induced CHOP and ATF4 expression [61,62]. These data reveal that complex positive or negative crosstalk loops between the UPR and NF-κB also operate through secreted cytokines (TNFα, IL-1β) or TLR agonists (LPS) that are all well known for strongly activating the canonical NF-κB pathway [63]. The IRE1α-dependent mechanism seems to cooperate with the PERK pathway for regulating NF-κB. DTT- or thapsigargin-induced NF-κB activity (as determined by EMSA) does not depend on IKKα, but rather on IKKβ, as revealed by the analysis of Mefs lacking IKKβ or IKKα/β or re-expressing a kinase-inactive mutant of IKKβ. Thapsigargin-mediated IκBα decay is abrogated in cells reconstituted with an IκBα super repressor mutant (SS32/36AA), suggesting that IκBα decay depends on IκBα phosphorylation. Puzzling though, neither IκBα phosphorylation, nor IKK activity, were directly triggered by thapsigargin or DTT through PERK. Rather, NF-κB activity and IκBα degradation were reduced in IRE1α-deficient cells. The half-life of IκBα is around 5 h in wild type cells but IκBα is very stable in IRE1α -/- cells. Both, reduced basal phosphorylation of IκBα and diminished NF-κB activity in IRE1α-deficient cells were restored by IRE1α, but also by IKKβ (but not IKKα) [64]. These results suggest that basal IKK activity, maintained by IRE1α, is critical for the activation of NF-κB when PERK-induced translation inhibition (by thapsigargin) occurs. Tam et al. also showed that the extent of DTT- or thapsigargin-induced activation of NF-κB correlates with the amount of IRE1 re-expressed in IRE1α-deficient cells. NF-κB activity inversely correlates with CHX-mediated translational inhibition, showing a proportional correlation with the level of NF-κB activation. Based on these genetic experiments, IRE1α-dependent regulation of basal IKK activity is necessary for effective activation of NF-κB by PERK [64]. Such a link may also operate in disease, as demonstrated in a model of dextran sulfate-induced colitis in mice, where IKKα suppressed ER stress through the TNFR and nucleotide-binding oligomerization domain-containing protein (NOD)1/2 receptor-mediated pathways. Mice expressing a non-activatable IKKα mutant show increased IRE1α-dependent ER stress [65]. Thus, these data show that upon chemical ER stress, the PERK and IRE1α branches of the UPR may, in fact, work in concert to modulate the NF-κB pathway. However, it remains an open question how exactly IRE1α regulates basal or inducible IKK activity and vice versa. In conclusion, multiple lines of evidence suggest that strong activation of the canonical NF-κB pathway (by IL-1, TNFα, LPS) is accompanied by (moderate) activation of the IRE1α and PERK branches of the UPR. The various levels of interplay between NF-κB and the UPR are schematically displayed in Figure 1.

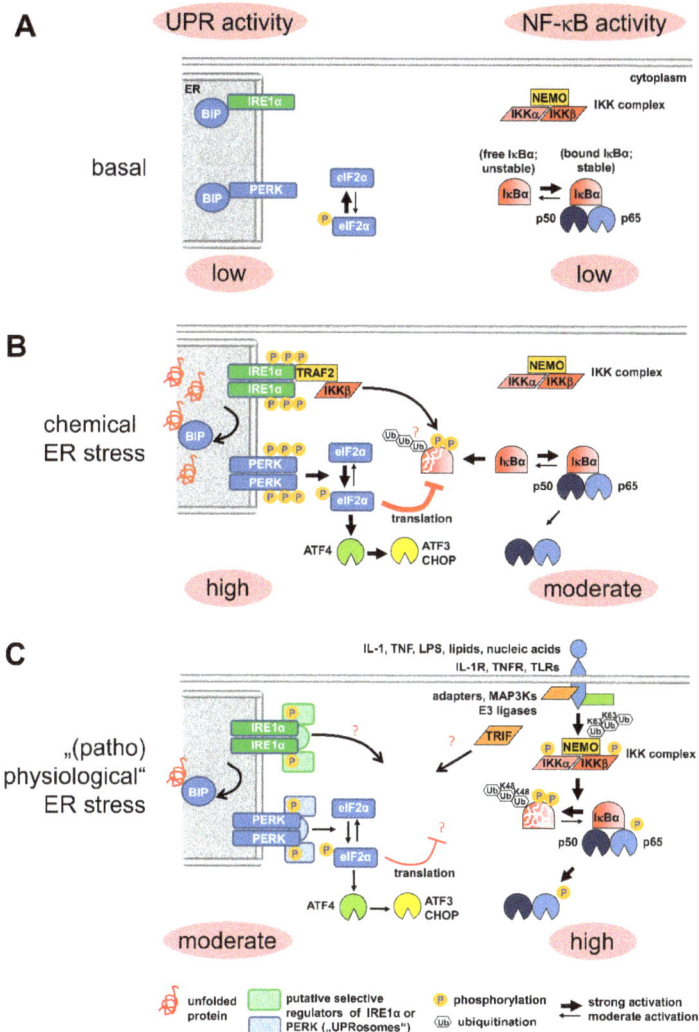

Figure 1. Integration of unfolded protein response (UPR) and NF-κB signaling. (**A**) In the absence of signals, most cells have no or very low basal activity of the UPR and of the canonical NF-κB pathway. The protein kinases IRE1α and PERK are kept inactive by binding to the chaperone BIP/GRP78 within the endoplasmic reticulum (ER) lumen. Most of the IκBα inhibitor is bound to NF-κB subunits retaining them in the cytoplasm. (**B**) Different classes of chemical stressors (tunicamcyin, thapsigargin or dithiotreitol (DTT)) increase the unfolded protein load in the ER causing massive auto-phosphorylation and activation of both IRE1α and PERK. IRE1α binds to the adapter protein TRAF2 and (indirectly) to the protein kinase IKKβ. PERK phosphorylates the eukaryotic translation initiation factor eIF2α causing translational shut-off for multiple proteins including the free unbound IκBα. Destruction of free IκBα also requires phosphorylation by IRE1α-associated IKKβ. (**C**) During infection, inflammation or cancer, strong activation of NF-κB by cytokines or toll-like receptor (TLR) agonists occurs in parallel to ER stress. In this case, the formation of specific signaling complexes at PERK and IRE1α sensors by still putative "UPRosomes" is suggested to restrict maximal ER stress, thereby contributing to context-specific gene activation or repression.

9. Cross-Interference of ER Stress and NF-κB at the Level of Transcription Factors and Gene Regulation

Arguably the most striking but least discussed aspect of ER stress interactions with the NF-κB system involves the chromatin response within the nucleus. Early studies showed that thapsigargin strongly induces the protein levels of the transcription factors ATF3 and ATF4 and CHOP (also called growth arrest and DNA damage-inducible protein (GADD)153 or DNA damage-inducible transcript 3 protein (DDIT3)). Besides transcriptional induction, this occurs by a unique mechanism that allows preferential translation from downstream open reading frames (ORFs) of ATF4 or CHOP (and also GADD34) [66–68]. ATF3 seems to be mainly regulated by transcriptional mechanisms involving ATF4, CHOP and Jun family members [69,70]. ATFs belong to the bZIP family of TFs, while CHOP is a member of the C/EBP family of TFs. Thapsigargin strongly induces the expression of all three of them [27]. This effect is absent in PERK-deficient cells, while during amino acid starvation, GCN2 is employed for their expression. Importantly, cells lacking ATF4 do not express ATF3 or CHOP, whereas cells lacking ATF3 normally induce ATF4 and CHOP. ATF4 and ATF3 are also required for the regulation of GADD34, thus forming a negative feedback loop for eIF2α phosphorylation. These landmark studies defined the PERK/GCN2–ATF4–ATF3–CHOP pathway and clearly established a hierarchy of the three TFs within this signaling cascade [27,71]. Upon amino acid starvation, ATF3 and ATF4 bind to the amino acid response elements (AARE) of the ATF3 promoter and numerous metabolic genes, such as asparagine synthetase (ASNS), sodium-coupled neutral amino acid transporter-2 (SNAT2), and the γ-glutamyl cyclotransferase (CHAC1) but also to the vascular endothelial growth factor (VEGF) promoter [72–75]. Interestingly, some of the genes regulated by nutritional stress are well known inflammatory NF-κB target genes, such as *IL8* and *CXCL2* [74,76–78]. Thus ATF4 and ATF3, together with the NF-κB subunits, form a transcription factor network that coordinates metabolic gene expression programs during nutritional and ER stress [79]. As nutritional and metabolic changes are major hallmarks of cancer, these observations form a natural link between ER stress, the UPR and the NF-κB system in malignant disease that warrants further investigation.

Less is known about the regulation of ER stress target genes by NF-κB subunits. In one report, the thapsigargin-inducible expression of CHOP and ATF4 was unchanged in cells lacking p65 [36]. Expression of BIP/GRP78 upon thapsigargin treatment is normal in PERK-deficient cells suggesting that this crucial ER chaperone is regulated independently from ATF3, ATF4 or CHOP [27]. Interestingly, another observation from the study of Tam and coworkers is that thapsigargin-induced expression of BIP/GRP78 is reduced in p65-deficient Mefs, and p65 binds to the BIP/GRP78 promoter, suggesting a new direct link from NF-κB subunits to ER stress target genes [64].

The connection of ATF3/ATF4/CHOP to the NF-κB system also became apparent with the advent of bioinformatics approaches and genome-wide assays correlating DNA-binding profiles (obtained by chromatin immunoprecipitation sequencing (ChIP-seq)) with mRNA expression. In macrophages, a combination of microarray analyses and motif searches resulted in the identification of ATF3 as an early LPS-induced gene that bound in the proximity of p50 (REL) NF-κB sites. ChIP-qPCR confirmed the binding of NF-κB and ATF3 to the *IL6* and *IL12b* (*IL12p40*) promoters [80]. ATF3 is required for the recruitment of HDAC1 and negative regulation of eleven LPS-target genes [80]. This was confirmed by a systems biology approach showing that ATF3 attenuated LPS-induced *IL6* as part of a regulatory circuit that involves sequential activation and cooperative activity of NF-κB and C/EBPδ. Depletion of ATF3 or C/EBPδ or pharmacological inhibition of NF-κB by the IKK inhibitor SC-514 disrupted the kinetics and amplitude of LPS-induced *IL6* expression [81]. ATF3 was also identified as a high-density lipoprotein-inducible repressor of TLR-induced proinflammatory cytokines. This first ATF3 ChIP-seq study confirmed the inducible binding of ATF3 to the promoters of the *IL6*, *IL12p40* and *TNFA* genes [82]. ATF3 is also a type I interferon-inducible negative regulator of multiple interferon response genes (ISGs) and binds to the *IFNβ* promoter [83]. These studies assigned a broad and largely negative regulatory role to ATF3 in regulating the NF-κB response and provided links to immunity and cancer (also reviewed in [84].

Han et al. systematically identified the target genes of ATF4 and CHOP in response to tunicamyin by a combination of ChIP-seq and RNA-seq approaches using ATF4 or CHOP-deficient Mefs. They showed that ATF4 and CHOP co-occupy many genomic regions and identified around 3000 binding sites within the genome. This analysis also revealed several hundred common, but also unique, target genes of both factors. Upregulated gene sets were clearly enriched for ER stress target genes involved in protein folding, amino acid synthesis and protein transport, while downregulated genes were involved in proliferation, wound healing, anti-apoptosis, and steroid and lipid synthesis [85]. We interrogated this data set for factors that may represent the crosstalk of ER stress with (NF-κB-dependent) immunoregulatory genes. Indeed, as illustrated by the selection shown in Figure 2, tunicamycin also regulates multiple genes with annotated functions in immune responses such as various cytokines and their receptors (e.g., *Il23a*, *Il1a*, *Il6*, *Ifnar1*, *Il17ra*, *Il6ra*), chemokines (*Ccl2*, *Ccl9*), adhesion molecules (*Icam1*) and prostaglandin synthetases (*Ptgs2*, Cox-2). In many cases these are suppressed in an ATF4- or CHOP-dependent manner, while *Il23a* is induced as previously reported (Figure 2), [86,87].

The significant overlap in target genes of NF-κB and the UPR transcription factors, ATF4- and CHOP, was also seen by the comparative analysis of gene expression sets visualized in Figure 3. These data show ATF4- and CHOP-dependent regulation of 58 genes that have a documented role in regulation of the NF-κB signaling pathway, including three IκBs, (IκBα, IκBe, IκBz). Data sets of this kind, therefore, not only highlight how closely ER stress and the NF-κB response are intertwined, they also provide a rich resource for further analyses of the exact mechanisms of crosstalk between these two systems. In physiological settings, this crosstalk between ER stress transcription factors is not necessarily always negative or repressive. For example, it was recently shown that ATF4 is a positive regulator for LPS-induced *Ccl2* expression in the endothelium and mediates leukocyte infiltration within the retina, while ATF3 can support breast cancer metastasis [88,89]. *Ccl2* is a prototypical NF-κB target gene that also requires additional transcription factors from the c-Jun/Fos family for full activation [90]. Likewise, all three canonical ER stress TFs, ATF4, ATF3 and CHOP, engage in multiple further interactions with other TFs that also shape their influence on the NF-κB response, as reported in several reviews [91,92]. More generally, the main ER stress TFs are likely always integrated into large protein–protein interaction (PPI) networks and co-occupy their target genes, as shown recently by large studies that combined multiple ChIP-seq and mass spectrometry profiling approaches [93,94].

gene	alternative names	wt +Tm / wt	CHOP-/- +Tm / wt	ATF4-/- +Tm / wt	function	category
		fold change in mRNA expression (log$_2$) in response to Tunicamycin (Tm)				
Ddit3	Chop Chop10 Gadd153	5.2	1.4	2.5	ER stress	UPR
Atf3		4.9	3.1	0.8	ER stress	UPR
Hspa5	Grp78 BIP	4.0	3.3	2.8	chaperone	UPR
Atf5	Atfx Nap1	3.0	1.5	-2.0	ER stress	UPR
Hspa4	Apg2 Hsp110	2.4	2.9	3.1	chaperone	UPR
Atf4		2.0	2.1	-0.7	ER stress	UPR
Hspa9	Grp75 Hsp74 Hspa9a	2.0	1.6	1.3	chaperone	UPR
Xbp1	Treb5	1.8	1.2	1.3	ER stress	UPR
Atf6		1.6	0.8	0.6	ER stress	UPR
Hspa1a	Hsp70-3 Hsp70A1	-1.1	-0.5	0.8	chaperone	UPR
Igf2bp3	Vickz3	-1.3	-0.8	-0.4	growth factor	UPR
Igfals	Albs Als	-2.5	-1.9		growth factor	UPR
Igf1	Igf-1	-5.7	-5.0	-8.7	growth factor	UPR
Il23a		4.1	0.3	1.8	cytokine	immune response
Ifnar1	Ifar Ifnar	3.8	3.8	3.4	interferon recptor	immune response
Il1a		2.5	2.3	0.6	cytokine	immune response
Ik	Red Rer	2.4	2.4	2.5	unknown, cytokine (?)	immune response
Il5	Il-5	2.3	1.6	0.4	cytokine	immune response
Ifit1	Garg16 Ifi56 Isg56	2.3	1.7	1.0	ISG	immune response
Il17ra	Il17r	2.3	2.1	2.3	cytokine receptor	immune response
Cxcl4	Cxcl4 Scyb4	1.6			chemokine	immune response
Ifit3	Garg49 Ifi49 Isg49	1.5	-0.4	-0.5	ISG	immune response
Irf6		1.3		-0.8	ISG regulator	immune response
Irf7		1.3	-0.5	1.7	ISG regulator	immune response
Il12rb2		1.1	2.4	1.8	cytokine receptor	immune response
Il21r	Nilr	1.0			cytokine receptor	immune response
Cxxc4	Idax	-1.0		4.0	chemokine	immune response
Il12rb1	Il12rb	-1.0	-1.1	-0.2	cytokine receptor	immune response
Cxcl15	Scyb15	-1.2	-3.6		chemokine	immune response
Il13ra2		-1.3	-2.9		cytokine receptor	immune response
Ifitm3		-1.3	-2.1	-3.4	ISG	immune response
Ccl2	Je Mcp1 Scya2	-1.5	-4.1	-5.7	chemokine	immune response
Igtp		-1.6	-2.7	-2.6	ISG	immune response
Cxcl5	Scyb5	-1.7	-2.9	-7.6	chemokine	immune response
Il6ra	Il6r	-1.8	-0.9	-2.2	cytokine receptor	immune response
Il13ra1	Il13r Il13ra	-1.8	-3.7	-4.3	cytokine receptor	immune response
Ptgs2	Cox-2 Cox2 Pghs-b Tis10	-1.9	0.3	-1.2	COX-2	immune response
Il6	Il-6	-2.0	-7.0	-8.0	cytokine	immune response
Ccl9	Mrp2 Scya10 Scya9	-2.1	-3.5	-7.4	chemokine	immune response
Il20ra		-2.2	-3.1		cytokine receptor	immune response
Cxcl12	Sdf1	-2.3	-2.7	-4.7	chemokine	immune response
Ifitm2		-2.3	-2.8	-3.0	ISG	immune response
Il4ra	Il4ra	-2.3	-4.4	-3.9	cytokine receptor	immune response
Icam1	Icam-1	-2.5			adhesion molecule	immune response
Il11		-2.5	0.8	0.8	cytokine	immune response
Il18rap		-2.8	-0.4	-3.4	cytokine receptor	immune response
Il16		-2.8	-4.5		cytokine	immune response
Ifitm1		-3.8	-6.0	-5.0	ISG	immune response
Il1rn	Il-1ra	-4.0	-4.0	-9.0	cytokine receptor antagonist	immune response

fold change (log$_2$)

-4   4

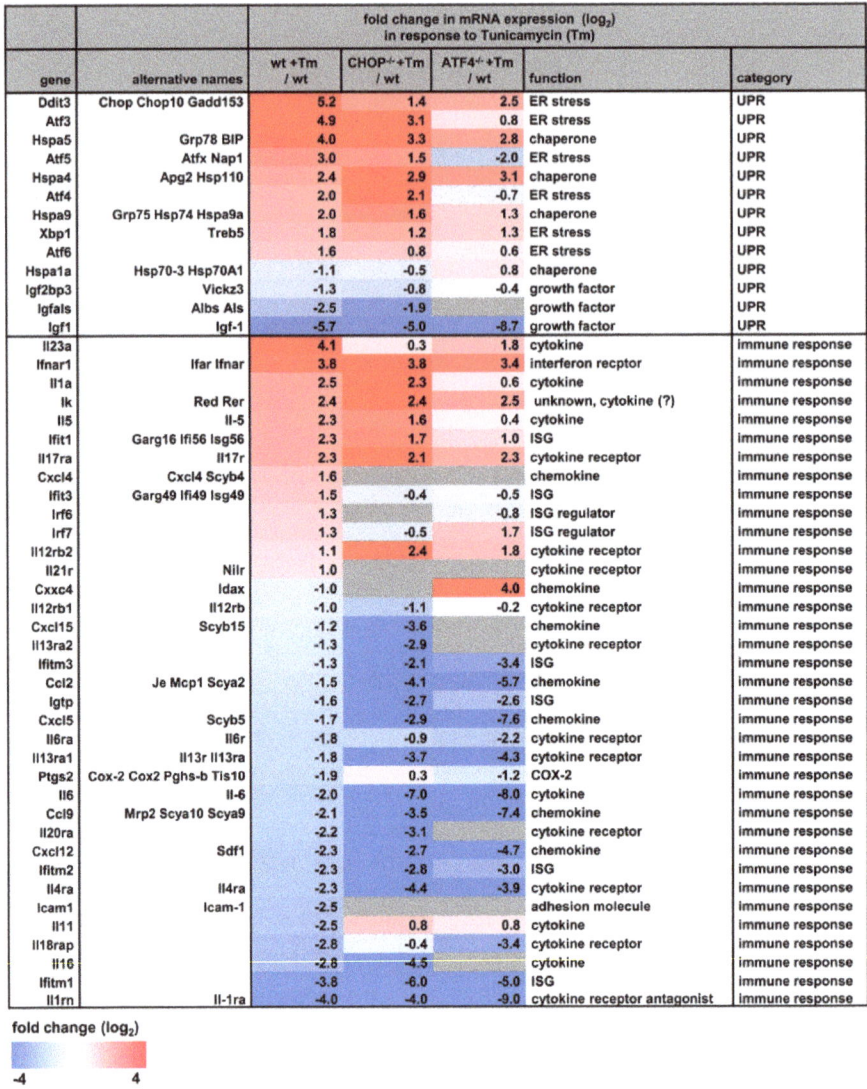

Figure 2. ATF4- or CHOP-dependent regulation of the UPR and of immune modulators by chemical ER stress. Published RNA-seq data sets from wild type murine embryonic fibroblasts or cells deficient in CHOP (CHOP-/-) or ATF4 (ATF4-/-) treated with tunicamycin (Tm) for 10 h or left untreated were extracted from GEO (GSE35681). Data were filtered for genes regulated by at least two-fold. The gene list is sorted by fold change compared to untreated wild type cells. Shown is a selection of prototypical target genes of the UPR and of immune responses. Gray colored boxes indicate the absence of expression.

NF-κB pathway component	gene	fold change in mRNA expression (log₂) in response to Tunicamycin (Tm)					requirement for basal expression		requirement for Tm-regulated expression	
		wt+Tm / wt	CHOP-/- / wt	CHOP-/- +Tm / wt	ATF4-/- / wt	ATF4-/- +Tm / wt	CHOP	ATF4	CHOP	ATF4
atypical	Rbck1	2.4	2.0	1.6	1.8	1.5				
atypical	Csnk2a1	2.0	1.9	1.9	2.7	2.5				
atypical	Prkcd	1.7	1.2	1.0	1.4	1.2				
atypical	Map3k14	-1.2	-0.5	-1.4	-0.3	-0.1				
atypical	Zc3h12a	-2.0	-1.7	-2.6	-2.8	-4.1				
atypical	Prkch	-2.3	-2.6	-2.1	-0.7	-1.6				
cofactors	Bag1	4.0	3.7	3.8	4.3	4.2				
cofactors	Ppp4c	2.6	3.1	2.7	3.7	3.8				
cofactors	Cdk5rap3	2.4	1.3	2.2	0.5	1.7				
cofactors	Ank2	1.8	-0.5	1.7	-1.6	-1.6				
cofactors	Lyl1	1.7	1.9	1.9	3.0	2.8				
cofactors	Tsc22d3	1.3	0.2	2.0	-0.5	-0.5				
cofactors	Mtdh	1.0	0.7	1.4	0.6	0.7				
cofactors	Scn1a	-1.0	0.0	-1.0	2.5	1.6				
cofactors	Zbp1	-1.0	-2.4	-3.9		-3.9				
cofactors	Cul2	-1.1	-1.2	-1.1	-0.7	-1.2				
cofactors	Bcl3	-1.3	-2.7	-3.9	-3.1	-3.3				
cofactors	Col2a1	-1.4	4.8	4.3	2.2	1.7				
cofactors	Hspb1	-1.5	-4.6	-6.2	0.8	0.3				
cofactors	Cdk6	-1.8	-1.5	-2.0	-1.6	-0.9				
cofactors	Uxt	-1.8	-2.5	-1.6	-2.3	-2.1				
cofactors	Errfi1	-2.1	-1.0	-1.0	-4.3	-4.8				
cofactors	Cdk2	-2.4	-1.4	-2.2	-0.6	-0.5				
cofactors	Nr4a1	-2.4	0.2	-0.4	-0.7	-0.6				
cofactors	Il33	-2.8	2.7	0.9						
cofactors	Brca1	-3.1	-2.0	-3.2	-0.7	-0.3				
cofactors	Ctnnal1	-3.3	-2.1	-3.0	-1.0	-1.9				
cofactors	Peli2	-4.1	-8.7	-10.0	-7.9	-9.5				
cofactors	Muc1	-4.4	-4.9	-6.1	-4.7	-5.8				
core	Cul1	3.7	3.5	3.8	3.8	3.8				
core	Relb	1.0	-1.1	-0.8	-0.4	0.4				
core	Nfkbiz	-1.2	-3.0	-3.9	-3.5	-5.5				
core	Nfkbia	-1.3	-1.2	-2.2	0.0	-0.2				
core	Nfkbie	-1.9	-2.5	-3.2	-2.1	-2.3				
ireceptors	Ddx21	1.0	1.2	1.6	1.5	2.0				
ireceptors	Nlrp3	1.0		1.0		1.6				
kinases	Raf1	3.5	3.0	3.3	3.8	4.0				
kinases	Prkca	2.7	2.9	2.8	1.8	2.1				
kinases	Tbk1	-1.6	-1.1	-1.4	-1.2	-1.1				
kinases	Camk2a	-2.4	-1.1	-3.2	-3.6	-4.6				
kinases	Rps6ka5	-4.5	-4.5	-4.6	-3.4	-4.0				
negative	Itch	3.4	2.7	3.0	3.1	3.4				
negative	Ier3	2.2	2.4	1.9	2.3	0.5				
negative	Noc3l	1.0	0.2	1.1	0.5	1.0				
negative	Pdlim2	-1.3	0.2	-0.5	-3.2	-3.8				
negative	Socs3	-3.5	-2.2	-3.3	-3.3	-4.1				
non-canonical	Traf3	-1.5	-1.0	-1.4	-0.9	-0.8				
rpr	Traf2	2.0	0.9	0.8	1.4	2.3				
rpr	Tram1	1.2	-0.1	1.2	0.2	1.2				
rpr	Mkl	-1.0	4.9	4.6	3.9	4.3				
rpr	Fyb	-5.9	-3.2	-5.0		-10.3				
Tbcell	Pik3ca	2.4	1.5	1.6	2.6	2.5				
Tbcell	Grb2	2.1	1.9	2.0	2.2	2.4				
Tbcell	Malt1	-1.2	-1.0	-0.7	-0.8	-1.2				
Tbcell	Card10	-1.3	1.4	0.3	4.3	4.1				
Tbcell	Ripk3	-1.8	-2.0	-2.3	-0.5	-0.5				
Tbcell	Map3k8	-2.5	-2.5	-3.4	-3.0	-5.6				
Tbcell	Gad1	-2.7								

function of CHOP or ATF4: repressor / activator / no effect

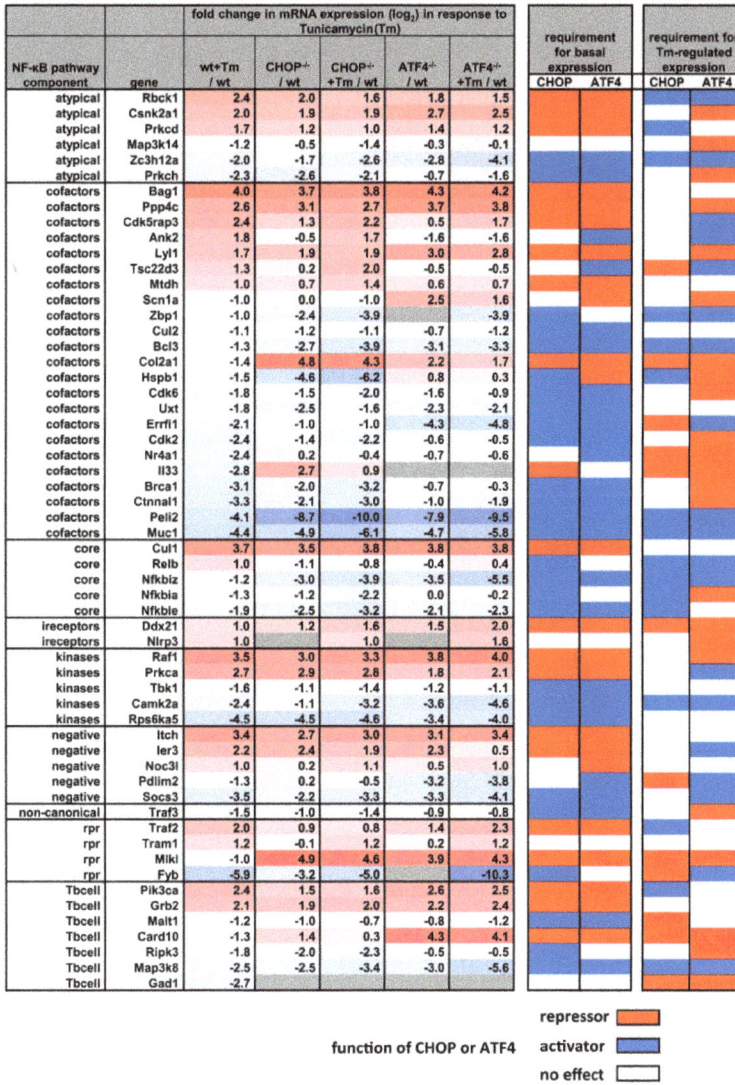

Figure 3. ATF4- or CHOP-dependent modulation of NF-κB regulators by chemical ER stress. The same data sets described in Figure 2 (GSE35681) were filtered for mRNA expression of 262 genes with a documented role in the regulation of NF-κB in diverse systems (as revealed by searching public data bases). The left column shows their role in regulation of the canonical (core), non-canonical, or atypical NF-κB pathways and some additional information on their established functions according to (Perkins, 2007). Two hundred and thirty-two of these factors were found to be expressed in all conditions. The left heatmap summarizes 58 components (25% of all components) that were deregulated by at least two-fold by tunicamycin (Tm) in wild type cells and the corresponding changes in CHOP- or ATF4-deficient Mef cells. The right color map categorizes the effects of loss of CHOP or ATF on basal and Tm-inducible expression of the NF-κB components based on log₂ differences of more than 0.5. Gray colored boxes indicate the absence of expression. Abbreviations are as follows: ireceptors, intracellular receptors; rpr, receptor-proximal component; Tbcell; NF-κB pathway components mainly characterized in T- or B-lymphocytes.

10. The Impact of Pharmacological Inhibition of PERK or IRE1α on NF-κB Function

As outlined above, much of the mechanistic knowledge on the ER stress–NF-κB relationship has been derived from genetically altered mouse fibroblasts. Recently, fast acting pharmacological compounds have become available, facilitating dissection of ER stress pathways in more complex models. In 2012, a novel, small molecule, ATP-competitive PERK inhibitor (GSK2606414) was reported to inhibit thapsigargin-induced PERK autophosphorylation in intact cells within the nM range. This inhibitor is highly selective and only affected 20 out of 294 kinases tested by >85% at 10 μM and suppressed tumor growth in a mouse xenograft model [95]. An optimized version of this compound (GSK2656157) was 1000-fold more active on PERK against a panel of 300 kinases, including HRI, PKR and GCN2 [96,97]. Again, this compound suppressed a range of human xenograft tumors and inhibited PERK autophosphorylation, eIF2α phosphorylation and the induction of UPR target genes in cell lines and tumors [96]. However, the inhibitor was also cytotoxic to the exocrine/endocrine pancreas tissues in non-tumor controls consistent with the results from PERK-deficient mice and loss of function PERK mutations in humans [98,99]. These and other data suggest tissue-specific functions of PERK in the development and exocrine function of the pancreas that may also partly operate through P-eIF2α independent effects [100]. The latter is supported by the observation that the PERK inhibitor caused cell death in human HT1080 tumor cells engineered to express the eIF2α S51A mutant, thus bypassing the P-eIF2α pathway [101]. Like for any protein kinase inhibitor, this may simply be the result of off-target effects and indeed, there is a single study suggesting that both PERK inhibitors also inhibit RIPK1 in the TNFα pathway [102]. However, recent studies have reported suppressive effects of PERK inhibitors on prototypical NF-κB target genes in diverse models. Using transcriptome analyses, Iwasaki et al. found that saturated fatty acids acting as proinflammatory factors induce nuclear translocation of ATF4 and p65 in macrophages. As demonstrated by ChIP, both TFs are co-recruited to the *IL6* promoter and activate *IL6* transcription. A functional interaction between p65 and ATF4 was found by reduced nuclear translocation and promoter binding of p65 in ATF4-deficient cells [103]. Iwasaki et al. also showed that the expression of ATF4 was suppressed by the PERK inhibitor, GSK2656157. These data confirm, in physiological settings, that ATF4 links various metabolic stresses to the NF-κB chromatin response [103]. In a rat model of intervertebral disc degeneration, silencing of PERK or ATF4 or application of GSK2606414 also suppressed inducible *Il6* expression in addition to *Tnfa* [104].

A further inhibitor acting on the PERK–eIF2α pathway is salubrinal. This compound suppressed eIF2α dephosphorylation through the GADD34/PP1 and CReP/PP1 complexes [105]. Recently, salubrinal was shown to inhibit TNFα (but not IL-1)-induced activation of NF-κB and the expression of *Ccl2/Mcp-1*. This effect was not abrogated in cells with siRNA-mediated suppression of eIF2α, suggesting that the salubrinal effect occurred through PP1, independent from eIF2α. In support of this, the same effect was seen with guanabenz, another PP1 inhibitor [106].

Another newly developed inhibitor is based on the finding that oligomerization of the IRE1α kinase domain controls the catalytic activity of the adjacent endoribonuclease domain. Ghosh et al. developed an optimized small molecule inhibitor called kinase-inhibiting RNase-attenuator (KIRA)6. KIRAs disrupt IRE1α oligomerization thereby suppressing the RNase activities of IRE1α [107]. IRE1α is often mutated in cancer and it will be interesting to learn if KIRA compounds also modulate NF-κB activities in malignant diseases [9,107]. Consistent with the observations cited above, Keestra et al. found that thapsigargin induces *IL6* mRNA and protein levels through NOD1/NOD2 receptors—two cytosolic sensors of bacterial peptidoglycans. Interestingly, IL-6 secretion was suppressed by KIRA6 but not by GSK2656157 [108]. Together, these pharmacological approaches provide independent support for a link between ER stress and NF-κB, although the underlying mechanisms and the specificity of the effects require further investigation.

11. Concluding Remarks and Open Questions

In this review we have summarized studies that connect the ER stress-mediated activation of the UPR to activation of the NF-κB pathway. So far, the evidence has been mainly derived from genetic experiments and chemical activators of ER stress, with a focus on cytosolic signaling pathways. However, as schematically depicted in Figure 4 these two pathways can also be viewed as a combinatorial system of ten transcription factors that cooperate to shape gene activation patterns and protein folding capacities. In the future, it will be important to test the concept of specific signaling complexes assembling at IRE1α and PERK during infection, inflammation and in malignant processes. The idea of such "UPRosomes" is attractive as it provides ample opportunity to adapt the extent and kinetics of UPR activation to the activity of other signaling pathways, such as NF-κB [5]. Another important area of research will be the systematic investigation of combined UPR and NF-κB activation on transcriptional programs and chromatin structure in disease.

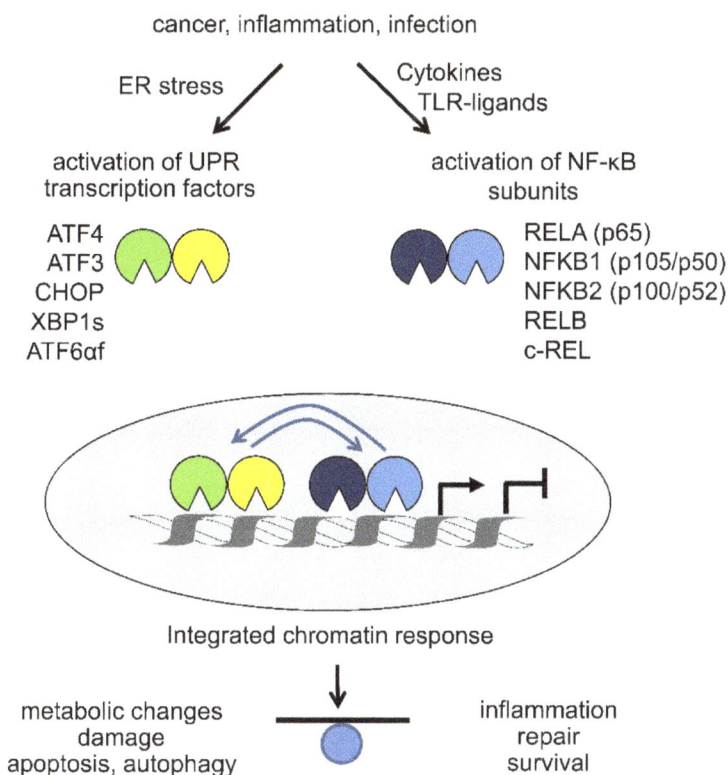

Figure 4. Integration of UPR and NF-κB signaling at chromatin. Despite representing unique signaling cascades, new data indicate that UPR and NF-κB pathways converge within the nucleus through ten major transcription factors. The combinatorial occupancy of numerous genomic regions (enhancers and promoters) coordinates the transcriptional activation or repression of hundreds of genes that collectively determine the balance between metabolic and inflammatory phenotypes and the extent of apoptosis and autophagy or repair of cell damage and survival.

Author Contributions: M.K. wrote the first draft of the manuscript; all other authors provided substantial help in the correction of the manuscript.

Acknowledgments: This work was supported by grants from the Deutsche Forschungsgemeinschaft Kr1143/7-3, KR1143/9-1 (KLIFO309), TRR81/2 (B02), SFB1213 (B03) and SFB1021 (C02) (to MK). The work from M.L.S. was supported by grants from the Deutsche Forschungsgemeinschaft (SCHM 1417/8-3), SFB/TRR81, SFB1021, SFB1213, the Deutsche Krebshilfe (111447) and the IMPRS program of the Max-Planck Society. Work from both laboratories was further supported by the Excellence Cluster Cardio-Pulmonary System (ECCPS) and the LOEWE/UGMLC program.

Conflicts of Interest: The authors declare no conflict of interest.

References

1. Zhang, Q.; Lenardo, M.J.; Baltimore, D. 30 Years of NF-kappaB: A Blossoming of Relevance to Human Pathobiology. *Cell* **2017**, *168*, 37–57. [CrossRef] [PubMed]
2. Baud, V.; Karin, M. Is NF-kappaB a good target for cancer therapy? Hopes and pitfalls. *Nat. Rev. Drug Discov.* **2009**, *8*, 33–40. [CrossRef] [PubMed]
3. Ben-Neriah, Y.; Karin, M. Inflammation meets cancer, with NF-kappaB as the matchmaker. *Nat. Immunol.* **2011**, *12*, 715–723. [CrossRef] [PubMed]
4. Taniguchi, K.; Karin, M. NF-kappaB, inflammation, immunity and cancer: Coming of age. *Nat. Rev. Immunol.* **2018**, *18*, 309–324. [CrossRef] [PubMed]
5. Hetz, C.; Papa, F.R. The Unfolded Protein Response and Cell Fate Control. *Mol. Cell* **2018**, *69*, 169–181. [CrossRef] [PubMed]
6. Wang, M.; Kaufman, R.J. Protein misfolding in the endoplasmic reticulum as a conduit to human disease. *Nature* **2016**, *529*, 326–335. [CrossRef] [PubMed]
7. Hanahan, D.; Weinberg, R.A. Hallmarks of cancer: The next generation. *Cell* **2011**, *144*, 646–674. [CrossRef] [PubMed]
8. Wang, M.; Kaufman, R.J. The impact of the endoplasmic reticulum protein-folding environment on cancer development. *Nat. Rev. Cancer* **2014**, *14*, 581–597. [CrossRef] [PubMed]
9. Chevet, E.; Hetz, C.; Samali, A. Endoplasmic reticulum stress-activated cell reprogramming in oncogenesis. *Cancer Discov.* **2015**, *5*, 586–597. [CrossRef] [PubMed]
10. Urra, H.; Dufey, E.; Avril, T.; Chevet, E.; Hetz, C. Endoplasmic Reticulum Stress and the Hallmarks of Cancer. *Trends Cancer* **2016**, *2*, 252–262. [CrossRef] [PubMed]
11. Li, S.; Ye, L.; Yu, X.; Xu, B.; Li, K.; Zhu, X.; Liu, H.; Wu, X.; Kong, L. Hepatitis C virus NS4B induces unfolded protein response and endoplasmic reticulum overload response-dependent NF-kappaB activation. *Virology* **2009**, *391*, 257–264. [CrossRef] [PubMed]
12. Poppe, M.; Wittig, S.; Jurida, L.; Bartkuhn, M.; Wilhelm, J.; Muller, H.; Beuerlein, K.; Karl, N.; Bhuju, S.; Ziebuhr, J.; et al. The NF-kappaB-dependent and -independent transcriptome and chromatin landscapes of human coronavirus 229E-infected cells. *PLoS Pathog.* **2017**, *13*, e1006286. [CrossRef] [PubMed]
13. Claudio, N.; Dalet, A.; Gatti, E.; Pierre, P. Mapping the crossroads of immune activation and cellular stress response pathways. *EMBO J.* **2013**, *32*, 1214–1224. [CrossRef] [PubMed]
14. Janssens, S.; Pulendran, B.; Lambrecht, B.N. Emerging functions of the unfolded protein response in immunity. *Nat. Immunol.* **2014**, *15*, 910–919. [CrossRef] [PubMed]
15. Fung, T.S.; Liu, D.X. Coronavirus infection, ER stress, apoptosis and innate immunity. *Front. Microbiol.* **2014**, *5*, 296. [CrossRef] [PubMed]
16. Fung, T.S.; Huang, M.; Liu, D.X. Coronavirus-induced ER stress response and its involvement in regulation of coronavirus-host interactions. *Virus Res.* **2014**, *194*, 110–123. [CrossRef] [PubMed]
17. Grootjans, J.; Kaser, A.; Kaufman, R.J.; Blumberg, R.S. The unfolded protein response in immunity and inflammation. *Nat. Rev. Immunol.* **2016**, *16*, 469–484. [CrossRef] [PubMed]
18. Wu, H.; Ng, B.S.; Thibault, G. Endoplasmic reticulum stress response in yeast and humans. *Biosci. Rep.* **2014**, *34*, e00118. [CrossRef] [PubMed]
19. Cui, W.; Li, J.; Ron, D.; Sha, B. The structure of the PERK kinase domain suggests the mechanism for its activation. *Acta Crystallogr. D Biol. Crystallogr.* **2011**, *67*, 423–428. [CrossRef] [PubMed]

20. Carrara, M.; Prischi, F.; Nowak, P.R.; Ali, M.M. Crystal structures reveal transient PERK luminal domain tetramerization in endoplasmic reticulum stress signaling. *EMBO J.* **2015**, *34*, 1589–1600. [CrossRef] [PubMed]

21. Holcik, M.; Sonenberg, N. Translational control in stress and apoptosis. *Nat. Rev. Mol. Cell Biol.* **2005**, *6*, 318–327. [CrossRef] [PubMed]

22. Proud, C.G. eIF2 and the control of cell physiology. *Semin. Cell Dev. Biol.* **2005**, *16*, 3–12. [CrossRef] [PubMed]

23. Novoa, I.; Zeng, H.; Harding, H.P.; Ron, D. Feedback inhibition of the unfolded protein response by GADD34-mediated dephosphorylation of eIF2alpha. *J. Cell Biol.* **2001**, *153*, 1011–1022. [CrossRef] [PubMed]

24. Connor, J.H.; Weiser, D.C.; Li, S.; Hallenbeck, J.M.; Shenolikar, S. Growth arrest and DNA damage-inducible protein GADD34 assembles a novel signaling complex containing protein phosphatase 1 and inhibitor 1. *Mol. Cell. Biol.* **2001**, *21*, 6841–6850. [CrossRef] [PubMed]

25. Brush, M.H.; Weiser, D.C.; Shenolikar, S. Growth arrest and DNA damage-inducible protein GADD34 targets protein phosphatase 1 alpha to the endoplasmic reticulum and promotes dephosphorylation of the alpha subunit of eukaryotic translation initiation factor 2. *Mol. Cell. Biol.* **2003**, *23*, 1292–1303. [CrossRef] [PubMed]

26. Jousse, C.; Oyadomari, S.; Novoa, I.; Lu, P.; Zhang, Y.; Harding, H.P.; Ron, D. Inhibition of a constitutive translation initiation factor 2alpha phosphatase, CReP, promotes survival of stressed cells. *J. Cell Biol.* **2003**, *163*, 767–775. [CrossRef] [PubMed]

27. Jiang, H.Y.; Wek, S.A.; McGrath, B.C.; Lu, D.; Hai, T.; Harding, H.P.; Wang, X.; Ron, D.; Cavener, D.R.; Wek, R.C. Activating transcription factor 3 is integral to the eukaryotic initiation factor 2 kinase stress response. *Mol. Cell. Biol.* **2004**, *24*, 1365–1377. [CrossRef] [PubMed]

28. Silvera, D.; Formenti, S.C.; Schneider, R.J. Translational control in cancer. *Nat. Rev. Cancer* **2010**, *10*, 254–266. [CrossRef] [PubMed]

29. Olden, K.; Pratt, R.M.; Yamada, K.M. Role of carbohydrates in protein secretion and turnover: Effects of tunicamycin on the major cell surface glycoprotein of chick embryo fibroblasts. *Cell* **1978**, *13*, 461–473. [CrossRef]

30. Yoo, J.; Mashalidis, E.H.; Kuk, A.C.Y.; Yamamoto, K.; Kaeser, B.; Ichikawa, S.; Lee, S.Y. GlcNAc-1-P-transferase-tunicamycin complex structure reveals basis for inhibition of N-glycosylation. *Nat. Struct. Mol. Biol.* **2018**, *25*, 217–224. [CrossRef] [PubMed]

31. Quynh Doan, N.T.; Christensen, S.B. Thapsigargin, Origin, Chemistry, Structure-Activity Relationships and Prodrug Development. *Curr. Pharm. Des.* **2015**, *21*, 5501–5517. [CrossRef] [PubMed]

32. Treiman, M.; Caspersen, C.; Christensen, S.B. A tool coming of age: Thapsigargin as an inhibitor of sarco-endoplasmic reticulum Ca(2+)-ATPases. *Trends Pharmacol. Sci.* **1998**, *19*, 131–135. [CrossRef]

33. Oslowski, C.M.; Urano, F. Measuring ER stress and the unfolded protein response using mammalian tissue culture system. *Methods Enzymol.* **2011**, *490*, 71–92. [PubMed]

34. Pahl, H.L.; Baeuerle, P.A. A novel signal transduction pathway from the endoplasmic reticulum to the nucleus is mediated by transcription factor NF-kappa B. *EMBO J.* **1995**, *14*, 2580–2588. [PubMed]

35. Pahl, H.L.; Baeuerle, P.A. Activation of NF-kappa B by ER stress requires both Ca^{2+} and reactive oxygen intermediates as messengers. *FEBS Lett.* **1996**, *392*, 129–136. [CrossRef]

36. Jiang, H.Y.; Wek, S.A.; McGrath, B.C.; Scheuner, D.; Kaufman, R.J.; Cavener, D.R.; Wek, R.C. Phosphorylation of the alpha subunit of eukaryotic initiation factor 2 is required for activation of NF-kappaB in response to diverse cellular stresses. *Mol. Cell. Biol.* **2003**, *23*, 5651–5663. [CrossRef] [PubMed]

37. Jiang, H.Y.; Wek, R.C. GCN2 phosphorylation of eIF2alpha activates NF-kappaB in response to UV irradiation. *Biochem. J.* **2005**, *385*, 371–380. [CrossRef] [PubMed]

38. Deng, J.; Lu, P.D.; Zhang, Y.; Scheuner, D.; Kaufman, R.J.; Sonenberg, N.; Harding, H.P.; Ron, D. Translational repression mediates activation of nuclear factor kappa B by phosphorylated translation initiation factor 2. *Mol. Cell. Biol.* **2004**, *24*, 10161–10168. [CrossRef] [PubMed]

39. Wu, S.; Tan, M.; Hu, Y.; Wang, J.L.; Scheuner, D.; Kaufman, R.J. Ultraviolet light activates NFkappaB through translational inhibition of IkappaBalpha synthesis. *J. Biol. Chem.* **2004**, *279*, 34898–34902. [CrossRef] [PubMed]

40. Wu, S.; Hu, Y.; Wang, J.L.; Chatterjee, M.; Shi, Y.; Kaufman, R.J. Ultraviolet light inhibits translation through activation of the unfolded protein response kinase PERK in the lumen of the endoplasmic reticulum. *J. Biol. Chem.* **2002**, *277*, 18077–18083. [CrossRef] [PubMed]

41. Kato, T., Jr.; Delhase, M.; Hoffmann, A.; Karin, M. CK2 Is a C-Terminal IkappaB Kinase Responsible for NF-kappaB Activation during the UV Response. *Mol. Cell* **2003**, *12*, 829–839. [CrossRef]
42. Zhang, K.; Kaufman, R.J. From endoplasmic-reticulum stress to the inflammatory response. *Nature* **2008**, *454*, 455–462. [CrossRef] [PubMed]
43. Laszlo, C.F.; Wu, S. Old target new approach: An alternate NF-kappaB activation pathway via translation inhibition. *Mol. Cell. Biochem.* **2009**, *328*, 9–16. [CrossRef] [PubMed]
44. Sen, R.; Baltimore, D. Inducibility of kappa immunoglobulin enhancer-binding protein Nf-kappa B by a posttranslational mechanism. *Cell* **1986**, *47*, 921–928. [CrossRef]
45. Renner, F.; Schmitz, M.L. Autoregulatory feedback loops terminating the NF-kappaB response. *Trends Biochem. Sci.* **2009**, *34*, 128–135. [CrossRef] [PubMed]
46. Schmitz, M.L.; Mattioli, I.; Buss, H.; Kracht, M. NF-kappaB: A multifaceted transcription factor regulated at several levels. *Chembiochem* **2004**, *5*, 1348–1358. [CrossRef] [PubMed]
47. Ghosh, S.; Hayden, M.S. Celebrating 25 years of NF-kappaB research. *Immunol. Rev.* **2012**, *246*, 5–13. [CrossRef] [PubMed]
48. Griffin, B.D.; Moynagh, P.N. In vivo binding of NF-kappaB to the IkappaBbeta promoter is insufficient for transcriptional activation. *Biochem. J.* **2006**, *400*, 115–125. [CrossRef] [PubMed]
49. Hinz, M.; Arslan, S.C.; Scheidereit, C. It takes two to tango: IkappaBs, the multifunctional partners of NF-kappaB. *Immunol. Rev.* **2012**, *246*, 59–76. [CrossRef] [PubMed]
50. Sun, S.C.; Ganchi, P.A.; Ballard, D.W.; Greene, W.C. NF-kappa B controls expression of inhibitor I kappa B alpha: Evidence for an inducible autoregulatory pathway. *Science* **1993**, *259*, 1912–1915. [CrossRef] [PubMed]
51. Perkins, N.D. Integrating cell-signalling pathways with NF-kappaB and IKK function. *Nat. Rev. Mol. Cell Biol.* **2007**, *8*, 49–62. [CrossRef] [PubMed]
52. Krappmann, D.; Wulczyn, F.G.; Scheidereit, C. Different mechanisms control signal-induced degradation and basal turnover of the NF-kappaB inhibitor IkappaB alpha in vivo. *EMBO J.* **1996**, *15*, 6716–6726. [PubMed]
53. Krappmann, D.; Scheidereit, C. Regulation of NF-kappa B activity by I kappa B alpha and I kappa B beta stability. *Immunobiology* **1997**, *198*, 3–13. [CrossRef]
54. Pando, M.P.; Verma, I.M. Signal-dependent and -independent degradation of free and NF-kappa B-bound IkappaBalpha. *J. Biol. Chem.* **2000**, *275*, 21278–21286. [CrossRef] [PubMed]
55. O'Dea, E.L.; Barken, D.; Peralta, R.Q.; Tran, K.T.; Werner, S.L.; Kearns, J.D.; Levchenko, A.; Hoffmann, A. A homeostatic model of IkappaB metabolism to control constitutive NF-kappaB activity. *Mol. Syst. Biol.* **2007**, *3*, 111. [PubMed]
56. Mathes, E.; O'Dea, E.L.; Hoffmann, A.; Ghosh, G. NF-kappaB dictates the degradation pathway of IkappaBalpha. *EMBO J.* **2008**, *27*, 1357–1367. [CrossRef] [PubMed]
57. Kaneko, M.; Niinuma, Y.; Nomura, Y. Activation signal of nuclear factor-kappa B in response to endoplasmic reticulum stress is transduced via IRE1 and tumor necrosis factor receptor-associated factor 2. *Biol. Pharm. Bull.* **2003**, *26*, 931–935. [CrossRef] [PubMed]
58. Hu, P.; Han, Z.; Couvillon, A.D.; Kaufman, R.J.; Exton, J.H. Autocrine tumor necrosis factor alpha links endoplasmic reticulum stress to the membrane death receptor pathway through IRE1alpha-mediated NF-kappaB activation and down-regulation of TRAF2 expression. *Mol. Cell. Biol.* **2006**, *26*, 3071–3084. [CrossRef] [PubMed]
59. Lalani, A.I.; Zhu, S.; Gokhale, S.; Jin, J.; Xie, P. TRAF molecules in inflammation and inflammatory diseases. *Curr. Pharmacol. Rep.* **2018**, *4*, 64–90. [CrossRef] [PubMed]
60. Shenderov, K.; Riteau, N.; Yip, R.; Mayer-Barber, K.D.; Oland, S.; Hieny, S.; Fitzgerald, P.; Oberst, A.; Dillon, C.P.; Green, D.R.; et al. Cutting edge: Endoplasmic reticulum stress licenses macrophages to produce mature IL-1beta in response to TLR4 stimulation through a caspase-8- and TRIF-dependent pathway. *J. Immunol.* **2014**, *192*, 2029–2033. [CrossRef] [PubMed]
61. Woo, C.W.; Cui, D.; Arellano, J.; Dorweiler, B.; Harding, H.; Fitzgerald, K.A.; Ron, D.; Tabas, I. Adaptive suppression of the ATF4-CHOP branch of the unfolded protein response by toll-like receptor signalling. *Nat. Cell Biol.* **2009**, *11*, 1473–1480. [CrossRef] [PubMed]
62. Woo, C.W.; Kutzler, L.; Kimball, S.R.; Tabas, I. Toll-like receptor activation suppresses ER stress factor CHOP and translation inhibition through activation of eIF2B. *Nat. Cell Biol.* **2012**, *14*, 192–200. [CrossRef] [PubMed]
63. Gaestel, M.; Kotlyarov, A.; Kracht, M. Targeting innate immunity protein kinase signalling in inflammation. *Nat. Rev. Drug Discov.* **2009**, *8*, 480–499. [CrossRef] [PubMed]

64. Tam, A.B.; Mercado, E.L.; Hoffmann, A.; Niwa, M. ER stress activates NF-kappaB by integrating functions of basal IKK activity, IRE1 and PERK. *PLoS ONE* **2012**, *7*, e45078. [CrossRef] [PubMed]
65. Diamanti, M.A.; Gupta, J.; Bennecke, M.; De Oliveira, T.; Ramakrishnan, M.; Braczynski, A.K.; Richter, B.; Beli, P.; Hu, Y.; Saleh, M.; et al. IKKalpha controls ATG16L1 degradation to prevent ER stress during inflammation. *J. Exp. Med.* **2017**, *214*, 423–437. [CrossRef] [PubMed]
66. Lee, Y.Y.; Cevallos, R.C.; Jan, E. An upstream open reading frame regulates translation of GADD34 during cellular stresses that induce eIF2alpha phosphorylation. *J. Biol. Chem.* **2009**, *284*, 6661–6673. [CrossRef] [PubMed]
67. Dey, S.; Baird, T.D.; Zhou, D.; Palam, L.R.; Spandau, D.F.; Wek, R.C. Both transcriptional regulation and translational control of ATF4 are central to the integrated stress response. *J. Biol. Chem.* **2010**, *285*, 33165–33174. [CrossRef] [PubMed]
68. Palam, L.R.; Baird, T.D.; Wek, R.C. Phosphorylation of eIF2 facilitates ribosomal bypass of an inhibitory upstream ORF to enhance CHOP translation. *J. Biol. Chem.* **2011**, *286*, 10939–10949. [CrossRef] [PubMed]
69. Chen, B.P.; Wolfgang, C.D.; Hai, T. Analysis of ATF3, a transcription factor induced by physiological stresses and modulated by gadd153/Chop10. *Mol. Cell. Biol.* **1996**, *16*, 1157–1168. [CrossRef] [PubMed]
70. Fu, L.; Kilberg, M.S. Elevated cJUN expression and an ATF/CRE site within the ATF3 promoter contribute to activation of ATF3 transcription by the amino acid response. *Physiol. Genom.* **2013**, *45*, 127–137. [CrossRef] [PubMed]
71. Harding, H.P.; Novoa, I.; Zhang, Y.; Zeng, H.; Wek, R.; Schapira, M.; Ron, D. Regulated translation initiation controls stress-induced gene expression in mammalian cells. *Mol. Cell* **2000**, *6*, 1099–1108. [CrossRef]
72. Pan, Y.; Chen, H.; Siu, F.; Kilberg, M.S. Amino acid deprivation and endoplasmic reticulum stress induce expression of multiple activating transcription factor-3 mRNA species that, when overexpressed in HepG2 cells, modulate transcription by the human asparagine synthetase promoter. *J. Biol. Chem.* **2003**, *278*, 38402–38412. [CrossRef] [PubMed]
73. Pan, Y.X.; Chen, H.; Thiaville, M.M.; Kilberg, M.S. Activation of the ATF3 gene through a co-ordinated amino acid-sensing response programme that controls transcriptional regulation of responsive genes following amino acid limitation. *Biochem. J.* **2007**, *401*, 299–307. [CrossRef] [PubMed]
74. Shan, J.; Lopez, M.C.; Baker, H.V.; Kilberg, M.S. Expression profiling after activation of amino acid deprivation response in HepG2 human hepatoma cells. *Physiol. Genom.* **2010**, *41*, 315–327. [CrossRef] [PubMed]
75. Crawford, R.R.; Prescott, E.T.; Sylvester, C.F.; Higdon, A.N.; Shan, J.; Kilberg, M.S.; Mungrue, I.N. Human CHAC1 Protein Degrades Glutathione, and mRNA Induction Is Regulated by the Transcription Factors ATF4 and ATF3 and a Bipartite ATF/CRE Regulatory Element. *J. Biol. Chem.* **2015**, *290*, 15878–15891. [CrossRef] [PubMed]
76. Kilberg, M.S.; Pan, Y.X.; Chen, H.; Leung-Pineda, V. Nutritional control of gene expression: How mammalian cells respond to amino acid limitation. *Annu. Rev. Nutr.* **2005**, *25*, 59–85. [CrossRef] [PubMed]
77. Bobrovnikova-Marjon, E.V.; Marjon, P.L.; Barbash, O.; Vander Jagt, D.L.; Abcouwer, S.F. Expression of angiogenic factors vascular endothelial growth factor and interleukin-8/CXCL8 is highly responsive to ambient glutamine availability: Role of nuclear factor-kappaB and activating protein-1. *Cancer Res.* **2004**, *64*, 4858–4869. [CrossRef] [PubMed]
78. Jurida, L.; Soelch, J.; Bartkuhn, M.; Handschick, K.; Muller, H.; Newel, D.; Weber, A.; Dittrich-Breiholz, O.; Schneider, H.; Bhuju, S.; et al. The Activation of IL-1-Induced Enhancers Depends on TAK1 Kinase Activity and NF-kappaB p65. *Cell Rep.* **2015**, *10*, 726–739. [CrossRef] [PubMed]
79. Kilberg, M.S.; Balasubramanian, M.; Fu, L.; Shan, J. The transcription factor network associated with the amino acid response in mammalian cells. *Adv. Nutr.* **2012**, *3*, 295–306. [CrossRef] [PubMed]
80. Gilchrist, M.; Thorsson, V.; Li, B.; Rust, A.G.; Korb, M.; Roach, J.C.; Kennedy, K.; Hai, T.; Bolouri, H.; Aderem, A. Systems biology approaches identify ATF3 as a negative regulator of Toll-like receptor 4. *Nature* **2006**, *441*, 173–178. [CrossRef] [PubMed]
81. Litvak, V.; Ramsey, S.A.; Rust, A.G.; Zak, D.E.; Kennedy, K.A.; Lampano, A.E.; Nykter, M.; Shmulevich, I.; Aderem, A. Function of C/EBPdelta in a regulatory circuit that discriminates between transient and persistent TLR4-induced signals. *Nat. Immunol.* **2009**, *10*, 437–443. [CrossRef] [PubMed]
82. De Nardo, D.; Labzin, L.I.; Kono, H.; Seki, R.; Schmidt, S.V.; Beyer, M.; Xu, D.; Zimmer, S.; Lahrmann, C.; Schildberg, F.A.; et al. High-density lipoprotein mediates anti-inflammatory reprogramming of macrophages via the transcriptional regulator ATF3. *Nat. Immunol.* **2014**, *15*, 152–160. [CrossRef] [PubMed]

83. Labzin, L.I.; Schmidt, S.V.; Masters, S.L.; Beyer, M.; Krebs, W.; Klee, K.; Stahl, R.; Lutjohann, D.; Schultze, J.L.; Latz, E.; et al. ATF3 Is a Key Regulator of Macrophage IFN Responses. *J. Immunol.* **2015**, *195*, 4446–4455. [CrossRef] [PubMed]

84. Thompson, M.R.; Xu, D.; Williams, B.R. ATF3 transcription factor and its emerging roles in immunity and cancer. *J. Mol. Med. (Berl.)* **2009**, *87*, 1053–1060. [CrossRef] [PubMed]

85. Han, J.; Back, S.H.; Hur, J.; Lin, Y.H.; Gildersleeve, R.; Shan, J.; Yuan, C.L.; Krokowski, D.; Wang, S.; Hatzoglou, M.; et al. ER-stress-induced transcriptional regulation increases protein synthesis leading to cell death. *Nat. Cell Biol.* **2013**, *15*, 481–490. [CrossRef] [PubMed]

86. Goodall, J.C.; Wu, C.; Zhang, Y.; McNeill, L.; Ellis, L.; Saudek, V.; Gaston, J.S. Endoplasmic reticulum stress-induced transcription factor, CHOP, is crucial for dendritic cell IL-23 expression. *Proc. Natl. Acad. Sci. USA* **2010**, *107*, 17698–17703. [CrossRef] [PubMed]

87. Kim, S.Y.; Hwang, J.S.; Han, I.O. Tunicamycin inhibits Toll-like receptor-activated inflammation in RAW264.7 cells by suppression of NF-kappaB and c-Jun activity via a mechanism that is independent of ER-stress and N-glycosylation. *Eur. J. Pharmacol.* **2013**, *721*, 294–300. [CrossRef] [PubMed]

88. Wolford, C.C.; McConoughey, S.J.; Jalgaonkar, S.P.; Leon, M.; Merchant, A.S.; Dominick, J.L.; Yin, X.; Chang, Y.; Zmuda, E.J.; O'Toole, S.A.; et al. Transcription factor ATF3 links host adaptive response to breast cancer metastasis. *J. Clin. Investig.* **2013**, *123*, 2893–2906. [CrossRef] [PubMed]

89. Huang, H.; Jing, G.; Wang, J.J.; Sheibani, N.; Zhang, S.X. ATF4 is a novel regulator of MCP-1 in microvascular endothelial cells. *J. Inflamm. (Lond.)* **2015**, *12*, 31. [CrossRef] [PubMed]

90. Wolter, S.; Doerrie, A.; Weber, A.; Schneider, H.; Hoffmann, E.; von der Ohe, J.; Bakiri, L.; Wagner, E.F.; Resch, K.; Kracht, M. c-Jun controls histone modifications, NF-kappaB recruitment, and RNA polymerase II function to activate the ccl2 gene. *Mol. Cell. Biol.* **2008**, *28*, 4407–4423. [CrossRef] [PubMed]

91. Arensdorf, A.M.; Diedrichs, D.; Rutkowski, D.T. Regulation of the transcriptome by ER stress: Non-canonical mechanisms and physiological consequences. *Front. Genet.* **2013**, *4*, 256. [CrossRef] [PubMed]

92. Pakos-Zebrucka, K.; Koryga, I.; Mnich, K.; Ljujic, M.; Samali, A.; Gorman, A.M. The integrated stress response. *EMBO Rep.* **2016**, *17*, 1374–1395. [CrossRef] [PubMed]

93. Garber, M.; Yosef, N.; Goren, A.; Raychowdhury, R.; Thielke, A.; Guttman, M.; Robinson, J.; Minie, B.; Chevrier, N.; Itzhaki, Z.; et al. A high-throughput chromatin immunoprecipitation approach reveals principles of dynamic gene regulation in mammals. *Mol. Cell* **2012**, *47*, 810–822. [CrossRef] [PubMed]

94. Xie, D.; Boyle, A.P.; Wu, L.; Zhai, J.; Kawli, T.; Snyder, M. Dynamic trans-acting factor colocalization in human cells. *Cell* **2013**, *155*, 713–724. [CrossRef] [PubMed]

95. Axten, J.M.; Medina, J.R.; Feng, Y.; Shu, A.; Romeril, S.P.; Grant, S.W.; Li, W.H.; Heerding, D.A.; Minthorn, E.; Mencken, T.; et al. Discovery of 7-methyl-5-(1-{[3-(trifluoromethyl)phenyl]acetyl}-2,3-dihydro-1*H*-indol-5-yl)-7*H*-pyrrolo[2,3-*d*]pyrimidin-4-amine (GSK2606414), a potent and selective first-in-class inhibitor of protein kinase R (PKR)-like endoplasmic reticulum kinase (PERK). *J. Med. Chem.* **2012**, *55*, 7193–7207. [CrossRef] [PubMed]

96. Atkins, C.; Liu, Q.; Minthorn, E.; Zhang, S.Y.; Figueroa, D.J.; Moss, K.; Stanley, T.B.; Sanders, B.; Goetz, A.; Gaul, N.; et al. Characterization of a novel PERK kinase inhibitor with antitumor and antiangiogenic activity. *Cancer Res.* **2013**, *73*, 1993–2002. [CrossRef] [PubMed]

97. Axten, J.M.; Romeril, S.P.; Shu, A.; Ralph, J.; Medina, J.R.; Feng, Y.; Li, W.H.; Grant, S.W.; Heerding, D.A.; Minthorn, E.; et al. Discovery of GSK2656157: An Optimized PERK Inhibitor Selected for Preclinical Development. *ACS Med. Chem. Lett.* **2013**, *4*, 964–968. [CrossRef] [PubMed]

98. Delepine, M.; Nicolino, M.; Barrett, T.; Golamaully, M.; Lathrop, G.M.; Julier, C. EIF2AK3, encoding translation initiation factor 2-alpha kinase 3, is mutated in patients with Wolcott-Rallison syndrome. *Nat. Genet.* **2000**, *25*, 406–409. [CrossRef] [PubMed]

99. Harding, H.P.; Zeng, H.; Zhang, Y.; Jungries, R.; Chung, P.; Plesken, H.; Sabatini, D.D.; Ron, D. Diabetes mellitus and exocrine pancreatic dysfunction in *perk-/-* mice reveals a role for translational control in secretory cell survival. *Mol. Cell* **2001**, *7*, 1153–1163. [CrossRef]

100. Harding, H.P.; Zyryanova, A.F.; Ron, D. Uncoupling proteostasis and development in vitro with a small molecule inhibitor of the pancreatic endoplasmic reticulum kinase, PERK. *J. Biol. Chem.* **2012**, *287*, 44338–44344. [CrossRef] [PubMed]

101. Krishnamoorthy, J.; Rajesh, K.; Mirzajani, F.; Kesoglidou, P.; Papadakis, A.I.; Koromilas, A.E. Evidence for eIF2alpha phosphorylation-independent effects of GSK2656157, a novel catalytic inhibitor of PERK with clinical implications. *Cell Cycle* **2014**, *13*, 801–806. [CrossRef] [PubMed]

102. Rojas-Rivera, D.; Delvaeye, T.; Roelandt, R.; Nerinckx, W.; Augustyns, K.; Vandenabeele, P.; Bertrand, M.J.M. When PERK inhibitors turn out to be new potent RIPK1 inhibitors: Critical issues on the specificity and use of GSK2606414 and GSK2656157. *Cell Death Differ.* **2017**, *24*, 1100–1110. [CrossRef] [PubMed]

103. Iwasaki, Y.; Suganami, T.; Hachiya, R.; Shirakawa, I.; Kim-Saijo, M.; Tanaka, M.; Hamaguchi, M.; Takai-Igarashi, T.; Nakai, M.; Miyamoto, Y.; et al. Activating transcription factor 4 links metabolic stress to interleukin-6 expression in macrophages. *Diabetes* **2014**, *63*, 152–161. [CrossRef] [PubMed]

104. Fujii, T.; Fujita, N.; Suzuki, S.; Tsuji, T.; Takaki, T.; Umezawa, K.; Watanabe, K.; Miyamoto, T.; Horiuchi, K.; Masaya Nakamura, M.M. The unfolded protein response mediated by PERK is casually related to the pathogenesis of intervertebral disc degeneration. *J. Orthop. Res.* **2017**. [CrossRef] [PubMed]

105. Boyce, M.; Bryant, K.F.; Jousse, C.; Long, K.; Harding, H.P.; Scheuner, D.; Kaufman, R.J.; Ma, D.; Coen, D.M.; Ron, D.; et al. A selective inhibitor of eIF2alpha dephosphorylation protects cells from ER stress. *Science* **2005**, *307*, 935–939. [CrossRef] [PubMed]

106. Nakajima, S.; Chi, Y.; Gao, K.; Kono, K.; Yao, J. eIF2alpha-Independent Inhibition of TNF-alpha-Triggered NF-kappaB Activation by Salubrinal. *Biol. Pharm. Bull.* **2015**, *38*, 1368–1374. [CrossRef] [PubMed]

107. Ghosh, R.; Wang, L.; Wang, E.S.; Perera, B.G.; Igbaria, A.; Morita, S.; Prado, K.; Thamsen, M.; Caswell, D.; Macias, H.; et al. Allosteric inhibition of the IRE1alpha RNase preserves cell viability and function during endoplasmic reticulum stress. *Cell* **2014**, *158*, 534–548. [CrossRef] [PubMed]

108. Keestra-Gounder, A.M.; Byndloss, M.X.; Seyffert, N.; Young, B.M.; Chavez-Arroyo, A.; Tsai, A.Y.; Cevallos, S.A.; Winter, M.G.; Pham, O.H.; Tiffany, C.R.; et al. NOD1 and NOD2 signalling links ER stress with inflammation. *Nature* **2016**, *532*, 394–397. [CrossRef] [PubMed]

MDPI

Article

Deubiquitinylase USP47 Promotes RelA Phosphorylation and Survival in Gastric Cancer Cells

Lara Naghavi [†], **Martin Schwalbe** [†], **Ahmed Ghanem** and **Michael Naumann** *

Institute of Experimental Internal Medicine, Otto von Guericke University, 39120 Magdeburg, Germany;
lara.naghavi@med.ovgu.de (L.N.); martin.schwalbe@med.ovgu.de (M.S.); ahmed.ghanem@med.ovgu.de (A.G.)
* Correspondence: Naumann@med.ovgu.de; Tel.: +49-391-671-3227
† These authors contributed equally to this work.

Received: 13 February 2018; Accepted: 12 May 2018; Published: 22 May 2018

Abstract: Every year, gastric cancer causes around 819,000 deaths worldwide. The incidence of gastric cancer in the western world is slowly declining, but the prognosis is unpromising. In Germany, the 5-year-survival rate is around 32%, and the average life span after diagnosis is 6 to 9 months. Therapy of gastric cancer patients comprises a gastrectomy and perioperative or adjuvant chemotherapy. However, resistance of gastric cancer cells to these agents is widespread; thus, improved chemotherapeutic approaches are required. Nuclear factor kappa B (NF-κB) transcription factors are associated with anti-apoptosis, carcinogenesis, and chemoresistance, and thus, constitute attractive targets for therapeutic intervention. In immunoblots, we show that ubiquitin specific protease 47 (USP47) promotes β-transducin repeat-containing protein (βTrCP) stability and phosphorylation of RelA. Furthermore, after knockdown of USP47 by RNA interference, we analyzed in gastric cancer cell lines metabolic activity/viability in an MTT assay, and apoptotic cell death by Annexin V staining and poly(ADP-Ribose) polymerase (PARP)-1, caspase 3, and caspase 8 cleavage, respectively. We found that USP47 contributes to cell viability and chemoresistance in NCI-N87 gastric carcinoma cells treated with etoposide and camptothecin. Inhibition of USP47 might be a suitable strategy to downregulate NF-κB activity, and to overcome chemoresistance in gastric cancer.

Keywords: deubiquitinylases; NF-κB signaling; apoptotic cell death; gastric carcinoma; therapeutic targets; chemoresistance; ubiquitin-specific proteases

---

1. Introduction

With 1.3 million incident cases and 819,000 deaths recorded globally, gastric cancer was the fifth most common malignancy, and the third leading cause of cancer deaths in 2015 [1]. Surgical gastrectomy remains the only curative therapy. However, relapse occurs in 40–60% of cases [2]. Diagnosis of gastric cancers often occurs when tumors are inoperable and patients have median survival times of three to five months. First-line or adjuvant chemotherapy (or chemoradiation) extends patient survival times by 6–7 months [3]. The high number of deaths from gastric cancer, low cure rates, and tumor relapse after gastrectomy therefore demand the development of better chemotherapeutic agents to improve patient survival. In particular, overcoming chemoresistance in tumor cells that otherwise could result in relapse or metastasis poses a major challenge [4,5].

Nuclear factor kappa B (NF-κB) constitutes a family of transcription factors (RelA, RelB, c-Rel, p105/p50 and p100/52) that regulate many target genes involved in inflammation, immunity, cell proliferation, or cell survival. Without stimulation, NF-κB molecules are sequestered to the cytoplasm by inhibitors of kappa B (IκB) molecules. Stimulation of the canonical NF-κB pathway results in the activation of the IκB kinase (IKK) complex, which in turn phosphorylates IκBα, promoting its ubiquitinylation and subsequent degradation. After release from IκBα, NF-κB factors translocate to the nucleus stimulating transcription of their target genes. Transcriptional activity of NF-κB is further regulated by posttranslational modifications like phosphorylation, ubiquitinylation, or acetylation [6]. Importantly, dysregulation of NF-κB signaling promotes carcinogenesis [5,7,8] and chemoresistance [9–11].

As novel therapeutic targets, deubiquitinylases are receiving increased attention, due to their association with cancer and neurodegenerative diseases [12,13]. Deubiquitinylases (DUBs) as antagonists to E3 ubiquitin ligases are an integral part of the ubiquitin-proteasome-system. E3s conjugate ubiquitin moieties to substrate proteins, and thus, regulate their stability, protein interactions, or subcellular localization. In contrast, to the more than 600 E3 ligases [14] in humans with distinct substrate specificities, less than 100 DUBs reverse the conjugation and cleave ubiquitin moieties from substrate proteins. In particular, within the NF-κB pathway, E3s and DUBs regulate IKK complex activation, IκBα degradation, or NF-κB activity [15–17]. Because the NF-κB pathway constitutes an attractive target for therapeutic intervention in gastric cancer [5,18], we performed an siRNA screen aiming to identify potential DUB targets that are essential for NF-κB activity. As one such target, we identified the ubiquitin specific protease 47 (USP47), that was previously shown to be essential for the proliferation of gastric carcinomas [19]. Furthermore, depletion of USP47 sensitized osteosarcoma and breast cancer cell lines towards drug and UV radiation-induced apoptosis [20].

In this study, we addressed the molecular details of NF-κB regulation using USP47, and examined the viability of USP47 as a promising target for drug intervention to enhance the action of current drugs, and to overcome chemoresistance in gastric cancer cells. Our results show that depletion of USP47 in AGS gastric cancer cells results in decreased protein levels of phospho-RelA and β-transducin repeat-containing protein (βTrCP). Even though USP47 depletion failed to increase apoptotic cell death in AGS cells treated with camptothecin (CPT) and etoposide (Eto), it overcame chemoresistance in NCI-N87 gastric carcinoma cells. Therefore, targeting of USP47 represents a suitable strategy to overcome drug resistance in gastric carcinomas.

2. Experimental Section

2.1. Cell Culture and siRNA Transfection

Gastric carcinoma cell lines AGS and NCI-N87 (ATCC, LGC Standards GmbH, Wesel, Germany) were cultured in an RPMI 1640 medium, supplemented with 10% fetal calf serum (FCS) and 100 U/mL penicillin/streptomycin, and then incubated at 37 °C in a 5% CO_2 humidified incubator. AGS and NCI-N87 cells were seeded at a density of 1.2 or 1.5×10^6, respectively, per 100 mm culture dish (CellStar, Greiner Bio-one GmbH, Frickenhausen, Germany). For transient knockdown of USP47, cells at 30% confluency were transferred into serum-free OptiMEM medium (Thermo Fisher Scientific, Waltham, MA, USA), and transfected with 50 nM siRNA that had been mixed with 10 μL/10 cm culture dish SiLentFect Lipid Reagent (Bio-Rad, Hercules, CA, USA). USP47 siRNA 5'-CUAUAACUGUUCGUGCUUA-3' was obtained from Eurofins Genomics, Ebersberg, Germany, and AllStars negative control siRNA from Qiagen, Hilden, Germany. Six hours after transfection, a fresh RPMI 1640 medium with 10% FCS was added, and incubation continued for 42 h. Sixteen hours prior to TNF treatment, cells were serum starved with an RPMI 1640 medium supplemented with 0.2% FCS; 10 ng/mL TNF (in PBS) (PeproTech, Hamburg, Germany) was used, as described previously [17]. Camptothecin or etoposide (all Sigma-Aldrich, St. Louis, MO, USA) were used at concentrations of 1 μM (in DMSO) or 10 μM (in DMSO), respectively.

2.2. Subcellular Fractionation

Cells were washed twice with PBS and scraped into chilled 1 mL PBS. After removing the PBS by centrifugation (2 min, $1000 \times g$), the cell pellet was suspended in 400 μL chilled buffer A (20 mM Tris pH 7.9, 10 mM NaCl, 1.5 mM $MgCl_2$, 10% glycerol, 1 mM DTT, 0.5 mM AEBSF, 1 mM Na_3VO_4,

1 mM Na$_2$MO$_4$, 10 mM NaF, 10 mM K$_2$HPO$_4$, 20 mM 2-Phosphoglycerate, 7.5 mM N-Ethylmaleimid (NEM), 5 mM ortho-phenanthroline (OPT), 1× cOmplete EDTA-free protease inhibitor cocktail (Roche Diagnostics, Mannheim, Germany)), and incubated for 10 min on ice. Subsequently, 4 µL NP40 was added to lyse the cells. After 5 min incubation on ice, nuclei were separated from cytosolic supernatants by centrifugation (10 min, 2000× *g*). Cytosolic supernatants were clarified by centrifugation (10 min, 13,000× *g*), and nuclear pellets washed with 100 µL buffer A. Nuclear pellets were suspended in 50 µL buffer C (20 mM Tris pH 7.9, 420 mM NaCl, 1.5 mM MgCl$_2$, 10% glycerol, 0.2 mM EDTA, 0.5 mM DTT, 0.5 mM AEBSF, 1 mM Na$_3$VO$_4$, 1 mM Na$_2$MO$_4$, 10 mM NaF, 10 mM K$_2$HPO$_4$, 20 mM 2-Phosphoglycerate, 7.5 mM NEM, 5 mM OPT, 1× cOmplete EDTA-free protease inhibitor cocktail), and incubated for 30 min on ice, with occasional vortexing. Soluble nuclear proteins (N1) were obtained after centrifugation (10 min, 13,000× *g*). The BCA assay (Thermo Fisher Scientific) was used to determine protein concentration.

2.3. SDS-PAGE and Immunoblotting

Proteins were separated by SDS-PAGE in Tris-Glycine buffer gels and transferred onto PVDF membranes (Merck, Darmstadt, Germany), followed by 1 h blocking at room temperature using 5% skim milk in TBS containing 0.1% Tween (TBS-T). The PVDF membranes were incubated overnight at 4 °C, with the primary antibodies in either 5% BSA or 5% skim milk/TBS-T. The membranes were washed thrice with TBS-T, and incubated with the appropriate HRP-conjugated secondary antibody diluted 1:5000 in 5% skim milk/TBS-T for 1 h at room temperature. All antibodies used in this study are listed in the Supplementary Materials Table S1. After washing the membranes thrice with TBS-T, the blots were developed using a chemiluminescent substrate (Merck), and protein bands were visualized using the ChemoCam Imager (Intas, Göttingen, Germany).

2.4. Apoptosis Detection

Forty-eight hours after transfection, cells were treated for 24 h with either camptothecin or etoposide. Control cells were treated with equal volumes of DMSO. Externalization of phosphatidylserine was determined 24 h after treatment using the Annexin V-FITC kit (Biotool, Houston, TX, USA). Cells were labeled with Annexin V-FITC according to instructions, and, apoptotic cells (Annexin V-FITC positive) were counted by flow cytometry (CyFlow space, Partec, Görlitz, Germany). Whole cell extracts for immunoblotting were prepared by suspending the cells in 100 µL RIPA buffer (50 mM Tris, pH 7.5, 150 mM NaCl, 10 mM K$_2$HPO$_4$, 5 mM EDTA, 10% glycerol, 1% Triton X100, 0.05% SDS, 1 mM AEBSF, 20 mM 2-phosphoglycerate, 20 mM NaF, 1 mM $_3$VO$_4$, 1 mM Na$_2$MoO$_4$, 1× cOmplete EDTA-free protease inhibitor cocktail), followed by 10 min incubation on ice, and subsequent clarification of the extract (10 min, 13,000× *g*).

2.5. Cell Viability Assay

Cells were seeded in 96-well plates with 50,000 cells per well, and transfected as described above. Twenty-four hours after camptothecin or etoposide, treatment viability was assessed using the MTT Cell Viability Assay Kit (Abnova, Taipei, Taiwan). Assays were performed according to instructions, and absorption was measured at 570 nm on a Spectramax M5 plate reader (Molecular Devices GmbH, Biberach an der Riss, Germany). The average absorption value for the control cells was set to 100% viability.

2.6. Statistics

All quantitative data were presented as mean ± S.D (standard deviation). Statistical analysis was performed using Student's *t*-test (Excel, Microsoft Office Plus 2010). $p < 0.05$ was regarded as significant.

3. Results

3.1. Depletion of USP47 Decreased Phospho-RelA and βTrCP Protein Levels

Based on the observation that USP47 promotes the proliferation of gastric carcinomas [19], we performed a USP47 knockdown in AGS gastric carcinoma cells. Stimulation with tumor necrosis factor (TNF) for up to 20 min in USP47 knockdown cells resulted in decreased Ser536-phosphorylated RelA protein levels in the cytosol, and to a minor extent, also in the nucleus (Figure 1A). In contrast, IκBα degradation appeared unaffected, even though a minor decrease in IκBα phosphorylation was noticed after USP47 knockdown (Figure 1A). Analysis of regulatory NF-κB pathway components revealed that USP47 depletion constitutively decreased βTrCP levels (Figure 1A). βTrCP is the substrate adaptor of the E3 ligase complex that facilitates IκBα ubiquitinylation after phosphorylation. Inhibition of protein translation through cycloheximide (CHX) treatment showed that USP47 depletion decreased the basal expression levels of βTrCP (Figure 2). Similar to AGS cells, knockdown of USP47 in gastric cancer cell line (NCI-N87) constitutively decreased βTrCP levels, and reduced nuclear translocation of RelA (Figure 1B).

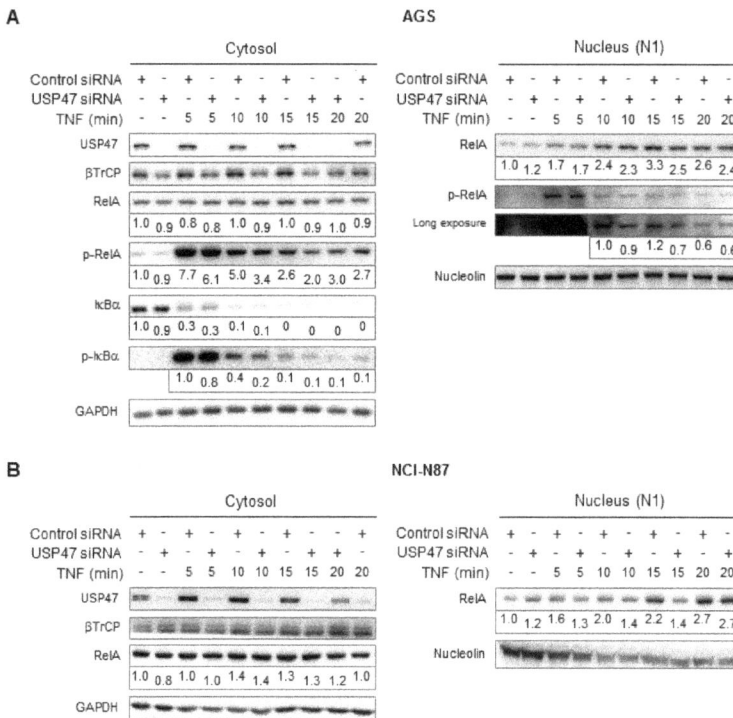

Figure 1. USP47 depletion downregulates protein levels of βTrCP and Ser536-phosphorylated RelA. (**A**) AGS or (**B**) NCI-N87 cells transfected with USP47 or a negative control siRNA were stimulated with TNF for up to 20 min. Subcellular fractions were prepared at the indicated time points and NF-κB pathway components analyzed in cytosolic and soluble nuclear fractions (N1). An additional immunoblot with increased signal acquisition time is shown for p-RelA in (**A**). Detection of GAPDH and nucleolin served as control for equal protein load. For densitometric analysis, band intensities of key proteins were normalized to the band intensities of the respective cytosolic or nuclear loading control. The normalized band intensities are shown below the corresponding blots.

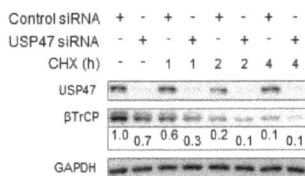

Figure 2. USP47 depletion downregulates basal βTrCP protein levels. AGS cells transfected with USP47 or a negative control siRNA were treated with CHX for up to 4 h. Subcellular fractions were prepared at the indicated time points and βTrCP stability was analyzed in the cytosolic fractions.

3.2. USP47 Promotes Chemoresistance and Cell Viability in NCI-N87 Gastric Carcinoma Cells

TNF can induce diverse cellular responses, including cell survival and apoptosis [21]. Gastric cancer patients have low cure rates after chemotherapy; we therefore asked whether USP47 knockdown could enhance apoptosis in gastric cancer cells when treated with chemotherapeutic drugs. We selected the topoisomerase I and II inhibitors CPT and Eto to treat different gastric carcinoma cell lines, because their mode of action is similar to drugs used in current chemotherapy regimens [3], and because they are known to also promote NF-κB-dependent apoptosis resistance [22,23]. Eto or CPT treatment of AGS cells increased apoptotic cell death by ca. 18% or 38% in a concentration dependent-manner, respectively, (Figure 3A–C), promoted caspase 8, caspase 3, and poly(ADP-Ribose) polymerase (PARP)-1 cleavage (Figure 3D), and decreased cell viability by ca. 20% (Figure 4). Combining USP47 knockdown with Eto or CPT treatment did not further enhance apoptotic cell death, and resulted in comparable levels of apoptotic cells and cell viability (Figure 3C,D and Figure 4).

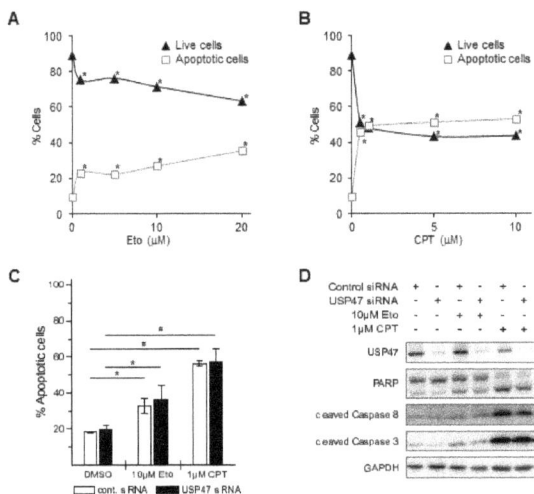

Figure 3. Etoposide (Eto) or camptothecin (CPT) induce apoptosis in AGS cells. 24 h after incubation of AGS cells with increasing amounts of (**A**) Eto (1–20 μM) or (**B**) CPT (0.5–10 μM) live (black triangles) and apoptotic cells (open squares) were counted by flow cytometry using Annexin V staining. Standard deviations are in the range of the marker size. * $p < 0.05$. USP47 depletion does not enhance Eto or CPT-induced apoptosis in AGS cells. AGS cells were incubated with 10 μM Eto or 1 μM CPT for 24 h alone or in combination with USP47 knockdown; (**C**) Apoptotic cells were counted by flow cytometry using Annexin V staining. * $p < 0.05$, # $p < 0.001$; (**D**) Whole cell extracts were subjected to immunoblotting and analyzed for apoptotic markers. Detection of GAPDH served as control for equal protein load.

Analysis of additional gastric carcinoma cell lines showed that the NCI-N87 cell line was resistant to Eto and CPT treatment, independent of the concentration (Figure 5A,B). Interestingly, after USP47 knockdown, the NCI-N87 cell line displayed increased apoptotic cell death (Figure 5C), increased cleavage of PARP, caspase 3, and caspase 8 (Figure 5D), and reduced cell viability (Figure 6).

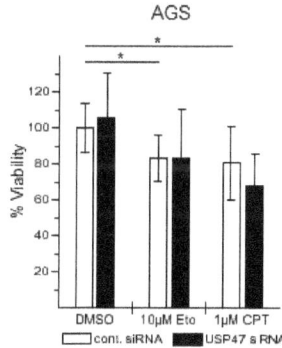

Figure 4. USP47 promotes cell viability in AGS cells. AGS cells were seeded in 96 well plates, transfected with control or USP47 siRNA and incubated for 24 h with either with 10 μM Eto or 1 μM CPT. Cell viability was measured using the 3-(4,5-Dimethylthiazol-2-yl)-2,5-diphenyltetrazoliumbromid (MTT) assay. * $p < 0.05$.

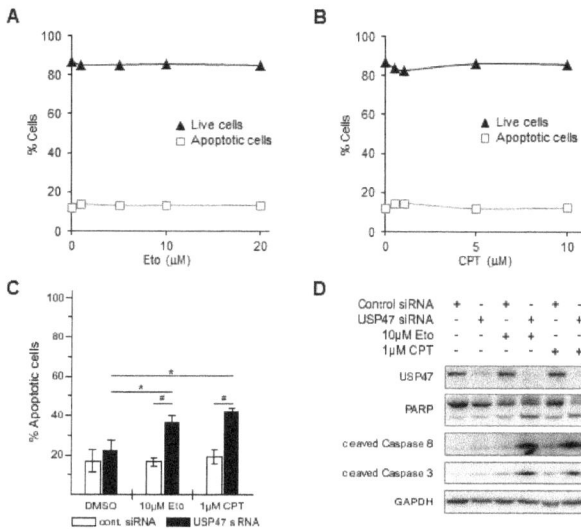

Figure 5. USP47 attenuates etoposide (Eto) or camptothecin (CPT)-induced apoptotic cell death in NCI-N87 cells. NCI-N87 cells were incubated with increasing concentrations of (**A**) Eto (1–20 μM) or (**B**) CPT (0.5–10 μM) for 24 h. Live (black triangles) and apoptotic cells (open squares) were counted by flow cytometry using Annexin V staining. Standard deviations are in the range of the marker size; (**C**) To analyze the effect of USP47 on cell survival, NCI-N87 cells were transfected with USP47 or control siRNA, treated for 24 h with 10 μM etoposide or 1 μM CPT and apoptotic cells were counted by flow cytometry using Annexin V staining. * $p < 0.05$, # $p < 0.001$; (**D**) Whole cell extracts of NCI-N87 cells treated as in (**C**) were subjected to immunoblotting and analyzed for apoptotic markers. Detection of GAPDH served as control for equal protein load.

NCI-N87

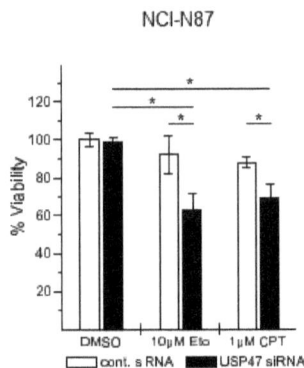

Figure 6. USP47 promotes cell viability in NCI-N87 cells. NCI-N87 cells were seeded in 96 well plates, transfected with control or USP47 siRNA and incubated for 24 h with either with 10 μM Eto or 1 μM CPT. Cell viability was measured using the 3-(4,5-Dimethylthiazol-2-yl)-2,5-diphenyltetrazoliumbromid (MTT) assay. * $p < 0.05$.

4. Discussion

After cardiovascular diseases, cancer is the second leading cause of death worldwide, and with an increasing and ageing population, it is expected that patient numbers will rise [1]. Chemoresistance constitutes a major problem of chemotherapy, because it promotes relapse and thus increases morbidity [4]. Therefore, it is necessary to improve current chemotherapy regimens and identify new drug targets, especially for malignancies like gastric carcinomas with low cure rates [3]. Deubiquitinylases represent promising novel targets for drug discovery in cancer therapy, due to their involvement in metastasis and their regulatory role in apoptosis pathways [24]. In this study we could identify USP47 as a regulator of the NF-κB pathway. Therapeutic intervention of NF-κB signaling is an intensively studied research field [18,25], because aberrant NF-κB activation in the context of cancer promotes carcinogenesis, tumor progression, metastasis, or chemoresistance [5].

Our results demonstrate that USP47 stabilizes the E3 ligase complex substrate adaptor protein βTrCP, and promotes phosphorylation of RelA at Ser536 (Figure 1). The residual βTrCP levels after USP47 depletion suffice to allow for IκBα degradation. Within the chosen time points, differences between USP47 and control cells did not become apparent, probably due to the rapidness of the process. Nonetheless, nuclear translocation of RelA is diminished after USP47 knockdown (Figure 1), which is consistent with a delayed release from IκBα.

Apart from its regulatory role in the NF-κB pathway, USP47 constitutes an interesting target for gastric cancer chemotherapy, because it promotes gastric carcinoma cell proliferation [19], and its depletion sensitize cancer cells to chemotherapy, probably due to the upregulation of Cdc25a [20]. We therefore addressed the question of whether USP47 inhibition could affect cell survival/proliferation of gastric cancer cells. For instance, USP47 inhibition could be beneficial in attenuating NF-κB-dependent chemoresistance in response to DNA double strand break inducing agents [22]. AGS cells displayed pronounced apoptotic cell death when treated with CPT or Eto (Figure 3A–D), but no further enhancement was observed after USP47 depletion. It is likely that the already strong response masks additional contributions to the apoptosis caused by the USP47 knockdown. Interestingly, in the CPT- and Eto-resistant gastric carcinoma cell line NCI-N87 (Figure 5A,B) USP47 depletion resulted in pronounced apoptosis induction after CPT and Eto treatment (Figure 5C,D and Figure 6). The molecular mechanisms responsible for the drug-resistance are currently unknown. However, it is likely that in addition to the attenuation of NF-κB activity, other factors involved in DNA double strand break repair might play a role. For instance, USP47 stabilizes DNA polymerase β, and hence regulates base excision repair after DNA damage [26]. Similarly, USP47 might stabilize protein factors involved in DNA double strand break repair, and hence, depletion of USP47 could sensitize cancer cells to the action of DNA damaging agents.

USP47 depletion on its own caused no increase in apoptotic cell death (Figures 3–6), suggesting that loss of USP47 activity is not cytotoxic. This together with its potential to overcome chemoresistance in gastric carcinoma cells suggests that USP47 could represent a promising therapeutic target structure. Furthermore, a combination of current chemotherapy regimens together with USP47 inhibition might allow the reduction of the concentration of chemotherapeutic agents without impairing the cytotoxic effect on cancer cells, while reducing side effects. We propose that the development of USP47 inhibitors and their combination with current chemotherapy regimens might reduce relapse after gastrectomy, and extend overall patient survival times.

5. Conclusions

Here, we could show that USP47 regulates NF-κB activity by promoting the phosphorylation of RelA through the stabilization of βTrCP and subsequent degradation of IκBα. In addition, USP47 contributes to chemoresistance and viability in NCI-N87 gastric carcinoma cells. Inhibition of USP47 activity could represent a viable strategy for gastric cancer chemotherapy by downregulating transcription of NF-κB-regulated pro-survival genes, and overcoming NF-κB-dependent chemoresistance.

Supplementary Materials: The following are available online at http://www.mdpi.com/2227-9059/6/2/62/s1, Table S1: List of primary and secondary antibodies: Uncropped immunoblots corresponding to data shown in this study.

Author Contributions: M.N. conceived the study; M.N., L.N. and M.S. designed the experiments; L.N. performed the flow cytometry and cell viability assays of AGS and NCI-N87 cells; M.S. performed subcellular fractionations and subsequent immunoblot analysis of NF-κB components; A.G. contributed to RNAi optimization and immunoblotting; M.N., M.S. and L.N. wrote the paper.

Acknowledgments: The work was supported by the Ministry of Economy, Science and Digitalisation (Förderung von Wissenschaft und Forschung in Sachsen-Anhalt aus Mitteln der Europäischen Struktur- und Investitionsfonds in der Förderperiode 2014-2020, ZS/2016/04/78155) and a stipend from the Medical Faculty of the Otto von Guericke University to L.N.

Conflicts of Interest: The authors declare no conflict of interest.

References

1. Global Burden of Disease Cancer Collaboration. Global, regional, and national cancer incidence, mortality, years of life lost, years lived with disability, and disability-adjusted life-years for 32 cancer groups, 1990 to 2015: A systematic analysis for the global burden of disease study. *JAMA Oncol.* **2017**, *3*, 524–548. [CrossRef]
2. Orditura, M.; Galizia, G.; Sforza, V.; Gambardella, V.; Fabozzi, A.; Laterza, M.M.; Andreozzi, F.; Ventriglia, J.; Savastano, B.; Mabilia, A.; et al. Treatment of gastric cancer. *World J. Gastroenterol.* **2014**, *20*, 1635–1649. [CrossRef] [PubMed]
3. Wagner, A.D.; Syn, N.L.; Moehler, M.; Grothe, W.; Yong, W.P.; Tai, B.-C.; Ho, J.; Unverzagt, S. Chemotherapy for advanced gastric cancer. *Cochrane Database Syst. Rev.* **2017**. [CrossRef] [PubMed]
4. Zheng, H.-C. The molecular mechanisms of chemoresistance in cancers. *Oncotarget* **2017**, *8*, 59950–59964. [CrossRef] [PubMed]
5. Sokolova, O.; Naumann, M. NF-κB Signaling in Gastric Cancer. *Toxins* **2017**, *9*, 119. [CrossRef] [PubMed]
6. Neumann, M.; Naumann, M. Beyond IκBs: Alternative regulation of NF-κB activity. *FASEB J.* **2007**, *21*, 2642–2654. [CrossRef] [PubMed]
7. Maeda, S.; Omata, M. Inflammation and cancer: Role of nuclear factor-kappaB activation. *Cancer Sci.* **2008**, *99*, 836–842. [CrossRef] [PubMed]
8. Kwon, H.-C.; Kim, S.-H.; Oh, S.Y.; Lee, S.; Lee, J.H.; Jang, J.S.; Kim, M.C.; Kim, K.H.; Kim, S.-J.; Kim, S.-G.; et al. Clinicopathologic significance of expression of nuclear factor-κB RelA and its target gene products in gastric cancer patients. *World J. Gastroenterol.* **2012**, *18*, 4744–4750. [CrossRef] [PubMed]

9. Manu, K.A.; Shanmugam, M.K.; Li, F.; Chen, L.; Siveen, K.S.; Ahn, K.S.; Kumar, A.P.; Sethi, G. Simvastatin sensitizes human gastric cancer xenograft in nude mice to capecitabine by suppressing nuclear factor-kappa B-regulated gene products. *J. Mol. Med.* **2014**, *92*, 267–276. [CrossRef] [PubMed]

10. Zhou, W.; Fu, X.-Q.; Zhang, L.-L.; Zhang, J.; Huang, X.; Lu, X.-H.; Shen, L.; Liu, B.-N.; Liu, J.; Luo, H.-S.; et al. The AKT1/NF-kappaB/Notch1/PTEN axis has an important role in chemoresistance of gastric cancer cells. *Cell Death Dis.* **2013**, *4*, e847. [CrossRef] [PubMed]

11. Kinoshita, J.; Fushida, S.; Harada, S.; Makino, I.; Nakamura, K.; Oyama, K.; Fujita, H.; Ninomiya, I.; Fujimura, T.; Kayahara, M.; et al. PSK enhances the efficacy of docetaxel in human gastric cancer cells through inhibition of nuclear factor-κB activation and survivin expression. *Int. J. Oncol.* **2010**, *36*, 593–600. [CrossRef] [PubMed]

12. Clague, M.J.; Barsukov, I.; Coulson, J.M.; Liu, H.; Rigden, D.J.; Urbé, S. Deubiquitylases from Genes to Organism. *Physiol. Rev.* **2013**, *93*, 1289–1315. [CrossRef] [PubMed]

13. Harrigan, J.A.; Jacq, X.; Martin, N.M.; Jackson, S.P. Deubiquitylating enzymes and drug discovery: Emerging opportunities. *Nat. Rev. Drug Discov.* **2017**, *17*, 57. [CrossRef] [PubMed]

14. Li, W.; Bengtson, M.H.; Ulbrich, A.; Matsuda, A.; Reddy, V.A.; Orth, A.; Chanda, S.K.; Batalov, S.; Joazeiro, C.A.P. Genome-Wide and Functional Annotation of Human E3 Ubiquitin Ligases Identifies MULAN, a Mitochondrial E3 that Regulates the Organelle's Dynamics and Signaling. *PLoS ONE* **2008**, *3*, e1487. [CrossRef] [PubMed]

15. Wertz, I.E.; Dixit, V.M. Signaling to NF-κB: Regulation by Ubiquitination. *Cold Spring Harb. Perspect. Biol.* **2010**, *2*, a003350. [CrossRef] [PubMed]

16. Schweitzer, K.; Bozko, P.M.; Dubiel, W.; Naumann, M. CSN controls NF-κB by deubiquitinylation of IκBα. *EMBO J.* **2007**, *26*, 1532–1541. [CrossRef] [PubMed]

17. Schweitzer, K.; Naumann, M. CSN-associated USP48 confers stability to nuclear NF-κB/RelA by trimming K48-linked Ub-chains. *Biochim. Biophys. Acta* **2015**, *1853*, 453–469. [CrossRef] [PubMed]

18. Erstad, D.J.; Cusack, J.C. Targeting the NF-κB Pathway in Cancer Therapy. *Transl. Cancer Res. Surg.* **2013**, *22*, 705–746. [CrossRef] [PubMed]

19. Zhang, B.; Yin, Y.; Hu, Y.; Zhang, J.; Bian, Z.; Song, M.; Hua, D.; Huang, Z. MicroRNA-204-5p inhibits gastric cancer cell proliferation by downregulating USP47 and RAB22A. *Med. Oncol.* **2014**, *32*, 331. [CrossRef] [PubMed]

20. Peschiaroli, A.; Skaar, J.; Pagano, M.; Melino, G. The ubiquitin-specific protease USP47 is a novel β-TRCP interactor regulating cell survival. *Oncogene* **2010**, *29*. [CrossRef] [PubMed]

21. Varfolomeev, E.E.; Ashkenazi, A. Tumor Necrosis Factor: An Apoptosis JuNKie? *Cell* **2004**, *116*, 491–497. [CrossRef]

22. Habraken, Y.; Piette, J. NF-κB activation by double-strand breaks. *Biochem. Pharmacol.* **2006**, *72*, 1132–1141. [CrossRef] [PubMed]

23. Choi, Y.S.; Jeong, S. PI3-Kinase and PDK-1 Regulate HDAC1-mediated Transcriptional Repression of Transcription Factor NF-κB. *Mol. Cells* **2005**, *20*, 241–246. [PubMed]

24. He, M.; Zhou, Z.; Wu, G.; Chen, Q.; Wan, Y. Emerging role of DUBs in tumor metastasis and apoptosis: Therapeutic implication. *Pharmacol. Ther.* **2017**, *177*, 96–107. [CrossRef] [PubMed]

25. Hoesel, B.; Schmid, J.A. The complexity of NF-κB signaling in inflammation and cancer. *Mol. Cancer* **2013**, *12*, 86. [CrossRef] [PubMed]

26. Parsons, J.L.; Dianova, I.I.; Khoronenkova, S.V.; Edelmann, M.J.; Kessler, B.M.; Dianov, G.L. USP47 Is a Deubiquitylating Enzyme that Regulates Base Excision Repair by Controlling Steady-State Levels of DNA Polymerase β. *Mol. Cell* **2011**, *41*, 609–615. [CrossRef] [PubMed]

MDPI
St. Alban-Anlage 66
4052 Basel
Switzerland
Tel. +41 61 683 77 34
Fax +41 61 302 89 18
www.mdpi.com

Biomedicines Editorial Office
E-mail: biomedicines@mdpi.com
www.mdpi.com/journal/biomedicines

MDPI
St. Alban-Anlage 66
4052 Basel
Switzerland

Tel: +41 61 683 77 34
Fax: +41 61 302 89 18

www.mdpi.com

MDPI

ISBN 978-3-03897-118-4

www.ingramcontent.com/pod-product-compliance
Lightning Source LLC
Chambersburg PA
CBHW051712210326
41597CB00032B/5452